GEOLOGY OF
NATIONAL PARKS

GEOLOGY OF NATIONAL PARKS

Third Edition

Ann G. Harris
Youngstown State University

Esther Tuttle
Science Editor, Iowa City, Iowa

KENDALL/HUNT
PUBLISHING COMPANY
Dubuque, Iowa

Consulting Editor

Sherwood D. Tuttle
University of Iowa

Cover: Waves erode Arch Rock and stacks at the eastern tip of An-acapa Islands in Channel Islands National Park, California. The bedrock is volcanic. Photograph courtesy Channel Islands National Park Collection.

The drawings by Genevieve Shimer on pages 1, 111, 215, 319, 429, and 503 are from *Field Guide to Landforms in the United States* by J. A. Shimer. © 1972 J. A. Shimer; used by permission. The Mac-millan Company.

The map reproductions appearing on pages 3, 36, 77, 113, 217, 243, 275, 285, 296, 372, 398, 410, 464, 475, and 497 are maps of the national parks from *A Guide to the National Parks—Their Landscape and Geology* by William H. Matthews III. Copyright © 1968, 1973 by William H. Matthews III. Reprinted by permission of Doubleday & Company, Inc.

Copyright © 1975 by Ann G. Harris
Copyright © 1977, 1983 by Kendall/Hunt Publishing Company

Library of Congress Catalog Card Number: 82–83166

ISBN 0–8403–2810–9

Second Printing, 1984

Printed in the United States of America

C   402810   02

To Dale Noel

Acknowledgments

Typists: Marion Beam, Malcena Mooney, Jean
Peterson
Editorial assistance: Carole Hallenbeck
Library assistance: Hildegard Schnuttgen, Louise Zipp
National park location maps drafted by Robert Tynal are
reproduced in figures 2.2, 4.2, 8.2, 9.2, 11.2, 13.2, 14.2,
15.2, 16.2, 20.2, 20.3, 21.2, 25.2, 26.2, 27.2, 28.2, 29.2,
31.2, 34.3, 36.2, 37.2, 40.2, and 42.2. Kay Chambers
drafted additional maps and diagrams.

Contents

x   Contents

Foreword

National Parks are sources of challenging and interesting illustrations of geologic concepts. That is not surprising because the presence of "geologic features worthy of preservation and protection" is one of the criteria for elevation of an area to national park status.

Like the previous editions, this third edition of *Geology of National Parks* is developed around the idea that students and others who use the book like to "get right into the parks" and learn about geology as they go along. Many of them have visited national parks and are surprised, as they read about the geology, to find out how much they did not see when they were there.

Because geology has a strong visual component, helping students to "see" natural surroundings and then relate what they observe to dynamic processes has been a major concern of those who have helped to compile this book. For this kind of learning, the national parks offer a multitude of spectacular examples. Explanations of actual examples of rocks and landscapes in the parks are more vivid to beginners in geology than abstract discussions of geologic processes and earth materials.

The design of this book groups national parks in terms of basic geologic processes; that is, stream erosion and weathering, glacial action, igneous activity and volcanism, mountain-building, and ground water action. This approach provides a pattern for organizing data and concepts about the parks' landforms, rocks, structures and tectonics and relates geologic topics emphasized in individual parks to other parks with similar themes within each group. Using the chapters in another sequence is possible, however, because each chapter can stand alone and is not necessarily dependent on chapters that precede or follow. Thus an instructor—or any user of the book—has considerable latitude in the selection and arrangement of topics basic to geologic study.

Geology of National Parks presents basic elements of physical and historical geology in a nontraditional format. A background in the earth sciences is not required because terms and concepts are defined and explained as they are introduced. Within each chapter are discussions of significant geologic features and processes and a summary of the geologic history of the park, as well as information about the geographic setting and the local history. Also included for each park are a location map, illustrations, and a geologic column. The geologic columns are not stratigraphically comprehensive. They are designed, rather, to help readers grasp essential aspects of a park's geology; that is, the most important rock units and the relative times during which geologic events occurred. The references at the end of each chapter acknowledge the sources from which the chapter content was drawn and also provide suggestions for further reading by those who desire more detailed information about a particular park. With a few exceptions, the references given are not highly technical. Most of the titles are both accessible to and understandable by general readers.

As this third edition goes to press, there are 48 formally established national parks in the National Park System, 12 of them added since publication of the second edition. Five of the new parks are described in chapters in this book; namely, Badlands National Park, Theodore Roosevelt National Park, Denali National Park and Preserve (formerly Mount McKinley National Park,

now redesignated and more than doubled in size), Channel Islands National Park, and Biscayne National Park. The geology of the seven other new parks, all in Alaska, is discussed briefly in chapter 42 of Part VI. They could not be given more complete coverage in this edition because of constraints of time and space. One former park, Platt National Park, which was enlarged and redesignated Chickasaw National Recreation Area, has been omitted from this edition. The national seashores, many of the national monuments, and some of the national historical parks are of geologic interest but could not be included within the scope of this volume.

National parks such as Great Smoky Mountains, Grand Canyon, Yellowstone, and Yosemite attract a million or more visitors annually. Because of their remoteness or inaccessibility, other parks, such as Voyageurs, Channel Islands, Haleakala, and Gates of the Arctic, are less familiar to the general public and may not have been as thoroughly studied. But each park in the National Park System has its own beauty and uniqueness, and its own geologic story to be told. All are treasures of our national heritage. They are ours to learn about and appreciate.

Sherwood D. Tuttle

Preface

"You have taken on all the great problems of American geology!" a sympathetic friend remarked as we were working on this book. What our friend referred to was the fact that applications and implications of the still relatively new, composite theory of sea-floor spreading and plate tectonics have impelled geologists to rethink nearly all of the tectonic aspects of geologic history—including, naturally, the geologic histories of the national parks. Furthermore, a number of the parks happen to be in the midst of areas that geologists regard as "controversial."

Many nongeologists are unaware of the "revolution" that has taken place in the earth sciences due to a general acceptance of the unifying theory of plate tectonics. They might well ask, what does this have to do with national parks? After all, the rocks and the magnificent scenery haven't changed. That is true, but geology is a dynamic, speculative science. The "facts" may be the same; but as new insights are perceived, interpretations and explanations that once sufficed are no longer valid. The challenge in preparing the third edition of *Geology of National Parks* has been to provide scientific explanations for geologic phenomena that are current and accurate as well as being clear and understandable to readers who may not have a geologic background. Essentially what this has meant is that in this third edition, each chapter contains new material and has been completely revised.

We recognize the hazards of trying to synthesize incomplete and sometimes controversial geologic information, but we used what seemed to us to be the best explanation for each park in light of what is known at this time. Some may fault us for not including enough detail about a significant process or feature in what may be their favorite park. On the other hand, some readers may be overwhelmed by the large number of geologic terms and too much complexity. In all cases, when we have had doubts about what to include and what to omit, we have made the decisions on the basis of what we felt would help users of the book build on and expand their general knowledge and appreciation of the national parks and their geologic features.

In bringing this project to completion, we have had the help of numerous colleagues, several of whom authored chapters. These contributors are Rodney M. Feldmann (chapters 9 and 41), Ann Budd Foster (chapters 33, 34, and 36), Arthur N. Palmer (chapters 39 and 40), Donald F. Palmer (chapter 29), Lisa A. Rossbacher (chapter 26), and Sherwood D. Tuttle (chapters 24, 25, 32, 38, and 42). Introductions to Parts I through VI were also written by Sherwood D. Tuttle.

Other geologists who are knowledgeable about particular regions have guided us when we were uncertain, especially in regard to dating and interpreting large-scale tectonic events with respect to specific national parks. The assistance, also, of those who reviewed chapters or sections of this book in manuscript form was invaluable. They not only made comments, corrections, and criticisms that we could hardly have done without; they also suggested additional sources of information, sent us unpublished materials, and, in some instances, lent us their diagrams and photographs. We owe them all a great deal, and we appreciate their help. Their names are listed below:

Philip Bjork, Museum of Geology, South Dakota School of Mines and Technology
Robert S. Carmichael, University of Iowa
Robert R. Churchill, Middlebury College
R. David Dallmeyer, University of Georgia
Ann Budd Foster, University of Iowa
Charles T. Foster, University of Iowa
Michael O. Garcia, University of Hawaii
Robert E. Gernant, University of Wisconsin—Milwaukee
Brian F. Glenister, University of Iowa
Richard A. Hoppin, University of Iowa
Robert P. Larkin, University of Colorado
Gwendolyn W. Luttrell, U.S. Geological Survey
Walter L. Manger, University of Arkansas
Richard W. Ojakangas, University of Minnesota—Duluth
Willard H. Parsons, Wayne State University (emeritus)
Charles C. Plummer, University of California, Sacramento
Weldon W. Rau, Department of Natural Resources, State of Washington
John C. Reed, Jr., U.S. Geological Survey
Dallas D. Rhodes, Whittier College
Lisa A. Rossbacher, Whittier College
Holmes A. Semken, University of Iowa
Richard J. Stewart, University of Washington
Rowland W. Tabor, U.S. Geological Survey
Paul Tychsen, University of Wisconsin, Superior
Robert E. Wallace, U.S. Geological Survey
Bradford Washburn, Museum of Science, Boston

Instructors using this book as a text will, of course, draw on their own scientific interests and geologic experience to meet the special needs of their students. A number of instructors who have used the previous editions of *Geology of National Parks* have made suggestions for this revision, calling our attention to errors and pointing out significant items that were overlooked. We are indebted to them and hope their interest and helpfulness continues.

We cannot adequately express our gratitude for their assistance to National Park Service personnel and to those who staff the affiliated Natural History Associations. Invariably, superintendents, naturalists, rangers, interpretive specialists, technicians, and secretaries responded efficiently and helpfully to our requests. Those who work in the national parks are very special people. We admire their dedication to the philosophy and ideals of the National Park Service. As we ourselves have learned more about the irreplaceable wonders of the national parks, our own commitment to the preservation of America's parklands has been strengthened. We hope that users of this book will share this experience so that all of us, as citizens, will help to protect the national parks for future generations to enjoy.

Ann G. Harris
Esther Tuttle

PART I
Stream Erosion and Weathering

© by G. Shimer

Weathering and erosion over a prolonged period of time have produced the Grand Canyon of the Colorado River. Grand Canyon National Park, Arizona.

Spectacular cliffs, canyons, arroyos, pinnacles, and pedestals are some of the striking landforms in this group of national parks, all of which are in the drier regions of the United States, and all but two on the Colorado Plateaus. Despite the dry climate of these areas, running water has been the dominant agent of erosion in the shaping of the land surface. Weathering, which prepares rocks for erosion, operates wherever moisture is present; stream erosion—the most important external geologic process—sculptures the slopes and valleys. Resisting the *external,* erosive processes and influencing their activities are the various characteristics of the rocks and the underlying geologic structures. They compose the *internal* framework of the land. Because vegetation is generally sparse, the dynamic interactions between topography (the surface features) and the geologic framework (rocks and structures) can be examined more easily in these parks than in humid areas where forests tend to obscure the geology.

1

Grand Canyon National Park

Location: Northwest Arizona
Area: 1,218,375.24 acres; 1903.7 square miles
Established: February 26, 1919

Figure 1.1 Inner Gorge of the Grand Canyon, showing an unconformity (a break in the geologic record) between Precambrian rock units and Lower Paleozoic beds. The unconformity represents an erosional surface on which sediments were deposited by Cambrian seas. The narrow, V-shaped gorge is cut in ancient metamorphic and igneous rocks, which are resistant to erosion. Above the Inner Gorge, the canyon widens out in a series of slopes and cliffs carved in flat-lying sedimentary beds. National Park Service photograph.

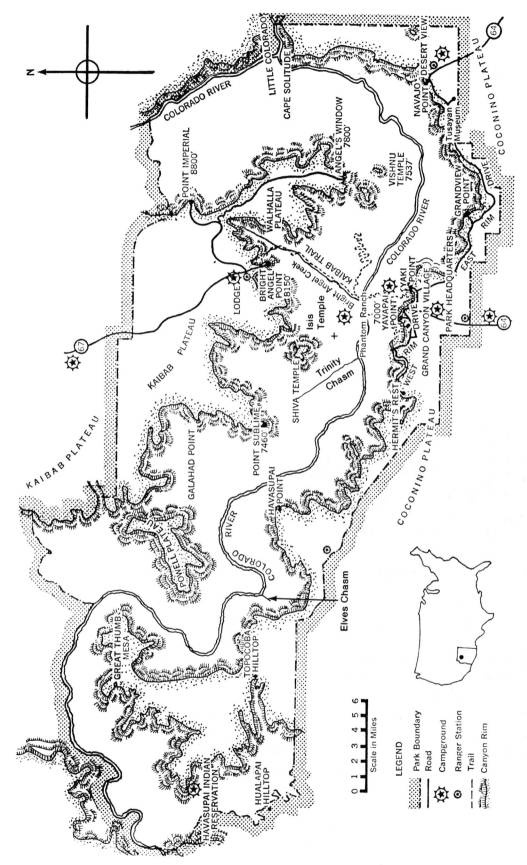

Figure 1.2 Grand Canyon National Park, Arizona.

3

Table 1.1. Geologic Column, Grand Canyon National Park

Time Units		Rock Units			Geologic Events
Era	Period	Super-group	Group	Formation	
Cenozoic				Slump and slide deposits, terrace gravels, river deposits, travertine; lavas	Development of Grand Canyon by uplifting and tilting, accompanied by downcutting. Minor lava flows
Mesozoic	Cretaceous				Uplift, minor deformation, erosion.
	Jurassic				
	Triassic			Moenkopi	Flood-plain deposition Withdrawal of seas; erosion
Paleozoic	Permian			Kaibab Limestone	Transgression of shallow seas; regression and erosion.
				Toroweap	Transgression of shallow seas, flattening sand dunes
				Coconino Sandstone	Flood-plain deposits covered by migrating sand dunes
				Hermit Shale	Deposition on flood plain Erosion
	Pennsylvanian		Supai	Esplanade Sandstone	Flood-plain deposition Retreat of seas Marine sedimentation Erosion
				Wescogame	
				Manakacha	
				Watahomigi	
	Mississippian			Redwall Limestone	Deposition in a warm shallow sea; uplift; formation of karst topography. Erosion
	Devonian			Temple Butte Limestone	Deposition of sediments in sinkholes and channels. Erosion
	Cambrian		Tonto	Muav Limestone	Deposition in a transgressing sea, forming a facies change between these formations.
				Bright Angel Shale	
				Tapeats Sandstone	
Late and middle Precambrian time (Proterozoic)		Grand Canyon	Chuar	Sixty Mile	Stream deposition
				Kwagunt	Volcanic activity; erosion
				Galeros	Grand Canyon orogeny, block faulting, thrust faulting, folding
				Nankoweap	Deposition in a shallow sea. Erosion.
			Unkar	Cardenas Lava	Faulting, igenous intrusion, lava flows, erosion
				Dox Sandstone	Return of a shallow sea
				Shinumo Quartzite	Stream deposition and delta deposits
				Hakatai Shale	Temporary retreat of the sea.
				Bass Limestone (including Hotauta Conglomerate)	Deposition in warm, shallow sea Transgression of seas Major unconformity
Early Precambrian time (Archean)				Zoroaster Plutonic Complex	Intrusion of batholith. Erosion of Mazatzal Mountains.
				Trinity and Elves Chasm Gneisses	Uplift of Mazatzal Mountains Mazatzal orogeny, folding, faulting, intrusions, metamorphism
			Vishnu	(formerly called Vishnu and Brahma Schists)	Deposition of marine sediment, volcanic activity, additional deposition.

Source: modified after Huntoon et al. 1976; Billingsley and Breed 1980.

Silurian Ordovician missing (handwritten annotation, left margin near Devonian/Cambrian)

Local History

The early inhabitants of the Southwest had legends to explain the origin of the Grand Canyon. The Indians' stories have a certain logic and show an awareness of natural causes and effects, even though these processes were not fully understood by the Indians, of course. The legends also indicate that the Indians were as much in awe of the Grand Canyon as were the white men who arrived much later.

The Navajos believed that continuous rain for many days and nights brought about a great flood that covered the land with water that rose higher and higher. Finally an outlet formed, and as the rushing waters drained away, the Grand Canyon was cut deeply into the earth. According to the legend, the Navajos survived because they were temporarily changed into fish.

Another legend tells of a great chief who could not cease from mourning the death of his beloved wife. Finally the gods offered to take him to visit his wife so that he could see she was contented in the happy hunting ground. In exchange, he was to stop grieving when he returned to the land of the living. When the chief had promised this, the gods made a trail through the mountains, creating the Grand Canyon. After the chief returned, the gods sent a river to conceal the trail forever so that no one else could use it. This is how the Colorado River was formed.

Nomadic hunters are believed to have been the first inhabitants of the Grand Canyon region between 3,000 and 10,000 years ago. Twigs shaped into forms resembling animals have been found in caves located in Grand Canyon National Park; and it is believed that they were left by the early hunters (fig. 1.3). A more advanced culture, "the basket makers," followed the hunters; and in turn, it was succeeded by the Pueblos, the builders of "apartment houses." More than 500 Indian sites in the park indicate that as recently as 600 years ago, large numbers of Indians lived in the area.

Three tribes still call the Grand Canyon region their home. They are the Navajo and Hopi, who live east of the park, and the Havasupai, who live in a valley watered by Havasu Creek in the western section of the park. (The name *Havasupai* means "people of the blue-green water.")

In 1540, a band of Coronado's conquistadores, led by Don Lopez de Cardenas, were the first white men to see the Grand Canyon. After being directed by Hopi Indians to the Gorge, they tried for three days to descend to the Colorado River but managed to get only a third of the way down. Discouraged, they gave up; they climbed back to the rim and left.

More than 300 years later, Lt. Joseph C. Ives, a U.S. Army surveyor and his party managed to reach Vegas Wash in the lower Grand Canyon, southeast of the

Figure 1.3   Natural caves and a flowing spring (to the left) in Redwall Limestone in the Grand Canyon. Ground water (underground water seeping downward through the rock layers) dissolved some of the soluble limestone, forming openings, or caves, in the rock. In prehistoric time, humans, as well as animals, found shelter in some of these caves. From *Geology of Arizona* by D. Nations and E. Stump. © 1981 Kendall/Hunt Publishing Company.

Shivwits Plateau, and then crossed over to Diamond Creek. From this point, after an unsuccessful attempt to reach the rim, they concluded that much of the Grand Canyon of the Colorado River would never be visited.

John S. Newberry, a member of the Ives party and a geologist, thought otherwise. He convinced Major John Wesley Powell, a fellow geologist who had lost an arm in the Civil War, that an expedition by boat down the Colorado River through the Grand Canyon would be worth the risk and that a survey could then be accomplished. With scientific help from the Smithsonian Institution, the Powell expedition of four boats and nine men left Green River, Wyoming, in May 1869, and began their hazardous journey. One boat was smashed; scientific instruments and food were lost. Most of the time the party were wet, tired, hungry, discouraged, and convinced they would never get out of the canyon alive. Three of the men climbed out of the canyon and started cross-country, but they were killed on Shivwits Plateau by hostile Indians who mistook them for hunters who had killed some squaws. The men who stayed with Powell survived. Two years later Powell went down the river again in order to collect more data for his geographical and geological report on the Colorado River region.

Theodore Roosevelt, after a trip to the Grand Canyon in 1903, was determined that it should be preserved. Unfortunately, some people had other plans for the area and they blocked establishment of a national park until after World War I (although Roosevelt was able to have part of the area placed under government protection in 1908). Some park opponents wanted to mine the minerals; another group wanted to charge a toll on Bright Angel Trail and make cable cars available for descents into the canyon. Even today, some people advocate damming the Colorado River so that as the water level in the park rises, tourists will be able to take easy boat trips to isolated spots which are now inaccessible.

Geologic Features

As it passes through the Grand Canyon, the Colorado River (which has often been described as "too thick to drink but too thin to walk on") carries about a half million tons of sediment past any one point each day. The rate of erosion is not constant throughout the course of the river, but it is estimated that about every thousand years the drainage basin of the Colorado River system is lowered by an average of 6 1/2 inches. It was his observation of the sediment load that convinced Powell he would not be washed over high falls (as some people had predicted) on his initial expedition. He reasoned correctly that a river carrying that much silt would have scoured its bed down to a more or less even grade, or slope.

The Grand Canyon is about a mile deep, with the distance across from rim to rim ranging from 9 to 18 miles. The elevation of the North Rim (8,900 feet) averages some 1,200 feet higher than the South Rim (6,900 feet). Two hundred eighty miles of the total length of 1,450 miles of the Colorado River flows through the park.

Dimensions of such magnitude have biological as well as geological significance. Life forms on the two canyon rims have evolved differently because of their physical separation over time. At the higher altitude of the North Rim, under conditions of cooler average temperatures and somewhat greater annual precipitation, plants and animals characteristic of the Canadian and sub-Arctic Hudson zone of life became established. Trees and shrubs such as blue spruce, pine, fir, and aspen, that are rather dark in color, took over. A typical animal here is the Kaibab squirrel, which has a black body, black stomach, and white tail. On the South Rim, vegetation that is lighter in color and typical of the Upper Sonoran life zone (junipers, piñon, etc.) predominates. Here is found the Abert squirrel, with its grizzled gray body, white stomach, and gray tail. The common ancestor of these two modern-day species was a black and white squirrel with tufted ears that lived throughout the whole area before the Grand Canyon was

eroded. When the squirrels were no longer able to migrate back and forth across the physical barrier of the canyon, and as the differences in life zones on the North Rim and South Rim became pronounced, the new species developed due to evolutionary adaptation—the darker Kaibab squirrel on the North Rim and the lighter Abert on the South Rim.

Around 9 million years ago, uplift of the Colorado Plateau began, and this caused the Colorado River to start carving the Grand Canyon. Although mountain-building took place to the east and west, the Plateau itself remained relatively undeformed as it was gradually raised, and the Colorado River eroded and cut down about as fast as the area was being uplifted. Exposed by this downcutting is one of the most nearly complete geologic columns on earth, encompassing some two billion years of geologic history. The layers of sedimentary rock seen in the Canyon are now, as when they were deposited, essentially flat-lying; and many of these horizontal beds extend throughout most of the Colorado Plateau.

Geologists have theorized that the ancestral Upper Colorado flowed along Marble Canyon and turned south along the edge of the Kaibab uplift (where the Little Colorado River flows today), finally terminating in a basin in the mountains. At the same time, on the other side of the Kaibab uplift, the Hualapai drainage system drained to the northwest. Both were rivers with stream valleys that were not especially deep.

The region underwent uneven but relatively rapid uplifting, and by the time the movement ceased, a major change had occurred. The headwaters of the Hualapai system had cut back into and through the Kaibab uplift, thus capturing the headwaters of the ancestral Colorado River forcing it to swing across the uplift and flow somewhat westerly. The Little Colorado, which had drained to the south, changed its direction and drained into the Colorado (fig. 1.2).

An alternate theory (but one not widely accepted) suggests the possibility that at one time the Colorado River drained toward the east. The Kaibab Plateau acted as a divide, preventing the river from draining to the west and into the Pacific. Uplift in the Arizona-New Mexico border region dammed the river so that it formed a large lake. Eventually, continued cutting back by streams breached the Kaibab divide and began to drain the lake to the west. This was when the cutting of the canyon began.

Igneous Features Associated with the Geology of the Grand Canyon—Some Definitions and Explanations

Igneous rocks are formed from solidification of magma: *magma* is molten rock below the earth's surface. A *magma chamber* (the space occupied by a body of molten rock) usually forms under conditions of high temperature and pressure 35 to 40 miles down within the earth's crust.

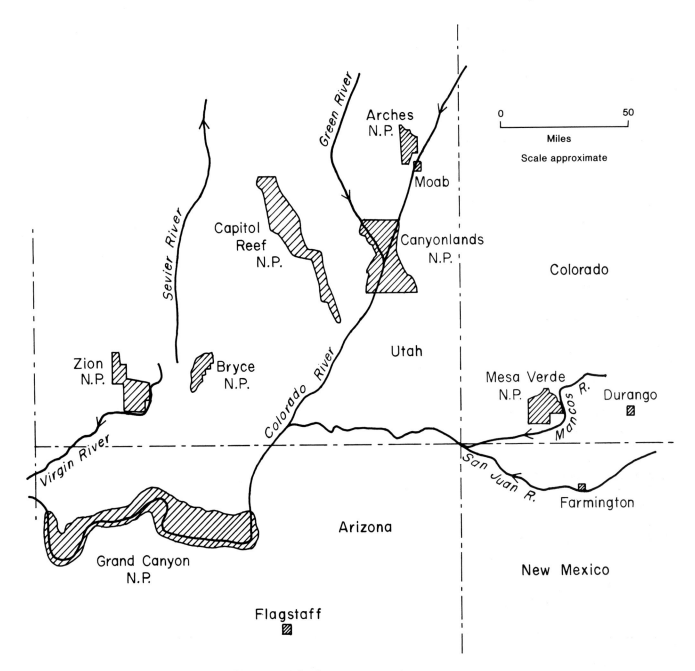

Figure 1.4 Location on the Colorado Plateaus of the first seven national parks described in Part I. The Colorado Plateaus, a large physiographic province, is about 130,000 square miles in area. Distinctive characteristics of the region are (1) flat-lying sedimentary rocks; (2) relatively high elevations, mostly between 5,000 and 11,000 feet; (3) remarkable canyons dissecting the plateau surfaces; (4) retreating escarpments produced by the erosion of rock layers of varying resistance; and (5) semiarid and arid climatic conditions.

Box 1.1
The Colorado Plateaus—an Unsolved Puzzle

Why are the rock units exposed in the Colorado Plateaus—except for the Precambrian rocks in the bottom of the Grand Canyon—essentially flat-lying, when sedimentary layers of the same age and type all around the Colorado Plateaus have been repeatedly folded and faulted, forming great mountain ranges? Does plate tectonic theory have an explanation? (The essential idea of *plate tectonic theory* is that large, thick crustal plates make up the earth's surface and that as these plates move slowly over the face of the globe, intense geologic activity occurs along plate margins.)

The spectacular scenery of western North America is largely the result of tectonic activity—plus erosion—that occurred during relatively recent geologic time; that is, within the Cenozoic Era, or roughly the last 60 million years. By the time the Mesozoic Era came to a close, the major structural features of the North American Cordillera (the mountain belts of western North America) had been formed by the Rocky Mountain orogeny. This vigorous mountain-building was followed by long periods of uplift and erosion. The dominance of uplift as the mountains were worn down has meant that continents have remained generally high since the Mesozoic.

The uplift that has characterized the Cenozoic was preceded by a long, interesting, but complicated sequence of events. As the Paleozoic landmass of Pangaea broke up (beginning in the Triassic Period), with the various fragments drifting in different directions, the large plate carrying what is now North America moved westward in response to sea-floor spreading. The plates diverged from a rift that opened and eventually became the Atlantic Ocean (fig. 1.5B). Meanwhile, as the Mesozoic Era went on, an upwarp of considerable length (the East Pacific Rise) developed in the Pacific, moving a large oceanic plate eastward.

As the plates approached each other, a trench formed parallel to the western margin of the North American plate and the oceanic plate was subducted, with some of its material being scraped off the top, thus adding to the size of the continent. (The Coast Ranges of California are thought to be made up largely of material from the subducting plate.) With pressure at the plate boundary continuing, the North American plate was forced to override the Pacific plate (fig. 1.5C). This overriding is thought to account for the orogenic activity of the Mesozoic that resulted in the building of the North American Cordillera.

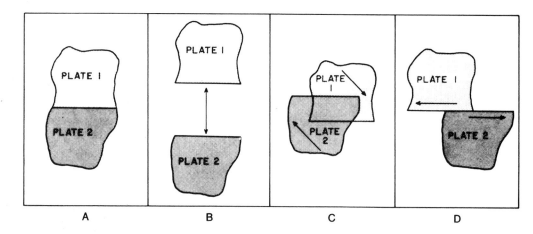

Figure 1.5   Schematic representation of possible types of boundary zones along the margins of tectonic plates. *A.* Plates in contact. *B.* plates moving apart, or *diverging*, from a spreading center. *C.* One plate overriding another as they come together, or *converge. D.* Plates sliding past each other. From *Physical Geology Text and Manual* (2nd ed.) by R. D. Dallmeyer. © 1978 Kendall/Hunt Publishing Company.

Subduction apparently ceased when the North American plate reached the flank of the East Pacific Rise. Compressive forces gave way to shearing stress at the plate boundary as the Pacific plate began to move more to the north so that the two converging plates were sliding by each other. This deduction has been made from observations of California's San Andreas fault, which is thought to be the present location of the plate boundary (fig. 1.5D).

As the compressive forces causing subduction ceased, the plates could begin to sink or rise in order to attain equilibrium. Thus isostatic activity could be the explanation of the uplifting and block-faulting in the Cordillera that has been the pattern during the Cenozoic.

This brings us back to the puzzle we started with. How have the Colorado Plateaus remained virtually undeformed as a sort of stable shelf, moving up and down like an elevator platform, when on all sides similar sedimentary beds have been folded and mangled, faulted and overthrust, and built and rebuilt into great mountains? We can say that the Colorado Plateau acted as a sort of buttress, with its perimeter subjected to the folding and faulting that produced highlands and mountains all around. Why this is so has not been worked out with any certainty as yet.

We do know that deep faults cracked the Colorado Plateaus during uplift, permitting volcanic materials to rise to the surface. The San Francisco Mountains, resting on the level plateau just south of the Grand Canyon, were formed in this way. Upwarping or arching and some normal faulting also occurred during uplift; but in comparison to the troughs and heights of blocks in the Basin and Range Province to the west, displacement of faults on the Plateau has been minor.

Some geologists who have studied heat flow data on the North American plate have suggested that the high heat values in the Basin and Range Province, indicating an area of heat deep in the earth, may be the result of a plume, or hot spot, rising from a convection current in the mantle. This could account for the recent volcanism and earthquake activity along the Colorado Plateau's perimeter, and may foretell the eventual reopening of an ocean basin in this region. If this should happen (in another million years or so), it seems probable that the Colorado Plateau will again escape being folded or compressed.

In subsequent chapters dealing with national parks where the relationships between geologic structure and plate tectonic theory are more clear cut, we take a closer look at plate movements and the accompanying activities.

When magma forces its way upward into the confining bedrock through any cracks or fissures it can find, it is called an *intrusion*. Hence, *intrusive rocks* are igneous rocks that have hardened from magma forced up into surrounding rock, typically far below the surface. A rock body that is the result of an igneous intrusion, regardless of its shape or size, is termed a *pluton* (fig. 1.6).

The intrusive igneous rocks exposed in the bottom of the Grand Canyon are among the oldest on this continent. They were originally solidified many miles below the surface but are now exposed in a *batholith* which is a large intrusive body (or pluton) that crops out over an area of at least 40 square miles. A *stock* is a pluton smaller than a batholith (i.e., cropping out in an area less than 40 square miles). A pluton that is tabular in shape is either a dike or a sill. A *dike* is a tabular intrusive structure that is *discordant*; i.e., it cuts *across* the surrounding rock, or "country rock." A *sill* is a *concordant* tabular intrusive structure that has spread into country rock between or parallel to bedding planes (fig. 1.6).

Extrusive Igneous Rocks. Magma that reaches the surface, either through fissures or volcanic vents is called *lava.* Any igneous rock that has solidified at the earth's surface, whether from cooling of molten lava or from

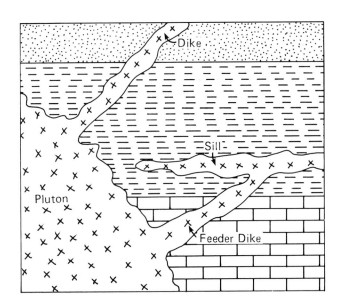

Figure 1.6 Igneous structures shown schematically in relation to country rock (the pre-existing rock) that is being intruded by magma far below the surface. Plutons, which crystallize deep underground, tend to have irregular shapes. Sills are igneous bodies that were intruded parallel to the bedding of the country rock. Dikes cut across the bedding.

fragments of volcanic explosions is classified as *extrusive rock*. Particles of dust and ash that have been explosively ejected tend to pack together, particularly when covered by subsequent layers of ash or solidified lava; and the extrusive rock resulting from this is called *tuff*. Usually lava flows out over a land surface, but sometimes it comes out through an underwater fissure forming *pillow lava*, which, when solidified, resembles a series of flour sacks.

Geologic Structures Associated with Grand Canyon Rocks

Layered sedimentary rock units, such as those exposed in the canyon walls above the igneous and metamorphic rocks of the Inner Gorge, also have interesting implications about geologic conditions in the past. The presence of marine fossils leads to the obvious conclusion that these rock layers have been uplifted more than a mile from their original position beneath an ancient sea. Uplift must have been uniform over a large area to keep them so nearly horizontal.

What happens to layered bedrock when the uplifting forces are not uniform? Keep in mind that most tectonic activity takes place at considerable depth when rocks are subjected to great heat and pressure. Under such conditions, layered rock that would fracture or break in a surface environment may become *plastic*, or malleable, and bend into *folds* rather than break. The sides, or arms, of a fold are called *limbs*. Most folds are either *anticlines* (upfolds) or *synclines* (downfolds) (fig. 1.7,B,C). If the limbs are parallel (as with a hairpin), the fold is called an *isocline* or *isoclinal fold*, and we know it must have been formed by intensely compressive forces (i.e., squeezed together). Folds of this type are found in the Precambrian metamorphic rocks in the lowest parts of the Grand Canyon.

A type of fold much more characteristic of the Colorado Plateau is the monocline, which is usually found in otherwise flat or slightly tilted beds. *Monoclines*, or *monoclinal folds*, have one limb and a gentle bend, or flexure, connecting rock layers at one level (elevation) with the same layers at another level (fig. 1.7A).

A *fault* is the result of *movement* along a break or fracture in bedrock, causing *displacement*. Faults are named and identified according to the type of movement or displacement that occurred. Faulting may or may not be related to the topography of an area. In some regions the surface features give no indication of the geologic structure of the bedrock below; but in the Basin and Range Province of the southwestern United States, for example, great mountains and valleys display evidence of faulting on a grand scale.

These faults (fig. 1.8E) separate a series of blocks that have been either raised or dropped, mainly as a result of vertical forces. The uplifted blocks are called *horsts*; the

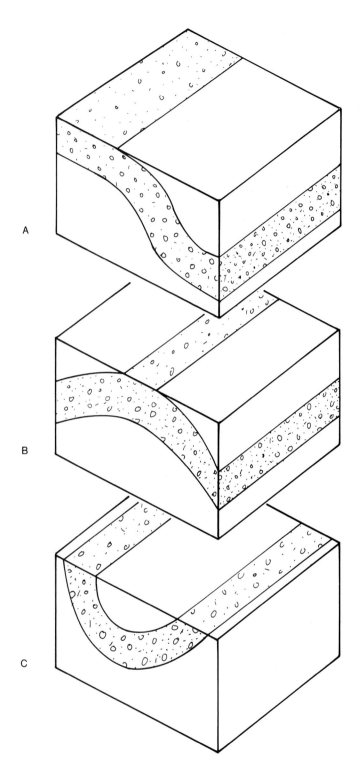

Figure 1.7 Block diagrams of simple folds.
A. Monocline. Flat-lying beds (left) are folded downward (middle) and become flat-lying again (right). **B. Anticline.** An upfold, opening downward, with older beds in the center of the fold. **C. Syncline.** A downfold opening upward, with younger beds in the center of the fold.

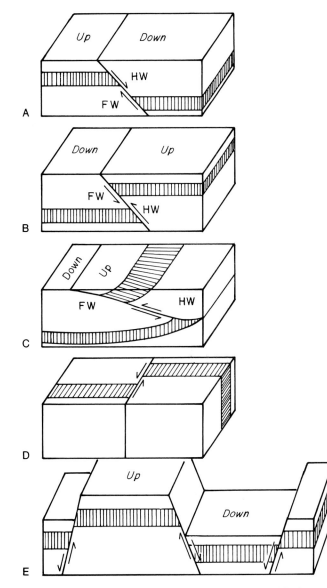

Figure 1.8 Block diagrams of common types of faults and a horst-and-graben structure bounded by faults. Stresses that bring about fault movement may cause one block to move up, down, obliquely, horizontally, or even to rotate, in relation to another block. **A. Normal fault.** The hanging wall side of the fault has moved *down* in relation to the footwall side. **B. Reverse fault.** The hanging wall side of the fault has moved *up* in relation to the footwall side. **C. Thrust fault.** A type of reverse fault in which the fault plane dips at a low angle. **D. Strike-slip fault.** One side of the fault has moved horizontally by the other side. **E. Horst and graben structure.** The upfaulted block, bounded by normal faults, is a horst. The downfaulted block, also bounded by normal faults, is a graben. The surface of this block diagram represents topographic features resulting from faulting. Topography is not shown on the other block diagrams because a land surface may or may not reflect the presence of a fault (or faults) in the underlying structures.

downdropped blocks, *grabens*. Faults of this type have also been found in the Inner Gorge of the Grand Canyon. Here, of course, they indicate block-faulting during a much earlier geologic period and are unrelated to the present surface features.

Faults involve relative movement between two blocks, one of which is called the footwall and the other the hanging wall. If a person were able to stand on a fault surface within the earth (as if in a mine), his feet would be supported by the *footwall*, while the *hanging wall* would be the rock surface over his/her head.

In a *normal fault*, the hanging wall has moved *down* in relation to the footwall (fig. 1.8A). If the hanging wall has moved *up* in relation to the footwall, it is called a *reverse fault* (fig. 1.8B). A special type of reverse fault, in which the fault plane has a low angle (about 10° or less), is referred to as a *thrust fault* (fig. 1.8C).

Note that the faults described above indicate largely vertical movement. When two blocks slide past each other, resulting in lateral displacement, the fault is called a *strike-slip fault* (fig. 1.8D).

What Causes Gaps in the Sedimentary Rock Record?

The marine sedimentary rocks making up the Grand Canyon geologic column cover a fairly extensive portion of geologic history, but some periods are not represented. Were the marine beds of the missing geologic periods completely eroded away? Or did marine sedimentation not occur during those periods? Reconstructing the sequence of events can be quite complex.

We know that ocean levels have fluctuated throughout geologic time. During glacial episodes, for example, so much water is frozen in the great continental ice sheets that world-wide sea level may be lowered considerably. In time, warming climates melt the glaciers and sea level rises again. Moreover, because of large-scale tectonic activity, continents have stood at times higher and at times lower during geologic history. The resulting changes in sea level have occurred very gradually over millions of years of time.

When sea level rises, the oceans *transgress* onto the land; and when sea level falls, they *regress* from the land. During the long slow process of transgression or regression, streams carrying sediment down to the ocean continue to erode. If the ocean is regressing, the streams may strip and redeposit (at lower elevations) most or all of the sediment layers that were put down during an earlier period of transgression. Then when the seas finally come back and again cover an area that has been stripped of sediment, a gap occurs in the geologic record. Such a gap is called an *unconformity*, and the rock above it is always considerably younger than the underlying rock. Geologists find unconformities to be useful indicators when they are trying to reconstruct ancient changes in elevation leading to either erosion or deposition. Typically an unconformity can be recognized by the presence of a *basal conglomerate* (resting on the old erosion surface) made up of fragments of old eroded sediment incorporated with younger sediment deposited by a transgressing sea.

As a shoreline shifts landward or seaward (depending on whether the sea is transgressing or regressing), beach deposits accumulate in this narrow depositional environment. Eventually, a *time-transgressive rock unit* (e.g., the Tapeats Sandstone), with the rocks becoming younger in the direction in which the sea was moving, may show the shifting location of the old shoreline. The principle that rock units or unconformities may vary in age from place to place within an area of deposition (or erosion) is called *time (temporal) transgression.*

Unconformities, or breaks in a sequence of deposition, can be produced in several ways. A *nonconformity* is a type of unconformity that developed between sedimentary rocks and older plutonic or metamorphic rocks that had been exposed to erosion before the overlying sediments covered them. A *disconformity* is a surface that indicates missing rock layers, or strata, but is parallel to beds above and below that surface. In other words, a definite time gap exists in the sequence (fig. 1.9A, B).

In an *angular unconformity* the bedding planes of the rock units above and below the erosion surface are *not* parallel; specifically, the younger overlying sediments rest upon the eroded surface of tilted or folded older rocks (fig. 1.9C).

Base Level

Base level is a theoretical limiting surface below which a stream (or a network of streams) cannot erode. When large areas of land have been reduced nearly to base level, an *erosional surface* of low relief is produced. Here and there, knobs of resistant bedrock remain as isolated hills on the erosional surface. If such an erosional surface is later covered by sediment deposited by a transgressing sea, the former plain appears in the rock record as a relatively smooth unconformity.

Stream Erosion

A stream whose *gradient*, or slope, is low or nearly flat does not have much erosive power. But if an area is uplifted, thus lowering the base level and increasing the gradient, then a stream begins to downcut. Sometimes during uplift, a stream maintains its original course but cuts a narrow trench in the plain (or upland). While continuing to cut down through bedrock, an entrenching stream may encounter different geologic structures or rock units of varying resistance and yet still maintain its course. The stream would then be described as *superimposed.* On the other hand, if a structure develops (such as an arch)

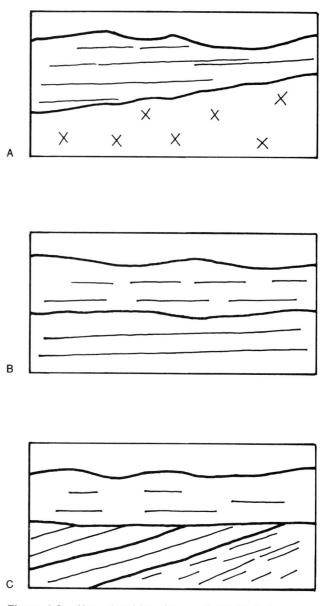

Figure 1.9 Unconformities shown schematically in cross section. **A. Nonconformity.** The contact between older igneous rocks (below) and the younger sedimentary rocks (above) is an erosional surface that represents a break in the geologic record. When igneous rock has intruded sedimentary rock, altering the surrounding rocks by heat and pressure, no unconformity exists because there is not a gap in the geologic record. **B. Disconformity.** When an erosional surface forms the contact between parallel units of sedimentary rocks, this shows that some units have been eroded or were never deposited in the particular location. Thus a disconformity is an erosional unconformity. **C. Angular unconformity.** The older beds below the unconformity were deformed (folded) and subsequently eroded before younger beds were deposited on the erosional surface. Beds above and below the unconformity are not parallel.

while the stream is downcutting and the stream continues to incise at approximately the same rate as the land is rising, then the stream is described as *antecedent* because it existed prior to the present topography.

Earlier in this chapter, the Colorado River is described as downcutting and eroding as the Colorado Plateau was being uplifted. As the main river was entrenching itself, its tributaries lengthened themselves by *headward erosion*, which means that they cut back into the upland at the head of each valley. Meanwhile (according to one theory of the Grand Canyon's development), the tributaries of the neighboring Hualapai drainage system, which were apparently more favorably situated, were able to erode headward at a faster rate until they intersected the headwaters of the Colorado and, as a result, captured the Colorado drainage system. This process is called *stream capture*, and the stream that loses its headwaters is called a *beheaded stream*.

Mass Wasting

Downcutting deepened the Grand Canyon, but it was the important process of *mass wasting* that widened the canyon, created the "steps," and separated the rims. *Mass wasting* involves the various types of movement (from very slow to very fast) by which rock debris and soil, under the influence of gravity, go downslope. As material is loosened and moved downward by gravity, more and more of it is picked up by running water and transported by streams. Thus mass wasting, in conjunction with stream flow, is probably the most important element in landscape development.

Looking for Clues about Ancient Environments in Sedimentary Rocks

Most (but not all) of the common sedimentary rocks are made up of sediment that has been lithified (changed into rock). *Sediment* is a collective term for loose solid particles that may be (1) weathered or eroded from pre-existing rocks, (2) precipitated from solution, or (3) secreted by organisms. Therefore, studying the size, type, arrangement, etc., of sediments can often reveal a great deal about the origin of sedimentary rocks and the environmental conditions that prevailed at the time the sediment was deposited.

For example, a geologist may be able to tell what type of weathering produced the sediment that makes up a sedimentary rock. (*Weathering* refers to destructive processes that change or break down rock at or near the earth's surface.) Physical weathering, or *disintegration* (also called mechanical weathering) involves the breaking up of rock into smaller pieces by frost action, abrasion, etc. Chemical weathering, or *decomposition*, refers to the changes in rock resulting from exposure to water and atmospheric gases. Usually these weathering processes work

in conjunction so that rock is altered both mechanically and chemically.

In the Grand Canyon many interesting examples of weathering processes can be seen. *Brecciated* deposits of rock that has been broken up and crushed into angular fragments are found where a cave ceiling has fallen in or in fault zones where movement along fault surfaces has crushed rock. Rock broken up in this way weathers more rapidly because more surfaces are exposed to water and the atmosphere. Tributaries of the Colorado River have tended to form along fault zones where the rock is more susceptible to weathering and erosion, an example being Bright Angel Canyon.

Sandstones may be formed from old dunes, such as the *eolian* (wind-blown) deposits that make up the Coconino Sandstone. Other sandstones were formed from water-laid deposits, either in fresh-water lakes or in marine environments. The degree of sorting and rounding of sand grains, the presence or absence of fossils, the type of bedding, and the kind of cementing (or matrix) holding the grains together can all provide clues as to the geologic history of an area.

Geologic History

1. Deposition of the sediment that formed the Vishnu Schist early in Precambrian time.

During the Precambrian, more than 2,000 million years ago, sands, silts, and muds were deposited in a shallow marine basin, along with debris from nearby active volcanoes. Igneous intrusions that were mafic in chemical composition forced their way into these beds, and some lava flows may have poured out into the ancient ocean. The basin sank under the weight of accumulating material as more and more sediment and volcanic fragments built up great thicknesses of sedimentary rock. Evidence of life (i.e., fossils) has not been found in these marine beds. Later tectonic activity converted the layers of sedimentary rock into the metamorphic Vishnu Schist.

2. Volcanic activity and the formation of the Brahma Schist.

The pillow structure of many of the lava flows indicates that they were extruded from fissures in the ocean floor (rather then issuing from a central volcanic vent on land). Since the fissures were extensive, great thicknesses accumulated. At present the 15,000 feet of lava flows are tilted almost to vertical. The presence of interbedded tuff implies that some of the lava flows erupted on the land surface, evidently during times of ocean regression. Metamorphism of the lava and tuff formed the Brahma Schist, which is a mixture of quartzite and a quartz mica schist. Together, the Vishnu Schist and the Brahma Schist form the Vishnu Group.

Figure 1.10   This drawing of the north wall of the Grand Canyon originally appeared in John Wesley Powell's 1875 report of his explorations of the Colorado River. At the bottom, next to the river, are the ancient crystalline rocks of the Vishnu Group. The massive, vertical spires, portrayed so well by the artist, were produced by the weathering and erosion of these resistant rocks. In the middle of the view, between two angular unconformities, are the dipping beds of the Unkar Group, deposited later in Precambrian time. Above the higher unconformity, erosion of Paleozoic rock units has formed stairstep topography (box 1.2). Because of the steepness of the slopes in this part of the Grand Canyon, only the early Paleozoic formations are visible from the river level. Reproduced from *The Colorado River Region, John Wesley Powell.* 1969. U.S. Geological Survey Professional Paper 669.

3. The formation of the Mazatzal Mountains by folding, faulting, intrusion, and metamorphism.

Between 1,500 and 2,000 million years ago the members of the Vishnu Group were folded into isoclinal folds. Intrusion of granites into these beds metamorphosed the sands into quartzite, and the shales and mudstones into schists. The heat, pressure, and fluids of contact metamorphism formed garnets, staurolite, and sillimanite schists in the Vishnu along with tourmaline, epidote, and hornblende schists in the Brahma. The accompanying orogeny built the Mazatzal Mountains.

4. Faulting, vein-filling, and additional faulting.

During folding of the rock layers, a series of intersecting faults developed, trending northwest-southeast and northeast-southwest. Quartz veins, fed by hydrothermal (hot water) solutions that were squeezed out of the intruding granite, worked their way up into the fractures and joints opened by the faulting. In Trinity Chasm near Isis Temple, and Elves Chasm, near Explorer Monument, these uplifted beds and veins can be seen as light-colored felsic gneisses containing appreciable amounts of quartz, feldspar, and micas.

The Zoroaster Plutonic Complex, the granite that intruded the faulted beds, slowly crystallized, forming a batholith. Faulting continued after the intrusion took place.

5. The erosion of the Mazatzal Mountains and the formation of the Mazatzal erosional surface.

Even as the Mazatzal Mountains were being raised up, they were subjected to the processes of weathering and erosion; and as uplift ceased, erosion proceeded at a fairly rapid rate until the mountains were worn down. Eventually a broad, relatively featureless erosional surface (peneplain) developed where the mountains once stood. Apparently the region was arid during the last stages of erosion.

6. Transgression of the seas and deposition of the Unkar Group 1,000 million years ago. (figs 1.10, 1.11).

From the west a sea gradually transgressed onto the land. As the sea advanced, waves worked over and smoothed out the old land surface with larger fragments accumulating as gravel in some areas, creating a basal conglomerate known here as the Hotauta. The Hotauta Member of the Bass Limestone (which marks the beginning of the Unkar Group) consists of igneous and metamorphic fragments derived from the weathering of the older rocks. The Bass Limestone, with its basal conglomerate indicating the unconformity, is subdivided into five units. Ranging from 120 to 340 feet thick, the formation was deposited in a shallow sea as a mixture of sandstone and shale. Since much of the calcium in the limestone has been replaced by magnesium, the rock should be called dolomite rather than limestone. (The variation in lithology is the basis for dividing the Bass Limestone into several units.) In the arid climate of the Colorado Plateau, the Bass Limestone is resistant to weathering and is, therefore, a cliff-former.

The Hakatai Shale, the next younger member of the Unkar Group, is also divided into several units since it is a mixture of thin-bedded shales, mudstones that have no bedding (or very thick bedding), and sandstones. In the beds are some depressions that may represent salt crystal molds. Since most of the shale, ranging in thickness from 580 to 830 feet, is nonmarine, the formation represents a temporary retreat of the sea. The Hakatai Shale beds are very susceptible to erosion, so this rock is a slope-former in the Grand Canyon walls.

The Shinumo Quartzite, originally a marine sandstone that later was metamorphosed, indicates the return of the sea to the region. The sand must have been deposited in shallow water because the cross-bedding, which can be plainly seen, would have been the result of wave action. The resistant Shinumo beds, 1,100 to 1,560 feet thick, form impressive cliffs.

The Dox Sandstone, also subdivided into several units, is mostly sandstone but contains some shale layers. This member, which has a thickness of 3,000 feet, was also deposited in a shallow sea. The rock has a tendency to be a cliff-former with steep slopes.

7. Renewed faulting and volcanic activity.

Movement recurred along the intersecting fault system. Later on flows of the Cardenas Lava pushed up through joints and fractures associated with the faults.

Part of the Cardenas Lava intruded any available openings or zones of weakness in the rock, such as bedding planes and fault systems. The dikes and sills that formed had largely diabasic or ophitic texture (lath-shaped crystals in the basalt). The remaining Cardenas unit is made up of basaltic lava flows up to 1,000 feet thick, which erupted from fissures. Cardenas Lava is the youngest formation of the Unkar Group.

8. Deposition of the Nankoweap Formation and the Chuar Group.

Once again the region was covered by a shallow sea, in which was deposited approximately 300 feet of Nankoweap beds lying unconformably over the eroded surface of the Cardenas Lava.

Also deposited in a shallow sea, but accumulating to a thickness of 5,000 feet, was the Chuar Group, consisting of three formations—Galeros, Kwagunt, and Sixty Mile, from oldest to youngest. These formations are in turn divided into several members.

The Precambrian units of the Unkar Group and the Chuar Group, make up the Grand Canyon Supergroup (table 1.1).

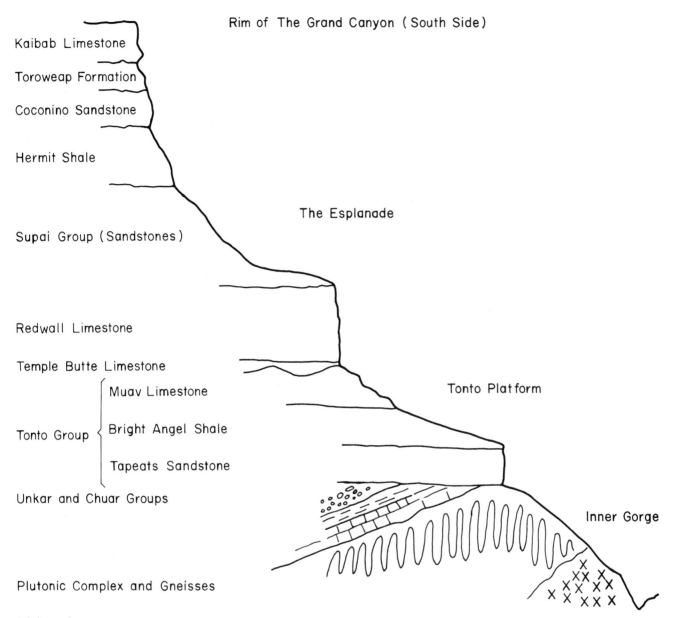

Figure 1.11 at the top shows:

Rim of The Grand Canyon (South Side)

Kaibab Limestone

Toroweap Formation

Coconino Sandstone

Hermit Shale

The Esplanade

Supai Group (Sandstones)

Redwall Limestone

Temple Butte Limestone

Tonto Group {
 Muav Limestone
 Bright Angel Shale
 Tapeats Sandstone
}

Tonto Platform

Unkar and Chuar Groups

Inner Gorge

Plutonic Complex and Gneisses

Vishnu Group

Figure 1.11 Schematic profile of the south wall of the Grand Canyon and the rock units exposed. Relative thicknesses of the formations are approximately to scale. Slope angles are greatly exaggerated since the Grand Canyon is about a mile deep here and over ten miles wide from rim to rim. In the semiarid climate of the Grand Canyon, most of the limestones are cliff-formers and the shales tend to form gentle slopes. The sandstones, depending on their purity and cementation, are resistant in varying degrees. Weathering and erosion are much less effective on the Precambrian crystalline rocks of the Inner Gorge. Adapted from Billingsley and Breed 1980.

9. **Fault-block mountain-building during the Grand Canyon orogeny with accompanying thrust-faulting and the formation of monoclines.**

The block-faulting occurred in stages along the old fault lines of the northeast-southwest fault system, forming first a horst-and-graben topography (fig. 1.8E). Next, the faulted beds were folded in a monocline (fig.1.7A), with enough stress being built up to produce thrust-faulting along the base of the monocline in a northeast-southwest direction. Thirdly, normal faulting occurred along the original set of faults (fig. 1.8A). This movement offset some of the folded monoclinal beds and also permitted further downdropping of several of the graben blocks. Last of all, renewed thrust-faulting parallel to the monocline occurred.

10. **The wearing down of the Grand Canyon Mountains to an erosional surface.**

During the very long erosional period that ensued, much of the Chuar beds and some of the Unkar Group were stripped from the mountain slopes. This exposed some of the more resistant beds that formed isolated hills on the erosional surface. Long-time weathering in an arid climate is suggested by the depth of weathering in rock, the degree of oxidation of some minerals, and the decomposition of feldspar into clay minerals.

Remnants of the Unkar and Chuar Groups are preserved in the down-faulted blocks. The contact between the Precambrian beds and the Paleozoic beds is close to the horizontal because only a few resistant mountains broke the ancient worn-down landscape.

11. **Deposition of Cambrian beds by a transgressing sea, forming an angular unconformity on top of the Precambrian beds.**

The transgressing sea came from the west about 550 million years ago and deposited three formations, each representing a different environment. The way in which the sediment was laid down illustrates the principle of time transgression (fig.1.12).

Imagine yourself standing on a sandy beach (that would eventually become the Tapeats Sandstone). Lying on the sand are cobbles (2.5 to 10 inches in diameter) and a few boulders (over 10 inches in diameter) that had been loosened from the bedrock by pounding waves and tossed up on the shore in storms. These coarse fragments mixed with sand formed the basal conglomerate at the bottom of the Tapeats. As you wade out into the water, the sea bottom becomes muddy. These muds formed the Bright Angel Shale. Farther out, in deeper water, lime (calcium

carbonate, $CaCo_3$) is precipitating out, forming a limy ooze on the bottom. From this ooze was formed the Muav Limestone.

Because sea level is very gradually rising, the shoreline and the beach are gradually shifting eastward, and the pattern of sediment deposition is also moving eastward without a break in continuity. Thus when the Tapeats Sandstone and the Bright Angel Shale were first laid down, it was Lower Cambrian time; but by the time the last grains were being deposited some distance to the east, the time was Middle Cambrian. Therefore, these beds can be considered time-transgressive rock units.

That the sea was transgressing is shown by finer-grained shale deposited over coarser-grained sandstone. Had the sea been regressing, coarser sediment would have overlain the finer.

The Tapeats Sandstone, Bright Angel Shale, and Muav Limestone belong to the Tonto Group and are the beds that make up the Tonto Plateau. These beds lie in their original horizontal position, virtually undisturbed since they were deposited on top of the old erosional surface formed on rocks that had been folded and faulted. Thus an angular unconformity occurs between the bottom of the Tonto Group and the top of the Precambrian beds (figs. 1.9C, 1.10).

12. **Disconformity between the Cambrian and Devonian beds.**

Beds from the Ordovician and Silurian periods are missing from the Grand Canyon geologic column. There is no evidence to indicate whether beds were deposited in the region during these time periods and subsequently removed by erosion or whether deposition did not occur. At

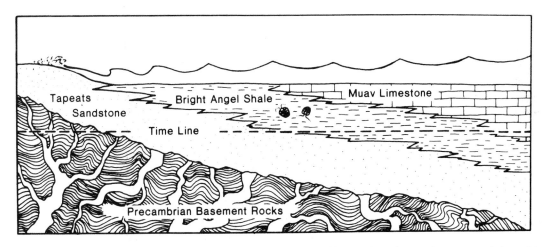

Figure 1.12 Sedimentary rock layers illustrating the principle of time transgression. The Tapeats Sandstone formed as a beach deposit over the Precambrian basement rocks. The Bright Angel Shale was deposited in shallow water, and the Muav Limestone in deeper water. As sea level rose and the land subsided, the shoreline shifted, along with the changing environments of deposition, so that the age of the Tapeats Sandstone, for example, is younger above the time line than below the time line. From *Geology of Arizona* by D. Nations and E. Stump. © 1981 Kendall/ Hunt Publishing Company.

any rate, deep channels were carved by streams (or possibly by marine scour) on top of the Muav. In these depressions was deposited the Devonian Temple Butte Limestone, which is not present in some areas but is up to 100-feet thick in other areas. The surface of this disconformity between the Muav Limestone and the Temple Butte Limestone represents a break in time in the sequence of sedimentary rocks.

13. Erosion of the Muav and Temple Butte Limestones.

As soon as the Devonian Temple Butte was deposited, it began to undergo erosion. Most of it was removed, as was part of the Muav. The only remnants of the Temple Butte that can be found are where the sediment was deposited in some type of hollow and protected by the surrounding Muav. Thus another unconformity is found between Cambrian-Devonian beds and the overlying Mississippian beds.

14. Deposition of the Mississippian Redwall Limestone.

The Redwall, a bluish-gray limestone containing chert nodules, was deposited in a shallow sea. Thickness averages 500 feet. Following the deposition of the Redwall Limestone, the region was gradually uplifted and the beds underwent erosion. Because limestone is highly susceptible to solution by ground water, some natural caves developed. The visually striking cliffs of the Redwall in the Canyon today have been stained red by rust streaks dripping down from the overlying beds as they have weathered back.

15. Deposition of the Supai Group during Pennsylvanian and Permian time.

The Supai, which is mostly nonmarine, contains footprints of amphibians and perhaps primitive reptiles. These footprints, along with the plant fossils, indicate that the region was once a vast flood plain. The lower formations (Watahomigi, Manakacha, and Wescogame) may be marine in origin and are probably Pennsylvanian in age. The upper beds (Esplanade Sandstone) are nonmarine and Permian in age and have been oxidized to a bright red color. The total thickness is around 1,000 feet.

16. Deposition of Permian beds.

The Hermit Shale, also red in color and nonmarine in origin, contains ripple marks, mudcracks, and footprints which indicate an environment of deposition similar to that of the underlying Supai Group. The main fossils here are plants, but insect remains can also be found.

The flood plain on which the Hermit Shale accumulated was gradually buried by migrating sand dunes. This formed the Coconino Sandstone, the whitish beds near the top of Grand Canyon. The uniformly sized, perfectly rounded, frosted quartz grains, arranged in a cross-bedded pattern characteristic of dunes, testify to a wind-blown

(eolian) origin. The dunes must have been fairly large as the Coconino is 400 feet thick. Footprints preserved in the sandstone show that animals wandered around the dunes while the sediment was accumulating.

The marine Toroweap Formation, with its nearly 300-foot thickness of sandstone and limestones, indicates that the sand dunes were smoothed over by transgressing seas and gradually buried.

Forming the rim of the Grand Canyon and the edge of the Plateau is the massive, thick-bedded Kaibab Limestone, more than 300 feet in overall thickness. Mostly limestone, the Kaibab also contains chert layers. Some zones have sand grains and chert nodules. Fossils are middle Permian in age. The formation is divided into three members.

17. Withdrawal of the Kaibab Sea; erosion and deposition of the Moenkopi about 220 million years ago; possible deposition of other Mesozoic sediments.

During the Triassic Period, which began the Mesozoic Era, the continents were rising. As the seas regressed, streams carved wide low valleys on the newly exposed land surface. Sediment was transported from higher land to lower slopes and basins, forming continental deposits that became the Moenkopi Formation, a mixture of brightly colored sandstones and shales with interbedded gypsum layers. The number and variety of fossils indicate an abundance of life forms. Fish swam in the streams and ponds; reptiles and amphibians lumbered across the land.

Because Jurassic, Cretaceous, and early Cenozoic beds have been identified in adjacent areas, there is the possibility that beds of the same age were deposited in the Grand Canyon region and later removed by erosion. For example, only a small remnant of the Moenkopi has been preserved in the Cedar Mountain area in a graben.

18. Monoclinal folding, thrust faulting, and normal faulting during Mesozoic mountain-building.

Even though beds were being folded, faulted, and uplifted, and mountain-building was going on all around, the Colorado Plateaus remained relatively stable and undisturbed (box 1.1).

Monoclinal folds with thrust-faulting at the base occurred again, but in a direction perpendicular to the earlier folds formed during the Precambrian. Normal and strike-slip faulting offset sections of the monoclines (fig. 1.8D). Many of the tributary canyons have been eroded along these strike-slip faults, an example being Bright Angel Canyon.

19. Ancestral Colorado River and the development of the present drainage system; uplift of the Colorado Plateaus during the Miocene Epoch.

The ancestral Colorado River formed during Miocene and Pliocene time and may have originally drained eastward. The loss of its headwaters by stream capture

Box 1.2
The Terraced Walls of the Grand Canyon

To a discerning observer who first looks down into the Grand Canyon from either the South Rim or the North Rim, the most obvious feature of the Canyon is the "stairstep" appearance of the sidewalls (figs. 1.13, 1.14). The upper part of the Canyon is terraced while the lower part, or Inner Gorge, is narrower and V-shaped.

Four geological formations are responsible for the most prominent cliffs or terrace risers. They are, from the top, the Permian Kaibab Limestone and the Supai Group, the Mississippian Redwall Limestone, and the Cambrian Tapeats Sandstone. Each cliff has at its base blankets of loose rock fragments, or *talus,* overlying gentle slopes. On top of the Tapeats Sandstone a broad terrace top, the Tonto Platform, has been created by removal of less resistant rock units. A similar platform, called the Esplanade, has developed above the resistant Esplanade Sandstone downriver near the western part of the Park. These different slope segments, formed as the Canyon was sculptured downward through successive geological formations, illustrate how in the same climate some rock units are more resistant to weathering and erosion than others. This phenomenon is called *differential erosion.*

The steeper-walled, narrower Inner Gorge is not terraced or stepped, because the metamorphic and intrusive igneous rocks at that level are of fairly uniform hardness and are more resistant to weathering and mass wasting than the Paleozoic rocks above, which form the series of cliffs and slopes.

One's first impression of the Grand Canyon is of the immensity of erosional downcutting. However, inasmuch as the Canyon is more than 10 miles wide from rim to rim in most places and about a mile deep, it becomes evident that the processes of valley widening and cliff recession are more significant than the downcutting. Everywhere in the Canyon, except on the faces of the steepest cliffs, great sheets of broken rock lie on the valley sides, testifying to the effectiveness of weathering—which produces the fragments—and mass wasting, which is moving the loose rock debris downslope.

Figure 1.13 The stairstep topography of the Grand Canyon is the result of erosional processes acting differently on resistant and less resistant sedimentary rock layers. The Tonto Platform, a nearly flat surface underlain by resistant Tapeats Sandstone, is in the center of the photograph. (The Inner Gorge, cut into resistant Precambrian rock units, is below the Tapeats.) Weathering and erosion of the weak Bright Angel Shale have formed the Tonto Platform. Farther up the canyon walls, the Muav Limestone has formed a second set of cliffs. National Park Service photograph.

North of Grand Canyon National Park, stepped topography on a larger scale is produced by the retreat of the edges of Mesozoic and Cenozoic formations (Chocolate Cliffs, Vermilion Cliffs, White Cliffs, and Pink Cliffs) on the Colorado Plateau. The cliffs trend generally east-west and step up toward the north.

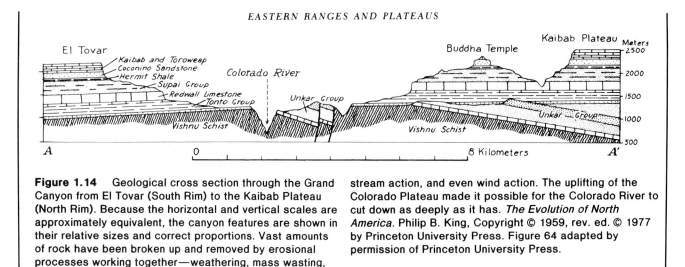

Figure 1.14   Geological cross section through the Grand Canyon from El Tovar (South Rim) to the Kaibab Plateau (North Rim). Because the horizontal and vertical scales are approximately equivalent, the canyon features are shown in their relative sizes and correct proportions. Vast amounts of rock have been broken up and removed by erosional processes working together—weathering, mass wasting, stream action, and even wind action. The uplifting of the Colorado Plateau made it possible for the Colorado River to cut down as deeply as it has. *The Evolution of North America.* Philip B. King, Copyright © 1959, rev. ed. © 1977 by Princeton University Press. Figure 64 adapted by permission of Princeton University Press.

may be what changed the river's flow to the southwest. Perhaps a lake formed and the river began as an overflow outlet. At any rate, the early Colorado River flowed on a fairly low, gently sloping surface of soft sediment (such as valley fill) and weak rocks (easily eroded sandstone and shales). As the Colorado Plateaus were uplifted and warped, the development of the drainage system was accelerated. During uplift, some local faulting and volcanic activity also took place.

20. Uplift and tilting of the Colorado Plateaus and superposition of the Colorado River during the Pliocene Epoch.

Because the gradient of the Colorado River was increased by uplift, the stream began to downcut rapidly, becoming entrenched while maintaining its course. With continued uplift, the Colorado River cut down to the buried geologic structures in the Precambrian rocks, thus becoming superimposed on these structures.

In the eastern portion of the park, some modification of the course took place where the Colorado River encountered some of the monoclinal folds and faults. By middle Pliocene time the present stream pattern was well established. With the entrenchment and later superposition of the Colorado River, the development of the Grand Canyon had begun.

During late Pliocene time, volcanic activity covered parts of the plateau with lava flows; and some of the flows poured down side valleys, especially in the western section of Grand Canyon. These lava flows are being eroded by the Colorado River today.

21. Deepening of the Grand Canyon by the Colorado River; widening of the Canyon by mass wasting.

Weathering loosens particles of bedrock by both mechanical and chemical processes and thus enables mass wasting, powered by gravitational energy, to move the materials downslope. In an arid climate these processes operate slowly except when flash floods or landslides occur, at which times quantities of loose sediment are transported very rapidly. Thus by processes that are mostly very slow or very fast, weathering and mass wasting deliver sediment to streams that carry it away. Since the Colorado River still has a steep gradient, downcutting of the Grand Canyon continues, but most of its sediment load is the result of mass wasting.

Geologic Maps and Cross Sections

Billingsley, G. H., and Breed, W. J. 1980. Geologic cross section along Interstate 40, Kingman to Flagstaff, Arizona, Petrified Forest National Park: Petrified Forest Museum Association.

Huntoon, P. W., Billingsley, G. H., Breed, W. J., et al. 1976. Geologic map of the Grand Canyon National Park, Arizona. Flagstaff, Arizona: Grand Canyon Natural History Association.

Maxson, J. H. 1962. Geologic map of Bright Angel quadrangle. Flagstaff, Arizona: Grand Canyon Natural History Association.

Bibliography

Beal, M. D. 1978. *Grand Canyon, the story behind the scenery.* Las Vegas, Nevada: KC Publications.

Breed, W. J. (ed.) 1974. *Geology of the Grand Canyon.* Flagstaff, Arizona: Museum of Northern Arizona and Grand Canyon.

Collier, M. 1980. *An introduction to Grand Canyon geology.* Flagstaff, Arizona: Grand Canyon Natural History Association.

Hamblin, W. K., and Rigby, J. K. 1969. *Guidebook to the Colorado River,* Part I (v. 15, part 5) and Part II (v. 16, part 2). Provo, Utah: Brigham University Geology studies.

Hunt, C. B. 1969. Geologic history of the Colorado River *in Colorado River Region: John Wesley Powell.* U.S. Geological Survey Professional Paper 669, pp. 59–130.

————. 1976. *Geology of the Grand Canyon.* Flagstaff, Arizona: Museum of Northern Arizona and Grand Canyon; Grand Canyon Natural History Association.

Karlstrom, T., Swann, G. A., and Eastwood, R. L. (eds.) 1974. *Geology of northern Arizona with notes on archeology and paleoclimate,* part I, regional studies. Geological Society of America Rocky Mountain meeting, Flagstaff, Arizona.

King, P. B. 1977 (rev. ed.) *The evolution of North America.* Princeton, New Jersey: Princeton University Press (chapter VII).

McKee, E. D. 1938. *The environment and history of the Toroweap and Kaibab Formations of northern Arizona and southern Utah.* Carnegie Institution of Washington Publication 492.

————. 1969. Stratified rocks of the Grand Canyon *in Colorado River Region: John Wesley Powell.* U.S. Geological Survey Professional Paper 669, pp. 23–58.

Nations, D., and Stump, E. 1981. *Geology of Arizona.* Dubuque, Iowa: Kendall/Hunt Publishing Company.

Redfern, Ron. 1980. *Corridors of time.* New York Times Books.

Rigby, J. K. 1977. *Southern Colorado Plateau,* K/H Geology field guide series. Dubuque, Iowa: Kendall/Hunt Publishing Company.

Stegner, Wallace. 1953. *Beyond the hundredth meridian.* Boston: Houghton Mifflin Company.

Address

Grand Canyon National Park
P.O. Box 129
Grand Canyon, Arizona 86023

2

Zion National Park

Location: Southwest Utah
Area: 146,546.97 acres; 228.96 square miles
Established: November 19, 1919

Figure 2.1 Downcutting by the North Fork of the Virgin River, mass wasting, and erosion sculptured Zion Canyon in Zion National Park. National Park Service photograph by G. A. Grant.

Key

———————— Roads

– – – – – – – Trails

. . . . Intermittent streams

~~~~~~~~~ Rivers and creeks

Wayne Canyon

Horse Range Mt.
8,740

Lava
Field

Taylor
Creek Road

Lee Pass

Nagunt
mesa
7,803

Timber
Top Kolob
Shuntavi   Mt. Arch
Butte

Gregory Butte
WILLIS 7,705'
VERKIN   Burnt Mt.
Neagle Ridge 7,669
LA

Hop Valley

CREEK

Bullpen Mt.
7,100

Langston
Mtn.

Little
Creek
Sinks

Jobs Head

Firepit Knoll
7,274

Spendlove Knoll

Lee Valley

Tabernacle
Dome

Pocket
Mesa
Pine
Valley
Peak
7,428

North
Guardian Angel
7,408'

Guardian
Angel Pass

South
Guardian
Angel
7,164'

Wildcat Canyon

Great West Canyon

Ivans Mt.
7,019

Kalob Creek

Lava Point
7,890

West Rim Trail

Horse Pasture Plateau

Goose Creek

Kalob Creek

Deep
Creek

Virgin River

The Narrows

North Fork

West Rim

Orderville
Canyon

Bullock
Gulch

Mt. of
Mystery
6,545

Temple of
Sinawava

Echo Canyon

Angels Landing

The Great
White Throne

Echo Canyon

Echo Canyon
Trail

Three Patriarchs

Court of the Patriarchs

The Sentinel
7,157'

Towers
of the Virgin

The Beehives

The West
Temple

Crater Hill

ZION   CANYON

The
East Temple

The
Great Arch

(Tunnel)

White Cliffs

Checkerboard Mesa

Springdale

North Fork

The
Watchman

East Fork Virgin River

Parunuweap Canyon

VIRGIN R.

N

Park boundary

0   ½   1   1 ½   2 mi.

Scales in miles

(All mountain elevations in feet)

Tynal

Figure 2.2   Zion National Park, Utah.

Table 2.1. Geologic Column, Zion National Park

Era	Period	Epoch	Group	Formation	Member	Geologic Events
Cenozoic	Quaternary	Holocene Pleistocene		Basalt lavas		Weathering, canyon-cutting, cliff retreat. Volcanic eruptions ~~~ Erosion ~~~
Cenozoic	Tertiary — Neogene	Pliocene		Sevier River		Flood deposits Block faulting; uplift ~~~ Erosion ~~~
Cenozoic	Tertiary — Neogene	Miocene		Sevier River		
Cenozoic	Tertiary — Paleogene	Oligocene		Lava flows		Episodes of volcanic eruption; uplift
Mesozoic	Cretaceous					~~~ Erosion ~~~
Mesozoic	Jurassic	Upper	San Rafael	Carmel		Silts deposited in shallow sea. ~~~ Erosion ~~~
Mesozoic	Jurassic	Middle and Lower	Glen Canyon	Navajo Sandstone	Temple Cap	Area buried by migrating sand dunes; arid climate; Units form White Cliffs and part of Vermilion Cliffs. ~~~ Erosion ~~~
Mesozoic	Triassic	Upper	Glen Canyon	Kayenta		Intermittent stream deposits; Exposed in Vermilion Cliffs.
Mesozoic	Triassic	Upper	Glen Canyon	Moenave	Springdale Sandstone	Stream deposits; makes up sand portion of Vermilion Cliffs.
Mesozoic	Triassic	Upper	Glen Canyon	Moenave	Dinosaur Canyon Sandstone	Lake deposits ~~~ Erosion ~~~
Mesozoic	Triassic	Lower		Chinle	Petrified Forest	Stream, swamp deposits, plus volcanic ash; wood fossils; makes up part of Vermilion Cliffs.
Mesozoic	Triassic	Lower		Chinle	Shinarump Conglomerate	Basal conglomerate ~~~ Erosion ~~~
Mesozoic	Triassic	Lower		Moenkopi	Upper Red	Inland stream deposits
Mesozoic	Triassic	Lower		Moenkopi	Shnabkaib Shale	Coastal plain; gradually changed from marine deposition. Units eroded to form Chocolate and Belted Cliffs
Mesozoic	Triassic	Lower		Moenkopi	Virgin Limestone	
Paleozoic	Permian			Kaibab Limestone		Deposition in shallow sea.
Paleozoic				Older units not exposed.		

Source: modified after Eardley and Schaack 1971; Gregory 1956; Rigby 1976.

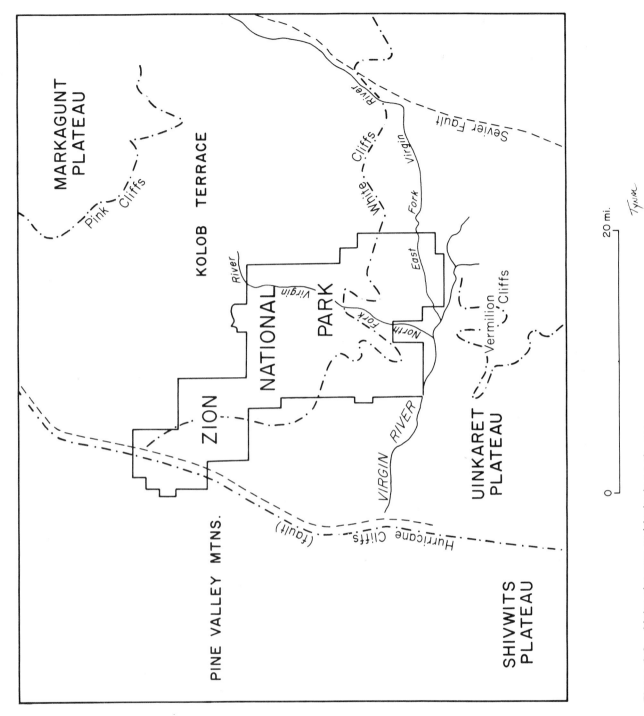

Figure 2.3 Major plateaus and faults in southwestern Utah in the vicinity of Zion National Park.

25

Box 2.1
How Rocks Are Classified

Rock classification is basically genetic; that is, how a rock originated determines the category it belongs in, the three main divisions being *igneous, sedimentary, and metamorphic.* However, identification of rocks depends upon further classification by *texture and mineral composition.* When you look at a hand specimen, for instance, you may not be able to tell how the rock originated; but you can determine the texture, and with the aid of a hand lens you may be able to recognize some of the mineral constituents by their crystal form or other physical characteristics.

Rocks are made up of grains of one or more varieties of minerals. In the geological sense, a *mineral* is a naturally occurring, inorganic, crystalline solid with a definite chemical composition and characteristic physical properties that are either uniform or variable within definite limits. Therefore, because the grains in rock represent the constituent minerals and because each mineral has a specific crystalline structure, the mineral grains, their size, and how they combine provide the basis for identifying and classifying rocks.

Rock texture, referring to the grain size of the component minerals, ranges from extremely fine to extremely coarse. For example, most *extrusive igneous rocks* have fine-grained texture with crystals too small to be seen with the unaided eye. This is because at surface temperatures and pressures, lava cools too fast for crystals to attain much size while the rock hardens. A rock that cools so quickly that *no* crystals have time to form becomes a natural glass or *obsidian.*

Intrusive igneous rocks tend to be coarse-grained in texture because magma generally cools at great depth under confining pressure over a long period of time—thousands, perhaps millions, of years. As the mineral crystals separate from the melt, they grow in size, interlocking and filling all the available space.

Sometimes as a magma or lava cools, conditions change or reheating and cooling occurs with the result that large crystals are enclosed in a mass of finer-grained crystals, like raisins in a loaf of bread. The texture of such rocks is described as *porphyritic.*

Igneous rocks (as discussed in chapter 1) originate from magma (or lava); that is, the rock material passed from very hot liquid to a solid state as cooling took place. A classification of common igneous rocks is shown in box 12.1, p. 143.

When igneous rocks and any other rocks are exposed at the earth's surface, they are subjected to processes of *weathering* that bring about chemical changes *(decomposition)* and physical changes *(disintegration).* The debris produced by weathering is transported and deposited, usually in layers, and may eventually become *sedimentary rock,* the second major category; that is, rock made up of sediment that has been lithified, or consolidated.

Sediment is a collective term for loose, solid particles of rock that can originate in several ways. *Clastic* sediment refers to any accumulation of unconsolidated fragments freed by weathering and erosion of pre-existing rock. Sizes of the fragments range from large boulders to gravel, sand, silt, and even to clay particles that are microscopic. During transportation by running water, glaciers, or other agents, these rock particles become rounded and broken up further by processes of *abrasion,* which is the wearing away of rock by friction.

During weathering, decomposition of rock involves chemical alteration by the addition or subtraction of various materials. Some minerals, for example, go into *solution* when exposed to water and are transported in dissolved form. Later, the minerals may be *precipitated out* by evaporation or by organic secretion (by marine organisms, etc.). Sediment formed by chemical precipitation or by organic secretion is referred to as *nonclastic.*

A sedimentary rock made up of grains or fragments of clastic sediment is described as having *clastic texture.* Examples are *conglomerate,* a coarse-grained rock made up of consolidated rounded gravel; *sandstone,* a medium-grained rock formed by cementation of sand grains; and *siltstone* or *shale,* fine-grained rocks made of silt, wet mud, etc., lithified by compaction and cementation.

The most common sedimentary rock having a *nonclastic texture* (i.e., made up of nonclastic sediment) is *limestone*, which is composed mainly of interlocking crystals of calcite (calcium carbonate) that has been precipitated by inorganic or organic means. Table 3.2 (p. 39) shows classification of the more abundant sedimentary rocks.

Metamorphic rocks, the third major category, are derived from pre-existing rocks that might have been igneous, sedimentary, or other metamorphic rocks (table 2.2). However, the term metamorphic—meaning "changed form"—indicates that the original, or "parent" rock has been converted by heat, pressure, and chemically active fluids into another type of rock, generally (but not always) having a different texture and composition. For example, sandstone changes into *quartzite*, shale into *slate*, and limestone into *marble*. If metamorphism persists or intensifies, slate may change into *phyllite*, which, in turn, may become a *schist*, and then a *gneiss*. The texture of metamorphic rocks is described as *foliated* if some alignment of minerals is present, or *nonfoliated* when the arrangement of minerals is random. In general, foliated rocks (such as gneiss, schist, phyllite, and slate) have been subjected to higher temperatures and pressures than rocks with nonfoliated texture (such as quartzite, marble, and some greenstones). During metamorphism, if the melting point of a rock is exceeded, the rock becomes molten. The resulting magma may eventually be "recycled" as an igneous rock.

This continual reworking of the earth's crust over geologic time can be depicted graphically by a theoretical *rock cycle* (fig. 2.4). A rock cycle model shows the interrelationships among weathering, erosion, sedimentation, deposition, burial, tectonic forces, melting, volcanism, the major rock types, and so on. What is evident is that with the passage of time, new rocks are formed from old rocks under many different conditions, and that minerals combine and recombine to produce the infinite variety of rocks found in nature.

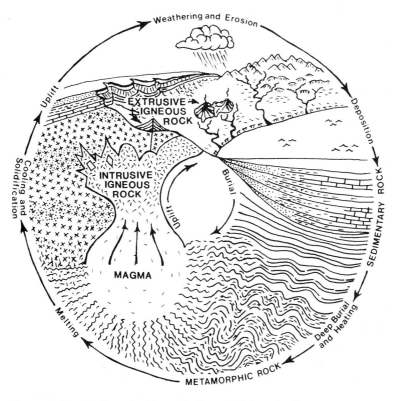

Figure 2.4   Rock cycle. From *Geology of Arizona* by D. Nations and E. Stump. © 1981 Kendall/Hunt Publishing Company.

Table 2.2
Metamorphic Rock Classification

Individual Mineral Grains May Be Seen	Mineralogy	Special Characteristics	Rock Name	Metamorphic Grade
Foliated Textures				
No	Chlorite, muscovite, quartz, sodium-plagioclase	Dark-colored, breaks into thin plates with generally dull surfaces	Slate	Very low
Difficult	Biotite, muscovite, quartz, sodium-plagioclase	Breaks into thin plates with a lustrous sheen	Phyllite	Moderately low
Yes	Biotite, muscovite, quartz, sodium-plagioclase are common; garnet, staurolite, kyanite, and orthoclase may occur.	Irregular, often contorted foliation surfaces	Schist	Moderately high
Yes	Quartz, plagioclase, orthoclase are common; biotite, muscovite, garnet, sillimanite, kyanite, and hornblende may occur	Generally has a banded appearance due to alternating light- and dark-colored layers	Gneiss	Very high
Nonfoliated Textures				
Difficult	Recrystallized quartz	Light-colored and massive with sharp edges; may have a glassy or sugary appearance	Quartzite	Variable
Yes	Recrystallized calcite	Light-colored, generally coarse-grained	Marble	Variable
Yes	Mostly hornblende and plagioclase; biotite, garnet, and quartz may occur	Massive varieties have a salt and pepper appearance (resembling an igneous texture). With increasing biotite content may become slightly foliated	Amphibolite	Variable
No	Variable	Dark-colored, massive, dense	Hornfels	Very low

(From R. D. Dallmeyer, *Physical geology laboratory text and manual,* 2nd ed. © 1978 Kendall/Hunt Publishing Company.)

Local History

Traces of the earliest inhabitants of the Park area have been found in Parunuweap Canyon. Artifacts of the Basket Maker Indian culture have been found there. They were followed by the Cliff Dwellers, or Pueblos, who lived in Zion Canyon and Parunuweap Canyon.

The next group to come in were the Piutes, who built no permanent dwellings but had settlements in the Virgin River Valley. They called the Virgin River *Pahroos*, which means "muddy, turbulent water;" and Zion Canyon *loogoon*, meaning "arrow-quiver" or "come out the way you came."

The first recorded venture of white men into the area was the Escalante-Dominquez Expedition of 1776. Escalante and Dominquez were Spanish priests who crossed the Virgin River while trying to find a direct route between Santa Fe, New Mexico, and Monterey, California.

In 1825, Jedediah S. Smith, a mountain man, left the American Fur Company outpost at Great Salt Lake with a party of sixteen. In following some of the streams, he encountered a river which he named Adams River after President John Quincy Adams. During a second expedition in 1827, Smith renamed it the Virgin River after Thomas Virgin, a member of the expedition who was wounded by the Indians.

The Fremont Expedition of 1843/44 passed through the region also. However, none of the members of the expedition saw Zion Canyon.

Mormons started settling the Virgin River region in 1847 with plans to farm and grow cotton. The Mormon search for farm land was responsible for the Piute Indians leading Nephi Johnson to the canyon in 1858. Joseph Black explored it thoroughly in 1861 and found several locations suitable for farming. The settlers called the region "Little Zion" and stayed in the area until 1909 when it became a national monument.

In 1872 Major John Wesley Powell's party crossed over Parunuweap Canyon ("water that roars") and Little Zion Canyon, which he named Mukuntuweap ("straight canyon"). Descriptions and photographs by J. K. Hillers brought the park its fame.

About the same time (1872), Captain George M. Wheeler and G. K. Gilbert (on separate expeditions) mapped the region. Additional mapping was done in 1880 by Major C. E. Dutton, with sketches made by W. H. Holmes.

Public pressure led to the establishment of a national monument in 1909, called Mukuntuweap (after Powell). In 1918 the name was changed to Zion, and in 1919 it became a national park.

A railroad spur was built to the park in 1930. In addition, a wagon road was built into the canyon. This was improved later to accommodate automobiles. The Pine Creek tunnel is over 5,000 feet long and has six large windows cut into it for views.

Geologic Features

Zion Canyon, a deep chasm with nearly vertical walls, has been and still is being cut by the North Fork of the Virgin River, a stream with great erosive power whose gradient, or slope, ranges between 50 and 70 feet per mile. In addition to the high velocity (caused by the steep gradient), the Virgin River has great volume during much of the year and is able to transport some three million tons of rock and sediment downstream annually. This debris is derived from products of mass wasting picked up by the stream as well as from the downcutting, widening, and scouring of the canyon as the water rushes through it (fig.2.5).

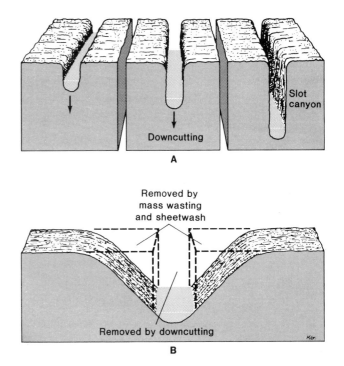

Figure 2.5 *A.* Rapid downcutting in resistant rock may create slot canyons, especially in arid regions where weathering processes operate at a slow rate. *B.* Downcutting in conjunction with mass wasting and weathering eventually widens most canyons into V-shaped valleys. From *Physical Geology* (2nd ed.) by Charles C. Plummer and David McGeary. © 1979, 1982 Wm. C. Brown Company Publishers, Dubuque, Iowa. Reprinted by permission.

Since the Virgin River is downcutting more rapidly than it is widening, its tributaries, with smaller volumes of water, are literally "left hanging" because they are unable to downcut as fast as the main stream. The tributaries have smaller drainage basins and some of them are intermittent (i.e., they do not flow year-round). Because the tributary valley floors are at a higher level than the floor of the main stream, they are referred to as *hanging valleys*. Waterfalls connect the hanging valleys with the river below. Angels Landing is a typical hanging valley, as are the valleys between Twin Brothers mountain peaks and Mountain of the Sun (located on the eastern side of the canyon north of Pine Creek).

It is natural when one sees Zion Canyon to wonder how much more deeply the Virgin River can cut before it "hits bottom" (i.e., reaches base level). Geologists estimate that the Virgin can cut down another thousand feet and still have enough energy to transport its sediment to the Colorado River!

The erosive power of sediment-laden water running at high velocity is well shown in the Pine Creek Narrows and in the Narrows near the head of the Virgin River, where the floor of the valley is only a few feet in width but the canyon walls rise more than 1,500 feet nearly straight up. The walls are so steep that sunlight reaches the valley floor only at high noon when the sun is directly overhead. The canyon is so narrow that the water level may rise 25 feet in 15 minutes during a heavy rain or flash flood. The scouring effect of sediment in transport helps to widen the valley. This also explains why the valley walls show evidence of scouring high above normal water levels.

Mass wasting in Zion Canyon is gradually widening the valley, and in many places the canyon rims are developing a scalloped look as they retreat unevenly. Sapping by springs and the undermining of sandstone cliffs by removal of parts of the softer underlying shales due to gravity has had a twofold effect. First the sandstone cliffs become overhanging, and then, due to lack of support, large blocks break off and fall to the valley floor (fig. 2.5).

The variety of the features is also due in part to the many bedding planes, joints, and faults, some of which are curved, others horizontal or vertical, some close together, and still others widely separated. Thus both resistant and nonresistant types of bedrock are increasingly exposed to the ravages of weathering. Frost action has a

significant effect in the Park, which receives an average of 22 inches of snow each year. Rainwater, seepage, tree roots, etc., also are effective agents of weathering, even in the arid to semiarid climate of southern Utah.

The generally rectangular stream pattern is the result of closely spaced *joints* (fractures in bedrock without displacement) and vertical faults that permit broken or exposed rock to weather and erode more easily, thus forming valleys. The "temple formations" are the result of resistant, unfractured rock standing as columns or pyramids after the highly fractured rock in between has been eroded and transported away. Examples of these are East Temple, Checkerboard Mesa, Sentinel Mountain, Twin Brothers, Temple of Sinawava, West Temple, the Three Patriarchs, Towers of the Virgin, and the Beehives.

The rock unit that dominates Zion National Park is the Navajo Sandstone. This massive formation, 1,500 to 2,000 feet in thickness, is famed for its *cross-bedding* that developed as sand dunes migrated to and fro on the shores of an ancient shallow sea (fig. 2.6). Parts of the formation appear to have been deposited in shallow water with the

Figure 2.6   Recent interpretations suggest that the Navajo Sandstone was deposited as a series of dunes in a nearshore marine environment as well as in a coastal dune environment. Thus some cross-bedding may have been formed by water currents in shallow seas as well as by wind reworking coastal dunes. See figure 2.7. Strong winds and a dry climate prevented vegetation from anchoring or stabilizing the dunes. National Park Service photograph.

cross-bedding caused by currents. The eolian cross-bedding formed as shore dunes were impelled by the prevailing winds. In this process wind picks up sand grains on the gentle back slope of a dune and deposits them on the (steeper) downwind side. The loose grains cascade down the face of the dune tending to keep the slope at about 34°, the "angle of repose." As sand underneath becomes compacted and the top of the dune becomes oversteepened, the face of the dune shifts direction slightly as more sand grains keep tumbling down. Gradually the entire dune is overridden as it moves in a downwind direction. Any shift in wind direction intensifies the cross-bedding effect (fig. 2.7). The source area of the sand for the Navajo formation lay to the north and west, and the coastal dunes gradually spread southward and westward. Checkerboard Mesa is one of the locations in Zion National Park where these "fossil" dunes can be seen.

The Great White Throne, a natural feature sculptured by erosion of the Navajo Sandstone, is a large monolith rising approximately 2,450 feet from the valley floor. Despite its name, the throne is not completely white. The color of the Navajo's upper part is actually a light tan; but from a distance, particularly if sunlight is striking it, the rock appears to be a brilliant white. The base of the throne is stained a deep red, the result of iron oxide leaching out from higher rock layers as they were weathered

and eroded. The Great White Throne is also an example of how the lower beds of the Navajo Sandstone tend to form vertical cliffs, while the upper portion has a tendency to weather as rounded domes.

The highest point in the Park is at West Temple, the top of which is 7,795 feet above sea level and 3,802 feet above the canyon floor. The lowest point is near South Campground and is about 3,900 feet above sea level. Thus the *relief* in the Park (i.e., the difference in elevation between the highest and lowest points) is approximately 3,900 feet.

Around the Pine Creek area the Chinle Formation can be seen. At a location near the highway the beds are a mixture of sandstone and shale. The Chinle was deposited by streams and contains fossils of a large aquatic reptile, the *phytosaur,* quite similar in appearance to the crocodile. Phytosaurs became extinct at the end of the Triassic Period.

The geologic history of Zion Canyon begins where the history of Grand Canyon ends. The rock sequence of Grand Canyon is primarily Precambrian and Paleozoic and covers about two billion years, whereas Zion formations are Mesozoic and represent only about 150 million years of geologic time.

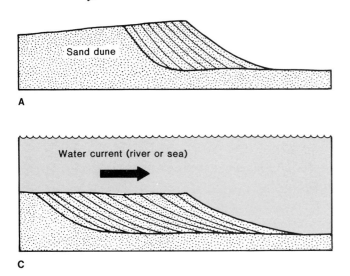

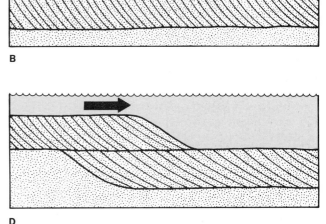

Figure 2.7   How cross-bedding may develop in wind-blown sand (*A,B*) and in sand deposited by currents in water (*C,D*). *A.* Sand accumulates in slanting layers on the downwind side of a dune. *B.* With a change in wind direction, a second dune, with a different orientation, covers the first dune. *C.* Currents deposit layers of sand over a depression in the sea floor or a river bottom. *D.* Continued sedimentation may cover first set of cross-beds with another. From *Physical Geology* (2nd ed.) by Charles C. Plummer and David McGeary. © 1979, 1982 Wm. C. Brown Company Publishers, Dubuque, Iowa. Reprinted by permission.

Geologic History

1. Formation of the Kaibab Limestone during the Permian Period.

Although the Kaibab Limestone is not actually exposed within Zion National Park boundaries, it is found just beyond the southwest border. The Kaibab, which is the same formation that caps the rim of Grand Canyon, is a marine limestone that was deposited in a shallow sea.

2. Deposition of the Moenkopi Formation (the Belted Cliffs) during Lower Triassic.

The Moenkopi, the oldest rock within the park boundaries, is a mixture of sandstone, shale, limestone, and gypsum. The different lithologies produced bright bands of color, which accounts for their name, the Belted Cliffs.

Not all of these deposits were marine. Some of the sediment was deposited in and along the streams that flowed on the coastal plain adjacent to an ancient sea. Along the shoreline in shallow lagoons, the water evaporated rapidly without replenishment, precipitating out the mineral *gypsum* which, when lithified, becomes a sedimentary rock, also called *gypsum* or gyprock. (Some of the gypsum is a very dense variety called alabaster.) Fossils in the beds help geologists to date the formation as having been deposited approximately 225 million years ago.

The Moenkopi, primarily a cliff-forming formation, accumulated to a thickness of 1,900 feet. Its three members, from oldest to youngest, are the Virgin Limestone, Shnabkaib Shale, and Upper Red Member. The fossiliferous Virgin Limestone is a mixture of oil-bearing sandy limestones, shales, and gypsum-bearing shales. The Shnabkaib Shale is primarily a gypsum that overlies the sandy shales. The Upper Red Member is a mixture of gypsum-bearing shales and sandstone.

3. Uplift and erosion of the Moenkopi during Triassic time; deposition of the Shinarump Conglomerate at the bottom of the Chinle Formation.

During a period of uplift and erosion that followed the deposition of the Moenkopi, a surface formed that had a relief of 100 feet or more. Rivers picked up sands and gravels and washed down trees that became buried in sand or gravel bars during floods. Later the sands and gravels were lithified into a basal conglomerate while the wood became petrified. Because the Shinarump is a conglomerate, it is fairly resistant to weathering and is thus a cliff-former even though it is only 100 feet thick.

The name *Shinarump* means "wolf's rump." Indians in the region where the Shinarump crops out thought the gray beds (which have a tendency to weather to rounded hills) looked like the back end of a wolf.

4. Deposition of Chinle Formation by streams and in shallow lakes and ponds, followed by some uplift and erosion.

The sediment making up the Chinle is a mixture of sandstone, shale, limestone, and volcanic ash. The latter is of great interest because it supplies the silica necessary for wood petrification in the Chinle Formation (chapter 38). The brilliant red colors of the 1,000-foot-thick Chinle are caused by the variation in the lithology and the amount of iron oxide present. Other colors of the Chinle formation include shades of yellow, pink, red, purple, blue, gray, and brown. These contribute to make it one of the most brightly colored formations in the American West.

Since the deposits were mainly laid down by streams, the fossils are for the most part those of river dwellers such as fish, amphibians, phytosaurs, and freshwater clams and snails. The region was a nearly flat plain with lakes scattered over its surface. Because of the low gradient, only fine-grained sediment and material in solution were transported. The volcanic ash was wind-deposited. At a much later time, uranium salts, dissolved and transported in percolating ground water, were redeposited in parts of the Chinle and Shinarump.

Some uplift and erosion occurred after the deposition of the Chinle Formation.

5. Deposition of the Moenave Formation (Vermilion Cliffs) in lakes and streams.

Two members make up the Moenave (Mo-en-ah-vee) Formation: the Dinosaur Canyon Sandstone, which was deposited in a lake that eventually silted in; and the Springdale Sandstone, which was stream-deposited.

The Dinosaur Canyon Member is a slope-former, ranging from 140 to 375 feet thick. It is a mixture of interbedded mudstones, thin-bedded siltstones, and sandstones. Fish fragments, the main fossils present, date the beds as being Upper Triassic.

The Springdale Sandstone Member, a light-colored, pale reddish-brown sandstone, is a cliff-former, 75 to 150 feet in thickness. The stream deposits of the Springdale contain fish fossils; and in the flood plain are deposited the bones of dinosaurs and tritylodonts (crocodile-like reptiles).

6. Deposition of the Kayenta Formation as a flood plain and stream deposit containing dinosaur tracks and cross-bedding.

During late Triassic time, the adjacent areas were uplifted at a fairly rapid rate. As sediments became coarser, beds of maroon sandstone called the Kayenta Formation were formed. The beds were deposited as sands and silts by streams meandering back and forth over the flood plain. Cross-bedding was produced by shifting currents and the building of sandbars. Dinosaurs coming down to the stream for water left footprints in the damp sediment along the banks. The Kayenta has a tendency to be a slope-former.

7. Formation of the Navajo Sandstone as migrating sand dunes during early Jurassic time.

As the climate became more arid, streams dried up or became intermittent. Strong dry winds from the north and west picked up sand grains and piled them in dunes that migrated because no vegetation could grow there to anchor them. Dunes formed both in the shallow waters of the nearshore environment and as coastal dunes that spread gradually inland. Some sand was "recycled" from the increasingly dry lands to the east as grains were picked up by onshore winds (figs. 2.6, 2.7).

At first the dunes interfingered with the Kayenta stream deposits, but finally they covered the entire drainage basin. Migration of the dunes and shifting winds produced the pronounced cross-bedding characteristic of the Navajo Sandstone, although it should be noted that cross-bedding in the nearshore environment was probably formed by currents in the shallow sea.

Because of differences in the type and amount of cement, the Navajo Sandstone varies in color from white to tan to red. A cliff-former, the Navajo is over 1,000 feet thick and makes up such well-known features as the White Cliffs, Checkerboard Mesa, the Great White Throne, etc. (fig. 2.8). Dinosaur footprints have been found in the Navajo as well as in the Kayenta.

The Moenave, Kayenta, and Navajo are all members of the Glen Canyon Group. (In other regions the Wingate is included as a lower member.)

8. Deposition of the Carmel Limestone in a shallow sea that covered the sand dunes during late Jurassic time.

About 170 million years ago a shallow, warm sea gradually spread over the area and wave action flattened the sand dunes. A calcareous silt, characteristic of a shallow sea environment, buried the sand dunes. Massive beds

Figure 2.8   View of Zion Canyon from the east side of Observation Point. The nearly vertical cliff in the foreground is made up of resistant Navajo Sandstone. Part of the stream bed and the slopes at the base of the cliffs are in the Moenkopi Formation. The lowest part of the stream bed cuts into the Kaibab Limestone. National Park Service photograph by G. A. Grant.

of compact limestone from one to four feet thick separated by thin sandy limestone beds containing fossils of shallow water organisms make up the Carmel Limestone. The composition and the type of bedding confirm the shallow sea environment in which the Carmel was deposited.

The Carmel Limestone, which is light-colored and between 200 and 300 feet thick, caps many of the mesas on the Kolob Terrace, as well as the Altar of Sacrifice and some of the Temples.

Within the boundaries of Zion National Park, the Cretaceous beds that overlie the Carmel Limestone have been removed by erosion. However, north of the Park the Cretaceous beds make up the Gray Cliffs.

9. Lava flows, uplift, and erosion.

After the Mesozoic Era came to a close, erosion continued into Cenozoic time, but it ceased in the Eocene Epoch as deposition took place along streams and in lakes outside the present boundaries of the Park.

Terraces on the Kolob Plateau were formed by a series of basaltic lava flows mixed with pyroclastic debris from nearby volcanoes. The older flows, which are latite or andesite, are Oligocene/Miocene (30 to 15 million in age); and the younger flows, which are basaltic, are Pleistocene/Holocene (one million to a few hundred years ago).

Between these flows, the Sevier River Formation, which may be Pliocene in age forms terraces along the Virgin River. The Sevier gravels (formerly called Parunweap Gravel) are flood deposits that formed either during a period of exceptionally heavy rainstorms or as the result of gradient change when uplift occurred along the Hurricane fault. As the uplift developed, the gravel terraces and lava beds were removed from the Park area by erosion, but they are still found in areas surrounding the Park.

10. Block-faulting with uplift of 10,000 feet creating the Markagunt Plateau.

Beginning about 13 million years ago, all of southern Utah and the adjacent areas underwent intermittent uplift. This continued for several million years until the entire region was elevated about 10,000 feet. This uplifting rejuvenated the streams, and, with new base levels, they rapidly began to downcut and form canyons.

Because uplifting was uneven, the beds broke into a series of blocks, some of which were miles across. One of these, the Markagunt Plateau block, which is the site of Zion National Park, is bounded by Hurricane fault on the west and Sevier fault on the east (fig. 2.3). Although faulting tilted the block eastward only slightly (the slope being one to two degrees), the uplift was as much as 2,000 feet. This is the reason why the same beds that cap Elkhart Cliffs are found in the Virgin River below.

11. Dissection by the Virgin River and mass wasting of the Markagunt Plateau forming Zion Canyon.

As faulting and tilting again increased the gradient of the stream, downcutting carved the canyon. The differences in the lithology of the beds, their resistance or lack of resistance to weathering, and the location of the faults and joints produced the features that we now view.

Geologic Map and Cross Section

American Association of Petroleum Geologists. 1967. *Geological highway map no. 2 of the southern Rocky Mountain region.* Tulsa, Oklahoma: American Association of Petroleum Geologists.

Eardly, A. J.; and Schaack, J. W. 1971. *Zion-the story behind the scenery.* Las Vegas, Nevada: KC Publications. p. 9.

Bibliography

Eardley, A. J.; and Schaack, J. W. 1971. *Zion-the story behind the scenery.* Las Vegas, Nevada: KC Publications.

Gregory, H. E. 1950. *Geology and geography of the Zion Park region, Utah and Arizona.* U.S. Geological Survey Professional Paper 220.

———— 1956. *Geological and geographic sketches of Zion and Bryce Canyon National Parks.* Springdale, Utah: Zion-Bryce Natural History Association.

Hamilton, W. L. 1980. *Guide to the geology of Zion National Park.* Springdale, Utah: Zion Natural History Association.

Redfern, Ron. 1980. *Corridors of Time.* New York: New York Times Books.

Rigby, J. K. 1976. *Northern Colorado Plateau,* K/H Geology Field Guide Series. Dubuque, Iowa: Kendall/Hunt Publishing Company

---

Address

Zion National Park
Springdale, Utah 84767

3

Bryce Canyon National Park

Location: Southern Utah
Area: 36,010 acres; 56.28 square miles
Established: February 25, 1928

Figure 3.1 The Wall of Windows (as seen from Peek-a-Boo Trail) shows variations in the degree of weathering because of differences in resistance of rock types in the Wasatch Formation. The intersecting joints control the location and distribution of the erosional features. Spheroidal weathering and exfoliation of the rock produced the windows. National Park Service photograph.

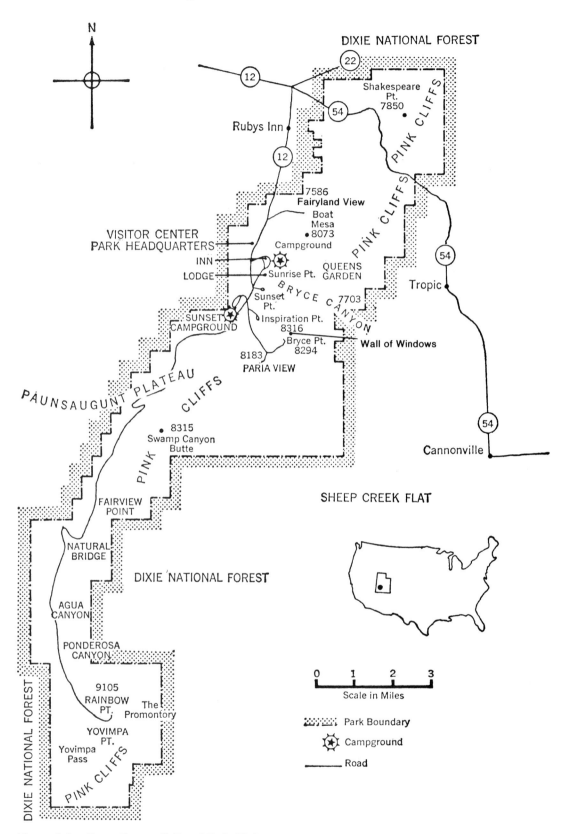

Figure 3.2   Bryce Canyon National Park, Utah.

Table 3.1. Geologic Column, Bryce Canyon National Park

Time Units			Rock Units		Geological events
Era	Period	Epoch	Group	Formation	
Cenozoic	Quaternary	Holocene Pleistocene			Weathering and erosion of Wasatch beds to form present landscape ～～Uplift and erosion～～
Cenozoic	Tertiary / Paleo-gene	Eocene		Wasatch	Streams depositing into inland lakes. Erosion.
Mesozoic	Cretaceous	Upper		Kaiparowits	Deposition in meandering streams. Erosion
Mesozoic	Cretaceous	Upper		Wahweap Sandstone	Deposition along a shifting shoreline.
Mesozoic	Cretaceous	Upper		Straight Cliffs Sandstone	Deposition in a shallow regressing sea and brackish lagoons Erosion
Mesozoic	Cretaceous	Upper		Tropic Shale	Deposition in a shallow sea that periodically regressed and in coal swamps.
Mesozoic	Cretaceous	Upper		Dakota Sandstone	Deposition of a beach deposit in a transgressing sea, as a blanket sand.
Mesozoic	Jurassic	Upper	San Rafael	Winsor	Deposited as sand dunes and intertidal mud flats. Erosion
Mesozoic	Jurassic	Upper	San Rafael	Curtis	Return of the sea, deposition of gravel, sandstone, shale.
Mesozoic	Jurassic	Upper	San Rafael	Entrada Sandstone	Accumulation of dune sands; erosion Marine deposition; then retreat
Mesozoic	Jurassic	Upper	San Rafael	Carmel	Seas flattened out dunes, then retreated. Erosion.
Mesozoic	Jurassic	Lower	Glen Canyon	Navajo Sandstone	Migration of sand dunes over an eroded landscape.
Mesozoic				Older units not exposed	

Source: modified after Gregory 1956; Rigby 1976.

Local History

The first inhabitants of this area were the Basket Maker Indians, and they were followed by the Pueblo Indians. The Pueblos apparently had no long-term settlements here. Both cultures lived in the Paria Valley. They were followed by the peaceful Piutes who farmed and hunted, and they did not build permanent settlements here either. The Piute name for Bryce was *Unka-timpe-wa-wince-pock-ich*, which translates to "red rocks standing like men in a bowl-shaped canyon."

Few white men had visited the Bryce Canyon area before Major J. W. Powell initiated a survey to map much of southern Utah in 1871 after his second trip down the Colorado River. Under Powell's direction, Almon H. Thompson, a geographer, surveyed and explored the Bryce Amphitheater. He was responsible for the first scientific description of Bryce. His explorations were followed by those of Edwin E. Howell, Grove Karl Gilbert, and Lieutenant W. L. Marshall, who were members of a survey party under the direction of Captain George M. Wheeler from 1870 to 1876. Howell mapped the Pink Cliffs for 100 miles and studied them in detail at Table Cliffs. Gilbert went as far as Paunsaugunt Plateau and Paria Valley. John Weyes, an artist with this expedition, made the first sketches of Bryce.

Captain C. E. Dutton surveyed the area in 1880; and J. K. Hiller, of his party, took the first photographs.

The few settlers, mostly Mormons from 1874 to 1891, were unimpressed by the scenery. They were more concerned about the lack of soil and water and the loss of cows in the terrain.

Ebenezer Bryce (for whom the canyon is named) kept his cows there in the late 1870s. He is credited with saying it was "a hell of a place to find a cow."

Paunsaugunt Plateau or "home of the beaver" became a national forest in 1905. When the wagon road was improved (1915-1917), the area became more popular. On June 8, 1923, it became a national monument, and on February 25, 1928, a national park.

Box 3.1
Sedimentary Rocks, a Record of the Past

By volume, sedimentary rocks make up only a small percentage of the earth's crust; and yet because they cover most of the continental areas of the earth like a thin skin, they are the common rocks with which most people are familiar. Sedimentary rocks are widespread because they are derived from the processes and products of weathering. All rocks weather when exposed to the earth's atmosphere, and water intensifies the weathering process.

By studying sedimentary rocks, geologists can learn a good deal about the climate, weathering agents, and other environmental conditions that existed during the geologic time in which the sediment was deposited and the rocks were lithified. Moreover, fossils found in sedimentary rocks reveal much about the development of life forms on earth, when some organisms became extinct, and when others first appeared.

The first seven national parks discussed in this book are all located in the Colorado Plateaus Province of western United States, and all seven parks are famous for their sedimentary rock exposures. In this region the beds of sedimentary rocks are essentially undeformed; that is, they lie one above the other, almost horizontally, in the same order they were deposited. Numerous episodes of uplift have raised the beds high above their original positions; then erosion and downcutting have dissected the beds, exposing them to view (see box 1.1). Moreover, vegetation is relatively sparse in this region because of the generally arid climate. For these reasons, the sedimentary rock record can be "read" with relative ease as the interested traveler goes from one park to the next (fig. 1.4, p. 7).

Some sedimentary rocks are also important because of their economic value. Limestone, for example, is the main ingredient of cement. The three most abundant sedimentary rocks are shale, sandstone, and limestone, in that order. These and other common sedimentary rocks (most of which can be found in the parks of the Colorado Plateaus Province) are listed in the Table 3.2.

Geologic Features

In the near-desert climate of Bryce Canyon National Park (less than 12 inches of precipitation a year), the processes of weathering and mass wasting, rather than downcutting by streams, have been mainly responsible for the remarkable scenery that visitors see. Bryce Canyon, a horseshoe-shaped basin (Paria Basin) rather than a true canyon, is about 12 miles wide, 3 miles long, and 800 feet deep. The arches and windows in Bryce, its sculptured terraces, and its intricate pinnacles have all been weathered and eroded out of the pink and white Wasatch Formation—a mixture of limestone, limy siltstone, shale, sandstone, and a few remnants of conglomerate. The famous pink color is due to the presence of iron oxide.

The first beds of the Wasatch were laid down in an Eocene inland sea some 54 million years ago, and deposition ceased about 40 million years ago. Uplift and faulting formed blocks that streams carved into plateaus. The joint systems and cracks produced by uplift control the location of arches, windows, and natural bridges. The pinnacles are the result of differential weathering acting along vertical joints in the flat-lying beds that have varying degrees of hardness (figs. 3.3, 3.4).

Because the Wasatch beds vary in hardness, they weather at different rates. A feature called Queen Victoria, located in the Queen's Garden area, is one of the many notable examples of differential weathering. Some beds have been completely removed. For example, the conglomerate capping two buttes at Bryce Point is about all that remains of that particular layer in the entire park.

Most of the arches found in Bryce Canyon, such as the famous Natural Bridge, are carved from the sandstone beds of the Wasatch. Despite their appearance of solidity, they are not permanent features; In the Fairyland area, an arch known as Oastler's Castle collapsed under its own weight in the late 1960s. However, as old ones are destroyed, new ones are being created by the unremitting processes of weathering (fig.3.5, table 3.3).

Weathering Processes

The fragmentation or *disintegration* of rock is a *physical* process that is inextricably linked to *decomposition* of rock which involves *chemical* processes. In other words, rock

Table 3.2
Classification of Common Sedimentary Rocks

Nonclastic Sedimentary Rock Classification

Principal Mineral Component	General Characteristics	Rock Name
Inorganic		
Halite	Massive, variably crystalline	Rock salt
Gypsum	Massive, variably crystalline	Gypsum
Calcite	Finely banded, crystalline or microcrystalline	Travertine
Finely divided quartz	Massive or layered, microcrystalline	Chert
Organic		
Calcite	Massive, variably crystalline	Limestone
	Massive, variably crystalline; firmly embedded fossil fragments	Fossiliferous limestone
	Fragmental rock with abundant, poorly cemented fossils	Coquina
	Massive, microcrystalline	Micrite
	Massive and soft; composed of abundant microscopic fossils	Chalk
	Massive rock with small, very well rounded calcite grains	Oolitic limestone
Chemical Alteration Product		
Dolomite	Dense, massive, crystalline or microcrystalline	Dolomite (Dolostone)

Clastic Sedimentary Rock Classification

Grain Size	Composition and/or Characteristics	Rock Name
Coarse-Grained (> 2 mm)	Composed of rounded rock or mineral fragments; usually poorly sorted	Conglomerate
	Composed of generally angular rock or mineral fragments; usually poorly sorted	Breccia
Medium-Grained (0.062–2mm)	Composed mostly of quartz; sorting and roundness are variable	Quartz sandstone
	Composed mostly of quartz and feldspar; usually poorly sorted, angular	Arkose
	Composed of clays, quartz, and rock fragments; poorly sorted, angular grains	Graywacke
Fine-Grained (0.062–0.005 mm)	Generally massive rock composed of quartz and clays (feels "gritty")	Siltstone
Very Fine-Grained (<0.005 mm)	Generally laminated rock composed of quartz and clays (feels "smooth")	Shale

(From R. D. Dallmeyer, *Physical geology laboratory text and manual,* 2nd ed. 1978 © Kendall/Hunt Publishing Company.)

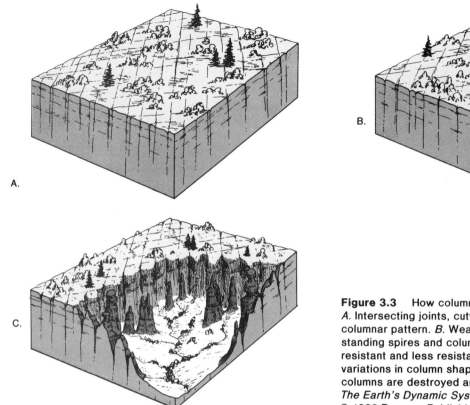

A.

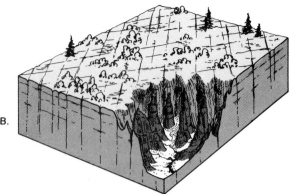

B.

C.

Figure 3.3   How columns may develop as a cliff retreats.
A. Intersecting joints, cutting across the rock layers, form a
columnar pattern. *B.* Weathering along joints produces
standing spires and columns. Differential weathering of
resistant and less resistant horizontal beds etches out
variations in column shape. *C.* As cliff retreat continues, old
columns are destroyed and new ones are created. From
The Earth's Dynamic Systems (3rd ed.) by W. K. Hamblin.
© 1982 Burgess Publishing Company; used by permission.

Figure 3.4   Bryce Canyon is being cut back into the
Paunsaugunt Plateau due to weathering along intersecting
joints, mass wasting, and erosion (see fig. 3.3). The rate of
cliff retreat is approximately one foot every 50 years.
Extensive talus slopes accumulate as columns and spires
are eroded from the rock layers of the Wasatch Formation,
which is made up of limestones, siltstones, shales, and
sandstones of varying resistance. National Park Service
photograph.

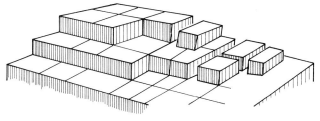

A.

B.

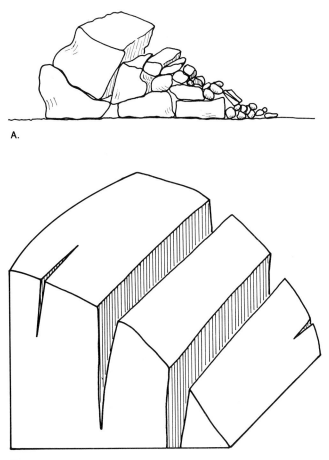

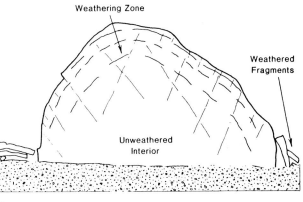

Weathering Zone

Weathered
Fragments

Unweathered
Interior

C.

D.

Figure 3.5 Examples of weathering processes. In nature, several processes of weathering are usually operating at the same time to break up rock. *A* **Granular disintegration.** Massive rocks that do not have bedding, joints, or foliation tend to break up into irregularly shaped blocks and fragments that, in turn, become broken up until eventually individual mineral grains are weathered out from the disintegrating rock. Whenever water is available, small fragments and particles are washed away. *B.* **Separation along joint blocks.** When a bedrock has bedding, joints, or definite foliation, weathering penetrates the fractures and, in time, causes the rock to come apart in blocks. Once loosened from the outcrop, blocks may move downslope or become rounded or broken up into smaller and smaller blocks. *C.* **Frost wedging.** Water freezing in cracks and fractures in rock expands and exerts pressure on the rock, enlarging the openings. Many cycles of freezing and thawing eventually cause slabs or blocks of the rock to split off. *D.* **Spheroidal weathering.** Processes of physical and chemical weathering operate more rapidly on the exposed edges and corners of boulders and outcrops than on flat surfaces. This causes cracks to form beneath the rock's exterior so that concentric shells or layers loosen and spall off.

weathers physically and chemically at the same time. However, in the environment of Bryce where the atmosphere is dry and rain is intermittent or scanty most of the time, the physical aspects of weathering tend to be dominant. In addition to the semiarid climate, the elevation of Bryce makes it one of the colder locations in Utah. Fairyland View, for example, is 7,586 feet above sea level; Bryce Point, 8,294 feet; and Rainbow Point, the highest spot on the rim, 9,105 feet.* At these altitudes frost action

*From Rainbow Point a person can view most of Bryce Canyon, the Table Cliffs, Plateau, White Cliffs, Vermilion Cliffs, and the Henry Mountains.

plays a significant role in the breaking up of bedrock, although the climate is generally warm and sometimes extremely hot at lower elevations (fig. 3.5C).

Such extremes of temperature tend to cause *granular disintegration* of coarse-grained rock masses such as conglomerate and coarse sandstone, both of which are found in Bryce. The mineral grains tend to separate from one another along their natural contacts to produce coarse sand or gravel (fig. 3.5A).

Granular disintegration brings about minor changes, however, in comparison to the processes of frost action. Moisture in cracks and joints freezes and expands, causing *frost wedging;* this enlarges cracks and breaks off rock fragments. Subsurface freezing of water in rock forces a

Table 3.3
Weathering Processes

A. Mainly physical (mechanical) weathering—*disintegration*

Type	Process
1. Granular disintegration	Separation of coarse mineral grains due to temperature extremes
2. Frost wedging	Grains and particles forced apart by freezing water
3. Frost heaving	Rock fragments raised up by ice pressure
4. Exfoliation	Stripping of concentric rock slabs from outer surface of rock mass
5. Spheroidal weathering	Rounding by weathering from an initial blocky shape
6. Organic weathering	Root pry, animal burrows, human activity, etc.

B. Mainly chemical weathering—*decomposition*

Type	Process
7. Oxidation	Reaction of mineral grains with atmospheric oxygen
8. Carbonation	Reaction of carbonic acid (from water and carbon dioxide) with mineral grains
9. Hydration	Minerals in rock combining with water
10. Solution	Soluble minerals go into solution; insoluble mineral grains transported as sediment

portion of rock upward, causing more fragmentation, in a process called *frost heaving*. As cracks are deepened, moisture penetrates farther into the rock and enables chemical weathering to attack the minerals, loosening and altering the grains (fig. 3.5C).

Spheroidal weathering is a physical process that has the effect of rounding off pinnacles and knobs in Bryce Canyon. Most of the features tend to be angular as they break up along the vertical joints (fig. 3.5B). However, a sharp corner has more surface area exposed than a smooth flat rock face, so weathering occurs faster on the corners. Thus angular shapes become rounded as the corners retreat. If enough moisture is present, chemical weathering penetrating into cracks may cause rounded layers to spall off in concentric shells (fig. 3.5D).

A similar process on a more massive scale is called *exfoliation*. Large curved sheets or slabs of rock peel off, or *exfoliate*, from a bedrock surface. As disintegration proceeds, decomposition by chemical processes leads to further rock decay.

Plants and animals (including man) cause fragmentation of rock by both physical and chemical processes. Plant roots pry open cracks and then plant acids react chemically with rock. Animals dig burrows, and men blast out rock in building roads and tunnels. All of these activities expose fresh rock surfaces to moisture and organic and atmospheric acids.

Most weathering processes are more effective in the presence of water. Although water is scarce in Bryce, it is present in the forms of frost, some snow and ice, and occasional severe thunderstorms that may cause the flash flooding characteristic of arid climates. The sudden heavy rainstorms do intensify (at least temporarily) the processes of weathering, but the storms' main effect is erosive as they wash down loose sediment and transport quantities of debris downslope to streams. Very little vegetation is available to hold the soil back or aid in absorption of the rainwater.

In the normally dry conditions of Bryce, the lack of moisture in the atmosphere and soil causes the chemical reactions of weathering to operate much more slowly than they do in humid climates. Nevertheless, the effects of chemical weathering are evident. *Oxidation,* which involves the reaction of certain mineral constituents in rock with the oxygen in the air, is the process responsible for the many shades of color characteristic of Bryce Canyon. The reds, pinks, and browns are produced by the iron oxide hematite (Fe_2O_3); the yellows by another iron oxide, limonite ($Fe_2O_3 \cdot nH_2$); and the purple by manganese oxide (MnO_2). Most of the white or light-colored beds have been leached out.

Hydration is a weathering reaction involving the combining of water with minerals in rock. It causes the grains to expand and the rock to be weakened. (Limonite, above, is an oxide that has been hydrated.)

The *carbonation* reaction begins with the combination of carbon dioxide in air and soil with water to form carbonic acid, which plays an important role in the weathering of many kinds of rock. Some feldspars, for example, weather by carbonation to produce calcite ($CaCO_3$), the main constituent of limestone.

Examples of *solution weathering* (i.e., the dissolving of the soluble minerals in rock) can be seen in such features as the Wall of Windows along Peek-a-boo Trail. Here again, physical and chemical processes of weathering interact since exfoliation also aids in the carving out of the windows. Many of the odd and unusual rock shapes in the Park are due in part to solution weathering of less resistant portions of the original bedrock.

Frost action on jointed rocks in a region of high relief accounts for the effectiveness of mass wasting in Bryce Canyon National Park. Evidence for this can be seen in the extensive accumulations of talus that overspread the base of slopes, pile up against the cliffs, and surround the columns (figs. 3.4, 3.7). Typically the steepness of a talus

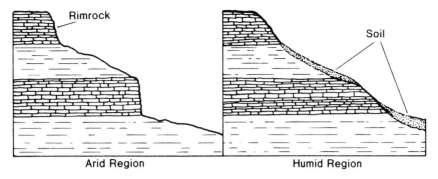

Figure 3.6 Land surfaces produced by weathering and erosion tend to reflect differences in climate from one region to another—even when the underlying bedrock is similar, as in the diagram above which shows rock layers of differing resistance being eroded. Slopes in arid regions are generally steeper and more angular than slopes in humid regions, which characteristically are smoother, more rounded, and mantled with soil. The major cause of the distinctly different slope profiles is the difference in the amount of moisture available for weathering processes. From *Landforms and Landscapes* (3rd ed.) by Sherwood D. Tuttle. © 1970, 1975, 1980 Wm. C. Brown Company Publishers, Dubuque, Iowa. Reprinted by permission.

FROST ACTION—TALUS

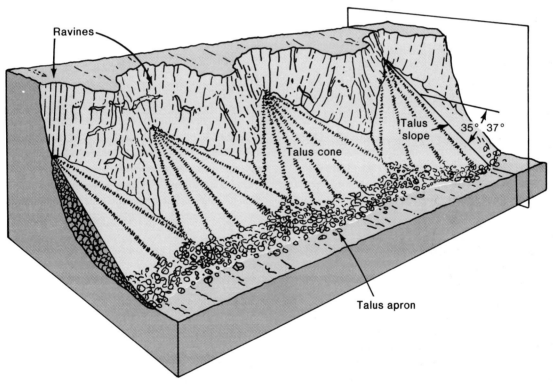

© WARD'S Natural Science Establishment, Inc. 1964. Rochester, N.Y.
Monterey, Calif.

Figure 3.7 Talus slopes developing in front of a retreating cliff due to weathering by frost action combined with mass wasting. Note that the steepness of the slopes is determined by the angle of repose of the accumulating sediments. © 1964 Ward's Natural Science Establishment; used by permission.

slope or cone is at the angle of repose (roughly 35°), and loose fragments continually roll downslope or shift position as they move downward.

On a landform scale, it can be seen that the terraces in the Bryce Canyon region are formed in the softer rocks, such as shales. The cliff-formers are the resistant sandstones and limestones. The plateaus are associated with the various formations: Pink Cliffs, the Wasatch; Gray Cliffs, the Dakota Sandstone; White Cliffs, the Navajo Sandstone; Vermilion Cliffs, the Wingate Sandstone; and Chocolate Cliffs, the Shinarump Conglomerate. The Bryce Canyon area itself has three topographic divisions: (1) the Paria amphitheater, a lowland at the base of the Paunsaugunt Plateau; (2) the Wall, which is the rim of the plateau; and (3) the Highland at the top of the plateau.

The Paria River and its headwaters drain Bryce Canyon, carrying away sediment and debris plus rock material in solution. The Paria flows southeastward, emptying into the Colorado River and eventually the Pacific Ocean. To the west a narrow drainage divide (barely five feet wide) runs along the Paunsaugunt Plateau, paralleling the Park's western boundary and separating the drainage basin of the Paria River system from that of a rather sluggish, slow-moving river system on the plateau. The streams on the Paunsaugunt, which have low velocity due to their low gradient, flow westward to the Great Basin and disappear in the desert. In terms of definition, a *drainage basin* can be described as the total area drained by a stream and its tributaries; while a *drainage divide* is the boundary between adjacent drainage basins (fig. 3.8).

As was noted earlier, Bryce is not a true canyon, but is rather a basin cut into a plateau made up of fairly thick, massive rock units. The very thin Dakota Sandstone, the oldest formation exposed in the Park, is an interesting exception and it can be seen in the Paria amphitheater. The Dakota Sandstone is lithified from *blanket sand*, that is, a sedimentary layer thinly covering a very large geographic area. In this region the deposit represents the beach area of a transgressing sea (fig. 3.9).

Geologic History

1. **Deposition of the Navajo Sandstone as migrating sand dunes; the Carmel Formation deposited in a transgressing sea.**

   During the Jurassic Period, the dunes that lithified into the Navajo Sandstone spread over a large area in what is now southwestern United States. The Carmel Formation was deposited in a transgressing sea that flattened the dunes as it advanced. The Navajo and the Carmel are not exposed within the boundaries of Bryce National Park but can be found in cores from drilled water wells.

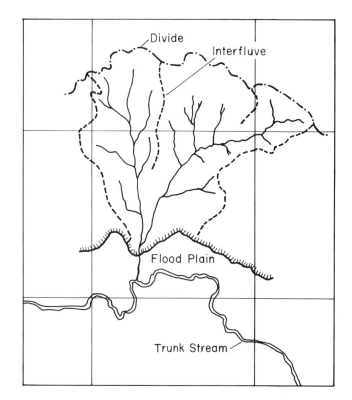

Figure 3.8 A small drainage basin, in map view, with a drainage divide separating it from neighboring drainage basins. An interfluve is a small ridge, or divide, that separates tributaries. Water from the drainage basin empties into a trunk stream flowing on a flood plain.

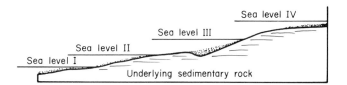

Figure 3.9 A blanket sandstone may form from beach sand as land subsidence or rise in sea level causes a shoreline to migrate inland. The rock formation shown here is the same lithologically from left to right but it becomes progressively younger toward the right as the sea encroaches on the land.

2. **Deposition of undifferentiated Jurassic beds.**

   These poorly defined beds, with indefinite contacts, are a mixture of sandstone and shale. They belong to the San Raphael Group, the oldest being the Entrada Sandstone, which was deposited in water in some regions, but gradually graded into land or terrestrial deposits and finally into sand dunes. None of these Jurassic beds are exposed in Bryce Canyon.

   The sea in which the Entrada was deposited retreated to the west as the region underwent uplift, forming the San Rafael swell. Some of the softer sediment was slightly folded. Streams meandered across the land surface, transporting sediment to the retreating shoreline. This created a landscape of gently rolling hills as the San Rafael swell continued to rise.

Later the seas returned, depositing a thin layer of gravel that lithified to conglomerate. Over the gravel, sand was laid down that became the Curtis Formation, a pale-colored deposit. Thin layers of shale containing the green mineral *glauconite*, an iron potassium-silicate, are mixed with the sandstone layers. Since glauconite can form only in a marine environment, its presence indicates that the beds were formed in a sea. The overlying Winsor Formation is a sandstone made up of dunes that accumulated along the intertidal mud flats of a Jurassic Sea.

3. Deposition of the Dakota Sandstone in a shallow sea during the Cretaceous.

The Dakota Sandstone represents the advancing shoreline of the last sea to cover the Colorado Plateaus region. The sediment was deposited as a blanket sand about 125 million years ago by a shallow sea that transgressed over the landscape. The Dakota is called a blanket sand because it is usually less than 50 feet thick, yet it covers a wide geographic area. Its composition is exceedingly varied. The initial sediment was a mixture of gravel and sand (conglomerate and sandstone eventually). The zones in which clay (shale) and coal are interbedded represent deposition in coastal marshes and lagoons. Deposits of pure sandstone indicate the location of ancient beaches. The age is also variable since the sea crossed time boundaries as it slowly encroached on the landscape (fig. 3.9).

4. Deposition of the Tropic Shale (mostly marine).

As they are incomplete and poorly represented, no sharp separation can be found between the Tropic Shale and Dakota Sandstone in exposures within the Park. The Tropic Shale was laid down in a shallow sea that occasionally regressed long enough for coal swamps to form along the coast. The coal beds are thick enough to mine on a commercial basis in areas adjacent to the Park.

5. Deposition of the Straight Cliffs and Wahweap Sandstones.

These sediments were deposited in a shallow regressing sea and in brackish lagoons. They have a total thickness of 1,500 to 2,000 feet. A buff-colored marker bed of sandstone 30 to 150 feet thick makes the formations easy to trace, because it forms a vertical cliff. Due to a shifting shoreline, the Straight Cliffs Sandstone consists of alternating layers interbedded with a portion of the overlying Wahweap. Each is easy to distinguish since the Straight Cliffs forms vertical cliffs and the Wahweap has a tendency to form slopes. Combined they form a series of giant steps.

6. Deposition of the Kaiparowits Formation as sandbars, deltas, and river flats during late Cretaceous time.

The Kaiparowits beds, ranging in color from a dark grayish-green to a yellowish-brown, were deposited by meandering streams in several environments. The beds are discontinuous, with a variation in composition all the way from pure white sandstones to arkoses containing sand-sized grains of micas, feldspars, and quartz. Argillaceous (clay) shales and calcareous (limy) shales are intermixed with the sand beds. Lenses of impure limestones, ironstone concretions, and sandballs, all of which are resistant to weathering, tend to form knobs on the eroding slopes of the Park as well as caprock on pinnacles and buttes. Within the beds fossils of crocodiles, turtles, dinosaurs, and freshwater invertebrates are found along with leaves and petrified wood. Deposition of this formation brought the Cretaceous to a close around 70 million years age.

7. Erosion of Kaiparowits beds at the beginning of Cenozoic time.

The Kaiparowits Formation varies considerably in thickness because much of it was removed by erosion. Therefore, the thickness ranges from less than 100 feet to over 1,000 feet. Because of the erosion, an unconformity formed with the overlying Eocene Wasatch beds.

8. Deposition of the Wasatch Formation along streams and in quiet water some 54 million years ago.

The Wasatch is a mixture of thick red limestones with thin layers of mottled, reddish shales and gray sandstones combined with discontinuous lenses of calcareous gravel. The lower portion of the Wasatch is a basal conglomerate containing rounded fragments of quartzite, quartz, igneous rocks, limestone, and sandstone. The gravels and sandstones were deposited in and along streams, and the limestones and shales formed in quiet water. Fossils of primitive mammals, clams, snails, turtles, and plants have been found in the Wasatch beds.

Thickness of the Wasatch varies from 800 feet to about 1300 feet. Because it is cemented mainly by iron oxide, most of the beds are striking reds and pinks in color. The formation has been covered in some areas outside the Park by basaltic lava flows. North of Bryce Canyon igneous rocks are abundant.

9. Elevation of the plateau around 2,000 feet by uplift and block-faulting.

About 13 million years ago, uplift of the High Plains of Utah caused the area to be broken into nine blocks, and a series of plateaus was formed. The western border of the Paunsaugunt Plateau is separated from the Markagunt Plateau by the Sevier fault along the Sunset Cliffs. To the east is the Aquarius Plateau, which is separated by the Paunsaugunt fault. A remnant of the Paunsaugunt Plateau, Table Cliffs, shows a displacement of 2,000 feet along the Paunsaugunt fault. Thus at Bryce Point the Wasatch

Formation has an elevation of 8,294 feet on the west side of the fault, while the same sedimentary layers on the east side are at an elevation of 10,000 feet. The earthquake that occurred in the late 1960s in Bryce Canyon indicated that movement along the fault has not entirely ceased.

10. Erosion of the plateau and development of Bryce Canyon during the Holocene Epoch.

The Sevier River, flowing in a shallow valley on the Paunsaugunt Plateau with a gradient of only 15 feet per mile, was relatively unaffected by uplift and has continued to drain toward the west. The Paria River, on the eastern side of the divide, was rejuvenated by the uplift and has a high gradient of 1000 to 1500 feet per mile as it flows over the rim and drains Bryce Canyon. Uplift has also caused some of the Paria tributaries (such as Bryce Creek, Yellow Creek, Willies Creek, and Riggs Creek) to capture the headwaters of streams that used to flow north. The Paria river system drains southward into the Colorado River.

The geologic processes that carved Bryce Canyon are, to summarize, (1) differential weathering of flat-lying sedimentary layers of varying resistance; (2) weathering (especially frost action) along vertical joint systems; (3) mass wasting; and (4) stream erosion. These processes in combination produced the unevenly eroded, or scalloped, edge of the plateau and carved out the bowl-shaped basin. Flash floods, which characterize dry climates, are responsible for some of the rapid changes that occur each year as violent rainstorms undermine and cause collapse of pinnacles, columns, and walls.

Cedar Breaks National Monument, located on the western edge of Markagunt Plateau resembles Bryce Canyon and is also eroded from the Wasatch Formation.

Geologic Map and Cross Section

American Association of Petroleum Geologists. 1967. *Geological highway map no. 2 of the southern Rocky Mountain region.* Tulsa, Oklahoma: American Association of Petroleum Geologists.

Bibliography

Follows, D. 1972. *Bryce Canyon.* Desert News Press, U.S.A. Originally published by Bryce Canyon Natural History Association.

Gregory, H. E. 1951. *The geology and geography of the Paunsaugunt region, Utah.* U.S. Geological Survey Professional Paper 226.

———— 1956. *Geologic and geographic sketches of Zion and Bryce Canyon National Parks.* Springdale, Utah: Zion-Bryce Natural History Association.

Rigby, J. K. 1976. *Northern Colorado Plateau,* K/H Geology Field Guide Series. Dubuque, Iowa: Kendall/Hunt Publishing Company.

Address

Bryce Canyon National Park
Bryce Canyon, Utah 84717

4
Canyonlands National Park

Location: Southeast Utah
Area: 337,570.43 acres; 558.7 square miles
Established: September 12, 1964

Figure 4.1 Looking east over the Land of Standing Rocks in Canyonlands National Park. The location of the fins (the rock ridges) is controlled by the vertical, closely spaced joint system of the Cedar Mesa Sandstone, a member of the Cutler Formation. Spheroidal weathering has rounded the exposed surfaces of the fins. National Park Service photograph.

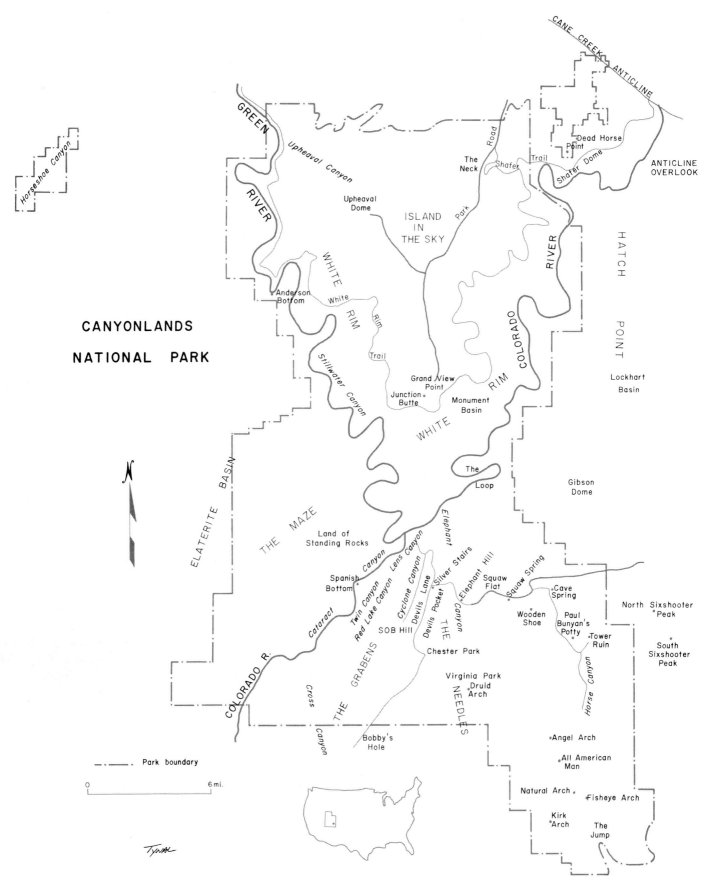

Figure 4.2   Canyonlands National Park, Utah.

Table 4.1. Geologic Column, Canyonlands National Park

Time Units			Rock Units			Geologic Events
Era	Period	Epoch	Group	Formation	Member	
Cenozoic						Arches and needles formed; entrenching of Colorado and Green Rivers. Migration and doming of salt beds. Continued uplift and block faulting.
Mesozoic	Cretaceous					Erosion, uplift, block faults
	Jurassic	Middle to Lower	San Rafael	Summerville		Deposition on tidal flats
				Curtis		Deposition in a transgressing sea.
				Entrada Sandstone	Moab Tongue	Wind-blown deposits
					Slick Rock	Deposited in a regressing shallow marine sea.
				Carmel	Dewey Bridge**	Deposited in a shallow transgressing sea.
	Triassic	Upper	Glen Canyon	Navajo Sandstone		Erosion Wind-deposited sand dunes
				Kayenta		Deposited by braided streams and in lakes
				Wingate Sandstone		Wind-deposited sand dunes
		Middle to Lower		Chinle	Church Rock	Deposited in streams and lakes
					Owl Rock	Deposited in lakes and rivers
					Petrified Forest	Deposited on fluvial flood plains. Nearby volcanic activity.
					Moss Back	Deposited as a fluvial complex in point bars, etc.
				Moenkopi	Upper Member	Erosion Tidal flats deposits
					Hoskinnini Tongue	Coarse-grained sandstone formed from the erosion of highlands.
Paleozoic	Permian			Cutler	White Rim Sandstone	Deposited in sublittoral marine environment.
					Organ Rock Tongue	Fluvial flood plain deposits on a coastal lowland.
					Cedar Mesa Sandstone	Sandbars forming in a transgressing sea, some wind-blown. Nearshore to coastal environment.
				Elephant Canyon } Halgaito Shale*		Fluvial to coastland to marine facies, shale grading into low coastlands from a shallow marine environment.
	Pennsylvanian		Hermosa	Honaker Trail		Cyclical deposition in a shallow sea.
				Paradox		Deposition in land-locked lagoons, evaporites
				Pinkerton Trail		Deposition in a transgressing sea, shallow water.
			Older rocks not exposed			

Source: modified after Mutschler 1969; Baars and Molenaar 1971; Lohman 1974.
*Referred to by some authors as Rico Formation.
**Some authors consider the Dewey Bridge as a member of the Entrada Formation.

Local History

People of the Fremont Culture lived in this area until A.D. 900. Their pictographs can be seen on Newspaper Rock, a state historical marker outside the boundaries of Canyonlands National Park. *Desert varnish*, a thin dark film of iron and manganese oxides that, with long exposure, forms on the surface of desert rocks, covers Newspaper Rock. The Indians discovered that they could make pictures on the rock by scratching through the coating to the lighter-colored, unweathered rock beneath.

Indians of the Anasazi (or Pueblo) Culture arrived in the region around A.D. 1075 and left during the 12th century. Modern Indians living in the area are the Utes and the Navajos. The early white settlers, as well as the Indian groups, left their marks on Newspaper Rock.

In 1859 Captain John N. Macomb led a group of people into the area and decided the region was totally worthless. However, John S. Newberry, a geologist in the party, was awed and delighted by the scenery.

Since this is desert country (only 5 to 9 inches of precipitation per year), water is precious and water holes are few and far between. Settlers were not attracted to the region, but in the late 1800s and early 1900s, cowboys and cattle and sheep herders passed through the area from time to time. Butch Cassidy and his "Wild Bunch" knew the region well and used their knowledge to evade posses of lawmen on numerous occasions. The area was ideal for outlaws and cattle rustlers who knew the location of the water holes and could come and go without getting lost in the confusing jumble of canyons, cliffs, and exceedingly rough terrain. Prospectors for gold, silver, and (more recently) uranium have also been lured to this barren land by the hope of "making a strike." Canyonlands is still a rough and primitive area that is best traveled by jeep, especially such sections as Elephant Hill, Sob Hill, Silver Stairs, Shafer Trail, and White Rim Trail.

Geologic Features

The rocks in Canyonlands are sandstones with uniformly sized, clear quartz grains cemented with iron oxide that gives the rock units their red color. Close, parallel vertical joints, differential weathering, and uplift are mainly responsible for the erosional features in the Park such as the fins (ridges), arches, and pinnacles. There are many collapse features, such as graben valleys. They are due to block-faulting, intrusion of salt beds, and the subsequent solution of salt by ground water, which results in the overlying beds sinking down from their own weight.

Upheaval Dome (fig. 4.3) is a very large, circular collapse feature. It exhibits intense deformation of beds and is surrounded by a ringlike depression. Geophysical studies indicate that a salt dome forcing its way up through sedimentary beds was responsible for Upheaval Dome. The salt came from the Paradox Formation, a member of the Hermosa Group, that was deposited during the Pennsylvanian Period. Above the Paradox Formation are many rock layers, a mile thick overall. Since salt is a substance that becomes *plastic* (i.e., able to flow without rupture) under pressure, the great weight of the overlying rock caused the salt to push upward as a huge plug that arched the overlying beds into a dome, or circular anticline. As the beds were pushed up, they became cracked and severely deformed. This made them more susceptible to weathering so that the top beds were removed by erosion. Then, since the beds exposed in the center of Upheaval Dome are softer than the resistant ones that form the rims, the less resistant rock eroded more rapidly, forming the depression in the center. A stream flowing to the northwest removed the sediment and carved a canyon in the rim called Upheaval Canyon. An intrusion of igneous rock below the salt beds and approximately 4,800 feet beneath the dome may have been in part responsible for the uplifting that occurred.

The same Paradox salt beds caused the collapse of the graben valleys between Cataract Canyon and Elephant Hill in Canyonlands. As the salt beds shifted and migrated, the overlying beds developed cracks from the stress and pressure, giving ground water easy access to the soluble salt.

The *fins*, which are narrow sandstone ridges, or walls, are a unique feature of the Park. The closely spaced vertical joints become enlarged by chemical and physical weathering as water seeping down through fractures dissolves iron oxide cement and loosens the sand grains. Frost action, with its alternate freezing, expanding, and thawing, also breaks up the sandstone; and the sediment is removed by streams. The joint spaces become gradually enlarged into narrow valleys that separate the narrow walls, called fins.

Weathering and the formation of arches separate the fins even more. As the joint spaces widen and the narrow valleys deepen, beds at the base of the fins are exposed to weathering. If the lower beds are slightly softer, they are removed more rapidly than the top beds. Because the fins are narrow, a hole, or *window*, appears where the rock is worn completely through. Sand blown by wind (sandblasting) polishes and rounds off the window edges. Snow, rain, and ice continue to attack the rock, dislodging fragments and causing rockfalls (a process of mass wasting) until the windows are enlarged to form *natural arches*.

Certain rock formations, because of their composition, type of cement, structure, etc., have a tendency to form arches. Most of the arches in Canyonlands are in the Cutler Formation, which tends to form standing rocks rather than arches. On the other hand, the Entrada Sandstone, exposed in nearby Arches National Park (chapter

Figure 4.3 Upheaval Dome, which is almost 1,600 feet deep and three miles across, was formed by the intrusion and subsequent collapse of salt domes. The inner rim of cliffs around the collapsed dome is formed of Wingate Sandstone. Beds of the Chinle and Moenkopi Formations are exposed in the dissected cone in the dome's center. The oldest exposed beds, the arkosic rocks of the Cutler Formation, can be seen only in the deep canyons of the innermost part of the dome. Ringing the outer part of the structure, like a doughnut, are prominent cliffs of Navajo Sandstone. The circular valley between the Navajo and Wingate rims is carved in the less resistant Kayenta Formation. National Park Service photograph.

5) tends to form arches very rapidly. This explains why Canyonlands has only 25 arches while Arches National Park has around 90, even though both areas are in sandstone formations. Eventually arches collapse, and only the buttresses remain standing as pillars.

The *needles* and *pillars*, or standing rocks, develop when two intersecting joint sets are spaced closely together. The second set of joints, which is at an angle to the first set, prevents fins from developing. Instead, pointed needles or more rounded pillars evolve as weathering and erosion sculpture the rock. The Cedar Mesa Sandstone, a member of the Cutler Formation, tends to form needles and pillars (fig. 4.1). Some are capped by resistant material that makes balancing rocks. The fins in the Cedar Mesa develop vertical fractures that permit more rapid weathering along these openings. Eventually, the fins break down into a line of columns, pinnacles, and spires. These features can be seen in Monument Basin.

The Colorado River flows through Canyonlands in a spectacular, deep, narrow-walled gorge. The Colorado is joined by the Green River (its largest tributary), which also has a deep and narrow canyon (fig. 4.4). Together they flow quietly for about 3 miles until they enter Cataract Canyon at Spanish Bottom. The turbulent cataracts in this 16-mile, appropriately named canyon exist because of changes in the bedrock of the stream channel.

Both the Green River and the Colorado River have cut down their channels to form *incised meanders* (figs. 4.4, 4.5). In the geologic past, these two rivers meandered back and forth on a fairly low, flat surface close to base level. As the region was gradually raised, creating new base levels, the rivers did not change their courses (as would have been the case if uplift had been rapid). Instead, the streams began downcutting their channels, the energy being supplied by the increased velocity. The streams still follow their old meandering courses but now flow in narrow, steep-sided canyons as the meanders have

Figure 4.4   Looking downstream (south) over the confluence of the Colorado River and the Green River, which joins the Colorado from the right. The water of the Colorado River, which carries a heavier load of sediment in suspension, is different in color from the water of the Green River. A bar of river gravel and sand has accumulated on the inside of the bend in the middle distance. The lower part of the Cutler Formation forms the canyon rim. Below the Cutler, the canyon is cut through the Halgaito Shale (also called Rico Formation) into massive beds of the upper part of the Hermosa Group. Aerial photograph courtesy Myron Agnew.

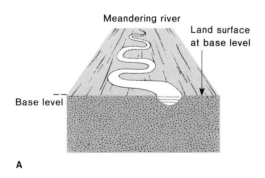

A

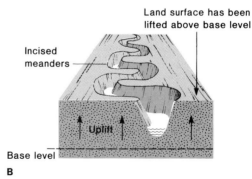

B

Figure 4.5   During and following regional uplift, incised meanders tend to develop. From *Physical Geology* (2nd ed.) by Charles C. Plummer and David McGeary. © 1979, 1982 Wm. C. Brown Company Publishers, Dubuque, Iowa. Reprinted by permission.

Figure 4.6 The Needles and Grabens, in the south central part of Canyonlands National Park, are part of a broad, dissected benchland east of the Colorado River. The Needles area is largely to the left of Chesler Park (foreground); the grabens, which are salt-trough valleys, are to the right. Chesler Park is a natural meadow, walled in by needles eroded from the Cedar Mesa Sandstone. The location and pattern of the canyons and mesas of the landscape is controlled by the geologic structure. Two sets of fractures, produced by tensional forces, have cut the bedrock, one trending northeast-southwest and the other northwest-southeast. Movement has not been perceptible along the northeast-southwest fractures, but vertical displacement has occurred along the northwest-southeast fractures. Between these fractures faulting has moved crustal blocks differentially several hundred feet up or down, forming horsts (uplifted blocks) or grabens (downdropped blocks). Although the grabens' average width is about 500 feet, they range in width from 7 to 8 feet to nearly 2,000 feet. Since the faulting occurred relatively recently, the processes of weathering and erosion have only just begun to modify the graben walls. From a U.S. Air Force oblique air photograph reproduced in Lohman 1974, U.S. Geological Survey Bulletin 1327.

been incised. The downcutting isolated an area known as Island in the Sky, a 1,400-foot high mesa that is 6,000 feet above sea level. The mesa stands between the canyons created by the Colorado and Green Rivers.

Island in the Sky is the highest of several surfaces (or levels) that can be seen in the Park. The Wingate Sandstone, which is a dune sandstone and very resistant to erosion, caps the mesa and has fairly effectively prevented the landform from being dissected and eroded away. The White Rim Plateau, the next surface lower, is a thousand-foot drop from Island in the Sky. The plateau is capped by the White Rim Member of the Cutler Formation, the White Rim being a white, cross-bedded sandstone similar to the Wingate, which explains why it is so resistant. The lowest surface is capped by the Cedar Mesa Sandstone, the lowest member of the Cutler Formation. Cedar Mesa is also a cross-bedded sandstone containing thin beds of cherty limestone. Its color varies from pale red to white. Many of the fins, needles, pillars, and arches for which the Park is noted are formed in the Cedar Mesa due to the joint systems discussed earlier. Angel Arch, for example, has a window that is 190 feet high, and David Arch is 250 feet tall, an indication of how thick this sandstone is. Both arches are located in the southeast corner of the Park and can be reached by four-wheel drive vehicles or on foot.

At the bottom of the canyon are the incised rivers, the Colorado and the Green. The Green River originates in Wyoming and flows southward to its junction with the

Colorado. Although both rivers are fresh water and flow year-round, they are not potable. This is because the local saline ground water seeps into the rivers from the salt beds.

People and animals in the Park are dependent for water on rock depressions in the White Rim Sandstone and Cedar Mesa Sandstone or on spring water that seeps from the contact between the Organ Rock Tongue and the White Rim Sandstone. The hollows in which water collects, locally referred to as "tanks" or "charcos," are formed by the combined process of solution of calcium carbonate cement and removal of loose rock grains by wind (deflation). Tanks are found by Junction Butte near Grand View Point in the central portion of the Park and also in the Needles area near Chesler Park in the south central portion (fig. 4.6).

Chesler Park derived its name from the fact that it is an expanse of a thousand acres of flat, grassy land surrounded by pinnacles of the Cedar Mesa Sandstone. To the west is the horst-and-graben topography and the so-called salt valleys, such as Devils Pocket, Devils Lane, Cyclone Canyon, Red Lake Canyon, Twin Canyon, and Lens Canyon.

The same sort of desert varnish that covers Newspaper Rock can be seen on many of the rocks in Canyonlands. It is believed to be the exudation of mineralized solutions from within the rocks that is hardened by evaporation. A different type of exudation is found in Elaterite Basin (west of the Maze) where a dark brown, elastic tar seeps out of the White Rim Member of the Cutler Formation and hardens to a brittle substance upon exposure to air. It is called elaterite and is composed of petroleum residues.

The Cutler Formation at the western end of Canyonlands is an *arkose*, a sandstone containing quartz and 25 percent or more feldspars. As it is traced eastward, the undifferentiated Cutler gradually changes from a fluvial (river) deposit to a coastal dune deposit, the Cedar Mesa Sandstone; then the Organ Rock Tongue, a river and flood plain deposit; and finally the White Rim Sandstone, which is a deposit of offshore sandbars. This type of sediment transition is called a *sedimentary facies*. That means that different rock types are deposited in different environments and geographic areas at the same time. The entire series represents a large *alluvial fan* deposit, a landform usually found where a stream's velocity decreases suddenly, as when a stream emerges from a narrow canyon onto a plain or basin at the foot of a mountain range. In this case, the fans formed in a graben, or fault-block valley created by the uplifting of adjacent blocks. The valley was probably a *closed drainage basin* (i.e., interior drainage) where the water evaporated or sank into the ground.

Geologic History

1. Beginning of the Cordilleran geosyncline during Cambrian time.

During the Cambrian Period, when most of what is now North America lay along the paleoequator, the Cordilleran mobile belt was a great elongate tract that stretched along the leading edge of the landmass as the plate moved slowly northward and westward. This mobile belt became the Cordilleran Geosyncline, which covered a wide area that stretched from Alaska down through Canada and the United States and into Central America. Throughout early Paleozoic time, seas covered most of what is now North America.

The Cordilleran geosyncline, which was the dominant feature of western North America throughout the Paleozoic and Mesozoic Eras, was finally destroyed by the Rocky Mountain orogeny, a period of severe crustal deformation that produced the Rocky Mountains.

2. During Ordovician/Silurian time, uplift and erosion; in the Devonian and early Mississippian, deposition.

The Cordilleran geosyncline evolved through stages of sedimentation, uplift, erosion, and renewed deposition as the seas periodically retreated and advanced over the mobile belt and the landmass. The seas regressed during the Ordovician; and uplift and orogeny followed in the Silurian, accompanied by erosion that wore down the mountains and highlands. During the Mississippian occurred the last great Paleozoic flooding of the North American landmass, although transgressions and regressions continued during late Paleozoic time.

3. Middle to late Mississippian time, regression of seas.

By the middle of the Mississippian Period, mountain-building and uplift to the west and southeast caused the seas to retreat from the Colorado Plateaus area.

4. Uncompahgre uplift in the Pennsylvanian Period; formation of adjacent Paradox Basin and deposition of salt beds.

Again the seas advanced, gradually covering the Paradox Basin and Canyonlands and reworking the red soils that had formed. This sediment was redeposited as the Pinkerton Trail Formation, which represents the first beds laid down by the transgressing sea in early Pennsylvanian time.

During this period, about 325 million years ago, the Uncompahgre (un-com-paa-grey) Range was developing as fault-block mountains along a system of faults that had formed much earlier. Eventually the region was divided by a series of northwest-trending faults into elongate ridges and troughs, which produced a closed drainage basin. In the troughs nearer the ocean, seawater became trapped.

In the arid climate the water evaporated rapidly, concentrating the salts and mineral matter in these lagoons. The beds in this entire sequence make up the Hermosa Group.

Ordinary seawater has 3.5 parts salt per 1000 parts water. When the ratio changes to approximately 10 parts salt per 1000 parts water (3x normal), the minerals anhydrite and gypsum precipitate as crystals. If no water is added to dilute this brine while evaporation continues, the ratio of salt to water increases. When salinity reaches 35 parts of salt per 1000 parts water (10x normal), the mineral halite (rock salt) begins to precipitate.

The lagoons in the Paradox Basin acted as large evaporation pans, and layer after layer of salt accumulated. Occasionally seawater would breach the lagoons for a time, depositing a layer of mud over the salt flats. Then the cycle would start all over again. The mud eventually became shale beds. By the time the sequence ended, there were 29 layers of salt and gypsum beds with intervening shale layers. The accumulated thickness of 4,000 feet became the Paradox Formation.

As the shoreline gradually shifted to the east, the water in this region became deeper, providing an environment in which the more usual sediment of sandstone, shale, and limestone accumulated to create the Honaker Trail Formation. This rock unit, over 1,800 feet thick, is exposed on the bottom of Cataract Canyon and forms the gray-colored inner gorge of the Colorado River.

As the Pennsylvanian Period closed and the Permian began, advancing seas deposited the Halgaito Shale that grades into the Elephant Canyon Formation, a rock unit deposited in coastal lowlands. The sequence ranges from 1,000 to 1,500 feet in thickness. (These beds are called the Rico Formation by some geologists.) In Canyonlands, beds of the formation are exposed at the Cane Creek anticline and in Cataract Canyon.

5. Deposition of the Cutler Formation sedimentary facies during the Permian Period.

While the salt beds and the overlying sedimentary layers were being deposited, the Uncompahgre Mountains to the east were being severely eroded. Streams stripped sediment off the slopes and dropped it on enlarging alluvial fans that gradually filled the basin at the foot of the range. The material in the fans lithified to arkose that became the Cutler Formation.

As the shoreline continued to shift eastward toward the mountains, sandbars, built from the plentiful supply of sediment, were shaped by waves and currents offshore. The sandbars became the Cedar Mesa Sandstone (a member of the Cutler Formation), the rock unit that is responsible for many of the striking erosional landforms in Canyonlands. Near the top, the Cedar Mesa contains a white cliff-forming sandstone that interfingers with the red arkosic beds in the vicinity of the Maze and the Needles. The red arkosic layers indicate deposition in shallow marine water. The Cedar Mesa Sandstone is between 900 and 1,200 feet thick.

Muds that oxidized and became the Organ Rock Tongue were laid down over the sandbars. The Organ Rock beds are the brightly colored red, orange, and brown layers that add color to the Park. This formation is also a slope-former that separates cliff-forming beds above and below. The Organ Rock shale beds are equivalent to the Hermit Shale of Grand Canyon National Park. In Canyonlands this shale sequence is 100 to 200 feet thick.

As noted earlier, the Cedar Mesa Sandstone tends to weather more rapidly along its numerous close-set parallel joints. Fracturing during crustal uplifts in fairly recent geologic time probably caused the jointing. The Needles area, where some of the columns, spires, etc., are more than 300 feet high, is a good place to look at the Cedar Mesa Sandstone and the erosional features that characterize it.

The Cutler Formation is more likely to erode as "standing rock," or fins. Some arches are found in the Cutler but not so many as in the Upper Jurassic Entrada Sandstone in Arches National Park (chapter 5). The best known examples in the Cutler are Druid, Angel, and Washer Woman Arches. Other distinctive features are Tower Ruin, Newspaper Rock, Wooden Shoe, Doll House, Cane Spring, and Confluence Overlook.

6. Late Permian time, deposition of the White Rim Sandstone, and then erosion.

Exposed on the western side of the Colorado River in Canyonlands is the marine White Rim Sandstone that has an average thickness of 250 feet. Outcroppings that exhibit cross-bedding are well exposed at Anderson's Bottom (northwest section of the Park). Here they are probably "fossil" sand dunes, but west of the Green River they are probably large offshore marine sandbars. Outside of Canyonlands' borders these same beds contain oil (e.g., Elaterite Basin). While these beds were being formed, the Cutler Formation was still being deposited in the east as the Uncompahgre Mountains were undergoing erosion by streams.

Northwest of the Needles region, west of the Green and Colorado Rivers, is the Land of Standing Rocks. Narrow fins and balanced rocks on a broad, extensive bench characterize this landscape. The area includes the Maze, a canyon network of interlocking gullies and buttes. The White Rim Sandstone, which forms the intermediate plateau, can also be seen in exposures at Stillwater Canyon and along the Green River, Dead Horse Point, and Monument Basin.

7. Deposition of the Mesozoic formations.

After the Cutler beds were laid down, the area underwent erosion; and deposition did not begin again until the Triassic, 230 million years ago. The first Triassic sedimentary unit consists of the brownish-red beds of the Moenkopi Formation that were deposited on shallow tidal flats. In some sections, the 300- to 400-foot-thick Moenkopi contains the Hoskinnini Tongue. These rocks are overlain by the brightly colored Chinle beds that are divided into four members in this region. The Moss Back Member (here replacing the Shinarump) is a channel sand and point-bar deposit (i.e., sandbars deposited on the inside of a stream curve). The Petrified Forest Member, noted for its petrified wood as the name implies, is a flood plain deposit. Last of all are the Owl Rock and Church Rock Members which formed in the lakes and ponds of the flood plains. Collectively the beds are 400 to 600 feet thick.

As the climate became drier, sand dunes from the Wingate desert migrated across the region, burying the rivers and their flood plains. The dune sediment became the cliff-forming Wingate Sandstone, a reddish rock about 300 feet thick. The Wingate and the overlying members of the Glen Canyon Group encircle the western portion of Canyonlands National Park. Exposures can be found near Shafer Trail, Lockhart Basin, and the buttes of the Cross Range Canyon. The two remaining members of the Glen Canyon Group are the Upper Triassic Kayenta Formation and the Lower Jurassic Navajo Sandstone. The Kayenta, a fluvial sandstone and siltstone that has a tendency to form gentle, ledgy slopes, is 200 to 250 feet thick. The Navajo, another cliff-forming dune sandstone with well developed cross-bedding, is 350 to 450 feet in thickness. Canyon View Point Arch and Millard Canyon Arch are in the Navajo.

Rock units of the San Rafael Group overlie the Glen Canyon Group. The Dewey Bridge Member of the Carmel Formation is a sandstone and siltstone unit laid down in a transgressing sea that spread over the Navajo Sandstone. The Entrada Sandstone, overlying the Carmel Formation, has two members, Slick Rock and Moab Tongue. The Slick Rock Member is a pinkish, cross-bedded sandstone and Moab Tongue at the top is a white, cross-bedded sandstone, probably wind-deposited.

8. Laramide phase of the Rocky Mountain orogeny; uplift; building up of the Uinta and San Juan Mountains.

Most of the folds and faults in the Rocky Mountains of today are the result of the Laramide phase of the Rocky Mountain orogeny, which brought to a close the Mesozoic history of the great Cordilleran geosyncline. As the North American plate moved westward against the Pacific plate and then overrode it, the sediments of the geosyncline were caught up in the intense compression and mountain-making upheavals that ensued. These continuing deformations began along the Pacific margin in the Jurassic and

shifted eastward, culminating with the Laramide phase in the Rocky Mountain region near the end of Cretaceous time.

An east-west-trending fold produced the Uinta Mountains, and the associated uplift tilted up the beds of the San Juan Mountains. All this movement arched the sedimentary rock layers and formed the joint systems so crucial to the development of the Canyonlands landscape. Nevertheless, the uplift that deformed the beds in the adjacent areas and created mountains barely tilted the beds in Canyonlands as they rose to their present elevation. The dip of the beds is so slight that they appear flat.

Although the joints are the controlling factor in the location of the weathered and eroded features of the Park, another force was necessary to produce some of the spectacular scenery, such as that seen in the Needles area. This force was provided by the weight of the overlying Cutler Formation, and it produced a curious effect. Becoming plastic because of the pressure, the salt beds below the Cutler began to flow slowly to the southwest where the overlying arkose beds were not so thick and hence not so heavy. When the salt in its plastic state encountered the walls of fault blocks, it pushed upward along the fracture, following the path of least resistance. This produced a horst-and-graben effect with narrow upthrown blocks that formed fins and narrow, down-dropped blocks that became north-south-trending "race-track valleys" (fig. 4.6).

As the salt beds underneath the Needles area dissolved due to percolating ground water, additional collapse occurred, complicating the fault structure even more. The results of this process can be seen from the Needles section to Cataract Canyon, but not west of the Colorado River because little salt bed solution takes place on that side.

9. Weathering and erosion of joints creating fins and arches; graben faulting; intrusion of salt domes; incising of Colorado and Green Rivers.

Exposure of the jointed beds accompanied by gentle uplifting helped the rivers to incise their canyons. Continued weathering and erosion in the arid climate has produced the geologic wonder that is Canyonlands.

Geologic Map and Cross Section

American Association of Petroleum Geologists. 1967. *Geological highway map no. 2 of the southern Rocky Mountain region.* Tulsa, Oklahoma: American Association of Petroleum Geologists.

Huntoon, P. W., Billingsley, G. H., and Breed, W. J. 1982. *Geologic Map of Canyonlands National Park and Vicinity, Utah.* Moab, Utah: Canyonlands Natural History Association.

Lohman, S. W. 1974. *The geologic story of Canyonlands National Park.* U.S. Geological Survey Bulletin 1327.

Bibliography

Baars, D. L.; and Molenaar, C. M. 1971. *Geology of Canyonlands and Cataract Canyon.* Four Corners Geological Society, sixth field conference.

Lohman, S. W. 1974. *The geologic story of Canyonlands National Park.* U.S. Geological Survey Bulletin 1327.

Mutschler, F. E. 1969. *River runners guide to the canyons of the Green and Colorado Rivers with emphasis on geologic features, volume II, Labyrinth, Stillwater, and Cataract canyons.* Denver, Colorado: Powell Society, Ltd.

Redford, R. 1978. *The outlaw trail.* New York: Grosset and Dunlap.

Rigby, J. K. 1976. *Northern Colorado Plateau.* K/H Geology Field Guide Series. Dubuque, Iowa: Kendall/Hunt Publishing Company.

Rigby J. K., et al. 1971. *Guidebook to the Colorado River,* Brigham Young University geology studies, Volume 18, Part 2. Provo, Utah: Brigham Young University Department of Geology.

———. 1971. *Moab to Hite, Utah through Canyonlands National Park,* Brigham Young University geology studies, Volume 18, Part 3. Provo, Utah: Brigham Young University Department of Geology.

Address

Canyonlands National Park
446 S. Main St.
Moab, Utah 84532

5
Arches National Park

Location: Southeast Utah
Area: 73,378.98 acres; 114.65 square miles
Established: November 12, 1971

Figure 5.1 Devil's Garden in Arches National Park. The uplifted Entrada Sandstone eroded along parallel vertical joints, forming fins, arches, and pedestals. National Park Service photograph.

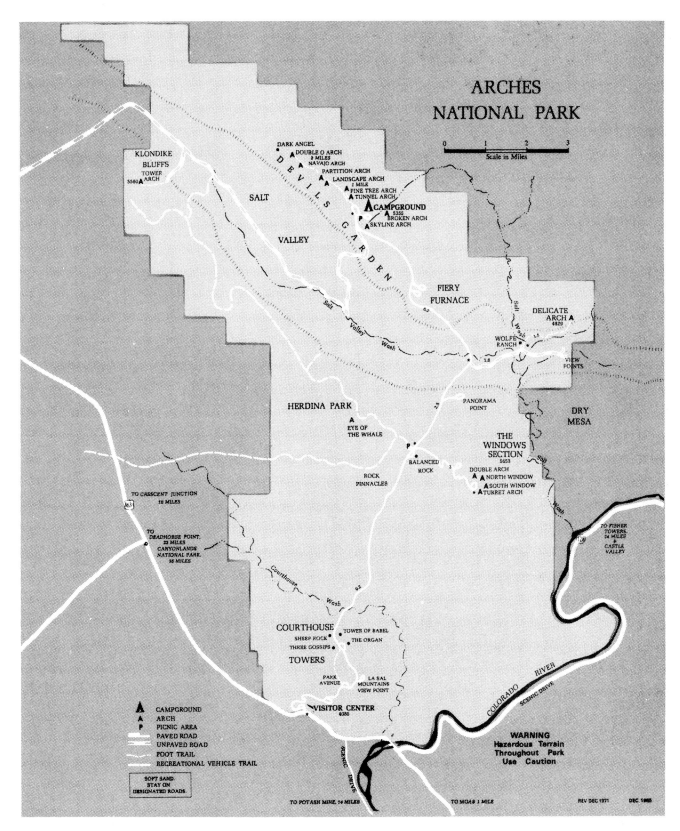

Figure 5.2   Arches National Park, Utah. National Park Service map.

Table 5.1. Geologic Column, Arches National Park

Era	Period	Epoch	Group	Formation	Member	Geologic Events
				Time Units	Rock Units	Geologic Events
Cenozoic						Weathering and erosion forming arches, etc. Solution along faults forming salt valleys. Block faulting. Laramide orogeny; uplift
Mesozoic	Cretaceous	Upper		Mancos Shale	Blue Gate	Blue-gray marine shale deposited in a transgressing sea.
					Ferron Sandstone	Deposited as lenticular, deltaic and barrier island sands.
					Tununk Shale	Deposited as marginal marine beds; some coal formed; regressing sea
				Dakota Sandstone		Formed as a beach deposit along the edge of a transgressing sea.
	Jurassic	Upper		Burro Canyon or Cedar Mountain		Deposition as bright green marine beds with some interbedded gray units, reflecting periodic exposure to the air.
				Morrison	Brushy Basin Shale	Contains dinosaur bones, deposited by meandering streams on flood plains; beds greenish-gray color, some maroon-gray.
					Salt Wash Sandstone	Deposited in meandering streams with ample water; maroon color; contains point bars.
			San Rafael	Summerville		Reddish rock, formed on tidal flats and as sand dunes.
				Curtis		Pale-colored marine sediments occurring in shallow, transgressing sea.
				Entrada Sandstone	Moab Tongue	White sandstone formed as wind-blown dune deposits.
					Slick Rock	Red and gray units formed in a shallow, regressing sea
					Dewey Bridge*	Red and gray units formed in a shallow, transgressing sea; has crinkly appearance
		Middle to Lower	Glen Canyon	Navajo Sandstone		Light yellow and gray sandstone deposited by the wind as dunes.
	Triassic	Upper		Kayenta		Slabby rock deposition by braided streams and deposition in lakes
				Wingate Sandstone		Sandstone formed from sand dunes
				Chinle	Church Rock	Red sandy facies, stream and lake deposits
					Owl Rock	Reddish-brown lake and river deposits
					Moss Back	Greenish silt units with a few pebbles; some gray beds (deposited in braided stream channels) containing uranium ore.
		Middle to Lower		Moenkopi	Undifferentiated	Reddish-brown marine deposit formed in a tidal flat environment. Erosion.
Paleozoic	Permian			Cutler	Undifferentiated	Brick-red unit deposited as a large alluvial fan. Erosion.
	Pennsylvanian		Hermosa	Honaker Trail		Cyclic deposition of limestone, sandstone; shale and salt in a shallow sea.
				Paradox**		Cyclic sequences of sandstone, shale, limestone, evaporites in land locked lagoons. Also arkose.
				Pinkerton Trail**		Marine limestones and shales deposited in a transgressing sea.
				Molas**		Formed from red soils that developed on a lowland.
				Older units not exposed		

Source: modified after Baars 1972; Rigby 1976.
*Regarded by some authorities as part of the Carmel Formation.
**Units not exposed.

Local History

Prehistoric Indians used the area around Landscape Arch for making arrowheads, points, and the like. It may have served as a winter tool-making campsite. The Indians were, no doubt, attracted to this place by the presence of chalcedony, a variety of quartz that could be chipped (painstakingly) to make a sharp edge.

The Arches National Park area was first discovered in historic time by Alex Ringhoffer, a prospector. Later, in 1859, Captain John Macomb guided an exploring party in the region and judged the land to be worthless. Certainly the lack of water and the barren, forbidding appearance of the terrain were not encouraging to settlers. Modern prospectors have been attracted to this part of Utah by deposits of oil, natural gas, potash, and uranium; and some dryland ranching is carried on.

The climate of the region is harsh. Temperatures range from 110° F in summer to −20° F in winter. It is always dry and windy, with precipitation averaging 5 to 9 inches per year and shelter from the wind being hard to find.

Part of the present Park area was proclaimed a national monument by President Hoover on April 12, 1929. The boundaries were changed three times—in 1938, 1960, and 1969; and on November 12, 1971, President Nixon signed the proclamation making it a national park.

Geologic Features

Since Arches National Park is only a short distance up the Colorado River from Canyonlands National Park (chapter 4), the geology of the two parks is similar in many ways; but yet, significant differences are also evident. Both Canyonlands and Arches have several of the same sandstones—the Cutler Formation of Permian age, the Triassic Wingate, and the Jurassic Navajo. Moreover, the Jurassic Entrada Sandstone and the Morrison Formation, as well as the Cretaceous Dakota Sandstone, are exposed in Arches National Park. With all of these sandstones in the same arid climate, why are not arches scattered throughout all of these beds? Why is it that most of the arches are concentrated in Arches National Park?

To find answers we must look at the processes by which arches are made, the differences in the origin of the sandstones, variations in composition, and differing reactions to uplift and weathering. The Cutler, for example, produced huge fins in Canyonlands because the formation tends to weather along vertical joints closely spaced together. However, in Arches National Park most of the formation is covered by several thousand feet of overlying beds. With very few exposures in the Park, the Cutler has not formed fins in this area.

The Wingate and Navajo Sandstones were lithified from sand dunes; consequently they are massive and cross-bedded. As a result—except where they are buried under other sediment—they weather in large rounded masses, as in Capitol Reef National Park (chapter 6). The Entrada, however, mainly because it is highly susceptible to weathering by exfoliation, is noted for its arches. *Exfoliation* (as was discussed in the examples of weathering in Bryce Canyon National Park, chapter 3) refers to the peeling off from bedrock of slabs or shells, usually in a series of concentric layers. Certain massive rocks, such as quartzite, granite, and some sandstone varieties, have a tendency to weather in this manner. The curved surfaces tend to spall off sheet by sheet, like the layers of an onion (fig. 5.3).

The arid climate of southern Utah is probably a key factor in the weathering of the Entrada by exfoliation. In a very humid climate, natural bridges or arches might have been formed by streams, but the features would not have been so numerous or long-lasting. Hence in this region stream erosion is not an important process in arch formation, although it may have had a role during the somewhat wetter climate of Utah in the fairly recent geologic past. Of course, much of the loose sediment resulting from weathering processes is washed down toward the Colorado River during flash floods, occasional snow melting, etc., so the processes of stream erosion do play a minor part.

Several weathering processes help exfoliation to get started. The Entrada Sandstone, originally deposited in a shallow marine environment, was raised and arched slightly during the uplift of the San Rafael Swell. The arching produced vertical linear joints in the massive sandstone, 10 to 20 feet apart; and these joints became zones of weakness through which weathering agents, such as water, ice, and frost, could attack the rock. In this way fins were developed. The process of exfoliation, working on both sides of the fin, eventually perforated the rock, forming windows. At the same time, water and frost enlarged these holes, and the wind, carrying fine sand, abraded and polished them. At the base of the fins, more moisture is available from seepage that tends to dissolve cement holding the sand grains together. Also, the wind doesn't dry out the base as fast; hence, the rocks at the base tend to weather more rapidly. This undercuts the fin, causing small rockfalls due to mass wasting. These processes are probably significant in starting many of the Entrada arches because the Dewey Bridge Member (regarded as part of the Entrada Formation here rather than the Carmel Formation) is less resistant than the overlying Slick Rock Member. Thus the Dewey Bridge has weathered more intensely, forming hollows or niches. Left unsupported, sections of the Slick Rock Member collapse, forming (or enlarging) openings. A few windows

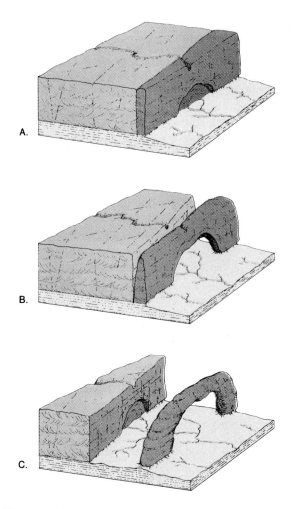

Figure 5.3 How natural arches develop in massive sandstone in an arid region. From *The Earth's Dynamic Systems* (3rd ed.) by W. K. Hamblin. © 1982 Burgess Publishing Company; used by permission. *A.* Water from infrequent precipitation tends to move downward through porous sandstone until it reaches an impermeable rock layer and then flows laterally along the bedding planes below the surface. Water also moves underground through stream gravels in dry stream channels. Where the subsurface flow emerges as a seep or spring along the base of a cliff, moisture becomes concentrated and gradually dissolves the cement that holds the sand grains together. Loosened sand grains fall away, leaving a niche or recessed area at the base of the cliff. *B.* Weathering enlarges the vertical joints in the sandstone, separating a slab or fin from the main body of rock. Meanwhile, the processes of exfoliation and abrasion by wind-blown sand enlarge the recessed area in the cliff until an arch is formed. *C.* All the exposed surfaces of the arch are subjected to weathering and erosion. The arch thins out and eventually collapses while a new arch is being formed in the retreating cliff.

appear to have been started as potholes on the sides of cliffs where the abrasion of rushing water formed hollows. This process also enlarges the "sills" of the windows. It becomes apparent then, that a number of processes in various combinations have formed arches; and it would be a mistake to assume that every arch in the Park has exactly the same erosional history. What is important is to realize that these features are impermanent, that they are constantly changing, and that new ones are being formed as old ones are being destroyed (fig. 5.3).

Most of the arches in the Park are in Devils Garden, which has 64 of them in all stages of development—from the faintest suggestion of a depression to the collapsed remains of those that have all but weathered away. Landscape Arch (fig. 5.4), the longest natural span in the world, is 291 feet long and 105 feet high. One part of the span is only 6 feet wide, an indication that by geologic time, the arch is in the last stage of its existence and may soon be no more.

Unique in its isolation, Delicate Arch stands alone at an elevation of 4,829 feet northeast of Wolfe Ranch. Estimations are that approximately 70,000 years were required for its sculpturing by water, wind, frost, and sun. Because of a narrowed segment on one side, this arch, too, may collapse in the near geologic future. Delicate Arch, 85 feet high and 65 feet wide, is also an Entrada feature. The cowboys had another name for it not so complimentary as its present name; they called it "Old Maid's Bloomers."

In the Devils Garden area, Skyline Arch last changed its profile in November, 1940, when a large section fell from the arch to the ground below, noticeably increasing the size of the opening. This event is another reminder that arches continue to change until they collapse under their own weight.

The uplift and deformation that created structural conditions favorable to arch formation are evident in the Salt Valley and Cache Valley anticlines. Both were flexed upward over migrating salt beds of the Paradox Formation (p. 54), and both have collapsed as grabens due to solution of salt by ground water. Salt Valley, the larger feature, trends northwest-southeast and is about 30 miles long. To the southwest the beds become part of the Courthouse syncline, locally called Courthouse Wash.

Near the Visitor Center, the effects of movement along the Moab fault, about 6 million years ago, are plainly visible because the beds on either side of U.S. Highway 163 do not match. The Navajo Sandstone and the Entrada Sandstone, which were dropped down, can be seen on the northeast side of the road. Across the highway, in the upthrust block, Paleozoic beds of the Hermosa Group

Figure 5.4   Landscape Arch is the longest natural span in the world (291 feet). National Park Service photograph.

and the Cutler Formation are exposed at the bottom of the canyon. Prominent in the high cliff above are beds of the Moenkopi, and Chinle Formations and, at the top, the Wingate Sandstone. Vertical movement along the fault amounted to 2,600 feet.

Most of the formations in this region are fairly thick and massive; but the Dakota Sandstone is an exception here, as in Bryce Canyon (chapter 3), inasmuch as it crops out as a thin, blanket sandstone.

Ham Rock, a balanced rock near The Windows, is a striking example of a feature shaped mainly by wind erosion (fig. 5.5). The rock is part of the Moab Tongue that overlies the Dewey Bridge Member of the Entrada Sandstone. A *balanced rock* is a large rock resting more or less precariously on its base, formed by weathering and erosion in place. Normally the rock on top is a cap rock that is more resistant than the rock below. The dominant process in the development of the balanced rocks in Arches National Park (and in most desert regions) is *sandblasting*, or abrasion by wind-blown sand. Wind velocity is at its greatest about 20 inches above the ground, so the most

intense erosion occurs at this level. First a pedestal is formed. As it becomes thinner and thinner, the weight of the caprock causes the top part to tilt very slowly to one side, thus compressing the rock on that side making it more difficult to erode. The opposite side, not under as much compression, weathers away slightly faster until this causes the rock to tilt back, compressing *that* side, and the process starts all over again. The rock keeps tilting very slightly back and forth until some force finally removes it or it falls from its own weight.

Balanced rocks formed in this manner are an example of *differential weathering*; that is, the weathering of rocks at varying rates in an area due to differences in resistance or differences in the intensity of weathering. Evidence of differential weathering can also be seen in the caves, pinnacles, and windows in the Park.

The *chalcedony* in the Park, from which the Indians made stone tools, is a cryptocrystalline variety of quartz (i.e., having no visible crystals) that is sometimes found in chert or as concretions in sedimentary rocks. Because chalcedony can be chipped or made to break with a smooth, curved surface, the Indians used the stone for

Figure 5.5 Ham Rock, a balanced rock 55 feet high, weighs 3,500 tons. Sandblasting (wind abrasion) has left this huge sandstone block (an outlier of the Moab Tongue Member) perched on a pedestal of crinkly beds of the Dewey Bridge Member of the Entrada Sandstone. From *Northern Colorado Plateau* by J. K. Rigby. © 1976 Kendall/Hunt Publishing Company; used by permission.

making tools and points with sharp, scalloped edges. This property or characteristic of some minerals to break along curved surfaces is called *conchoidal fracture*.

As one would expect, with all the sandstones in this area, most of the loose sediment that accumulates in wind-blown piles or is transported by streams is almost entirely fine sand grains that are clean, well rounded, and very smooth. In fact, they are so spherical that they do not cling together well, even when wet. When these sand grains become saturated with water, as at the mouth of a river, they make *quicksand*. Such a deposit is found by Salt Creek in Arches National Park. Water flowing upward due to hydrostatic pressure beneath the surface keeps the grains separate and neutralizes their frictional strength. The soft, semi-liquid mass of quicksand instantly shifts under the weight of an object, an animal, or a person and has a tendency to suck anything down. However, few of

the quicksand beds are more than waist-deep to an adult. The important thing to remember if you find yourself in quicksand is not to struggle, because the sand is reluctant to release anything. The best method for extricating yourself is to move slowly in a breast-stroke motion so as to distribute body weight more evenly. Then gradually work your way out of the quicksand. If you have to cross a quicksand area, walk slowly and steadily without stopping or standing in one place.

For "rockhounds," this region yields chalcedony, apache tears, almandite garnet, quartz crystals, geodes, flint, and petrified wood. However, *specimens must be collected outside Park boundaries*. It is illegal to take specimens from a national park.

Geologic History

1. **Deposition of Paleozoic beds; erosion.**

    The same Paleozoic sedimentary beds that make up the bedrock in nearby Canyonlands were also deposited in the Arches National Park area, but they underwent extensive erosion at the end of the Permian Period. Therefore, the uppermost Paleozoic beds have an unconforming contact with the overlying Triassic beds. The Paleozoic beds that crop out in Arches National Park are upper rock layers of the Hermosa Group and beds of the Cutler Formation. The Elephant Canyon Formation did not extend this far east.

2. **Deposition of the Moenkopi and Chinle Formations and the Glen Canyon Group in Triassic/Jurassic time, followed by erosion.**

    The geologic history of these formations is essentially the same in Arches National Park as in Canyonlands (chapter 4) because of the close proximity of the two areas. In the Arches vicinity the Triassic/Jurassic sedimentary layers make up the upper surface of the present plateau and the canyon rim. The plateau surface consists of weathered remnants of broken rock formed during earlier geologic periods. Collectively, the formations range from 1,550 to 1,700 feet in thickness.

3. **Transgression of the seas, deposition of the Entrada Sandstone during late Jurassic time.**

    The Entrada Sandstone was lithified from wind-blown sands deposited along the shores of shallow seas. The wind dropped the coarser sand close to the shoreline; and the finer particles fell nearer the center of the basins. The Entrada Sandstone, which was derived from the weathering and erosion of older sandstone formations, lies unconformably over earlier beds. The Entrada belongs to the San Rafael Group and is divided into three members that together are about 300 feet thick in this area. (The oldest, the Dewey Bridge Member, is considered by many geologists to be part of the Carmel Formation rather than a member of the Entrada Sandstone.)

A significant characteristic of the two lower members, the Dewey Bridge and the Slick Rock, is their "waviness," or large-scale, contorted bedding. Deformation must have occurred after deposition because the entire rock unit was affected rather than individual layers alone. A possible cause for this slumping on such a large scale may have been some minor rejuvenation of a pre-existing structure (since then destroyed) so that tilting was localized; i.e., crinkling of as much as 50 feet of beds, with but minor disruption of beds above and below. The odd-shaped spires near the entrance to the Park are eroded outcrops of the contorted beds. The Moab Tongue is the third member of the Entrada.

4. Deposition of the Curtis, Summerville, and Morrison Formations.

The Curtis and Summerville Formations (the upper members of the San Rafael Group) range from 345 to 575 feet thick. The Curtis was deposited in shallow water, and the Summerville in a tidal flat environment, as indicated by the presence of ripple marks, mud-cracks, etc. In the Arches area the Summerville has a very complex interfingering relationship with the Moab Tongue Member of the Entrada Sandstone. The Moab Tongue beds were sand dunes on beaches along the edge of the tidal flats.

The Morrison Formation has four members, but only two of them (between 550 and 650 feet thick) occur in Arches National Park. The lower member, the Salt Wash Sandstone, was stream-deposited and contains large enough amounts of uranium in point bars to be of commercial value. The upper member, the Brushy Basin Shale, was laid down as a flood-plain deposit. These beds contain most of the dinosaur bones, especially in localities farther north such as Dinosaur National Monument.

The Burro Canyon (or Cedar Mountain) Formation, ranging from 80 to 250 feet thick, is made up of bright green and gray marine beds that were deposited in fluctuating water levels, each bed reflecting its specific environment and depth of water.

5. Deposition of Cretaceous beds of sandstone, shale, and limestone.

The Dakota Sandstone represents the beach of the last major sea to cover the western continental area. Apparently this sea crept very slowly over the land, depositing a thin blanket sand that became the Dakota Sandstone (which averages 200 feet thick in this region) (fig. 3.9, p.

44). As the waters deepened in the Arches area, the sandy sediment changed to mud that lithified as the Mancos Shale. The Mancos is divided into three members here: the marine Tununk Shale, the deltaic Ferron Sandstone, and the Blue Gate, also marine shale.

Whenever the lower shale member becomes wet (where it is near or at the surface) it becomes soft, viscous, and very sticky. When dry, it hardens again into a concretelike substance. The asphalt runways for the town of Moab (outside Park boundaries) are laid over this shale, and in the occasional very wet seasons they tend to sink down into this muddy deposit. In the Park, however, these beds have been removed by erosion.

6. Laramide phase of the Rocky Mountain orogeny; formation of the Uinta and San Juan Mountains.

The uplifting of the Rockies, the San Juan Mountains, and the Uintas, as explained earlier, put stress and strain on all the rock layers and consequently formed the joints that initiated the development of most of the landforms in Arches National Park.

"Salt valleys," as in Canyonlands (chapter 4), formed along fault trends and later collapsed after the intruding salt was dissolved by ground water. The 30-mile-long Salt Valley is partially within Park boundaries and parallels Devils Garden, in which are found most of the major arches. Devils Garden also contains the younger beds in the Park. Summerville, Morrison, Dakota, and Mancos are in the down-dropped blocks, or grabens, that are bounded on the west by the Moab Fault.

A second salt valley, Cache Valley, lies outside the Park and is the site of the town of Moab.

7. Formation of arches and other erosional features in Arches National Park.

The cliff-formers in the Park are the Wingate Sandstone, the lower portion of the Kayenta Formation, the Navajo Sandstone, the upper portion of the Entrada Sandstone, the Salt Wash Sandstone (a member of the Morrison Formation), and the Dakota Sandstone. Other formations form slopes, such as the upper portion of the Kayenta, Dewey Bridge (or Carmel) Formation, Summerville, the upper part of the Morrison, and the Mancos Shale. Valley sides in the area tend to be steplike because of differential weathering of the rock units.

The lithology and structure of the Entrada Sandstone, along with the jointing, made it vulnerable to weathering by a combination of processes that produced the famous arches. The bright red color of the Entrada makes them all the more spectacular. The balanced rocks, formed mainly by sandblasting, or wind abrasion, add interest to the Park's scenery.

Geologic Map and Cross Section

American Association of Petroleum Geologists. 1967. *Geological highway map no. 2 of the southern Rocky Mountain region.* Tulsa, Oklahoma: American Association of Petroleum Geologists.

Baars, D. L. 1972. *Red Rock country-the geologic history of the Colorado Plateau,* New York: Doubleday/Natural History Press; p. 73.

Bibliography

Baars, D. L. 1972. (see above).

Doolittle, J. 1974. *Canyons and mesas,* the American Wilderness series. Alexandria, Virginia: Time-Life Books.

Hoffman, J. F. 1981. *Arches National Park, an illustrated guide and history.* San Diego, California: Western Recreational Publications.

Lohman, S. W. 1975. *The geologic story of Arches National Park.* U.S. Geological Survey Bulletin 1393.

Rigby, J. K. 1976. *Northern Colorado Plateau,* K/H Geology Field Guide Series. Dubuque, Iowa: Kendall/Hunt Publishing Company.

Address

Arches National Park
446 S. Main Street
Moab, Utah 84532

6

Capitol Reef National Park

Location: Southeast Utah
Area: 241,865.48 acres; 377.91 square miles
Proclaimed a National Monument: August 2, 1937
Established as a National Park: December 18,1971

Figure 6.1 Looking northward over the Waterpocket Monocline and Circle Cliffs. The less resistant beds of the Kayenta and Wingate Formations are exposed to the left in the broad valley. Beds of Navajo Sandstone, dipping slightly to the right, make up the resistant rimrock. National Park Service photograph.

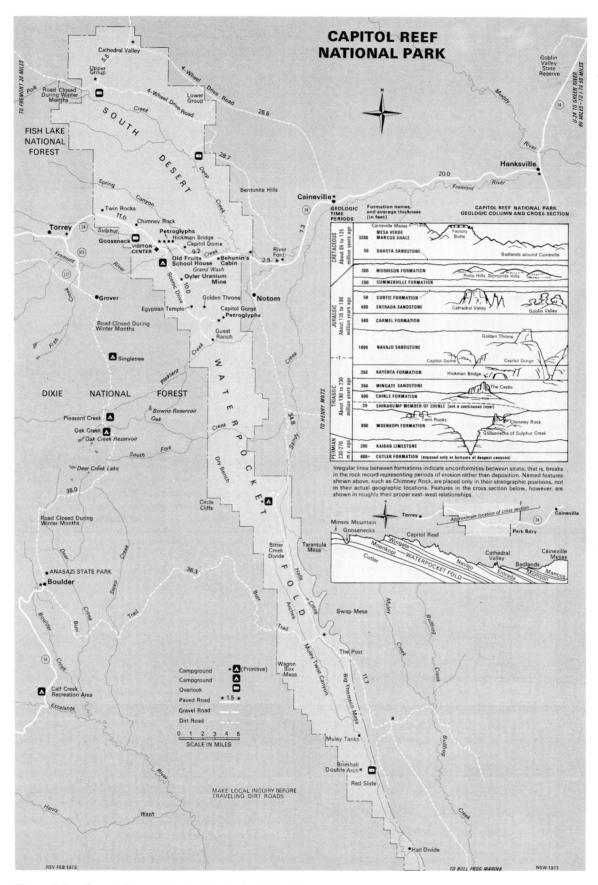

Figure 6.2 Capitol Reef National Park, Utah. National Park Service map.

Table 6.1. Geologic Column, Capitol Reef National Park

Time Units			Rock Units			Geologic Events
Era	Period	Epoch	Group	Formation	Member	
Cenozoic				Various igneous intrusions and eruptions		Boulders transported by glaciers and left as erratics. Dikes, sills, igneous plugs intruded into sediments; mostly basalts with interbedded tuff. Uplift, folding, erosion.
Mesozoic	Cretaceous	Upper	Mesa-verde	Undifferenti-ated		Deposited as a yellow marine sandstone
				Mancos Shale	Masuk	Sandy marine shale, deposited in shallow water.
					Emery Sandstone	Deposited on tidal flats; sandstone with thin coal beds.
					Blue Gate	A blue-gray shale, deposited in a transgressing sea.
					Ferron Sandstone	A lenticular sandstone with thin coal bed, deltaic and barrier island deposits.
					Tununk	Marginal marine beds deposited by a regressing sea.
				Dakota Sandstone	Upper	Beach deposits in a transgressing sea.
					Middle	Shallow water and beach deposits formed in a regressing sea.
					Lower	Oyster beds deposited in a transgressing sea
	Jurassic	Upper		Morrison	Brushy Basin Sandstone	Gray-green and maroon clays deposited by meandering streams as flood-plain units. Contains dinosaur bones.
					Salt Wash Sandstone	Mixture of maroon sandstone, conglomerate and siltstone deposited by meandering streams; some point bars.
		Middle	San Rafael	Summerville		Chocolate-brown shales deposited in a tidal flat region; a mixture of mudflats, sand dunes, and shallow lakes
				Curtis		Pale-gray shale with gypsum layers, deposited by regressing sea.
				Entrada Sandstone		Reddish-brown sandstone, deposited in shallow seas and land-locked basins
				Carmel		Shales, claystones, limestone and gypsum deposited in transgressing and regressing sea.
		Lower	Glen Canyon	Navajo Sandstone		White sandstone deposited as water-rounded, fine-grained sand in sand dunes, by the wind.
	Triassic	Upper		Kayenta		Interbedded red shales with pale sandstones, honeycomb development, flaggy; also red-brown lenticular sandstone deposited by streams.
				Wingate Sandstone		Bright red sandstone stained from weathering of overlying rocks. Desert varnish and petroglyphs; a cross-bedded dune deposit.
				Chinle	Upper	Gray slope and cliff-former units deposited on flood plains and alluvial fans.
					Middle	Reddish-brown slope-former; lake, river and flood-plain deposits including bentonitic ash.
					Shinarump	Gray-green slope-former with bentonitic ash and basal conglomerate. Contains uranium, lenticular beds, channel filling, point bars, and petrified wood.
		Lower		Moenkopi	Upper	Red, slabby, highly jointed, tidal flats deposits with ripple marks, mud cracks, raindrop impressions and cross-bedding
					Sinbad Limestone	Tan, ledge-former; deposited in warm, shallow marine water.
					Lower	Red slope-former with a light-colored sandstone; a tidal flat deposit.
Paleozoic	Permian			Kaibab Limestone	Undifferentiated	Light-colored cliff-former, deposited in a warm shallow sea.
				Cutler	Undifferentiated	A cliff-former, cross-bedded unit deposited as dunes.
				Older rocks not exposed		

Source: modified after Olson and Olson 1972; Rigby 1976.

Box 6.1
How Mass Wasting Operates in Arid Regions

Finding effects of mass wasting on landforms in the national parks of the Colorado Plateaus Province is not difficult. Loose pieces or chunks of bedrock that fall from cliffs form piles of *talus* that lie at the foot, and talus slopes usually surround the base of a mesa or butte. When roadcuts in dry regions are made through bedrock, this may bring about *rockslides*, especially along surfaces of weakness, such as bedding planes. On some steep slopes, scars of old slides can be seen above a block or mass of hillside material that has separated and slipped down along a curved surface, slumping at the bottom so that the base bulges out as a hummocky toe. Sudden torrential rains, perhaps accompanied by pelting hail, may loosen, wash down, or saturate so much debris that a desert stream turns into a thick *mudflow* that continues down a canyon picking up or burying everything in its path. When the rain stops, the mudflow keeps on going of its own momentum until the water dissipates, leaving the channel or the canyon mouth choked with debris.

Thus with the driving force of gravity supplying the necessary energy, loosened rock or debris may fall, slide, slip, slump, or flow from higher locations to lower. Controlling the rate and type of mass wasting processes are such factors as climate, degree of slope, relief, and vegetation cover, with climate usually being the key factor.

Relief and slope angle are important because steep slopes are more susceptible to mass wasting than gentle slopes, and slopes in areas of high relief tend to be less stable than slopes in areas of low relief. Climate influences what types of weathering occur and governs how much and what type of vegetation grows in an area. In cold climates freezing and thawing contribute to downslope movement in both humid and arid regions. The infrequent but heavy rain pelting down on slopes unprotected by vegetation in desert country aids the processes of mass wasting. Depending upon the amount of runoff, debris may wash down or become saturated, forming a debris slide or a mudflow.

In contrast to the more gently rounded hills and valleys typical of humid regions, the desert landscape looks angular with steep slopes and walled canyons, as in Canyonlands. This kind of topography tends to increase the effectiveness of mass wasting. Due to the thin soil and lack of moisture in the desert, one type of mass wasting characteristic of humid areas is extremely slow or ineffectual here, however. This is *creep*, the very slow, continuous downslope movement of soil or unconsolidated debris. In temperate regions, with water in the soil and seasonal freezing and thawing, even slopes covered with thick vegetation are subject to the continual action of creep.

Local History

The first inhabitants of the area are believed to have been the Anasazi, or "Ancient Ones." Their petroglyphs in Grand Wash Canyon have been dated at approximately A.D. 700. Traces of pre-Columbian Indians of the Fremont culture are also found. Known as the Basket Maker III Indians, they lived in open caves and raised corn. They also left petroglyphs, some of which were stained with colors derived from powdered minerals, instead of being simply scratched through the desert varnish. The stone granaries they built for storing corn were thought by the later Paiute Indians to be the huts of a race of tiny people, or Moqui, the name given them by the Paiutes.

Some of the basket makers lived in pit houses with rock walls built on the foundation of basaltic boulders. Posts were placed in the corners to support the roof. About all that is left of the huts today are the pits and the foundation stones. It is believed that the basket makers occupied the area for about 250 years, from A.D. 950 to A.D. 1200.

The Paiutes, who called this area "The Land of the Sleeping Rainbows," moved into the territory around 1600 and remained until white people came in the middle and latter 1800s.

John C. Fremont, leader of the Fremont expedition of 1854, was the first white man to see the region. A few years later Mormons passed through. A member of the

Powell expedition (1875) studied the region in some detail. The first permanent settler, Nels Johnson, arrived in 1880. He and a few other settlers lived here in a community they called Fruita. His cabin survived until the 1920s; and a school and community house built around 1890 have been restored by the National Park Service. The cabin built by another settler, Elijah Cutler Behunin, still stands. He established the route that Scenic Drive follows today.

Cohab Canyon, a hanging gorge, was used as a hideout by Mormons with more than one wife who wished to evade being charged with polygamy. Mormons who were polygamists were called "polygs" or "co-habs;" hence the canyon name. Outlaws also hid in the Park area. Cassidy Arch is named for one of the most famous ones, Butch Cassidy.

The old lime kilns in Sulfur Creek and near the campgrounds on Scenic Drive are all that remains of about the only industry that ever developed here. The early settlers used the lime for whitewash, fertilizer, and cement.

The name Capitol Reef was applied to the rounded white domes weathered from the Jurassic Navajo Sandstone in the area. Early visitors were reminded of the white, domed buildings of Washington, D.C., and many state capitals. The term *reef* was originally applied by Australians to the gold-bearing ridges of Australia and Africa that resembled marine coral reefs. The meaning changed through usage then to refer to any rocky barrier or escarpment that made traveling difficult. Hence, reef in this sense has nothing to do with the rocky masses built by marine organisms.

Originally designated a national monument in 1937, Capitol Reef was declared a national park by Richard M. Nixon on December 18, 1971. Acquisitions of land in 1958, 1969, and 1971 added Waterpocket Fold and Cathedral Valley to Capitol Reef National Park.

Geologic Features

The Waterpocket Fold, a large monoclinal fold dipping eastward and trending north-south for about 100 miles, forms the geologic framework of Capitol Reef National Park (fig. 6.1). All but about 30 miles of this dominant structure lies within Park boundaries. Its resistant sandstones have eroded to form the cap rock of ridges, cliffs, domes, mesas, and buttes, as well as smaller features, such as pinnacles, towers, arches, etc. Valleys and narrow canyons are carved in the softer rocks. A *monocline* is a fold having a local steepening in an otherwise uniform gentle dip (see fig. 1.7A, p. 10). In the Park, the fold forms the escarpment, or "reef," that for many years was a formidable barrier to travelers and settlers going west.

The Navajo Sandstone forms the impressive ridge cliff, with the Carmel Formation making up the low ridges at the base of the cliff to the east. Underlying the broad valley are the Entrada Sandstone and the Curtis Formation, while the Summerville and Morrison Formations and the Dakota Sandstone form a series of low ridges beyond. Still farther to the east, the severely dissected Mancos Shale makes a *badlands topography*, an area of horizontal clay beds that is vegetation-free, extremely rough, and steeply gullied. When torrential rains wet these unstable beds, they disintegrate rapidly, so the landscape changes with every storm. The shale beds in the Capitol Reef badlands are mainly composed of *bentonite*, an exceedingly fine-grained rock derived from weathering of igneous material such as tuff and volcanic ash. The rock is greasy and soaplike to the touch and has great ability to absorb quantities of water, with the result that its volume can increase enormously. Thus whenever water is available, the rock weathers and erodes rapidly, and badlands topography is created.

Capitol Reef is an ideal place to see large-scale examples of differential erosion of harder and softer beds because of the way they are exposed by the monoclinal fold. It is interesting, also, to note the effect of structure on the stream patterns of the area. With the exception of the Fremont River and a few others which are *consequent*, or slope-controlled, most of the streams are *subsequent*. A *consequent stream* flows along a course determined by the initial slope and configuration of the land surface; i.e., the direction of stream flow is "in consequence of the slope." Where geologic structure is the controlling factor, a stream is termed *subsequent*. Thus a *subsequent stream* is a tributary that has adjusted its course to the underlying rock structure. Deep Creek, a south-flowing, subsequent tributary of the Fremont, has carved Chimney Rock Canyon along the *strike*, or trend, of a belt of weak rock in the Waterpocket monocline. The Fremont River is consequent because it flows in an easterly direction *across the strike* of the beds, following the regional slope and eventually joining a tributary of the Colorado River. The regional slope coincides approximately with the slight tilt, or dip, of the beds—*dip* being defined as the vertical angle that beds form with a horizontal surface.

The flat-topped hills in the Park, capped by resistant rock layers, are erosional remnants carved from the nearly horizontal sedimentary rock layers. These landforms, called *mesas*, have steeply sloping erosion scarps on all sides. Buttes are very similar but smaller in size with less summit area than mesas. Most buttes were probably mesas originally (fig. 6.3).

Alluvial fans (p. 54) have been built up where streams loaded with sediment come out of steep canyons onto an open, flat floor or a basin or valley. The sudden loss of

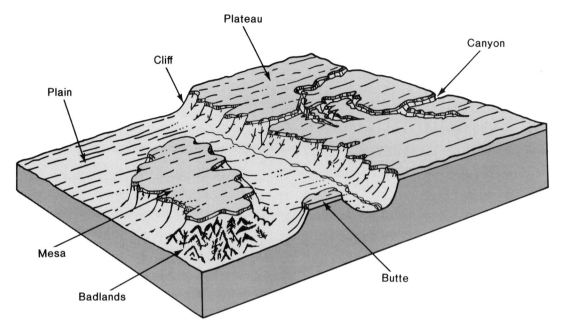

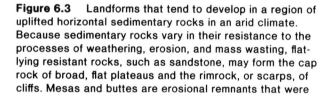

Figure 6.3   Landforms that tend to develop in a region of uplifted horizontal sedimentary rocks in an arid climate. Because sedimentary rocks vary in their resistance to the processes of weathering, erosion, and mass wasting, flat-lying resistant rocks, such as sandstone, may form the cap rock of broad, flat plateaus and the rimrock, or scarps, of cliffs. Mesas and buttes are erosional remnants that were once part of a plateau. Steep-walled canyons indicate rapid downcutting by streams during uplift. As cliffs retreat, less resistant shale beds become exposed and are eroded more rapidly. Badlands may develop in the least resistant shales and mudstones. © 1964 Ward's Natural Science Establishment; used by permission.

velocity causes the streams to unload as water spreads out and dissipates. Sometimes fans from several streams grow so large that they join together, forming a more or less continuous apron of sediment along the foot of a plateau or a group of mesas. This landform, called a *bajada*, or alluvial slope, is composed of detrital material of various sizes (see fig. 7.4, p. 81).

Some small but rather unique features in Capitol Reef National Park are the basaltic boulders. Their source rock was the result of igneous activity to the north of the Park and within its northern section. Lava flows capped the mountains, and igneous intrusions formed dikes, sills, and plugs. (A *volcanic plug, or neck*, is a vertical, pipelike body of magma that represents the conduit to a former volcanic vent. Dikes and sills were described on p. 9.) During the Pleistocene Epoch, when the climate was colder and wetter, small valley glaciers formed on the higher mountain slopes, such as on Thousand Lake Mountain, and plucked boulders from the basaltic rock. As the climate warmed and the glaciers melted back, about 25,000 years ago, meltwater streams draining off the ice spread the glacial boulders throughout the Park area. The Indians used these boulders for making their stone dwellings.

The "waterpockets" for which the monocline is named are really *potholes* that have been carved in bedrock by the abrasive action of fine and coarse sediment whirling in turbulent water and deepening a hole that may have started as a small depression. Some may owe their origins to the meltwater torrents of waning Pleistocene time, but many are still being actively eroded whenever flash floods fill narrow gorges and canyons in the Park with rushing water. Boulders and cobbles, trapped in low places and swirled around by the water may act like a drill boring holes in the rock. Some waterpockets are found at the foot of a cliff in the bed of an *intermittent stream*, that is, a temporary or seasonal stream. The action of falling water when the stream is in flood may create bowl-shaped potholes. The water stored in these natural pockets is the source of most of the drinking water for the area's inhabitants. Some of the pockets are filled with sand and gravel, but water can usually be found by digging.

Sudden flash floods that come with little warning are dangerous to persons who become trapped in narrow, steep-walled gorges. The Valley of Decision, for example, is a narrow gap that is quite subject to flash flooding.

Rockhounds may find rock layers in the Park and outside its boundaries that exhibit *primary structural features*, such as ripple marks, footprints of dinosaurs and amphibians, etc., which were put there during deposition but before consolidation. Some ripple marks are "mined" outside the Park and sold as "ripple rock".

Geologic History

1. Deposition of the Cutler Formation and Kaibab Limestone during Permian time; erosion.

From 270 to 230 million years ago, a shallow sea deposited about 800 feet of the Cutler Formation and 300 feet of the Kaibab Limestone. The Cutler is best exposed in Canyonlands National Park (chapter 4), while the Kaibab is well known as the rim of Grand Canyon (chapter 1). The Cutler is the equivalent of the Coconino Sandstone of the Grand Canyon.

In Capitol Reef National Park, the Cutler Formation is exposed only at the bottom of some of the deeper canyons, such as in the Goosenecks (incised meanders) of Sulfur Creek. Since the Cutler beds are overlain by more than 7,000 feet of other sedimentary beds, fins and arches like those in Canyonlands could not be formed. The Kaibab Limestone, also exposed in the deep canyons of Capitol Reef, marks the shore in western Utah of the Cordilleran sea. To the east the sediments change to continental deposits laid down by rivers and streams.

2. Deposition of the Triassic Moenkopi and Chinle Formations on an alluvial flood plain.

During Mesozoic time a huge upland area was developing to the east, in what is now Colorado, that eventually became the great Mesocordilleran *geanticline*. This is an anticlinal structure of regional extent that upwarps due to lateral compression in geosynclinal sediments. The Mesocordilleran geanticline separated the Cordilleran geosyncline into two components—the shallow *miogeosyncline*, in which sedimentation was nonvolcanic, and the deeper *eugeosyncline*, in which volcanism is associated with clastic sedimentation. During this major upwarp, minor uplifts continued which are sometimes referred to as the "ancestral Rockies" because of their geographic location, but actually they are unrelated to the development of the present Rocky Mountains. The names of these uplifts are Defiance, Kaibab, Monument, San Rafael, and Zuni. The San Rafael Swell was located northeast of Capitol Reef National Park. It was the source of sediment for the early Triassic beds, which were deposited on land and formed an unconformity, or break in the sequence of deposition. The northward tilting of the land about 230 million years ago caused the seas to retreat, transforming the area into a flood plain and initiating sedimentation on land.

The Triassic beds were deposited as a series of alluvial fans that were built up at the base of the mountains by outflowing streams. As the fans grew larger and larger, they coalesced, forming a continuous sloping alluvial plain along the foot of the mountain range. First laid down was the Moenkopi, made up of red and brown layers of mudstone, siltstone, and sandstone to a depth of 800 feet. In these beds are primary structural features, such as mud cracks and ripple marks that formed on the tidal flats and the flood plain.

The Moenkopi Formation was uplifted and underwent extensive erosion as is shown by the overlying but discontinuous beds of basal conglomerate of the Shinarump Member of the Chinle Formation. The yellow-colored Shinarump ranges in thickness from zero to 20 feet. Uranium salts that accumulated in this sedimentary unit were much later taken out as ore from the Oyler Mine in Grand Wash. The first claim was filed in 1904, many years before the uranium boom of the 1950s. The early roads that were built for the miners are rapidly disappearing because the Moenkopi, in which most of the roads were cut, erodes rapidly (fig. 6.4).

As marshes and flood plains formed once again, 500 feet of the Chinle Formation beds were deposited. In this region the Chinle consists of siltstone, sandstone, volcanic ash, and breccia, the colors being mainly shades of brown,

Figure 6.4 Chimney Rock, which towers over the highway between Torrey and Fruita, is capped by the resistant rock of the Shinarump Conglomerate, a member of the Chinle Formation. The Shinarump protects the softer, underlying rock of the Moenkopi Formation from erosion. National Park Service photograph.

gray, and green. Different depositional environments explain the variation in color since the large alluvial fans represented by the Chinle had on their surface rivers, lakes, and swamps.

The Chinle is a slope-former, and differential weathering and erosion of these beds produced the buttes known as Chimney Rock, Twin Rocks, and Egyptian Temple. The buttes are capped by the resistant Shinarump Member.

3. Uplift, erosion; formation of the Wingate Sandstone, the Kayenta Formation, and the Navajo Sandstone.

An uplift toward the end of the Triassic changed the region into a desert. Migrating sand dunes (that became the Wingate Sandstone) advanced over the Chinle surface, burying it with 250 feet of cross-bedded white sandstone and also creating an unconformity between the Chinle and the Wingate. The Castle is carved from these two layers, with the Wingate forming the highest part of the so-called reef.

As the region was tilted, streams deposited the Kayenta Formation in channels and on the flood plains to a depth of 350 feet. At places where dinosaurs and tritylodonts (crocodilelike animals) came to drink or cool off on the banks of streams, their footprints have been preserved in the rock. The Kayenta has a tendency to form ledgy slopes where it crops out between two massive sandstones above and below. Hickman Natural Bridge, which spans a dry stream bed, is carved from Kayenta rock. The bridge is 133 feet long and 72 feet high.

At the close of the Triassic (about 195 million years ago) and as the Jurassic began, the region was once again arid. Migrating dunes covered the area, accumulating to a thickness of 1,000 feet. These remarkably clean, cross-bedded sands became the well known Navajo Sandstone that dominates the modern landscape in many parts of the western states. The cross-bedding causes the Navajo to weather into rounded, dome-shaped hills, and its white color above the reddish-brown Kayenta beds inspired fanciful names for such landforms as the Golden Throne and Capitol Dome.

4. In Jurassic time, tilting; erosion; deposition of the Carmel Formation and the Entrada Sandstone in a sea.

When the land tilted to the south, erosion intensified, and the seas advanced and retreated periodically, forming sandstones, siltstones, and mudstones. In the sheltered lagoons limestone and gypsum were precipitated out, forming the Carmel Formation to a depth of 500 feet. Gradually the area was filled in, producing landlocked basins with interior drainage. This was the environment of the Entrada Sandstone as its beds accumulated to a thickness of 600 feet. The reddish-brown color of the Entrada is the same as in Arches National Park (chapter 5), but since the joint systems in the rock are different, arches are not formed in Capitol Reef. Cathedral Valley in the northern end of the Park has been eroded in the sandier sections of the Entrada. It can be visited only by walking or four-wheel drive vehicle. The Carmel Formation caps the Golden Throne, located south of Grand Wash.

5. Deposition of the Curtis, Summerville, and Morrison Formations.

After deposition of the Carmel and Entrada sedimentary units, the region was again tilted to the south. In some areas, the semisoft beds slumped and were wrinkled; and in other areas they underwent erosion so that an unconformity exists between the Entrada Sandstone and the overlying Curtis Formation. The Curtis, which is only 50 feet thick, gray in color, and resistant, contains a basal conglomerate and glauconite, indicating that it was deposited in a shallow sea. Numerous features in Cathedral Valley are capped by Curtis bedrock.

With the retreat of the Curtis Sea, the brightly colored beds of the Summerville Formation were laid down on tidal mud flats. The beds are a mixture of siltstone and mudstones that have a tendency to form cliffs and ledges. The mud flats existed long enough to accumulate 200 feet of sediment.

As the region rose, alluvial fans were deposited at the base of the newly created highlands. Here and there in low spots scattered over the land surface, swamps and lakes persisted. Winds picked up fine pyroclastic debris and volcanic ash from the nearby volcanoes, spreading the sediment over the area. This fine sediment which accumulated in lakes and was worked over by streams, eventually weathered into an unstable bentonite clay that forms badlands topography wherever it is exposed.

The Morrison Formation is best known for its dinosaur remains. The bones of *Brontosaurus, Diplodocus, Allosaurus, Stegosaurus, and Brachiosaurus* are concentrated along sandbars where the reptiles were buried during times of flood about 140 million years ago.

6. During the Cretaceous Period, erosion; deposition of Dakota Sandstone, Mancos Shale, and the undifferentiated beds of the Mesaverde Group.

Other sediments may have been deposited at the beginning of the Cretaceous, but their record has been lost due to the extensive erosion that occurred between 140 and 80 million years ago. Then the Cretaceous sea crept slowly over the land, depositing the thin blanket sand (rarely exceeding 50 feet in thickness) that became the Dakota Sandstone. This deposit is a mixture of reworked sands and gravel with abundant fossil shells, especially oysters.

During the advance and retreat of the sea, the Mancos Shale (and some limestone) formed in the shallow water while the Mesaverde Group beds accumulated on beaches and deltas. The Mancos and the Mesaverde, which have a combined thickness of over 3,200 feet, are the youngest beds in the Park. Because the Mesaverde Group is a mixture of beds, some more resistant than others, they have tended to form mesas and buttes in this region. Factory Butte is one example. Some of the Mesaverde beds are coal-bearing. The late Cretaceous was a time of warm, humid climate and lush vegetation that flourished in the swampy lowlands. Over a long period the plant materials that accumulated were buried and compacted and gradually converted to low-sulfur bituminous coal. The Cretaceous coals of the western states are an important energy resource. Of course, commercial mining is not permitted in the national parks.

7. Uplift and formation of the Waterpocket Fold.

On Miners Mountain, just west of the Park, the beds are essentially horizontal. They start to fold at the Goosenecks and once again level out by the Caineville Mesas, thus forming a monocline, the Waterpocket Fold. The end result is a series of horizontal beds, a cliff where the fold is flexed, and then another series of horizontal beds. Since uplift was very gradual, streams were able to cut through the fold as it formed; and a narrow, winding canyon called the Grand Wash was created. A number of the waterpockets were eroded in this area by boulders, cobbles, and gravel swirled around by high water rushing through the canyon.

8. Volcanic activity and igneous intrusions during Miocene time.

During the Miocene Epoch, about 20 million years ago, lava pushed up through vents forming volcanic domes in South Desert and Cathedral Valley. Dikes and sills intruded along joints and bedding planes in the sedimentary beds.

9. Development of the Fremont River and its tributaries; erosion, weathering, and glaciation during late Cenozoic time

The uplifting that created the Waterpocket Fold also changed the course of some of the streams so that they became subsequent and now flow parallel to the strike, or trend, of the beds. The streams carved steep-sided canyons because of uplift and also because they were able to erode the soft beds rapidly. The Fremont River, a consequent stream, and its tributaries (most of which are subsequent) have done most of the erosion of the present landscape in the Park, but the processes of weathering and mass wasting have played a significant part as well. Mountain glaciation (outside the Park area) during the Pleistocene and the accompanying cooler, wetter climate must have kept the streams well supplied with water for a considerable time. Glacial ice also plucked the basaltic boulders from the mountain slopes; meltwater streams distributed them over the Park.

Capitol Reef has been described as a barrier to transportation, which indeed it was, with its formidable cliffs and canyons. The canyon of the Fremont River, the largest stream that crosses the Park by cutting through the beds against their north-south trend, provides the main route. A second permanent stream, Pleasant Creek, which also flows east, crosses the Park about ten miles south of the Fremont. Pleasant Creek downcut Capitol Gorge Canyon. The only other permanent stream in the Park is Oak Creek in the area south of the domes. Polk Creek at the northern end of the Park cannot be considered a permanent stream since it sinks into South Desert. Other streams in the Park are intermittent.

The present landscape is remarkably varied with dry canyons, plateaus, mesas, high cliffs, some forested slopes, badlands, magnificent scenery, and very few people.

Geologic Map and Cross Section

American Association of Petroleum Geologists. 1967. *Geological highway map no. 2 of the southern Rocky Mountain region.* Tulsa, Oklahoma: American Association of Petroleum Geologists.

Olson, V. J.; and Olson, H. 1972 (1979 printing). *Capitol Reef-the story behind the scenery.* Las Vegas, Nevada: KC Publications, p. 9.

Bibliography

Olson, V. J.; and Olson H. 1972 (1979 printing). *Capitol Reef-the story behind the scenery.* Las Vegas, Nevada: KC Publications.

Rice, L. 1982. Journeying into the Fold. *National Parks* 56(9–10): 4–8, September/October.

Rigby, J. K. 1976. *Northern Colorado Plateau,* K/H Geology Field Guide Series. Dubuque, Iowa: Kendall/Hunt Publishing Company.

Address

Capitol Reef National Park
Torrey, Utah 84775

7
Mesa Verde National Park

Location: Southwest Colorado
Area: 52,036.24 acres; 81.31 square miles
Established: June 29, 1906

Figure 7.1 The Chapin Mesa Museum in Mesa Verde National Park is near the ruins of the unfinished Sun Temple, constructed on top of the mesa by the Pueblo people more than 700 years ago. A small cave dwelling in an alcove of the cliff is visible next to the lower right margin, to the right of the tree. National Park Service photograph.

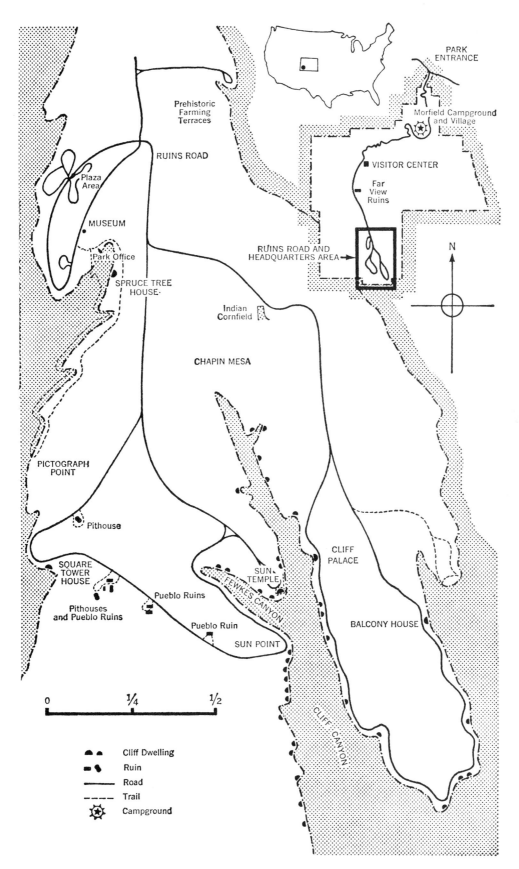

Figure 7.2 Map of Mesa Verde National Park (upper right) and Chapin Mesa (insert enlarged), showing location of major features. National Park Service map.

Table 7.1. Geologic Column, Mesa Verde National Park

Era	Period		Epoch	Group	Formation	Member		Geological Events

Era	Period	Epoch	Group	Formation	Member		Geological Events
	Time Units		Rock Units			Geological Events
Cenozoic	Quaternary	Holocene Pleistocene		Stream and mass wasting deposits Florida Gravel Bridgetimber Gravel			Deposition by streams; accumulation of sediment at the base of cliffs. Glacial outwash, mostly in valleys Pediment gravels, mostly on plateaus and mesas.
	Ter-tiary / Neo-gene	Pliocene					
Mesozoic	Cretaceous	Upper	Mesa-verde	Cliff House Sandstone	Upper Tongue		Marine, nearshore sandstone. Top of formation removed by erosion
					Barker Dome Tongue		Cross-bedded sandstone interbedded with shale. Deposited in a transgressing sea.
					Lower Tongue		Deposited in a transgressing sea
				Menefee*	Upper Coal Member	Upper Split	Carbonaceous clay shale deposited on a flood plain
						Middle Split	Bone coal and coal formed in swamps; interbedded above and below.
						Lower Split	Lenticular coal beds interbedded with sandstone and siltstone; formed in a coastal swamp
					Middle Barren		Carbonaceous, cross-bedded shale with sandstone concretions. Filled-in swamps.
					Lower Coal		Formed in swamps as the seas retreated.
				Point Lookout Sandstone	Upper (massive sandstone)		A cross-bedded beach deposit, massive cliff-former; deposited in a regressing sea.
					Lower (sand-stone and shale)		Alternating sandstone and shale in fluctuating water depths.
				Mancos Shale*	Masuk		Sandy marine shale deposited in shallow water.
					Emery Sandstone		Tidal flats deposits; sandstone with thin coal beds
					Blue Gate		A blue-gray shale deposited in a transgressing sea.
					Garley Canyon Sandstone		Shoreline continued to shift eastward with the water getting deeper
					Ferron Sandstone		Lenticular sandstone with coal beds; deposited on deltas and barrier islands.
					Tununk		Marginal marine beds deposited by a regressing sea.
				Dakota Sandstone			Blanket sands deposited in an advancing sea; contains shale and coal; cross-bedded.
				Older rocks not exposed			

Source: modified after Baars 1972; LeRoy 1960; Wanek 1959.
*Interbedded with formations above and below

Local History

In the middle 1700s Spanish traders called this large mesa *La Mesa Verde*, or "The Green Table." Father Escalante camped along the Mancos River a few miles to the east in 1776 during his explorations of the Colorado River area. Apparently he did not see the cliff dwellings. The first American to visit the area was William H. Jackson, the famous photographer with the Hayden Survey, who took photographs of a cliff dwelling in Mancos Canyon in 1874. The following year W. H. Holmes, an artist who became a geologist, sketched dwellings in the canyon.

The dwellings in the walls of Mesa Verde were first seen by Richard Wetherill and Charles Mason on December 18, 1888, when they were out searching for cattle. The five Wetherill brothers (Richard, Al, Win, Clayton, and John) and their brother-in-law Charles Mason set about exploring these dwellings during the winter when the chores on their nearby 160-acre ranch weren't quite so heavy. In the winter of 1889–90, their explorations brought them to Mug House and the other major ruins. Then in the summer of 1891, John Wetherill helped Gustaf Nordenskiold (the 23-year-old son of a Swedish nobleman) to excavate several of the cliff dwellings. Nordenskiold published his findings in 1893 in a book entitled *The Cliff Dwellings of Mesa Verde*.

In 1900 the Colorado Cliff Dwellings Association, a women's organization, was founded, and the members began their task of preserving the dwellings. As a result, Mesa Verde National Park was established in 1906 in order to protect the archeological sites so that they could be scientifically excavated and studied.

Periods of Cultural Development in the History of the Cliff Dwellers

1) *The Basket Makers* (A.D. 1–550). Originally hunters, these people probably settled down and became farmers here because of the reliable water supply and the favorable conditions on top of the mesa for raising corn and squash. They lived in shallow caves on the side of the mesa and stored their food in stone bins in the floor. They made tools of shell, bone, stone, and wood. The *atlatl* (a primitive throwing stick used to give a spear greater propulsion) was used for hunting. Their skill in weaving baskets, bags, sandals, etc., was remarkable. Their yucca baskets were woven so tightly that they held water, and the baskets they coated with clay and baked were fireproof. They made blankets of fur (mostly rabbit) and bags of leather. Most of the remains of this early culture were covered over by the cliff dwellings of later inhabitants.

2) *The Modified Basket Makers* (or Basket Maker III Culture) (A.D. 550–750). These Indians used the bow and arrow, made baskets and plain pottery, and raised turkeys in addition to crops of corn, beans, and squash. It appears that the turkeys were more important to them for feathers than for food; the feathers were used for arrows and for lining blankets and cloaks for warmth.

The people of this culture lived on the mesa top in pit dwellings, each consisting of a shallow round pit with a flat roof made of poles and mud. The huts were clustered in villages on the mesa. As time went on, these homes became more elaborate as rooms of various shapes (square, rectangular, and D-shaped) and ventilation shafts were added to the pit dwellings. Benches surrounding the main floor areas were also added by these people. Later four corner timbers supported the flat roof, which had a hole in the center for people to come in and for smoke from the central fire to escape. An antechamber with bins or pits in the ground served as a storage area. In some cases entry was through the antechamber rather than the roof.

"Front orientation architecture" is characteristic of these dwellings, in that the northwest portion of a site was for surface rooms and storage units. In front of these surface dwellings, in the southeast section, were the underground kivas, or secret club rooms for men. The trash pile was farther to the southeast beyond the homes and kivas.

Corn was ground in a hollowed-out stone called a *metate* (meh-taa-tay), using another stone, a *mano*, as a grinding tool. With the coarse flour they made gruel and various breads. Pot-making evolved, and eventually white pots with black designs became standard. The finding of an occasional clay or stone pipe indicates that tobacco was used, probably mainly in ceremonies. Jewelry made of shells from the Pacific coast has been found and makes it obvious that trading with other tribes was carried on. The dead were buried with their knees drawn up and arms close to the body. However, the finding of bodies is rare because the dead were buried some distance from the homes.

3) *Developmental Pueblo Culture* (A.D. 750–1100). The pit dwellings were abandoned for above-ground living quarters by this culture. These Indians built rectangular houses of adobe mud with straight sides and flat roofs. As time passed, more and more stonework was added, and houses eventually were built entirely of stone. Individual houses were joined together around a central court, with the front orientation arrangement persisting. Within the central court, the pit houses gradually evolved into larger kivas, the men's secret ceremonial rooms. A major change during this period was burial of the dead in the trash piles, with the bodies still in the flexed position, and the heads pointed in any direction except to the east.

Pottery became more advanced, although pots and jars were still made by the "coiled-rope" method and the vessels were still white with black designs. Neck-banding developed; unsmoothed "ropes" were coiled around the necks of cooking vessels and jars. Toward the end of this stage, pottery was corrugated by pinching the coils with the fingers.

Baskets were still used by this culture but were not so important as before. The women used a type of baby cradle that tended to flatten the back of the infants' heads, resulting in broadening of the faces. With the appearance of cotton, true cloth was woven on looms. In modern times the Pueblo men are the weavers, but whether men or women were the weavers in this earlier period is not known.

The Indians in the Mummy Lake area on Chapin Mesa dug a series of ditches to collect and store water. The central basin had no outlet, but a bypass ditch could send water into an irrigation ditch system 3.5 miles away. The Indians devised "check dams" of rocks to catch silt and moisture for their small gardens.

In the last part of this cultural stage, the villages consolidated further and dwellings were built as high as four stories. Farview House is an example. The sites selected were between 7,000 and 7,500 feet in elevation and afforded unobstructed views. Kivas were moved closer to the dwelling rooms, inside small courts. Many of the outer doors were sealed off with only inner doors to the courtyard remaining open. Outer walls were made thicker. All of this suggests a strong need for protection from outside intruders.

4) *Classic Pueblo* (A.D. 1100–1300). This was the period of maximum development. At the beginning of this last phase, the Indians lived in *pueblos* (a Spanish term for "small villages") on top of the mesa. About A.D. 1200 they began building towers, some separate, others part of the kivas. Whether the towers were symbolic or used as lookouts is not known. An example is Cedar Tree Tower on Chapin Mesa.

Early in this period the Indians abandoned the compact, well fortified pueblos on the mesa and returned to the cliff dwellings of their ancestors. For about 100 years they lived there; and then they abandoned the cliff dwellings, moved out of the area, and apparently settled elsewhere. Archeologists are not certain why the Indians left, but it is known that the cliff dwellings have not been lived in for nearly seven centuries. Possible explanations for the Indians' abrupt departure include a 24-year drought that began in 1276, constant war with neighboring tribes, and depletion of the soil by poor farming methods. It is also possible that they no longer wanted to live in dwellings that were hot in summer and cold and damp in winter. Some of the cliff houses look as if the inhabitants simply got up and walked away, taking little if anything with them. Obviously, the cliff houses were not so comfortable as the mesa-top houses were.

The interiors of the cliff homes were plastered and decorated. This made them more pleasing to the eye and also hid the sloppier masonry work; it appears as though the cliff dwellings were put up in a hurry. Retaining walls were added to prevent children from falling off the cliffs.

What were these people like? Archeologists have determined that the men averaged 5 feet 4 inches in height, and the women, about 5 feet one inch. They had an average life span of about 34 years, and about half of the children died before they were four years old. Adults had poor teeth, showing signs of pyorrhea and tooth decay, and their teeth were badly worn; these are indications of a poor diet that is very gritty. (Poor food and poor health may also have been a reason for their leaving.) Their build was stocky, and their faces were broad with prominent cheekbones and short noses. They had dark eyes, dark straight hair, and dark to light skin.

Culturally, the Indians of the Classic Pueblo period were considerably more advanced than the earlier peoples. Their pottery was magnificent. Their clothing was woven from both cotton and yucca fibers, and they made yarn and cord. They used juniper bark for baskets, bags, and sandals; and juniper poles were used in making walls and roofs. Jewelry was fashioned of many different materials. Children's toys have been found as well as games with playing pieces. Since the making of pottery came to be considered more important than basketry, the workmanship of baskets and sandals declined somewhat during this period.

The buildings of this period show skill and ingenuity. Spruce Tree House was constructed between A.D. 1230 and 1274, according to radiometric dating of the logs. The best preserved of the cliff dwellings, it has 114 living rooms and 8 kivas. It is 216 feet in length and 89 feet deep. Between 100 and 150 people lived there. With a spring at the corner of the cliff, obtaining water was not a problem. Each clan had its own kiva, set of living rooms, granaries, and food preparation area. When a male married, he moved to his wife's clan since family lines were carried through the females.

Cliff Palace, probably the best known dwelling, has 200 rooms and 23 kivas, and 8 floor levels. The buildings were constructed, added onto, and repaired between A.D. 1073 and 1273. The residents numbered between 200 and 250 people (figs. 7.5, 7.6).

Balcony House was built from A.D. 1190 to 1272 and gets its name from the balconies on many of the houses. The dwelling units are very well protected. A visitor entering Balcony House must climb a 30-foot ladder and squeeze through a crawl space. Exit is by chain and toehold climb.

The three cliff dwellings described above can be toured, but Square Tower House, New Fire House, and Oak Tree House can only be viewed from the outside. It should be pointed out that the cliff dwellings can be visited only on guided tours.

Sun Temple is located on top of Chapin mesa (fig. 7.1), 2,000 feet above the surrounding region. Built near Fewkes Canyon about A.D. 1200, it was left unfinished.

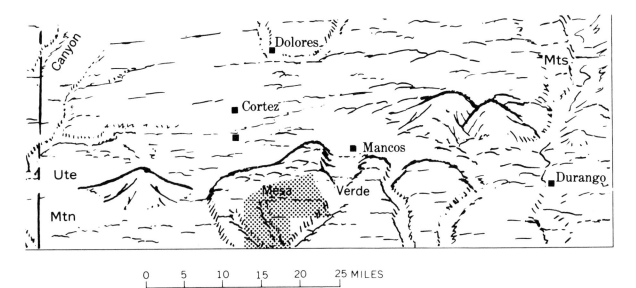

0   5   10   15   20   25 MILES

Figure 7.3 Physiographic diagram of the Mesa Verde plateau and neighboring landforms. Mesa Verde National Park (shaded area) is located on the gently dipping slope of a large cuesta formed of Cretaceous sandstone. North-facing cliffs rim the cuesta north of the Park boundary. On the south side the cuesta is cut by canyons eroded by the south-flowing Mancos River and its tributaries. Modified from Hunt 1969, U.S. Geological Survey Professional Paper 669.

The structure is laid out in a D-shape and has walls which are 4 feet thick. The length is 121 feet; the width, 64 feet.

Step House, Mug House, Long House, and Kodak House are all located on the Wetherill Mesa. Mug House, constructed around 1066, is named after the numerous mugs found there.

At the north end of Chapin Mesa is Cedar Tree House—with its tower connected to a kiva. Nearby are Far View Ruins, which are about the same age as the cliff dwellings. Far View may have been a trade center from A.D. 1100 to 1200. The living units are consolidated into thick-walled, multi-storied buildings, and the kivas are located within protected courtyards. One kiva is 32 feet in diameter.

Geologic Features

Mesa Verde is a *cuesta* that slopes gently to the south, conforming to the dip of resistant beds, and has on the north side a steep scarp (2,000 feet high) formed by the outcrop of the resistant beds (fig. 7.3). The Mesa Verde cuesta is an erosional remnant of an ancient pediment that once connected Mesa Verde with the La Plata and San Juan Mountains. A *pediment* is a broad, flat (or gently sloping) bedrock erosional surface that develops at the base of a mountain front. Usually a veneer of sediment called a *bajada*, covers the pediment and extends along the front edge, built from sediment that is washed down from the mountains and across the pediment surface (fig. 7.4). These landforms are characteristic of arid regions.

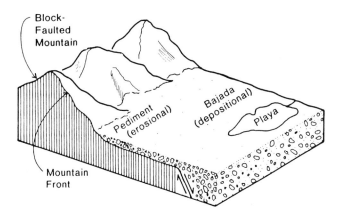

Figure 7.4 The development of a pediment and bajada in front of a block-faulted mountain range. The playa is a dry lake bed in the desert. From *Landforms and Landscapes* (3rd ed.) by Sherwood D. Tuttle. © 1970, 1975, 1980 Wm. C. Brown Company Publishers, Dubuque, Iowa. Reprinted by permission.

When the Mesa Verde region was uplifted, streams incised themselves into the pediment and divided it into its separate components. McElmo Creek, which flows west outside the Park *captured* (p. 13) the headwaters of streams that used to flow south along the north and west sides of the cuesta. The stream capture occurred because the south-flowing streams had lower gradients even though they were at higher elevations than McElmo Creek. When headward erosion by the McElmo tributaries cut back far enough, they encountered the south-flowing streams and captured their headwaters, thus changing their direction of flow from south to west.

The Mancos River, which drains the gentle slope of the cuesta, has cut a 1000-foot canyon along the mesa's eastern and southern edge. Streams that are now intermittent (but in an earlier time flowed year-round) have dissected the mesa into 15 long, parallel, steep-sided canyons, all of which drain into the Mancos. It is along these canyon walls and on the intervening mesa tops that the pueblo dwellings were built.

The upper member of the Point Lookout Sandstone is described as *massive* because it has no bedding planes, is very resistant, and forms cliffs. In the canyon walls, resistant sandstone layers, the cliff-formers, alternate with weak shale areas that weather back rapidly, forming the shallow caves that the Indians enlarged and made into cliff dwellings.

Some of the rock units, such as the Mancos Shale, contain *concretions*. These are hard, compact, rounded masses of mineral matter, of varying size and composition, that have been precipitated out from solution, usually around nucleii of some type (such as fossils). Because a concretion is more resistant than the sedimentary rock it forms in, it sticks out in an exposure or weathers out separately.

The springs in the caves that the Indians used for drinking water were fed by ground water seeping out between permeable and impermeable layers of bedrock. *Ground water* is the the subsurface water that fills the cracks, crevices, and pore spaces of rock. Sedimentary rocks that have many pore spaces (openings) can hold a considerable quantity of ground water and are said to have high *porosity*. If a porous rock is also *permeable*, then ground water can pass through it. Most sandstones and conglomerates are both porous and permeable. Shales, on the other hand, are usually porous, but they tend to be *impermeable* because the pore spaces are so small that water is trapped and cannot pass through. Thus when ground water coming down through a sandstone layer hits a shale layer, the water is forced to move laterally along the bedding plane contact and eventually seeps out as a spring, typically at the back of a cave.

Geologic History

1. Deposition of the Mancos Shale over the Dakota Sandstone in late Cretaceous time.

The Dakota Sandstone that underlies most of Colorado was deposited in an advancing shallow sea that gradually crept over the area that now forms the Colorado Plateau. On a shoreline in the vicinity of Mesa Verde National Park, wave action in the shallow water produced cross-bedded sand deposits that became a yellowish-orange sandstone. Layers of shale and coal indicate the presence of swamps just inland from the beach. This Upper Cretaceous formation is called the Dakota Sandstone, which is about 100 feet thick in this region.

As the shoreline shifted eastward, the water over the area that is now the Park became deeper. Mud layers were deposited that became the Mancos Shale, which is now 2,000 feet thick and subdivided into six members. Since these muds were dark gray and black in color, we know that the water was low in oxygen and probably stagnant. The oxygen deficiency prevented the iron in the sediment from oxidizing. The lack of circulation helps to account for the types of fossils found: swimmers, such as sharks and cephalopods (squid family); and floaters, such as foraminifers (single-celled marine animals with a calcium carbonate shell.) These life forms could only exist near the surface where wave action supplied oxygen. Thick-shelled clams that could stand muddy water were the only bottom dwellers. The Mancos Shale, which interfingers with the overlying formation, is the oldest formation exposed in Mesa Verde National Park. Concretions and tinges of a yellow-orange limestone are found in the Mancos.

2. Beach and shallow water deposits of the Point Lookout Sandstone.

The Mesaverde Group, made up of three formations (the Point Lookout Sandstone, Menefee Coal and Shale, and Cliff House Sandstone), reflects the changes in depositional environment that accompanied the upwarping and downwarping that was occurring at that time. The Point Lookout Sandstone, which forms the lower cliffs at the entrance of the Park, is interbedded with the Mancos Shale, indicating that the shoreline shifted back and forth. The Lower Member of the Point Lookout Sandstone, composed of alternate layers of sandstone and shale with a cumulative thickness of 60 to 140 feet, interfingers with the overlying beds of the Upper Member. Since the shale was deposited in deep water and the sandstone in shallow water, the change in depth of the sea is reflected in these beds. The cliff-forming Upper Member, a yellow-orange massive sandstone that was originally a cross-bedded beach deposit, protects the less resistant Lower Member and the underlying Mancos Shale from erosion.

3. Swamps along the shore producing the Menefee Formation, following regression of the sea.

The Menefee, the middle formation of the Mesaverde Group, contains two coal layers and varies in thickness from 230 to 800 feet. The lower coal layer interfingers with the massive sandstone of the Point Lookout; the upper coal layer, varying in color from gray to orange, is mixed with a cross-bedded, lenticular (lens-shaped) sandstone. A carbonaceous shale separates the upper and lower coal beds of the Menefee. Such a sequence indicates a swampy environment near the beaches.

The plants that grew and died and were buried in the swamps were unable to decay completely because of the toxic character of the water. As they were buried more and more deeply by accumulating sediment, the increasing pressure gradually changed the organic material into coal beds interbedded with the shale and sandstone. Outside Park boundaries, the Menefee coal is mined commercially.

The Upper Member of the Menefee Formation interfingers with the overlying Cliff House Sandstone. Differential weathering of this combination of a sandstone, shale, and coal in the canyon walls produces natural caves wherever the shale and coal occur, and ledges where the sandstone is located.

4. Return of the seas with several advances and retreats; deposition of the Cliff House Sandstone.

The Cliff House Sandstone (uppermost formation of the Mesaverde Group) is subdivided into three tongues (Lower Tongue, Barker Dome Tongue, and Upper Tongue) that interfinger with Menefee. The massive Cliff House Sandstone, youngest bedrock in the Park, rims the plateau and averages 400 feet in thickness. It was one of the last formations to be deposited in a sea in this region and is unconformable with the overlying gravels. The color of the sandstone ranges from pale to dark yellow-orange. The shale zones in the Cliff House Sandstone determine the location of the cliff dwellings. Weathering, due mainly to water seepage and frost action, plus mass wasting hollowed out the caves in the less resistant shale (figs. 7.5, 7.6).

5. Domelike uplift of the San Juan and La Plata Mountains and creation of the Mesa Verde pediment during the Laramide phase of the Rocky Mountain orogeny.

Upwarping and doming in the San Juan Mountains caused the region to slope toward Mesa Verde; and the combination of uplift and erosion produced the Mesa Verde pediment along the base of the mountains. On this erosional surface, meandering streams flowed toward the southwest, depositing gravels brought down from the mountains. The presence of tuff and other volcanic sediment records a relatively short period of volcanic activity that accompanied the uplift.

Figure 7.5 Cliff Palace, one of the largest cliff dwellings known, was the home of between 200 and 250 people. Located in a cave 325 feet long, 60 feet high, and 90 feet deep, the complex has 23 kivas and over 200 rooms. National Park Service photograph.

Figure 7.6   This view looks northeastward over Cliff Canyon from the Sun Point Overlook on Chapin Mesa. Cliff Palace (fig. 7.5) is in the deep alcove in the upper sandstone member of the Cliffhouse Sandstone (middle distance). The Cliffhouse Sandstone here is approximately 400 feet thick. From *Northern Colorado Plateau* by J. K. Rigby. © 1976 Kendall/Hunt Publishing Company; used by permission.

6. **Separation of Mesa Verde, La Plata, and the San Juan plateaus by stream erosion at the beginning of Tertiary time.**

Further uplift caused the streams on the pediment to downcut rapidly—removing most of the sediment, creating canyons and valleys, and isolating erosional remnants such as the Mesa Verde cuesta. Headward erosion created the 15 canyons.

7. **How seepage and weathering produced shallow caves.**

Because the Cliff House Sandstone is both highly porous and permeable, some of the melting snow and rain soaked in and readily percolated down as ground water until reaching the impermeable shale beds associated with the Menefee coalbeds. Then the ground water moved laterally along the contact until it seeped out at the canyon wall. Seasonal freezing and thawing broke up the shale. Mass wasting and erosion removed the loosened material, thus enlarging the hollows that the Indians later found and lived in, sheltered from the elements and protected from enemies. The alternate layers of permeable and nonpermeable rock also permitted springs to seep into the caves, bringing in a reliable water supply except during protracted droughts.

8. **Deposition of gravels during Pliocene and Pleistocene time.**

Quantities of pediment gravels, which make up the Bridgetimber Gravel, were deposited on top of the Mesa Verde cuesta in Pliocene time. Younger terrace gravels, consisting of glacial outwash from the LaPlata Mountains, filled the bottoms of the canyons during the cooler, more humid climates of Pleistocene time. These deposits formed the Florida Gravel.

Geologic Maps and Cross Sections

American Association of Petroleum Geologists. 1967. Geological highway map no. 2, Southern Rocky Mountain region. Tulsa, Oklahoma: American Association of Petroleum Geologists.

Baars, D. L. 1972. *Red Rock country, the geologic history of the Colorado Plateaus.* New York: Doubleday, p. 199.

Wanek, A. A. 1959. Geology of fuel resources of the Mesa Verde area, Montezuma, and LaMata counties, Colorado. U.S. Geological Survey Bulletin 1072M (map in pocket).

Bibliography

Devoto, R. H. 1972. Paleozoic stratigraphy and structural evolution of Colorado. *Colorado School of Mines Quarterly,* vol. 67, no. 4.

Hunt, C. B. 1969. *Geologic history of the Colorado River.* U.S. Geological Survey Professional Paper 669.

LeRoy, L. A. 1960. Generalized composite stratigraphic sections of western Colorado. Colorado School of Mines Department of Publications.

Rigby, J. K. 1977. *Southern Colorado Plateau,* K/H Geology field guide series. Dubuque, Iowa: Kendall/Hunt Publishing Company.

Rohn, A. H. 1971. Wetherill Mesa excavations, Mesa Verde National Park. National Park Service.

Wanek, A. A. 1959. Geology of fuel resources of the Mesa Verde area, Montezuma, and LaMata counties, Colorado. U.S. Geological Survey Bulletin 1072M.

Yardell, M. D. (ed.) 1972. *National Parkways, a photographic and comprehensive guide to Rocky Mountain and Mesa Verde National Parks.* Casper, Wyoming: World Wide Research and Publishing Company.

Address

Mesa Verde National Park
Mesa Verde National Park, Colorado 81330

8
Badlands National Park

Location: Southern South Dakota
Area: 380.16 sq. miles; 243,302.23 acres
Proclaimed a National Monument: March 4, 1929
Redesignated a National Park: November 10, 1978

Figure 8.1 Castle Butte, a prominent feature in Badlands National Park, towers above the Lower Prairie in the foreground. Resistant beds of sandstone form the pinnacles on the skyline. National Park Service photograph.

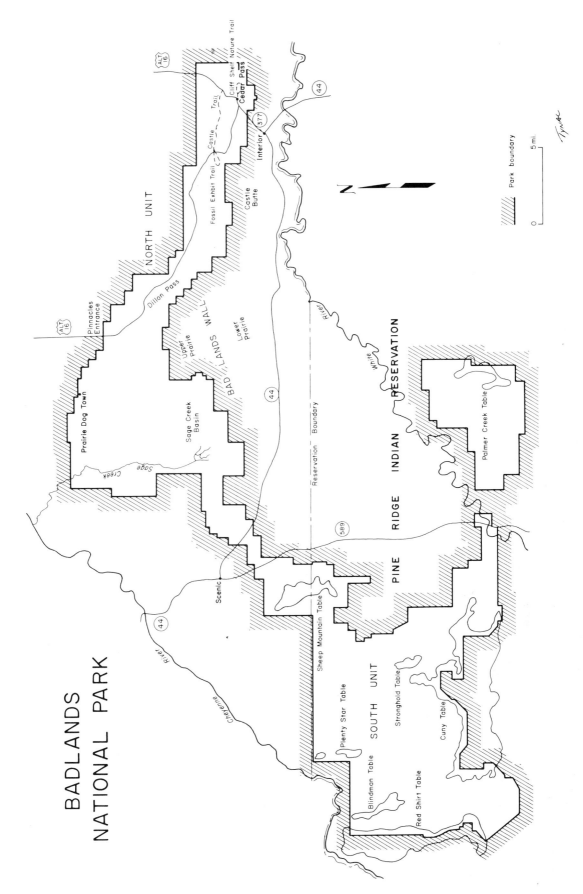

Figure 8.2 Badlands National Park, South Dakota.

Table 8.1. Geologic Column, Badlands National Park

Time Units			Rock Units			Geologic Events
Era	Period	Epoch	Group	Formation	Member	
Cenozoic	Quaternary	Holocene		Alluvium Landslide deposits Wind-blown sand		Deposits in streambeds, etc. Slumped masses of shale, silt, ash Stabilized dunes, yardangs
Cenozoic	Quaternary	Pleistocene		Older alluvium		Channel fills, pediment deposits
Cenozoic	Tertiary (Neogene)	Pliocene	//////////	//////////	//////////	Erosion
Cenozoic	Tertiary (Neogene)	Miocene	Arikaree	Sharps (Rosebud)	Rockyford Ash	Volcanic ash, silt, pebbles, concretions, sand; poorly consolidated beds
Cenozoic	Tertiary (Paleogene)	Oligocene	White River	Brule	Poleslide	Arid climate; sheet flow, mudflows, and flash floods; ashfalls, concretions
Cenozoic	Tertiary (Paleogene)	Oligocene	White River	Brule	Scenic	Arid climate; sheet flow, mudflows, and flash floods; ashfalls, concretions
Cenozoic	Tertiary (Paleogene)	Oligocene	White River	Chadron	Peanut Peak	Interfluve deposition; climate changing from humid to semiarid; sandstone, clay, conglomerate.
Cenozoic	Tertiary (Paleogene)	Oligocene	White River	Chadron	Crazy Johnson	Interfluve deposition; climate changing from humid to semiarid; sandstone, clay, conglomerate.
Cenozoic	Tertiary (Paleogene)	Oligocene	White River	Chadron	Ahearn	Flood-plain deposition; channel fill.
Cenozoic	Tertiary (Paleogene)	Eocene	—?—	Slim Buttes —?—	—?—	Channel-fill sandstone deposits on old erosion surface.
Cenozoic	Tertiary (Paleogene)	Paleocene	//////////	//////////	//////////	Unconformity Regional uplift
Mesozoic	Upper Cretaceous		Montana	Pierre Shale	"Interior Zone"	Soil layer formed by weathering in warm, wet climate. Weathered shale Black marine shale deposited in regressing sea.

(After Raymond and King 1976; Clark et al. 1967; Harksen and Macdonald 1969; O'Harra 1920)

Local History

The Badlands terrain is beautiful and desolate; fascinating, yet formidable. To the Dakota Indians the region was *Ma-koo-si-tcha*, or "bad land." The French trappers who had to cross it to get to their hunting grounds called the area *Les mauvaises terres à traverser*, meaning "a bad place to travel across." The early settlers avoided these parts—a land that was obviously no good for farming or hunting and a barrier to travel besides.

And yet from this hostile terrain has come much of what we know about the history of mammals that lived in North America between 40 and 25 million years ago. The strange bones in the walls of the canyons and cliffs and along stream banks aroused the curiosity of 19th-century naturalists. As it turned out, this region was probably the richest storehouse of vertebrate fossils in all of North America. Among the first scholars to come was Dr. Hiram Prout of St. Louis, who in 1846 discovered the jawbone of a titanothere, a large, extinct mammal. In the following year Dr. Joseph Leidy of Philadelphia found the bones of animals clearly related to camels, another astonishing discovery. From that time on, paleontologists came to the area

in increasing numbers. The U.S. Geological Survey, the great natural history museums, and universities of America and Europe sent expeditions and established "digs."* One can open practically any textbook on historical geology or vertebrate paleontology and find pictures of fauna

*Some of the more notable scientists and collectors are listed below, along with locations of several outstanding collections:
1839 John Evans of the David Dale Owen Geological Survey
1850 Thaddeus A. Culbertson, Smithsonian Institution
1853–1866 F. V. Hayden, leader of the Hayden Survey (forerunner of the USGS), made several visits to study and map the area.
1897 N. H. Darton, U.S. Geological Survey
1870–1874 O. C. Marsh, Yale University; additional specimens were collected for Yale between 1887 and 1908.
1886–1888 J. B. Hatcher, U.S. Geological Survey
1882–1893 W. B. Scott, Princeton University (J. B. Hatcher also collected for Princeton in 1894.)
1892–1916 H. F. Osborn, American Museum of Natural History
1894–1896 J. E. Todd, University of South Dakota
1899 (and subsequently) South Dakota School of Mines and Technology (Rapid City); one of the best collections of Badlands fossils is on display in its Museum of Geology.
1902 to the 1920s J. B. Hatcher and later O. A. Peterson, Carnegie Museum, Pittsburgh
1903, 1907 F. B. Loomis, Amherst College
1904 O. C. Farrington, Chicago Museum of Natural History

from the Brule and Chadron Formations of the White River Series—famous names, these, in American geology.

Unfortunately, as time went on, private collectors and amateurs—perhaps more enthusiastic than discriminating—began collecting on their own in alarming numbers. Congress decided in 1929 that it was high time this record of ancient life forms, set in the most remarkable and extensive badlands of the world, should be preserved as a national monument. After several additions to the protected area, the South Dakota Badlands were proclaimed a national park in 1978.

Today a 28-mile scenic highway, with numerous overlooks, winds through the Park's North Unit. The best time to take this drive in early summer is early in the morning (before it gets hot). At that time of day, the sun's rays slant into the deep gorges and gullies, lighting up hues of pink, yellow, green, maroon, cream, gray, and red on the spires and pinnacles. Buttresses and cliffs stand out in relief. After every rainstorm—the short, infrequent, heavy rains typical of a dry climate—the colors and shapes are different.

In the Park's South Unit (which lies within the Pine Ridge Indian Reservation) is Stronghold Table, a large butte where "ghost dances" were performed by Ogallala Sioux Indians late in the 19th century. They believed that the ceremony would bring back the buffalo and cause the white men to disappear from the Indian lands. In December 1890, after Sitting Bull was killed at Standing Rock, Chief Big Foot tried to elude the 7th Cavalry by leading his people through the Badlands toward the Pine Ridge Reservation. However, four troops of cavalry caught up with the Indians on December 28 and escorted them to Wounded Knee Creek (south of the Park) where they camped for the night. In the morning when the troops were searching and disarming the Indians, an Indian rifle discharged. Soldiers began firing, and indiscriminate shooting followed until more than 200 Sioux—men, women, and children—were dead. Twenty-five soldiers died and thirty-nine were wounded, most of them hit by their own fire. In the bitter cold, as a blizzard approached, the wounded soldiers and about fifty wounded Indians were taken in wagons to Pine Ridge. The bodies of the dead Sioux were later interred in a mass grave above the creek. This was the last major event of the Indian Wars.

Geologic Features

The Prehistoric Animals of the Badlands

During the Oligocene Epoch, between 37 and 25 million years ago, the region that is now the Badlands supported many kinds of animals. Some were the direct ancestors of modern horses and camels. Strangely enough, although most of the evolution of the camel and the horse took place in North America, both had become extinct on this continent before the white men came. Why horses survived and multiplied on the western plains after the Spaniards reintroduced them to the New World remains a mystery. Certainly the habitat was suitable in North America after the 1500s.

Some groups, the saber-toothed cats for one, have died out completely. The turtle, on the other hand, whose shell is the most common fossil found in the Badlands, belongs to a group that is virtually unchanged since early in the Mesozoic, some 200 million years ago. Other Oligocene reptiles also lived in this lush, well watered land; alligators and lizards, for instance. Many groups of birds were present, including eagles, owls, gulls, and pelicans.

Most significant, however, were the mammals. They began to take over after the dinosaurs (the rulers of the Mesozoic) disappeared from the face of the earth. The Cenozoic Era is called the Age of Mammals; and by the Oligocene Epoch, due to their rapid expansion into the habitats left by the dinosaurs, most of the modern mammalian orders had become established. Carnivores (flesh-eaters), for example, diversified rapidly as their main source of food, the ungulates (hoofed mammals), proliferated in the forests, marshes, and flood plains. Not only was this a time of great diversity among mammals; several species attained impressive size.

Largest of all mammals at that time were the *titanotheres,* an extinct order of *perissodactyls,* or odd-toed ungulates, that attained their peak in the Oligocene. *Brontotherium,* a giant titanothere the remains of which are found in the sandstones of the Brule Formation, stood about 12 feet high and had a pair of bony protuberances that grew hornlike above the snout. Like other ungulates, the titanotheres were herbivorous (eaters of foliage, etc.).

Most perissodactyl groups had three toes to begin with and tended to carry most of their weight on the middle toe, from which the single hoof of modern horses developed. *Mesohippus,* an early horse of the region that lived in the forests and ate foliage, still had three toes, but the middle one was large. *Pliohippus,* his successor and a grass-eater, had one great toe with two tiny ones on either side. The bones and teeth of the perissodactyl *Metamynodon,* an aquatic rhinoceros, have been collected from a sandstone member of the Chadron Formation. The rhinoceros, like the horse and camel, died out on this continent, but his descendants evolved in Africa. A third group of odd-toed ungulates, the tapirs, were present in the Badlands. Their descendants migrated to South America during the Pleistocene and have survived to the present.

The even-toed ungulates, the *artiodactyls,* were more successful mammals than the perissodactyls in terms of abundance, variety, and survival. Most of the Oligocene artiodactyls had four toes and tended to carry their weight on the third and fourth digits. Bones of camels *(Tylopoda)* at varying evolutionary stages have been found in the South Dakota Badlands, ranging from a four-toed animal the size of a jack rabbit to a species similar to modern camels. Although camels disappeared from this continent, a close relative, the llama, migrated to South America early in the Pleistocene and eventually became domesticated.

Many bones of piglike animals, called *entelodonts,* were preserved in rocks of the Badlands. These animals, which became extinct in the Oligocene, resembled peccaries or wild hogs, but they were not the ancestors of domestic pigs.

Primitive flesh-eaters, called *creodonts,* were forerunners of the true carnivores, who were more efficient hunters. *Hyaenodon,* about the size of a small bear, was an Oligocene carrion-eating creodont that was displaced by the carnivores. Early in the Oligocene, some carnivores became doglike and others became catlike; but they retained features in common for some time after they differentiated from their common ancestor. For example, several of the Badlands dogs had curved, hooded, catlike claws rather than the curved cylindrical cones for claws that are more typical of the dog family. Most of the dogs were about the size of foxes, but some were as large as black bears. (Bears, incidentally, evolved from dogs in Miocene time and have tended to become omniverous; i.e., eating all kinds of food.)

The most impressive members of the Oligocene cat family in the Badlands were the saber-toothed cats. Two species have been found in the White River beds, both belonging to an extinct group called the "stabbing cats." They had a unique ability to swing their lower jaw out of the way and stab their prey with their large, curved, daggerlike upper fangs. The saber-toothed cats were large and not so fast as most cats, and they preyed on large, slow-moving mammals.

The *insectivores* are represented in the Badlands fossil record by the golden mole, hedgehogs, and shrews. Other rodents found as fossils were the ancestors of modern beavers, rabbits, squirrels, and rats, with squirrels and rabbits being the most common.

Most of the modern ruminants (cud-chewers), the most diverse and abundant of all mammal orders today, evolved in Eurasia and Africa. However, a large group of primitive ruminants, unrelated to modern species, did thrive in the Oligocene. These animals are called *oreodonts* (fig. 8.3), and their fossilized bones are found only in North America. The cud-chewing oreodonts, about the size of foxes, roamed the plains in large herds. Of all the mammals in the Badlands region, the oreodonts were the most numerous. Many of their skeletons are found in huddled groups as if they were overcome by a natural disaster such as a sandstorm, an ashfall, or a blizzard. They became extinct well before the beginning of the Ice Age.

Figure 8.3   An articulated oreodont skeleton on display in the Visitors Center at Badlands National Park. A common fossil in the Park, the oreodont is an extinct vertebrate mammal, unrelated to modern species. Oreodonts were about the size of foxes. National Park Service photograph.

The Animals' Environment.

The Great Plains of the Dakotas are part of a vast mid-continent area overlying the North American *craton;* that is, a portion of the earth's crust that has been stable for a very long time. Nevertheless, the Great Plains in the Dakotas and elsewhere have been strongly affected by the intense tectonic activities that took place to the west at the end of the Mesozoic. During the Laramide orogeny, the regional tilt of the Great Plains was increased from west to east, and the dome of the Black Hills bulged upward (along the South Dakota-Wyoming border) only about 70 miles west of the Badlands. Beyond the Black Hills and another stretch of the Great Plains, the Rocky Mountains begin with the Bighorns (north central Wyoming and southern Montana), the first range of the Northern Rockies.

As the land rose, streams on the steepening slopes began eroding vigorously. Coarser sediment was dropped by mountain streams in great alluvial fans along the mountain fronts; finer sediment was spread out over the plains by overloaded rivers that built extensive flood plains. Over the millions of years that ensued, thousands of feet of gravel, sand, and silt accumulated and consolidated into conglomerates, sandstones, and mudstones. That the Black Hills were once much higher than now is indicated by the vast scale of the erosion that took place.

The climate was benign, much warmer than now, and on these flood plains both plants and animals flourished. During Oligocene time, the remains of successive generations of animals were covered over by the relatively rapid sedimentation, and their bones and teeth were preserved in the rocks. Sometimes animals perished in seasonal floods, were quickly buried, and thus whole skeletons became fossilized.

Gradually the climate became cooler and drier. This was the beginning of a long cooling trend that coincided with the slow northward movement of the North American plate away from the equatorial regions of the globe. Moreover, on the Great Plains, the mountains rising to the west blocked some of the moisture-bearing winds. The record in the rocks tells us that the mammals who lived on the plains either adapted to these changing conditions or migrated to more favorable habitats, or they became extinct.

Above the Oligocene rocks not much remains of the fossil record. Late in Oligocene time, volcanic ash, borne on the prevailing winds, began drifting over the Dakota plains. Most of the ash came from volcanoes erupting in the vicinity of Yellowstone National Park. Ash deposits show up as light-colored, very fine sediment, much of it mixed with silt. The Miocene beds, in which ash is much more extensive, are very poorly consolidated and fossils are not well preserved.

How the Badlands Developed

When renewed uplift occurred late in Cenozoic time, conditions were favorable for the development of badland topography on the old flood plains. Although the South Dakota Badlands are much more extensive than the badland topography in Capitol Reef National Park (p. 71), some similarities are apparent.

Again we have a region of sparse vegetation, semiarid climate, and seasonal or infrequent heavy rains. When precipitation does occur, the rain or snow falls on poorly consolidated sedimentary deposits containing minerals that dissolve or swell in contact with moisture. A fine drainage network of many small streams and rills develops short, steep slopes with narrow interfluves.

The eastern Dakota Badlands are being eroded into the "Wall," a south-facing, retreating escarpment (figs. 8.4, 8.5). From the top of the Wall, sloping gently to the north is the Upper Prairie, a nearly flat grassland, only slightly incised by north-flowing tributaries of the Bad River. The Wall itself is a 60-mile-long escarpment, about 3 to 5 miles wide, of deeply dissected badlands terrain. It is drained by tributaries that flow southward into the White River. The top of the Wall is thus a drainage divide. Below the Wall is an abrupt transition to the Lower Prairie, a partly grassy pediment area that slopes down to the flood plain of the White River. Between the top of the Wall and the distinct boundary where the Lower Prairie begins, relief averages about 150 feet but rises to more than 200 feet in some places. Incised, intermittent stream channels and sod-covered "tables" (erosional remnants) break up the surface of the Lower Prairie. The Park Visitor Center is on the Lower Prairie, which is about 3 miles in width. The White River flood plain, below the Lower Prairie, is about a half-mile wide (fig. 8.5).

Once much closer to the White River, the badlands area of the Wall is steadily eroding northward as the escarpment is dissected. The White River has cut all the way down through the Cenozoic rock units into the Mesozoic formations below. It is this downcutting and escarpment retreat that has exposed the fossils in the steep and barren slopes of the badland topography.

Erosional Features of the Badlands

Erosion rates in the Badlands are among the highest known.* When photographs of Badlands landforms taken early in the century are compared with recent photographs of the same features, it is apparent that marked

*During World War II, planes from a U.S. Army Air Force base near Rapid City sometimes discharged surplus ammunition in the Badlands on the way back to base after target runs. Thirty-some years ago a Park ranger found a projectile in the ground, its top flush with the surface on a sloping hillside. Today the projectile sticks up a foot above the ground! *R. R. Churchill, personal communication.*

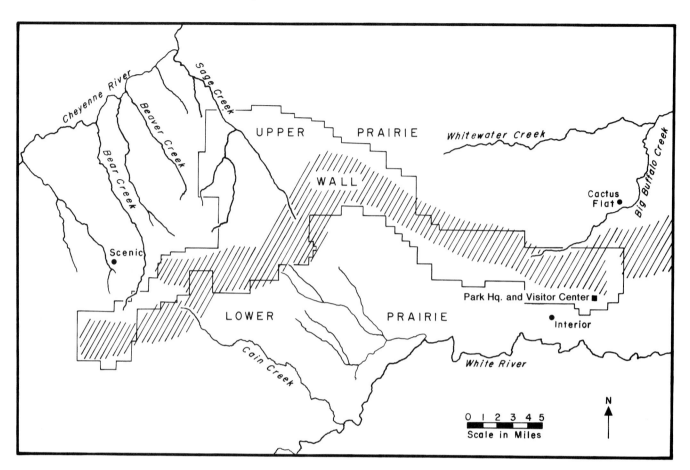

Figure 8.4   Map of the Wall in Badlands National Park showing how streams drain toward the northwest and southeast, with the Wall itself serving as a drainage divide. From Churchill 1979; courtesy R. R. Churchill.

Figure 8.5   Looking northward at the Wall, an erosional escarpment that is cutting back into the Upper Prairie. The Lower Prairie and the flood plain of the White River are in the foreground. The more rounded ridges and spurs along the lower part of the Wall are developed in the Chadron Formation; the steeper pinnacles and spires on the skyline are cut in the Brule Formation. Badlands topography is characterized by intricate stream dissection with many gullies and steep slope facets. National Park Service photograph.

changes in form and height have taken place. The rate of erosion varies with location and rock type. Spires in loosely consolidated ash may lose a half-foot of height per year. Tops of mudstone mounds may be lowered about an inch per year. Resistant sandstone caprock, on the other hand, may show an erosion rate of an inch in 500 years.

The dominant geologic agent of weathering and erosion in the Badlands is running water. During heavy rains, the pelting effect of raindrops loosens grains and particles, and running water washes the loose material downslope. Miniature gullies, eroding headward, enlarge rapidly. When infiltration occurs, the slope material may become saturated and move downward by mass wasting processes, slowly as creep or faster by sliding and slumping—sometimes an entire slope face at one time. Cliff Shelf, near Cedar Pass, the largest slump area in the Park, continues to slip downward from time to time, especially during unusually wet seasons.

The lithology of the rocks strongly influences the erosional processes. The high clay content and the fine-grained texture of the shale, siltstone, and ash of the Brule rocks, for example, limits infiltration of rain water, making runoff the dominant process. Thus many spires and pinnacles have formed. In the Chadron rocks, which have more sandstone and weaker cementation, a higher rate of infiltration makes mass wasting processes and downwastage more significant.

Frost action, which can be severe in the South Dakota climate, speeds the breaking up of the poorly consolidated rocks. Repeated wetting and drying in warm weather also causes them to crack, with each crack or tiny fissure providing more access for water.

In the Badlands, topographic position seems to be a factor in slope development of tables, or buttes. The north-facing slopes, being protected to some extent from the sun's radiation, retain moisture longer than south-facing slopes. Water has more time to react with calcium carbonate cement that holds the rock together, so that chemical weathering penetrates farther into the rock. Thus creep or slump of weathered material is more likely to occur on north-facing slopes; and when rain comes, the loose, weathered material is easily washed downslope. North-facing slopes tend to have more risers or slope segments and are usually not so steep as south-facing slopes.

South-facing slopes, on the other hand, tend to have high, steep slopes and angular dropoffs. Weathering on south-facing slopes is extremely limited because they dry very quickly when storms end and the sun comes out. Once dried out, any loosened fragments or blocks fall rapidly, collecting as debris at the bottom of the slope face (Churchill 1979).

Some of the strange-looking erosional effects in the Badlands are caused by layers or patches of resistant caprock. The rounded tops of the rock "toadstools" found here and there in the Park are usually made up of well cemented sandstone. The clay pedestals supporting these sandstone caps erode faster, becoming thinner and thinner until eventually the whole structure topples over (fig. 8.6). Arches have formed where weakly consolidated beds have washed out faster than beds above that are more firmly consolidated. These arches have a short life and usually cave in after a few years. Meanwhile, new ones are being carved out by the never-ending processes of erosion.

A ground water process called *piping* (a term from engineering geology) speeds mass wasting in the Badlands, particularly in the weakly cemented sandstones of the Chadron Formation. Percolating ground water, infiltrating soil and rock and dissolving cement in sandstone, initiates the formation of little tunnels, or pipes, in sediment and rock. These conduits enlarge as rivulets drain into them or as storm waters rush into them and scour out tunnels. Eventually a valley slope, an arch, or a slump body becomes undermined by the pipes and may collapse suddenly during or right after a hard rain. Landslide masses that have temporarily dammed tributary valleys may give way or wash out after being undermined by water flowing in such pipes or tunnels beneath the surface.

Clastic dikes stand out in relief against adjacent rocks that have been eroded back. These features were apparently formed by inwash of clastic sediment (i.e., sand, silt) that filled cracks and fissures in the sedimentary rock. Chalcedony (silica) precipitated by circulating ground water cemented the sandy sediment into hard, resistant rock. Little veinlets of chalcedony also filled joints and fractures in the rock. Once the less resistant enclosing rock has been removed by erosion, the dikes and veinlets are left as thin, resistant ridges, walls, and "shingles," some a foot or more wide and others only an inch or two wide (fig. 8.7).

Round, compacted *mudballs* accumulate in dry streambeds below the Wall. Ranging in size from that of apples to that of basketballs, these mudballs, consisting of clay mixed with pebbles, are formed by abrasion when a stream undercuts a bank and chunks of rock material are rolled around by rushing water and rounded by stream action.

When streams are running in the Badlands, they are always loaded with sediment. However, even when streams slow down or water stands so that material in suspension drops, the water remains cloudy and white in color. This is because submicroscopic particles of clay are in *colloidal dispersion;* that is, the particles carry a slight charge of electricity that causes them to repel each other and prevents them from settling to the bottom. The White River was given its name by early visitors to the region because of its thick, white water (fig. 8.8).

Figure 8.6   The pedestal rock shown is an erosional remnant with a resistant sandstone caprock overlying less resistant siltstone and mudstone. Note the pieces of sandstone lying at the base of the pedestal. Eventually the sandstone cap will be completely undermined and will tumble down, too. National Park Service photograph.

Figure 8.7   The resistant clastic dike now stands higher than the more rapidly eroding enclosing rock. The clastic dike formed when volcanic ash, blown in or washed in, filled a fissure in the sedimentary rock and hardened. National Park Service photograph.

Figure 8.8   The White River drains the Lower Prairie and the southern part of the South Dakota Badlands. The river water, which has a milky look because of its high "clay" content (both in suspension and colloidal dispersion), is undrinkable. National Park Service photograph.

Pioneers in the Badlands region discovered that pouring a pail of "alkali water" (taken from a waterhole in shale) into a barrel of white water would cause it to settle so that it could be used. Later, when the railroad was built through the Badlands in 1907, providing drinking water for the many teams of horses used in construction work was a serious problem until someone discovered that throwing slices of cactus pads into a horse trough or waterhole would clear the white water enough so that the animals could drink it. Obtaining a supply of clear water in the Badlands region can be a problem even today; however, the National Park Service has drilled shallow wells in gravel deposits near the White River that furnish enough clear water for Park facilities.

Faults

A linear series of normal faults (associated with a structural trough) crosses the Badlands area. The amount of displacement, a few feet in most cases, shows up in the alternating bands of light- and dark-colored sediment that can be seen on either side of a fault. The fault at Dillon Pass, however, shows that movement there was as much as 49 feet. The faults are related to movement in the Precambrian-Paleozoic basement complex far below the sedimentary rocks of the Badlands.

Concretions, Geodes, and Sand Crystals

Like the fossils, these curious objects are to be seen but *not collected* within the Park. The calcareous concretions, found in all the formations of the Badlands, are more numerous and more varied in size, shape, and type than those in the shale beds in Mesa Verde National Park (p. 82). The nodular zones of the Brule Formation have the most concretions. These zones, which are probably flood plain deposits, show up in the sedimentary sequence as red beds with a lumpy appearance. The nodules (concretions) that have weathered out tend to have a rusty brown color due to oxidation. Many of the concretions contain a fossil fragment as a nucleus, around which are tightly cemented layers of clay particles.

Like the calcareous concretions, geodes are present in all of the Park's formations and are also most numerous in the nodular zones of the Brule.* The nodules that weather out are apt not to show by their outward appearance whether they are calcareous concretions or geodes. But geode nodules that have been broken can be identified as such by their hollow interiors with small crystals projecting from the cavity walls. By definition, *geodes* are hollow (or partly hollow), globular-shaped bodies that occur in sedimentary rock and that weather out as discrete nodules or concretions. (Geodes are a type of concretion.) A geode's outer layer is made of dense chalcedony, and crystals project from its lining into a central cavity. The crystals, most commonly quartz or calcite, are usually perfectly formed and are not the same minerals as those of the enclosing rock (box 8.1). Sometimes the chalcedony shell is further enclosed by coatings of cemented clay.

*Dr. Philip Bjork, Director, Museum of Geology, South Dakota School of Mines and Technology. Personal communication.

Box 8.1
How Are Geodes Formed?

Geodes are like snowflakes; no two are exactly alike. They may be large or small, hollow or nearly solid. Some (called "rattle rocks") contain loose crystals that rattle when the geode is shaken; some are filled with connate water (water trapped in sediment). In the Badlands, geodes have even been found clustered together *inside* a calcareous concretion or, more surprising yet, encased in the long bone of a mammal fossil!

Geodes do not form in all sedimentary rocks; in fact, they are relatively uncommon. Their most typical occurrence is in limestone, but they occasionally occur in marine shales as well. The origin of geodes is not clear, because no one theory seems to account satisfactorily for all of the kinds of geodes and the different sedimentary environments in which they formed. The puzzle involves getting what was inside *out* and putting what is now inside *in.*

A widely accepted theory that appears to fit a good many geodes suggests that the beginning of a geode was the burial of a shelled animal, or other object, in the limy ooze of an ocean bottom, with the shell, for example, providing the original cavity. Silica gel accumulated in the void, surrounding and trapping a small amount of sea water. Later, after the sedimentary unit had been uplifted, ground water began circulating through the rock. Osmotic pressure between the trapped sea water inside the silica gel and the fresh water outside caused the membrane of silica to expand and eventually dry out, becoming the chalcedony layer. Minute cracks in the chalcedony permitted mineral ions in circulating ground water to enter the geode and gradually crystallize out of solution. Layers of clay particles may have become cemented to the exterior wall because they could not pass through the wall.

Obviously, some modifications in this theory must be made in order to account for the geodes in the Badlands. The only Badlands marine sequence is the Cretaceous Pierre Shale (table 8.1), with all the overlying beds formed in a terrestrial environment. Most of the rock layers are shales, siltstones, and sandstones; there is no marine limestone, and there are only a few lenses of fresh-water limestones here and there. Yet geodes are found in all formations (but not in every layer).

Here is an example of what is meant when it is said that "geologists deal in ambiguity." A geologist *expects* to find in the natural world things that don't fit the standard theory, or that deviate from it in one way or another. Some of these deviations, incidentally, turn out to be very useful clues in adding to our body of scientific knowledge. If you were a geologist intending to make a definitive study of the Badlands geodes—and so far as we know, none has been done—what questions would *you* ask?

A natural question is, what was happening during and after lithification of the sediment to make such a varied assortment of both calcareous concretions and geodes in the Badlands formations? Ground water must have been plentiful and it was probably close to the surface (high water table). But did it just percolate downward, or was it also forced back upward (by hydrostatic pressure) carrying mineral ions from rocks farther down? Silica in soluble form must have been readily available over a long period of time. Did abundant silica come from the highly siliceous volcanic ash that time after time settled over the Badlands region? The "thunder eggs" of central Oregon, which are geodelike features but not *true* geodes, formed from silica precipitated as chalcedony, opal, and agate, in welded tuff (p. 226), a type of volcanic deposit. Might there be a connection between how thunder eggs were made and the Badlands geodes?

In the Badlands, geodes are sometimes associated with calcareous concretions and sometimes not. Why is this so? Rockhounds who find geodes and regard them as fascinating natural objects can doubtless add many more questions!

The *sand crystals,* which are most apt to be found in the sandstones of the Sharps Formation, may occur singly or in a rosette structure. They range in size from a quarter of an inch to 15 inches. Although sand crystals are mainly calcite, they do contain inclusions of sand grains and they probably formed during cementation of the sandstone. Some are penetration crystals; that is, one crystal growing through another crystal.

Interesting and remarkable specimens of calcareous concretions, geodes, and sand crystals, as well as fossils, minerals, etc., are on display in the Museum of Geology at the South Dakota School of Mines and Technology in Rapid City, and in the Badlands National Park Museum.

Geologic History

1. Deposition of Pierre Shale in a shallow inland Cretaceous sea.

The geologic record in the rocks exposed in Badlands National Park begins in Late Cretaceous time, approximately 80 million years ago, with the deposition of silt and organic material in a shallow inland sea. Giant sea turtles and mesosaurs (swimming reptiles, similar to lizards) lived in its waters, as did several varieties of ammonites (marine mollusks). The mesosaurs and ammonites, although flourishing at that time, became extinct at the end of the Cretaceous. The giant sea turtles have survived in other oceans to the present.

As the land rose and the sea retreated, the water became brackish, and more and more sand and silt were washed down from the land. Uplift continued as the Black Hills to the west were domed up, and sedimentation in the Badlands region became entirely nonmarine.

2. Prolonged erosion during early Cenozoic time.

Still the land rose, and extensive drainage systems carved a rolling terrain of hills and valleys. In the Badlands region, the nonmarine layers were stripped off, and the black layers of Pierre Shale (pronounced "peer") became exposed. Intense weathering in a warm, wet environment turned some of the upper beds to a reddish-yellow soil (or laterite). In the Dillon Pass and Sage Creek areas, this ancient land surface is now exposed again. The bright-colored weathered beds contrast sharply with the black shale below. The deposits are referred to as the "Interior Zone," being named for the nearby town of Interior. Marine fossils have been found in the locality.

3. The Oligocene environment.

The mountain-building and uplift that began at the end of the Cretaceous apparently affected the Great Plains in a series of waves or pulses of activity, rather than by a continuous process. Periods of uplift were followed by long spans of erosion and deposition, and then the stress would begin again. At least twice the stress of upwarping produced faulting in the sedimentary beds. Relief was not

great, but enough of a structural trough resulted from the continuing isostatic adjustments so that a clastic (i.e., sedimentary) wedge buried the old erosional surfaces.

The environment of sedimentation was apparently a mixed one. The animals whose fossilized bones are preserved in the rocks roamed over open plains and along flood plains. Some preferred the extensive marshes. Ash falls may have been severe enough at times so that animals were buried by hot ash, or perhaps they starved when their food supply was destroyed. Floods were apparently a more or less regular occurrence, and probably dust storms were as well.

4. Deposition of the Slim Buttes Formation.

Slim Buttes rocks are mainly pale greenish to olive-white sandstones with some basal conglomerate. The sequence has an unconformable relationship with the Interior Soil Zone of the Pierre Shale and occurs as discontinuous lenses several miles long, with thicknesses of up to 40 feet. (Some geologists consider the Slim Buttes Formation to be part of the Ahearn Member of the Chadron because of the discontinuous distribution of Slim Buttes beds in the Badlands.) The rocks are assumed to be lithified channel fill deposits and are probably Eocene to Oligocene in age.

5. Fluvial deposits of the Chadron Formation.

Three members, Ahearn, Crazy Johnson, and Peanut Peak, make up the Chadron Formation. The Ahearn was deposited as channel fill in an ancient drainage network that antedated the drainage patterns of today. At the beginning of Ahearn time, the central valley was well developed and had a deep soil layer. Lenses of Slim Buttes sediment were scattered along the length of the valley. Due to increases in volume and gradient, several streams joined and enlarged, bringing in a fresh flood of material and forming a basal conglomerate. Above the basal beds, deposits of red clays from the older sedimentary rocks of the Black Hills alternated with greenish muds to produce a mottled sandstone. Because the river system's conditions changed frequently, cross-bedding and cut-and-fill deposits are common.

The Crazy Johnson and Peanut Peak Members* are made up of stream deposits. By this time, rainfall had decreased somewhat and the streams had lost some volume and erosive energy. The main source of sediment had shifted from the southern part of the Black Hills toward the northern part. Crazy Johnson Member is a mixture of mudstone and channel fill averaging 20 to 40 feet in thickness. The color ranges from green-gray conglomerate to gray-bluish or greenish mudstone. Oxidation lightened the color of the sediments later laid down.

*The names of these Chadron members refer to an early settler in the Badlands named Johnson, dubbed "Crazy Johnson" by his neighbors because he tried to grow peanuts on the flat summit of one of the large tables (still called "Peanut Peak").

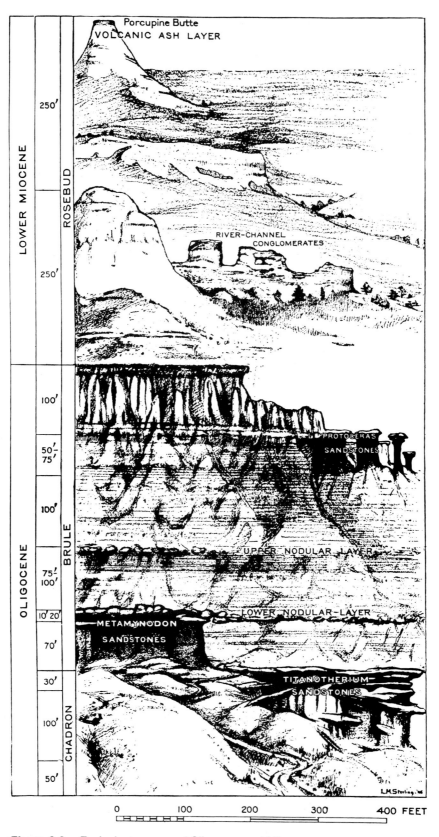

Figure 8.9 Typical exposures of Oligocene and Miocene rock units of the South Dakota Badlands are shown in this composite diagram. From Osborn 1909, U.S. Geological Survey Bulletin 361.

Peanut Peak Member is similar to Crazy Johnson but thinner and lighter in color, having more tan and orange mudstone and less of the greenish and gray. As the rainfall decreased, so did the amount of vegetation, and this brought about changes in faunal habitat.

The titanotheres were approaching their peak during this time, and many of their bones are found in the Chadron Formation, along with those of other mammal orders. Chadron rocks tend to weather to low, rounded "haystack hills" (fig. 8.5).

6. Deposition of the Brule Formation.

A thin, discontinuous nonmarine limestone marks the boundary between the Chadron and Brule Formations, suggesting the existence of shallow ponds at that time. Overlying this limestone sheet is the Lower Nodular Zone which contains, in addition to concretions, many fossils of turtles and oreodonts (primitive ruminants) (fig. 8.3). Although the overall trend during Brule time was toward a cooler, more arid climate, cycles of warm-wet climate with widespread flooding did occur. The flooding laid down sheets of thick, gooey mud that preserved fossil bones as it dried. About 75 to 100 feet above the Lower Nodular Zone is the Upper Nodular Zone. Since both of these are continuous across the Badlands, as well as distinctive in appearance, they serve as excellent markers (fig. 8.9).

The Scenic and the Poleslide are the two members of the Brule Formation, with the upper member, the Poleslide, being the thicker (about 270 feet). The Scenic Member, which includes the nodular zones, is between 80 and 100 feet in thickness. Interbedded clays, silts, sand, and volcanic ash, as well as channel fillings, make up the Scenic rocks. In the Poleslide Member are increasing amounts of volcanic ash, some of it fluvially reworked and mixed with silts, fine sands, and clays. The color of the rocks is mostly yellow or tan. The Poleslide channel-fill deposits are called "Protoceras sandstones" because of the large number of bones from this species of oreodont found in the rocks (fig. 8.9).

The Brule rocks, especially the upper layers, tend to form steep cliffs, knife-edge ridges, and pinnacles (fig. 8.1). On Brule slopes are many steps, overhangs, flutings, etc., all reflecting various responses to weathering processes. Some Brule beds are made up of loose clay; others of firm clay. Some are finer, some coarser. These differences suggest rapidly changing conditions of deposition and consolidation, with the result being variations in slope as the rock weathers and particles are washed down by running water.

7. Deposition of the Arikaree Group during the Miocene Epoch.

The Rockyford Member of the Sharps Formation is made up of white volcanic ash, somewhat mixed with mud (fig. 8.9). Because of its distinctive appearance, this member serves as a marker for the beginning of the Miocene.

Above the Rockyford are siltstones with a high ash content. They are gray or tan in color and form a flat, barren, dissected highland at the top of the Badlands.

8. Development of badlands topography.

Since Miocene time, the processes of erosion have been dominant over the processes of deposition, although the latter are by no means absent from the picture. Large deposits of river gravels are associated with the general disruption of drainage that went on during late Pleistocene and post-Pleistocene time; high winds created sand dunes, most of which later stabilized. (The Badlands were not covered by continental glaciers, but they were affected by the severe climate.)

Even during times of greater moisture than the present, vegetation never became established well enough to control the erosion of the poorly consolidated beds of ash and clastics. As long as these conditions continue, the Badlands will continue to erode.

Geologic Map and Cross Section

O'Harra, C. 1910. *The Badlands Formation of the Black Hills region.* South Dakota School of Mines Bulletin 9.

Raymond, W. H., and King, R. U. 1976. *Geologic map of the Badlands National Monument and vicinity, west central South Dakota.* U.S. Geological Survey, Miscellaneous Investigations Series Map I-934.

Bibliography

Brown, Dee. 1970. *Bury my heart at Wounded Knee.* New York: Holt, Rinehart & Winston.

Churchill, R. R. 1979. Topoclimate as a controlling factor in Badlands hillslope development. Ph.D. Thesis. Iowa City, Iowa: University of Iowa.

————. 1979a. The importance of mass movement and piping in Badlands slope development. In *Proceedings,* Iowa Academy of Science 86(1):10–14.

Clark, C. 1978. *The Badlands,* American Wilderness Series. Rev. ed. Alexandria, Virginia: Time-Life Books.

Clark, J., Beerbower, J. R., and Kietzke, K. K. 1967. *Oligocene sedimentation, stratigraphy, paleoecology, and paleoclimatology in the Big Badlands of South Dakota.* Fieldiana, Geology Memoirs, vol. 5. Chicago Museum of Natural History.

Fleisher, P. J. 1975. *Geology of selected national parks and monuments.* Dubuque, Iowa: Kendall/Hunt Publishing Company.

Harksen, J. C., and Macdonald, J. R. 1969. *Type sections for the Chadron and Brule Formations of the White River Oligocene in the Big Badlands, South Dakota.* South Dakota Geological Survey, Report of Investigation 99.

Hauk, J. K. 1969. *Badlands, its life and landscape.* Bulletin no. 2. Interior, South Dakota: Badlands Natural History Association.

Kennehan, R. W. 1980. Badlands National Park. Term paper. Youngstown, Ohio: Youngstown State University.

Kummel, B. 1970. *History of the earth.* 2nd ed. San Francisco: W. H. Freeman and Company.

Kurten, Bjorn. 1972. *The age of mammals.* New York: Columbia University Press.

Levin, H. L. 1978. *The earth through time.* Philadelphia: W. B. Saunders Company.

Madson, J. 1981. South Dakota's castles in clay. *National geographic* 159(4):524–539, April.

O'Harra, C. 1920. *The White River badlands.* South Dakota School of Mines Bulletin 13.

Osborn, H. F. 1909. *Cenozoic mammal horizons of western North America.* U.S. Geological Survey Bulletin No. 361.

Scheidegger, A. E., Schumm, S. A., and Fairbridge, R. A. 1968. Badlands. In *The encyclopedia of geomorphology,* ed. R. A. Fairbridge. New York: Reinhold Book Corporation.

Schumm, S. A. 1956. The role of creep and rainwash on badland slopes. *American journal of science* 254: 693–706.

Smith, K. G. 1958. Erosional processes and landforms in Badlands National Monument, South Dakota. *Geological Society of America Bulletin* 69: 975–1008.

Toepelman, W. C. 1922. *The paleontology of the area, the Badlands as a national park.* South Dakota Geological Survey Bulletin 6.

Wanless, H. R. 1922. Lithology of the White River sediments. American Philosophical Society 61:184–203.

———. 1923. The stratigraphy of the White River Badlands, South Dakota. American Philosophical Society 62:190–269.

Ward, F. 1922. *Geology of a portion of the Badlands.* South Dakota Geological and Natural History Survey Bulletin 11.

Address

Badlands National Park
P.O. Box 6
Interior, South Dakota 57750

9
Theodore Roosevelt National Park

by Rodney M. Feldmann
Kent State University

Location: North Dakota
Size: 70,344.64 acres; 109.91 square miles
Established: April 25, 1947; redesignated: November
10, 1978

Figure 9.1   Badlands along Little Missouri River in
Theodore Roosevelt National Park, western North Dakota.

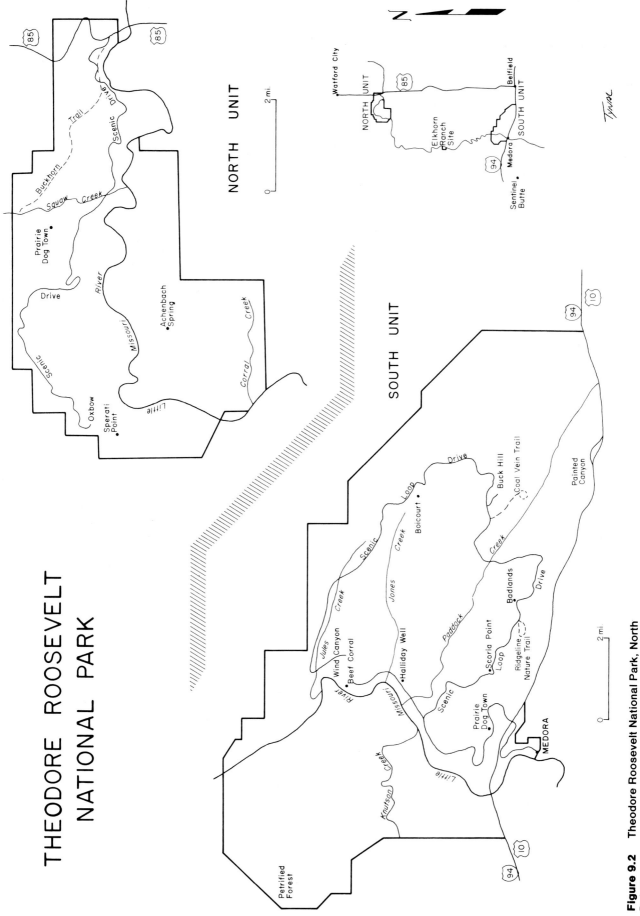

Figure 9.2 Theodore Roosevelt National Park, North Dakota.

Table 9.1. Geologic Column, Theodore Roosevelt National Park

Era	Period	Epoch	Group	Formation	Marker Bed	Geologic Events
	Quaternary	Holocene	Alluvial deposits, landslide features, slump features, "scoria" formation			Development of present badlands topography
	Quaternary	Pleistocene	Cut-and-fill structures and terraces along Little Missouri River and tributaries, glacial erratics, slump features			Glaciation reroutes Little Missouri River and Missouri River; possible early Pleistocene glaciation
Cenozoic	Tertiary	Paleogene / Paleocene	Fort Union Group	Sentinel Butte Formation	Upper sand	Grayish to brownish siltstone, clay, and sandstone deposited in habitats similar to Tongue River sediments; lignites less common; "Blue" bed represents weathered volcanic ash marker.
					Bullion Butte lignite	
					Upper "yellow" bed	
					Lower "yellow" bed	
					"Blue" bed	
					Basal sand	
				Tongue River Formation	HT lignite	Yellowish, pastel siltstone, clay, and sandstone deposited in fluvial, lacustrine, and overbank environments with numerous lignitic beds formed in swampy, flood plain setting.
					Meyer lignite	
					Garner Creek lignite	
					Harmon lignite	
					Hanson lignite	
					H lignite	
					Basal sandstone bed	

Modified from Royse 1970.

Local History

In 1883, when he was 24 years old, Theodore Roosevelt made his first visit to the badlands region of Dakota Territory. He traveled westward on the newly constructed Northern Pacific railroad for the purpose of hunting buffalo. He was so impressed with the badlands adjacent to the Little Missouri River that he soon purchased the Maltese Cross Ranch, located six miles south of Medora, and began a 15-year-long career as a rancher. He expanded his enterprise by developing the Elkhorn Ranch site several miles north of the Maltese Cross Ranch but still in the badlands along the Little Missouri River. (These badlands are some 300 miles north of the South Dakota badlands discussed in chapter 8.)

Although the badlands were referred to as *les mauvaises terres à traverser* ("the badlands to cross"), they possessed all of the characteristics necessary for sustaining successful ranching operations in the northern Great Plains. In an article written by Roosevelt and illustrated by Frederic Remington for *Century* magazine in February 1888, Roosevelt observed that the success of a ranching venture in the northern Great Plains was dependent upon three essential elements—food, water, and shelter. During the hot summer months, food and water were the two essentials and, because the Little Missouri River was one of the few perennial streams in the area, proximity to it assured a continuous, abundant water supply for cattle. Additionally, the badlands region of the Little Missouri River supported adequate vegetation to permit grazing. During the long harsh winter, availability of a water resource was less essential because cattle could derive adequate moisture from the snow cover. During the winter months, the most important element for survival of cattle was shelter, Roosevelt pointed out. The numerous canyons and gulleys of the region, as well as the cottonwood thickets along the Little Missouri River, provided the animals with excellent protection against the harsh winter winds that severely limited ranching enterprises in the bleak, exposed country of the Missouri Plateau.

Thus, the badlands were a paradox; they were both help and hindrance to development of the West. Wagon travel across the area was extremely difficult. Wagons had to be disassembled and hand-carried across many of the more rugged parts of the terrain. On the other hand, that very ruggedness provided the shelter from the elements that permitted development of a ranching industry in a climate that was otherwise too severe to permit anything but summer grazing.

Even at its best, the badlands did not support a large number of animals, and by the mid 1880s the number of cattle in the area had increased to the point where overgrazing of the range land resulted. The winter of 1886–1887 was extremely severe and dealt an almost killing blow to the ranching industry along the Little Missouri. The ranchers did survive, however, and Roosevelt maintained his ranching interests in the area until 1898.

Before the ranchers brought cattle into the northern Great Plains, herds of buffalo (bison) had thrived on the grazing lands for thousands of years. Hardier than cattle, the buffalo were able to survive periods of drought and the most severe winters. The buffalo were a mainstay of the Indians, who used the meat for food and hides for clothing and shelter. Most of the hostilities between Indians and the white ranchers were the result of Indian resistance to the loss of their hunting lands and the slaughtering of the buffalo herds. Organized resistance by the Indians had ended only a few years before Theodore Roosevelt came to the region.

Upon being elected President of the United States, Theodore Roosevelt initiated and supported numerous policies to assure conservation of lands in the National Park System. His interest in protecting these areas was based in large part upon the influence of his years as a cattleman in the badlands. Therefore, it is fitting that on April 25, 1947, two large separate tracts of land (North Unit and South Unit) along the Little Missouri River were established as the Theodore Roosevelt National Memorial Park. A part of Elkhorn Ranch, between the two units, is also under Park jurisdiction (fig. 9.2). In 1978, the area was redesignated the Theodore Roosevelt National Park.

Geologic Features

The topographic expression of any area is a function of structure, stratigraphy, and climate. In the badlands region of Theodore Roosevelt National Park, the rocks have been elevated subsequent to the time of their deposition but have not been substantially deformed tectonically. Therefore, with the exception of local mass wasting features, the beds are essentially horizontal (fig. 9.1). As stream erosion dissects the area, the drainage pattern tends to be dendritic, and no obvious "grain" related to structural deformation has developed. The majority of the park lies within the deeply dissected valley walls of the Little Missouri River, which meanders across the southwestern corner of North Dakota, and minor ephemeral tributaries that feed it. (*Ephemeral streams* flow only after heavy rains.) The valley is bounded by an upland surface, the Missouri Plateau, which represents a stripped structural surface.

Much of the interest in this particular badlands region lies in the variety of exposed sediments. The sedimentary types include sandstone, siltstone, shale, *lignite* (brown coal intermediate in coalification between peat and subbituminous coal), and occasionally limestone deposited in a rhythmic sequence during the Paleocene Epoch

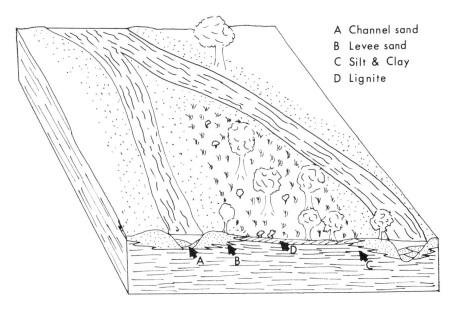

A Channel sand
B Levee sand
C Silt & Clay
D Lignite

Figure 9.3 Schematic representation of the environments in which the more common Fort Union sediments were deposited. Modified from Jacob 1976.

(table 9.1). These sediments are just a part of a very thick wedge of fluvial, lacustrine, and flood plain deposits formed during the uplift and erosion of the Rocky Mountains in Cretaceous and Paleogene (early Cenozoic) time. Faunal and floral evidence suggests that the climate then was generally warm and humid so that lush vegetation covered the land. Sluggish streams meandered across the region, depositing a variety of sediments now referred to as the Fort Union Group. (These sediments are perhaps analogous to those now accumulating in the lower reaches of the Mississippi Valley.) The streams deposited sand as channel fill and in *natural levees,* or embankments, paralleling the channels (fig. 9.3). During episodes of flooding, finer-grained material, silt and clay, accumulated as overbank deposits. Low-lying regions adjacent to the river valley were swampy, and in these areas organic material was buried and compacted, eventually becoming lignite. Thin beds of nonmarine limestone accumulated in lakes. Finally, occasional bursts of volcanic activity in the Rocky Mountains to the west yielded sufficient volcanic ash to deposit layers several inches thick over the region. These volcanic ashes have either been mixed with fine-grained flood plain sediments or remain as discrete layers. In either case, weathering of the material has yielded bentonitic clays with a lithology distinct from other units in the Fort Union.

Perhaps the most distinctive rock unit in the Theodore Roosevelt National Park, as contrasted to the sediments of the Badlands National Park, South Dakota, is lignite. The badlands in South Dakota do not contain lignitic beds and, therefore, the general appearance of the

terrain is quite different. The lignite in Theodore Roosevelt National Park is prominently exposed throughout the area and additionally has produced another interesting rock type, locally called "scoria." The term *scoria* usually refers to vesicular rocks of volcanic origin (see p. 143), the North Dakota scoria is baked clay (with a consistency much like that of pottery), which has resulted from the *in situ* burning of lignite. The incinerated bright red to black clinker beds are diagnostic of the Fort Union Group (fig. 9.4).

Figure 9.4 Small knolls capped by "scoria," baked clay resulting from burning lignite. National Park Service photograph.

The variation in lithology seen in the Fort Union Group is expressed as sediments with a wide range of colors and weathering characteristics so that the terrain is studded with erosional remnants capped by resistant rock units in a myriad of forms and sizes (figs. 9.5, 9.6). The variation in lithology also produces an interesting distribution of vegetation types. Because of the present-day climate, most of the vegetation is typical of a semiarid region dominated by short prairie grasses and sagebrush. In areas where particularly permeable rocks are exposed, however, stands of cedar are common. On lowland surfaces adjacent to the Little Missouri River, where ground water and surface water are plentiful, thick groves of cottonwood are the dominant vegetation.

The climate of the area can best be described as severe. Temperatures during the year fluctuate over a range of 150°F.; and rainfall, which averages about 15 inches per year, typically occurs in short, violent bursts. This climatic setting results in a relatively sparse vegetal cover. The absence of dense vegetation, coupled with periodic heavy rainfalls, makes an ideal setting for development of the high drainage density of badlands topography (fig. 9.7).

The geologic features of interest are, therefore, related to the stratigraphic setting as well as to the climate and its effect on the landscape. Some of the more obvious and interesting forms are described below.

Figure 9.5   Erosional remnants capped by ledgelike pieces of well cemented sandstone. National Park Service photograph.

Figure 9.6   Small erosional remnants capped by sandy concretions. National Park Service photograph.

Figure 9.7   Panoramic view of Painted Canyon illustrating the high density drainage pattern. The flat surface on the horizon is the Missouri Plateau.

Figure 9.8   Petrified log preserved in vertical position.
National Park Service photograph.

Petrified wood is relatively common throughout the Park. In most places it is preserved as tiny bits and pieces that have weathered out from the slopes and accumulated in low spots. In other places, however, nearly complete stumps (fig. 9.8) and logs of petrified wood can be found in place. Most of the petrified wood occurs in the silty and clayey sediments representative of the flood plain environment during the time of deposition of the Fort Union Group.

The occurrence of lignite in the Fort Union Group has yielded another interesting combination of geological events. The lignite beds exposed throughout the region are subject to spontaneous combustion. If beds begin to burn, they not only bake the overlying sediments but also affect the topography of the area. Coal seams tend to burn from the area of exposure back into the hillside, and this results in collapse of the overlying material and development of a hummocky terrain. Burning continues until such time as the lignite has burned so far back into the hillside that additional cracking and slumping do not provide access to oxygen for further combustion. One of the major coal seams that has recently been burning is in the area of Buck Hill in the South Unit of the park. It smoldered and burned until 1978. The effect on the topography is, however, quite evident. Coal seams have burned in many places and at many times throughout recent geologic history as evidenced by the widespread occurrence of "scoria," the baked clay resulting from the process. Additionally, the

areas in which scoria has been produced are particularly lacking in the lignite beds that should occupy that stratigraphic position.

Although the area is structurally simple, that is, the beds typically dip at angles of less than 1°, there are several places where much greater dips, on the order of 45°, can be observed. Many of these tilted blocks are tens of meters in length and width. Invariably, however, they occur along slopes, and investigation of the rock on the slopes suggests that the dips are of local significance, the result of mass wasting processes. Huge slump blocks (fig. 9.9) have moved downslope and rotated under the influence of gravity.

Geologic History

1. Pre-Paleocene deposition of sediments in the Williston Basin.

Throughout the Paleozoic Era and most of the Mesozoic, the area comprising western North Dakota, eastern Montana, and southern Saskatchewan and Manitoba was one of continued, slow subsidence and accumulation of sediments, mostly marine. Although none of the Paleozoic or Mesozoic rock sequence is exposed in the vicinity of Theodore Roosevelt National Park, the sequence is

Figure 9.9   Large slump block of Sentinel Butte material in North Unit. National Park Service photograph.

of interest. The Williston Basin is a significant oil-producing region, and the area surrounding the Park is dotted with pumps and drill rigs extracting petroleum from the Paleozoic rock sequence. Until near the close of the Mesozoic Era, the entire area was either slightly below sea level or about at sea level.

2. Uplift of the Rocky Mountains and the end of marine conditions.

Beginning in the Cretaceous Period and continuing into Paleogene time, the Laramide orogeny resulted in the development of the Rocky Mountain chain. One of the effects of the uplifting of the Rocky Mountains, hundreds of miles west of the Park, was that erosion of these newly formed mountains produced hundreds of cubic miles of sediment that was carried eastward from the mountains and deposited in great clastic wedges. This event is recorded in a sequence of terrestrial sediments ranging in age from Late Cretaceous through Oligocene time in many parts of the Dakotas and Montana. The Fort Union Group represents the Paleocene element of that clastic wedge. During the time of deposition of the Fort Union Group in the badlands area, the last vestige of the midcontinental seaway in North America was rapidly being filled in to the east of the park in central North Dakota. These rocks, comprising the Cannonball Formation, are well exposed along the Missouri River and some of its tributaries.

Scattered remnants of post-Paleocene sediments are found throughout western North Dakota in very isolated remnants capping butte tops. These units, the Golden Valley Formation and the White River Formation, span Eocene and Oligocene time.

3. Uplift and erosion of sediments following the Oligocene.

There are no extensive rock units in western North Dakota younger than the Oligocene White River sediments. Instead, the area had apparently been uplifted and was in the process of being dissected by rivers ancestral to the modern Little Missouri River system. The only evidence of any kind of deposition subsequent to the Oligocene is in the form of gravel and sand accumulations that are difficult to date.

4. Pleistocene glaciation and development of the modern drainage system.

During the Pleistocene, glaciers advanced southward from Manitoba and Saskatchewan at least as far as the course of the present Missouri River. Prior to that time, the Little Missouri River flowed northward into Canada. With the onset of glaciation, however, north-flowing streams were blocked. The drainage of this river, along with that of the Yellowstone River to the west, was diverted eastward and southward along the edge of the ice front. This ice-marginal drainage eventually developed into the present-day Missouri River, to which the Little

Missouri is tributary. The rate of erosion and downcutting of the Little Missouri River was not constant, and as erosion and deposition continued, the stream cut a series of terraces, remnants of which can still be seen in several places throughout the park.

Geological Map

Clayton, L., Moran, S. R.; Bluemle, J. P.; and Carlson, C. G. 1980. *Geologic map of North Dakota:* U.S. Geological Survey.

Bibliography

Bluemle, J. P.; and Jacob, A. F. 1973. *Geology along the South Loop Road.* North Dakota Geological Survey Educational Series 4.

Clark, C. 1978. *The Badlands,* American Wilderness series. Revised ed. Alexandria, Virginia: Time-Life Books.

Clayton, L.; Carlson, C. G.; Moore, W. L.; Groenewold, G.; Holland, F. D., Jr.; and Moran, S. R. 1977. *The Slope (Paleocene) and Bullion Creek (Paleocene) Formations of North Dakota.* North Dakota Geological Survey Report of Investigation 59.

Eliot, J. L. 1982. Roosevelt Country: T. R.'s Wilderness Legacy. *National Geographic* 162(3):340–363, September.

Holtzman, R. C. 1978. *Late Paleocene mammals of the Tongue River Formation, western North Dakota.* North Dakota Geological Survey Report of Investigation 65.

Jacob, A. F. 1976. *Geology of the upper part of the Fort Union Group (Paleocene), Williston Basin, with reference to uranium.* North Dakota Geological Survey Report of Investigation 58.

Laird, W. M. 1950. *The geology of the South Unit Theodore Roosevelt National Memorial Park.* North Dakota Geological Survey Bulletin 25.

————. 1956. *Geology of the North Unit Theodore Roosevelt National Memorial Park.* North Dakota Geological Survey Bulletin 32.

Roosevelt, T. 1978. *Ranch Life in the Far West.* Olympic Valley, California: Outbooks.

Royse, C. F., Jr. 1967. *Tongue River-Sentinel Butte Contact.* North Dakota Geological Survey Report of Investigation 45.

————. 1970. A sedimentologic analysis of the Tongue River-Sentinel Butte interval (Paleocene) of the Williston Basin, western North Dakota. *Sedimentary Geology* 4:19–80.

Schoch, H. A. 1974. *Theodore Roosevelt; the story behind the scenery.* Las Vegas, Nevada: KC Publications.

Ting, F. T. C., ed. 1972. *Depositional environments of the lignite-bearing strata in western North Dakota.* North Dakota Geological Survey Miscellaneous Series 50.

Address

Theodore Roosevelt National Park
Box 7
Medora, North Dakota 58645

PART II
Glaciers and Glacial Processes

© by G. Shimer

Glacial erosion has produced this bedrock landform, called a roche mountonnée, in the Kern River valley. Sequoia National Park, California (chapter 27, Part IV).

Pointed horns of rock, sharp ridges, amphitheaterlike cirques, and deep troughs form a distinctive skyline in high mountains that have been sculptured by glacier systems. Valley glaciers are still at work in the mountains of all but one of the national parks in this group, although glacial activity is less intense now than it was during the Ice Age. The continental glaciers that shaped the landscape of Acadia National Park in Maine have been gone since the Ice Age ended about 10,000 years ago.

10
Glacier National Park

Location: Northwest Montana
Area: 1,013,594.67 acres; 1583.74 square miles
Established: May 11, 1910; proclaimed part of
    Waterton - Glacier International Peace Park: June
    30, 1932

Figure 10.1 View looking west across St. Mary Lake toward the intensely glaciated ranges of Waterton-Glacier International Peace Park. Glacial horns and cols are outlined against the sky. St. Mary Lake occupies a glacial trough dammed by a moraine. The gravel in the left foreground is glacial outwash being reworked by waves and currents. National Park Service photograph by G. A. Grant.

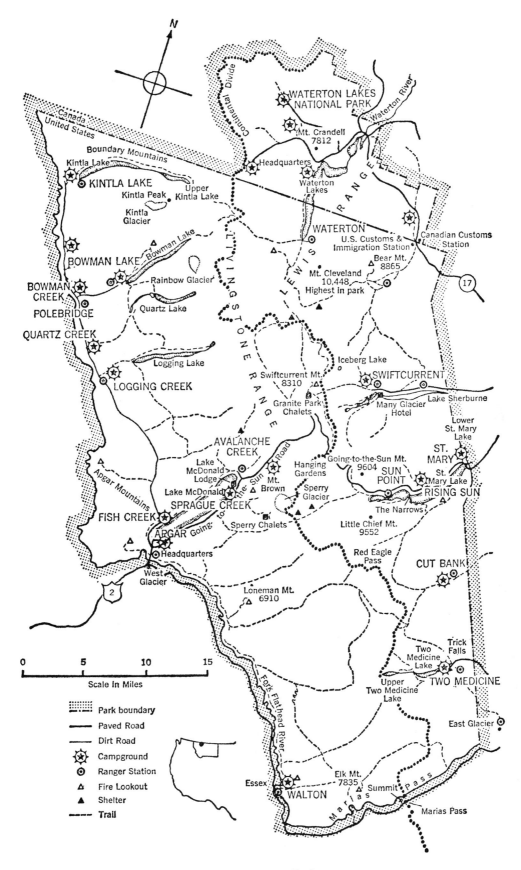

Figure 10.2 Waterton-Glacier International Peace Park, Montana and Alberta.

Table 10.1. Geological Column, Glacier National Park

Time Units			Rock Units		Geologic Events
Era	Period	Epoch	Group	Formation	
Cenozoic	Quaternary	Holocene		Alluvium; Glacial deposits	Erosion and deposition by streams. Return of glaciers in Neoglacial time. Present glaciers between 6,000- and 9,000-foot levels on north or east slopes
		Pleistocene		Glacial deposits	Valley and piedmont glaciation with several advances and retreats; intense glacial erosion.
	//////// Major erosional interval				Major erosional interval
Mesozoic	Cretaceous		Montana Group		Return of seas depositing Cretaceous beds; Laramide orogeny, folding, uplift, thrust faulting, vertical faulting
Paleozoic	//////// Major erosional interval				Major erosional interval
Late Precambrian (Proterozoic Y)	Belt Supergroup	Missoula Group		Mount Shields Formation	Red argillites with salt-crystal casts accumulated in shallow lagoons.
				Shepard Formation	Lava interbedded with sediment in shallow sea; siltstone near shore, limestone and dolomite farther out.
				Purcell Lava	Submarine extrusion of pillow lava
				Snowslip Formation	Major change in basin shape; beds thinning and coarsening to the east
		(Middle Belt Carbonate)		Helena (Siyeh) Dolomite	Deposition of marine limestone with algal layers. Intrusion of diabase dikes and sills associated with emplacement of copper; contact metamorphism.
		Ravalli Group		Empire Formation	Contains green beds, carbonate beds, and anomalous copper. Continued warping to form upper Ravalli basin.
				Spokane Formation	Stream-deposited beds of red clays, white sandstone, green muds containing fossils, mud cracks, ripple marks.
				Greyson Formation	Stream-deposited shales and argillites. (Sand layers form white strips in rock unit.)
		(Pre-Ravalli)		Altyn Formation	Alternating bands of conglomerate and sandstone at base, interfingered with overlying limestone containing stromatolite fossils.
				Waterton Formation of Canada	Deposition of limestone in a warm, shallow sea; later altered to dolomite.

Based on Harrison (1972) and information supplied by G. W. Luttrell (personal communication).

Local History

The Indians who lived in this region for at least 10,000 years found the fishing and hunting to be excellent. St. Mary Lake and Lake McDonald, the two largest of the more than 200 glacier-fed lakes in the Park, were favorite areas of Indian habitation. The Indians who lived here most recently were the Blackfoot Indians, who displaced the Kootenai (Flathead Indians) about 200 years ago.

The first white visitor was apparently a Canadian trapper, Hugh Monroe from Hudson Bay, who was at St. Mary Lake in 1816. He was called "Rising Wolf" by the Blackfoot. Father DeSmet, the Jesuit missionary and explorer, named St. Mary Lake in 1846.

In 1885, George Bird Grinnell, a naturalist and conservationist, vacationed in the region. He fell in love with the place and returned many times. As the long-time editor and publisher of *Forest and Stream,* he campaigned for preservation of the area for many years and was influential in the legislation that established Glacier National Park in 1910. Establishment was difficult because of opposition from private interest groups and congressional committees; two bills were rejected by Congress before one finally passed.

Marias Pass, which had been regarded as merely an Indian legend, was discovered by Major Baldwin in 1889 and surveyed in December of that year by John Stevens. The Great Northern Railroad constructed its line to the region in 1892, and settlers built a road to Lake McDonald in 1895. This road eventually became Going-to-the-Sun Highway. The road, as well as Going-to-the-Sun Mountain, is named after a legendary Blackfoot chief who was sent down by the Great Spirit to help the tribe regain its former greatness. After helping them, he departed, climbing up the side of the mountain during a terrible storm. According to the legend, when the storm broke, the snow on the mountain formed a profile of the chief going to the sun.

In 1932, largely through the efforts of Rotary International, the U.S. Glacier National Park and the Canadian Waterton Lakes National Park were joined to form the Waterton-Glacier-International Peace Park.

Geologic Features

The glaciation that produced the striking landscapes and the spectacular, ice-sculptured landforms in Glacier National Park was at its maximum extent during the Great Ice Age of the Pleistocene Epoch. Even in today's warmer climate more than 50 active glaciers exist in the Park.

How Glaciers Form and Move

Glaciers, which are exceedingly powerful agents of erosion, form in regions where more snow falls than melts each year. As more and more snow accumulates, a perennial snowfield gradually builds up, at the same time packing down and settling under its increasing weight. The individual snowflakes, formed from water vapor by direct crystallization, lose their characteristic six-pointed shape and become granules due to being compacted ever closer together in the snowfield. At the stage of transition from flakes of snow to glacial ice, the granular snow is called *firn*. Continued compaction forces the granules more tightly together and expels the air. Individual grains melt slightly and recrystallize, filling the voids and transforming the whole mass into ice. As soon as the ice mass becomes large enough and heavy enough so that gravity causes it to move or flow downward or outward, it is considered, by definition, a *glacier*.

Evidence of several types of glaciers can be found in Glacier National Park. During the Pleistocene, both piedmont and continental glaciers existed in the region, although continental glaciers never quite overrode this part of Idaho and Montana. The glaciers now in the Park are

cirque glaciers and valley glaciers. A *cirque glacier* occupies a hollow or basin at the head of a mountain valley and is usually the remnant of a much larger glacier. A valley (or alpine) glacier is confined to a valley (almost always a former stream valley) and flows from higher to lower elevations. *Piedmont glaciers* are formed when several valley glaciers come down and join together at lower elevations, making a wide bulbous ice mass where the valleys open out. Largest of all are the *continental glaciers,* or *ice sheets,* which cover vast areas of land and may be several miles thick. Ice sheets flow from a central high point, downward and outward.

The higher part of a glacier, the area of the perennial snowfield (or firn field), is called the *zone of accumulation.* Farther down, where melting or loss of ice and snow exceeds inflow, lies the *zone of wastage.* Separating the two zones is an irregular, shifting band, called the *firn line,* that marks the highest level to which the winter snow cover retreats during the summer season.

A glacier's motion is not uniform throughout its mass as it moves downslope. The ice on top is brittle and tends to fracture, forming *crevasses,* or cracks. This part of the glacier is the *rigid zone.* In the bottom part of the glacier, called the *plastic flow zone,* the ice becomes plastic, or bendable, under pressure from the overlying ice, and this increases the rate of movement of the ice near the bottom (fig. 10.3).

Examples of Glacial Erosion

Much of glacial erosion is done by *abrasion,* or grinding and scouring, as pieces of rock frozen in the ice are pushed, scraped, and dragged over underlying rock. The great weight and force of the ice mass also crushes and cracks the bedrock, so that chunks of rock are *plucked,* or quarried out, as the ice moves along (figs. 10.4, 10.5). At the

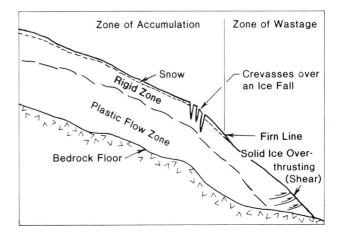

Figure 10.3 Mechanisms of glacier motion. Glacial ice flows differently in different parts of a glacier. Note that the firn line marks the transition between the zone of accumulation and the zone of wastage. From *Landforms and Landscapes* (3rd ed.) by Sherwood D. Tuttle. © 1970, 1975, 1980 Wm. C. Brown Company Publishers, Dubuque, Iowa. Reprinted by permission.

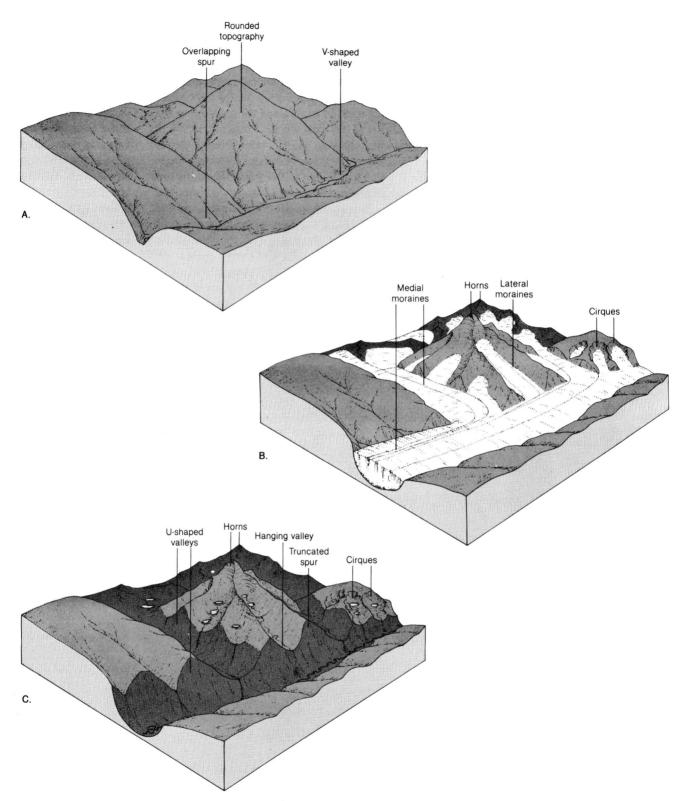

Figure 10.4   Erosion by valley glaciers may greatly modify the preglacial topography of a mountainous region. From *The Earth's Dynamic Systems* (3rd ed.) by W. K. Hamblin. © 1982 Burgess Publishing Company. *A.* Before glaciation, streams have rounded ridges and slopes and eroded V-shaped valleys in a mountain landscape. *B.* Glacial ice accumulating on the higher slopes sharpens peaks and ridges and carves cirques. Valley glaciers move down the preglacial stream valleys, eroding and gouging out bedrock. Intense frost action operates on the bare rock ridges that stick up above the ice surface, producing rock fragments that are carried down on and in the glaciers. *C.* After glaciers have melted away, an alpine topography is left, characterized by glacial erosional landforms: a serrated skyline, cirques, truncated spurs, hanging valleys, and U-shaped glacial troughs.

Figure 10.5 View along the trail leading to Sperry Glacier. The bedrock in the foreground has been smoothed by glacial abrasion, and the irregular, steplike ridges on the slope were produced by glacial plucking and quarrying. The scattered boulders were left when the ice melted, and talus, produced by frost action, has accumulated since the area was uncovered. Note the folded rock layers visible above the highest snowbank (upper left). National Park Service photograph.

head of a valley glacier, the continual scouring and plucking done by the ice mass produces a hollow or basin, called a *cirque.* Many of the lakes in Glacier National Park (such as Avalanche, Cracker, Ellen Wilson, Helen, Hidden, Iceberg, Kennedy, Pocket, Ptarmigan, and Upper Two Medicine Lakes) are *cirque lakes* that lie in bowl-shaped depressions from which the glacial ice has retreated.

Where two cirques have eroded headward until they are back to back or adjacent, the thin, blade-shaped wall of rock separating them is an *arête,* examples being Garden Wall and Ptarmigan Wall. A saddle-shaped depression, termed a *col,* results when glacial ice breaks down an arête between cirques. Jackson and Blackfoot Glaciers are separated by a col, and another can be found behind Haystack Butte. Some cols in the Park are also mountain passes, such as Logan Pass, Piegan Pass, and Gunsight Pass. A glacial *horn,* a feature usually having a pyramidal shape, is produced when cirques cut back into a mountain summit from three or four sides. Some of the horns in Glacier National Park are Clements Mountain, Kinnerly Peak, Little Matterhorn, Reynolds Mountain, Sinopah Mountain, Split Mountain, Mount Wilbur, St. Nicholas, and Haystack Butte.

Valleys that have been heavily glaciated usually look like troughs and have a characteristic U-shape. This is because valley glaciers tend to straighten bends or curves cut by preglacial streams, smooth and widen the side slopes, and deepen the valley floors. Typical U-shaped glacial valleys in the Park are McDonald Valley, St. Marys Valley, Swiftcurrent and Bowman Lake Valley, Belly River Valley, and Cataract Creek Valley (fig. 10.4).

Many beautiful hanging valleys can be seen in Glacier National Park. The mouth of a *hanging valley* opens out high on the steep side of a larger, glacial valley, many feet above the valley floor. The larger valley was eroded by a trunk glacier, which had much greater erosive power than its smaller tributary glaciers and thus was able to erode its valley more deeply. When the glaciers melted back, the tributary valleys were left hanging far above the main valley. Streams in the hanging valleys join the main stream as waterfalls, Avalanche Falls being one example. Other waterfalls, such as St. Mary, Rockwell, Baring, and Florence Falls, cascade down over glacial steps in the bedrock walls. Trick Falls are unique because at the top, the falls flow over resistant beds; when stream volume is low, the bottom falls (normally hidden from view beneath the top falls) issue out from a cave formed by the enlargement of joints in limestone.

Some glacial valleys have on their floor a series of *rock-basin lakes* connected by a single stream. The lakes occupy depressions scoured out of highly fractured or weaker sections of bedrock. This type of glacial feature (sometimes called "beaded lakes" or "paternoster lakes") can be seen between Many Glacier Hotel and Swiftcurrent Pass. Bullhead Lake and Swiftcurrent Lake are two of the rock-basin lakes in the string.

In late summer natural *ice caves* can usually be found at the base of either Grinnell or Sperry Glaciers, where streams come out from beneath the glacial snouts. The characteristic milky white color of the water is due to the heavy concentration of *rock flour* (finely pulverized rock debris) being carried off in suspension.

Processes of Glacial Deposition

Great quantities of sediment in all sizes are produced by glacial erosion—rock flour, sand, gravel, cobbles, even huge boulders. Debris is picked up, flattened down, pushed, or carried along on top of, beneath, beside, and in front of a glacier. This largely unsorted material, called *till,* accumulates in *moraines* of various types and sizes. Lakes dammed by moraines are typical of glaciated topography; examples of this in the Park are Bowman, Josephine, and Kintla Lakes.

Streams work over and sort till and other debris, redepositing the sediment as *glacial outwash,* alluvial fans, etc. In the bottom of glacial troughs, glacial outwash usually spreads over till that was brought down by the ice. Eventually the scoured bedrock downvalley from a retreating glacier may be buried by many feet of sediment. Sometimes streams back up, forming lakes held in by natural dams built up by a mixture of outwash and morainal material. Both St. Mary Lake and Lake McDonald, which lie in glacial troughs, were dammed in this way.

Avalanche Chutes

One type of mass wasting occurs in the Park that is only indirectly related to the glacial topography. During the winter, snow collects in pockets on steep slopes or in steep-sided, narrow trenches, called *avalanche chutes.* Because the snow is unstable, an avalanche may be triggered by a sudden vibration or sharp noise. Snow, ice, and debris rushing down the mountainside—with a wall of air in front of the slide—can build up such force that trees are torn out by the roots.

Features of the Rocks

The fact that the mountains in the Waterton-Glacier International Peace Park are made up of layered sedimentary (and metamorphic) rocks gives the glaciated topography a unique "architectural" character that is not seen in the Rockies farther south, where the rocks are largely granitic. The sculptured peaks in the international boundary area are often referred to as "mountains without roots" because they consist of ancient Precambrian rocks resting on top of much younger sedimentary beds. The reason for this is that a huge, low-angle thrust fault (known as the "Lewis Overthrust") transported the sequence of Precambrian rocks eastward for many miles over the younger rocks. The trace of the fault surface, which separates the older beds on top from the younger beds beneath, can easily be seen on the face of Wynn Mountain and at other places in the Park. The fault surface also crops out along the base of the vertical face of Appekunny Mountain.

Chief Mountain, a peak that stands alone astride the Park's eastern boundary, is an erosional outlier, or *klippe* (i.e., a remnant) of the thrust fault, which was of much greater size and extent before being eroded. The Lower and Upper Members of the Altyn Formation (pre-Ravalli rocks of late Precambrian age) are exposed in Chief Mountain. Making up the top third of Chief Mountain are the horizontal limestone beds of the Upper Member. The lower two-thirds of the mountain consist of the more dolomitic beds of the Lower Member, which have been slightly tilted and show minor thrusting. Underneath the mountain are the young Cretaceous beds separated from the overlying Precambrian formation by the fault surface of the Lewis Overthrust, visible at the mountain's base (fig. 10.6).

Within the Park, clearly exposed on the monumental glacial features, are many evidences of the marine origin of these ancient rocks. Ripple marks and mud cracks in

Figure 10.6 Drawing of Chief Mountain, which is located on the Park boundary, about 15 miles east of the Continental Divide. Chief Mountain is a striking example of a klippe, a feature formed by the erosion of a thrust sheet. The trace of the Lewis Overthrust (a fault), which is almost horizontal here, lies about at the top of the line of trees. The rocks above the fault surface are units of the Precambrian Belt Supergroup; those below the fault are Cretaceous in age. From Bailey Willis (1901) *Bulletin of the Geological Society of America,* vol. 13, p. 334.

the sandstones indicate deposition in shallow water. Fossilized calcareous algae (sometimes called "cabbage heads" because of their shape) record the presence of primitive, single-celled plants in the Precambrian seas. Collectively known as *stromatolites,* the blue-green algae lived only in shallow water where they could receive abundant sunlight. By secretion of calcium carbonate, the tiny plants formed broad, massive reefs of organic limestone, up to 60 feet high, called *biostromes.* These reefs—rather than being in the shape of mounds or lenses—had a bedded structure and spread blanketlike over wide areas.

In several exposures in the Park, basalt flows containing *pillow structures* indicate that lava was extruded over the floor of a Precambrian sea. Because the lava cooled suddenly, it is filled with bubble holes, called *vesicles.* At a much later time in geologic history, when the rocks were part of the land instead of beneath the sea, ground water percolating through the rocks filled the holes with calcite, forming *amygdules.*

Salt-crystal casts are another type of cavity that can be found in the rocks. When because of evaporation, salinity in the shallow lagoons reached 10 times normal, salt crystallized out, forming salt beds. Later, some of the salt crystals were dissolved and replaced by calcite from ground water. Some of the cavities became filled with fine silt or mud.

Erosion by sediment-laden streams draining off glaciers has brought about the development of rather large, deep *potholes* in some of the less resistant rocks in streambeds. Rock debris that collects in any slight depression or weak spot in bedrock or boulders acts as a drill, swirling around in the water and enlarging the pothole.

Drainage Divides

The map of Glacier National Park (fig. 10.2) shows the Continental Divide running from north to south through the region. Streams that drain the western side of the Continental Divide carry water that may eventually reach the Pacific Ocean. Water falling on the eastern side enters streams that end up in either Hudson Bay or the Gulf of Mexico. Triple Divide Peak is unique, inasmuch as water drains in three directions instead of two; i.e., west toward the Pacific, northeast toward Hudson Bay, and southeast toward the Gulf of Mexico.

Geologic History

1. Deposition of the Belt Supergroup during late Precambrian time (Proterozoic Y)*.

Ripple marks, mud cracks, and an abundance of calcareous algae in the rocks indicate that shallow seas covered the region most of the time although the land was occasionally above sea level. Biostromes as high as 60 feet

were built by the algae. All of the Precambrian rock units described below belong to the Belt Supergroup.

The basin in which the Belt Supergroup rock units were deposited held an arm of a sea that advanced over the North American craton along its western edge. The size, shape, and outline of the basin, plus the information derived from the sediment types and facies changes all indicate multiple source areas rather than a single source area. The sediments in general thin eastward. The extent of the Belt basin can be determined only approximately because parts of the boundaries have been obscured by younger rocks and by deformation.

2. Deposition of the Waterton Formation of Canada (pre-Ravalli).

The dolomitic Waterton beds are exposed in Waterton National Park but not on the U.S. side of the border in Glacier National Park. Thickness is unknown since the base is not exposed. Although a fresh surface is red-brown or gray, the dolomite weathers to gray, tan, or reddish brown.

3. Deposition of the Altyn Formation (pre-Ravalli), some 2,300 feet of sandy dolomite and limestone.

A ridge-former, this rock weathers to a distinctive light buff. The layers of sandstone and conglomerate at the base, interbedded with limestone, show that the clastic sediments came from a granite source bordering the sea. Lower layers contain stromatolite fossils, exposures of which can be seen at Appekunny Falls, Rising Sun Camp Ground, and Divide Mountain fire tower. Thin layers of quartz that filled in minute fractures have weathered out in sharp relief, making them easy to recognize.

Some of the Altyn beds contain mud cracks and ripple marks. These rocks form The Narrows of St. Mary Lake, the ledge over which Swiftcurrent Falls tumble, and the rim upon which Many Glacier Hotel is built.

4. Deposition of the Greyson Formation (Ravalli Group), 3,000 feet of greenish shales and argillites.

The difference between the shales and argillites is mainly one of degree of metamorphism, with the argillites being more indurated (compacted and hardened by heat, pressure, and so forth). Most of the beds are green mudstones that were deposited in water and never oxidized. Reddish layers indicate exposure to air and oxidation. Fossils have not been found in the sequence, but mud cracks, cross bedding, and ripple marks are common. Occasional beds of well sorted clean sand now form white stripes in the Greyson. Such features suggest that sediment-bearing streams traveled some distance before depositing their load in the shallow Precambrian sea.

McDonald Falls flow over Greyson beds, and Aster Creek carved its gorge in them. The beds can also be seen

---

*For an explanation of Precambrian chronology, see box 22. 1, p. 278.

in Old Squaw and Yellow Mountains. Because the rock is hard and splits easily, it is used for flagstone walks in the Park. Soil weathered from Greyson rocks usually supports a heavy forest cover.

5. Deposition of the Spokane Formation (Ravalli Group), 3,000 feet of red argillites interbedded with white quartzite.

The upper portion of the Greyson grades upward into the Spokane Formation without a sharp contact. The Spokane is predominantly red in color but contains layers of white sandstone and green mudstone. Since these beds formed in the same environment as the Greyson, mud cracks, raindrop impressions, and ripple marks have been found but few fossils. The dominance of red color indicates that the Spokane deposits had greater exposure to air. The iron has been oxidized mostly in the form of hematite. Both because of its thickness and its location, the Spokane Formation makes up the bulk of Precambrian rocks in the Park. Avalanche Gorge and Red Rock Canyon are cut into this formation; and McDonald Creek, noted for its many potholes, flows through it. Sperry Glacier, Grinnell Lake and Falls, Mount Henkel, Ptarmigan Creek, Rising Wolf Mountain, Ptarmigan Tunnel, and many other features lie within its domain.

6. Deposition of the Empire Formation containing anomalous copper.

Argillites, mudstones, and carbonates (mostly limestones) deposited in a shallow sea make up the Empire Formation, the uppermost rock unit of the Ravalli Group. The green carbonate rock layers of the Empire Formation contain "anomalous copper," that is, higher-than-normal concentrations of copper minerals, such as chalcopyrite, chalcocite, and bornite. Anomalous copper (at least 100 parts per million, and much higher in some places) has been found in nearly all parts of the Belt basin and in most of the formations. Silver and mercury, also in anomalous amounts, is usually associated with the copper. The occurrences of anomalous copper in the Belt rocks are similar to those that have been discovered in Precambrian sedimentary rocks of equivalent age in Canada, Africa, Australia, and Russia. Exploration geologists refer to such copper deposits as "stratabound" because the copper minerals tend to be concentrated along bedding planes, in sedimentary structural features (e.g., mud cracks), and in certain rock layers. Some of the anomalous copper in the Belt basin was deposited *with* the sediments (syngenetic), but probably some copper migrated or was reconcentrated *subsequent* to deposition (that is, by diagenetic processes). Concentrations of ore grade copper are mainly outside the Park.

The Empire Formation is considered the equivalent of the Werner Peak Formation, which is the uppermost Ravalli rock unit in the Whitefish Range west of Glacier National Park (Harrison 1972).

7. Deposition of the Helena Dolomite, 4,000 feet of carbonates containing algal zones, dikes, and sills (Middle Belt Carbonate).

The Helena was deposited in a northwest-trending basin that received clastic sediments from the east, south, and southwest. Carbonate deposits along the eastern shelf area were extensive.

Numerous algal zones, that accumulated in shallow water, are found in the Helena Dolomite. Of the three main zones, the lowest is *Collenia symmetrica*, which grew on a mud flat. These fossiliferous beds can be seen on the Garden Wall, near Lake McDonald, and along Going-to-the-Sun Highway, east of Logan Pass. *Conophyton* beds in the second zone are what most people notice in the Park because they show up as a light-colored band in the bedrock on either side of Logan Pass and along the Garden Wall. The algal colonies formed a concentric series of cones. The third zone, which consists of *Collenia multiflabella*, is also exposed at Logan Pass.

Intruded near the top of the Helena are a series of dikes and sills, ranging in thickness from 10 to 200 feet. Their composition is metagabbro or diabase. Some copper was emplaced in the enclosing limestone beds but not enough to be commercially mined. Contact metamorphism during intrusion bleached the beds above and below, transforming them into marble. These beds are easily traced all over the Park, especially near the top of Clements Mountain and Mount Gould and in the Natural Amphitheater. Because the beds are less resistant, differential weathering has formed deep grooves, some of which are avalanche chutes.

8. Tectonic adjustments in the Belt basin and deposition of the Snowslip Formation (Missoula Group).

Major tectonic adjustments in the Belt basin marked the end of the Middle Belt Carbonate accumulations and the beginning of Missoula Group deposition. Gentle upwarping and downwarping brought about the development of a long, low dome in the western part of the basin (Idaho panhandle area) and the sinking of the carbonate shelf along the eastern side (Glacier National Park area). Clastic sediment spread over the Helena Dolomite from two source areas, one to the northeast and one to the southwest. The argillites, shales, mudstones, and siltstones that accumulated in this north-northwest-trending trough make up the Snowslip Formation. Hematitic particles, an indication of deep weathering in the source areas, gave some of the Snowslip rock layers a reddish color. Greenish beds in the Snowslip contain anomalous copper that apparently came from the eastern source area.

9. Extrusion of Purcell Lava.

Slight uplift produced an erosional surface on the Snowslip Formation that sloped westward and cut down into the Middle Belt Carbonate rocks. Basaltic lava, ranging in thickness from 50 to 275 feet, spread over this uneven surface. Most of the lava was extruded under water, causing pillow structures to form. The exterior of the lava cooled very quickly producing a glossy skin. Trapped gas underneath formed bubble holes, or vesicles. Much later, calcite filled the holes, forming amygdules in the rock.

The Purcell Lava, which has been dated by radiometric procedures at 1,080 million years before the present (box 13.1, p. 159), is the only extrusive Precambrian rock that has been found in the Belt basin. The best place to see the lava is at Granite Park, which was misnamed by someone who did not know the difference between granite and basalt. There is no granite in Glacier National Park.

10. **Deposition of the Shepard Formation (Missoula Group), 600 feet of buff-colored carbonates.**

The Shepard Formation was deposited unconformably over the Purcell Lava. The rocks of the Shepard, which become more clastic toward the east, are a mixture of limestone and siltstone with some dolomite. Fossil algae, ripple marks, mud cracks, and some channel fillings suggest that the inland sea was becoming shallower.

11. **Deposition of the Mount Shields Formation, some 860 feet of bright red argillites containing salt-crystal casts.**

The Mount Shields, the youngest Precambrian rock unit exposed in the Park, formed in shallow lagoons where the water became very salty. Sometimes the lagoons were exposed as mud flats, other times they were covered with water. The bright red color is the result of oxidation occurring when the beds were exposed to air. The salt (halite) crystallized out when salinity was 10 times normal. The rock sequence also contains algal zones. Outcroppings can be seen on the peaks of many mountains in the Park, including Running Rabbit Mountain, Mount Custer, Boulder Pass, and Hole-in-the-Wall Basin.

The remaining members of the Missoula Group do not crop out inside Park boundaries, but are exposed to the west in the Whitefish Range.

12. **Uplift, withdrawal of seas, erosion of some of the Belt rocks; deposition of Paleozoic beds.**

As the region of Glacier National Park was gradually uplifted, the seas withdrew. Stream systems developed and rapidly stripped the land of sediments that were soft, unconsolidated, and easily eroded. Geologists surmise that in time the seas returned to this area because mountains both to the north and south of the Park are composed of Paleozoic beds deposited in shallow water. Paleozoic rocks are not exposed at the surface in the Park but might be found by drilling below the Cretaceous beds.

13. **Deposition during the Cretaceous about 100 million years ago.**

A shallow inland sea extended into the Glacier National Park region as the Cretaceous Period was drawing to a close. Beds deposited in this sea contain marine fossils, although in other areas beds of the same age contain dinosaur fossils. Because the Cretaceous beds were never deeply buried, they are not so strongly lithified as the Precambrian rocks and thus are more susceptible to weathering. Erosion of the Cretaceous rocks has formed a gently rolling landscape in contrast to the rugged mountains and steep slopes of the Precambrian Belt Supergroup. The Cretaceous beds cropping out along the eastern side of the Park contain oil, and those in North Fork Valley have natural oil seeps. Where the beds extend into Montana and Alberta, Canada, commercial oil fields are being worked.

14. **The Laramide phase of the Rocky Mountain orogeny.**

As the Cretaceous Period was ending, the beds deposited in the vast Cordilleran Geosyncline during the Paleozoic and Mesozoic Eras were caught in the compression of the North American plate moving westward against the Pacific plate. In the mountain-making upheavals that ensued, a mountain system was built that reached from Mexico to Alaska, with the Northern Rockies being a part of this system. In the Park region major folds that were produced can be seen along some trails. (Along the roads most of the beds appear horizontal.) Since the uplifting that occurred increased the stream gradients, large stream valleys were eroded.

15. **Late in Cretaceous and early in Eocene time, tectonic compression caused Precambrian rocks to override Cretaceous shales.**

Deformation was especially severe in the Northern Rockies in Late Cretaceous and early Eocene time, and massive thrust faulting occurred. In the Glacier National Park region, a broad arch, cresting about 100 miles west of the Park, was uplifted several thousand feet. As the beds tilted eastward, some of the rocks near the weaker zones began to slide and slip toward the east and northeast, creating the Lewis Overthrust. The force of the thrusting moved the Precambrian rocks some tens of miles over the weaker Cretaceous beds, which were only slightly compacted at that time. The Lewis Overthrust can be traced along mountain fronts for many miles north and south of the Park; but just how much eastward movement occurred has not been clearly determined.

During the slippage, as the Cretaceous beds were stretched and torn, *pull-apart valleys* formed, North Fork Valley being an example. The principle behind the formation of pull-apart valleys is similar to that of a snowslide on a steeply slanting roof. As the snow starts to slide down the roof, cracks appear because not all of the snow cover moves at the same rate due to differences in friction between the roof and the snow. The cracks are analagous to the pull-apart valleys. When the huge block of Precambrian Belt rocks was sliding over the soft (and possibly water-saturated) Cretaceous beds, the sliding was uneven. One section remained stationary while another continued to move, thus creating a valley. In all probability, Flathead Valley was formed in this way, with the Whitefish Range acting as a center block between Flathead and North Fork Valleys.

16. Vertical faulting in the Livingston Range; erosion, uplift, more erosion.

After the North Fork Valley was formed, its southern end was dropped down during continued tectonic activity. Late Eocene sediments were deposited in the newly created valleys; and erosion isolated sections of the overthrust sheet, creating klippes (such as Chief Mountain) completely surrounded by younger beds.

This series of events completely changed the drainage system. A single river once flowed to the northwest from Bear Creek, through Middle Fork Valley, over McGees Meadow, and then on to Canada. Today, two branches of the Flathead River flow to the southeast, suddenly swing west, and finally join southwest of the Apgar Mountains. With the change in drainage, swamps developed in some of the poorly drained valleys. A few coal beds were formed in the swamps, but the coal is not high grade. The beds deposited at this time in the Park region have been largely removed by erosion.

17. Pleistocene glaciation, covering all but the highest peaks in the Park with 3,000 feet of ice.

At least four separate episodes of intense and prolonged glaciation over North America and Europe are recognized as having occurred during the Pleistocene Epoch, which began some 2 million or more years ago. In Glacier National Park, evidence indicates extensive glaciation during the middle Pleistocene, two glacial substages during the Wisconsin (the last) glacial stage, and at least two episodes of ice accumulation and advance during post-Pleistocene time. Glacial erosion and deposition in the Park was mainly by valley and piedmont glaciers. The continental ice sheets from Canada came up to the boundaries of the Park on the northern and eastern sides, but never quite overrode the Park.

Plate tectonic theory suggests that the long cooling trend in global climates over the past 55 million years was concurrent with the gradual shifting of North America and Eurasia toward and North Pole while Antarctica drifted toward the South Pole. As more and more land surface moved into high latitudes (closer to the poles), reflection of solar radiation from land surfaces increased. Mean annual temperatures were lowered, and small glaciers began to form at high elevations or high latitudes.

The appearance of the continental ice sheets in the Northern Hemisphere about 2 million years ago marks the onset of the Pleistocene. Thus began a long series of fluctuations with climatic cycles of warming and cooling and the corresponding retreats and advances of glaciers. When glaciation was at its maximum extent, firn lines were several thousand feet lower on slopes; and after the last continental glaciers left North America about 9,000 years ago, mountain glaciers continued to advance and retreat, reflecting climatic shifts of shorter duration.

Glaciers are such powerful agents of erosion that they modify landscapes and landforms much more rapidly and drastically than streams can. However, what glaciers work on is the pre-existing topography of a region—in other words, the valleys and divides already molded by streams, which, in turn, are mainly controlled by the underlying rock structure.

In Glacier National Park (as throughout most of North America) the initial Pleistocene glacial stage was more extensive and severe than later glaciations. One indication of this can be seen on the crests of Two Medicine Ridge, Cut Bank Ridge, and Milk River Ridge, all of which have deeply weathered till, obviously older than the blankets of till laid down on the floors of the intervening valleys by later glacial stages. Evidence relating to the earlier glaciation also shows that the two large piedmont glaciers—Two Medicine Glacier, that flowed east, and St. Mary Glacier, which flowed northeastward—filled valleys with ice 500 to 1,000 feet higher than in succeeding glacial stages and also advanced farther eastward onto the plains of Montana. Both piedmont glaciers encountered the southward advances of the Canadian ice sheets. The location of moraines and glacial outwash show that at its maximum Two Medicine Glacier was 20 miles wide and extended some 10 miles out from the mouth of its valley.

Along Piegan Pass Trail are a bench (flat surface) above the trail and an enclosed sloping basin that represent traces of an older glacial valley. Another of these older valleys can be seen near Logan Pass. Granite Park Chalet is built on a bench that was a valley floor during an earlier glacial stage.

18. Final retreat of Pleistocene glaciers about 10,000 years ago; evolving of the present landscape.

Near the end of the Pleistocene, global climates gradually became warmer, with the result that the Park was denuded of ice—except for a few small, stagnant glaciers and snowfields persisting at high altitudes and on north-facing slopes. No major changes in the landscape have come about since the ice disappeared, but the effects of weathering and erosion continue. Small gorges cut by streams after ice retreat are Sunrise Gorge and the gorge at Hidden Falls, both formed as the result of enlargement of the weak zones in the bedrock joint system. The potholes in Avalanche Gorge have also developed over the past 10,000 years. Mass wasting is active under the prevalent freeze-and-thaw conditions, especially on steep slopes. Talus slopes are building up at the foot of cliffs and along the base of the mountain ranges. Avalanches, rockfalls, mudflows, and soil creep all continue to modify the landforms.

Since the melting back of the glaciers took place over a much shorter time than their growth and spread, enormous quantities of debris and sediment were released when the climate warmed. Streams were greatly overloaded and responded, as streams will, by dumping the excess wherever they could—in alluvial fans and outwash features, on top of till beds, moraines, bedrock, etc. The valleys holding Waterton Lake, St. Mary Lake, and Lower Two Medicine Lake (as well as other valleys in the Park) remain clogged, with the lakes held in by dams of glacial sediment.

19. Origin of glaciers that exist today.

The worldwide postglacial climatic optimum that brought about the disappearance of glaciers from the temperate zones and their retreat to the polar regions was a warm interval that peaked about 7,000 years ago. Average temperatures were warmer then and rainfall was greater. This warm interval, which ended the Great Ice Age, was followed by a global cooling trend that has continued into the present. Within this long-term cooling trend, several climatic fluctuations have occurred. A period cooler than now that prevailed about 5,000 years ago was probably the cause of renewed glacial activity in the Park that resulted in the development of the valley glaciers of the present day. The most recent significant advances of the valley glaciers within historic time occurred during the "Little Ice Age," roughly between A.D. 1450 and A.D. 1850. Since that time the glaciers have gradually receded. In 1910 about 90 glaciers were active; today between 50 and 60 might be called active. However, in the last 30 years or so, the global cooling trend appears to have resumed and the recession of the valley glaciers has slowed down. That global climate will continue to fluctuate seems certain.

The Park's active glaciers are found between elevations of 6,000 and 9,000 feet. The perennial snowfields lie in protected areas, usually on slopes that face north or east. Wind-drifted snow from over the peaks is a main source of their nourishment. Movement of glacial ice may range from 6 to 8 feet per year to as much as 12 to 50 feet per year, depending upon conditions of precipitation and temperature.

Geologic Maps and Cross Sections

Alt, D. D., and Hyndman, D. W. 1972. *Roadside geology of the Northern Rockies.* Missoula, Montana: Mountain Press Publishing Co. Pp. 72, 159, 165.

American Association of Petroleum Geologists 1972. *Geological highway map of the Northern Rocky Mountains region.* Map no. 5. Tulsa, Oklahoma: American Association of Petroleum Geologists

Ross, C. P. 1959. *Geology of Glacier National Park and the Flathead region, Northwestern Montana.* Geological Survey Professional Paper 296. Plates I and II (in pocket).

Bibliography

Alden, W. C. 1932. *Physiography and glacial geology of eastern Montana and adjacent areas.* U.S. Geological Survey Professional Paper 174.

Alt, D. D., and Hyndman, D. W. 1972. *Roadside geology of the Northern Rockies.* Missoula, Montana: Mountain Press Publishing Co.

———. 1973. *Rocks, ice, and water: the geology of Waterton-Glacier Park.* Missoula, Montana: Mountain Press Publishing Co.

Dyson, J. L. 1957. *The geologic story of Glacier National Park.* Kallispell, Montana: Glacier Park Natural History Association, special bulletin no. 3.

———. 1971. *Glacier National Park.* National Park Service.

Harrison, J. E. 1972. Precambrian Belt basin of northwestern United States: its geometry, sedimentation, and copper occurrences. *Geological Society of America bulletin* 83: 1215–1240.

Robbins, M. 1981. *High country trail: along the Continental Divide.* Washington, D.C.: National Geographic Society. Pp. 143–195.

Roberts, A. E. 1972. *Cretaceous and early Tertiary depositional and tectonic history of the Livingston area, southwestern Montana.* U.S. Geological Survey Professional Paper 526-C.

Ross, C. P. 1959. *Geology of Glacier National Park and the Flathead region, northwestern Montana.* U.S. Geological Survey Professional Paper 296.

———. 1962. The Precambrian of the northwestern United States—the Belt Series. In *The Precambrian,* ed. K. Rankama, vol. 4, pp. 145–251. New York: Interscience.

Ruhle, G. 1972. *Roads and trails of Waterton-Glacier National Parks.* Minneapolis: John W. Forney.

Salop, L. J. 1977. *Precambrian of the Northern Hemisphere and general features of early geological evolution.* New York: Elsevier Scientific Publishing Co.

Willis, B. 1901. Stratigraphy and structure, Lewis and Livingston Ranges, Montana. *Bulletin Geological Society of America* 13:305–52, November 5.

Address

Glacier National Park
West Glacier, Montana 59936

11

Denali National Park and Preserve

Location: South Central Alaska
Area: 5,695,493 acres; 8,899.2 square miles
Established as Mount McKinley National Park:
    February 26, 1917
Enlarged and redesignated Denali National Park and
    Preserve: December 2, 1980

Figure 11.1 Muldrow Glacier on the north side of Mount McKinley in Denali National Park and Preserve. The first climbers to reach North Peak on Mount McKinley ascended the mountain in 1910 by way of the Muldrow Glacier, a route still used by mountaineers. South Peak, the actual summit, is 20,320 feet in elevation. Turbulence in glacier flow has produced striking fishhook curves in the medial moraines in the foreground. Photograph by Bradford Washburn.

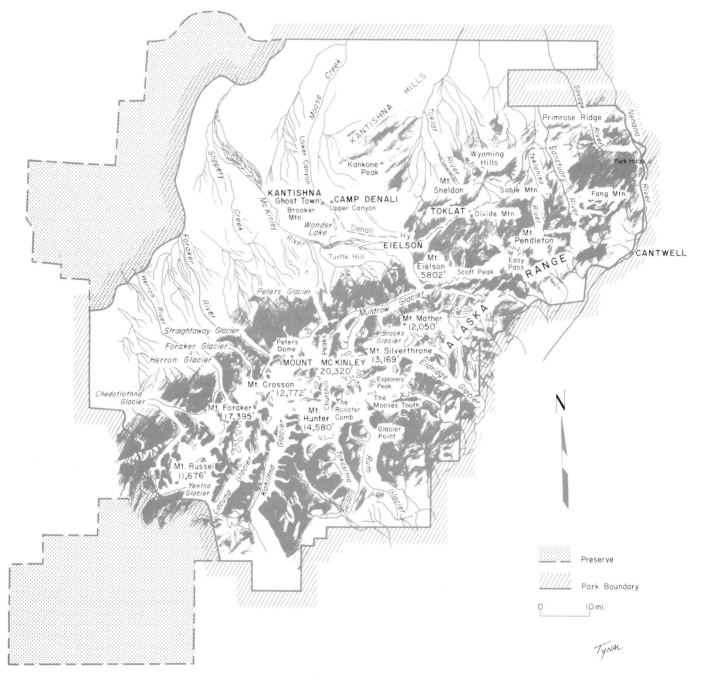

Figure 11.2 Denali National Park and Preserve, Alaska.

Table 11.1. Generalized Geologic Column, Denali National Park and Preserve

Time Units			Rock Units		Geologic Events	
Era	Period	Epoch				
Cenozoic	Quaternary	Holocene	Glacial, alluvial, and landsliding deposits		Glacial and periglacial activity Stream erosion and deposition Mass wasting Maximal glacial stages	Vertical uplift, strike-slip faulting, earthquakes
		Pleisto-cene	Extensive glacial deposits Nenana Gravel			
	Tertiary (Neogene)				Sedimentation in valleys Erosion	Uplift, block faulting
			Coal-bearing Group			
	Tertiary (Paleogene)		Mt. Galen Volcanics Teklanika Formation Cantwell Formation	Granitic rocks	Ashfalls Lava flows Nonmarine deposition in lowlands	Deformation Episodes of plutonic intrusion
Mesozoic	Cretaceous/ Jurassic				Intense orogenic activity accompanying plate convergence: thrust faulting, uplift, folding, metamorphism	
			Sandstones, shales (S) Pillow basalts (S) Slates, marble (S) Slates, marble (S)	Basement complex	Deposition in deep ocean basins	
	Triassic					
Paleozoic			Metavolcanic quartz schists, marble (N)		Orogenies, volcanism, plutonism, metamorphism	
			Quartzites, mica-quartz schists (N)		Accumulation of marine sediments	

Source: Modified after Gilbert 1979; Wahrhaftig 1958.
S = South side of Hines Creek fault
N = North side of Hines Creek fault

Local History

The areas added to the redesignated Denali National Park and Preserve surround the original park on three sides and ensure the protection of the entire Mount McKinley massif. The Kichatna Mountains on the southern flanks of the massif and most of the great glaciers on the southeast side are now in the Park and Preserve, as is the historic Kantishna mining district on the northwest side of the original park. Extension of the boundaries also protects the natural habitat of wolves, grizzly and black bears, caribou, and other wildlife within the mountain ecosystem.

Mount McKinley, in the central part of the Alaska Range, is the highest peak in North America. *Denali* is the Athapascan Indian name for the mountain and means "the Great One." William A. Dickey, who prospected for gold in the Susitna River valley in 1896, recognized the importance of the mountain and guessed its height to be about 20,000 feet. He named Mount McKinley for the 25th President of the United States, who was elected in that year.

In 1898 George H. Eldridge, a geologist, and Robert Muldrow, a topographer, were sent by the U.S. Geological Survey to explore the southern approaches to the range. Eldridge and Muldrow determined the map location and calculated the altitude of Mount McKinley with an accuracy close to currently accepted figures, an amazing achievement considering the difficulties of mapping a region without (at that time) topographic control points. The two men also plotted the route for the Alaska Railroad, built 25 years later from Seward to Fairbanks. The railroad, which skirts the eastern boundary of the Park and Preserve, is still a principal means of access to the Mount McKinley area.

Lt. Joseph S. Herron led a U.S. Army expedition in 1899 that was the first to cross the Alaska Range from south to north. On their way from Cook Inlet to the Yukon River valley (north of the range), they discovered and named Mount Foraker, the second highest peak of the Alaska Range—the mountain the Indians referred to as "Denali's Wife."

The first expedition to reach the base of Mount McKinley was led by Alfred H. Brooks, a pioneer geologist who took a U.S. Geological Survey party to the northern side of the Alaska Range in 1902. Brooks climbed a shoulder of Mount McKinley but was forced to turn back at about the 7,500-foot level because of icy slopes and steep cliffs. He placed a note in a cartridge shell in a rock cairn at the spot. The note and shell were recovered in 1954 by geologists of the U.S. Geological Survey and are now on display in the Park Museum (J. C. Reed 1955).

The scaling of Mount McKinley is a hazardous undertaking even today and can be attempted only by experienced climbers who have Park Service permission. North Peak is 19,470 feet high and South Peak, the actual summit, has an elevation of 20,320 feet. The peaks can be reached only by many hours of continuous ice climbing. The weather on the mountain is unpredictable. On a clear, bright day a blizzard may come up within 15 minutes, or dense fog and freezing mists may suddenly overtake climbers. Temperatures as low as -90° F. and winds as high as 87 miles per hour have been recorded. Normal temperatures are around 0° F., with winds 25 to 30 miles per hour. Even in July, the best month to attempt a summit ascent, conditions are often perilous. From time to time, climbers have died of exposure on Mount McKinley or have been fatally injured by falls (fig. 11.3).

In 1903 Judge James Wickersham and a party of four from Fairbanks made the first attempt to reach the top of Mount McKinley. Unfortunately, the route they chose was by way of Peters Glacier to the north face and a sheer escarpment, now called Wickersham Wall, that is still almost impossible to climb. The highest elevation they reached was 10,000 feet. A map of the Kantishna Hills which they filed with a mining claim after their return was the impetus for a gold rush. Rich placer diggings were discovered in 1903 in the Kantishna mining district. Later in life, when he was the territorial delegate to Congress, Judge Wickersham introduced the bill that provided for the establishment of the original Mount McKinley National Park.

Dr. Frederick Cook, starting from the south, tried three times in 1903 and 1906 to find a route to the summit of Mount McKinley, but he failed to get higher than about 6,000 feet. Foolishly he claimed to have reached the summit, but few believed him.

Then miners in the Kantishna district boasted that they could get the job done. Saloonkeepers put up the money, and the "Sourdough Expedition"—whose members were Thomas Lloyd, Charles McGonagall, Peter Anderson, and William Taylor—took up the challenge in 1910. Having prospected in the foothills, the miners knew the approaches to McKinley well and had already decided that the Muldrow Glacier on the north side offered the best access to the summit. They discovered a pass, now called McGonagall Pass, leading to the upper Muldrow Glacier, and established a base camp on a ridge at 12,000 feet. From here, Anderson and Taylor climbed to the top of North Peak, believing it to be the higher of the two summits. Once there, they realized their mistake, but they planted a pole to mark their achievement nonetheless. They descended North Peak to the saddle, and Anderson started toward South Peak but turned back because of lack of time.

The Muldrow Glacier route the Sourdough Expedition used is still one of the two principal routes by which climbers approach Mount McKinley's summit (fig. 11.1). The other, the West Buttress route, is shorter and safer

Figure 11.3 The east face of Mount McKinley showing South Peak (left), the highest point on the North American continent, and North Peak (right; elevation 19,470 feet). Denali Pass separates the two peaks. Photograph by Bradford Washburn.

and ascends by way of Kahiltna Glacier. The West Buttress route was pioneered in 1951 by the distinguished explorer-scientist Bradford Washburn, who led a number of expeditions into the high areas of the McKinley massif. In 1947, his wife, Barbara Washburn, became the first woman to climb to Mount McKinley's summit (fig. 11.6).

The Stuck-Karstens Expedition of 1913 achieved the honor of making the first successful ascent of Mount McKinley's South Peak. The leader was Archdeacon Hudson Stuck, an Episcopal missionary and an advocate of the native name Denali, both for the mountain and for the national park that was being proposed. The co-leader, Harry Karstens, had been a packer for the naturalist Charles Sheldon and had spent many months with him, roaming the region that is now Denali National Park and Preserve. Others on this first ascent were Robert Tatum, another missionary, and Walter Harper, whose father had been one of the first white men to explore the Alaskan interior in the 1870s.

Harry Karstens later became the first superintendent of Mount McKinley National Park when it was established in 1917. It was established largely through the efforts of Charles Sheldon, who was well known for his studies of Alaskan wildlife. In his original proposal for the national park, Sheldon had urged that it be named Denali.

Because Mount McKinley is such a famous mountain and a national park has been in existence there for many years, the region is more accessible and more visitor facilities are available than in most of the national parks in Alaska. Visitors can reach Denali National Park and Preserve via the Alaska Railroad, by the Denali and Alaska No. 3 Highways, and by private plane and charter aircraft. Private vehicles are not allowed on the park roads (except to get to the campgrounds), but shuttle buses operate daily during the summer season (fig. 11.4).

Summer temperatures in the Denali area are cool, but they may range from 40° F. to 80° F. The weather is often wet and windy. During the winter season, visitor facilities are closed, but the Park and Preserve is open to skiers and dog-sledders. Winter temperatures are generally between 30° F. and -50° F.

Geologic Features

Denali National Park and Preserve lies like a saddle across the central part of the Alaska Range, a curved mountain chain about 600 miles long that forms a topographic barrier between the coastal lowlands around Cook Inlet and

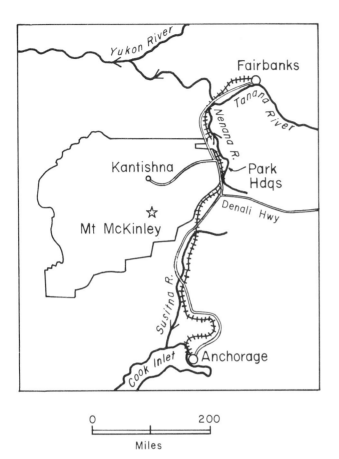

```
0              200
|_____|_____|
        Miles
```

Figure 11.4 Sketch map of access to Denali National Park and Preserve from Anchorage and Fairbanks. The Alaska Railroad and a paved all-weather highway follow essentially the same route from Anchorage to Fairbanks. The Denali Highway, which approaches the Park and Preserve from the east is open only from June through September. The road through the Park to Kantishna has limited access only. Drainage is to the north via the Nenana River and to the south along the Susitna River. The Alaska Range, which trends east-west in this area, forms a drainage divide between north-flowing and south-flowing streams in the region.

the Yukon lowlands of the interior. In the McKinley massif section, where the Alaska Range is about 60 miles wide, the mountains trend in a generally northeast-southwest alignment. Fewer than 20 peaks in this highest part of the range are over 10,000 feet in elevation, and elsewhere in the Alaska Range, the mountains are mostly between 7,000 and 9,000 feet high. "The Great One," Mount McKinley, at 20,320 feet, towers over them all. On clear days its majestic, snow-covered summit can be seen from ships in the Gulf of Alaska. From the north, its height is even more impressive. The steep north face of Mount McKinley rises 15,000 feet from its base.

Stuctures and Bedrock

Mount McKinley's height and location are the result of a lengthy and complex series of tectonic events that have been and are going on in the Alaskan mobile belt. The Aleutian Islands and most of southwestern, south central, and southeastern Alaska are within this tectonically active zone (see box 42.1, p. 510). The great Denali fault system, which cuts through Denali National Park and Preserve, is the dominant geologic structure of the region. Together, the fault system and the southern mountain ranges form an arcuate belt, with the Alaska Range lying approximately along the center of the arc (fig. 11.5).

In Denali National Park and Preserve, the two major segments of the fault system are the Hines Creek fault and the McKinley Strand, both of which are north of Mount McKinley. Movement along the fault system has been mainly lateral (i.e., strike-slip displacement); but locally, offset has also been vertical, especially along the currently active McKinley Strand where it skirts the north side of the McKinley massif. Here the rocks on the south side of the fault have been raised many thousands of feet and uplift may be continuing today. The Wickersham Wall on the northwest side of Mount McKinley is the result of this relatively recent movement. The trace of the McKinley Strand is marked by a linear, topographic depression where the rocks in the fault zone have been crushed, pulverized, and eroded. Between Gunsight Pass (below Wickersham Wall) and Anderson Pass to the northeast (where the glacier bends toward the northwest), the Muldrow Glacier follows the fault zone.

For more than 300 miles along its strike, the Denali fault system separates rock units that are grossly dissimilar in their lithologies, ages, and degrees of metamorphism (table 11.1). In the Park and Preserve this contrast is clearly evident along the Hines Creek fault, which crosses the Alaska Railroad near the Park Headquarters and runs westward through a series of fault valleys for about 50 miles. The oldest rocks of the region (rocks of the basement complex) are north of the Hines Creek fault. They are highly deformed and metamorphosed micaceous schists, quartzites, and marbles that were originally sedimentary and volcanic rocks deposited in Paleozoic and probably Precambrian time.* They crop out in an east-west band across the northern part of the Park and Preserve. Mount Healy, Mount Wright, the Wyoming Hills, and the Kantishna Hills are made up mostly of these old crystalline rocks. Good exposures can be seen near Park Headquarters and the Savage River Campground. The schistose rocks (Birch Creek Schist), which are highly foliated and cut by numerous fractures, tend to weather into thin slabs parallel to the orientation of the mica flakes.

*Rock descriptions from Gilbert 1979.

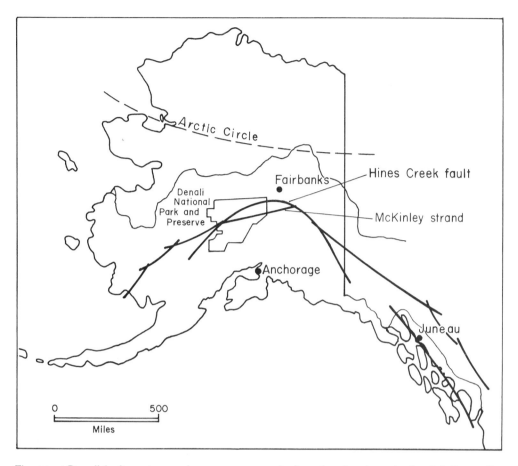

Figure 11.5 The great Denali fault system arches across southern Alaska, extending from south of Juneau to the western coast. The two major segments in Denali National Park and Preserve are the Hines Creek fault and the McKinley strand. The principal movement along the Denali fault system has been horizontal, the south and southwestern side moving west and northwest relative to the north and northeastern side. Vertical movement has also occurred on several of the fault segments. Both types of movement are still going on.

South of the Hines Creek fault are three groups of somewhat younger basement rocks, also metamorphosed and deformed, but of obviously different origin than those north of the fault. One group of gray slates and marbles, dark gabbros, and altered basalts is exposed along the crest of the Alaska Range between Windy Pass and Mount McKinley (fig. 11.6). They can be seen more easily from the park road on the slopes above the Toklat River. A second group consists of interbedded marbles and slates extending from Savage River westward to Stony Creek. The third group, made up of weakly metamorphosed sandstones and shales, is found in a band between the other two groups.

Unconformably overlying the basement rocks of the Park and Preserve, on both sides of the Hines Creek fault, are several young formations that were deposited in downfaulted basins as the mountain blocks were uplifted and shifted around. The oldest of these, the Cantwell Formation, consists of sandstones, shales, and conglomerates that have been tilted and folded by compressive forces. Deformed beds show up clearly on slopes above the East Fork River, and outcrops can also be seen from the park road. Overlying the Cantwell rocks are the rhyolitic and basaltic lava flows of the Teklanika Formation and the andesitic Mount Galen Volcanics. The bright hues of ash layers interbedded with clastic sediments and lava flows are strikingly displayed in Polychrome Pass and on Polychrome Mountain, on the south side of the park road.

The highest and most rugged peaks in the Park and Preserve, such as those on Mount McKinley and Mount Foraker, are carved from granitic rocks of great strength and resistance that were emplaced in plutons of varying ages. On the southeast side of Mount McKinley in the area around Ruth Amphitheater and along the Great Gorge enclosing the upper part of Ruth Glacier, great spires and walls of granite soar thousands of feet above the ice. And in the Kichatna Mountains at the southwest end of the Park and Preserve are the granite Cathedral Spires, probably the highest stand of vertical rock in North America and famed as a challenge to the most intrepid and skillful mountain climbers (fig. 11.7).

Figure 11.6   Contact between light-colored McKinley granite and dark-colored metamorphic rock on the flank of North Peak. Barbara Washburn is the climber in the foreground. Bradford Washburn, who was the leader of a scientific expedition that made the first geologic studies of the upper part of Mount McKinley in June, 1947, took this photograph at an elevation of 18,400 feet just south of Denali Pass.

Figure 11.7   Eight major glaciers and several smaller ones radiate from the vicinity of Cathedral Spires in the southwestern part of Denali National Park and Preserve. Kichatna Spire (elevation, 8,985 feet), which is rated as North America's greatest single stand of vertical rock, was first climbed in 1966. National Park Service photograph by M. Woodbridge Williams.

The youngest rocks in the Park and Preserve are the units of the Coal-bearing Group and the poorly consolidated Nenana Gravel. At one time coal was mined near Sable Pass in the northeastern part of the Park and Preserve. The Nenana Gravel (4,000 feet thick in some places) is widespread in the northern and northwestern parts of the Park and Preserve.

Glacial Features

The alpine scenery of Denali National Park and Preserve is large-scale and spectacular. When weather is favorable, charter aircraft fly visitors over the peaks of the Alaska Range so that they may see the glacier systems and the mighty sculpturing of rock by ice. The granite spires (mentioned earlier), the arêtes, huge cirques, and gorges are impressive examples of glacial erosion.

The more extensive glaciers and snowfields are on the southeastern side of the range because the moisture-bearing winds from the Gulf of Alaska drop more snow on this side. The five largest south-facing glaciers are the Yentna (20 miles long), Kahiltna (30 miles long), Tokositna (23 miles long), Ruth (31 miles long), and Eldridge (30 miles long). On the drier north side, except for the Muldrow Glacier, which is 32 miles long, all the glaciers are smaller and shorter. The Muldrow Glacier's length is due to the fact that it follows the linear topographic depression of the Hines Creek fault zone. The Muldrow's right-angle bend near Anderson Pass represents a drainage offset. Before the glacier formed, the McKinley River occupied and eroded this course.

The glaciers of the Alaska Range carry enormous amounts of rock debris on, in, and beneath the ice as they move downslope. Loose fragments and blocks that fall from cliffs or are scraped from the rock walls build up as low ridges of till that ride along on the edges of the glaciers. These are *lateral moraines*. Where tributary glaciers merge with each other or the trunk glacier, the adjacent lateral moraines join, forming *medial moraines* that are also carried down on the surface of the moving ice. Some of the large trunk glaciers, like the Yentna (fig. 11.8), have accumulated lateral and medial moraines that, seen from above, look like lane markers on a superhighway.

Figure 11.8   Air view of the Yentna Glacier, which flows southward from the McKinley massif. Ice-filled cirques feed the tributaries of this large trunk glacier, and frost-loosened talus blocks, fallen from the peaks, accumulate in lateral moraines along the outer edges of the moving glaciers. From both sides of the group of peaks in the center of the picture, lateral moraines have joined to form a wide band of multiple medial moraines being carried downglacier. The tributary glacier joining the Yentna from the left is cascading over a rock rim, and the curved crevasses on the glacier surface suggest that ice is flowing faster in the center and dragging on the sides as the glacier descends. National Park Service photograph.

From the snouts of the glaciers, braided meltwater streams, heavily loaded with rock debris, continually shift and intertwine their channels over the valley floors. New channels open up as old channels are filled. *Valley trains* build up as the streams drop quantities of poorly sorted sediment. Valley trains are long, narrow accumulations of glacial outwash, confined by valley walls. In some places coalescing outwash fans at the mouths of valleys have built up outwash plains over lowlands on both sides of the range.

Sometimes when glaciers are retreating and melting is rapid, blocks or masses of ice buried by till and outwash may persist for many years far beyond a glacier's terminus. If the ice finally does melt, the enclosing sediment slumps in, forming depressions called *kettles*. Filled with water, these depressions become *kettle lakes*. Wonder Lake (reached via the park road) was formed by the melting many years ago of a massive block of ice that was stranded between end moraines left by the retreating Muldrow Glacier. The presence on the outwash plain of old moraines, kettles, and the like are indications that the glaciers extended to lower altitudes in the past. The fact, also, that the present glaciers are "underfit" in their valleys shows that many of the glacial erosional and depositional features that we see in the Park and Preserve today are the result of earlier, more extensive glacial action.

Permafrost and Periglacial Activity

During the colder Pleistocene climates, all the ground in the Park and Preserve area became so solidly frozen that much of it, especially on the north side of the range, is still frozen because average annual temperatures in the region have not risen enough since the Ice Age for the ground to melt. Permanently frozen ground, or *permafrost*, is *discontinuous* in the Denali area. The mean surface ground temperature there of $-1°$ to $-5°C$. (a few degrees below freezing) has allowed some thinning of the permafrost and a great deal of local variability. (North of the Arctic Circle permafrost is *continuous* and generally more than 500 feet in thickness.)

Permafrost is covered by an active layer that freezes and thaws seasonally. The depth of the active layer is controlled by many highly variable factors, such as the vegetation, surface material, topography, amount of moisture, and so on. In the Denali area, the active layer ranges from less than an inch to 10 feet or more, depending upon local conditions. Thicknesses of permafrost below the active layer have been measured between 30 and 100 feet at various locations around the Alaska Range.

The *periglacial* milieu is the active layer above the permafrost where freeze-thaw oscillations are more or less continual. Water changing back and forth between its liquid and solid states produces periglacial phenomena. In the active layer, water is available (and usually plentiful) because it cannot seep down into the impermeable permafrost below. (Voids and pores in permafrost are filled with ice.)

Simple frost action—a process of mechanical weathering that breaks up rock—is not usually dominant in temperate regions. In subpolar climates, however, the freeze-thaw cycles are predominant over other processes most of the time. Periglacial effects are noticeable around the margins of melting glaciers, on slopes, in river valleys, and on the tundra plains. Under periglacial conditions, *frost splitting* of wet rocks produces pinnacles (such as Cathedral Spires) in massive, resistant rocks and thick accumulations of talus in fissured rocks. Sometimes talus masses in draws or gullies retain moisture, freeze, and become *rock glaciers* that move slowly downslope.

Frost heaving disrupts unconsolidated material, both sorted and unsorted. Soil's volume increases irregularly when it freezes and then decreases irregularly when thawing occurs. Thus the surface of the soil keeps heaving up and falling back a little differently each time. As this process goes on, grains, pebbles, and cobbles on a flat surface tend to rearrange themselves in polygonal patterns and strips. Underneath the surface, lenses of coarse and fine material slump and sag chaotically.

On slopes, the active layer may become saturated during summer melt seasons due to the interaction of rain, meltwater, and low evaporation. The saturated soil (usually carrying a cover of vegetation) slides and slumps down the slope, exposing fresh permafrost underneath. This process is called *solifluction*, or soil flow. A stand of white spruce growing on a lower slope of Mount McKinley is called the Drunken Forest because the oddly leaning trees appear to be staggering as the sliding soil slowly transports them down the mountainside.

Many of the shallow ponds in the Park and Preserve, such as those in the tundra plains northwest of the range, are *thaw lakes* and *cave-in lakes*. They form where sun-warmed water has melted basins in the underlying permafrost. The lakes gradually deepen from summer to summer, as warmth penetrates the frozen ground, and they enlarge as the rims collapse.

In winter, the freeze-thaw cycles slow and then become suspended, but ice (which has 10 percent greater volume than water) continues to exert pressure. Thermal expansion and contraction at depth can cause cracks to develop in the ice-cemented permafrost. During the melt season, water gets into the cracks, forming veins of ice called *ice wedges*, that enlarge with successive seasons of freezing and thawing. In periglacial regions, excavations or landslides sometimes reveal very large ice wedges that have been buried in sand and gravel for centuries.

Periglacial activity and the presence of permafrost cause engineering problems for highways, railroads, landing fields, and buildings because structural failures tend to occur whenever the natural thermal regime is disturbed. Uneven thawing, for example, causes roadbeds and foundations to tilt or collapse.

Geologic History

1. Basement complex of crystalline Paleozoic and possibly Precambrian rocks.

The oldest rocks in Denali National Park and Preserve, the metamorphosed units north of the Hines Creek fault, may originally have been deposited on the outer slope and shelf of a continental margin along the landmass that is now North America. The rocks lie at the southernmost edge of formations extending to the north and east which some geologists believe have long been part of North America.

The somewhat younger Paleozoic rocks on the south side of the Hines Creek fault were probably deposited in deep ocean basins unrelated to the North American landmass. Recent studies suggest that the rocks originally formed far from their present location, possibly south of the equator (see box 42.1, p. 510). All of the basement rocks have been altered by episodes of orogeny, plutonism, and volcanism that have caused them to be highly deformed and metamorphosed.

2. Late Mesozoic and early Cenozoic plate convergence.

In late Mesozoic time, an oceanic plate converging against the North American plate was subducted. The Hines Creek fault may represent a boundary between the two plates. Older rock units on both sides of the fault were further deformed by compression and uplift. Great volumes of granitic rock were emplaced in the Alaska Range by a series of plutonic episodes that lasted through most of Cenozoic time. Lava flows poured out at the surface and volcanoes erupted. Sedimentation was controlled by tectonic events. As the land rose, sediments washed down into basins between the highlands and were interbedded with ash and lava. In this way the Cantwell and Teklanika Formations and the Mount Galen Volcanics accumulated. Continued plate convergence caused the sedimentary rocks to be tilted and folded (fig. 11.9A).

3. Deposition of the Coal-bearing Group in middle Cenozoic time.

As erosion wore down the mountains, plant material accumulated in swamps and muddy lakes. Sand and gravel buried the plant remains, and eventually they became compacted to a low grade of coal.

4. Blockfaulting and uplift in Pliocene time.

Uplift caused renewed erosion. The Nenana Gravel, which is poorly consolidated, filled the downfaulted basins and spread over the lowlands north of the Alaska Range.

5. The pushing up of the McKinley massif.

As plate motion changed at the end of Pliocene and the beginning of Pleistocene time, strike-slip displacement occurred all along the Denali fault system. In the Park and Preserve region, rocks from southeastern Alaska moved westward along the McKinley Strand, the new plate boundary. However, near Mount McKinley, the boundary bends sharply southwest toward the Aleutian Range (fig. 11.9B). Subduction is now occurring under western Alaska and the Aleutian Islands (box 42.1).

Some geologists have suggested that continental crust riding on the leading edge of the Pacific plate, on the south side of the McKinley Strand, is too lightweight and too bulky to sink. In other words, the igneous and metamorphic rocks on the south side are low in density and tend to pile up rather than sink. Thus plate movement here tends to keep pushing Mount McKinley upward (Gilbert 1979).

6. Pleistocene glaciation.

Deciphering the record of past glaciations in the Denali area has been complicated by the influence of the relatively rapid tectonic uplift occurring simultaneously. At the time of the first glaciation (Browne) in late Pliocene time, the mountains in the McKinley part of the range were only of moderate height. But apparently by the time the second glaciation (Dry Creek) began, deep stream valleys had been cut in the rising massif. Glaciers flowed down these valleys, widening and deepening them more. On the snowier south side, ice covered nearly all the land between the Alaska Range and the Gulf of Alaska. North and west of the range, glaciers filled valleys and flowed outward but did not cover the whole area. (The interior of Alaska between the Alaska Range and the Brook Range was never glaciated.) The third glaciation (Healy) and the fourth (Riley Creek) followed the same pattern of ice advance, but they were not so extensive as the second (Wahrhaftig 1958).

An *icecap* (small ice sheet) that developed on the Alaska Range during each of the last three glaciations apparently did not connect with the great Canadian ice sheets to the east. The Alaskan icecap originated locally and had its own supply area, direction of movement, and persistence. Several minor advances and retreats of valley glaciers have occurred since Pleistocene time, but even the warmest interglacial period did not greatly reduce the size of the glaciers.

7. Continuing tectonic activity.

Along the Denali fault system the Pliocene Nenana Gravel has been folded and faulted. Moraines, river terraces, and outwash features have been offset. Some of these

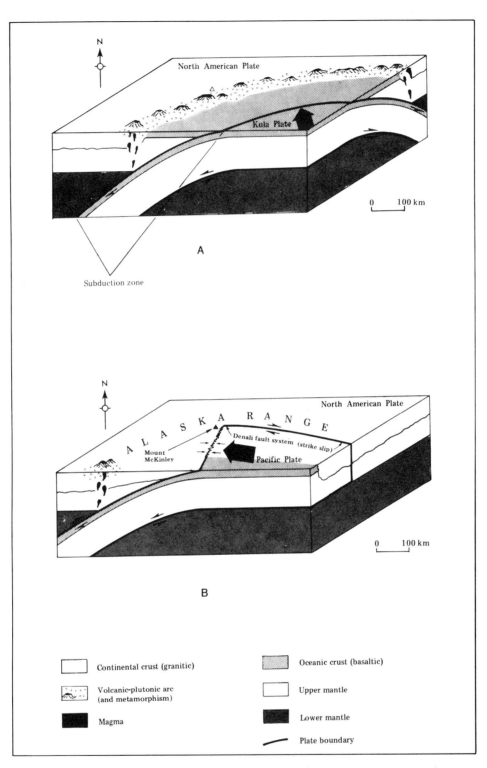

Figure 11.9   Block diagrams illustrating models for possible plate movements during (*A*) late Mesozoic and early Cenozoic time and (*B*) during late Cenozoic time. From *A Geologic Guide to Mount McKinley National Park* by W. G. Gilbert. © 1979 Alaska Natural History Association; used by permission. *A.* The oceanic Kula plate (i.e., forerunner of the Pacific plate) is subducting as it converges with the North American plate. Heat and pressure causes melting of rock, forming magma which works its way toward the surface and erupts, building a line of volcanoes. The small triangle (slightly off-center on the North American plate) is the future site of Mount McKinley. *B.* The oceanic plate—now the Pacific plate—is subducting beneath the North American plate farther to the west. Along the Denali fault system, a change in plate motion through a series of strike-slip displacements is causing the plates to slide by each other. Where the plate boundary bends south, Mount McKinley is being shoved upward.

displacements can be seen from the Alaska Railroad and the north-south highway skirting the Park and Preserve. These are all indications of recent fault movement. Frequent earthquakes in southern Alaska are also a sign that tectonic activity in the region is ongoing.

Geologic Maps and Cross Sections

American Association of Petroleum Geologists. 1974. *Geologic highway map of Alaska and Hawaii, map no. 8.* Tulsa, Oklahoma: American Association of Petroleum Geologists.

Gilbert, W. G. 1978. Geologic map of northern Mount McKinley National Park. In Gilbert 1979 (below).

Reed, J. C., Jr. 1961. Geology of the Mount McKinley quadrangle. U.S. Geological Survey Bulletin 1108A

Special Topographic Map

Washburn, Bradford. 1963. *Map of Mount McKinley.* Boston, Massachusetts: Museum of Science.

Bibliography

Alaska Travel Publications Editors. 1976. *Exploring Alaska's Mount McKinley National Park.* 2nd ed. Anchorage, Alaska: Alaska Travel Publications, Inc.

Brown, Dale. 1976. *Wild Alaska,* American Wilderness Series. New York: Time-Life Books.

Buskirk, Steve. 1978. *Mount McKinley, the story behind the scenery.* Las Vegas, Nevada: KC Publications

Churkin, M., Jr., Carter, C., Trexler, J. H., Jr. 1980. Collision—deformed Paleozoic continental margin of Alaska—Foundation for microplate accretion. *Geological Society of America bulletin* 91:648–654, November.

Gilbert, W. G. 1979. *A geologic guide to Mount McKinley National Park.* Anchorage, Alaska: Alaska Natural History Association.

Péwé, T. L. 1975. *Quaternary geology of Alaska.* U.S. Geological Survey Professional Paper 835.

Reed, B. L., and Lanphere, M. A. 1974. Offset plutons and history of movement along the McKinley segment of the Denali fault system, Alaska. *Geological Society of America bulletin* 85:1883–1892, December.

Reed, J. C., Jr. 1955. Record of the first approach to Mt. McKinley. *American alpine journal* 9:78–83.

————.1961. Geology of the Mount McKinley quadrangle, Alaska. U.S. Geological Survey Bulletin 1108A

Tricart, J. 1968. Periglacial landscapes. In *Encyclopedia of geomorphology,* ed. R.W. Fairbridge, pp. 829–32. New York: Reinhold Book Corporation.

Wahrhaftig, Clyde. 1958. *Quaternary geology of the Nenana River and adjacent parts of the Alaska Range.* U.S. Geological Survey Professional Paper 293A

————, and Black, R. F. 1958. *Engineering geology along part of the Alaska Railroad.* U.S. Geological Survey Professional Paper 293B.

Williams, Howel, ed. 1958. *Landscapes of Alaska, their geologic evolution.* Berkeley: University of California Press.

Address

Denali National Park and Preserve
Box 9
McKinley Park, Alaska 99755

12
Rocky Mountain National Park

Location: Northern Colorado
Area: 263,790.69 acres; 412.16 square miles
Established: January 26, 1915

Figure 12.1 Upper Cascade Creek, in Rocky Mountain National Park, meanders over an alpine meadow underlain by glaciofluvial deposits. The valley was once filled with glacial ice that flowed from the U-shaped glacial trough in the background. National Park Service photograph.

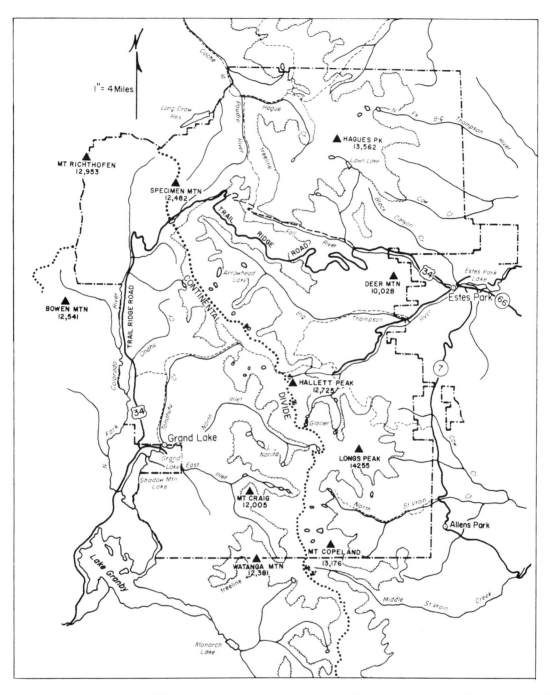

Figure 12.2 Rocky Mountain National Park, Colorado. Map reproduced from the *Southern Rocky Mountains* (K/H Geology Field Guide series) by R. P. Larkin, P. K. Grogger, and G. L. Peters. © 1980 Kendall/Hunt Publishing Company.

Table 12.1 Generalized Geologic Column, Rocky Mountain National Park

Era	Period	Epoch	Rock Units	Geologic Events
Cenozoic	Quaternary	Holocene	Neoglacial deposits, alluvium, talus	Development of modern glaciers; active mass wasting; periglacial processes; stream erosion Melting away of earlier glaciers during post-Pleistocene warming

		―?―
		Pleistocene
			Glacial tills, moraines, etc., of successive ages; early deposits largely buried or destroyed by later advances	Pinedale Glaciation
				Intermediate Glaciation
				  Late Bull Lake
				  Early Bull Lake
				Early Glaciation (at least two major advances and retreats)

	Tertiary	Paleogene / Neogene	Volcanic ash Alluvial deposits Rhyolite, welded tuff, obsidian, pyroclastic debris, mudflow deposits	Development of modern drainage patterns Extensive erosion exposing basement rocks on summits, hogbacks on flanks Block-faulting forming basins Uplift along old faults raising Front Range to present height Volcanism (mainly to west and south of Park) Mudflows, lava flows, ash falls
Mesozoic	Cretaceous Jurassic Triassic		Shales	Intermittent uplift, faulting, tilting of sedimentary beds; stripping of beds producing erosional surface Rocky Mountain orogeny (Laramide phase) ―――――――――――― Last transgression and withdrawal of seas Reduction of relief
Paleozoic	(Cambrian to Permian)		Remnants of sedimentary rock units exposed on eastern and western flanks of Front Range (none in Park)	Uplift and erosion of Ancestral Rockies Long deposition, first marine, then terrestrial (due to uplift)
			/////////////	Long erosion exposing granite batholiths
	Late Precambrian (Proterozoic Y) Middle Precambrian (Proterozoic X)		Basaltic dikes, granites, pegmatites Granites, schists, gneisses (basement complex)	Successive intrusions and emplacement of batholiths; uplift Uplift, orogeny, metamorphisms, intrusions Long accumulation of sediment interbedded with volcanics

(After Larkin, Grogger, and Peters 1980; Richmond 1974)

Local History

The area of Rocky Mountain National Park, in the Southern Rockies, was Indian country for thousands of years before Europeans came to North America. As settlers pushed westward, displaced Indians who were also migrating westward came into conflict with tribes who had earlier occupied the hunting grounds and meadows of the Rocky Mountains. More than 40 Indian campsites have been found in and around the Park, some of the locations being Forest Canyon Pass, Milner Pass, and meadows at lower elevations.

Like the white men, the Indians found the Rocky Mountains, especially the Front Range, to be a barrier to travel. One of the Indian trails over the mountains was called *Taleonbaa*, or "the Children's Trail," because it was so steep that even the children had to dismount from horses and walk.

Lieutenant Zebulon M. Pike (for whom Pikes Peak, near Colorado Springs, is named) explored the region that is now Rocky Mountain National Park in 1806–07. He referred to the tallest summit as "Great Peak" but did not get to the mountain. The peak was located and renamed Longs Peak in 1820 by the Colonel Stephen H. Long expedition, which was commissioned to map the region by President Madison.

In 1859 Joel Estes and his son Milton found the valley that now bears their name (Estes Park) and the following year built a cabin along Willow Creek. They left in 1866, and their claim came into the possession of Griff Evans (1867), who later transferred it to the Earl of Dunraven who wished to make the area into a game preserve. Dunraven paid people to file individual claims and then transfer them to him. This method of land acquisition kept out enterprises that might have ruined the region, but the claims were later declared invalid. Dunraven did, however, keep a ranch in Estes Park and built the region's first hotel.

The first person to climb Longs Peak was William N. Beyer, who got to the top on August 23, 1868, after a failed earlier attempt in 1864. Included in his party were Major John Wesley Powell (the geologist who explored the Grand Canyon in 1869), N. E. Farrell, S. Gorman, S. W. Keplinger, and J. C. Sumner. An early guide in the region was E. J. Lamb, an archdeacon, who first climbed Longs Peak in 1871 and led a number of parties to the peak in the years that followed.

The person who was most involved in the effort to preserve the region as a national park was Enos Mills. He built a home in Longs Peak Valley in 1866 when he was only 16 years old. After helping a survey party map Yellowstone in 1891, he was convinced that his homesite should be preserved as Yellowstone was. He fought long and hard for this until success was achieved in 1915 with the establishment of Rocky Mountain National Park. A skilled mountain climber, Mills was the first to go up the east face of Longs Peak, a very difficult route, in 1903.

The famous Trail Ridge Road, built in 1932, starts near Estes Park and climbs to the so-called Roof of the Rockies, a broad, relatively flat upland that extends over most of the central area of Rocky Mountain National Park. Above the timberline, the upland is a treeless, rolling terrain of alpine tundra with a short summer season and nearly continual high winds. Passing through this remarkable area, Trail Ridge Road crosses the Continental Divide and then follows the Colorado River south to Grand Lake.

Box 12.1
How Igneous Rocks Are Classified

The brief discussion of ancient igneous features exposed at the bottom of Grand Canyon (chapter 1) suggested that igneous activity and mountain-building might be closely associated. Again, in the description of the origin of the very young mountains of Denali National Park and Preserve (chapter 11), a relationship is evident between plate tectonic activity, igneous intrusions, volcanism, and mountain-building.

In previous chapters you have become familiar with the terms referring to igneous bodies of various sizes and shapes (dikes, sills, batholiths, stocks, etc.). Moreover, you have already learned that igneous rocks are formed from the solidification of magma, or molten rock, that originates under conditions of intense pressure and heat far below the earth's surface. Some magma rises to the surface and is ejected (as in a volcanic explosion) or erupts as lava. Rocks formed in this way are called *extrusive igneous rocks*. They normally cool so rapidly that the individual mineral grains making up the rock cannot be identified in most specimens without the aid of a microscope. *Intrusive igneous rocks*, on the other hand, tend to be medium- or coarse-grained because they solidify from magma that cools slowly at depth over a long period of time.

Thus in classifying igneous rocks, determining the texture, or grain size, is one of the most important steps (fig. 12.3). The major grain size categories are (1) *phaneritic*, which includes rocks containing individual mineral grains visible to the unaided eye (larger than 0.062 mm); and (2) *aphanitic*, those rocks with grains too small to be seen without the aid of a microscope (generally smaller than 0.062 mm).

Sometimes changes in a magma's environment during cooling result in the formation of mineral crystals of distinctly different size within a rock, creating *porphyritic texture*, an effect that is sometimes described as similar to rice pudding with raisins (i.e., the rice representing the smaller grains and the raisins the larger ones). Other less common textures, mostly associated with volcanic activity, are (1) *glassy*, or noncrystalline texture, describing rocks (for example, obsidian) that cooled too rapidly for mineral grains to form; (2) *vesicular*, referring to rocks with holes due to gas bubbles trapped by the cooling process; (3) *pyroclastic*, made up of fragments, such as ash, pumice, and other *ejecta*, thrown up into the air during a volcanic eruption or explosion (table 12.2).

Igneous rocks are also classified on the basis of their mineral composition, or mineralogy. What this is for any particular rock is determined by the chemical composition of the parent magma. From the *felsic* magmas, rich in aluminum, silicon, sodium, and potassium, come the light-colored rocks that mainly contain the minerals orthoclase, sodium-plagioclase, and quartz. At the other extreme are the *mafic*, or iron-rich magmas, that are the source of the dark-colored rocks, made up primarily of the ferromagnesian minerals. In table 12.2, which is a graphical summary of igneous rock classification, the principal igneous rocks are shown, ranging in color from light to dark (according to their mineral composition) and from coarse to very fine texture.

Observe that a phaneritic (coarse-textured) granite may have essentially the same mineralogical composition as an aphanitic (very fine-textured) rhyolite, with both being light in color. Similarly, a dark, coarse gabbro may contain the same minerals as a dark, aphanitic basalt. What is surprising, however, is that the intrusive granite and the extrusive basalt are the most abundant igneous rocks on earth, while rhyolite (an extrusive rock) and gabbro (intrusive), the respective mineralogical equivalents, are relatively rare. For this paradox of nature, involving the most abundant extrusive and intrusive igneous rocks, geologists have as yet no wholly satisfactory explanation. However, in many respects, the distribution patterns of igneous rocks and their great variety can be accounted for by plate tectonic interpretations, based on field data; studies of composition and texture of rocks; and experimental work on artificial magmas.

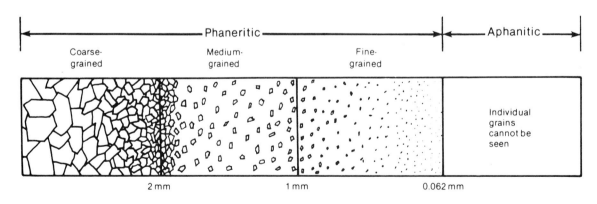

Figure 12.3 Diagram illustrating grain sizes of phaneritic and aphanitic textures in igneous rocks. From *Physical Geology, Laboratory Text and Manual* (2nd ed.) by R. D. Dallmeyer. © 1978 Kendall/Hunt Publishing Company.

Many of the national parks are noted for their exposures of igneous rock. The immense batholiths of the High Sierras underlying Yosemite, Sequoia, and Kings Canyon National Parks serve to illustrate the abundance of granitic rocks in the earth's crust. In the Hawaiian parks, for example, vast outpourings of mafic lava have formed large areas of basaltic rocks. Igneous rocks of intermediate composition (andesite, etc.) are found in Mount Rainier and Crater Lake National Parks. In these and other parks that have igneous rocks, their variety is noticeable to anyone who gets off the road and does some hiking around. Many igneous rock specimens found in nature have mineral grains large enough to be identified by eye using the characteristic physical properties of the constituent minerals. Students ordinarily learn these identification techniques in laboratory classes and on field trips. For convenience, diagnostic physical properties of the common rock-forming minerals are given in table 12.3. In subsequent chapters, where appropriate, the tables and information about igneous rock classification can be referred to.

Table 12.2
Igneous Rock Classification*

General Color	Light				Intermediate				Dark			
20% Abundance	20%	40%	60%	80%	20%	40%	60%	80%	20%	40%	60%	80%

Mineralogy / Possible Ranges in % Volume Abundance of Constituent Minerals

Orthoclase				
Quartz				
Muscovite				
Plagioclase				
Biotite				
Hornblende				
Pyroxene				
Olivine				

Textures

Texture	Light	Intermediate	Dark
Phaneritic	Granite	Diorite	Gabbro
Porphyritic Phaneritic	Porphyritic Granite	Porphyritic Diorite	Porphyritic Gabbro
Aphanitic	Rhyolite	Andesite	Basalt
Porphyritic Aphanitic	Porphyritic Rhyolite	Porphyritic Andesite	Porphyritic Basalt
Glassy	Pumice		Scoria
Pyroclastic	Tuff, Breccia		

Ultramafic rocks (typically phaneritic):
If mostly plagioclase = anorthosite.
If mostly olivine = dunite.
If mostly pyroxene = pyroxenite.
If a combination of olivine and/or pyroxene and/or plagioclase = peridotite.

*From *Physical Geology, Laboratory Text and Manual* Second ed. by R. D. Dallmeyer. © 1978 by Kendall/Hunt Publishing Company, Dubuque, Iowa

Geologic Features

Rocky Mountain National Park lies on the crest of the Front Range, the highest part of the American Rockies. The Front Range has 42 peaks exceeding 12,000 feet in elevation; among them are Longs Peak (14,256), Mount Meeker (13,911), Ypsilon Mountain (13,314), Taylor Peak (13,153), Flattop Mountain (12,342), Snowdrift Peak (12,274), Mount Adams (12,121), and Battle Mountain (12,044). Extending more than 200 miles from central Colorado northward into southeastern Wyoming, the Front Range borders the Great Plains and is the first major range encountered by travelers coming from the east.

The Development of the Rocky Mountains

Although geologists have studied the Front Range and the Rocky Mountains for many years, the relationships of plate movement, subsidence, and episodes of orogeny have not been entirely worked out. One indication of a very long history of tectonic activity in this region is the elevation of basement rocks (Precambrian gneisses, schists, and granites) exposed in the Front Range. Here they are on the Continental Divide, thousands of feet above sea level. In the Grand Canyon, you will recall, basement rocks are found in the depths of the Inner Gorge.

The present topography of the Front Range is due to a broad upwarping that took place during the Cenozoic Era, when vertical uplift brought Precambrian basement rocks to heights far above their surroundings (fig. 12.4). However, the mountains that we see today are but the most recent manifestation of plate motion and plate convergence that have gone on in this part of the earth's crust for many millions of years.

The basement rocks were formed when the earth's crust was considerably thinner than at present, particularly in its sial layer. The term *sial* refers to the earth's continental crust and means rocks high in *si*licon and *al*uminum; i.e., granitic rocks. Oceanic crust, or *sima*, is made up largely of basaltic rocks high in *si*licon and *magne*sium. Since the granitic rocks are lighter in weight (density) than the basaltic rocks, the continental plates tend to ride higher than the sima material that makes up the ocean floors and is also believed to extend under the sial layer.

When downwarps and crinkles developed in the thin primitive crust, basaltic lavas pushed upward and accumulated in troughlike structures. Eroded sediments were washed down from higher areas, with the added weight causing further subsidence and widening of the depression. As the trough sank, sedimentary rock layers were formed that eventually became metamorphosed by compressive stresses and magmatic intrusions. This produced the Precambrian gneisses, schists, and granites. Because

Table 12.3

Characteristic Appearance in Igneous Rocks of Common Rock-Forming Minerals. *

Mineral	Diagnostic Physical Properties
Orthoclase feldspar	1. Generally equidimensional grains. 2. Usually white or pink. 3. Two directions of cleavage nearly at right angles.
Plagioclase feldspar	1. Lath-shaped grains. 2. Usually white or gray, may be slightly greenish. 3. Two directions of cleavage nearly at right angles. 4. Parallel markings on cleavage faces.
Quartz	1. Usually interstitial, poorly defined grains. 2. Colorless and glassy appearance. 3. No cleavage.
Olivine	1. Small, round-shaped grains. 2. Pale green and glassy appearance. 3. No cleavage.
Hornblende (amphibole)	1. Lath-shaped grains. 2. Black or dark green. 3. Two directions of cleavage not at right angles.
Augite (pyroxene)	1. Blocky or lath-shaped grains. 2. Black or dark green. 3. Two directions of cleavage nearly at right angles.
Biotite	1. Flat, flaky grains. 2. Black or dark brown. 3. One perfect direction of cleavage.
Muscovite	1. Flat, flaky grains. 2. Light green or silvery white. 3. One perfect direction of cleavage.

*From Physical Geology, Laboratory Text and Manual, Second ed., by R. D. Dallmeyer. © 1978 by Kendall/Hunt Publishing Company, Dubuque, Iowa.

of subsequent geologic activities that occurred, details of these early tectonic events may never be completely deciphered.

During Paleozoic time both marine and terrestrial sedimentation continued in an extensive linear trough, or geosyncline, as crustal downwarping persisted in the region that is now Western North America. Intermittent tectonic activity in the Pennsylvanian Period deformed the sedimentary rock layers in the orogenic belt and elevated them into the "Ancestral Rockies." Subsequent erosion

Figure 12.4   Longs Peak (elevation 14,256 feet) is about eight miles air distance from Beaver Meadows in the foreground. Note the upper part of a cirque headwall on the face of Longs Peak, which rises above the eroded uplands in Rocky Mountain National Park. National Park Service photograph.

stripped away large amounts of the earlier Paleozoic beds and some of the Precambrian rocks.

Throughout this time the segment of crust that is now called the North American plate was part of a supercontinent, far to the south and east of the plate's present location. But as the Paleozoic drew to a close, drastic geographic changes were taking place—changes involving withdrawal of the shallow continental seas, mountain-building, and regional warping. The continents became fragmented, and the drifting began that ultimately led to the broad, deep ocean basins and high continents that we know today.

During the Mesozoic Era, as the North American plate moved westward and northwestward, it encountered the Pacific plate (see box 1.1, p. 8) that was moving slowly eastward. Beginning in the Triassic, the entire Cordilleran mobile belt along the western side of the North American plate began to buckle under compressive stresses, with the most intensive activity gradually moving from the outer edge (California) toward the east as mountain-building and deformation proceeded. By the end of the Cretaceous, the Rocky Mountain orogeny—which is

the term designated for this long and complicated episode—had reached eastern Colorado, and it was entering what is called the Laramide phase, a major part of this ongoing sequence of events. The Laramide phase of the Rocky Mountain orogeny is characterized by powerful plate motions that broke, uplifted, squeezed, and tilted the rocks of the crust, enabling younger plutonic and volcanic rocks to be formed.

As the mountains were raised higher and higher, erosion intensified, with sediments from summits and mountain slopes filling the valleys and lowlands. A period of crustal adjustment, called the isostatic phase, ensued as the orogenic pressures subsided. The principle of *isostasy* proposes a condition of equilibrium, comparable to floating, of units of brittle crust on the plastic zone of the *mantle* (the thick rock shell of the earth's interior between the crust and the core). Thus blocks of crust that are not in equilibrium, due to tectonic forces raising them up or the weight of accumulated sediment pushing them down, tend to compensate by moving up or down in the direction toward equilibrium. Block-faulting and other vertical adjustments of crustal units can sometimes be explained in terms of isostatic compensation. However, this principle cannot account for the much more powerful mechanisms of plate subduction and plate convergence that clearly involve forces of much greater intensity.

In this chapter the structural features of Rocky Mountain National Park are depicted as the culmination of an immeasurably long sequence of geologic events, the most recent being Cenozoic upwarp. But in one way or another, all the national parks of the western United States were affected by the Rocky Mountain orogeny and the Mesozoic/Cenozoic tectonic activities associated with mountain-building. Some of the aspects that you should note as being of particular significance are block-faulting in the Grand Tetons (chapter 25); volcanism in Mount Rainier, Crater Lake, and Yellowstone National Parks (chapters 17, 19, 24); thrust-faulting in Glacier National Park (chapter 10); plutonic activities in Yosemite, Sequoia, and Kings Canyon National Parks (chapters 13, 27, 28); and in the Olympic, Denali, and North Cascades National Parks (chapters 16, 11, 14), combinations of compression and uplift accompanied by plutonic and volcanic activity. The rocks of the Colorado Plateaus, on the other hand, show only moderate faulting and uplift in response to the tectonic stresses of the Mesozoic and Cenozoic Eras (chapters 1 through 7).

Volcanic Activity, Pegmatites, and Hydrothermal Deposits

The faults on both sides of the Front Range, as well as most of the other faults and folds in the Park region, resulted from Laramide tectonism. In the faulted areas, eruptions and intrusions of various kinds occurred repeatedly from the Cretaceous Period on into the early Pleistocene, with the greatest intensity being in Tertiary time. The earlier episodes of mountain-building and faulting (such as the tectonic activity that produced the Ancestral Rockies) were also associated with igneous intrusives and extrusives, but much of this evidence was obliterated by subsequent geologic events and long periods of erosion.

Plate movement and increasing volcanic activity during the Tertiary was characteristic not only of the Rocky Mountain areas; it was widespread throughout much of western North America. In the Southern Rockies the more violent eruptions produced great quantities of pyroclastic dust, ash, and other debris that consolidated as tuff. Lava flows were mainly andesitic (of intermediate composition), but some were basaltic or rhyolitic (table 12.2). Some of the lava solidified as obsidian, or natural glass; and where lava flows overran streams, the cold water shattered the volcanic glass so that it consolidated as breccia. Stream gravels mixed in with the lava, which then hardened and formed andesitic conglomerate. Remnants of these lavas can be seen in the Never Summer Range

along the Park's western border and in the northwest corner of the Park near where Mt. Richthofen now stands. Lava flows on Specimen Mountain, which may be an extinct volcano, can be viewed from Trail Ridge Road.

Some of the igneous material solidified below the surface in dikes, sills, and plutons that were later exposed by erosion and uplift. Among these bodies are the pegmatite dikes and pods (small lenses). *Pegmatites* are extremely coarse-grained igneous rocks, generally granitic in composition and made up mainly of large quartz and feldspar crystals. Most pegmatites are found around the edges of the main batholiths where the pegmatite magma intruded faults and folds in the country rock. Sometimes large sheets of muscovite mica and deposits of rare minerals, such as beryl, can be mined from pegmatite bodies.

Hydrothermal veins of quartz, some of them associated with pegmatites, are the source of much of the mineral wealth of the Front Range. These veins formed when hydrothermal solutions containing mineral ions escaped from the hot magma and forced their way upward through cracks in the rock. As the water cooled, the silica precipitated as quartz and the other minerals were deposited in the quartz matrix. Most of the mineral-bearing pegmatites and veins are located in the Colorado Mineral Belt, south of Rocky Mountain National Park.

Old Erosional Surfaces

The early Rocky Mountain ranges, uplifted by Laramide mountain-building, were separated by intermountain valleys that rapidly filled with sediment. Interbedded with fluvial deposits washed down from the uplands were deposits of volcanic ash and lava from the extensive eruptions. The whole region underwent a lowering of relief during early Tertiary time, and a sloping erosional surface with low hills and wide shallow valleys only a few thousand feet above sea level was produced. The relatively flat summits of Deer Mountain and Flattop Mountain are remnants of this old erosional surface.

The broad upwarping that began in the Cenozoic was apparently related to the continuing subduction of the Pacific plate. This affected a wide area that today stretches from Denver to Reno. Episodes of strong vertical uplift were particularly intense in the Front Range region. As a result, streams began intensive downcutting, stripping away sedimentary beds and incising canyons in the resistant granitic rocks.

A second erosional surface of moderate relief developed late in Tertiary time as uplift slowed. Remnants of this surface (called the Rocky Mountain pediplain or peneplain) are found on Giant Track Mountain, Old Man Mountain, Castle Mountain, Needles, and Twin Owls. The highest peaks stood as isolated knobs, or *monadnocks,* above the erosional surface. Again, alluvial deposits built fans, filled basins, and spread over the plains to the east.

The increasing height of the mountain ranges to the west cut off some of the moisture supply from the Pacific and the climate became more arid.

Inferences that can be drawn from the old erosional surfaces suggest that (1) thousands of feet of rock and sediment were removed from the high areas during and after uplift, and (2) periods of uplift can be related to the erosional history of the region.

A last stage of uplift, near the end of the Pliocene Epoch, brought the mountains of the Front Range to their present height and grandeur. Streams cut deep canyons in the resistant rocks. Along the eastern side of the Front Range, much of the weaker sedimentary rock was eroded away, accenting the barrier appearance of the range. This was differential erosion on a grand scale.

For all their toughness, however, the basement rocks that make up the "roof" of the Park are not impervious to erosive forces. The extensive joint systems, initiated or intensified by up-arching, provide conduits for weathering processes. The Needles and Twin Owls are examples of joint weathering. Exfoliation domes, such as McGregor Mountain, show the weathering of massive granite. The results of frost action at high altitudes are clearly evident along mountain trails and highways.

Glacial Features

A long worldwide cooling trend that began in Tertiary time culminated in the severe glaciation of the Pleistocene. In the Rocky Mountains the effect of climatic cooling was intensified by the increasing elevation of the region, and conditions favorable for glacier development were brought about very early in the Pleistocene. Glacial ice, instead of running water, became the dominant agent of erosion, and mountain glaciers did the major shaping of the alpine scenery that we see in Rocky Mountain National Park today (fig. 12.5). On the high peaks are prominent cirques, cols,

Figure 12.5   View of Longs Peak from the Forest Canyon Overlook, looking south. A wide glacial trough slopes from left to right across the picture, its walls grooved by avalanche chutes. The horns and cirques along the skyline were sculptured from Precambrian granite by Pleistocene glaciers. Intensive frost action, suggestive of ongoing periglacial processes, produced the jumble of loose boulders in the foreground. National Park Service photograph.

and horns. The crest of the Front Range itself is a serrated ridge. At their maximum extent, the glaciers of the Front Range coalesced, forming a small icecap with the high peaks jutting up through the perennial snowfields and glacial ice.

Four major periods of glaciation in the Park have left great quantities of *glacial drift* (glacial deposits, debris, outwash, etc.) in the basins, valleys, and foothills of the region. Segments of moraines of various sizes, ages, and types are scattered over the eastern and western sides of the Park in the mountain meadows and the deep glacial troughs. Some of the trails and roads in the Park follow moraines, such as a section of the road west of the Park entrance in Fall River valley that was constructed on an end moraine. Most of the moraines are forested. Since the lodgepole pine prefers to grow on glacial drift, a good stand of these trees usually indicates the location of a moraine.

Rock glaciers, left by retreating ice, form continuous strips of talus for thousands of yards along the walls of high valleys. Rock glaciers also formed on some of the cirque floors, enlarging themselves as they pushed out over the rim and downward.

Glacial lakes appeared on the landscape as the glaciers departed. Grand Lake, just outside the southwest corner of the Park, is dammed by an end moraine. Lakes in morainal kettle holes are Burstadt, Copeland, Bear, Dream, and Sheep Lakes. Cirque Lakes are Blue, Black, Shelf, Lawn, Crystal, Fern, Odessa, Tourmaline, and Chasm Lakes, as well as Lake Mills. Chains of rock-basin lakes (paternoster lakes) are found on the "stair treads" of several rocky gorges that rise like staircases. Examples are the Gorge, Loch Vale, Cub Lake, East Inlet (including Lake Verna), and Ouzel Creek Lake. Some lakes, such as Nanita, Nokoni, and Bench Lakes, lie in hanging valleys. Streams that occupy hanging valleys are Roaring River, Chiquita Creek, Sundance Creek, and Fern Creek.

Roches moutonnées—characteristic of glacial erosion on massive rocks—are common in the Park. A roche moutonnée is an elongate knob or hillock of bedrock that has been smoothed and scoured by moving ice on the up-glacier (stoss) side so that the rock is gently inclined and rounded. On the downglacier (lee) side, the rock is steep and hackly from glacial plucking or quarrying. Glacial Knobs is a well-known roche moutonnée area in the Park.

Many rock fragments and exposed bedrock surfaces in the Park show glacial *striations*. These are furrows or lines that were inscribed on rock surfaces as rock fragments embedded in ice were dragged along with great force by a moving glacier. On outcrops, striations are usually oriented in the direction of glacier movement.

A bedrock surface that has been planed down, striated, and polished by glacial abrasion until it is relatively smooth is called *glacial pavement*. The area around North Inlet has particularly good examples of glacial pavement.

Glacial erratics are abundant at Boulder Field and on the sides of Fern Creek canyon. An *erratic* is a relatively large rock fragment that has been transported by ice from its place of origin, sometimes a considerable distance away.

In the Park are some good examples of V-shaped stream valleys that can be compared with glacially sculptured U-shaped valleys (all of which were originally V-shaped). Fox Creek, Red Creek, and Cow Creek have characteristic V-shapes. Fall River valley, Forest Canyon, Fern Lake valley, and North St. Vrain valley are U-shaped and display other evidence of glacial erosion and deposition.

Mass Wasting Features

Periglacial conditions in the Park have brought about some modification of glacial topography due to mass wasting on both steep and gentle slopes. On Sundance Mountain and along Trail Ridge Road, *solifluction,* or flowage of soil, has caused slumping and development of hummocky lobes on the gentle upland slopes above the timberline and above the glacial valleys. Because water cannot percolate down through the frozen ground that exists at this altitude, the soil becomes saturated during the melt season and begins to slide on even a slight slope. On steeper slopes, especially at lower elevations, landsliding and avalanching may occur seasonally. The abundance of talus debris, talus cones, and rock glaciers is also an indication of active mass wasting.

Where there are loose rock fragments or soil cover on high, flat surfaces above the timberline, various forms of *patterned ground,* such as stone rings, polygons, stripes, and nets, have developed. Again this is due to periglacial conditions, intense frost action, and very slow mass movement, which in combination tend to assemble stones in symmetrical patterns or designs.

Geologic History

1. Accumulation of sediments during Precambrian time.

The oldest rocks in the Park, now metamorphics, were deposited in a primitive trough, or geosyncline, over 1,800 million years ago. Many thousands of feet of sediment formed shales, sandstones, and carbonate rocks interbedded with volcanic rocks, probably basalt (fig. 12.6).

2. Folding, intrusion of igneous rock, and metamorphism.

As the geosyncline was compressed and uplifted by tectonic activity, the sedimentary beds were intensely folded, metamorphosed, and intruded by plutonic material from a magma chamber far below the surface. (fig. 12.7). As the orogeny continued, increased heat and pressure intensified the metamorphic processes, changing the rocks further into schists and gneisses. Prominent exposures of these ancient rocks are in outcrops along Trail Ridge Road and in cliffs along valley walls in the western and central parts of the Park. Schists and gneisses can

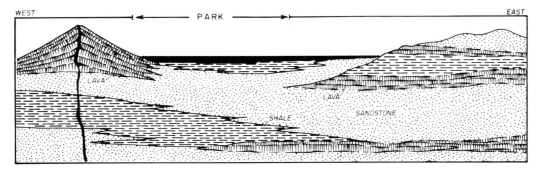

°Figure 12.6 Volcanic and clastic sediments accumulated on the land and in the ocean early in Precambrian time.

*Figures 12.6 through 12.19 are reproduced from *Raising the Roof of the Rockies* by G. M. Richmond. © 1974 by Rocky Mountain Nature Association, Inc.; used by permission.

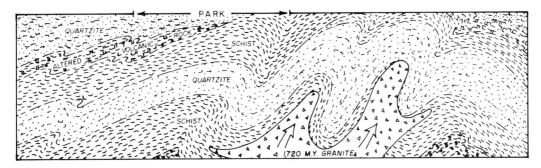

Figure 12.7 The layers of sediment were buried and lithified. Then during mountain-building episodes, over a long period of time, the Precambrian rocks were folded, metamorphosed, and intruded by rising magma.

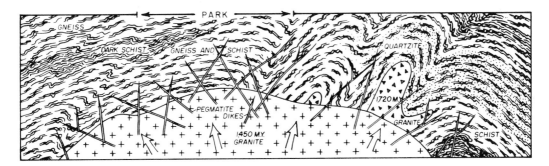

Figure 12.8 Later in Precambrian time, a very large granitic pluton was emplaced during an extensive period of mountain-building. This intrusion, in combination with intense deformation and metamorphism, produced a basement complex of crystalline Precambrian rocks.

also be found along the Continental Divide in the cirque headwalls.

3. Precambrian uplift and intrusions.

While the earlier magma was cooling, a larger intrusion pushed its way up into the schists and gneisses, eventually creating a granitic mass about 30 miles across. Intensely hot hydrothermal solutions penetrated the country rock surrounding the batholith, and pods and lenses parallel to the banding of the schists and gneisses were formed. In some places the confining pressure was so great

that pegmatites were formed during the long, slow cooling. Radioactive minerals in the intruded rocks have provided an absolute age (p. 159) figure indicating that the event took place about 1,450 million years ago (fig. 12.8).

Exposures of these granitic intrusions can be seen in the western part of the Park, in the area east of Shadow Mountain and Grand Lake, and also in the eastern part of the Park, north and south of Estes Park. Additional exposures are on Hagues Peak, near the northern boundary, and on the east face of Longs Peak, near the southeast boundary. The contact between the older schists and gneisses and the granitic intrusions shows up on canyon walls east of the Continental Divide.

4. Erosion of the Precambrian highlands.

The mountains that were formed in the Precambrian orogenies were worn down during a span of about a thousand million years until the tops of the granite batholiths were exposed in the region of the Park. The sands and other sediment that were the products of erosion were lithified as sandstones overlying the gneisses, schists, and granites. A few remnants of these ancient sandstones have been found overlying granite in the Colorado Springs area southeast of the Park, but none within the Park itself (fig. 12.9).

A basaltic dike, called "Iron Dike," suggests that a possible third intrusion occurred in late Precambrian time. The dike outcrops near Mount Chapin and Storm Pass north of Trail Ridge Road.

5. Development of the Cordilleran geosyncline west of the Park during the Paleozoic; uplifting of the Ancestral Rockies.

During early and middle Paleozoic time, shallow inland seas advanced and withdrew, at times covering most of the area that is now Rocky Mountain National Park.

For 200 million years Paleozoic gravels, sands, muds, and oozes accumulated and were lithified. None of the sedimentary rocks deposited in this period are present in the Park today.

Gradual uplift in the Pennsylvanian Period, about 300 million years ago, created the Ancestral Rockies in the region west of the Front Range. At their maximum the mountains may have been about 2,000 feet high. Along with the uplift came increasing erosion and downcutting by streams, processes that stripped away the previously deposited marine sediments, along with some of the Precambrian igneous and metamorphic rocks (fig. 12.10).

6. Mesozoic reduction of relief and transgressions of seas.

By the end of the Paleozoic, in Permian time, the Front Range region was a low, swampy area with only low hills remaining where the Ancestral Rockies had stood. Amphibians, reptiles, and dinosaurs roamed over the land. Their footprints and fossilized remains have been found in sedimentary rocks not far from the Park. In the swamps some Jurassic coal beds formed. Extensive coastal dunes migrated along the shores of the inland sea, but neither the dunes nor the sea overspread the Park area. Remnants of the dunes can be found in the sedimentary rocks east of the Park along the mountain front.

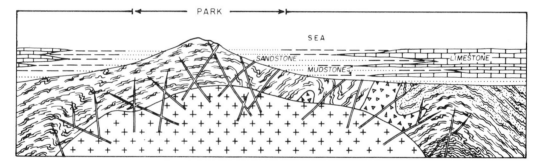

Figure 12.9   Following a long period of erosion that lowered relief in the area, a sea moved over the land, submerging the flanks and possibly the summits of the Precambrian mountains. Layers of Paleozoic sedimentary rocks, deposited in the sea, covered the Precambrian rocks.

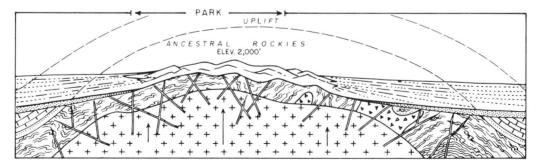

Figure 12.10   In middle to late Paleozoic time, the region was uplifted, forming mountains that are called the Ancestral Rockies. Erosion followed.

7. Maximum advance of seas in Cretaceous time, followed by uplift.

Eventually (about 100 million years ago), the seas covered the region completely (fig. 12.11). Many feet of sand and clay accumulated on the shallow sea floor. Remains of some of the shale beds that were deposited at this time are in the northwest part of the Park, east of Lead Mountain.

About 70 million years ago, the region that was to become the Front Range began to rise again, forming a low island in the sea. Uplift continued for about 16 million years until by the end of the Cretaceous, all of Colorado was above sea level. As the land rose, stream systems de-

veloped, eroding the marine rocks and redepositing the sediment in new layers over the surrounding areas.

8. Laramide phase of the Rocky Mountain orogeny; uplift and faulting; stripping of sedimentary beds.

At the end of the Cretaceous Period and the beginning of Tertiary time, a major event in the geologic history of North America took place. In the Laramide phase of the Rocky Mountain orogeny, the earth's crust was pushed up with such force that widespread faulting occurred, creating block mountains from New Mexico northward into Canada. In the Rocky Mountain National Park region, large blocks were uplifted along deep fractures in the Precambrian basement rocks (figs. 12.12, 12.13).

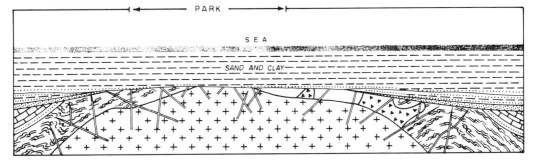

Figure 12.11 During the Cretaceous Period, accumulations of marine sediments became widespread as the sea advanced over the region for the last time.

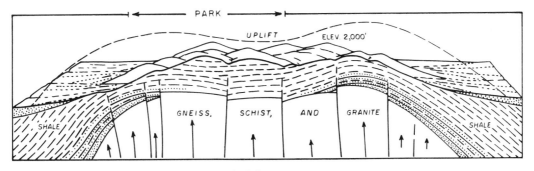

Figure 12.12 During Laramide mountain-building, differential uplifting produced fault blocks. Some thrust-faulting occurred.

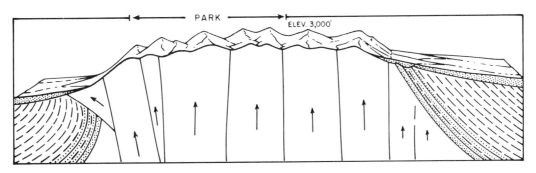

Figure 12.13 Continued uplift and erosion caused the stripping off of the sedimentary layers, exposing the crystalline basement rocks in early Tertiary time.

Periods of volcanism ensued. In the Park the lava flows at Mt. Richthofen mark an episode. Near Specimen Mountain, mudflows, lava flows, and obsidian are indications of a later and different type of volcanic activity of Oligocene age. Evidence of volcanic ash can be seen near Iceberg Lake along Trail Ridge Road (fig. 12.14).

During and between the episodes of volcanism and uplift, the processes of erosion went on, lowering relief, stripping away the sedimentary rocks, and cutting into the Precambrian schists, gneisses, and granites (fig. 12.15). In all, about 5,000 feet of sedimentary rocks were removed. The wearing down of the highlands and the filling of valleys produced the erosional surfaces described earlier.

9. Late Tertiary upwarping and erosion.

In late Tertiary time, successive uplifts that were part of a broad regional up-arching brought the mountains of the Park to their present dramatic height. The uplifts occurred primarily along the old faults, increasing elevations as much as four to six thousand feet (fig. 12.16). The Front Range appeared to rise as a series of steps, particularly on the east side of the Continental Divide. Volcanic activity, mainly the eruption of ash and pyroclastics, was intermittent (fig. 12.17). (Most of the intrusive and extrusive activity and accompanying mineralization took place south of the Park region.)

Because of the Tertiary uplifts, the rocks in the Park are almost entirely very old or very young; that is, either

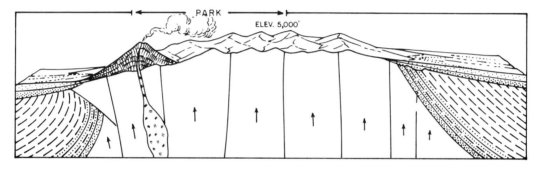

Figure 12.14   Volcanism accompanied uplift and erosion in early Tertiary time.

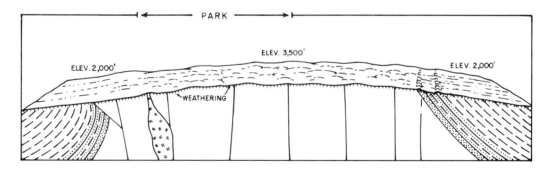

Figure 12.15   By middle Tertiary time, vast amounts of rock had been removed, summits had been lowered, and relief was subdued.

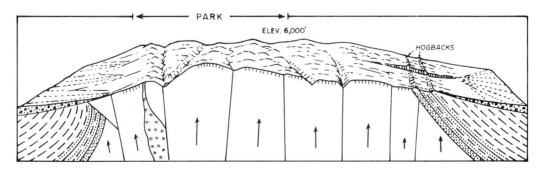

Figure 12.16   Beginning in middle Tertiary time and continuing in late Tertiary time, renewed uplift raised the mountains again.

Precambrian basement rocks or the remains of young volcanics deposited in valley-fill areas or on protected slopes that "rode up" with the uplands. Virtually all of the rocks of intermediate age, the Paleozoic and Mesozoic sedimentary beds, that once covered the area were stripped away. Erosional remnants of these rocks form the *hogbacks* of the foothill ranges that make up a narrow transitional belt east of the Park between the plains and the Front Range (fig. 12.18).

By the end of the Tertiary, parts of the land surface had been markedly displaced by vertical movement of the large mountain blocks. Basins such as Estes Park were down-dropped as much as a thousand feet while parts of the upland along the Continental Divide were raised to their present height of over 12,000 feet (fig. 12.19).

Drainage patterns were disrupted or altered with each episode of uplift. In the last stage of faulting and uplift, waterfalls formed, and streams incised their valleys, carved deep canyons, and dissected old surfaces. In this way the principal drainage patterns that exist today were established.

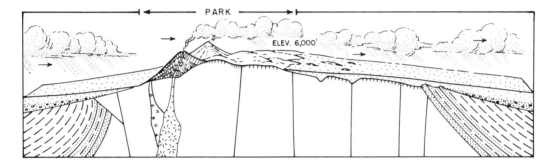

Figure 12.17 Volcanism accompanied uplift and faulting.

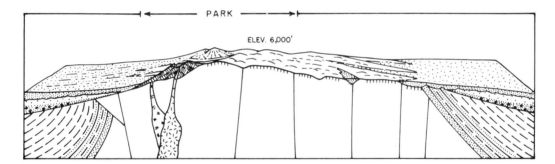

Figure 12.18 Erosional remnants of the Paleozoic and Mesozoic sedimentary rocks formed hogback ridges in the foothill ranges to the east.

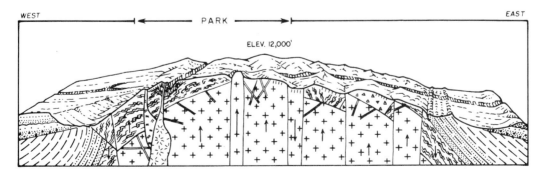

Figure 12.19 By the beginning of Quaternary time, the Front Range had attained approximately its present elevation. Vertical fault movements raised some blocks and dropped others down, causing disruptions of drainage and in some cases significant displacement of the land surface.

10. Pleistocene glaciation.

In Rocky Mountain National Park the glacial chronology is generally divided into four periods of ice advances and retreats:

(1) Early Glaciation
   Probably two major advances, beginning around one and a half million years ago

(2) Intermediate Glaciation, between 500,000 and 87,000 years ago
   Buffalo Glaciation (or Early Intermediate)
   Bull Lake Glaciation (or Late Intermediate), from about 127,000 years ago to 70 to 80 thousand years ago
      Early Bull Lake, ending about 105,000 years ago
      Late Bull Lake, from about 100,000 to around 70,000 years ago

(3) Pinedale Glaciation, maximum extent about 15,000 years ago
   Warm (hypsithermal) interval, from about 9,000 to 2,500 years ago; disappearance of glaciers about 7,500 years ago

(4) Neoglaciation (or Little Ice Age), since about 3,800 years ago

Since within the Park virtually all traces of the early glaciation were obliterated by later glaciations, the ice advances of that period are inferred from evidence remaining in other regions. The oldest glacial deposits in or adjacent to the Park are from Buffalo Glaciation and are found near the eastern boundary, some in roadcuts along Colorado Highway 7 in the southern Tahosa Valley and some near the Fall River entrance to the Park. These deposits are around 160,000 years old.

The Bull Lake Glaciations were extensive and severe, and evidence of their erosion and deposition is distributed throughout the Park and adjacent to Park boundaries. Both the early and late Bull Lake advances deepened the same cirque basins, enlarged canyons, joined together tributary glaciers, and left prominent end moraines that are usually not far apart. Since the moraines tend to be similar in size, shape, and type of material, the amount and depth of weathering is what distinguishes them. Examples of Bull Lake end moraines can be seen in Aspenglen Campground and also near the Beaver Meadows entrance. On the west side of the Park, Bull Lake glaciers that occupied the Colorado River valley left end moraines near Shadow Mountain Lake.

Pinedale Glaciation is notable for the sculpturing and quarrying that is evident in the Park's present alpine scenery. The steep cirque headwalls along the Continental Divide and the nearly vertical east face of Longs Peak are the result of Pinedale Glaciation, although the ice in the valleys did not attain the thickness of the Bull Lake glaciers.

On the east side of the Park, the Pinedale glaciers began in the perennial snowfields and cirques of the Mummy Range and then flowed down the valleys that trend eastward. Other glaciers, such as the one 13 miles in length occupying Big Thompson valley, flowed down from the eastern side of the Continental Divide. Another group of glaciers originated in the cirque basins of the Never Summer Mountains in the northwest part of the Park, flowed eastward to the Colorado River valley, and joined westward-flowing glaciers from the Continental Divide upland to form the Park's longest glacier, which extended for about 20 miles from an area west of La Poudre Pass southward to the morainal islands in Shadow Mountain Lake.

In the process of scouring and deepening earlier glacially eroded valleys, the Pinedale glaciers gouged out chains of rock-basin lakes, such as the Gorge Lakes and those in Tyndall Gorge. Large basins were dug out by the glacial snouts, such as Horseshoe Park by Falls River Glacier, Moraine Park by Thompson Glacier, and Bartholf Park by Bartholf Glacier.

End moraines and lateral moraines left by Pinedale glaciers show less evidence of weathering than the older Bull Lake morainal landforms. In some places, as at Horseshoe Park, Pinedale moraines overlie Bull Lake moraines. The Pinedale moraines tend to be prominent ridges with numerous boulders scattered over the surface, as, for example, the end moraines west of the Park entrance in Fall River valley.

11. Post-Pleistocene glaciers and erosion.

Although all the Pleistocene glaciers melted away during the post-Pleistocene warming period, new glaciers reformed in the old cirques several thousand years ago during the Little Ice Age, or Neoglacial advances. The modern glaciers have shrunk back to little more than perennial snowfields and glacierets (miniature alpine glaciers). In recent years, a few, such as Sprague, Andrews, Tyndall, and Taylor Glaciers, have increased somewhat in volume.

The moraines built by the post-Pleistocene glaciers are the easiest to identify because they are the least weathered and dissected. Near the cirques are the best places to see the Neoglacial moraines because the glaciers that built them did not extend very far down the valleys. The most accessible recent moraines are those associated with Andrews, Chiefshead, Mills, and Taylor Glaciers.

With glaciers restricted to small areas at high altitudes, the processes of stream erosion and deposition have again become dominant in the shaping and modification of the Park landscapes. However, the processes of mass wasting, particularly those associated with periglacial conditions, continue to play a significant role in landform evolution in the Park.

Geologic Map and Cross Section

Clark, T. H.; and Stearn, C. W. *The geologic evolution of North America.* New York: Ronald Press. P. 212, figs. 10–18.

American Association of Petroleum Geologists. 1967. *Geological highway map no. 2, southern Rocky Mountains.* Tulsa, Oklahoma: American Association of Petroleum Geologists.

Richmond, G. M. 1974. *Raising the roof of the Rockies.* Estes Park, Colorado: Rocky Mountain Nature Association, Inc.

Bibliography

Dickinson, W. R., and Snyder, W. S. 1978. Plate tectonics of the Laramide orogeny. In *Laramide folding associated with basement block-faulting in western United States,* ed. V. Matthews, III, Geological Society of America Memoir 151, pp. 255–56.

Kenofer, L. 1970. *Rocky Mountain trails-Rocky Mountain National Park.* Pruett Publishing Co.

Larkin, R. P.; Grogger, P. K.; and Peters, G. L. 1980. *Southern Rockies,* K/H geology field guide series. Dubuque, Iowa: Kendall/Hunt Publishing Company.

Peterson, M. S.; Rigby, J. K.; and Hintze, L. F. 1980. *Historical geology of North America.* 2nd ed. Dubuque, Iowa: Wm. C. Brown Company Publishers.

Richmond, G. M. 1974. *Raising the roof of the Rockies.* Estes Park, Colorado: Rocky Mountain Nature Association, Inc.

Yardell, M. D., ed. 1972. *National parkways-a photographic and comprehensive guide to Rocky Mountain and Mesa Verde National Parks.* Vols. III/IV. Casper, Wyoming: Worldwide Research and Publishing Co., Division of National Parks.

Address

Rocky Mountain National Park
Estes Park, Colorado 94956

13
Yosemite National Park

Location: Central California
Area: 760,917.18 acres; 1,188.93 square miles
Established: October 1, 1890

Figure 13.1 Yosemite Falls in Yosemite National Park drop 2,425 feet from crest to base. Upper Fall, shown here, drops 1,430 feet. Lower Fall and the cascades between the falls are hidden from view by the projecting rock spur. The flat floor of Yosemite Valley in the foreground is an old lakebed. U.S. Geological Survey photograph by R. E. Wallace.

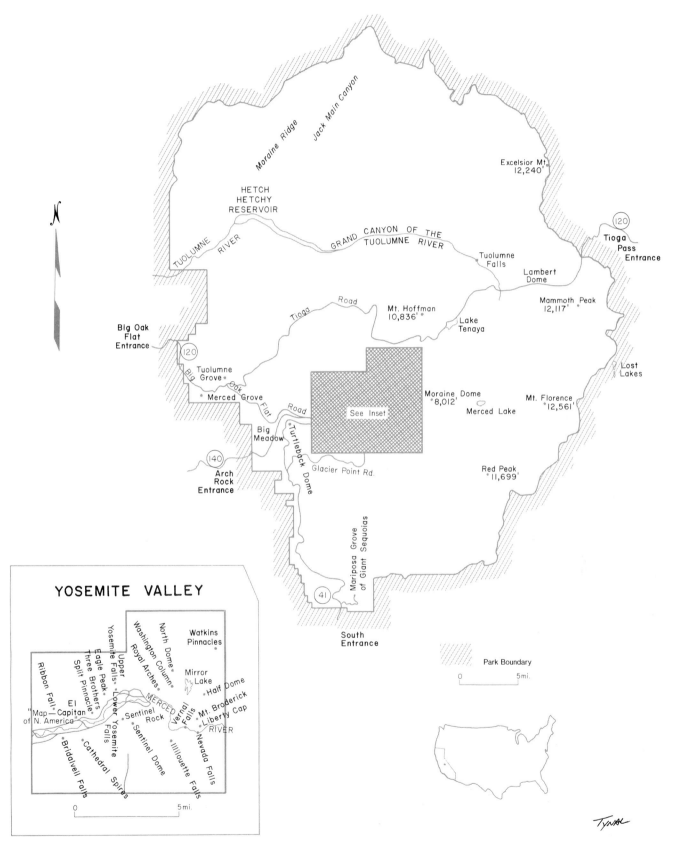

Figure 13.2 Yosemite National Park, California.

Table 13.1 Generalized Geologic Column, Yosemite National Park

Time Units			Rock Units		Geologic Events
Era	Period	Epoch	Hetch Hetchy Valley	Yosemite Valley	
Cenozoic	Quaternary	Holocene	Glacial deposits, alluvium, landslides, rockslides, talus		Erosion and deposition by glacial, fluvial, and mass wasting processes
		Pleistocene	Glacial tills, moraines, etc.; some early deposits buried or destroyed by later ice advances		Severe erosion during glacial advances (Sherwin, Tahoe, Graveyard, Tioga stages). Sculpturing of giant staircases, hanging valleys, basins, etc.
	Tertiary — Neo-gene	Pliocene Miocene	Mudflows (lahars) Andesite lava		Volcanism (northern part of Park) Series of uplifts with faulting and tilting of Sierra Nevada block.
	Tertiary — Paleo-gene		/////////////		
Mesozoic	Cretaceous		*Tuolumne Intrusive Series*		
			Aplite Yosemite Creek Granite Mt. Gibson Diorite	Johnson Granite Porphyry Cathedral Peak Granite	Major intrusions into country rock and igneous rock of the composite batholith Dikes, veins
			Half Dome Quartz Monzonite Sentinel Granodiorite		
			Minor Intrusive Series		
				Diorite of "Map of North America" Quartz-mica diorite Bridalveil Granite Leaning Tower Quartz Monzonite	Minor intrusions of composite batholith
			Western Intrusive Series		
			Ten Lakes Alaskite Rancheria Mt. Granodiorite; Double Rock and Mt. Hoffman Granodiorites	Taft Granite	Major intrusions into country rock and emplacement of composite batholith
			El Capitan Granite		
			Tamarack Creek Quartz Diorite	Granodiorite of the Gateway Arch Rock Granite Diorite of the rockslides	
	——?—— Jurassic	——?——	South Fork of Tuolumne River Quartz Diorite	Mariposa Slates	Deformation, metamorphism, beginning of intrusions Interbedded sediment and volcanics
	Triassic		Diorite		
Paleozoic			Metasediments and metavolcanics (mostly outside Park; some preserved in roof pendants)		Prolonged sedimentation followed by deformation, metamorphism, erosion

Source: modified after Smith, 1962; Peck, Wahrhaftig, and Clark, 1966; Harbaugh, 1975

Box 13.1
Time and Geology

The geologic columns in this book are a way of looking at the "record of the rocks" in each park in relation to the standard geologic time scale (see inside front cover). Using well established methods of field observation and map study and applying stratigraphic principles, geologists have worked out the rock sequences shown in the geologic columns in order to show which units are older, which younger, where gaps occur in the record, and so on. What we are mainly concerned with in the geologic columns is *relative time*; that is, the chronology or sequence indicating, for example, that lava flowed over a sandstone and that the hardened lava was later buried by volcanic ash. This statement does not tell us the number of years involved, but it does tell us the age of the lava flow *relative* to earlier and later episodes of deposition.

The concept of relative time is based on the principle of *uniformitarianism,* which is the conviction that geologic processes operating in the present also operated in the past. A familiar restatement of the principle says, "The present is the key to the past." Acceptance of the concept of uniformitarianism leads to awareness of the immensity of geologic time as we visualize changes on the earth's surface that took place over the passage of millions of years, in contrast to the chronicles of human history that are based on spans of hundreds or thousands of years. Before the modern era, people tended to think of earth time in terms of catastrophic events, such as volcanic eruptions, earthquakes, great floods, etc., because gradual geologic changes were difficult to perceive.

Absolute Age. Frequently in this book ages in millions or billions of years are attached to events, rock sequences, etc. These "dates" are based largely on a group of techniques that are used to determine *absolute age* of rocks and earth materials, expressed in units of time. Notice that absolute age is not the same as exact age. For example, we say that a certain rock may have been formed *between* 75 and 50 million years ago, or that it may be *about* 65 million years old.

Absolute age determinations are based on methods of a *radiometric dating;* that is, measuring how much time has elapsed in the process of radioactive decay since a particular rock or earth material was formed. *Radioactive decay* is the spontaneous disintegration or transformation of certain unstable atoms, called the parent material, to *daughter products* through the gain or loss of nuclear particles. The four most useful, naturally radioactive, long-lived elements that decay at a constant rate and form daughter products are uranium-235 and uranium-238, which decay eventually to lead-207 and lead-206; potassium-40, which decays to argon-40; and rubidium-87, which decays to strontium-87. By carefully determining the ratio of daughter products to parent material in a given sample, the radiometric age of the rock can be calculated. This value is expressed in years before the present, or B. P.

Igneous rocks containing mineral grains of radioactive elements usually yield the most precise radiometric values because, having crystallized from a melt, the rocks are unlikely to be contaminated by previously decayed daughter products. Weathered rocks and sedimentary rocks are not suitable for radiometric dating because so much of the original parent material has been lost or altered. In dating metamorphic rocks, several techniques can sometimes be combined to yield reasonably accurate age determinations.

A method for dating quite recent geologic events, mainly those occurring within the last 50,000 years, uses short-lived radioactive atoms of carbon-14 produced by cosmic ray activity in the earth's atmosphere. The minute amounts of carbon-14 quickly combine with stable carbon atoms in atmospheric carbon dioxide and enter the earth's carbon cycle. As living things incorporate carbon into their tissues, the ratio of carbon-14 to other carbon atoms usually remains the same as the ratio in the atmosphere. When an organism dies and new tissue is no longer added, the carbon-14 decays radioactively at a *fixed* rate, however. By comparing the ratio of carbon-14 to the other carbon atoms in the organic remains, a radiometric date can be calculated that establishes the length of time that has elapsed since the death of the organism. Recent glacial and volcanic features can sometimes be dated by this method, and the technique is also useful in archeological research.

Local History

An Indian legend explaining how two of Yosemite's domes were made tells of quarrels between a squaw named Tis-sa-ack and her husband Nangas. Their constant bickering made the Great Spirit so angry that in order to separate them forever, he turned Tis-sa-ack into Half Dome, on one side of Yosemite Valley, and Nangas into North Dome on the opposite side of the valley. Supposedly, by looking carefully at Half Dome, one can see the outline of Tis-sa-ack's face on the cliff. Another legend tells that the Great Spirit was angry with a chief who became so lovesick that he neglected his tribe. As punishment, the Great Spirit aimed thunderbolts at the chief's throne on the mountaintop, splitting it in half and forming the two domes.

Yosemite Valley was first inhabited by the Ahwahneechee Indians. Their name for Yosemite was *Ahwahnee,* which means "deep grassy valley in the heart of sky mountain." They called Bridalveil Falls *Pohono,* meaning "puffing wind." Lake Tenaya, named for an Indian chief, was called *Pywiack,* or "lake of the shining rocks." The bedrock around the lake is shiny because of smoothing and polishing by glacial ice during the Ice Age. Three Brothers, with its gabled summits, was named for the three sons of Chief Tanaya who were captured at that location.

Chief Tenaya, who led his tribe of Uzumati, a subgroup of the Ahwahneechee Indians, into the area in the middle 1800s, tried to keep white men out of Yosemite Valley. However, Major James D. Savage, who came to California in 1845, explored the region and set up a trading post on the Merced River, about 15 miles downstream from Yosemite Valley. He married into five separate Indian tribes in an effort to make alliances with the tribes so that he could trade glass beads for gold nuggets. Nevertheless, the Indians got together and raided the trading post and nearby white settlements in 1850. By the following year, Major Savage had organized miners and former soldiers into a militia unit, called the Mariposa Battalion, consisting of three companies of men. Their plan was to force the Indians out of their stronghold without killing them, and then to convince them to give themselves up to the Indian commissioners.

Tenaya actually came into the battalion camp with some of his men, but most of the tribe stayed behind. The Mariposa Battalion decided to go in after them and thus saw Yosemite Valley for the first time. That night around the campfire the troops argued about a name for this impressive area, finally deciding upon *Yosemite,* meaning "grizzly bear," the name being an anglicization of the Indian tribe's name, *Uzumati.*

Members of the battalion wrote such enthusiastic letters about the valley to their friends that word began to get around. James M. Hutchings, publisher of *California Magazine,* brought a horseback party to Yosemite in 1855 to see its wonders. Sketches and accounts of the experience by Thomas Ayres, Albert Bierstadt, and Henry Pratt were published by papers all over the world. A year later, a camp on the south fork of the Merced was built and horseback tours were conducted through Yosemite Valley.

Horace Greeley, editor of the *New York Tribune,* was an early visitor to Yosemite. Unfortunately, he came in August when the water level in the streams was low and most of the waterfalls were dried up or merely trickles, so that he was somewhat disappointed. However, the beauty of the region enthralled him. In the glowing accounts of Yosemite published in his newspaper, he suggested that the region be preserved for future generations.

In the early 1860s, Josiah Whitney, head of the newly established California State Geological Survey, led the first mapping expeditions into the Yosemite area and published a guidebook at his own expense since the state survey lacked funds.

Congress granted Yosemite Valley and Mariposa Big Tree Grove to the state of California in June 1864 in a proclamation signed by President Lincoln. Thus Yosemite became the first public park under the administration of a state. Galen Clark, who had campaigned for preservation of the area, was put in charge of the Park.

John Muir, the Scottish-born naturalist and conservationist, came to California as a young man in 1868 and made the state his home. He wrote many articles about Yosemite and campaigned tirelessly for its preservation. He persuaded Ralph Waldo Emerson to visit the park in 1871. Emerson, then in his late sixties, spent ten days with Muir, camping and traveling on horseback throughout the Yosemite region. Emerson was "aghast with admiration" at what he saw. Years later in 1903, Theodore Roosevelt spent three days with Muir in Yosemite.

Largely through Muir's efforts, Yosemite National Park was established around the original Yosemite Grant in 1890. Then in 1905 the state of California returned the grant to the federal government, increasing the size of the Park. Several additions to the Park were made in subsequent years, bringing it to its present size of 1,189 square miles.

Wagon roads were laid out in the Park in 1878; before that, travel by horseback and on foot were the only methods of getting around in the Park. In 1914, the first year in which automobiles were admitted, 127 cars passed through the area. In recent years, with over two million people visiting the Park annually, automobile travel has had to be restricted because of traffic problems and pollution.

The popularity of the Park created problems in earlier years as well. Stephen T. Mather, an industrialist from Chicago who visited Yosemite in 1914, became so angered over the way the Park was being abused that he wrote a letter of complaint to the Secretary of the Interior. The Secretary, in turn, offered him the job of running the nation's parks. Mather accepted the position in 1915; and two years later when Congress created the National Park Service, Mather became its first director. He spent the rest of his life fighting for the preservation of the national parks.

The needs of the rapidly growing city of San Francisco for water and electric power resulted in plans being drawn for the construction of a dam on the Tuolumne River in Yosemite National Park. In spite of opposition by conservation groups, Congress gave permission for the flooding of the Hetch Hetchy Valley in 1913. The dam, which was completed in 1923 and enlarged in 1938, turned the Hetch Hetchy Valley into a nine-mile-long lake which is used for generating power and for supplying water by aqueduct to San Francisco. Unfortunately the lake flooded scenery that once rivaled Yosemite Valley in grandeur and beauty. Now during much of the year, the Hetch Hetchy Valley is a mud flat dotted with rotting tree stumps because of low water level.

Geologic Features

The museum at the Yosemite National Park Visitor Center displays a large, three-dimensional geologic map that shows both topographic features and outcropping rock units. A visitor who looks at the map can see that here in this small area the great Sierra Nevada batholith produced a complex of igneous rocks of varying composition that were intruded at different times. Because the granites, granodiorites, quartz diorites, and monzonites that make up the batholith have different properties, the rocks have responded differently to geologic processes since their emplacement. Thus the character of the igneous structures has influenced, and to a large extent controlled, the development of the magnificent and varied scenery that we see in the Park today. El Capitan, for example, which rises 3,604 feet above the valley, is made up of massive granitic rock with almost no joints or fractures that is highly resistant to weathering and erosion. Close by, however, are intricately jointed igneous rocks, also granitic, that form castlelike landforms of striking appearance.

Development of the Sierra Nevada Batholith

Yosemite National Park is located in approximately the central section of the Sierra Nevada Province. The Sierra Nevada is the largest continuous mountain range in the United States. This great block of the earth's crust trends north and south for some 400 miles, with its fault system lying along the steep east-facing slopes and the block itself tilted toward the west. The range is geologically young, and earthquake activity suggests that the block is still rising.

Long before the block-faulting and uplift occurred, however, the plutons that make up the Sierra Nevada batholith began to intrude the sedimentary rocks along the western margin of the North American plate. The intrusions were part of Mesozoic mountain-building that occurred as the westward movement of the plate subjected the whole coastal area to intense structural deformation when oceanic crust was subducted and thrust under the continent.

As the layers of older rock were squeezed and contorted by tectonic forces, the plutons rose from a source region far below the surface. Portions of the country rock apparently sank and were incorporated into the melt, a process that is believed to account for some of the variations in composition that resulted. Some chunks of the roof rock dropped into the magma and hardened as metamorphic inclusions within the igneous rock, while some remained attached to the chamber roof as pendants. Other sections of the country rock remained in place but were metamorphosed by the intense pressures and heat.

Radiometric dating (box 13.1) and lithologic studies of the igneous rocks in the Yosemite region indicate that as many as a dozen plutons of varying size, age, and composition were emplaced. Ages of the plutons tend to fall into three groups with large gaps of time between groups. This suggests that plutonic activity may have occurred in pulses with quiet intervals in between. The first group of dates, ranging from 190 to 210 million years ago, indicates that a period of plutonism occurred during late Triassic time. The second group, from roughly 126 to 136 million years ago, was in late Jurassic and early Cretaceous time; while the last group of dates, about 80 to 90 million years ago, shows that episodes of plutonism occurred in late Cretaceous time.

Following emplacement, the plutons must have been well insulated and confined by the surrounding host rocks over a long period of time because the texture of most of the igneous rocks is fairly coarse. Some of the plutonic bodies are *porphyritic,* which indicates cooling at different rates or possible reheating and cooling. While the plutons were still far below the surface, characteristic joints and joint systems began to develop in the solidified rocks as they were subjected to stresses.

Meanwhile, the mountain-building processes that brought the land up higher also caused accelerated erosion of the surface rocks. The mountains were worn down, and eventually (probably by mid-Tertiary time) the granitic rocks were exposed, revealing the regional extent of the batholith structure. With Cenozoic block-faulting and intermittent uplift, the Sierra Nevada range came into being.

How Joint Systems Affect Landform Development

The surficial processes that have shaped the features of this geologically young mountain range are weathering, stream erosion, mass wasting, and glaciation. But the distinctiveness and uniqueness of the features was predetermined by the geologic history. In Yosemite National Park the types of jointing and joint systems of the various granitic rocks have been a key factor in the development of the landforms. Examples illustrating this are as follows:

a. *Massive, unjointed rock*—El Capitan, Cathedral Rocks, Mount Broderick, Liberty Cap.
b. Rocks with *vertical joints*—Glacier Point, Upper Yosemite Falls, Ribbon Falls, Washington Column, Sentinel Rock, part of Half Dome.
c. *Joint sets*—Cathedral Spires, Three Brothers.
d. *Sheet jointing*—Sentinel Dome, Turtle Back Dome, North Dome, Royal Arches, and part of Half Dome.

Some of the Yosemite granites have intersecting or complex jointing with major trends in several different directions. Where joints are fairly closely spaced, pillars, columns, and pinnacles have formed—as, for example, Watkins Pinnacle, Split Pinnacle, Washington Column, and Cathedral Spires. A larger feature with joints more widely separated is Three Brothers, with its asymmetrical, "gabled-roof" summits. At a number of locations, rock with widely spaced vertical joints is beside rock with close, intersecting joints, an example being Cathedral Rocks (massive features) next to Cathedral Spires (columns and pinnacles).

Sheet jointing, or *exfoliation,* is a type of physical (mechanical) weathering that is related to *pressure release* or unloading. As the great weight of rock overlying a batholith is removed by erosion, the outer part of the batholith is able to expand slightly because of the reduced confining pressure. Cracks, or sheet joints, develop parallel to the outer surface as a result of expansion. With continued weathering, the thin layers of rock begin to spall off or break loose in concentric slabs from the granite mass, a process usually aided by gravity (mass wasting). In massive, unjointed granites such as those in Yosemite, *exfoliation domes* have developed. Exfoliation of concentric shells of rock can be readily observed at Sentinel Dome and Turtleback Dome.

Half Dome developed mainly due to vertical jointing and exfoliation, with some help from glacial activity. This famous landform was never a whole dome that was cut in two by a tremendous glacier, as it may appear; in fact, the "other half" was never there. The face of Half Dome began to take shape as thin slabs of rock between parallel vertical joints were loosened by weathering and mass wasting and fell away, layer by layer—a process that is continuing at the present time. In the Pleistocene, the Tenaya Glacier moved down the valley, reaching within 500 feet of the top of Half Dome but never overriding it. The glacier cleared away the exfoliation debris and talus; and, because it was flowing in a direction parallel to the trend of the vertical joints, it removed more sheets of rock by plucking, thus leaving the precipitous face that we see today. The rounded back of Half Dome developed more gradually as curved shells of rock were exfoliated (fig. 13.3).

In some of the massive granites, exfoliation produced arches on the steep side of the domes—Royal Arches near the head of Yosemite Valley being the classic example. These are not "see-through" arches but are, rather, a series of recessed arches set back into the main arch that curves out over the dome's base, a thousand feet below the top of the overhang. Here the Tenaya Glacier probably removed the lower part of the concentric shells of rock, leaving the upper parts unsupported.

Effects of Glacial Erosion

Domes such as Liberty Cap, Lembert Dome, and Mount Broderick, which were overridden by glaciers, took on the characteristic profile of giant roches moutonnées; that is, a smooth, rounded, and polished stoss side and a steep, plucked lee side. One must conclude that on this highly resistant rock it was easier for the down-flowing ice to quarry out and carry away great chunks of bedrock loosened at vertical joints than to wear down massive, unjointed rock in its path.

Abrasion was severe, however, and on many of the rocks in Yosemite the glacial polish is remarkable (e.g., Lake Tenaya). The comparative freshness and sheen of the polish (i.e., the degree to which weathering has affected it) can sometimes be helpful in determining the extent of the various stages of glaciation. For instance, no polish remains on glacial pavement scoured by the earliest glaciers.

Glacial erratics, some quite sizable, are a common sight in the Park. Glacier Monument, which is 1,500 feet high, has erratics on top that were brought by the ice from a mountain 12 miles to the east! Erratics, along with patches of till, are found mostly on the uplands above the valleys. Morainal remnants are numerous in the Park. The ones that are deeply weathered and dissected are associated with earlier glacial stages. Nunataks, such as Half Dome and El Capitan, provide clues as to the size and extent of glaciers. (*Nunataks* are mountain peaks that projected above the glaciers and were not overridden by ice.)

In the high basins where the glaciers originated, cirque lakes formed following the retreat of the ice. Some still exist—Lost Lake, for example—but others have silted in, becoming subalpine marshes and meadows, such as Tuolumne Meadow (elevation, 8,600 feet).

Figure 13.3 Half Dome, the most famous landform of Yosemite National Park, was formed by exfoliation of the massive Half Dome Quartz Monzonite, by mass wasting along vertical joint planes, and by glacial plucking. National Park Service photograph.

Glacial Lake Yosemite, which was probably impounded by a moraine near El Capitan, was about 5.5 miles long when the ice was melting back. The lake rapidly filled with sediment and disappeared, leaving hundreds of feet of glacial drift and lake deposits over the bedrock valley floor. The floors of Little Yosemite and Tenaya Canyons are covered by similar deposits of drift and outwash. Most of the moraines that were left in Yosemite Valley by retreating glaciers were washed out by swiftly flowing meltwater or were covered over by lake deposits.

Waterfalls

The jointing of the bedrock, the fault-blocking and uplift, and Pleistocene valley glaciers all influenced the development and location of Yosemite's spectacular waterfalls. Most of the highest waterfalls in the country are in Yosemite National Park. Even the smallest of the falls listed below is nearly as high as the American Niagara Falls (186 feet).

> Yosemite Falls—Upper Fall, 1,430 feet (436 m); Lower Fall, 320 feet (98 m); cascades between falls, 675 feet (206 m); total drop between canyon rim and base of falls, 2,425 feet (740 m)
> Ribbon Fall—1,612 feet (491 m)
> Bridalveil Fall—620 feet (189 m)
> Nevada Fall—594 feet (181 m)
> Illilouette Fall—370 feet (113 m)
> Vernal Fall—317 feet (97 m)

Yosemite, Ribbon, Bridalveil, and Illilouette Falls all issue from hanging valleys that existed before glaciation and were heightened by glacial action. The regional uplift that raised the Sierra Nevada Range to its present height disrupted the drainage systems of the whole area and caused the main streams, such as the Merced River, to flow generally parallel to each other in a southwesterly direction. In the later stages the uplift was so rapid that the trunk streams eroded their valleys at a faster rate than the tributary streams. By the time the Ice Age began, the tributary streams of Yosemite were already entering the main valley in steep cascades many feet above the floor of the main valley. When glaciers formed and filled the valleys, the erosive action of the ice straightened and gouged out the main valley, making it even deeper and accentuating the "hanging" relationship of the tributary valleys. Some of the cascading falls then became free-leaping falls.

Nevada Fall and Vernal Fall have a somewhat different geomorphic history. They descend over giant glacial steps created as ice moving downvalley plucked out sections of closely jointed bedrock between strips of massive bedrock, thus forming a giant glacial staircase.

Geologic History

1. Origin of Paleozoic metamorphic rocks.

The oldest rocks in the Park occur as scattered remnants in the form of roof pendants (masses suspended from the roof of the batholith) and as separators between plutons. Just beyond the Park's western boundary, the batholith is flanked by a belt of metamorphosed sedimentary and volcanic rocks of the Calaveras assemblage (Paleozoic in age). Included are greenstones, siltstones, cherts, and quartzites. Some have been complexly folded and contorted.

On the Park's eastern boundary, which approximately follows the crest of the Sierra Nevada range, is the contact between the granitic rocks and an eastern metamorphic belt. The metamorphic rocks here are similar to those of the western metamorphic belt, but some are older. The eastern metamorphics occur as roof pendants in this area along the crest.

2. Metamorphic rocks of Jurassic age.

The Mariposa Slates, which crop out to the west of the Calaveras rocks, are part of the western metamorphic belt, which lies outside the Park. Jurassic rocks, mainly metavolcanics, are also found in the eastern metamorphic belt, close to the Park's eastern boundary. The Jurassic rocks are generally less severely metamorphosed than the Paleozoic rocks. If we assume that the Paleozoic and early Mesozoic rocks once covered the batholith, long periods of erosion must have been required to strip them away.

3. Mesozoic intrusions.

The overall description of the Sierra Nevada batholith as made up of diverse igneous bodies that were emplaced during pulses of plutonic activity probably does not do justice to the exceeding complexity of the structures, particularly in the Yosemite region. Even with years of careful study, both in field and laboratory, questions remain about the various plutons, dikes, etc., particularly with regard to their origin, their relationships to the country rocks, the processes of emplacement, etc. However, the geometric and age relationships have been fairly well worked out, although some of these are by no means definitive.

In the Yosemite Valley, in early Cretaceous time, the Jurassic and older sedimentary rocks were intruded by a sequence of magmas, making up what is called the Western Intrusive Series, the members of which have been generally named for prominent exposures. The igneous bodies are, from older to younger:*

> Diorite of the Rockslides—dark gray, generally coarse rock
> Granite of Arch Rock—medium light gray, medium-grained
> Granodiorite of the Gateway—dark gray, medium-grained
> El Capitan Granite—light gray, medium- to coarse-grained
> Taft Granite—very light gray, fine-grained

The most plentiful, or widespread, of this group is the El Capitan Granite, which crops out over a large area.

The sequence of the next group, the Minor Intrusives, is not so well worked out. However, the Leaning Tower Quartz Monzonite, a medium-gray, speckled, medium-grained rock, underlies the Bridalveil Granite at Bridalveil Fall. Abundant biotite in the Bridalveil Granite gives this fine-grained rock a salt-and-pepper appearance. Next in age is the medium-dark-gray Quartz-mica Diorite followed by the Diorite of the "Map of North America," a very dark-gray, fine-grained rock. The Map of North America appears as a dark mass and actually looks like a map on the light-gray cliff face of El Capitan. Some thin, light-colored pegmatite dikes, younger than the diorite cut across the maplike intrusion (fig. 13.4).

The youngest of the Cretaceous intrusions are the rocks of the Tuolumne Intrusive Series. These igneous bodies are:

> Sentinel Granodiorite—medium-dark-gray, variable in texture
> Half Dome Quartz Monzonite—light- to medium-gray rock that includes both porphyritic and nonporphyritic facies
> Cathedral Peak Granite—light-gray; contains large phenocrysts of potassium feldspar in medium-grained groundmass
> Johnson Granite Porphyry—light-gray, fine-grained quartz monzonite porphyry

*as described in Harbaugh (1975).

Figure 13.4   The face of El Capitan. Note that the massive El Capitan Granite has very few fractures or joints. The large, irregular, dark mass (slightly left center) is called the "Map of North America" because of its shape. This mass is a diorite dike, 2,000 feet in height, that intruded the El Capitan Granite. From *Field Guide to Northern California* by John W. Harbaugh. © 1975 Kendall/Hunt Publishing Company, Dubuque, Iowa.

The first two members of the Tuolumne Intrusive Series crop out in Yosemite Valley; but the latter two (Cathedral Peak Granite and Johnson Granite Porphyry) crop out east of the valley toward the Tuolumne Meadows. The Sentinel Granordiorite is an elongate igneous body with numerous exposures both north and south of Yosemite Valley. Half Dome Quartz Monzonite is a large mass that extends eastward.

To the north, in the Hetch Hetchy Valley of the Tuolumne River, extensions of several of the larger Yosemite Valley plutons (El Capitan Granite, Sentinel Granodiorite, Half Dome Quartz Monzonite) are exposed. The smaller igneous bodies of Hetch Hetchy are similar to their counterparts in the Yosemite Valley but show differences in age and composition. For the names and sequence of the Hetch Hetchy intrusions, see table 13.1.

4. Late Cenozoic extrusive activity.

Most of the Late Cenozoic volcanics are outside Park boundaries, mainly to the east in the Mono Craters vicinity. However, some volcanic activity occurred in the Park north of the Grand Canyon of the Tuolumne River. Lava flows of intermediate (andesite) composition as well as mudflows of volcanic ash (lahars) have been found, some of them overlain by scattered patches of Pleistocene glacial drift.

5. Uplift of the mountain range.

The present elevation of the crest of the Sierra Nevada—over 11,000 feet in most places—indicates the magnitude of the upfaulting of the Sierra block. Since the batholith rocks now exposed were formed at a considerable depth in the earth's crust and the eastern and western metamorphic belts are all that remain of the country rock that once covered the batholith, the amount of uplift must have been well over 11,000 feet. The uplift and westward tilting of the block began early in the Cenozoic and has continued spasmodically to the present time. At first the western slope was gentle and the eastern slope not so steep as today. But the tilting increased with successive episodes of uplift so that by the Pleistocene, the eastern face of the range was very steep and the western slope had become steeper than it was originally.

6. Development of the Merced River valley.

As the regional slope began to dip to the west due to uplift and tilting, the Merced River (and the other main streams) began to flow toward the west and south. During this time, called the broad-valley stage, the Merced valley lay between rolling hills several hundred feet high. With the gradient fairly low, the Merced meandered back and forth, widening its flood plain. As uplift continued, the gradient increased, causing the river to incise its channel in the broad valley floor.

By late in the Miocene, the westward tilt of the Sierra block had begun to have a major effect on the landscape. Side slopes became steeper as the incised channel of the Merced became deeper. The old valley floor was now an upland with gentle slopes on which tributaries flowed. Because of their lower gradient, the tributaries on the upland were not able to cut down so rapidly as the main stream. In this way, hanging valleys began to develop, and the water in the tributaries flowed down steep cascades in order to join the main valley. This was the mountain-valley stage.

As downcutting in the main stream slowed, some of the tributaries began to catch up, so that their channels became closer to the elevation of the main valley floor. However, the tributaries (such as Bridalveil Creek) that flowed on resistant, unjointed granite could cut down only at a slow rate. As a result, when uplift began again, later in the mountain-valley stage, the Merced River's downcutting was so rapid that waterfalls began to develop in the tributaries located on resistant rock.

Major faults forming on the east side of the range permitted the block to rise even higher, and Owens Valley was formed on the down-dropped block. The Merced River, by now in the canyon stage, eroded its valley to a depth of 2,000 to 3,000 feet. The narrow, inner, V-shaped gorge was perhaps as much as 1,500 feet deeper than the channel was during the mountain-valley stage. The relief of the hanging valleys was accentuated by the downcutting of the main stream. By the onset of the Ice Age, the highest part of the Sierra had reached an elevation of at least 14,000 feet.

7. Pleistocene glaciation.

For hundreds of millions of years geologic processes—igneous intrusions, uplift, stream erosion, and mass wasting—had all contributed to create what must have been a beautiful mountain landscape of considerable relief. At the beginning of the Pleistocene, bedrock cliffs were exposed in the stream valleys and gorges, but many of the mountain slopes were probably mantled by forests growing in largely residual soils. In the Yosemite Valley spurs narrowed the canyon and steep ridges formed divides between tributaries. The powerful force of great masses of moving ice was soon to change all that. In fact, none of the surface features of the region could escape alteration by glacial action in one way or another.

The glacial chronology of Yosemite National Park has four recognized stages, the last one being the least extensive. From oldest to youngest, the stages are designated: Sherwin, Tahoe, Graveyard, and Tioga. Ice during the Tioga stage came down the Merced River only as far as Nevada Fall. Tioga moraines are fresh and undissected.

During the Graveyard stage, the ice advanced to the foot of Bridalveil Fall; but Tahoe glaciers extended to a few miles west of El Portal, near the Park entrance. Most of the evidence for Sherwin glaciation was destroyed by subsequent advances.

The glaciers that advanced, retreated, and readvanced during the earlier stages lasted the longest and were the most erosive. The great Tuolumne Glacier, for example, was fed by the snowfields of such mountains as Dana, Gibbs, Lyell, Maclure, etc. Miles wide, it spread over the uplands and then, seeking lower levels, flowed around Mount Hoffman and divided into two trunk glaciers. Snows were so abundant that glacial ice accumulated in the high areas faster than the trunk glaciers could carry it away. One filled Big Tuolumne Canyon and Hetch Hetchy Valley. The other crossed a 500-foot divide that separated the Tuolumne and Merced basins and then flowed through Tenaya Canyon into the Yosemite Valley. Branch glaciers filled and coalesced until finally only the highest peaks and domes, such as El Capitan, Eagle Peak, Sentinel Dome, and Half Dome, stuck up through the ice. In the narrow canyons and gorges, spurs and projections were truncated by the moving ice, which also gouged out the bedrock of the valley floors.

The 2,000-foot-thick glaciers of Tenaya Canyon and Little Yosemite joined together and filled the main Yosemite Valley. Tenaya Glacier moved parallel to one set of bedrock joints, carving and deepening a canyon with a gently sloping floor. However, the ice flowing down Little Yosemite, which is underlain by massive granite with widely spaced joints, created a giant staircase by plucking away rock to form "risers" at the joints and smoothing "steps" between joints. The water of Vernal and Nevada Falls now descends over this staircase. The streams in the hanging valleys (carved in the unjointed rock) pour out over the canyon edge and fall precipitously to the floor of the valley several hundred feet below.

8. Deglaciation and the present landscape (fig. 13.5).

Glaciers are great excavators; but when the active digging stops and the melting processes take over, the basins and troughs that have been scooped out make convenient receptacles for water, glacial debris, and sediment. The bedrock floor of Yosemite Valley, for example, is covered in some places by more than 2,000 feet of alluvium and outwash. Streams are still filling in basins and lakes, as well as eroding their valleys. Exfoliation of domes and arches and active mass wasting continue. Landslides dammed Tenaya Creek near the mouth of the canyon, forming Mirror Lake. Near the inlet a delta is building up in the shallow water. Rockslides (possibly triggered by earthquakes) have formed long talus slopes in the valleys.

About 25 small glaciers exist in the Park, the two largest being Lyell and Maclure Glaciers. These produce glacial features today on a small scale.

Figure 13.5   View of Yosemite Valley looking upstream from Wawana Tunnel parking lot. Sediment deposited in glacial Lake Yosemite formed the flat valley floor during the time that the Merced River drainage was impounded by a series of moraines left by retreating glaciers. Bridalveil Fall (right) drops 620 feet to the valley floor from the rim of a hanging valley formed by a tributary glacier. National Park Service photograph.

Probably between 3,000 and 4,000 years ago, the giant sequoia trees reestablished themselves in scattered groves on the western slopes of the Sierras. Once widespread in temperate North America, the sequoias and redwoods were nearly exterminated by successive glaciations. Since in their native stands, giant sequoias grow only on sandy loams (or loamy sands), the soil derived from granitic rocks is favorable to their establishment. The sequoia groves in the Park are Mariposa (200 trees), Tuolumne (25 trees), and Merced (20 trees).

Geologic Maps and Cross Sections

California Division of Mines and Geology. 1967. *Geologic atlas of California, Mariposa sheet.*
Matthes, F. E. 1930. *Geologic history of the Yosemite Valley.* U.S. Geological Survey Professional Paper no. 160. Plate 29.

Bibliography

Bowen, Ezra. 1976. *The High Sierra,* American Wilderness Series. Rev. ed. New York: Time-Life Books.
Ditton, R. P.; and McHenry, D. E. 1973. *Yosemite road guide, Yosemite National Park, California.* Yosemite Natural History Association.
Geologic guide to the Merced Canyon and Yosemite Valley, California. 1962. California Division of Mines and Geology Bulletin 182.
Harbaugh, J. W. 1975. *Field guide to northern California,* K/H geology field guide series. Dubuque, Iowa: Kendall/Hunt Publishing Company.
Hill, Mary. 1975. *Geology of the Sierra Nevada.* Berkeley: University of California Press.
Jones, W. R. 1971. *Yosemite, the story behind the scenery.* Las Vegas, Nevada: K. C. Publications.

Keeney, N. T. 1974. The other Yosemite. *National geographic* 145:762–780.

Matthes, F. E. 1930. *Geologic history of Yosemite Valley.* U.S. Geological Survey Professional Paper 160.

———. 1950. *The incomparable valley,* ed. F. Fryxell. Berkeley: University of California Press.

———. 1958. *The story of the Yosemite Valley.* U.S. Geological Survey. Text on back of topographic map.

Muir, John. 1961 reprint (originally published in 1894). *The mountains of California.* Natural History Books.

Peck, D. L.; C. Wahrhaftig; and L. D. Clark. 1966. Field trip guide to Yosemite Valley and Sierra Nevada batholith. In *Geology of northern California,* ed. E. H. Bailey, pp. 487–502. California Division of Mines and Geology Bulletin 190

Russell, C. P. 1968. *100 years in Yosemite.* Yosemite Natural History Association.

Smith, A. R. 1962. *Petrography of six intrusive units in the Yosemite Valley area, California.* California Division of Mines and Geology Special Report 91.

Address

Yosemite National Park
Box 577
Yosemite National Park, California 95389

14

North Cascades National Park

Location: North Central Washington
Size: 504,780.34 acres; 788.72 square miles
Established: October 2, 1968

Figure 14.1 The Picket Range, North Cascades National Park. Icefalls descending from the horns, cols, and arêtes along the crest are eroding headward into the rock, enlarging the cirque. The face of the nearly vertical cliff (right center) clearly shows the strong foliation characteristic of metamorphic rocks. National Park Service photograph.

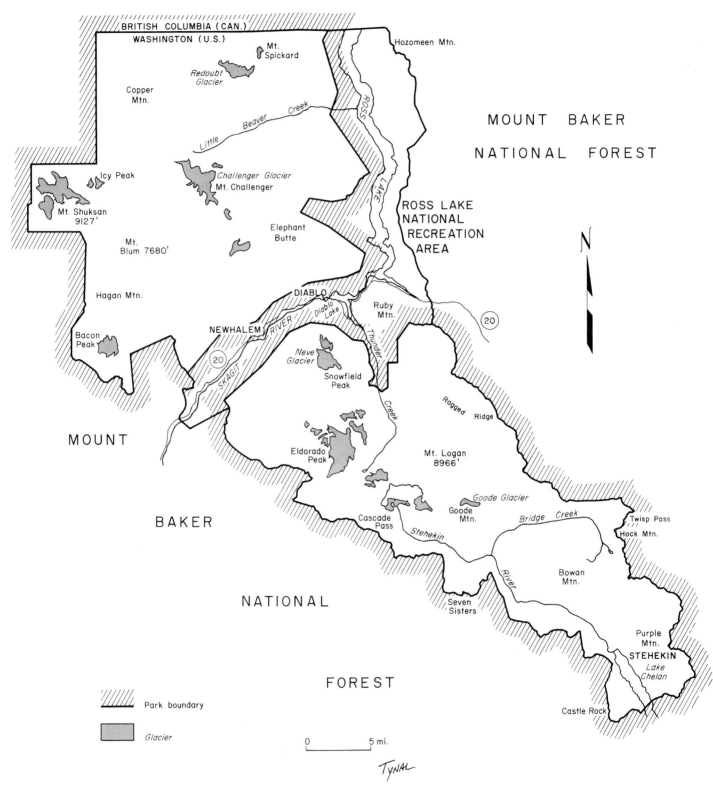

Figure 14.2   North Cascades National Park, Washington.

Table 14.1. Simplified Geologic Column, North Cascades National Park

Time Units		Rock Units (W＝West side; E＝East side)	Geologic Events
Era	Period		
Cenozoic	Quaternary	Glacial and alluvial deposits Cascade extrusives (W): (composite cones, Mount Baker and Glacier Peak)	Fraser Glaciation, latest of successive major episodes; severe glacial erosion, some deposition. Volcanism, construction of composite cones.
Cenozoic	Tertiary — Neogene	Tertiary Volcanics (W and E) Chuckanut Formation (W)	Uparching along north-south axis; block faulting; erosion, sedimentation, volcanism.
Cenozoic	Tertiary — Paleogene	Various late Cretaceous and Tertiary plutons and volcanics, including early Tertiary Chilliwack Composite Batholith (early, middle, and late phase) and the Eastern Intrusive Belt	Multiple intrusions (batholith, plutons, stocks); dikes, veins, mineralization; hydrothermal and metamorphic activity.
Mesozoic	Cretaceous	Skagit Metamorphic Suite Skagit Gneiss / Cascade River Schist Marblemount Metaquartz Diorite Shuksan Metamorphic Suite (W) Jack Mountain Phyllite (E) Shuksan Greenschist Darrington Phyllite	Major orogeny with intense compression, metamorphism, migmatism, and thrust faulting.
Mesozoic	Jurassic	Elijah Ridge Schists (E) (derived from North Creek Volcanics)	
Mesozoic	Triassic	Hozomeen Group (E)	
Paleozoic	Upper	Chilliwack Group (W) (not exposed in Park)	Interbedded sediments and volcanics deposited in eugeosynclinal basin.
Paleozoic	Middle	///////Profound unconformity///////	Long erosion
Paleozoic	Lower	Yellow Aster Complex (pre-Devonian crystalline basement)	Long sedimentation, volcanism, intrusions, metamorphism

Source: modified after Misch 1966, 1977; Armstrong 1980

Local History and Geographic Setting

The Indians of the Pacific Northwest had a legend accounting for the presence of the Cascade Range that is in some respects in accord with the scientific explanation. They believed that when the earth was young, the Cascades region was flat. Rain never fell, but forests and plants drew up the moisture they needed from ground water that kept coming to the surface. After a while, water stopped coming up in eastern Washington. In desperation, the tribe sent a delegation west to Ocean to plead with him to send rain. Ocean agreed to help the Indians and sent his children, Clouds and Rain, eastward with a good supply of rain. All went well for a while, but the Indians were greedy and would not let Clouds and Rain go back home. The members of the tribe kept digging large pits for Clouds and Rain to fill up with more and more water.

Ocean promised the Indians that if they would let Clouds and Rain come home, he would send water regularly for their crops, but still the Indians refused to let Ocean's children go. Ocean then appealed to the Great

Spirit to punish the Indians. So the Great Spirit reached down from heaven, scooped up a huge mass of dirt and rock and molded it into the Cascades. Ocean quickly flowed into the ditch where the earth had been, and this made Puget Sound. East of the Cascades the land dried up and has remained fairly dry to this day. The only water left to the Indians was what was in the pits, and the largest of these is Lake Chelan.

Presumably the Indians, who had great skill in naturalistic observations, were struck by the contrast between the always moist and often rainy character of the land on the western side of the range and the dryness of the land on the east side. They appear to have understood the principle of the *rain-shadow effect* of a high mountain range that lies in the path of prevailing winds that are moisture-laden from having passed over water.

Another tribal legend accounts for Lake Chelan in a somewhat different way. Again the myth begins with a flat grassy prairie on which game was abundant. Then a terrible monster came, and killed and ate so many animals that the Indians were starving. They beseeched the Great Spirit for help and he answered their prayers by killing the monster, who shortly came back to life, not once but three times. After killing the monster the third time, the Great Spirit was so angry that he pinned the terrible creature to the ground with a great stone knife. The earth quaked and a huge cloud of dust turned day into night. When the sky finally cleared, a high mountain range stood on the plain, and great canyons were left where rocks and dirt had been dug out to make the mountains. Into the longest and deepest canyon the Great Spirit threw the monster, then he filled the trough with water. This was Lake Chelan. The monster stayed dead this time—all but its tail, which kept thrashing around, creating great waves. This made it unsafe for the Indians to paddle their canoes on the lake. Even today small boats may capsize or be dashed on the rocky canyon walls when strong winds blow in certain directions, creating a wind tunnel effect. Lake Chelan is narrow, but it is 55 miles long (fig. 14.2).

Hunters and trappers were the first white men to explore the North Cascades, and they were followed by miners. Before the 1860s, gold, copper, molybednum, and zinc were mined here on a small scale. Today mining is not an important enterprise, although some commercial mining has been carried on in the North Cascades region in recent years. Logging is the main industry in the areas around the Park at present.

The early prospectors who looked for "strikes" in the North Cascades, especially near the Canadian border, must have endured great hardships. The same could be said for the geologic field parties who did the original reconnaissance work in these mountains. Travel was by foot, and pack mules carried equipment and supplies. Even today, since the only road that crosses the North Cascades

is the one going through the Ross Lake National Recreation Area between North Unit and South Unit, most geologists have to backpack their own supplies when working in remote areas. Mining companies and the U.S. Geological Survey have sometimes used helicopters for supplying field parties. Under the best of conditions, field work here is difficult. In the high areas some snow normally falls even during summer months. Rain and mist reduce visibility a great deal of the time. Below the snowline, the forest cover is heavy because of ample moisture. For those hardy ones who appreciate wilderness beauty, hiking or backpacking in the Park is truly rewarding.

North Cascades National Park was created to preserve the remaining wilderness areas and protect the watershed. Following the recommendations of a study team organized in 1963, the Park was established five years later in 1968. Four units are under Park jurisdiction: (1) North Unit, which includes the former Picket Range Wilderness area; (2) Ross Lake National Recreation Area, a reservoir comprising the Skagit River drainage and separating North and South Units; (3) South Unit, which includes the former Eldorado Peaks Wilderness area; and (4) Lake Chelan National Recreation area. The damming of the Skagit River to form Ross Lake provides hydroelectric power for Seattle and other populated areas of the region.

Geologic Features

Water, ice, and jagged peaks are the scenic elements for which the North Cascades are called "the American Alps." Adding interest to the skyline—although they are not inside Park boundaries—are two volcanoes of the Cascade volcanic chain; namely, Mount Baker and Glacier Peak.* Both are several thousand feet higher than the ranges of the North Cascades, which average around 5,000 feet in elevation south of the Park and rise to more than 8,000 feet in the Park.

Over 300 active glaciers within the Park—and as many more in the rest of the range—make up this country's largest concentration of glaciers outside Alaska. Also in this wild and rugged region are many spectacular arêtes, horns, cirques, and deep glacial troughs that were sculptured by earlier, more severe and extensive glaciers. Why, one asks, do these mountains of only moderate height bear so much ice? The answer is water—water as rain, fog, mist, snow, and more snow, all adding up to the average annual precipitation on the western side of the range of 110 inches. A winter's accumulation of 46 feet of snow is not unusual, and four or five feet of snow may pile up in a single storm. In the North Cascades the snowline on the western slopes is between 6,500 and 7,500 feet in elevation. (In the Swiss Alps, by way of contrast, the glacial

---

*A second Glacier Peak, not a volcano, is in the North Unit of the Park, very close to the Canadian border.

features are similar, but the snowline is between 8,000 and 10,000 feet in elevation.)

In the *rain-shadow zone,* east of the range, annual precipitation rates drop sharply, to only 35 inches, for example, around Lake Chelan, and a mere 12 inches in the Pasayten Wilderness area. What happens is that the prevailing winds, carrying moisture from the Pacific Ocean, are forced upward in order to pass over the North Cascades. Expanding and cooling, the wet air loses moisture as it rises. Descending air on the eastern side of the Cascades compresses and warms, holding more moisture and causing high evaporation. Thus the rain shadow zone on the downwind side of the range receives little rainfall.

Development of the North Cascades

The North Cascades Range is not only a region of great scenic beauty; it is also one that is geologically and geographically complicated and challenging. The present mountains are carved into a part of the earth's crust that was uplifted along a north-south axis. The uplifting, late in Cenozoic time, was accompanied by volcanism and plutonism. Although the mountains we see—that have been carved into the uparched bedrock—are spectacular, the uplift represents only the latest (and a relatively mild) event in a region that has been in the midst of intense geologic activity since the Precambrian. The record of the various orogenies, involving severe compression or crustal shortening, is exposed in rocks throughout the North Cascades. The latest uplift may be largely a matter of isostatic adjustments.

In the North Cascades we have in close proximity very old rocks and some of the youngest intrusive and extrusive rocks in North America. Mount Baker and Glacier Peak, for example, the two volcanoes just outside Park boundaries and only about 60 miles apart, were formed from Cenozoic lavas that came from great depths and worked up through many layers of much older crustal rocks.

Glacier Peak, a relatively inaccessible cone in the middle of the range, is built on top of a high ridge of pre-Cretaceous metamorphic rocks (slate, phyllite, and greenschist) between two plutons of quite different age. In other words, the andesite and later dacite lavas that make up the volcano have no apparent relation to the older rocks that compose the mountains of the range. The *dacitic* (i.e., containing more sodic plagiclase and quartz than andesite) lavas produced a great deal of ash and pumice late in the Pleistocene and in the Holocene because this type of lava tends to erupt more explosively than andesite. The most recent ash eruption was only about 200 years ago (Beget 1982).

Mount Baker, in the western foothills near the western edge of the range, has a much larger cone (elevation, 10,750 feet) that was built on top of older sedimentary and volcanic rocks (fig. 14.3). In shape and construction

Figure 14.3 Mount Baker, an active volcano, rises above the Chilliwack Range in the far distance. The somewhat lower peak in front of Mount Baker and slightly to the left is Mount Shuksan, which is located on an uplifted fault block (see fig. 14.4). The western boundary of North Cascades National Park passes between these two mountains. In the foreground, Silver Lake occupies a glacially eroded rock basin. National Park Service photograph by M. Woodbridge Williams.

Box 14.1
What Makes the Cascade Volcanoes Hazardous?

In March 1975, Mount Baker began belching out clouds of steam from Sherman Crater, located on the southeast side of the cone about 1,150 feet down from the glacier-covered summit. For several months thermal activity increased, melting some of the ice in the crater. Sulfurous ash came out with the steam and polluted the air downwind of the volcano.

Public campgrounds in the area were closed because of the possibility of a mudflow being triggered by melting snow mixed with hot ash and other debris. The level of Baker Lake, a reservoir at the foot of the mountain, was lowered because U.S. Geological Survey geologists feared that a mudflow might cause the dam to break, which would devastate the communities farther down the valley. Thermal activity subsided after a time, but the possibility of a dangerous eruption is still very real as recent events at Mount St. Helens (170 miles to the south, air distance) have demonstrated.

Composite volcanoes, like those in the Cascades, sometimes have explosive eruptions, depending upon the amount of gas in the lava and its *viscosity,* or resistance to flow. The viscosity of a lava controls the ease or difficulty with which the gas escapes into the atmosphere. Composite cones are built up by lava that is predominantly intermediate, or andesitic, in composition. Andesitic lava is ordinarily more viscous than basaltic lava; but if the temperature of an andesite lava is extremely hot, then it becomes more fluid, or nonviscous, and is able to flow easily from a crater or vent. If a vent becomes clogged and gas pressure builds up, then an explosion occurs which ejects pyroclastic debris, clouds of ash, etc. Usually eruptions of composite volcanoes are intermittent, with periods of possibly violent activity followed by many years of apparent dormancy. Thus volcanoes of this type are long-lasting and tend to construct very large, lofty cones over a long span of time.

Sudden release of intense heat is the immediate danger when a volcanic explosion occurs. All living things on the volcano's flanks may be incinerated by the blast. As hot ash, gas, and steam spew out, snow and ice turn to hot water and steam that mix with ash and pour down the mountain as a mudflow which buries everything in its path.

The water storage reservoirs that have been constructed on the lower slopes of most of the Cascade volcanoes may serve as catch basins for mudflows—providing enough warning has been given so that the water can be drawn down. Whether or not the dams can hold is uncertain. A large enough mudflow could overtop or destroy a dam, endangering all the communities downstream as hot mud and water race down a river valley. Such a flood would be catastrophic.

Hot falling ash carried rapidly downwind may start fires over a wide area, contaminate water supplies, collapse roofs, render highways impassable, clog storm drains, endanger aircraft, damage crops, etc.

Can eruptions be predicted? The monitoring program of the U.S. Geological Survey, in cooperation with state surveys, seeks to interpret data and issue warnings based on several types of indicators: (1) increases in thermal activity; (2) increases or changes in seismic activity; (3) gravity and magnetic measurements; (4) remote sensing and infrared studies; (5) gas and temperature monitoring; and (6) observations of actual changes (i.e., bulges, tilting, etc.) in the volcano's outward conformation through comparison of photographs or use of time-lapse photography.

In the Cascades, volcanic eruptions have been infrequent in terms of human life spans. In fact, before Mount Baker started heating up and steaming, many residents did not believe the geologists and other scientists who warned of potential hazards. However, Mount St. Helens's behavior since 1980 has succeeded in making believers out of skeptics. What is now recognized is that major evacuations are really the only adjustment that human beings can make when a volcano becomes threatening.

the cone is a typical *composite volcano,* (or *stratovolcano*) in that it was built up over a long period of time by alternating layers of andesite lava and pyroclastics. Because of recent emissions of steam and some ash, Mount Baker (along with other Cascade volcanoes) is being continuously monitored (box 14.1; see also chapters 17, 18, and 19).

Regional Structure of the North Cascades

In the North Cascades Range, several distinct structural blocks were formed by steeply dipping faults of late Mesozoic and early Cenozoic age. The faults trend north-northwest, roughly parallel to the trend of the range. Within each block the rocks are fairly similar, but they differ markedly from the rocks in the adjacent blocks. From west to east across the range there are four major zones of rock distribution (fig. 14.4). Most of North Cascades National Park is within the second zone, the Chilliwack Batholith, with small areas of the Park extending into the first zone, on the west, and the third zone, just east of the Chilliwack Batholith. The four rock distribution zones are described briefly below, but the emphasis throughout the rest of the chapter is on the rock units exposed within the Park boundaries.

Zone 1 is the western foothills region, bounded on the east by the Shuksan fault, a thrust fault. In this zone Mount Baker sits on top of folded Late Paleozoic and Mesozoic sedimentary and volcanic rocks.

In *Zone 2* the composite Chilliwack Batholith lies between the Shuksan fault and the Ross Lake fault. The granitic magmas of the batholith intruded older, severely deformed and metamorphosed schists and gneisses.

In *Zone 3* the Methow graben, a downdropped fault block, lies between the Ross Lake fault and the Chewack Pasayten fault. The Methow graben consists of Upper Mesozoic sedimentary beds that have been intruded by Cenozoic granites. Near the western edge of the block, the Jack Mountain thrust, a wedge of rock forming the tops of several peaks in the Park, rests on top of younger Upper Mesozoic beds. The Park's eastern boundary is in this zone.

Zone 4, the eastern highlands, is outside the Park on the east side of the Chewack Pasayten fault. This block consists mainly of schists, gneisses, and granites.

Regional metamorphism in the North Cascades produced a wide range of *metamorphic facies* indicating the physical and chemical conditions in the various subsurface environments in which metamorphism occurred. Each metamorphic facies represents a particular combination, or assemblage, of minerals that are associated with certain pressure and temperature relationships. By identifying the metamorphic facies of different outcrops of schists, for example, geologists can infer the approximate depth at which metamorphism took place. In a metamorphic facies series, a greenschist suggests an environment of relatively low temperatures and pressures, while an amphibolite schist indicates that the parent rock was subjected to higher temperatures and pressures at a greater depth.

Under some conditions of metamorphism, *migmatites,* which are interlayered or intermixed rocks, partly igneous and partly metamorphic, are formed. The North Cascades migmatites were probably formed at great depth

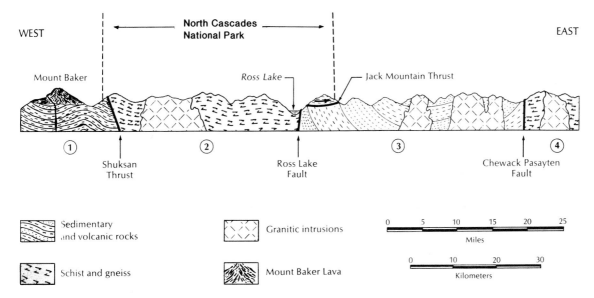

Figure 14.4 A schematic west-east cross section through the North Cascades. The major blocks are (1) the western foothills, made up of folded sedimentary and volcanic rocks and bounded on the east by the Shuksan fault; (2) the Chilliwack batholith intruding highly metamorphosed schists and gneisses; (3) the Methow graben, a down-faulted block between the Ross Lake and Chewack Pasayten faults; and (4) the eastern highlands. (Adapted from *Cascadia,* fig. 7.3, by Bates McKee. © 1972 McGraw-Hill Book Company; used by permission.)

during regional metamorphism as very hot water, bearing mineral ions, was forced at high pressure through planes of weakness (foliation) in the metamorphosed rocks, changing them to granite along the paths where the water traveled. (Elsewhere migmatites are more generally formed by partial melting, whereby granite magma "sweats" out of the metamorphic rock and collects in foliation planes.)

The North Cascades are not rich in mineral deposits, but some gold, copper, and silver grains were carried by mineral-bearing hydrothermal solutions into quartz veins during cooling of the intrusions. Some of the gold was subsequently freed by weathering and erosion from its quartz matrix and washed into stream gravels. Since gold has a high specific gravity, it tends to sink to the bottom of a stream and the grains ("gold dust") and nuggets become concentrated in sand bars, outwash, etc. Such accumulation is called a *placer deposit*. Before the filling of the Ross Lake Reservoir, placer gold was taken from glacial outwash and stream gravels in Ruby Arm, but this site is now under water.

Glacial Features

From the air, looking down over North Cascades National Park, one sees an awesome terrain—snowfields from peak to peak; ice bulging out over cirques and tumbling down valleys; sharp, jagged ridges that extend for miles; steep-sided horns; precipitous mountain slopes; deep U-shaped troughs; and other signs of glacial erosion past and present. Some of the names that prospectors and mountain men gave to the rugged features near the 49th parallel (the U.S.-Canadian boundary) evoke images of difficulties they must have had in trying to get around in this region on foot—names like Devils Dome, Terror Ridge, Fury Ridge, Ragged Ridge, and so on (fig. 14.5).

Excavation by Pleistocene glaciers provided some access into these formidable mountains. The one highway across the range, U.S. Highway 20, goes along glacial troughs now occupied by Diablo Lake and the southern end of Ross Lake and then follows the U-shaped valley of the Skagit River to the lowlands and the seacoast. A well known trail (used by Indians, explorers, pioneers, and

Figure 14.5 Arêtes and horns form the serrated crest of the Picket Range in the northern part of North Cascades National Park. The "sawtooth skyline" is characteristic of alpine topography in mountains that have been intensely eroded by glaciers. National Park Service photograph by M. Woodbridge Williams.

Figure 14.6 Forbidden Peak rises above the large compound cirque of the Boston Glacier near Cascade Pass in the southern part of North Cascades National Park. In the foreground, masses of glacial ice and thin streams of meltwater are cascading downward over the nearly vertical slopes. National Park Service photograph by M. Woodbridge Williams.

today by backpackers) goes from fiordlike Lake Chelan on the southeast side of the range, up the glacial valley of the Stehekin River, across Cascade Pass (a glacial col), down the Cascade River trough, and then on down the Skagit.

Above Cascade Pass (elevation 5,392 feet) towers Eldorado Peak, 8,868 feet high, a sharply eroded glacial horn rising above active glaciers. Part of the Boston Glacier, longest glacier in the Park, is on Eldorado (fig. 14.6). Other areas of numerous active glaciers in the Park are in the vicinities of the northern Glacier Peak (near the Canadian border), Mounts Challenger and Shuksan, Bacon Peak, and Colonial Peak. The two volcanoes outside the Park, Mount Baker and Glacier Peak (west of Lake Chelan), also have extensive glaciers. Glaciologists have pointed out that the Cascade volcanoes—even though they are higher and tend to have more glaciers—show the effects of glacial sculpturing less than the nonvolcanic peaks of the North Cascades. The inference is that since volcanic eruptions continued during and after the Pleistocene, the volcanoes were able to keep "repairing" their cones.

Mount Shuksan (fig. 14.7), near Mount Baker, is a glacial horn with numerous glacial cirques indenting its flanks. The Picket Range, with its thin, serrated ridge, was supposedly named for its resemblance to a picket fence extending along the skyline. Mount Johannesberg, at the headwaters of the Skagit River, is another knife-edged glacial arête. To the south, paralleling Lake Chelan, is the aptly named Sawtooth Ridge. Three Fools are glacial horns on the crest of a nearly vertical headwall that drops thousands of feet to the cirque floor. These are but a few examples of alpine erosional features in the Park.

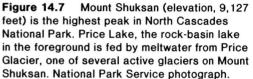

Figure 14.7   Mount Shuksan (elevation, 9,127 feet) is the highest peak in North Cascades National Park. Price Lake, the rock-basin lake in the foreground is fed by meltwater from Price Glacier, one of several active glaciers on Mount Shuksan. National Park Service photograph.

Of the glacial troughs, Lake Chelan (in the Lake Chelan National Recreational Area and Wenatchee National Forest) is the most remarkable. Over 2,000 feet deep in some places, the Chelan Lake trough was excavated by an 80-mile-long glacier that came down the Stehekin River valley from Cascade Pass. Many of the flat-floored glacial troughs, such as Ruth Creek valley and Nooksack valley, contain braided streams because of the low gradient and high sediment load. Hanging valleys, most of them containing waterfalls, are numerous throughout the Park. The most spectacular is a 1,500-foot waterfall from the cirque on Bacon Peak that holds Green Lake. Ann Lake and Rainy Lake are also examples of cirque lakes.

Except for moraines on valley floors and sidewalls, some outwash deposits, and similar accumulations, glacial depositional features in the Park are relatively minor. Most of the glacial debris was moved on southward by the Cordilleran ice sheet as it pushed down from Canada. Erratic boulders, however, from British Columbia were left sitting on some of the lower summits that were scoured by the ice sheet.

Geologic History

1. Origin of the oldest rocks, the Yellow Aster Complex.

The term *complex* is appropriate because the early geologic history of the Yellow Aster rocks (the "crystalline basement" rocks of the region) has been obscured by long erosion and by igneous and tectonic activity. The Yellow Aster Complex is pre-middle Devonian in age and is made up of early Paleozoic and Precambrian sedimentary and igneous rocks that have been intensively metamorphosed. Some Yellow Aster rocks crop out in the western foothills (Zone 1 in fig. 14.4).

2. Profound unconformity during middle Paleozoic time.

A very long period of erosion is indicated by this unconformity.

3. Late Paleozoic deposition and volcanism.

West of the Park, the metamorphosed rocks of the Chilliwack Group (not to be confused with the much younger Chilliwack Batholith) were deposited as alternating volcanics and sedimentary beds, mostly in shallow water. In some places the Chilliwack beds are associated with Mesozoic clastics (conglomerates, slates, etc.), suggesting that deposition was more or less continuous in a eugeosynclinal basin from the end of the Paleozoic Era into Mesozoic time.

4. Mesozoic deposition and volcanism.

Part of the Ross Lake glacial trough, Mount Hozomeen, and the eastern boundary of North Unit are within a belt of rocks of the Hozomeen Group that are believed to be Triassic (possibly Permian) in age. However, most of the thick sequences of Mesozoic volcanic and sedimentary rocks are outside Park boundaries in the western foothills and the eastern highlands. These rock units are typical eugeosynclinal deposits.

5. Late Mesozoic metamorphism, mountain-building, and thrust-faulting.

A major orogeny in the Cretaceous resulted in metamorphism and thrust-faulting that profoundly affected the rocks on the west and east sides of the range. During the movement of the Shuksan thrust fault on the west side of the range, rocks were sheared, shattered, and metamorphosed, producing the complex structures of the Shuksan Metamorphic Suite, which trends north-northwest in the western foothills and along the western edge of the Park. The unit consists of the Darrington Phyllite (parent rock, shales) overlain by the Shuksan Greenschist (parent rock, submarine basalt flows). Some unusual blue amphibole minerals (crossite and glaucophane) occur in the greenschist, which was apparently altered in a low-temperature but fairly high-pressure environment (i.e., greenschist metamorphic facies). In addition, tectonic slices of the ancient Yellow Aster rock were brought up from deep within the continental crust, along with slices of serpentized peridotite (ultramafic rock, largely olivine) believed to be derived from upper mantle rock (i.e., came from just below the earth's crust).

The Skagit Metamorphic Suite, which is apparently somewhat younger than the Shuksan Suite, crops out in the main part of the range and shows varying degrees of metamorphism (from greenschist metamorphic facies to amphibolite facies). The Cascade River Schist (part of the Skagit Metamorphic Suite) crops out mainly in the western part of the Park's South Unit. The Skagit Gneiss, which is the migmatized equivalent of the Cascade River Schist, crops out in both North Unit and South Unit. Spectacular exposures of Skagit Gneiss in the road cuts along the Skagit River and near the Ross Lake Dam are well worth stopping to see. Veins and dikes of permatites and other igneous rocks crosscut and interfinger the gneiss. This locality has been described as the *pièce de résistance* of North Cascades geology.

Several granitic plutons were emplaced before, during, and after metamorphism of the Skagit Suite and were themselves metamorphosed eventually.

On the east the Ross Lake fault separates the Skagit Gneiss from a long, thin belt of metamorphosed rocks. In this belt is the Elijah Ridge Schist (probably derived from the North Creek Volcanics), overlain by the Jack Mountain Phyllite. The Jack Mountain thrust fault moved older rock units eastward over younger beds (fig. 14.4). Parts of these fault zones have been destroyed by younger intrusions.

6. Late Mesozoic and early Cenozoic deposition; minor deformation.

In late Cretaceous and early Tertiary time, the highlands (precursors of the present mountains) that were elevated by the Cretaceous orogeny became a source of sediment for deposits in marine embayments on the west and east sides of the range. Then as the seas withdrew, continental beds (mixed sedimentary and volcanic) were laid down over the marine units. Patches of these beds, such as the Chuckanut Formation and Tertiary volcanics, exist in the Park, but mainly the sequences are found outside Park boundaries.

7. Multiple Cenozoic intrusions, including the Chilliwack Composite Batholith and the Eastern Intrusive Belt.

Emplacement of a large, composite batholith that, with associated older metamorphics, forms the core of the Cascade Range began with the intrusion of stocks having a diorite or gabbroic composition and was followed by a series of major intrusions, most of which were quartz diorite. Later in the sequence they tended to be granodiorite. Next, stocks of very light-colored granitic magma pushed up into the plutons placed earlier. Small intrusions of granitic composition continued through the rest of the Tertiary.

An Eastern Intrusive Belt, emplaced east of the Ross Lake fault, includes the Ruby Creek Heterogeneous Plutonic Belt, the Black Peak Batholith, the Golden Horn Batholith (alkali-granitic rock), and the Perry Creek intrusives of grandiorite and quartz diorite.

8. Late Tertiary block-faulting and uplift.

Uplift along a north-south axis was greater toward the north, with maximum uplift occurring in the area of the 49th parallel. Concurrent and subsequent erosion exposed more of the older metamorphic rocks and deep-seated intrusions in the North Cascades than in the South Cascades (southern Washington and Oregon).

9. Pleistocene and Holocene composite volcanoes.

The lavas that formed Mount Baker and Glacier Peak (the two Cascade volcanoes just outside the Park's western boundary) rose from great depths through thousands of feet of metamorphic and igneous rocks. Both volcanoes are andesitic composite cones, although recent eruptions from Glacier Peak have tended to be dacitic. Eruptions in the Cascade volcanic chain have continued into the present (see chapters 17, 18, and 19).

10. Pleistocene and Holocene glaciations.

Since before the beginning of the Peistocene, glaciers have been present in the North Cascades and they have been extensive during most of this time. Again and again, the glaciers in the mountains of western Canada grew until they formed a Cordilleran ice sheet that moved down over northern Washington, expanding and at times overrunning the mountain glaciers of the North Cascades until they enlarged into piedmont glaciers, accumulated great ice fields, and eventually merged with the ice sheet from the north. During warmer interglacial periods, the process was reversed as the ice sheet melted back, leaving alpine glaciers occupying the same valleys where they are found today.

In the high mountains, evidence of early and middle Pleistocene glaciations was destroyed by later glacial advances. Fraser Glaciation, the last major advance of the Cordilleran ice sheet, began about 25,000 years ago and attained its maximum stand about 50 miles south of Seattle between 15,000 and 13,500 years ago. The ice sheet's recession was followed by a readvance into northern Washington about 11,000 years ago, after which it disappeared. Since that time the North Cascades glaciers have made several minor advances and retreats in response to climatic changes, the most recent advances being in the Neoglacial period—one advance between 3,500 and 2,000 years ago and another about a thousand years ago.

The Cascade glaciers are an important water resource in the Pacific Northwest because they accumulate and store water (as ice and snow) in the winter season, and release it in the form of meltwater during the summer months when it is needed for irrigation and other purposes (fig. 14.8).

Figure 14.8 Trapper Lake and Trapper Magic valley in the southern part of North Cascades National Park. During Pleistocene time, glacial ice filled this valley and extended to a much lower elevation. The glacier in the center of the photograph has now retreated almost to the head of the valley. This view, taken during the summer melt season, shows water draining from snowfields among the peaks and from remnant glaciers. An end moraine is building up in front of the glacier terminus (center). Talus blocks and debris coming down the steep slopes on either side are raising the level of the valley floor. Trapper Lake, in the foreground, is probably dammed by both bedrock and glacial drift. National Park Service photograph by M. Woodbridge Williams.

Although glacial erosion is still a significant geologic process in the high mountains, stream erosion and deposition are more important in the valleys and lowlands at the present time.

Geologic Maps and Cross Sections

Misch, P. 1979. *Geologic map of the Marblemount quadrangle, Washington.* GM–23. Washington Division of Geology and Earth Resources.

Staatz, M. H., et al. 1972. *Geology and mineral resources of the northern part of the North Cascades National Park, Washington.* U.S. Geological Survey Bulletin 1359. Plate I.

Tabor, R. W.; and Crowder, D. F. 1969. *On batholiths and volcanoes—intrusion and eruption of late Cenozoic magmas in the Glacier Peak area, North Cascades, Washington.* U.S. Geological Survey Professional Paper 604. Figs. 2, 3, 12, 52 and p. 61.

Bibliography

Armstrong, R. L. 1980. Geochronometry of the Shuksan Metamorphic Suite, North Cascades, Washington. Abstracts with Programs, Geological Society of America (Cordilleran section). Vol. 12, no. 3, p. 94.

Beget, J. E. 1982. Recent volcanic activity at Glacier Peak. *Science* 215:1389–90, March 12.

Crandell, D. R. 1965. The glacial history of western Washington and Oregon. In *The Quaternary of the United States,* (VII INQUA Congress vol.), ed. H. E. Wright, Jr., and D. G. Frey. Princeton, New Jersey: Princeton University Press.

Darvill, F. T. 1973. *North Cascades highway guide.* Mount Vernon, Washington: F. T. Darvill.

Easterbrook, D. J.; and Rahm, D. A. 1970. *Landforms of Washington.* Bellingham, Washington: Department of Geology, Western Washington University.

Malone, S. D., and Frank, D. 1975. Increased heat emission from Mount Baker, Washington. Pp. 679–85 in *EOS Transactions.* American Geophysical Union.

May, A. 1973. *Up and down the North Cascades National Park.* Longmire, Washington: Mount Rainier Natural History Association.

McKee, Bates. 1972. *Cascadia.* New York: McGraw-Hill Book Company.

Misch, P. 1966. Tectonic evolution of Northern Cascades of Washington State. In *A symposium on the tectonic history and mineral deposits of the Western Cordillera,* ed. H. Gunning, special volume 8. Canadian Institute of Mining and Metallurgy.

———. 1977. Bedrock geology of the North Cascades. In *Geological excursions in the Pacific Northwest,* eds. E. H. Brown and R. C. Ellis. Bellingham, Washington: Department of Geology, Western Washington University.

Sato, M., Malone, S. D., Moxhan, R. M., and McLane, J. C. 1976. Monitoring of fumarolic gas at Sherman Crater, Mount Baker, Washington. Abstract. Vol. 57, no. 2 in *EOS Transactions.* American Geophysical Union.

Spring, R.; and Spring, T. 1969. *The North Cascades National Park.* Seattle, Washington: Superior Publishing Co.

Staatz, M. H., et al. 1972. *Geology and mineral resources of the northern part of North Cascades National Park, Washington.* U.S. Geological Survey Bulletin 1359.

Tabor, R. W., and Crowder, D. F. 1969. *On batholiths and volcanoes—intrusion and eruption of late Cenozoic magmas in the Glacier Peak area, North Cascades.* U.S. Geological Survey Professional Paper 604.

Vitaliano, D. B. 1973. *Legends of the earth.* Bloomington, Indiana: Indiana University Press.

Zoretich, F. 1976. Volcano alert in the Pacific Northwest. *Science digest,* January, pp. 39–47.

Address

North Cascades National Park
800 State Street
Sedro Woolley, Washington 98284

15
Acadia National Park

Location: Maine (mostly Mount Desert Island)
Area: 38,523.77 acres; 60.19 square miles
Established: February 26, 1919

Figure 15.1 A cliff of massive diorite at Great Head on the southeast shore of Mount Desert Island. Glacial plucking by a mile-high ice sheet flowing from the northwest produced the cliffs. Direct wave erosion has had little effect on this resistant bedrock since Pleistocene glaciers retreated and sea level rose to its present stand. However, as blocks of rock between joints are slowly but inevitably loosened by weathering, they fall into the ocean where they are smashed and rounded in the "wave mill." The cliffs here are within the "shatter zone" (p. 187) that encircles a granitic pluton, the large structure forming the highlands of Mount Desert Island. The narrow white streaks are dikes and veins of quartz and feldspar forced into the rocks during magmatic cooling. The rounded dark mass (upper left) is an inclusion of rock of different composition. (The discoloration just above the high water line is caused by weathering of organic material on the rock.) Nylander photograph.

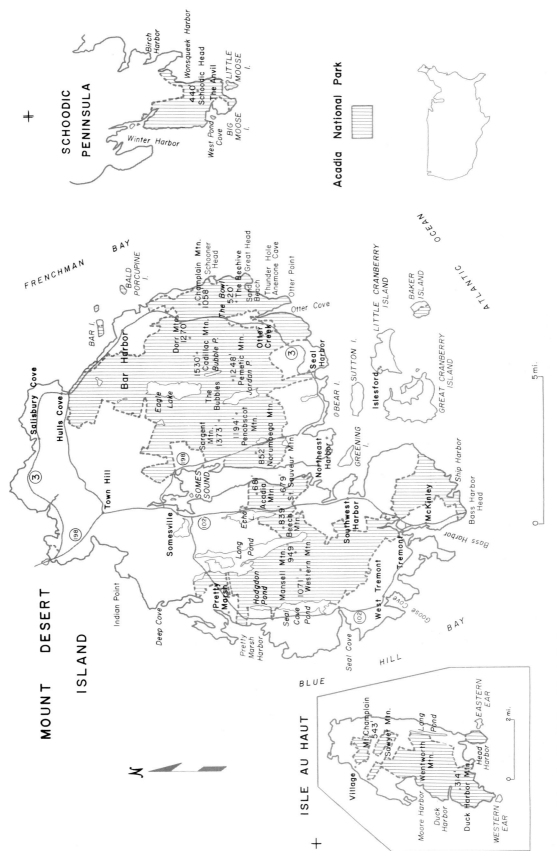

MOUNT DESERT ISLAND

SCHOODIC PENINSULA

Birch Harbor
Wonsqueek Harbor
440 Schoodic Head
The Anvil
LITTLE MOOSE I.
West Pond Cove
BIG MOOSE I.
Winter Harbor

Acadia National Park

FRENCHMAN BAY

BALD PORCUPINE I.
BAR I.
Bar Harbor
Salisbury Cove
Hulls Cove
Indian Point
Deep Cove
Town Hill
Somesville
SOMES SOUND
Dorr Mtn. 1270'
Champlain Mtn. 1058'
Schooner Head
The Bowl
The Beehive 520'
Sand Great Head Beach
Thunder Hole
Anemone Cove
Otter Point
Cadillac Mtn. 1530'
Bubble P.
Pemetic Mtn. 1248'
Jordan P
Eagle Lake
The Bubbles
Sargent Mtn. 1373'
Penobscot Mtn. 1194'
Norumbega Mtn. 852'
Otter Creek
Otter Cove
Seal Harbor
OBEAR I.
SUTTON I.
GREENING I.
Northeast Harbor
Acadia Mtn. 681'
St. Sauveur Mtn. 679'
Echo
Long Pond
Beech Mtn. 839'
Western Mtn
Hodgdon Pond
Seal Cove Pond
Mansell Mtn. 1071' 949'
Pretty Marsh
Pretty Marsh Harbor
Southwest Harbor
West Tremont
Tremont
McKinley
Bass Harbor Head
Ship Harbor
Bass Harbor
Goose Cove
Seal Cove
BLUE HILL BAY
Islesford
LITTLE CRANBERRY ISLAND
GREAT CRANBERRY ISLAND
BAKER ISLAND
ATLANTIC OCEAN

98
102
3
5 mi.

ISLE AU HAUT

Village
Moore Harbor
Duck Harbor
Mt. Champlain 543'
Sawyer Mtn.
Wentworth Mtn.
Long Pond
Duck Harbor Mtn 314'
Head Harbor
EASTERN EAR
WESTERN EAR
2 mi.

N

Figure 15.2 Acadia National Park, Maine.

183

Table 15.1. Geologic Column, Acadia National Park

Time Units		Rock Units	Geologic Events
Era	Period		
Cenozoic	Quaternary	Wave and stream deposits relatively minor Minor glacial deposition	Wave erosion relatively minor Crustal rebound Postglacial submergence Pleistocene glaciation; severe scouring, abrasion, excavation of troughs.
Cenozoic	Tertiary (or Paleogene/ Neogene)		Minor uplift; renewed erosion
Mesozoic		/////////	Prolonged weathering and erosion Unconformity
Paleozoic	? Mississippian ? Devonian	Miscellaneous granites Medium-grained granite Recrystallized granite Coarse-grained granite Fine-grained granite ///////// Diorite (in intrusive sheets and sills) ///////// Bar Harbor Series	Uplift and erosion Ring dikes, recrystallization Shattering, ring dikes, uplift Shattering, contact metamorphism, hydrothermal activity, pegmatites Uplift and erosion Dikes and veins throughout this and succeeding intrusions Contact metamorphism; volcanism Beginning of Acadian orogeny Uplift and erosion Deposition of siltstones, shales, conglomerates—mainly marine
Paleozoic	Silurian ? Ordovician	///////// Cranberry Island Series ///////// Ellsworth Schist	Deformation, metamorphism, erosion, Volcanism—ash, tuff, interbedded sedimentary rocks Deformation, uplift, metamorphism, erosion Marine sedimentation; volcanic ash

Source: modified after Chapman 1968, 1970

Local History

For more than 6,000 years prior to the arrival of the Europeans, Indians inhabited this part of the Maine coast. Relics of the earliest culture—spearheads, slate lancepoints, etc.—more closely resemble the artifacts of the Southwest Indians than they do New England Indian artifacts. Apparently these people came to this region to fish for tuna. The Red Paint People, so-named because they decorated their faces and bodies with red ocher, succeeded the tuna fisherman. The relics of this culture—tools of slate, pottery, and so forth—are similar to those of the Mound Builders of the Middle West. The Abnaki, who developed a birch-bark culture, apparently displaced or conquered the Red Paint People sometime before the Europeans came. The Penobscot and the Passamoquoddy Abnaki tribes, who summered in the Mount Desert region, were the Indians whom the white explorers and settlers encountered. These Indians continued to camp in the Somes Sound area each summer to fish, dig for clams, trap mink, etc., until the early years of the 19th century. Deep shellheaps indicate that they camped in the region for hundreds of years. They spent the winters in their permanent villages at the headwaters of the Penobscot and other rivers of the area, and traveled back and forth to the coast in their birch-bark canoes.

Perhaps the Norsemen and certainly the Portuguese explored this coast. A 1529 Portuguese map shows Mount Desert Island and the main estuaries of the region. Since Cadillac Mountain (1,530 feet) is the highest point on the Atlantic coast, it was an important landmark for sailing ships for the next several centuries.

To Samuel de Champlain, the father of New France, however, goes the honor of being the official discoverer of Mount Desert Island. Sieur de Monts, Champlain's patron, held a grant from Henri IV, King of France, that comprised the entire French claim to North America (i.e., all the land between latitudes 40° and 60°). The territory was called *La Cadie* (later anglicized to "Acadia"). After establishing a trading post on an island in the St. Croix River (between Maine and New Brunswick), de Monts sent Champlain in 1604 to explore and chart the nearby coast and to encourage the local Indians to do business with the French. In his journal of the voyage, Champlain described what is now Acadia National Park as: "an island, very high, and cleft into seven or eight mountains, all in a line. The summits of most of them are bare of trees, nothing but rock. I named it *l'Isle des Monts-deserts*" ("the Isle of Bare Mountains"). Attracted by the smoke from an Indian camp, Champlain put in to shore near Otter Cliffs and, guided by friendly Indians, made a reconnaissance of the locality. His 1607 chart of the coastline shows fairly accurately the islands, inlets, and estuaries.

Several years later the Marquise de Guercheville acquired the de Monts grant, which had been enlarged by Louis XIII to include the whole of North America to the Gulf of Mexico. At her own expense, she sent a group of Jesuit missionaries to convert the Abnaki to Christianity. At the invitation of the Penobscot chief, the Jesuits set up a mission near the entrance to the Somes Sound and the Penobscots' summer encampment, naming the colony Saint-Sauveur.

Unfortunately, a few weeks later the settlement was spotted by Samuel Argall of the English Virginia colony. He had been ordered by King James I, who disputed the French claim, to clear out any French colonies along the coast. In a surprise raid, Argall wiped out the settlement, sent the French sailors out to sea in a longboat, and carried the priests back to Virginia. Luckily the sailors met a French ship off the bay of Fundy and got home safely; and the Jesuits were taken to England eventually, from whence they made their way back to France. This was the beginning of more than a century of raids, skirmishes, and intermittent warfare between the French and the English over possession of North America.

The Sieur de Cadillac (Antoine LaMothe) did make one more try for the French side in the summer of 1688. He had obtained a grant for the territory from the Governor of Canada (New France). However, apparently fearing that his fate would be the same as the Jesuits', he went inland, became a fur trader, and founded Detroit. Cadillac Mountain bears his name.

Even after the Peace of Utrecht in 1713 awarded Maine to the English, the coast was not safe for permanent settlement. Both English and French warships used the safe harbors between Bar Harbor and Bangor for staging and refitting. Finally by 1760 the French and Indian wars ended, and this made the New England settlements free from the danger of Indian raids (organized by the French).

Governor Francis Bernard of the Massachusetts Bay Colony was awarded the grant for the Island of Mount Desert by King George III. In 1761 he helped the Somes and Richardson families to settle on the island, and this was the beginning of Somesville, the first permanent settlement. During the American Revolution, all of Governor Bernard's property was confiscated because he was a Loyalist. After the war, both Bernard's son and the granddaughter of Cadillac petitioned the General Court of Massachusetts (which included Maine until 1820) for possession of Mount Desert Island. The land was surveyed and the eastern half of the island was awarded to the French heirs and the western half to Bernard, who sold his share and returned to England. Cadillac's granddaughter, who settled on the island, sold off her land to other settlers a piece at a time.

Maine became a state in 1820, but the island remained an area of remote fishing hamlets for many years. Later on, some quarrying of the local granites for building stone was an important industry.

In the 1840s the charm and beauty of the island scenery was "discovered" by Thomas Cole, Frederick Church, and other artists of the "Hudson River School" of American painting. Soon college professors, wealthy summer visitors, and cruising yachtsmen began coming to the island and building "cottages" (many of them mansions). These summer folk were called "rusticators." Coastal steamers served the area from the 1850s until the advent of the automobile. Although the island was mainly a resort for the very rich, an increasing number of other people who appreciated the uniqueness of the region came to visit the island. By 1901 the wealthy people who owned most of the land began to fear that lumbering and commercial development would ruin the natural beauty of the island, so they formed a corporation to protect the area. By 1913 they had acquired approximately 6,000 acres of land on Mount Desert Island, the Schoodic peninsula (across the bay from Bar Harbor), and the Isle au Haut. President Charles Eliot of Harvard, John D. Rockefeller, and George Bucknam were among the leaders in this effort to transfer the land to the federal government for a national park. Proclaimed as Sieur de Monts National Monument in 1916, the area was established as Lafayette National Park in February of 1919. Ten years later the name was changed to Acadia National Park.

Geologic Features

Visitors to Acadia National Park see a mountainous, rockbound coast that has been severely abraded by ice sheets and submerged by an encroaching sea. Waves beat against an irregular shoreline with many bays, headlands, and islands. The role of water and ice in the shaping of this dramatic scenery is significant, but the origin of the rocks that have resisted these forces is perhaps equally dramatic.

An Ancient Mobile Belt

The emplacement of the Maine coastal plutons (one of which provided the dominant granites of Mount Desert Island) came in Paleozoic time at the very end of the Acadian orogeny. This orogeny is believed to be one of the most prolonged and severe episodes of mountain-building to which the North American landmass has been subjected. In retracing ancient plate movements, paleogeographers have concluded that in the early part of the Paleozoic, the crustal segments that eventually became North America and Europe were in low latitudes, in the equatorial regions of the globe. This means, of course, that at that time the climate of the region was tropical.

As a result of the development of an active, long-lasting mobile belt—that lay roughly along the east coast of North America but cut diagonally across New England—the Maine coastal region and the Canadian maritime provinces were a part of the European plate rather than the North American plate during most of the Paleozoic. Between the two continental landmasses lay a proto-Atlantic ocean, probably narrower than it is today. Extensive sedimentation was filling the eugeosynclinal belt. This belt, or trough, paralleled both the western landmass and an island arc dotted with volcanoes.

In the Silurian Period, the proto-Atlantic ocean was closing, bringing northern Europe and northern North America closer together. Scattered occurrences of volcanic rocks from Maine to Newfoundland suggest that the leading edge of the eastern landmass may have been a subduction zone with active volcanoes.

Fusion of the continents took place in Devonian time as the proto-Atlantic closed from the north southward, from Greenland to approximately New Jersey. Continental collision resulted in the folding and refolding of the geosynclinal rocks, along with crustal shortening. In this way a mountain chain bordering the former landmass margin was produced during the Acadian orogeny. In the thick sequence of volcanic and sedimentary rocks along the Maine coast, a large igneous complex was emplaced in a series of pulses, the earlier magma being gabbroic (mafic) or dioritic (intermediate in composition) and the later ones more felsic, or granitic (box 12.1, p. 142)

Then the stage was set for the intrusion of the Maine coastal plutons, which are made up principally of coarse-grained and medium-grained granites. In Acadia National Park these granites form the mountains and the core of Mount Desert Island, with older rocks surrounding the plutons in roughly circular patterns.

Isostatic adjustments apparently followed or may have accompanied the intrusive activity at the end of the Devonian. The area of intrusion (now the Maine coast) was uplifted, and a secondary geosyncline, or trough area, probably developed along the eastern side of the range. Meanwhile, off to the south, the African plate was moving northwestward to join the American-European landmass in forming the supercontinent Pangaea in late Paleozoic time.

When Pangaea started to split up in Triassic time, it had been in existence for around 150 million years. As rifts developed between Europe and North America during the Mesozoic, the Maine coastal region and adjacent parts of Canada stayed with the North American plate. The North Atlantic Ocean began to occupy its present basin, becoming wider and wider as the two continents moved apart. Long erosion stripped away most of the weaker rocks that once covered the Acadian mountains.

This exposed their granitic roots. In the Pleistocene Epoch, continental glaciers sculptured and abraded this rocky coast.

The Mount Desert Intrusives

The light-colored, barren summits that so impressed Champlain when he sighted them from his ship are made up of massive, coarse-grained granite that is highly resistant to erosion. This is the granite, pinkish in color, that makes up the Mount Desert range and all of the high points of the island. Wedged into the northwest section of the pluton is a somewhat younger body of medium-grained granite that is gray in color and speckled with biotite (black mica). From quarries in this granite (such as those between Somes Sound and Echo Lake), building stone of fine quality was shipped to cities along the eastern seaboard. (The Philadelphia Mint is one of the more famous buildings constructed of this stone).

Along the northwest side of the island, next to the granites, is an *intrusive sheet* of diorite, the third most abundant rock type of the Park. In the early intrusive phases of the Acadia orogeny (long before the emplacement of the plutons), the dioritic magma that formed the intrusive sheet came up very hot from great depths and worked its way through layers of crustal rock. Then the magma spread out horizontally between sedimentary bedding planes, forming sills in some places. The intrusive sheet is a larger, somewhat irregular body of diorite that probably spread out from an extensive fracture. The presence of the metamorphic rock *hornfels* in the contact zones indicates that the magma was extremely hot when it injected the stratified rocks.

A fine-grained granite that crops out between Southwest Harbor and Goose Cove (southwest part of the island) was emplaced after the diorite but before the coarse- and medium-grained granites. This fine-grained granite may be part of a large body of similar rock of the same age that makes up nearly all of the Schoodic Peninsula and also crops out on Isle au Haut.

The Forming of the Ring Dikes

By the time the magma that was the source of the coarse- and medium-grained granites began to rise through the crust, the roof rocks of the Mount Desert area had become seriously weakened, probably due to the length and severity of previous orogenic and intrusive activity. With renewed pressure and heat from the rising magma, the roof rocks began to fracture and subside. Eventually a large, pluglike mass of roof rock was incorporated into the melt, and some of it was preserved as inclusions in the slowly cooling magma that became the coarse-grained granite of the Mount Desert range. Into circular fractures

around the collapsed area, more magma welled up forming *ring dikes* of medium-grained granite. Curved fractures also developed within the pluton as the coarse-grained granite cooled and subsided. Intensely hot, magmatic liquids under high pressure permeated these curved fractures and recrystallized the rock. These patches of recrystallized granite have a fine-grained matrix with scattered large crystals of quartz and feldspar.

Following further subsidence and fracturing, a second pluglike mass of roof rock collapsed and sank on the western side of the pluton without becoming completely detached. Magma that flowed up over this block filled the curved fissure and solidified as medium-grained granite; this formed a large, bulging ring dike in the northwest section of the pluton.

A zone of severely shattered, fractured, and brecciated rock, up to a mile or more in width, lies between the coarse-grained granite and the metamorphosed country rock. Called the *shatter zone,* this band of rock that partially encircles the island (along the eastern shore and across the southern part) marks the edge of the old collapsed area. Shatter zone rock, which is now tough and coherent, is easy to spot because of its brecciated appearance. Angular blocks of diorite, chunks of old sedimentary rock, shattered fragments of various colors, veins, small dikes, pegmatites, partially blended rubble, etc., are all congealed in a matrix of granite.

Pegmatites, Veins, and Dikes

The pegmatites, quartz veins, and light-colored dikes of granite material are most numerous in the shatter zone, but they can also be found at other places in the Park where cracks, fractures, or joints in the rock gave access to hot hydrothermal solutions and magma forcing their way upward. Small amounts of minerals (epidote, hematite, etc.) are present in the quartz veins and pegmatites, but none are of economic value.

More numerous and more widely distributed than the granitic dikes are the basaltic (diabase) dikes, that range in width from less than an inch to more than 50 feet. Some dikes extend straight over mountaintops for thousands of feet. On bare outcrops the dark color of the basalt shows up against the lighter granites and the contact tends to be clear and sharp; this makes the dikes easy to follow. The Schoodic Point area (due east from Mount Desert Island, across Frenchman Bay) is the best place in the Park to see basalt dikes. Here large, dark basaltic dikes cut through light, fine-grained granite. In some places where the dikes come down to the shore, wave action has cut back straight-walled chasms. The less resistant basalt has been eroded, leaving the granite on either side. The basalt intrusions in the Park rocks apparently came from a very deep source. From time to time magma was released and moved up toward the surface. Radiometric dating techniques have yielded several sets of distinctly different ages for the basalt dikes.

Glacial and Wave-Action Features

The coarse-grained granite of Mount Desert Island formed a highly resistant rock ridge trending east and west. Before the time of Pleistocene glaciers, the processes of weathering and stream erosion had only slightly dissected the ridge. Possibly the summits of the present range could be distinguished, although reconstructing the preglacial drainage patterns is speculative because most of the evidence has been destroyed. What we see today, however, are six huge troughs, going north and south, cut through this very tough, resistant ridge. The amount of rock that was excavated and the force required to gouge it out and transport it are awesome to think about (figs. 15.3, 15.4). Not much of the rock debris is left on the island—in contrast to some of the shorelines of southern New England where prominent headlands are made up of glacial drift.

The Mount Desert ridge must have been a barrier temporarily to the great continental glacier moving down from the north, but eventually the mass of ice that was banked up against the north side of the ridge overrode the summits, scouring and rounding them off. At shallow gaps in the ridge or areas of weakness, the erosive effect of the ice was intensified and channeled as flowing ice currents probed for access to the south. As time went on and more and more ice pushed down from the north, troughs developed and became deeper and broader. The deepest trough, Somes Sound, which nearly bisects the island, is a fiord, the only true fiord on the U.S. Atlantic coast. A *fiord* is a long, deep arm of the sea, U-shaped, steep-walled, and always found on a mountainous coast that has been severely glaciated. A fiord represents the seaward end of a glacial trough, partially submerged after melting of the ice. Most of the embayments on the coast of Maine are drowned valleys, and many of them have been modified by glacial action. Somes Sound qualifies as a fiord because notches and grooves were cut into the rock walls by glacial ice, and because the deepest part of the sound is

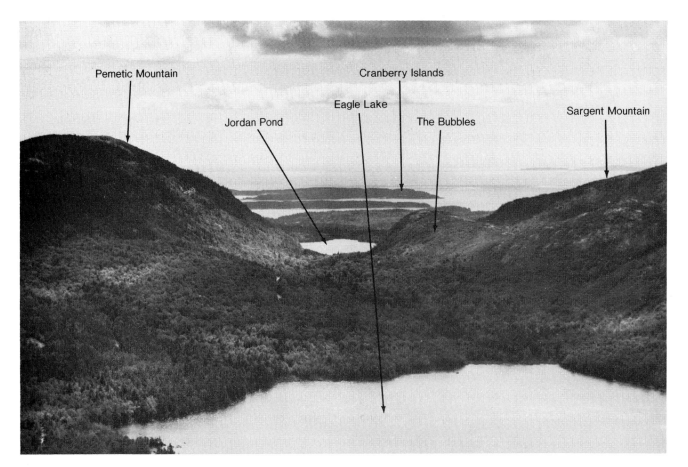

Figure 15.3   Ice-rounded topography in Acadia National Park. View is looking south toward the Atlantic Ocean. Preglacial stream erosion developed an east-west trending ridge on coarse-grained, massive granite. During the Pleistocene, south-moving ice scoured and excavated great quantities of bedrock, leaving large roches moutonnées, such as the Bubbles; smoothly rounded bare summits; and U-shaped, north-south glacial troughs. Eagle Lake and Jordan Pond occupy glacially scoured depressions in the trough shown here. Nylander photograph.

Sargent Mountain

Jordan Pond

The Bubbles

Pemetic Mountain

Cadillac Mountain

Figure 15.4 View looking north over the same glacially eroded trough (as in fig. 15.3) toward mainland hills in the far distance beyond Frenchman Bay. Summits of both north and south Bubbles are visible in this view. Notice the crisscross pattern of joints in coarse-grained granite exposed on the summit of Sargent Mountain. Compare this with the close-up view (fig. 15.5) of joints on Cadillac Mountain. Nylander photograph.

some distance back from the submerged mouth. Characteristically, a fiord becomes shallower near the mouth where glacial scouring was less severe.

Trough lakes, such as Seal Cove Pond, Long Pond, Echo Lake, Eagle Lake, and Jordan Pond, filled the low areas gouged out in the floors of the U-shaped valleys after the ice melted. In the valleys occupied by Long Pond and Jordan Pond, the water backed up to somewhat higher levels because of morainal dams at the south end.

The erosive power of the mile-high ice sheet also altered the shape of the mountains as it carved up the broad, previously unbroken, east-west ridge. Northern slopes became longer, smoother, and rounded; while southern slopes became steep and foreshortened, their surfaces rough and jagged from glacial plucking. Two large roche moutonnées that resulted from this type of glacial action are The Bubbles that stand between Cadillac Mountain and Penobscot Mountain.

Glacial grooves and striations on bedrock have a generally north-south orientation, indicating the direction of ice flow. *Chattermarks* (a series of small, close, curved

cracks made by vibratory chipping of ice-embedded rock fragments against a bedrock surface) show where some of the greatest pressures were exerted by the ice sheet. Many glacial erratics, large and small, are scattered over the island.

Jointing in the bedrock influenced both the glacial and wave action features. On the summits and in roadcuts at high locations, particularly around Cadillac Mountain, sheet-jointing, or exfoliation, of the granite is noticeable. These joints that run parallel to the surface are the result of unloading when the rocks covering the pluton had been eroded, long after the granite had cooled. On some of the long-exposed surfaces, *weather pits* have developed in the coarse-textured rock because the less-resistant mineral grains (or crystals) weather out first (fig. 15.5).

The geometric pattern of the broad, intersecting vertical joints in the granites is visible from the air and prominent on the ground wherever the bedrock is bare of vegetation. On the summits weathering and frost action have widened these joints, separating the rock into large blocks. On the south-facing slopes of the island, especially where high cliffs stand on the shoreline, the effect of the

Figure 15.5. Massive, glacially scoured granite on the summit of Cadillac Mountain (1,530 feet), highest point on the North American Atlantic coastline. Notice the weathering along joints and the pitting in this coarse-grained granite. National Park Service photograph.

glacial plucking along joint planes is most striking. Subsequent mass wasting of loosened blocks has steepened already jagged slopes and in some places undercut cliffs. Cliffs exposed to wave action are being further undercut as the sea continues the attack following the glacier's retreat. On the wave-cut bench formed at the base of a cliff, at and above the water level, rock fragments are rolled around, pounded by waves, and banged up against the cliff face during storms. Intensifying this mechanical action are the salt in the water, the spray, and the mist, which increase the rate of chemical weathering.

One of the many interesting features of the shoreline is the contrast between the effects of wave action on the granites, the shatter zone, and on the older metamorphic (originally sedimentary) rocks. At Bar Harbor on the eastern side of Mount Desert Island, the shoreline rock in the harbor area is within the Bar Harbor Series that was formed from gravel deposits, sand, silt, and some volcanic ash. Beds of purplish-gray rock dip toward the water, while the risers (tops of the beds) lap up in sequence against the shore. Loose pebbles and cobbles broken off from the beds tend to be flattened, or shingle-shaped. Most of the angles and corners of the beds have been smoothed and rounded by waves. Near the high water line is a small *stack* (free-standing rock column), an erosional remnant made up of layers of siltstone. Wave erosion (since the retreat of the ice and the rise of sea level) appears to be moderately effective on these rocks.

To the south of Bar Harbor as far as Great Head, the shoreline is mostly in the shatter zone. Along this stretch of jumbled ledges and cliffs, there are few places where one could safely land a boat at either high or low tide. Most of the rock is fairly resistant, but where weak spots exist, the waves have been able to erode. Anemone Cave is a large sea cave, facing the open ocean, that has been formed by wave action and frost action in a zone of fractured rock.

A short distance farther south, ledges of coarse-grained, resistant granite make up the shore. Here the erosive action of the waves has had little effect since the Pleistocene except along joint planes at exposed locations. Thunder Hole, which emits loud booming sounds when waves rush in and out, is a chasm eroded by wave action along large joints in the coarse granite. Frost action, by loosening slabs of rock, helps to enlarge the chasm. At low

Figure 15.6   Surf beating in on the exposed shoreline of the Schoodic Peninsula is gradually eroding this wave-cut platform on resistant bedrock. National Park Service photograph.

tide, between waves, one can see large, egg-shaped cobbles swirling and grinding away in the water at the head of the chasm.

One of the few places in the Park where post-Pleistocene wave erosion of granite is obvious is on the unprotected ledges and cliffs of the headlands at the extreme southern tip of Mount Desert Island near Bass Harbor Head Lighthouse. Here, where the rock is constantly exposed to the full force of waves from the open ocean, a *wave-cut platform* of considerable size is being cut back. A wave-cut platform (which includes the lower part of the wave-cut bench) slopes gently seaward from the base of a cliff. At low tide a wave-cut platform is partially exposed, but it is generally covered at high tide (fig. 15.6). On the northern Maine coast, the tidal range is fairly high, normally 8 to 10 feet of rise and fall.

Wave-Depositional Features

Most of the glacial drift and rock material scraped off and excavated from Mount Desert Island is probably still strewn over the ocean bottom out in the Gulf of Maine. Not much sand has accumulated since sea level rose after the Pleistocene. Thus wave-depositional features consisting of sand-sized material are minor except for a few sandbars in protected areas (such as the bar at Bar Harbor that connects the harbor shore with Bar Island at low tide) and small pocket beaches in coves. Most of the pocket beaches have more cobbles and shingles than sand. (Cobbles are generally spherical, while shingles tend to be disk-shaped.) Sand Beach (on the eastern side of the island between Great Head and Otter Cliff) was built across the head of a small bay and receives some sand from a small stream. However, Sand Beach is remarkable for the fact that the "sand" contains a high proportion of finely ground organic material (shells, bones, etc.) that has been washed up from the ocean and reworked by the waves.

Storm waves have built *seawalls* (natural beach-ridges) of cobbles and shingle along the high-water line on some of the less rugged parts of the shoreline (where the preglacial slopes were gentle). The seawalls were the source of smooth, even-sized paving stones (called "popplestones" in Maine) that were shipped down to cities on the eastern seaboard for street construction in the 19th century.

Geologic History

1. Ordovician deposition; the Ellsworth Schist.

Thousands of feet of marine sediment accumulated in a eugeosynclinal sea, and explosive volcanic eruptions added ash and pyroclastic debris. Late in the Ordovician Period and early in the Silurian, the sedimentary rocks were deformed, compressed, and metamorphosed, forming the Ellsworth Schist, the oldest rock exposed in the Park. These rocks, which are of two types—a mica-rich schist and a gneiss—crop out mainly along the shore at the northern end of Mount Desert Island. Following uplift, a long period of erosion took place.

2. The Cranberry Island Series; Silurian volcanism.

Later in the Silurian Period, after the Ellsworth Schist had been submerged, this rock unit was covered by the largely volcanic sediment (ash and pyroclastics) of the Cranberry Island Series. Some light-colored lava flows and sedimentary rocks are interbedded with volcanic sediment. Again deformation occurred, along with mild metamorphism, uplift and erosion. Felsite and tuff make up this rock unit.

3. Early Devonian deposition of Bar Harbor Series.

In the vicinity of Mount Desert Island, sand and gravel were washed into a shallow sea. This coarse sediment was overlain by many feet of finer sand and silt, the whole sequence eventually forming the Bar Harbor Series. The siltstone, which splits along bedding planes, is the source of some of the shingles of the cobble and shingle beaches.

4. The diorite intrusions; beginning of Acadian orogeny.

Great quantities of diorite (the intrusive equivalent of andesite) intruded the layers of the Bar Harbor Series as sills, dikes, and intrusive sheets. Some of the magma probably reached the surface and came out as lava flows. *Contact metamorphism,* the result of the intense heat of the magma, baked the country rock—especially the shales and the siltstones—into hornfels. Most of the diorite is fine-grained; but in the thicker intrusions, slower cooling in the center produced a coarser texture. Some of the sills in the Bar Harbor Series are 100 feet thick. The largest body, an intrusive sheet about 3,000 feet thick, formed between the Ellsworth Schist and the Bar Harbor Series.

5. Intrusion of fine-grained granite.

The granitic magma that rose after the diorite had solidified apparently hardened near the surface because the texture is fine-grained. Heat and pressure from the magma shattered the country rock, and many fragments of rock from the Bar Harbor Series and the diorite fell into the liquid magma and became inclusions or blended with the granite. Cracks and fractures in the country rock filled with magma that hardened as dikes, veins, and long stringers (fingers) within the older rock. Uplift of the land and long erosion followed. Sediments probably spread over and buried the Mount Desert region, but subsequent erosion has stripped away these beds.

6. Climax of Acadian orogeny; intrusions of coarse-grained and medium-grained granites.

From late Devonian time and probably extending into the Mississippian Period, the Mount Desert region was in the midst of intense crustal disturbances involving plate collisions (described earlier) and numerous pulses of intrusive and extrusive activity. The ring dikes, plutons, patches of recrystallized granite, etc. (explained in the geologic features section), are the roots of what was then part of a great mountain system, as high as the Rockies, or perhaps higher. A very long period of erosion followed, during which the mountains were worn down to ridges and hills.

7. Minor Tertiary uplift and renewed erosion.

In Tertiary time the Mount Desert range was a continuous ridge, trending east and west; and the Park region, including the offshore islands was part of the mainland. A minor uplift of the land, that tilted the region slightly seaward, was probably related to isostatic adjustment. Renewed erosion followed by minor subsidence with encroachment of the sea, produced an irregular shoreline, probably similar in outline to what we see today. Fluctuations of the shoreline since that time have generally been associated with glacial episodes that began in Pleistocene time.

8. Pleistocene glaciation.

The continental glaciers that covered New England and the Maritime Provinces were more than a mile thick. Ice extended southeastward over the Gulf of Maine, as far as the terminal moraine on Georges Banks. At that time sea level was much lower because water had been drawn from the oceans to build the enormous ice sheets, and their great weight was depressing the land. When the ice melted, sea level rose rapidly. The ocean came in over the land, drowning the shoreline, and filling the estuaries. Georges Banks, which had been a group of islands, was submerged and is now a famous offshore fishing ground. On Mount Desert Island sea level was probably 300 feet higher than it is today. Old beaches and deltas on the island mark the locations of earlier sea levels. Not long after the unloading of ice, however, the coast of Maine began to rise slowly, a process called crustal rebound. Sea level continues to fluctuate slightly, and has risen a few feet in the last several thousand years. Recent data also suggest a slight isostatic lowering of the crust along the east coast.

Geologic Map and Cross Section

Chapman, C. A. 1970. *The geology of Mount Desert Island, Maine.* Chatham Press.

Bibliography

Butcher, R. D. 1977. *Field guide to Acadia National Park, Maine.* New York: Reader's Digest Press.

Chapman, C. A. 1968. A comparison of the Maine coastal plutons and the magmatic central complexes of New Hampshire. In *Studies of Appalachian geology, northern and maritime,* eds. E. Zen, W. S. White, J. B. Hadley, and J. B. Thompson, pp. 385–96. New York: Interscience Publishers.

———. 1970. *The geology of Mount Desert Island, Maine.* Chatham Press.

Johnson, Douglas. 1925. *The New England-Acadian shoreline.* New York: John Wiley & Sons, Inc.

Morison, S. E. 1960. *The story of Mount Desert Island.* Boston: Little, Brown and Company.

Petersen, M. S.; Rigby, J. K.; and Hintze, L. F. 1980. *Historical geology of North America.* 2nd ed. Dubuque, Iowa: Wm. C. Brown Company Publishers.

Shaler, N. S. 1889. The geology of the island of Mount Desert, Maine. U.S. Geological Survey Annual Report 8:987–1061.

Address

Acadia National Park
Route 1, Box 1
Bar Harbor, Maine 04069

16
Olympic National Park

Location: Northwest Washington
Area: 901,216.23 acres; 1,408.15 square miles
Proclaimed a National Monument: March 2, 1909
Redesignated a National Park: June 29, 1938

Figure 16.1 Air view of the Hoh Glacier and its zone of accumulation on the summit of Mount Olympus, Olympic National Park. Intense glacial erosion has sculptured the resistant sandstones and conglomerates of the Olympic massif. National Park Service photograph.

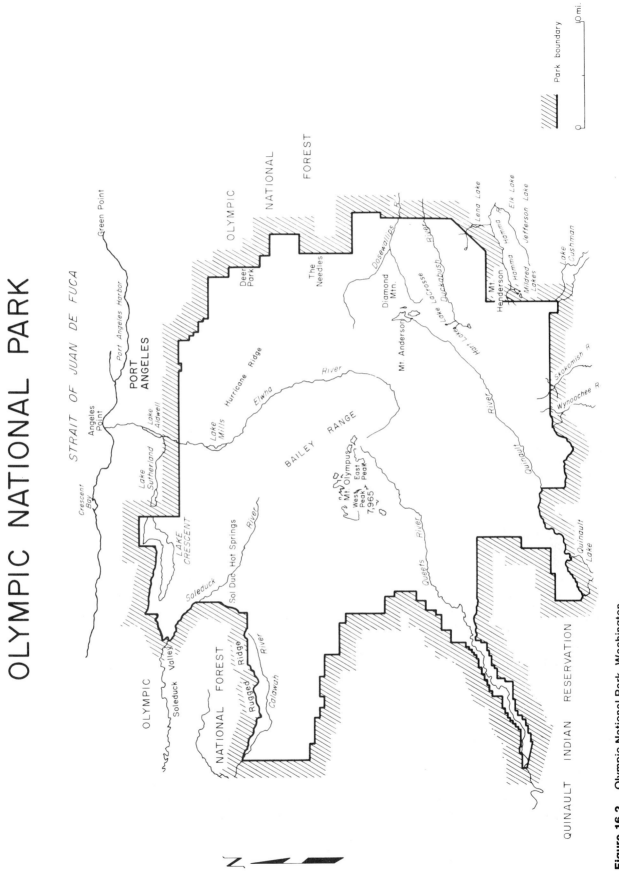

Figure 16.2 Olympic National Park, Washington.

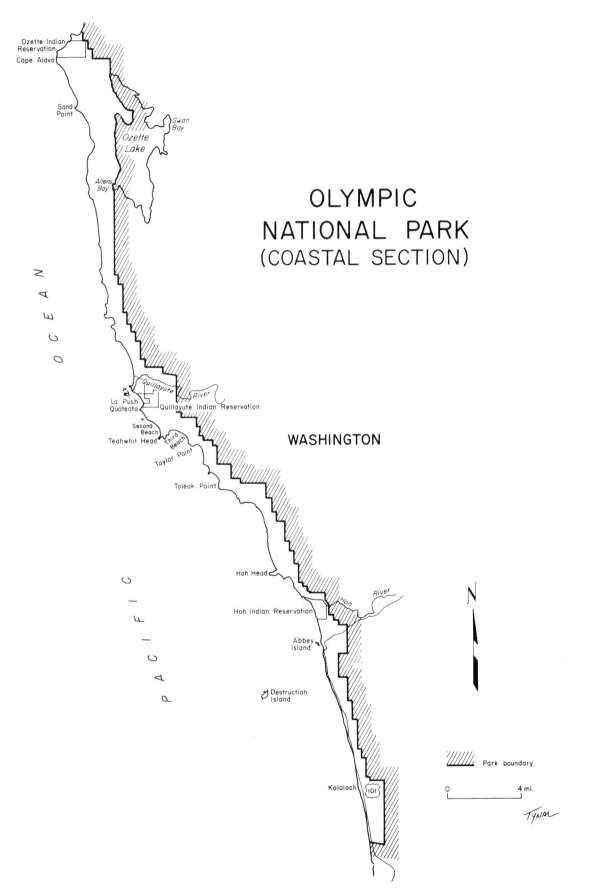

OLYMPIC
NATIONAL PARK
(COASTAL SECTION)

WASHINGTON

Ozette Indian Reservation
Cape Alava
Sand Point
Swan Bay
Ozette Lake
Allens Bay
O C E A N
Quillayute River
La Push
Quateata
Quillayute Indian Reservation
Second Beach
Teahwhit Head
Third Beach
Taylor Point
Toleak Point
Hoh Head
Hoh Indian Reservation
Hoh River
Abbey Island
Destruction Island
P A C I F I C
Kalaloch   101

N

Park boundary
0   4 mi.

TYNAL

Figure 16.2   Continued

Table 16.1. Generalized Geologic Column, Olympic National Park

Time Units			Rock Units	Geologic Events
Era	Period	Epochs		
Cenozoic	Quaternary	Holocene	Glacial drift, fluvial and landslide deposits, talus Beach materials	Erosion and deposition by streams, mass wasting, and wave action
			///// Unconformity /////	Unconformity
		Pleistocene	Younger glaciofluvial deposits and loess Older glaciofluvial deposits	Wind deposition on and near coast Fluctuation of sea level; uplift and warping Repeated episodes of alpine and continental glaciation
	Tertiary — Neogene		///// Unconformity /////	Unconformity
		Pliocene	Quinault and Quillayute Formations (western Olympic coast, not exposed in Park) Largely marine sedimentary rocks; slightly tilted.	Marine deposition in coastal areas Folding and upwarp
		—?—	///// Unconformity /////	Unconformity
		Miocene	Hoh rock assemblage (western Olympic coast), late Eocene to Miocene. Steeply tilted, overturned sedimentary rocks; turbidites; tectonic mélange rocks	Faulting and uplift Plate convergence and subduction accompanied by intense folding, shearing, slicing, partial metamorphism
	Tertiary — Paleogene	Oligocene	Core rocks (central and western) Eocene to Miocene highly folded sandstones, shales; some highly disrupted. Basaltic horseshoe (Crescent Formation); Eocene to Miocene peripheral sedimentary rocks	More or less continuous marine sedimentation
		Eocene	Inner Olympic rocks Eocene to Oligocene severely deformed sedimentary rocks and basalts	Submarine eruption of lavas
		Paleocene	///// Unconformity /////	Unconformity
			Pre-Cenozoic rocks of the old continent (not exposed in Park)	

Source: modified after Tabor 1975; Rau 1980

Local History

The natural features of Olympic National Park are astonishing in so many ways that it is difficult to avoid using too many superlatives when discussing the Park. Where else in the world can one find a wilderness area with such contrasting elements as—

—some sixty active glaciers on rugged mountains plus rain forests on the lower western slopes, both nourished by the highest annual precipitation in conterminous United States;

—a rain shadow zone on the northeastern slopes that is nearly as dry as the Baja California coast;

—a shoreline exposed to surf as violent and stormy as anywhere in the world;

—very young sedimentary rocks pushed up from the ocean floor by a plate collision that left them intermixed with submarine lavas in a chaotic, incredibly mangled heap;

—hot springs associated with faults;

—the largest remaining herd of Roosevelt elk left on earth;

—and in the offshore waters, seals, sea lions, and whales, plus thousands of marine birds nesting on rocks sticking up from the water.

Actually it was the elk for which the land was originally set aside and the forests preserved. A naturalist, C. Hart Merriam, discovered the Olympic elk, or wapiti, and named the new species for Theodore Roosevelt in 1897. In 1909, just before leaving office, President Roosevelt signed the proclamation that created Olympic National Monument. President Franklin D. Roosevelt, 29 years later, approved legislation that established Olympic National Park, protecting a much larger area (nearly a million acres in all) on the Olympic Peninsula.

Preservation of the wilderness habitat of the Olympic elk meant that the remarkable Olympic trees were also saved. In the rain forests of the Park are towering Douglas fir of record size, giant western hemlock and red cedar, Sitka spruce, and other kinds of trees. Some trees have diameters of 12 feet or more; some attain heights of 300 feet; many are between 100 and 200 feet tall. Beneath the umbrella of the giant trees, ferns, mosses, and smaller plants grow in profusion on the forest floor (fig. 16.3).

Higher up are forests of thin-spired, subalpine firs and alpine meadows near the treeline; then at the top, bare ledges, glaciers, snowfields, and the peaks of the massif.

The Olympic Peninsula, with its snow-capped mountains looming up, was a landfall for ships exploring the Pacific, perhaps as long ago as five centuries or more. Metal artifacts of great antiquity found at Indian village sites along the shore suggest possible contacts between the coastal Indians and voyagers of long ago from the Orient. A Greek pilot in Spain's service, Juan de Fuca, is said to have entered the strait that bears his name in 1592.

The first documented expedition was that of Juan Perez, a Spaniard, who was sent up the coast from California in 1774. The Spanish had heard rumors that Russians were moving down from the far north. In the following year Perez returned with Bruno Heceta. They explored the coast and went ashore at Grenville Bay (north of Grays Harbor) to lay claim to all of the land for the King of Spain.

The English navigator Captain James Cook searched these waters for a "northwest passage" in 1778. Although he failed to find the fabled route, he took on board a load of sea otter pelts that he sold for great riches in China, thus establishing the maritime fur trade. Other English fur traders followed. Captain Robert Gray (for whom Grays Harbor is named) was the first American trader to explore the "Northwest Coast." During his second voyage, from 1790 to 1793, he met Captain George Vancouver, who had been sent from England to chart the coast. Vancouver went on board Gray's ship "Columbia" in April 1792 near Destruction Island, and the two captains exchanged information about the coastal waters and harbors. In subsequent years American and English fur traders established trading posts along the Washington and Oregon coast, successfully keeping out both the Russians and the Spaniards.

Although the fur traders and the Indians had a thriving business relationship, hostile encounters between them did occur, especially after the Indians began demanding more goods for their pelts. These Indians were brave hunters and fierce fighters. Descendants of some of the tribes still live on reservations along the coast.

Greatly admired for their courage were the Makah Indians; they hunted whales in dugout canoes with spears and harpoons, the points fashioned from stone. Today much is being learned about the Makah culture because a few years ago at Cape Avala in the Ozette Indian Reservation, a hillside of clay slumped down, exposing a series of Makah villages that had been built on the site at different levels. Apparently when one village was covered by a slump or slide of clay, the Makahs would build a new village on the ruins of the old one. Many artifacts, such as paddles, baskets, weapons, carvings, and other items of the Indians' daily life, were sealed over and perfectly preserved by the clay. Anthropologists from Washington State University and Makah descendants living in the area have been excavating the site and cleaning and cataloguing the artifacts, many of which are now on display at a tribal museum in Neah Bay, a village on the Makah Indian Reservation. The museum and the area being studied are close to the northern end of the coastal strip belonging to Olympic National Park (fig. 16.2).

Figure 16.3 Mosses drape giant trees in the Hoh Rain Forest in Olympic National Park. Precipitation in this area may be as high as 150 inches a year. The couple standing on the trail (center) look tiny in comparison to the trees on either side. National Park Service photograph.

The interior of the Olympic Peninsula remained an unexplored wilderness for many years after the northern coast and the Puget Sound area were settled. On published maps the central mountainous area was left blank as late as 1890. Highway maps today show roads skirting the perimeter of the Peninsula but none crossing Olympic National Park from east to west or from north to south. Only logging roads and trails provide access to the Park's interior. A satellite photo or air view of the Olympic Peninsula shows great ridges in disarray, appearing to trend in every direction, with few passes through the mountains. Drainage is radial from the high area near Mount Olympus, the highest peak (elevation 7,965 feet). Streams rush down narrow valleys between steep interfluves. Getting from one drainage to another is obviously difficult, particularly where glaciers block the way. Below the treeline, dense forests impede one's passage.

The first official exploration of the mountains was led by Lieutenant Joseph O'Neill of the U.S. Army in 1885. He and his party managed to reach the top of Hurricane Ridge above the Elwha River, which flows north to the Coast near Port Angeles (now the site of Park Headquarters).

A newspaper, the *Seattle Press,* sponsored an expedition that set out in December 1889 to cross the Olympic Range. The five men in the party had planned to go up the Elwha River by boat, but that soon proved to be impossible. The sledges they built next to carry their supplies bogged down in the heavy wet snow. Then they tried mules, but one fell off a cliff and the other dropped from exhaustion. So they backpacked. After 14 weeks of climbing and searching for a divide that would get them to a southward-flowing stream, and with their supplies nearly gone, they finally found a river going to the southwest. Following this stream down, they met a white hunter with Indian guides, who took them by canoe to Lake Quinault. From there, they hiked to the Pacific shore and walked south along the beach to Aberdeen on the Chehalis River in the southern part of the Peninsula. The trip from Port Angeles to Aberdeen took them six months. What they had proved was that the mountains could be crossed. They had marked a trail of sorts, and had named many previously unknown peaks and rivers.

In the summer of 1890, Lieutenant O'Neill led a scientific expedition into the Olympic Mountains. In the party were a naturalist-cartographer and a botanist, as well as soldiers to assist them. With great effort, the party constructed a mule trail up the North Fork of the Skokomish River and down the Quinault. They also explored and mapped the South Fork of the Skokomish; the Wynoochee River, which joins the Chehalis; the Humptulips, which empties into Grays Harbor; and several minor streams. Finally they traveled down the Queets River (which flows through a section of the Park now called Queets Corridor) and reached the Pacific at the southern tip of the Park's coastal section.

The first group to reach the top of West Peak, the highest point on Mount Olympus, were eleven members of the Seattle Mountaineers Club, ten men and one woman, in August of 1907.

The Making of Geologic Maps

The geology of the Olympics remained obscure for a long time, even after the geographic explorations and much of the topographic mapping had been done. This was because of the difficulties of doing field work in such rugged and densely forested terrain, and also because the geologic structures and stratigraphic relationships are extremely complex. A noted geologist who in his youth was with a U.S. Geological Survey party in the rain forests described field work as "going sideways through the trees day after day, soaking wet." Another geologist who was with an early field party that was collecting hand specimens, measuring bed thicknesses, and taking strike and dip of beds, recalls "crawling up creek beds through the water, next to the bank and beneath overhanging bushes, because that was the only place where bedrock was exposed!" Field garb in those days consisted of heavy, long-sleeved shirts; "tin pants" (duckcloth impregnated with paraffin); hobnail boots; and leather gloves—the latter worn as protection against "devils club," a bush plant with poisonous barbs, very painful to the touch. By this kind of arduous, close-to-the-ground work, and hiking and climbing for months at a time, geologists finally accomplished the task of collecting field data necessary for the preparation of geologic maps of the Olympic Peninsula.

In making a geologic map, all outcrops visited are marked on a base map, and the type of rock, the strike and dip, and other pertinent observations are entered in a field notebook. Later, field notes are collated; and hand specimens are examined for fossils, bedding characteristics, size of grains, color, and the like. Eventually patterns of rock distribution are developed and their relative positions are studied. Using geologic maps based on field data and laboratory studies, together with scientific knowledge about nearby areas, geologists develop a model for the geologic history of the region. Such a model is by no means finite, for it is revised and refined from time to time as new data and new knowledge about the geologic history become available.*

Geologic Features

From early on, geologists have realized that the rocks of the Olympic Mountains are relatively young. For example, we might look at a sea stack (erosional remnant) of Miocene rock—only about 15 million years old—on the western Olympic shore and be awestruck by the fact that the beds are *vertical*. We may speculate on what force since Miocene time could have stood on end the beds that had been deposited and lithified on the sea floor in a horizontal position. Not far away on the shore at the Point of Arches in the Makah Indian Reservation are the oldest rocks of the Peninsula, some pre-Cenozoic outcrops of gabbro, at least 80 million years older than the oldest rocks in the Olympic Mountains. But even the oldest rocks of the Olympic National Park are still much younger than most of the rocks in the national parks that have been described in the preceding chapters.

A Plate Tectonic Explanation

What seems evident from the data available, is that most of the sedimentary rocks of the Olympics were deposited on basaltic oceanic crust, although some were laid down on an ancient erosional surface of submerged continental rock (i.e., igneous and sedimentary rock of the North American plate). According to plate tectonic theory, when an eastward-moving segment of oceanic crust (called the Juan de Fuca plate) encountered the westward-moving North American plate, the heavier basaltic oceanic crust was subducted or overridden by the lighter continental crust (fig. 16.4). In the process, during a period of serveral million years, layers of marine sedimentary rock along the edge were scraped off the plate, folded, compressed, thrust downward or squeezed upward, fractured, partially metamorphosed, sliced, and uplifted. The creation of the Olympic Mountains was one result of this powerful tectonic activity. Farther east the effects can be inferred (ac-

*See Tabor (1975), pp. 19–30, for an interesting account of how geologists developed theories for the origin of the Olympic structure.

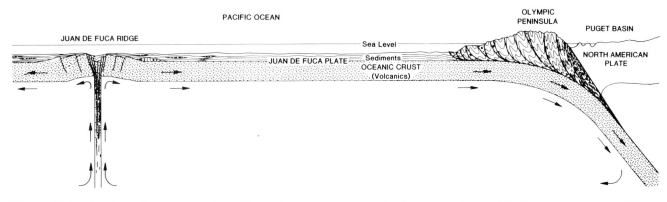

Figure 16.4 A schematic representation of how plate convergence may have produced the complex structures of the Olympic Peninsula. Oceanic crust, bearing marine sediments, moved eastward for millions of years from a spreading center along the Juan de Fuca ridge in the Pacific Ocean. When the heavier oceanic crust met the less dense continental crust of the North American plate, the oceanic plate was subducted. In the process, some of the sedimentary material was dragged down with the oceanic plate and some was scraped off and piled up against the margin of the North American plate, enlarging the continent by accretion. From Rau 1980, *Washington coastal geology between the Hoh and Quillayute Rivers,* Bulletin 72, State of Washington Department of Natural Resources.

cording to plate tectonic theory) in the upwarping of the North Cascade Range (chapter 14) and in later tectonic episodes involving the Cascade volcanoes (chapters 17, 18, 19).

While the sediments were still accumulating on the ocean floor, great quantities of basaltic submarine lava began welling up and spreading over or interfingering with the sedimentary layers. In some places lava flows piled up so high that offshore islands may have been formed. Most of the flows cooled quickly under water as pillow lavas or brecciated rocks, but some solidified more slowly in dikes and sills between sedimentary beds or in previously erupted lavas. The chemical composition of the basalt indicates that the lava rose from very deep in the earth and that it had not been previously mixed with surface or marine rocks. However, alteration of the basalt was not long in coming as the lavas were subjected to the same forces of deformation as the sedimentary rocks (fig. 16.5).

Between eruptions of pillow lava, during quiet intervals, a few beds of fossiliferous red limestone were deposited on the ocean floor. The fossils in the rock are Foraminifera (one-celled protozoa that secrete calcite), whose tiny shells can barely be seen with a magnifying glass. The red color of the rock is from iron leached out from the cooling pillow lava by hot sea water. Associated with the iron are manganese and copper. Some copper and manganese ores were mined early in this century, but the deposits were too small to be of much commercial value. The upper North Fork of the Skokomish River, one of the areas where the red limestones are associated with pillow lavas, was well explored by prospectors and miners.*

*A claim staked near Black and White Lakes yielded over a hundred tons of manganese ore that was carried out of the mountains by mule train during World War I. (The name of the lakes refers not to the color of the rocks but to a well known brand of whiskey.)

Because people were interested in the manganese and other minerals, the basalts of the Olympic Peninsula were mapped first. Geologists found that most of the basalts cropped out in an arcuate or horseshoe pattern, partly ringing the Peninsula, with the arc portion of the horseshoe on the east side and the arms extending westward. This feature, named the Crescent Formation (for exposures at Crescent Bay, not for the arcuate shape), is also referred to as the "basaltic horseshoe."

Since the fossil Foraminifera in the red interbedded limestone were dated as Eocene in age (Rau 1964), the associated lavas were also presumed to be Eocene. No older bedrock has been found on the Peninsula except for the small patch of rock (apparently from the old continent) on Point of Arches. In fact, the sedimentary layers between the Crescent Formation and the western coast become progressively younger from east to west.

The plate tectonic explanation is that the basalts peeled off and were wedged up against the older rocks of the continent as the ocean floor was underthrust and bent downward (fig. 16.5). Rocks that had been deposited seaward of the basalt pile were crammed accordionlike against the more rigid arch of basalt, and some beds were dragged down by the subducting plate. All the layers were severely deformed, fractured, and sliced by the process, as well as partially metamorphosed by heat and pressure. The younger rocks farther west were also deformed, but they were not so contorted and mashed up as the beds jammed inside the basaltic horseshoe. Microfossils (mainly Foraminifera) are more common in the younger beds. The lack of fossils in the older rocks is probably due to destruction of the shells by severe deformation and recrystallization.

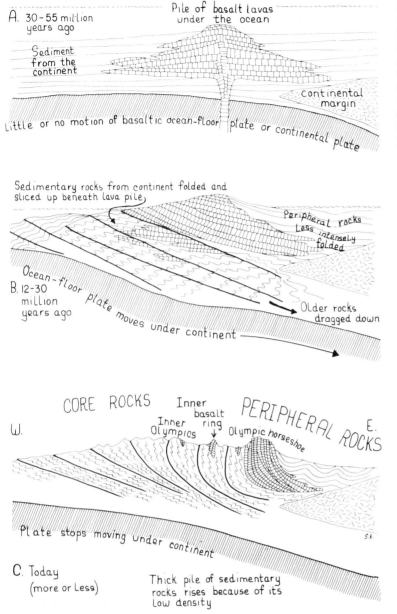

Figure 16.5 How the Olympic Mountains developed. *A.* Eruption and accumulation of basaltic lavas, interbedded with sediments on the ocean floor. *B.* Subduction of the oceanic plate, causing deformation of the sedimentary and volcanic rock layers. *C.* Uplifting of sedimentary rocks because of their lower density. From *Guide to the Geology of Olympic National Park* by R. W. Tabor. © 1975 University of Washington Press; used by permission.

Paleomagnetism

Study of the fossil Foraminifera helped geologists to better understand the structural and stratigraphic relationships of the Olympic sedimentary rocks, but, of course, no fossils existed in the lavas. Fortunately, the study of *paleomagnetism,* the record of ancient magnetic fields, which was developed in conjunction with the combined theories of sea-floor spreading and plate tectonics, proved to be helpful in discerning age relationships and interpreting structures.

Basalts (and numerous other igneous rocks) contain iron-bearing minerals, such as magnetite, that record the strength and direction of the earth's magnetic field *at the time* they crystallize in a cooling lava flow. The atoms in the crystals respond to the earth's magnetic field by "pointing" toward the location of the north magnetic pole. Unless the rocks are remelted, they permanently retain the magnetic alignment of the time of their crystallization, and this time can be ascertained by techniques of radiometric dating (box 13.1).

The polarity of today, that we depend upon for navigation, places the magnetic north pole close to the geographic North Pole. This is what we call *normal polarity.* However, paleomagnetic studies of ocean floor basalts and basalt flows stacked up on land have revealed that some lavas cooled with a magnetic orientation that is directly opposite of the earth's present magnetic alignment. This means that time after time during the geologic past *magnetic reversals* occurred when the magnetic north pole and the south pole exchanged positions so that the magnetic lines of force ran in the opposite direction; i.e., from north to south, *reversed polarity,* instead of from south to north, the normal polarity of the earth's magnetic field today.

Scientists have determined that the youngest basalts of the ocean floor are those most recently erupted from a spreading center between two oceanic plates. Outward from a spreading center, the basalts get progressively older on either side as the plates move like two conveyor belts taking the rocks farther and farther away from the point of eruption. Times of magnetic reversal show up in a banded pattern in the ocean-floor basalts, with each side (outward from the spreading center) being a mirror image of the other.

When it was discovered that the patterns of progressively older magnetic reversals recorded in the ocean basalts could be matched up with reversals of polarity shown in lavas on land, a powerful tool thus became available to geologists for determining ages and structural relationships of Olympic rocks. Much of the scientific basis for the concept that oceanic crust plunged beneath the North American plate near the present Olympic coast is derived from study of magnetic patterns in basalts on land and on the ocean bottom.

Faults and Hot Springs

Geologic mapping and interpretation of fault patterns in the Olympic core rocks appear consistent with the idea of rock layers—caught between converging crustal plates—being compressed and faulted so severely that the layers underwent slicing, thrusting, and stacking. In other words, rock layers that had been spread over a large expanse of sea floor were, during convergence and subduction, sliced up, stacked in a pile of great thickness, turned on edge, and curved into arcuate belts roughly concentric with the basaltic horseshoe. Faults separate the rock layers from each other and from the basaltic horseshoe, with the most intense disruption and metamorphic alterations occurring in the rocks in the center of the bend of the horseshoe (Tabor 1975, p. 30). Later, isostatic adjustments, or "rebounding," allowed the compressed and at least partially subducted rock layers to move upward. The sedimentary rocks, being less dense, tended to rise higher than the heavier basalt. Eventually, as fault movement continued, the Olympic core rocks rose to their present height.

The Calawah fault zone, a major fault on the Olympic Peninsula, follows a wide curving pattern within the basaltic horseshoe and shows evidence of strike-slip motion. The fault zone of sheared and shattered rock extends westward from Crystal Ridge (west of the Elwha River) to the Soleduck River and continues out to the ocean. To the east, the Calawah fault zone bends southward; it parallels the inner curve of the Crescent Formation (basaltic horseshoe), and then interlinks with a group of smaller faults and slate belts, called the "southern fault zone" (fig.

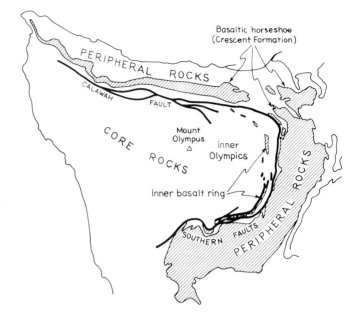

Figure 16.6 Location of the dominant rock groups and major faults of the Olympic Peninsula. From *Guide to the Geology of Olympic National Park* by R. W. Tabor. © University of Washington Press; used by permission.

16.6). In combination, the two fault zones clearly mark the separation of the simpler structure of the peripheral rocks from the highly deformed, complex rocks of the inner core.

The hot springs in the Park, which are associated with the fault zones, are not considered a sign of recent or current fault activity. Active faulting probably took place many millions of years ago when the rocks now exposed at the surface were deeply buried.

The water that issues from the hot springs is not noticeably different in chemical composition from surface water in the vicinity. The most likely explanation for the existence of the hot springs is that ground water percolates deep into the earth, is heated by the higher temperatures at depth, and then rises to the surface via the easiest, most direct route, which is usually through the shattered rock of the fault zones.

The Olympic Hot Springs flow from a hillside above Boulder Creek in the northern part of the Park near Olympic Hot Springs Campground. The hot water issues from rock in the Calawah fault zone. The Sol Duc Hot Springs (south of the Soleduck River) do not flow directly from the Calawah fault zone, but reach the surface by way of joints in sandstone beds adjacent to the fault. Elk Lick, which used to be a hot spring but is now a seep or boggy area used by elk as a wallow, is in the southeastern part of the Park where Elk Lick Creek joins the West Fork of the Dosewallips River. This lick, or spring, is near a branch of the southern part of the Calawah fault zone.

Turbidites, Tectonic Mélanges, and Piercement Structures

In spite of their rather forbidding names, these rock features are often so striking in appearance that a hiker seeing one is apt to stop and ask, "What's *that*?" To a geologist, however, these features—whether found singly or in association—may provide significant clues when he or she is studying a locality and attempting to reconstruct past geologic events.

Turbidites are rocks (or sediments) that were deposited in a marine environment by settling out from muddy or turbid water that flowed along a sloping ocean bottom. Such a sediment-laden flow is called a *turbidity current.* Under the influence of gravity, currents of this kind may flow quite swiftly. Storm waves may start them, or an earthquake that causes undersea slumping or sliding, or even perhaps a flooding river pouring huge quantities of sediment into the ocean. As a turbidity current slows down, the sediments tend to sort or grade themselves by size so

that the coarsest particles drop first and the finest are deposited last. Turbidite sequences are usually deposited one on top of the other, each sequence grading from coarse to fine, then to coarse again, and so on. When consolidated, the beds become siltstones, sandstones, and conglomerates.

In Olympic National Park, turbidite beds are well exposed in vertical and overturned positions on the western shoreline (fig. 16.7) and in the high mountains. Some have been fractured, as well as folded and overturned, and the fractures have been subsequently filled with white veins of calcite.

Since tectonic mélanges are associated with intense deformation and shearing, their occurrence in the Olympic rock sequences is not surprising. By definition, a *tectonic mélange* is a body of rock (large enough to be mapped) that includes fragments and blocks of numerous sizes and kinds of rock embedded in a generally sheared matrix of more malleable or tractable rock, such as a claystone or shale. Some of the Olympic mélange blocks

Figure 16.7 Turbidite beds of the Hoh Formation that have been overturned by compressive forces. These rocks in the coastal section of Olympic National Park are being eroded by wave action. The dashed line indicates an older wave-cut surface eroded at a time during the Pleistocene when sea levels were higher. Above the line are reworked glacial, stream, and beach materials that were deposited subsequently on top of the older surface. Compare with figure 16.10. Photograph courtesy W. W. Rau and State of Washington Department of Natural Resources.

Box 16.1
Oil on the Olympic Coast?

Oil seeps, traces of oil in coastal streams, and a natural gas seep on the western coast of Washington prompted efforts aimed at recovering petroleum in the early years of this century. None of these attempts was a commercial success. The test wells were drilled in the marine shales and sandstones of the Hoh rock assemblage. Oil well sites that are now within Olympic National Park will not be explored further, but several localities outside the Park are of potential interest to petroleum companies.

The early drillers had trouble with the high clay content of subsurface beds. They called such rock "heaving shale" because of its aggravating (to them) tendency of swelling up when wet, closing off drill holes, buckling pipes, and so on. Then there were horrendous problems in getting equipment to drill sites on a coast where an overland approach was impossible, safe natural harbors were few, and ships foundered in heavy surf on dangerous offshore rocks.

The first well drilled, around 1900, was at the top of a mélange rock bluff at the north end of Third Beach. A barge carrying equipment and supplies was destroyed during the landing attempt at the beach, but most of the machinery was salvaged. A donkey engine was dismantled and carried piece by piece up to the site where it was reassembled and used to haul the rest of the equipment from the beach to the top of the cliff. Things didn't get much better when the crew finally began to sink a hole. Drilling in the mélange rocks was fraught with difficulties. A slight "show" of petroleum must have buoyed their hopes for a while, but drilling conditions became worse and worse. Finally the site was abandoned, along with the machinery and equipment that had been so arduously brought in. An old steam boiler and other equipment are still rusting away in the underbrush, close to one of the Park trails.

The Jefferson Oil Seep, also within the present boundaries of the Park, was a more promising site. It was explored initially in 1913-14 and again in the 1930s. Being near the mouth of the Hoh River, the locality was much less difficult to get to. Eleven wells were drilled in the 1930s, and a land development called "Oil City" (now a ghost town) was begun. Some petroleum was recovered, but production was never high enough to be profitable. Today the wooden derricks have tumbled down, but oil still seeps out at the surface in small amounts and natural gas bubbles up.

Drilling has been tried since that time in the Hoh valley in areas outside the Park, but no wells have produced oil in commercial quantities.

What are the conditions that can make possible the recovery of petroleum from the ground? First of all, there has to be a *source rock* consisting of marine sediment containing organic matter buried during sedimentation. From the decayed organisms, droplets of petroleum (hydrocarbons) form and migrate through pores in the rock, generally upward because oil is lighter than water. Unless a geologic structure is present that can "trap" and retain the migrating droplets, they may rise all the way to the surface, seep out, and be lost forever. An example of a favorable structure might be an anticline with a layer of impermeable rock that could prevent upward migration of droplets. A third requirement is a permeable *reservoir rock* (below the impermeable rock) that has enough porosity to hold the accumulating petroleum. The Olympic coastal sedimentary beds have a source rock, but apparently the right combination of reservoir rock and appropriate geologic structure is yet to be found.

are as large as houses, which gives an indication of the tremendous force involved in their formation. In the Park's coastal section, along the western beaches, is the best area to see the mélange rocks. Since moisture causes the clay minerals of the matrix to swell, this part of the rock becomes crumbly, making it susceptible to wave erosion and separable from the resistant inclusions. Locally these rocks (which are part of the Hoh rock assemblage) are called "smell rocks" or "smell muds" because of their strong petroleum odor. Microfossils in the mélange siltstones indicate that the depositional environment was a deep-water ocean basin beyond the continental shelf.

Piercement structures are associated with tectonic mélange rocks. Under conditions of high pressures, certain rocks become "plastic," or able to flow; they are squeezed upward into overlying rocks, doming them or even breaking through them. In this way, older rocks may be forced up through younger beds, the result being what is called a piercement structure.

The Shaping of the Olympic Landscape

Running Water. All during the time that great masses of rock from the ocean floor were being pushed higher and higher above sea level, the agents and processes of erosion were actively shaping the surface of the newly risen land. Moisture-bearing clouds from the Pacific discharge rain and snow copiously on the western slopes; an average of 140 inches of precipitation falls per year at present, and it probably was at least that much in the past.

Water running off in every direction from the central "dome" initiated the radial drainage, and the swiftly descending water cut steep, narrow valleys. To some extent, the differing hardness of the rock types has made for modifications in the original radial drainage. The basalt ridges tend to be higher and sharper in relief than the less resistant sedimentary beds surrounding the basalt. Since the sandstone beds are harder than the shale beds, the tributaries in the latter tend to cut more deeply than the tributaries that flow over sandstone. As a result, some of the streams in the deeper valleys have captured the drainage of the streams on more resistant rocks.

Also affecting how the radial drainage is evolving is the contrast between the amount of moisture on the western slopes and the amount that falls in the rain shadow zone on the eastern slopes. In the city of Sequim, for example, outside the northeast corner of the Park, precipitation is only 16 to 18 inches per year, or a little more than a tenth of what falls on the western slopes. Tributaries on the east side of the Olympic Mountains generally have steeper gradients, are less numerous, and carry smaller volumes of water than those on the west side.

Glacial Ice. The Olympic landscape shows much evidence of severe glaciation in the past; and although glacial ice is still a significant shaping agent, it is secondary to running water at the present time. What is most remarkable about existing Olympic glaciers is their low elevation on the western slopes due to the abundance of snow that keeps them well nourished. Above the snowline, at around 6,000 feet, winter snows last through the summer season, and some of the glaciers extend down their valleys to as low as 4,500 feet. At three miles in length, the Hoh glacier is the longest.

Largest in size is the Blue Glacier, which has been closely monitored by glaciologists since 1938. A University of Washington glacier research station, built close to the Blue Glacier in 1957–58, collects data and makes observations continuously as part of a study to discern relationships between regional climate, accumulation and evaporation of snow and glacial ice, melting rates, velocity of ice flow, etc. Measurements show that in recent years the Blue Glacier has grown slightly.

Moraines, U-shaped valleys, cirques, rock-basin lakes, erratics, and other glacial features beyond the limits of the existing valley glaciers attest to the more extensive alpine glaciation of Pleistocene time. Because of rapid weathering and erosion in the Olympics, some of these features are not so long-lasting as similar glacial features in more resistant rock or a drier climate (as in Yosemite National Park, chapter 13).

Rock-basin lakes, for example, tend to have short lives in the high Olympics. As a lake silts in, vegetation spreads over the shallows. Soon the lake is a bog; and then it is a meadow, with perhaps a stream winding through it. If the basin is below the treeline, the forest eventually takes over the open space. Some of the high, glacially carved valleys have lakes in various stages of evolution, a good example being the Mary Ann Lakes in a hanging valley above the Quinault River. The highest basin holds snow a good part of the summer. The main lake a little farther down is beginning to fill in with moss. Next a meadow that was once a lake is giving way to trees. Below the meadow are boggy clearings among the trees.

Seven Lakes Basin on north-facing slopes at the headwaters of the Soleduck River will probably retain its glacial lakes longer, because the rock basins are largely the result of glacial plucking that removed blocks from the thick beds of sandstone. Not much silt is available from the sandstone ridges to fill in the lakes. Moreover, ice left the north-facing slopes more recently, so vegetation is not so well established.

Marmot Lake, Hart Lake, and Lake La Crosse are cirque lakes left by glaciers that once flowed down the Duckabush River. One of the most beautiful glacial troughs in the Park is the flat-floored Enchanted Valley, between steep mountain walls that lead up to hanging glaciers on either side of Mount Anderson. This is the headwaters region of the Quinault River. Farther down, just south of the Park boundary, is Lake Quinault (in the Quinault Indian Reservation), dammed by a glacial moraine. These are but a few examples of the many interesting alpine glacial features that can be visited in the Olympic wilderness.

Part of the Peninsula was covered by advances of continental ice. Numerous granitic and gneiss pebbles in the stream gravels on the north and east slopes are among the indications of continental glaciation. Presumably the pebbles were dumped by invasions of Cordilleran ice from British Columbia, because no granite or gneiss bedrock is exposed in the Olympic Peninsula. Also foreign to the Peninsula are some granitic and metamorphic boulders and cobbles dropped by icebergs in fiordlike lakes that at one time filled valleys on the north and east sides of the mountains. The Cordilleran ice sheet dammed these valleys after the alpine glaciers that carved them melted back.

Icebergs broke off ("calved") from the waning ice sheet and floated up the lakes, eventually melting and dropping their rock debris.

On the northern side of the Peninsula, continental ice sheets at least 3,500 feet thick scraped and excavated the weak peripheral rock (on the outer side of the basaltic horseshoe), destroying much of the preglacial topography and generally lowering relief along the coast.

Mass Wasting. Processes of mass wasting were most effective during times of glacial retreat when freshly excavated valley slopes, unprotected by vegetation, were unstable. Sometimes part of a ridge collapsed when it was no longer supported by glacial ice. Surface layers of rock on oversteepened cliffs or cirque walls slid down and became piles of rubble. Tilted beds of rock units are susceptible to rock sliding and avalanching, and many examples of this rapid type of mass wasting can be seen in the Park (fig. 16.8). A recent rock slide occurred near the terminus of the Hoh Glacier, which has retreated more

Figure 16.8   This view looks south toward Mount Olympus, about ten miles away (air distance). The Bogachiel River rises from the cirque valley in the foreground. The steep slopes of the ridge, which are underlain by tilted beds of late Tertiary sandstone, are mantled by thick sheets of talus resulting from frost action and mass wasting. A dense forest cover, supported by ample precipitation, has begun to stabilize some of the slopes. The valley of the Hoh River is between the ridge and Mount Olympus. National Park Service photograph by George A. Grant.

than a mile since the early 1800s. Some 20 or more years ago, cracks formed on the ridge crest, and the unsupported valley wall gave way; this left a fresh scarp on the valley side slope and a pile of rock debris on the moraine below the glacier. Throughout the Park blocks of rock fall from steep cliffs as weathering by frost action and chemical decomposition (aided by the plentiful supply of moisture) attack the bedrock.

A number of lakes in the Park have backed up behind natural dams which have resulted from landsliding. Three examples are Jefferson Lake, Elk Lake, and Lena Lake in the southeastern part of the Park near the Hamma Hamma River. On the Elwha River, in the fairly recent past, a large rockslide dammed a narrow canyon, forming a lake. In 1967 the dam gave way, releasing the impounded lake water, which washed out a trail bridge and spread gravel over the flood plain downriver. The river had caused the rockslide in the first place by undercutting the base of an unstable slope.

A great landslide mass (that probably came down soon after the last Cordilleran ice sheet retreated) separates Lake Crescent, the largest lake in the Park, from Lake Sutherland. The two lakes were originally one body of water that occupied a long glacial trough drained by Indian Creek, which flows eastward into the Elwha River. The landslide, which was caused by the collapse of the valley side slopes near Mount Storm King, dammed the valley and cut off Lake Sutherland. Lake Crescent cut a new outlet, forming the Lyre River which flows north into the Juan de Fuca Strait. Indians may have witnessed the catastrophic landslide because an Indian legend describes the spirit of Mount Storm King becoming so angry over tribal warfare going on in the valley that he threw down great rocks that blocked the valley and, naturally, put an end to the fighting.

Creep is another form of mass wasting that is actively, but slowly, moving slopes downward in the Olympic region. Counteracting the processes of mass wasting, however, is the thick vegetative cover on most slopes which tends to make them more stable. Areas outside the Park that have been logged over are less protected from erosion.

The Olympic National Park Coastal Strip and Its Shore Features

Between the west coast of Washington and the islands of Japan are 6,000 miles of open ocean. Impelled by westerly winds, waves travel unimpeded across this vast stretch of water. During storms, waves 15 to 20 feet high beat on the unprotected shore, often for days at a time in winter. This is a *high-energy coast* because large waves continually come in on the beaches and cliffs. In other words, a great deal of wave energy is available throughout the year

to erode the shore and transport beach material. Comparison of coastal surveys reveals that the Washington Pacific coast (near Hogsback) was cut back 225 *feet* in 60 years. At this rate, 375 feet of shore are being eroded back every 100 years!

The contrast between the high rate of coastal retreat on the Olympic coast and the relatively low rate of erosion of the coast of Acadia National Park (chapter 15) is remarkable. But the Acadia shore is by no means a *low-energy coast* (i.e., having little wave energy available). The difference between the North Atlantic waves that beat on unprotected parts of the northern Maine coast and the somewhat more powerful North Pacific waves is not that significant. What makes the difference is the character of the bedrock. The contrast we see is between the effects of wave erosion on very old, highly resistant, igneous and metamorphic rocks and wave action on very young, relatively nonresistant sedimentary rocks (fig. 16.9).

The coastal section of Olympic National Park is a narrow strip extending about 40 miles from the mouth of the Queets River on the south to Cape Alava and the Ozette Indian Reservation on the north. Two small Indian reservations are located within this section of the coast,

Figure 16.9 At Third Beach, near Taylor Point, blocks of conglomerate (Hoh Formation) have fallen to the beach from sea cliffs that are being cut back by wave action. Rounded pebbles and cobbles and fragments of a variety of rock types make up this poorly consolidated conglomerate. Photograph courtesy W. W. Rau and State of Washington Department of Natural Resources.

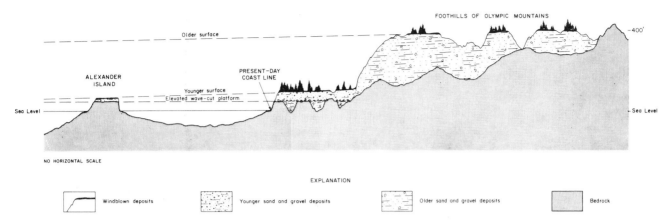

FOOTHILLS OF OLYMPIC MOUNTAINS

Older surface

ALEXANDER
ISLAND

PRESENT-DAY
COAST LINE

Younger surface
Elevated wave-cut platform

Sea Level

Sea Level

NO HORIZONTAL SCALE

EXPLANATION

Windblown deposits   Younger sand and gravel deposits   Older sand and gravel deposits   Bedrock

Figure 16.10   Simplified cross section of the Olympic coastal area showing the relationship of sediments to older wave-cut surfaces during Quaternary time. The position of the shoreline has changed several times to elevations both higher and lower than at present because of tectonic uplift of the Olympic Peninsula and also changes in sea level associated with Pleistocene glaciations. The older land surface (top), built by glacial and fluvial sediments from the higher parts of the Olympic Peninsula, extended seaward for many miles beyond the present shoreline. As the coast was uplifted, the sediments were eroded back and a younger wave-cut surface developed. This surface was, in turn, covered by a thinner series of sediments that extended westward of the present shoreline. Wave action following uplift has eroded the sediments back to the present shoreline, leaving Alexander Island as a remnant of the former mainland. Uplift, sea level change, and erosion continue on this tectonically active coast. From Rau 1980, *Washington coastal geology between the Hoh and Quillayute Rivers,* Bulletin 72, State of Washington Department of Natural Resources.

the Hoh Indian Reservation beside the Hoh River and the Quillayute Indian Reservation at the mouth of the Quillayute River. A coastal highway (Route 101) follows the shore from Queets to the Hoh Indian Reservation. The rest of the Park's coastal section is a wilderness area.

In the last 8 million years or so (since early in Pliocene time), the Olympic coast has been at times higher and at times lower than present sea level. Some wave-cut platforms are several hundred feet higher than sea level today, and some old wave erosional surfaces are covered by alluvial deposits that once extended to a shoreline hundreds of feet to the west. Because sea level fluctuated considerably, the shoreline has also shifted eastward and westward a number of times. The raising and lowering of sea level was partly due to Pleistocene glaciation that drew moisture from the oceans when the glaciers were enlarging and then released water as they melted back. But tectonic warping and minor uplift of this coast was also occurring during the Pleistocene, making interpretation of features of emergence and submergence difficult (fig. 16.10).

Rapid erosion also alters the position of the shoreline. Cliff retreat produces a widening, submerged wave-cut platform that extends several hundred feet out from the water's edge. Stacks and small islands of more resistant material that were recently part of the mainland now sit on the wave-cut platform surrounded by water. Sea arches collapse, becoming stacks. Caves in cliffs are eroded into sea arches (figs. 16.11, 16.12).

Figure 16.11   A sea stack of massive sandstone stands offshore from Taylor Point on the Olympic coast. The bedrock here is in the sedimentary sequences of the Hoh Formation. Intense weathering and wave action in the surf zone produced the wave-cut notch around the base of the stack. Photograph courtesy W. W. Rau and State of Washington Department of Natural Resources.

Figure 16.12   A sea arch eroded by wave action along a fracture zone in massive sandstone of the Hoh Formation. Second Beach, Olympic National Park. Photograph courtesy W. W. Rau and State of Washington Department of Natural Resources.

As the cliffs are cut back and landsliding occurs, the forests that grow right to the edge of the shore begin to sag and slump down. On the north side of the mouth of the Hoh River is a "drunken forest" where live trees are attempting to bend and grow vertically again after being slumped down and tilted by a slide. Many drunken forests can be seen along the coast.

Beaches are sometimes covered with dead trees that have fallen and great logs that have floated down the rivers to the ocean and then washed ashore. When storms come, these logs act as battering rams hurled up against the rocks by the waves.

The beaches between the headlands tend to be narrow and have fairly steep slopes. In some areas the steep cliffs come right down to the water. Nevertheless, the Olympic beaches have been the most convenient "pathway" from one place to another along the coast since prehistoric times because of the difficulty of traveling inland through the thick vegetation and the rain forests. Today beach hikers can follow well marked trails over the cliffs between stretches of beach. However, the warning still holds that hikers must guard against becoming trapped on a narrow beach at high tide, because most of the cliffs are too steep to climb. The beaches must also be avoided when heavy surf or storm waves are running. (In any wilderness area, taking normal precautions is sensible.)

From Captain George Vancouver on, what has fascinated visitors to Olympic beaches are the unusual landforms and rocks. In his ship's log (April 28, 1792), as he was sailing north up the coast, Vancouver wrote, " . . . The shores we passed this morning . . . were composed of cliffs rising perpendicularly from a beach of sand and stones [and] had many detached rocks of various romantic forms . . ." (quoted in Rau 1973).

Most of the rocks and cliffs in the Park's coastal section are in the Hoh rock assemblage of mostly Miocene age. The Kalaloch Rocks, offshore from the Kalaloch Ranger Station, are the southernmost exposure of the Hoh turbidites along the coast. They are mainly sandstones and conglomerates, but they are called "graywackes" because they contain fragments of other rocks (as well as individual grains). Most of the beach sand is derived from these rocks (or from sediment washed down from the mountains). Blocks of tectonic mélange Hoh rocks form many of the stacks and other strange-shaped erosional remnants sticking up in the water close to shore. The blocks from the mélange rocks are jumbled masses of resistant basaltic breccias, conglomerate, sandstone, and siltstone. Much of the original matrix of softer rock has been removed by weathering and erosion. Some of the mélange rocks, such as those in the bluffs at Jefferson Cove, are thought to be part of piercement structures.

Overlying the Hoh rocks along the shore are Pleistocene deposits of sand, gravel, and silt, most of them alluvial and derived from glacial outwash that poured out of the mountains as the valley glaciers retreated. The youngest Pleistocene deposits consist of fine, wind-deposited silt, called *loess*, that tend to form bluffs because the silt grains, being coherent, cling together. The loess at Quateata headland (First Beach) stands directly on an old wave-cut platform of bedrock and gravel; and at Taylor Point (Third Beach), the wind-blown deposits rest on older alluvial sediment. The loess accumulated near the end of the Pleistocene Epoch when the ice was melting back rapidly but sea level was still quite low. Large areas of land, unprotected by vegetation, were uncovered and left exposed to unusually high winds. Fine surface sediment, picked up by the winds, was redeposited over the coastal area.

The beaches between the headlands are made up of sand, gravel, and cobbles of many sizes and different composition, derived from the local bedrock, from glacial drift, and from sediment brought down by rivers from the Olympic Mountains. Here and there patches of red sand have accumulated. The red sand grains are tiny garnet crystals that have weathered out of erratics or been transported in glacial debris and then eroded by streams. Being heavier than quartz and other sand grains, the garnet crystals tend to be concentrated in stream beds and beach terraces. Presumably the red sands have been transported, concentrated, re-eroded, and reconcentrated several times on their journey from their places of origin to the shore.

Box 16.2.
How Waves Do Geologic Work

Wind-driven water waves erode, transport, and deposit material in a narrow coastal zone where land, water, and air meet. The source of wave energy is wind. In the forming of waves, energy is transferred from the moving air into wave shapes that move through the water. Note that the water itself does not move very far; rather it is the waves, impelled by the wind, that transmit the wind's energy, releasing it when they strike land. The size of waves is a function of (1) wind velocity, (2) the length of time that the wind blows in the same direction, and (3) the *fetch,* or distance, of open ocean over which the wind is blowing.

Great waves may be several tens of feet high and several hundred feet long. Once wave sets develop in an open ocean, they may move across hundreds of miles of sea before they break against a coast, their power and size unaffected by lighter, or changeable local winds.

When deep water waves begin to "feel" a shallowing bottom near shore, their lower parts begin to flatten due to the drag while their upper parts become shorter and steeper. Within a few seconds, the top of the wave falls forward, becoming a curling, breaking wave, or *breaker,* that slams against a cliff or rushes up a beach with great force and speed. With the wave's forward motion stopped, the water runs back down the beach face and slides under the curl of the next breaker. This upwash and backwash (called "swish and swash") breaks up and abrades rock and sediment, sorting coarser fragments from finer particles and rolling them back and forth. Coarser sediment tends to accumulate along the top of the beach while finer materials tend to stay in the *surf zone* (the strip of shore where the waves break and roll up the beach or beat against a cliff face). The surf zone shifts up and down the beach twice a day as the tides rise and fall. Thus the day-by-day erosion by waves is most intense on the *foreshore,* the area between mean low water level and mean high water level.

Storm waves, however, attack any part of the shore they can reach, often with devastating effect. Sizable rocks may be picked up by an incoming storm wave and tossed high into the air, often shattering as they hit. Tons of sand may be removed from a beach in a single storm.

Most beach sediment is only temporarily at rest, which is why beaches are sometimes described as "rivers of sand." Sand is moved along beaches by *beach drift.* When waves come in obliquely, as they often do, instead of head on, sand is carried up the foreshore at an angle. But the grains of sand roll back, under the influence of gravity, perpendicularly, thus moving along the beach in a series of looping movements.

Sometimes water "piles up" just outside the surf zone and starts moving in a *longshore current,* parallel to the shore, carrying sand from the bottom. When the current slows down or is deflected by a headland, the sand drops and may form a bar or *spit* (a fingerlike extension of a beach, with the far end terminating in open water). This type of sand transportation and deposition is called *longshore drift.* Sediment entering the ocean at the mouth of a river may encounter a longshore current and be carried up (or down) the shore, along with beach sediment. Meanwhile, the mouth of the river has to keep changing its location to get around the sandbar or spit built by the longshore current. Records show that the Quillayute River entered the Pacific at several different locations in the late 1800s and in this century before 1931, when the U.S. Army Corps of Engineers constructed a harbor at La Push and stabilized the mouth of the river with dikes.

Geologic History

1. The Pre-Cenozoic rocks of the old continent.

Little is known of these older rocks. Some were stripped off by erosion and probably served as a sediment source for the marine sedimentary rocks that make up the Olympic core. On the Olympic Peninsula, pre-Cenozoic rocks crop out only at Point of Arches just to the north of the Park's coastal section.

2. The Crescent Formation (basaltic horseshoe) and the peripheral rocks.

In early Eocene time the lavas of the Crescent Formation began to pile up on the ocean floor, probably fairly close to the old continent. Eruptions continued into middle Eocene time, even as subduction of the oceanic plate proceeded. As the plate moved down, the thick mass of basalt acted as a buttress, protecting the sandstones, shales, and conglomerates of the peripheral rocks deposited between the accumulated lavas and the continental margin, so that they were only moderately deformed. The sedimentary peripheral rocks are Eocene to Upper Miocene in age (fig. 16.13).

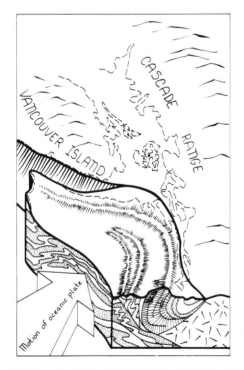

Figure 16.13 Diagram showing how opposing movement of oceanic and continental plates may have jammed the rocks that make up the Olympic Mountains into a tight arc between Vancouver Island and the Cascade Range. From *Guide to the Geology of Olympic National Park* by R. W. Tabor. © 1975 University of Washington Press; used by permission.

3. Eocene to Oligocene rocks of the inner Olympics.

While protecting the peripheral rocks, the Crescent Formation resisted the eastward movement of the sedimentary rocks that were deposited on the seaward side of the lavas. Therefore, the marine basalts, sandstone, phyllites, slates, and conglomerates within the arc of the horseshoe are most thoroughly metamorphosed, foliated, and recrystallized between the basalt buttress and the younger beds to the west. All the innermost rocks have been highly disrupted by slicing, shearing, and fracturing. These are the rocks that were scraped off and partially underthrust by the descending oceanic plate and subjected to heat and pressure in the process.

4. The core rocks from the inner Olympics westward.

Both the inner Olympic rocks and the core rocks are grouped in "lithic assemblages," rather than formations, because boundaries and specific rock types are not well defined. As is evident on a geologic map, the lithic assemblages tend to have an arcuate pattern, similar to that of the Crescent Formation; again there is the suggestion that they were jammed up against the basaltic horseshoe by the convergence of the plates.

The core rocks range in age from Eocene to Oligocene, with the beds closer to the inner Olympics rocks being more intensely folded and disrupted. Many of the core rock units show a well developed slaty cleavage, which indicates some degree of metamorphism. They consist of sandstones and shales with minor occurrences of conglomerate and basalt.

Mount Olympus and the other high peaks in the center of the Peninsula are in what is called the "Western Olympic lithic assemblage," a generally younger, less deformed group primarily made up of thick-bedded sandstone, with elongated patches of slate, phyllite, foliated sandstone, and semischist.

5. The rocks of the Olympic Peninsula western coast.

The greater part of the rocks on and near the Pacific coast are in the Hoh rock assemblage of mostly Miocene age. (Some Eocene volcanics interbedded with siltstones are exposed at Point Grenville, south of the Park's coastal section.) The Hoh turbidites and tectonic mélange rocks form many of the prominent shore features described earlier.

6. The Quinault and Quillayute Formations (not exposed in the Park).

The Quinault rocks, Pliocene in age and largely of marine origin, are separated from the Hoh rock assemblage by an unconformity. Quinault outcrops can be seen only on or near the shore in the Quinault Indian Reservation, just south of the Park's coastal section. However, for about 40 miles northward along the coast from the outcrops, the Quinault Formation extends seaward beneath the Quaternary sediments of the continental shelf for about 20 miles. The formation is of interest because it is regarded as a potential petroleum reservoir. The Quinault beds are only mildly tilted and were deposited quite some time after the Hoh rocks had been raised above sea level, eroded, and again partly covered by the ocean.

The Quillayute Formation, very late Miocene in age, is similar to the Quinault Formation in rock type and structure. The Quillayute Formation crops out in the Quillayute River valley (upstream from the coast), in a few places where thick Pleistocene deposits have been eroded away.

7. Pleistocene glaciation and deposition.

Alpine glaciers have probably existed on the highest mountains of the Olympic Peninsula since the onset of the Pleistocene and have advanced and retreated many times. At their maximum extent, the largest valley glaciers on the west side reached as far as the present coast (and possibly beyond), although parts of this coastal region were left unglaciated.

The far northwest coast and all of the north and east sides of the Peninsula were heavily glaciated, because the great Cordilleran ice sheet from British Columbia encountered the Olympic glaciers as it moved southward and then divided into two lobes. One lobe filled the Puget Sound lowland as far south as Tacoma, while the other spread over the Juan de Fuca Strait and the northern Peninsula coast. At least four times, possibly six, the ice sheet advanced and retreated. Cordilleran ice on the northern side of the Peninsula, at least 3,500 feet thick, scraped and excavated the weak peripheral rocks (on the outer side of the basaltic horseshoe), destroying much of the preglacial topography and generally lowering relief along the northern coast.

During warmer interglacial periods, when the glaciers melted back, great quantities of glaciofluvial deposits were laid down over the lowlands and coastal areas. The Pleistocene sands and gravels on the west side and part of the south side are better preserved, because they were not reworked (except in the glacial valleys) by later glaciations. Exposed on the Pacific coastal region are several sequences of Pleistocene and Holocene deposits. The oldest ones, which are semiconsolidated, originally extended for a considerable distance farther to the west than now, but they have been cut back by wave erosion.

Sea level has fluctuated on the Peninsula, both during and since the Pleistocene. Some fluctuations have been due to depression by glacial ice and subsequent release by melting; and some changes have been the result of tectonic uplift and warping (fig. 16.10).

8. Modern glaciers and erosional processes.

In recent years Olympic glaciers have shown slight increases in volume, but stream erosion in the Park at present is dominant over glacial erosion. Mass wasting is also a significant factor. In the Park's coastal section, wave erosion is unusually rapid compared to most other coasts.

Geologic Maps and Cross Sections

Brown, R. D., Jr.; Gower, H. D.; and Snavely, P. D., Jr. 1960. *Geology of the Port Angeles—Lake Crescent area, Clallam County, Washington.* U.S. Geological Survey Oil and Gas Investigations Map OM 203.

Rau, W. W. 1975. *Geologic map of the Destruction Island and Taholah quadrangles, Washington.* Washington Division of Geology and Earth Resources Geologic Map GM-13.

———. 1979. *Geologic map of the vicinity of the lower Bogachiel and Hoh River valleys and the Washington coast.* Washington Division of Geology and Earth Resources Geologic Map GM-24.

Tabor, R. W., and Cady, W. M. 1978. *Geologic map of the Olympic peninsula, Washington.* U.S. Geological Survey Miscellaneous Investigations Series Map I-994.

———, Yeats, R. S., and Sorenson, M. L. 1972. *Geologic map of the Mount Angeles quadrangle, Washington.* U.S. Geological Survey Geological Quadrangle Map GQ-958.

Bibliography

Cady, W. M. 1975. Tectonic setting of the Tertiary volcanic rocks of the Olympic peninsula, Washington. U.S. Geological Survey *Journal of research* 3(5):573–582.

Crandell, D. R. 1965. The glacial history of western Washington and Oregon. In *The Quaternary of the United States* (VII INQUA Congress), eds. H. E. Wright, Jr., and D. G. Frey, pp. 341–8. Princeton, New Jersey: Princeton University Press.

Danner, W. R. 1955. *Geology of Olympic National Park.* Seattle: University of Washington Press.

Fagerlund, G. O. 1965. *Olympic National Park, Washington.* Natural History Handbook Series, no. 1. National Park Service.

Kirk, R. 1962. *The Olympic seashore.* Olympic Natural History Association.

———. 1980 (3rd ed.). *Exploring the Olympic peninsula.* Seattle: University of Washington Press.

———, and Namkung, J. 1966. *The Olympic rain forest.* Seattle: University of Washington Press.

Leissler, F. 1973. *Roads and trails of Olympic National Park.* 2nd ed. Seattle: University of Washington Press and Olympic Natural History Association.

Rau, W. W. 1964. *Foraminifera from the northern Olympic peninsula, Washington.* U.S. Geological Survey Professional Paper 374–G.

———. 1973. *Geology of the Washington coast between Point Grenville and the Hoh River.* Washington Division of Geology and Earth Resources Bulletin 66.

———. 1977. General geology of the southern Olympic coast. In *Geological excursions in the Pacific Northwest,* ed. E. H Brown and R. C. Ellis, pp. 63–83. Bellingham, Washington: Department of Geology, Western Washington University.

———. 1980. *Washington coastal geology between the Hoh and Quillayute Rivers.* Washington Division of Geology and Earth Resources Bulletin 72.

Stewart, R. J. 1971. Structural framework of the western Olympic peninsula, Washington. Geological Society of America Abstracts with Programs 3(2): 201.

Tabor, R. W. 1975. *Guide to the geology of Olympic National Park.* Seattle: University of Washington Press.

———, and Cady, W. M. 1978. *The structure of the Olympic mountains, Washington—Analysis of a subduction zone.* U.S. Geological Survey Professional Paper 1033.

Williams, R. L. 1979. *The northwest coast,* The American Wilderness Series. Rev. ed. Alexandria, Virginia: Time-Life Books.

Note: Tabor (1975) and Rau (1980) are especially helpful guides for Park visitors interested in geology to take along on excursions in the Olympic Mountains and in the coastal section.

Address

Olympic National Park
600 East Park Avenue
Port Angeles, Washington 98362

PART **III**
Igneous Activity and Volcanism

© by G. Shimer

Aa lava from a recent eruption on Lassen Peak. Lassen Volcanic National Park, California.

Volcanic activity in relatively recent geologic time and ancient volcanism have been dominant processes in the forming of landscapes in the national parks of this group. Five parks have active (or recently active) volcanoes: Mount Rainier, Lassen Volcanic, Crater Lake, Hawaii Volcanoes, and Haleakala. Catastrophic, explosive volcanic action in the recent geologic past shaped the Yellowstone landscape. Ancient plateau eruptions were significant in forming the rocks and surface features of Voyageurs and Isle Royale National Parks.

17
Mount Rainier National Park

Location: South Central Washington
Size: 235,404.00 acres; 367.82 square miles
Established: March 2, 1899

Figure 17.1 The dissected volcanic cone of Mount Rainier as seen from the east. Gibraltar Rock, Steamboat Prow, and Willis Wall are erosional remnants of volcanic layers that extended much higher about 75,000 years ago when Mount Rainier's summit was at its maximum elevation. Emmons Glacier, the longest glacier on Mount Rainier, descends from the snow-covered summit to the White River valley. Morainal debris covers the surface of Emmons in the foreground. Glaciers on the mountain's east side extend to lower elevations because they are shaded from the afternoon sun and are nourished by wind-blown snow from the summit area. U.S. Geological Survey photograph by R. E. Wallace.

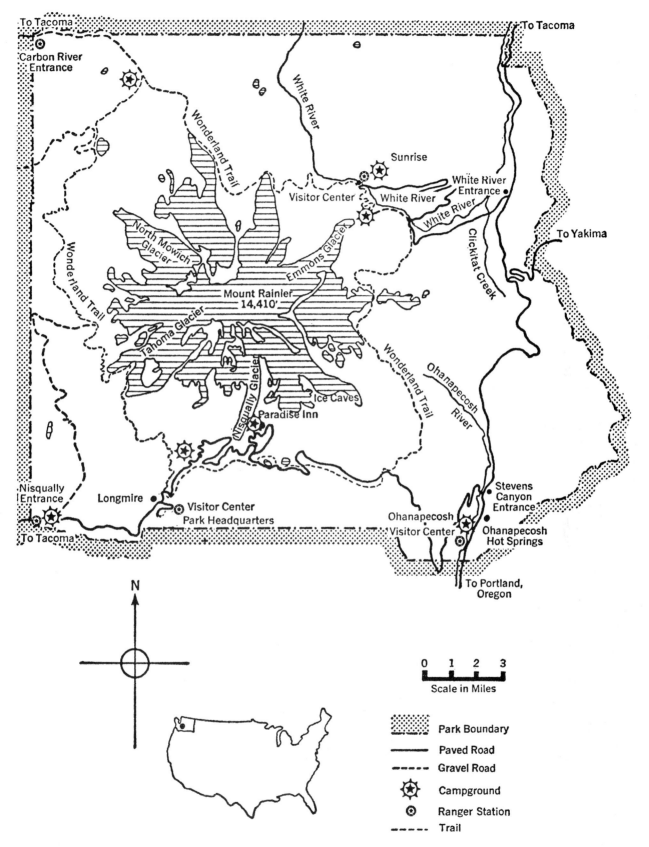

Figure 17.2   Mount Rainier National Park, Washington.

Table 17.1. Generalized Geologic Column, Mount Rainier National Park

Era	Period	Epoch	Rock Units	Geologic Events
Cenozoic	Quaternary	Holocene	Surficial deposits: avalanches, lahars, ash, pumice, glacial and alluvial material	Mudflows, debris flows, ash falls
				Growth and retreat of modern glaciers
				New cone constructed in old caldera
			Andesite lava: pumice, tuff	Osceola mudflow and lahars Explosive destruction of old summit
		Pleistocene	Ash, pumice, lahar deposits Interbedded andesite lavas and breccias Glacial deposits throughout epoch Andesite lavas Intracanyon lavas	Building of composite cone to maximum height
				Glacial and interglacial stages Beginning of Mount Rainier
				Filling of valleys by lavas
	Tertiary (Neogene)	Pliocene		Beginning of glaciation
				Uplift and erosion of Cascade Range
		Miocene	Tatoosh granodiorite	Intrusion of Tatoosh pluton with associated volcanics, dikes, sills
			Fifes Peak Formation	Andesite eruptions; basalt lava flows; pyroclastics
	Tertiary (Paleogene)	Oligocene	Stevens Ridge Formation	Extensive ash flows, ash falls; nuées ardentes; welded tuff
			/////////	Unconformity Folding, erosion; weathering of red soil
			Ohanapecosh Formation	Andesite lava eruptions; volcanic breccias, mudflows, ash falls deposited in shallow water and on land
		Eocene	Puget Group (not exposed in Park)	Sand, clay, swamp deposits; coal beds

Source: modified after Fiske et al. 1963; Crandell 1969

Local History

Back around the turn of the century, when Mount Rainier (elevation, 14,410 feet) was believed to be the highest mountain in the United States, the name "Columbia Crest" was given to what was thought to be the highest point on the crater's rim. More recent surveys have shown Register Rock to be the highest point at the summit. The grand old title of Columbia Crest remains, although the distinction of being "the highest" has long since passed to loftier peaks.

Few mountains anywhere, however, dominate their surroundings as does Mount Rainier. Viewed from Puget Sound on the west, the volcano appears to rise from the sea. And from any side, Rainier's cone towers over all other peaks of the skyline. Yet the hills and ridges around the base of Mount Rainier are not insignificant. Several of these lesser peaks are nearly 5,000 feet higher than Cadillac Mountain in Acadia National Park (chapter 15), a prominent landmark on the Atlantic coast.

To the Indians who inhabited the region for thousands of years before white men arrived, the mountain was *Tahoma*, or "Snow Mountain," a holy and fearsome place. Out of respect and awe, they hunted on the lower slopes only, not daring to approach the summit. A myth of the Nisqually Indians personifies Mount Rainier as a Monster Spirit, who long ago moved from the Olympic Peninsula to the east side of Puget Sound because the mountains on the Olympic side were growing too fast and crowding her out. In her new location, the Monster Spirit, who had an insatiable appetite, sucked in any animal or human who came too close. One day Changer, a sly spirit, challenged Tahoma to swallow him—after carefully preparing himself against her power by taking the form of a fox and tying himself with strong ropes to a neighboring mountain. Tahoma, the Monster Spirit, sucked and sucked, but could not swallow Changer, who continued to defy her. Finally she burst a blood vessel in the effort and died as torrents of blood poured down her flank. The event that in Indian memory was a "torrent of blood" may have been a volcanic mudflow (lahar) that roared down the mountain nearly 6,000 years ago, filling the White River valley for many miles and overspreading the lowlands north and east of Tacoma. Indian encampments were destroyed by the mudflow, and some of the inhabitants must have perished.

Certainly Mount Rainier must have seemed impressive to Captain George Vancouver of the British Navy when he sailed his ship into Puget Sound in 1792 on his mission to explore and chart the North Pacific coast. He named the mountain for his friend and superior officer, Rear Admiral Peter Rainier, who never saw the great volcano that bears his name.

The demand for furs for the China trade brought other sea captains, traders, and mountain men to "Oregon country," as it was known then. By 1818 a treaty between the United States and Great Britain provided for joint occupancy and the development of trading posts by fur companies. When the Hudson Bay Company established Fort Nisqually (near Tacoma's present waterfront) in 1833, William Fraser Tolmie, a young Scottish physician and botanist employed by the company, set about exploring the region and was apparently the first white man to enter the Park area. In that year, with five Indian companions, he went into the northwest section of what is now the Park and climbed the peak that bears his name, Tolmie Peak (5,030 feet).

Missionaries (such as Marcus Whitman) came into the region and were followed by an influx of settlers in the 1840s, who came over the "Oregon Trail." Trouble between the settlers and the British quieted down after the United States-Canadian boundary was established on the 49th parallel in 1846; but then Indian troubles began as the Indians resisted the encroachment on their land. Still settlement continued, especially in the lowlands and foothills. Adventurers hired Indian guides to help them scout the region and find promising sites for land and mineral claims. Mount Rainier, a difficult climb under the best of circumstances, seemed to be a challenge to some of these men. One account tells of two white men who ascended the mountain in 1855, leaving their Indian guide on the slopes and continuing upward on their own. They reported seeing a lake and steam vents at the top.

The first officially recorded attempt to reach the summit of Rainier was that of Lieutenant A. V. Kautz and his party in 1857. Kautz, Dr. R. O. Craig, four soldiers, and a Nisqually Indian guide almost made it to the top but were turned back by high winds and violent storms. Their approach was from the south and up the Kautz Glacier. Thirteen years later, in August of 1870, General Hazard Stevens and P. B. Van Trump carefully planned an approach to the summit up a narrow arête, the Cowlitz Cleaver, that separates the snowfields of Cowlitz and Nisqually Glaciers and leads to Gibraltar Rock, which is close to the top (fig. 17.1). This route to the summit via Gibraltar is still considered the safest and most convenient approach.

When Sluiskin, the Indian guide, realized that Stevens and Van Trump really meant to get to the top, he refused to accompany them because he believed that an evil spirit lived in a lake of fire at the summit. He promised to wait three days and asked for a letter absolving him of blame for their deaths if they did not make it back. The two climbers did reach the peak but had to spend the night there because of darkness and bad weather. They thought they would freeze to death on the mountain until they discovered steam vents and ice caves at the summit. They spent an uncomfortable night roasting by the vents and shivering in the caves, but they survived and descended the next day. When the anxious Sluiskin saw the two ice-encrusted figures emerging from the fog that had enveloped the mountaintop, he fled in terror, believing them to be ghosts. Finally they convinced him that they were alive, and the party returned to the settlements safely.

James Longmire, an early settler in the area, helped Stevens and Van Trump and also Samuel F. Emmons of the U.S. Geological Survey in his explorations of Mount Rainier and the surrounding region. Emmons did most of the early geologic reconnaissance of Rainier. Longmire himself finally climbed the mountain in 1883. On that trip he discovered mineral springs on the southwest side, and he later built a resort hotel at the site. The Park Headquarters and Visitor Center are now close to Longmire's springs, which was the first settlement within the Park area.

Considerable disagreement arose in the late 1880s over what to call this mountain area that was fast developing into a park—Tahoma, Tacoma, or Mount Rainier. The last name won out, and in 1899 the Mount Rainier area became the fifth national park to be established. At the present time, 80 miles of paved roads within the Park provide access to notable viewing points and campgrounds, as well as to many miles of trails into the wilderness areas.

Geologic Features

In addition to being the highest peak in the Cascade Range, Mount Rainier has the greatest single-peak glacial system in the Cascades. From its broad summit, glaciers radiate in all directions. In the million years of the volcano's existence, fire and ice (volcanism and glaciation) have continually interacted in molding, altering, shaping, and reshaping its form and appearance (fig. 17.3).

Matthes (1928) described Rainier's cone as "resembling an enormous tree stump with spreading base and irregularly broken top." Yet, lofty and massive as the summit is now, it was, as recently as 6,000 years ago, more than a thousand feet higher that it is today. A series of eruptions and explosions began about 6,600 years ago, culminating in an extremely violent explosion 5,700 years ago that destroyed the summit. Enormous blocks of rock and glacial ice were tossed up in the air and landed miles away. Great avalanches of ice and rock roared down the mountain. Burning, steaming mudflows plunged with great speed down the valleys. Dark clouds of ash and steam must have blotted out the sun, and torrential rains probably began to fall soon after the explosion. The Indians' fear of the mountain god Tahoma is not surprising since the stories of this disaster must have been passed down through many generations.

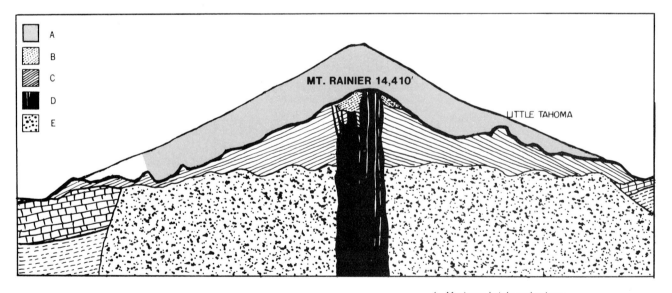

A. Maximum height and volume.
B. Holocene summit cone.
C. Lava flows.
D. Lava conduits.
E. Granodiorite.

Figure 17.3   Cross section diagram of Mount Rainier, past and present. The approximate profile of the volcano's maximum height and volume is superimposed over its present outline. From *Fire and Ice: The Cascade Volcano* by S. L. Harris. © 1976, 1980 by The Mountaineers; used by permission.

Box 17.1

The Cascade Volcanoes, a Part of the Pacific "Fire Zone"

Mount Rainier is one of more than a dozen geologically young volcanoes with composite cones (stratovolcanoes) that are perched on older rocks of the Cascade Range. In chapter 14, North Cascades National Park, you learned something of the geologic history of the Cascades and the development of two of these volcanoes, Mount Baker and Glacier Peak. In the chapters that follow—18, Lassen Volcanic National Park, and 19, Crater Lake National Park—we take the view that the general quietness of the Cascade volcanoes during the past century (except for Lassen Peak and, more recently, Mount St. Helens) is merely a lull in a period of fairly intense volcanic activity over the past few thousand years.

The Cascade volcanoes are remarkably aligned in a north-south direction, roughly paralleling the Pacific coastline (fig. 17.4). They are a small segment of the *circum-Pacific volcano belt*, or "Ring of Fire," that encircles the Pacific Ocean. This major volcano belt is located along converging plate boundaries where oceanic crust is subducted beneath overriding continental crust. Magma accumulates at depth (possibly as a result of friction and pressure heating the rocks to the melting point), works its way toward the surface, and erupts as lava, forming or adding to a volcano. Since plates converge at different rates in different areas, not all volcanoes in a major volcanic belt erupt at the same time. In recent geologic time, for example, the oceanic plate being thrust under the Pacific Northwest coastal region is apparently moving at a slower rate than the segments off Alaska and Mexico. However, any area of active subduction is considered a "fire zone" and should be regarded as potentially dangerous.

The volcanoes of major volcanic belts are almost entirely composite cones (stratovolcanoes) and are andesitic in rock composition. Composite cones are so named because they are built by alternating layers of lava flows and pyroclastic eruptions (ash falls, etc.) The Cascade volcanoes are typical composite cones in that the prevailing rock type is andesite and each volcano has grown,

over thousands of years of time, as brief eruptive episodes (lasting years or decades) interspersed with long periods of quiesence.

As explained in box 12.1 (p. 143), andesite is classified as an intermediate igneous rock because it is compositionally in between the light-colored, silica-rich rocks (e.g., rhyolite) and the dark, basaltic rocks that are relatively deficient in silica. Andesite, commonly dark gray or green in color, is made up of approximately equal amounts of light-colored minerals (plagioclase feldspars) and dark minerals (hornblende, olivine, pyroxenes). Chemically, andesites contain about 55 or 60 percent silica (SiO_2), with the remaining 45 to 50 percent mostly oxides of aluminum, calcium, sodium, iron, and magnesium. To geologists studying the Cascade volcanoes, the variations in chemical composition, texture, and crystal structure of the andesites are important inasmuch as very minor differences often enable investigators to identify and date lava flows and ash falls. Volcanic emissions are like fingerprints; that is, no two are exactly alike when samples are examined under a microscope and analyzed chemically.

Although andesite is clearly the predominant rock of the Cascade volcanoes, it is not the exclusive rock type; for at some time in its history, each volcano has had basalt and rhyolite eruptions. Several of the more explosive volcanoes have had mainly dacitic eruptions, dacite being a light andesite with a higher percentage of quartz (i.e., closer to rhyolite). Lavas containing higher percentages of silica are more viscous and tend to erupt more explosively. (See box 14.1, p. 174.)

Why andesitic rock is so prevalent in the Cascade volcanoes is a topic of debate among geologists. The oceanic crust that is being subducted is largely basalt. Where does the andesite come from? According to plate tectonic theory, the basalt originated from submarine eruptions along the mid-oceanic ridge in the Pacific and has moved like a conveyor belt away from the ridge on a long journey to the far boundary of the plate. (At a rate of 3 to 6 centimeters per year, it would have taken millions of years for the journey from the spreading center to the distant boundary.) Along the Pacific Northwest coastline, where the westward-moving North American plate is moving against a segment of eastward-traveling oceanic crust, the less dense continental crust overrides the denser oceanic crust. In other words, the oceanic crust is subducted beneath the continent. The subducted slab of basalt, along with some of the overlying sediment, descends at an angle to progressively greater depths, becoming hotter and hotter as it goes down—partly due to the geothermal gradient and partly because of the friction of the slab grinding past the overlying crust. (The *geothermal gradient* is the rate of temperature increase associated with increasing depth beneath the surface of the earth.) At some critical depth, the temperature is high enough so that some of the rock melts, forming magma. As pockets of magma coalesce, they tend to rise—perhaps as large blobs, shouldering aside rocks of the continental crust, or perhaps ascending fairly rapidly through fissures and erupting on the surface.

As to why this rising magma is andesitic, several lines of speculation have been suggested. Perhaps a silica-deficient basaltic melt is formed along the subduction zone and is contaminated by silica-rich rocks of the continental crust to produce an intermediate magma; or perhaps only part of the basaltic crust beneath the subduction zone melts, giving a melt higher in silica than the solid residue of oceanic crust remaining in the earth's interior. Or, possibly silica-rich sediment and sedimentary rock carried downward by the descending slab are incorporated into the molten basalt to form andesitic magma.

The remarkable alignment of the Cascade volcanoes can be explained because the plate convergence zone is parallel to the coastline in this area, and all along this zone the subducting oceanic plate goes downward at an angle of about 30°. Thus material on the descending slab reaches melting depth below the surface at about the same time and roughly the same distance inward from the coast. The line of volcanoes is believed to be located approximately above the zone of melting deep within the crust. They are supplied by magma rising from this zone and erupting at the surface as lava (fig. 17.4). Does it follow, then, that all the Cascade volcanoes have had a monotonously similar history? The fact is that they are more like a large family of brothers and sisters. They share a common background and have similar characteristics, but each grew and developed in a highly individual way.

Snow and glacial ice again accumulated and the glaciers rebuilt themselves, beginning again their work of erosion. Then, a mere 2,500 years ago, lava flows and ash falls began to repair the summit of Mount Rainier. Within the old *caldera* (a large basin produced by explosion and collapse) a new cone began to rise. Later more lava opened another smaller crater just to the east of the earlier vent. Where the two new craters intersect is Register Rock, the highest point of the volcano. Within historic time (which in the Pacific Northwest does not go back very far), minor events in the form of smoke, steam, and ash emissions were reported by Longmire, F. G. Plummer, and other observers in the late 19th century, but the mountain has been quiet for about a hundred years. The volcano is continuously monitored, however, by seismographs placed at various locations and elevations, infrared images that record variations in summit temperatures, and other observations, direct and indirect. Data from these sources have shown recent increases in thermal activity and some movement of magma at depth.

Volcano watchers emphasize that such indications (once believed to be signs of waning energy or even imminent extinction) might well be harbingers of renewed volcanic activity on a mountain that has had a remarkably varied history since its beginning early in the Quaternary Period.

Mount Rainier differs from other Cascade volcanoes in that about 90 percent of its eruptions have been in the form of lava flows. Having only a tenth of its bulk made up of pyroclastics (tephra) is abnormally low for a composite cone. Moreover, some of those pyroclastic layers were originally lava that was blasted apart by exploding gases and later reconsolidated as volcanic breccia.

By contrast, Mount St. Helens has had a more violent history with few lava flows and a high volume of pyroclastics. In fact, much of the ash and pumice on Rainier's slopes came from St. Helens during explosive episodes not unlike the 1980 explosions that littered the region with pyroclastics. (*Pumice* is a frothy rock resulting from the

The postglacial pumice and ash beds on Mount Rainier are identified by letter; i.e., the fall from Mount Mazama is called Pumice Layer O. The fall of maximum thickness, 20 inches, is Pumice Y from Mount St. Helens. The most recent fall,* Pumice Layer X, which came from Mount Rainier itself and accumulated between 100 and 150 years ago, has a maximum thickness of only one inch.

Glacial Features

The cone of Mount Rainier, from summit to base, has been deeply furrowed and scarred by severe and long-continued glacial erosion. Only remnants of the volcano's outer layers remain on the mountain flanks as sharp crags and ridges (the "cleavers") that jut through the ice and separate the glaciers descending on all sides. One of these residual rock masses, Mount Tahoma, is described by Matthes (1928) as "a sharp, triangular tooth on the east flank" of Rainier that rises to an elevation of over 11,000 feet. "In its steep, ice-carved walls one may trace ascending volcanic strata aggregating 2,000 feet in thickness that point upward to the place of their origin, the former summit of the mountain" (more than a thousand feet higher than the present top).

The caldera rim left by the explosion that destroyed the summit has been breached and the rock crumbled by overriding ice cascades. Even the new cone is beginning to show minor indications of erosion by glacial ice. Several hundred feet of snow fill the two young craters. Beneath the snow are the famous "steam caves," with the more extensive network of tunnels being in the east crater. The west crater has a smaller cave system but contains a pool of meltwater at the present time under a canopy of ice. If only a small increase in vent temperature occurred for any length of time, lakes might form at the surface of the snowpack in the craters (which has apparently happened from time to time in the past).

Because the storm clouds that carry most of the rain and snow drop a good part of their load at the lower and middle altitudes, these zones get much more precipitation than does the summit. These wet clouds are borne by the prevailing westerlies from the Pacific Ocean. Annual snowfall at middle elevations on the western slopes of the mountain averages 100 feet. Added to high snowfall at the middle levels are great volumes of snow that are blown off the summit or come down in avalanches. And since summer temperatures remain relatively low at these altitudes, the middle zone, between 8,000 and 10,000 feet, is the most favorable for glacier development.

Rainier's "atmospheric layering" is familiar to those who live near the mountain or come to climb it. On many days a climber on the higher ridges can be in bright sunshine and look back and see thick white clouds covering

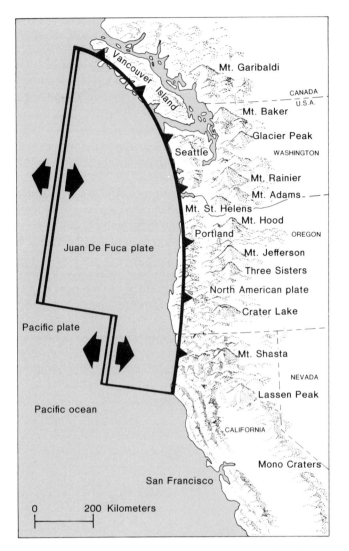

Figure 17.4 Map and schematic diagram showing the relationship of the Juan de Fuca plate to the North American plate and the chain of Cascade volcanoes. Heat and pressure at depth along the subducting Juan de Fuca plate causes magma to form, which works its way toward the surface. Eventually rising magma erupts as lava, building up the volcanoes. From *Physical Geology* (2nd ed.) by Charles C. Plummer and David McGeary. © 1979, 1982 Wm. C. Brown Company Publishers, Dubuque, Iowa. Reprinted by permission.

sudden explosion of lava containing a large amount of dissolved gas and steam. Due to the explosive release of pressure, tiny gas bubbles cause the solidified lava to have a spongy or frothy texture.)

About 6,600 years ago, the explosion of Mount Mazama in Oregon (chapter 19) spread several inches of ash over the Mount Rainier area and, in fact, the whole region. This was the most widespread postglacial ash fall. Ash deposits that fell before and during glacial stages are difficult to trace or date with accuracy because weathering and extensive glacial erosion and redeposition have left very few undisturbed patches.

---

*Most recent, that is, before the 1980 eruption of Mount St. Helens deposited fresh ash on the slopes of Rainier.

everything below; valleys and all the lower mountains have disappeared from view. The converse is true (as climbers should be so cautioned); sudden storms may envelop the summit while the valleys below are in sunshine.

The glaciers that originate at the summit—Kautz, Tahoma, Winthrop, Emmons, Ingraham, and Nisqually— are "refueled," so to speak, after they get 4,000 feet or so down from the top and reach the high snowfall zone. Another group of glaciers— among them Carbon and Russell, North and South Mowich with Edmunds in between, Puyallup, South Tahoma, Paradise, Cowlitz, Ohanapecosh, Fryingpan, and several others—start from cirques in the mountain flanks at the middle elevations where more snow accumulates. A few smaller glaciers in the Mount Rainier system still exist at around 6,000 feet on north-facing slopes of peaks at these altitudes. Here the winter snowfalls are heaviest, and snowfields on north slopes are protected from summer sunshine. In all, 26 named glaciers, occupying around 40 square miles of land surface, exist in the Park.

Periodic variations in climate affect the volume and length of the glaciers. Recently the glaciers have been advancing slightly after about a century of shrinking. The larger glaciers are surveyed and measured carefully at regular intervals because they are the source of major rivers (the Nisqually, Puyallup, White, Carbon, and Cowlitz Rivers) that supply hydroelectric power to the Northwest. Emmons Glacier, largest in the Park, supplies the White River. Flow in these rivers tends to be low in winter and high in the summer melt season. The effect of meltwater on the streams is perhaps most striking near the headwaters. In the early morning a small stream issuing from a glacier can be crossed easily on foot; but by midafternoon on a sunny summer day this same stream will have become a torrent of milky-white water, banging cobbles along its bed, far too swift and dangerous to walk or wade across. The white color is due to the pulverized fragments or grains of rock (*rock flour*) scoured and abraded by moving glaciers.

Glacial erosion on a large scale is most evident in the precipitous cirque headwalls, such as Willis Wall, the largest cirque on the mountain, with a headwall of 3,600 feet and a diameter of a mile and a half. In this great basin, opening to the north, is stored the snow and ice supply that enables the Carbon Glacier to descend to an altitude lower than any other glacier on the mountain (fig. 17.5).

Figure 17.5   So much fine, dark sediment clings to the wet ice of the Carbon Glacier's snout that it looks almost black. Great quantities of morainal debris also cover the top, sides, and bottom of the snout. A thick valley train of glacial outwash and debris extends for miles down the Carbon River valley from the end moraine being built here. The Carbon River was named for coal deposits of Eocene age that crop out in the valley walls many miles downstream from the Carbon Glacier. National Park Service photograph.

Conspicuous lateral moraines descending with the glaciers give us some idea of the quantity of rock being removed from the valley sidewalls, both by abrasion and by mass wasting. On the valley sidewalls, above the rock debris moving along with the ice, are the remains of earlier lateral moraines built when the climate was colder and the glaciers larger. Most of the present glaciers of any size have medial moraines as well as lateral moraines. The medial moraines trail down from a junction point where a branch joins the main glacier.

Drift and till deposits of various ages are found mostly at lower elevations in moraines built during earlier major glaciations. A large moraine from the last major glacial stage (about 11,000 years ago) dams Mystic Lake at the foot of the mountain on the north side. *Rock-glacier deposits* left by rock glaciers that are no longer active are also associated with the last glaciation. The fronts of these large talus accumulations are abrupt and steep and from 25 to 100 feet high. Also built of talus are the *protalus ramparts*, ridges 5 to 15 feet high of loose, angular rock fragments that accumulated along the bottom of large perennial snowbanks that lay at the base of a cliff. When the climate warmed and the snowbank finally melted, a protalus rampart was left skirting the toe of the talus slope.

Valley train deposits cover the floors of the narrow valleys that descend from the ends of the glaciers. *Valley trains* are outwash deposits dropped or spread within the valleys of the meltwater streams that drain the glaciers. Some of the valley trains open out onto outwash plains at lower elevations. All of the valley trains that come down from Mount Rainier have been modified to some degree by either ash falls or mudflows, or both.

A number of lovely cirque lakes, such as Tipsoo Lake, Crescent Lake, Mowich Lake, and others, can be seen in Mount Rainier National Park. Some tributary streams that flow in hanging valleys have waterfalls, Christine Falls and Comet Falls being examples. Ice caves, formed by meltwater, are usually visible at the bottom of Paradise, Carbon, and several other glaciers. Some ice caves have been enlarged by steam vents.

The Role of Mass Wasting Processes on Mount Rainier

Rockfalls, rockslides, avalanches, debris flows, mudflows, and other processes of downslope movement, singly or in combination, operate more or less continually in the Mount Rainier environment. Moreover, the interaction of processes of mass wasting, weather, volcanism, and glaciation has created a most powerful group of mechanisms in the eroding, transporting, and depositing of debris far and

wide. Their potential for seriously disrupting the environment is a matter of grave concern to volcano-watchers. Consider the following items:

. . . On almost any clear day, those who visit the Paradise Park or Sunset Park lookouts (or other vantage points) can see dust clouds rising from small rockfalls and avalances near the summit that occur almost hourly. And scanning the snowfields of the major glaciers with telescope or field glasses, one sees fresh, dark accumulations of debris that have fallen since the last snowfall—which may have been the previous night, or within the past few days.

. . . . Geologists have mapped more than 55 *lahars* (flows of pyroclasts, volcanic debris, mud, etc., usually mixed with hot water) that have flowed down Mount Rainier in Holocene (postglacial) time. The longest and largest of these, the Osceola Mudflow (± 5700 years ago) ran about 70 miles to the northwest, altered the course of the White River, and filled in an arm of Puget Sound.

. . . In October 1947, within a span of about 8 hours following heavy rains, several mudflows roared down Kautz Creek to the Nisqually River and left several cubic miles of muddy debris in the Kautz valley. Witnesses reported that large trees and boulders 13 feet in diameter were borne along by the mudflows, which had the consistency of wet concrete.

. . . In December 1963, a series of rockfalls and avalanches, apparently triggered by a steam explosion, fell in quick succession from the north face of Little Tahoma Peak. Plummeting downward, this great volume of broken rock struck the Emmons Glacier with great force and then bounced and flew down the White River Valley. While dropping some 6,200 feet in altitude, the avalanche traveled nearly 4 miles from its source. Some of the boulders that were transported are as large as buildings (S. Harris 1980).

These examples are by no means unusual events in the geologic story of Mount Rainier. Blankets consisting of mixtures of ash, pumice, fragmented bedrock, and solid blocks are found on all flanks of the mountain. These deposits also occur as thick fill in the major valleys that lead away from the cone. The action of hot water and steam on the lavas and ashfalls at the summit and on the slopes causes any consolidated material to break up both mechanically and chemically, producing quantities of debris for mass wasting processes, glaciers, and streams to carry away.

Continual freeze-and-thaw action at high altitudes on rock unprotected by soil or vegetation tends to break up and shatter the rock. Loosened fragments fall off cliffs and roll down slopes. Tremors or even slight explosions shake loose more rocks.

Glaciers are very efficient at breaking and grinding up rock, and Mount Rainier has plenty of glaciers to oversteepen slopes, expose fresh surfaces, and pulverize loose fragments.

The fact that andesite, the dominant rock, is more susceptible to chemical weathering than, say, granite is significant. The feldspars and ferromagnesian minerals that make up andesite tend to weather out as clay minerals and soluble salts, especially in the presence of hot water, which hastens the chemical action.

With all this fragmentation of rock going on, plus the intermittent addition of ash, lava, and pyroclastics, it's no wonder that it doesn't take much to trigger all kinds of slides and flows. Given the availability of great quantities of loose (or easily loosened) material, the dynamics of slides and flows can be explained as follows. (1) There is much rain and snow fall on Mount Rainier, and runoff is high. Rocks and debris are water-soaked much of the time. Material that is thoroughly wet tends to move, slip, and slide faster than dry material. (2) As noted above, the slopes at high altitudes are generally steep and unprotected by vegetation. Heavy snow accumulations tend to set off avalanching and slides. (3) Eruptions drop ash and pumice on snow accumulations. Mix in steam and hot water and all at once you have a mudflow. Add melting glacial ice and rock debris, and you have a disaster roaring down the valleys.

Note that the distinction between debris flows and mudflows is gradational; that is, a *mudflow* is a variety of *debris flow* in which the solid component is composed of at least 50 percent sand, silt, ash, clay, etc. The water content critically affects velocity and mobility. By gaining water along the way, a mudflow becomes more of a slurry; and if it runs into a good-sized river, then that river becomes an overloaded stream that floods lowlands and deposits the debris from the mountain on flood plains and, in the Pacific Northwest, in harbors.

According to estimates, lahars and mudflows travel at speeds of about 50 miles per hour. This doesn't allow much time for people in their paths to get out of the way. However, authorities at Mount Rainier National Park have taken sensible precautions by carefully monitoring daily conditions in the Park and setting up early warning systems, evacuation plans, and so forth. Campgrounds have been relocated on higher ground, for example. The Park would be closed immediately if thermal or seismic data indicated any possibility of an eruption. Meanwhile, visitors are able to enjoy the Park and its many natural wonders without being overly apprehensive.

Geologic History

1. Eocene lowland.

In the early part of the Cenozoic Era, the entire area presently occupied by the Cascades was a coastal plain along what was then the Pacific shore. Over several million years, sand, clay, and organic swamp deposits accumulated in subsiding tidal marshes. The accumulated material later became a 10,000-foot thick sequence of sandstone, shale, and coal (some of which has been mined) known as the Puget Group.

2. Eocene volcanism, continuing into Oligocene.

About 40 million years ago, volcanic activity began. This was millions of years before Mount Rainier's cone was built. The evidence for this volcanism is in the Ohanapecosh Formation, which adds another 10,000 feet of layered rock to the area and overlies the Puget Group. Arkose of the Puget Group interfingers with the Ohanapecosh Formation near the Park's western boundary, but none of the rocks of the Puget Group are exposed in the Park.

The Ohanapecosh Formation contains much breccia, which indicates volcanic explosions and accumulation in shallow water. The rock unit also includes lava flows, mudflows, and ash falls. Sinking continued to keep pace with the build up of deposits on the ocean floor.

3. Oligocene uplift.

Uplift accompanied by compression raised the Puget and Ohanapecosh rocks above sea level and gently folded them during early Oligocene time. It seems likely that the volcanic eruptions marked the beginning of subduction. The continued collision of the North American and oceanic plates could account for the subsequent folding and uplift of the layered rocks.

During this period, a warm, wet climate prevailed, resulting in deep weathering and extensive erosion. A thick red soil developed, and streams carved gulleys between low hills.

4. Oligocene volcanism; Stevens Ridge Formation.

Later in the Oligocene, perhaps 25 or 30 million years ago, volcanism began again. Ash falls and flows covered the area. The source vents were outside the present Park boundaries. Violent, exploding clouds of fine pumice (*nuées ardentes*) erupted. When these glowing avalanches settled, the ash particles with still molten edges bonded together to form a *welded tuff*. Ash flows filled the valleys, and eventually a plateau was built. This rock unit became the Stevens Ridge Formation. Outcrops of the light-gray welded tuff, containing darker inclusions of angular pumice fragments, can be seen in the southern part of the Park along the highway through Stevens Canyon.

5. Continued volcanism in Miocene time; Fifes Peak Formation.

The character of the volcanism changed as the Fifes Peak Formation was being deposited. Many flows of both andesite and basalt lava, ranging from 150 to 500 feet thick, piled up to a total thickness of 2,500 feet. The rocks of the Fifes Peak Formation make up many of the cliffs and peaks in the northwestern part of the Park.

6. Miocene deformation and uplift; intrusion of granodiorite.

As Fifes Peak volcanism ceased, the region was again uplifted and gently folded. (The already folded, underlying Ohanapecosh Formation was compressed into tighter folds.) Some faults, with displacements of thousands of feet, resulted from the compression.

A body of magma welled upward through the crust. Only a small amount of the magma reached the surface and erupted. Most of it cooled slowly and solidified underground, forming the Tatoosh pluton, composed largely of relatively coarse-grained granodiorite having a "salt-and-pepper" appearance; that is, mostly very light-colored rock speckled with large dark mineral grains. A number of dikes and sills are associated with the pluton, and some contact metamorphism occurred in the country rock. The Tatoosh pluton has been dated radiometrically as being about 12 million years old.

7. Pliocene uplift and erosion of the Cascade Range.

The Cascade Range continued to rise, and deep erosion carved the region into peaks and ridges separated by stream-carved valleys. The rocks overlying the Tatoosh pluton were eroded, and the pluton was "unroofed." It was on this mountainous terrain that the Mount Rainier volcano would be built. The granodiorite of the pluton, now exposed in the Park, underlies the Carbon and White River valleys and part of the upper Nisqually River valley. Plutonic rocks also form the central part of the craggy Tatoosh Range in the southern section of the Park, near the base of Mount Rainier.

8. Building of the composite cone, the Mount Rainier volcano, during repeated episodes of Pleistocene glaciation.

The climate was already very cold and the Tatoosh Mountains were undergoing glaciation when construction of the cone began with a series of relatively fluid lava flows early in the Pleistocene. These lavas flowed into the stream-carved canyons, eventually filling them. As the eruptions built a low, broad volcano, mudflow and glacial deposits were interlayered between flows, indicating that glaciations continued throughout this time. Then a large number of short lava streams from a central area began building up the cone, interspersed with ejections of breccia and ash. Eventually the cone rose to a height of perhaps 15,500 to 16,000 feet. A plug of solidified magma filled the central vent and dikes radiated from the center, like spokes in a wheel. The peak probably attained its greatest height about 75,000 years ago.

9. Erosion of the cone and destruction of the summit.

Intense glacial erosion steepened the mountain flanks, and by the end of the Pleistocene, perhaps a third of the mountain's bulk had been removed. However, steam and occasional eruptions at the summit prevented glaciers from eroding at the top of the mountain, so the summit remained high.

Beginning about 6,000 years ago, a series of violent explosions blasted away an enormous portion of the eastern side of the mountain and, later, the summit itself. It was probably this last explosion that caused the catastrophic Osceola Mudflow that traveled as far as Puget Sound. Rainier's lowered summit area was then a caldera about 2 miles across, the highest points on the rim barely 14,000 feet above sea level. Glacial ice filled the caldera and eroded the rim and caldera basin (fig. 17.6).

10. Reconstruction of the summit by lava flows and some pumice.

Reconstruction began about 2,500 years ago when pyroclastic eruptions, followed by lava flows, built a new cone within the caldera. In time, the eruptions shifted somewhat, creating a double crater for this "volcano upon a volcano." The mountain probably erupted at least once between 1820 and 1854 during the time that Mount St. Helens was having a series of pyroclastic eruptions. Minor eruptions may have occurred again on Mount Rainier in the 1870s and 1880s. *Fumaroles* (small vents emitting vapor and gas) and occasional steam explosions on the flanks of Mount Rainier indicate continued high temperature. *Microseisms* (very small earth tremors) beneath the mountain also tell us that eruptions may begin again.

Geologic Maps and Cross Sections

Crandell, D. R. 1969. *Surficial geology of Mount Rainier National Park, Washington.* U.S. Geological Survey Bulletin 1288. Map in folder.

Fiske, R. S., Hopson, C. A., and Waters, A. C. 1964. *Geologic map and section, Mount Rainier National Park, Washington.* U.S. Geological Survey Misc. Geological Investigations. Map I-432.

Crandell, D. R. 1969. *The geologic story of Mount Rainier, Washington.* U.S. Geological Survey Bulletin 1292.

Figure 17.6   This view of the southwest side of Mount Rainier (taken from Indian Henrys Hunting Ground) shows the extensive dissection and erosion of the volcanic cone by glaciers. The two high points, Liberty Cap and Point Success, are remnants of the old crater rim that was partially destroyed by violent eruptions. National Park Service photograph.

Bibliography

Barnett, J. 1969. *Mount Rainier National Park.* Fresno, California: Awani Press.

Crandell, D. R. and Mullineaux, D. R. 1967. *Volcanic hazards at Mount Rainier, Washington.* U.S. Geological Survey Bulletin 1238.

———. 1969. *Surficial geology of Mount Rainier National Park, Washington.* U.S. Geological Survey Bulletin 1288.

———. 1969. *The geologic story of Mount Rainier, Washington.* U.S. Geological Survey Bulletin 1292.

———. 1971. *Postglacial lahars from Mount Rainier volcano,* Washington. U.S. Geological Survey Professional Paper 677.

Fiske, R. S.; Hopson, C. A.; and Waters, A. C. 1963. *Geology of Mount Rainier National Park, Washington.* U.S. Geological Survey Professional Paper 444.

Harris, S. L. 1980. (rev. ed.). *Fire and ice, the Cascade volcanoes.* Seattle, Washington: The Mountaineers.

Hopson, C. A., Waters, A. C., Bender, V. R., and Rubin, M. 1962. The latest eruptions from Mount Rainier volcano. *Journal of geology* 70:635-647.

Matthes, F. E. 1915. The Mount Rainier National Park. Text on back of quadrangle map, Mount Rainier National Park, Washington. Scale 1:62,500 (revised 1955).

———. 1928. *Mount Rainier and its glaciers.* National Park Service.

Plummer, C. C.; and McGeary, D. 1982. *Physical geology.* 2nd ed. Dubuque, Iowa: Wm. C. Brown Company Publishers.

Plummer, F. G. 1899. *Mount Rainier Forest Reserve.* U.S. Geological Survey 21st Annual Report, part V.

Sigafoos, R. S.; and Hendricks, E. L. 1972. *Recent activity of glaciers, Mount Rainier, Washington.* U.S. Geological Survey Professional Paper 387-B.

Stagner, H. R. 1966. *Behind the scenery of Mount Rainier National Park.* Longmire, Washington: Mount Rainier Natural History Association.

Tuttle, S. D. 1980. *Landforms and landscapes.* 3rd ed. Dubuque, Iowa: Wm. C. Brown Company Publishers.

Address

Mount Rainier National Park
Tahoma Woods, Star Route
Ashford, Washington 98304

18
Lassen Volcanic National Park

Location: Northern California
Size: 106,372.22 acres; 166.21 square miles
Established: August 9, 1916

Figure 18.1   View of Bumpass Hell in Lassen Volcanic National Park. In this active thermal area, approximately 500 feet by 1,400 feet in extent, are bubbling mud pots, boiling springs, steam vents, and sulfurous vapors. Photograph by J. W. Harbaugh.*

*Figures 18.1, 18.4, 18.5, and 18.6 are from *Northern California* (K/H Geology Field Guide) by J. W. Harbaugh. © 1975 by Kendall/ Hunt Publishing Company.

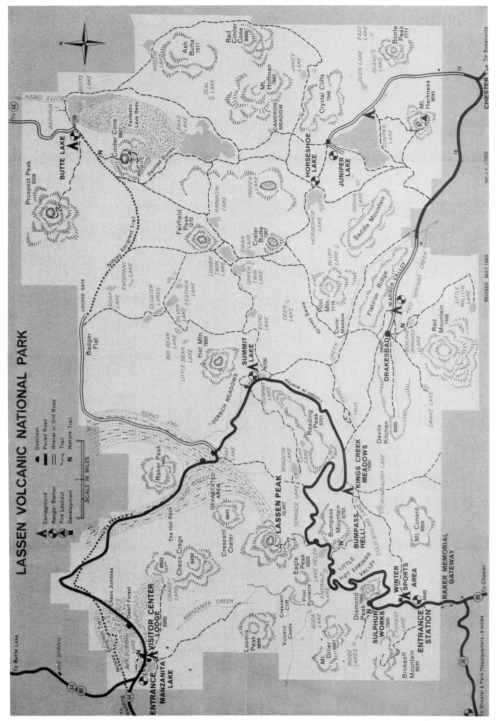

Figure 18.2 Lassen Volcanic National Park, California. National Park Service Map. (The Visitor Center Lodge is now located outside the Park.)

Table 18.1. Geologic Column, Lassen Volcanic National Park

Era	Period	Epoch	Rock Units	Geologic Events
Cenozoic	Quaternary	Holocene	1914–1921 dacite pyroclastics, lavas, mudlows	Lassen Peak eruptions
			Quartz basalt flows	Eruptions from base of Cinder Cone
			Pyroclastics, cinders	Construction of Cinder Cone
			Dacite breccias	Chaos Jumbles produced by avalanching from base of Chaos Crags
			Dacite domes	Eruption of Chaos Crags
				Volcanic domes extruded from older features, such as Raker Peak
				Eruptions of smaller domes near Lassen Peak
				Construction of Lassen Peak on flank of Brokeoff Mountain
		—?—	Pre-Lassen fluid dacite flows	Columnar dacite formed at base of Lassen Peak
				Erosion of Brokeoff caldera
				Decline and collapse Brokeoff Cone
		Pleistocene	Fluid andesite lavas; some basalt flows	Construction of shield-type volcanoes
			Glacial deposits interbedded with volcanics	Episodes of glaciation before, during, and after volcanic eruptions throughout epoch
			Brokeoff Andesite (and pyroclastics)	Construction of composite volcano, Ancient Brokeoff Cone (Mount Tehama)
			Eastern basalts and pyroclastics	Lava ridge; later eroded to rugged hills on east side of Park
	Tertiary / Neogene	—?—	Flatiron Andesite	Construction of Central Plateau and lava plain
		Pliocene	Juniper Lake Andesite; Twin Lakes Andesite	
			Willow Lake Basalt	Flowed across southern portion of Park
			Tuscan Formation (not exposed in Park)	Pyroclastics, lahars

Source: modified after Diller 1895; Williams 1932; S. Harris 1980

Local History

When President Theodore Roosevelt proclaimed Lassen Peak and Cinder Cone as national monuments in May 1907, the generally held assumption was that these little known areas in the northern California wilderness contained interesting volcanic phenomena, obviously extinct, but worthy of preservation for future generations to study and enjoy. Seven years later, in May 1914, Lassen Peak came to life again with a roar, opening up a new crater and emitting lava and ash. After several hundred intermittent eruptions of varying length and severity between 1914 and 1921, the volcano quieted down again. It was during the period of activity—not long after some particularly impressive outbursts—that Lassen Volcanic National Park was established, on August 9, 1916.

Actually, the Indians who lived in the area had told the settlers that the mountain was not "dead," but apparently the white people did not believe them. The Indians said that Lassen Peak was full of fire and water, and they called it "The Sweat of the Gods." They believed that one day the mountain would sweat so much it would blow itself into pieces, the fire and water would disappear, the old gods would return, and everyone would be happy.

The Lassen Peak region was the ancestral summer home of four Indian tribes—the Atsugewi, Mountain Maidu, Yahi, and Yana, all peaceful tribes who got along well together. They hunted and fished in the area during the summers and then spent their winters in sheltered valleys in the foothills of the southern Cascade Range. The Atsugewis (or Hat Creek tribe) had a more fanciful legend about Lassen Peak. Their story was that a brave warrior chief became so lovesick for a beautiful maiden that he neglected his hunting and his tribal leadership. To punish him, the fire and smoke spirits who lived in Lassen Peak abducted the maiden and hid her inside the mountain. When the chief saw her footprints disappearing toward the volcano, he knew what had happened and dug furiously until he reached the spirits' council chamber. The spirits were so impressed with his devotion for her and his strength that they forgave the lovers and invited them to make their home inside the mountain, too. Whenever smoke was seen coming from the top of Lassen Peak, the Atsugewis believed the chief was smoking his peace pipe.

Lassen Peak, the southernmost volcano in the Cascade chain, is near the northern end of the Sacramento Valley. The volcano, 10,457 feet high, was a landmark for immigrants coming to the Sacramento Valley in the mid-1800s. The peak was named for Peter Lassen, a Danish blacksmith who came to California in the 1830s. After receiving a land grant from the Mexican governor of California, Lassen started a ranch in the region and began guiding parties of immigrants to California. Unfortunately (so the story goes), he frequently got lost because he couldn't tell the difference between Lassen Peak (then called St. Josephs Mountain) and Mount Shasta (the volcano about 80 miles north of Lassen). One of the parties he was guiding forced him at gunpoint to climb the peak that bears his name so that he could locate himself! By the 1850s, word had gotten around not to take Peter Lassen's trail. Instead, the wagon trains from Nevada to Sacramento traveled via Nobles' Emigrant Trail, parts of which can be seen in the Park today as deep ruts and wagon tracks (fig. 18.2). Peter Lassen struck gold near Susanville, California, in 1853 and settled in a cabin along Lassen Creek. He was killed by Indians in 1859.

Although Nobles' Emigrant Trail winds along the base of Cinder Cone in the northeast corner of the Park, apparently only a few miners, homesteaders, and travelers witnessed the most recent eruptions of ash, steam, and fragmental material from Cinder Cone in 1850–51. Observers reported that fire on the top of the cone could be seen for many nights from points as far distant as 40 miles and from even a hundred miles away. Population in the region was still sparse in 1914 and 1915 when the most violent of Lassen Peak's eruptions occurred. Some property damage resulted from the explosions and mudflows, but people were able to get out of the way in time. It was, perhaps, sheer luck that enabled some of the photographers and observers who recorded these events at close range to survive with only minor injuries. At least during historic time, no lives have been lost due to volcanic activities in the Lassen region.

At the recommendation of the U.S. Geological Survey, the Visitor Center and accommodations near Manzanita Lake were closed in 1974. These facilities would be in the path of an avalanche from Chaos Crags should movement of the unstable slopes be triggered by seismic activity or renewed volcanism. A temporary Visitor Center at the Park's North Entrance Station is in use until a new one can be constructed in a safe location, probably in Lassen National Forest just outside the northwest corner of the Park.

Geologic Features

Lassen Peak, one of the major volcanoes of the Cascade Range, is not a composite cone (or stratovolcano) as the ones to the north are—including Lassen's near neighbor, Mount Shasta. Oddly enough, an even closer neighbor, Brokeoff Mountain, is an erosional remnant of a once mighty composite cone. It is more than a thousand feet higher than Lassen, and the second highest peak in the Park. Ancient Brokeoff Cone (also called Mount Tehama) exploded and collapsed late in the Pleistocene, leaving, besides Brokeoff Mountain, several other erosional remnants that outline the rim of the old caldera (i.e., Mount Diller, Pilot Pinnacle, and Mount Conrad) (fig. 18.3).

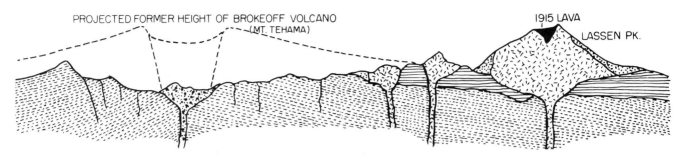

Figure 18.3 Generalized cross section through Lassen Peak and its mid-Pleistocene predecessor, Brokeoff Volcano (Mount Tehama). The collapsed caldera of Brokeoff was eroded by glaciers and has since been further destroyed by weathering of rocks softened by hot water and corrosive gases. From *Fire and Ice: the Cascade Volcanoes* (rev. ed.) by S. L. Harris. © 1980 by The Mountaineers; used by permission.

How Eruptive Activity Is Related to Type of Lava

After the collapse of Brokeoff Cone, Lassen Peak developed as a *volcanic dome* rising from a vent on the northeast flank of its predecessor, the earlier composite cone. A volcanic dome is characterized by an upheaved, plug-like conduit filling that forms when viscous lava is forced through a vent onto the earth's surface. Like toothpaste squeezed out of a tube, the viscous lava is unable to flow freely, so it builds a bulbous dome over its vent.

The rock type of the Lassen dome is *dacite*, a rock intermediate in composition between rhyolite and andesite. As dacite has a higher percentage of silica than andesite, it tends to flow less easily, hence its propensity for constructing domes. Dacite is also associated with more violently explosive eruptions because gases are not able to bubble out freely through the viscous magma. Instead they are trapped until enough pressure builds up so that they blast their way out. Lassen Peak is the world's largest volcanic dome, incidentally.

Volcano-watchers at Mount St. Helens, which has also had recent dacite eruptions, have noted many similarities between its renewal of eruptive activities (spring 1980) and the first year (1914) of Lassen's eruptions. Activity began at both volcanoes with a series of *phreatic eruptions* (i.e., explosive, steam-propelled ejections of ash and mud caused by heating and expansion of ground water) that shot cold ash and steam columns thousands of feet into the air (Stoiber et al., 1980).

Lassen Volcanism and Plate Tectonics

How does Lassen fit in with plate tectonic theory? As explained earlier (box 17.1, p. 221), a spreading center lies some 300 miles offshore from northernmost California, Oregon, and Washington. This spreading center adds about an inch of new crust each year to a small plate of oceanic crust, gently but inexorably pushing it toward the Pacific Northwest coast. As this small plate collides with the westward-moving North American plate, it bends down and slips beneath the leading edge of the continental plate. The Cascade volcanoes are apparently products of the subduction of the oceanic crust. To the south, the much larger Pacific plate and the North American plate *slide past one another* (the San Andreas fault is regarded as the plate boundary), and subduction is not occurring. That is why Cascade volcanic activity does not continue southward (at the present time). That movement and subduction of the small plate are relatively slow (around 3 cm per year) is a likely reason why the Cascade volcanoes erupt infrequently, as compared with some of the other sections of the circum-Pacific volcanic belt (South America, for example).

Why is dacite the rock type rather than andesite? One possible explanation is that being farther inland from the continental oceanic boundary, the magma comes from deeper along the subduction zone. Magmas that are more silica-rich seem to be associated with such sources. Alternatively, the relative age of volcanic activity may explain the dacitic composition. Late-stage eruptions of composite volcanoes seem to be more silica-rich. The fact that Lassen Peak formed on the flanks of Ancient Brokeoff Cone suggests that the construction of Lassen represents a late episode in long-continuing activity.

Eruptions and Volcanic Features

The volcanic features in the southwest corner of the Park, in or close to the old caldera, are associated with Brokeoff Cone. Sulphur Works, with its hot springs, fumaroles, and strong rotten egg odor (indicating the presence of hydrogen sulphide), is thought to be at the center of the former cone. Between Sulphur Works and Little Hot Springs Valley, another area of hydrothermal activity, is Diamond Point, an old conduit through which lava moved upward during the construction of Brokeoff Cone. As a result of hydrothermal alteration, many of the rocks at the Sulphur Works and Little Hot Springs Valley—which were formerly hard, gray-green andesite lavas—have been transformed into bright-colored clays. Other rocks are rapidly decaying and have become yellow, buff, and red in color from iron oxides. The hot water coming from the fissures contains both sulphurous acid and sulphuric acid which react with the minerals of the original rock to produce the clays and iron oxides. Nearly pure opal (but not gem quality opal) is formed where the acidity of the hot water is very high.

Just beyond Little Hot Springs Valley is Bumpass Hell, a 16-acre tract of boiling springs, steaming sulphur vapors, and bubbling mud pots, which is also associated with old caldera fissures. Bumpass Hell is named for a cowboy who worked in the area in the 1860s. He broke through the crust over a mud pot and scalded his feet. When his boss asked him where he had been, he replied, "In hell." Later when he was escorting a newspaper editor through the area, Bumpass broke through the crust again. This time he was so badly burned that he lost a leg. The superheated steam in the springs at Bumpass Hell and the fumaroles are an indication of the late stages of volcanism in the Brokeoff Cone area (fig. 18.1).

Devils Kitchen is another area of hot sulfurous springs. The springs are so strongly acidic that holes and pits have been eaten into the bedrock here. Note on the map (fig. 18.2) that these hot springs and the similar features described above are clustered along a northwest-southeast trend and appear to be associated with an old fault area.

The volcanic dome of Lassen Peak, with its surrounding talus, represents about a cubic mile of material (fig. 18.4). As noted earlier, this makes a very large summit area for a dome, but its bulk is small compared to the composite cone of Mount Rainier (chapter 17). An exceedingly unusual feature of Lassen Peak was its development of summit craters from which pyroclastics and lava were ejected. Most domes do not have summit craters because the explosions are emitted from the base of the plug. An example is Chaos Crags, a complex dome (on the northwest flank of Lassen), which is made up of several dacite plugs but does not have a crater.

In the crater bowl on Lassen's summit, a pre-1914 vent and a 1914–15 vent were filled and covered over by a large dacite lava flow in May 1915, and a third summit crater, or vent, was opened. A fourth crater was blasted out of the northwest corner of the summit by a series of steam, gas, and ash explosions in April, May, and June of 1917 (fig. 18.5). These were the last of the *major* eruptions of the 1914 to 1921 period of activity, although slight fumarolic activity at the summit (in the form of steam escaping occasionally from small fissures) has persisted to the present.

The May 19, 1915, dacite flow poured down the western side of Lassen for about a thousand feet before it hardened. Observers some 20 miles away to the west and south reported that beginning about 9 o'clock that night the mountain was "boiling over" for two hours. The photographer B. F. Loomis and the postmistress of nearby Manton had described a fire at the summit a few days

Figure 18.4   The summit area of Lassen Peak showing the 1915 dacite flow (foreground) and a small crater, formed by subsequent eruptive activity, between the dacite flow and Lassen Peak in the background. The people standing on the dacite flow (near middle of view) show scale. Photograph by J. W. Harbaugh.

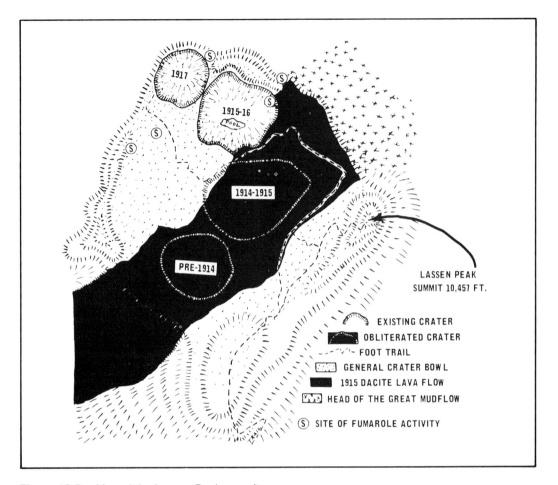

LASSEN PEAK
SUMMIT 10,457 FT.

EXISTING CRATER
OBLITERATED CRATER
FOOT TRAIL
GENERAL CRATER BOWL
1915 DACITE LAVA FLOW
HEAD OF THE GREAT MUDFLOW
(S) SITE OF FUMAROLE ACTIVITY

Figure 18.5 Map of the Lassen Peak summit area
showing sites of eruptions and fumaroles. National Park
Service.

earlier and then a large black mass beginning to rise at
the top, two days before the eruption. This "black mass"
was the plug, or column of lava, that was pushing up
through the conduit and breaking open the rock over the
new vent.

Some of the lava and ash also flowed down the steep
northeast side onto packed snow left from the previous
winter. Hot lava, ash, and snow quickly combined to form
a thick, pasty lahar that slid rapidly down the slope, burying a mile-wide swath of forest in its path, filling Lost
Creek, then gliding up over a divide and down into Hat
Creek. In these valleys, fences, bridges, farm buildings,
and fields were totally devastated. Fortunately, homesteaders were awakened by the roar of the oncoming mudflow, just in time to flee to high ground and safety. The
worst was yet to come.

A few days later, on May 22, 1915, at 4:30 in the
afternoon a violent explosion shot a great cloud of ash
thousands of feet into the air. Photographs of the event,
referred to as "the Great Eruption," show an enormous
cloud in the shape of a mushroom over Lassen Peak, very

much like an atomic bomb cloud. Prevailing westerly winds
carried fine ash particles for many miles to the east. However, much of the force of the blast was deflected downward across the northeast slope, snapping off many more
trees, some with trunks 6 feet in diameter, and throwing
them many feet from their stumps. Hot ash mixed with
additional melting snow reactivated the mudflow and it
again sped down the same valleys, split in front of Mount
Raker, and rushed on by, the two branches extending beyond the Park's northern boundary. Smaller mudflows
streamed down the north and northwest slopes. (By an odd
coincidence, the similar violent dacitic explosion of the
Mount St. Helens cone happened on May 18, 1980, just
four days short of the 65th anniversary of Lassen's Great
Eruption.)

The tract of greatest destruction at Lassen is about
4½ miles long and is called the Devastated Area. Park
visitors who drive through the area can see that the scars
on the landscape are being covered by natural reforestation.

Several small dacite domes south of Lassen Peak (namely, Eagle Peak, Vulcan's Castle, Mount Helen, and Bumpass Mountain) are apparently a continuation of the volcanic activity that built Lassen Peak on the ruins of Brokeoff Cone. White Mountain, a larger dome about 2 miles to the southeast of Lassen Peak, is nearly covered by banks of talus. Its summit and flanks were severely eroded by glaciers coming down from Lassen.

Chaos Crags (north of Lassen Peak) is younger than the other volcanic domes. It was formed by four or more dacite domes less than 1,200 years ago. Chaos Jumbles, which extends toward the northwest corner of the Park from the base of Chaos Crags, originated as a volcanic avalanche about 300 years ago. A series of explosions set off avalanching that transported large angular blocks a mile or more from their source. At speeds of around 100 miles per hour, the rock debris apparently traveled some of the distance on a cushion of compressed air that reduced friction (a mass-wasting phenomenon called "air-layer lubrication"). The damming of Manzanita Creek by the avalanche formed Lake Manzanita (fig. 18.6).

Four shield-type volcanoes, built up by successive flows of fluid andesite, are located in the Park. (Their construction is similar to that of the basaltic Hawaiian volcanoes described in chapters 20 and 21.) These volcanoes are Raker Peak, north of Lassen; Prospect Peak in the extreme northeast corner; Red Mountain, next to the south-central boundary; and Mount Harkness in the southeast corner. All are between 7,000 and 8,400 feet in elevation. They have been modified by glacial erosion and by the ash falls from Lassen Peak and Cinder Cone, which have erupted more recently. Although apparently constructed entirely of lava, without interbedded pyroclastic deposits, each of the four shield volcanoes is topped by a cinder cone rising from a central summit crater. The cinder cones and perhaps the top layers of lava are probably postglacial in age. On Raker Peak, a large dacite dome has pushed up on the southwest side, forming a line of remarkable dacite cliffs with columnar structure.

Cinder Cone, just southeast of Prospect Peak, has been built up during the past several centuries by a succession of ash eruptions. It rises as a symmetrical cone some 600 feet above the flatlands and is surrounded by a cinder sheet of varying thickness. Two circular craters are at the top (fig. 18.7). Slopes are steep on the cone, but the climb up the trail is not difficult. After cinder eruptions had built the cone, several lava flows of "quartz basalt" came out from the base of the volcano and spread north and south, damming streams and forming Butte Lake at the north and Snag Lake at the south. The quartz basalt is a "hybrid" lava. The quartz crystals, included in the basalt as partially assimilated *xenocrysts* (i.e., "foreign" crystals in an igneous body of rock), were picked up by the basalt as it rose toward the surface.

Figure 18.6   Chaos Jumbles (foreground) and Chaos Crags (background) in Lassen Volcanic National Park. The angular dacite blocks that make up Chaos Jumbles were transported by high-velocity avalanches from Chaos Crags several hundred years ago, following a series of explosions near the base of the volcanic dome. Photograph by J. W. Harbaugh.

Figure 18.7   Air view of Cinder Cone with its double-rimmed crater. The most recent eruptive activity at Cinder Cone probably occurred in 1851. National Park Service photograph.

Glacial Features

The Pleistocene ice sheets that covered the Park area during the time when the major volcanoes were being built up undoubtedly added a lot of glacial debris to the volcanic debris. Both kinds of material are found mixed together and reworked in mudflows, lahars, landslides, etc., and piled up in a few moraines. Some were buried by pyroclastics that fell after the moraines were deposited.

Some of the volcanoes, such as Red Mountain, Raker Peak, and Lassen Peak (a center of ice radiation), were more severely modified by glacial erosion than others. Small valley glaciers persisted at the higher altitudes and in protected locations well into post-Pleistocene time. Lakes formed in some of the cirque basins, examples being Emerald Lake and Lake Helen; the latter named in honor of Helen Tanner Brodt, the first woman to climb Lassen Peak (in 1864). Summit Lake is dammed by a moraine. Other lakes of glacial origin include Feather, Silver, Big Bear, Little Bear, Swan, Echo, and Twin Lakes.

Geologic History

1. The pre-Cenozoic rocks.

Inasmuch as Paleozoic and Mesozoic rock units, intruded by granite, crop out only 15 miles southeast of Lassen Volcanic National Park, it seems reasonable to assume that these Sierra Nevada basement rocks extend northward under Lassen Peak. This seems especially likely since similar rocks of the same age make up the Klamath Mountains that rise west of Lassen and exend northward up the coast into Oregon. Gravity studies have shown that the lighter (less dense) volcanic rocks occupy a deep basin and that very far below them are the basement rocks. Presumably, the rocks linking the Sierra Nevada and the Klamath Mountains are the ones that are buried beneath Lassen and its neighboring volcanoes.

Lassen Peak is, as we have said, the southernmost volcano of the Cascade Range, which is largely made up of relatively young extrusive and intrusive rocks. It is this southernmost segment of the Cascades that separates the Klamath Mountains on the west from the Sierra Nevada to the south, both of which are comprised of much older

Figure 18.8   The development of Lassen Peak. From *Fire and Ice: the Cascade Volcanoes* (rev. ed.) by S. L. Harris. © 1980 by The Mountaineers; used by permission. *A.* A dome of viscous dacite lava rises to the top of the volcano's crater. *B.* Lava from the still-rising dome overflows the old crater and pours down the sides of the volcano. *C.* As the dacite dome solidifies, fragments fall from the dome onto the flanks of the cone, forming aprons of breccia over the hardened flows.

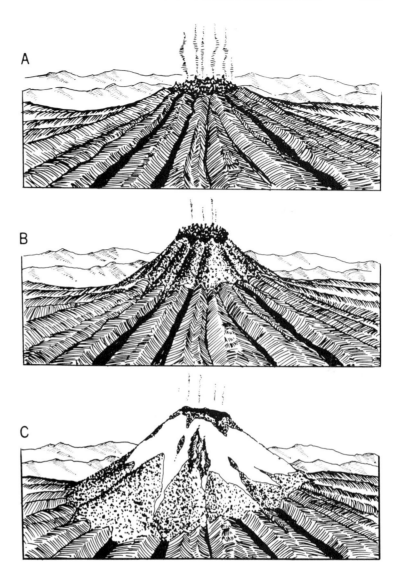

rock. For reasons that geologists can only speculate about, the portion of older rock between the Klamath and Sierra remained low instead of being uplifted and, consequently, was deeply buried by the volcanics.

Sierra Nevada and Klamath rocks began accumulating on an ocean floor in the early part of the Paleozoic Era. Included in the great thickness of rocks that formed are basalts typical of the ocean floors and a wide variety of sedimentary rocks—mainly shales, sandstones, and limestones. Perhaps some of the rocks formed as an "island arc," or chain of volcanic islands that marked an ancient collision between two plates.

Marine deposits continued to build through most of Paleozoic time and into the early part of the Mesozoic Era. Late in Jurassic time the area was subjected to intense compressive deformation. The layered rocks were folded, faulted, and metamorphosed (the shales to slates and phyllites, and the basalts to "greenstone"). Some of the faulting involved great vertical displacement because

material believed to have once been part of the mantle (underlying the basaltic oceanic crust) is presently exposed over wide belts in the Sierra and Klamath Mountains. This rock, originally a peridotite (composed entirely of ferromagnesian minerals), was metamorphosed to serpentinite (the California state rock) on its way up from the mantle.

While deformation was continuing, melting was probably occurring at a greater depth in the earth's crust, probably along a subduction zone. Great masses of magma were generated, but it took millions of years for the large blobs of granitic magma to work their way upward through the overlying rocks. Ultimately the magma reached higher levels in the crust and slowly solidified to form the numerous batholiths and plutons of the Klamath Mountains and, more especially, the Sierra Nevada. Toward the end of the Cretaceous Period, erosion had leveled these mountain ranges to low hills and a coastal plain along what is now the Sacramento Valley (west and south of Lassen Volcanic National Park). A marine sandstone, the Chico

Formation, was deposited along the shore of the coastal plain. This sandstone crops out in stream valleys west of the Park and may extend southward under volcanic rocks in the Park.

2. Origin of the placer gold deposits near the Park.

During early Cenozoic time, sand and gravel eroded largely from the Chico Formation and associated rock units were transported along the channels of somewhat sluggish streams. Included in some of these gravels was placer gold that had been eroded from quartz veins in metamorphic rocks. Some of this placer gold eventually became concentrated in the stream beds and terraces that attracted prospectors and adventurers to northern California from all over the world after the original strike in 1849 northwest of Sacramento. The Lassen area itself was never a promising section for the miners, but they traveled back and forth across the region looking for gold in the surrounding country.

3. Cenozoic uplift and the beginning of volcanism.

Later in the Cenozoic the region began to be uplifted, and the Sierra Nevada became a westward-tilted fault block (chapters 13, 27, 28). As uplift proceeded, extensive volcanic activity began. Apparently volcanic vents were located along what is now the crest of the Sierra. The volcanic debris washed down the gentle western slopes of the Sierra and along stream channels as huge mudflows (lahars). These became the Tuscan Formation of Pliocene age. Although the Tuscan rocks are not exposed in Lassen Volcanic National Park, they lie close beneath the surface in a number of places.

4. The Willow Lake basalt flows of Pliocene time.

Beginning in the southern portion of the Park, basaltic flows poured out of fissures and vents, covering an increasingly wide area.

5. Build-up of a plateau by extensive andesite flows in Pliocene and Pleistocene time.

The basalt flows were covered over by a thick series of very fluid andesite flows, the Jupiter lavas. The Twin Lakes lavas of black porphyritic andesite flooded a large area at about the same time. Later the Flatiron andesites erupted and spread over the area south and west of White Mountain's present location. By this time the landscape had become a fairly flat lava plain stretching over a wide area. No pyroclastic debris had accumulated.

A renewal of basalt flows, the Eastern basalts, built low hills along the Park's eastern boundary, which was later eroded to a rugged topography. With the advent of some pyroclastic eruptions, some cones began to appear in the northern part of the Park.

6. Development of Ancient Brokeoff Cone during the Pleistocene.

Meanwhile, in the southwestern corner of the Park, a great volcano was becoming larger and larger from successive andesite flows interbedded, toward the top of the cone, with a good deal of tuff and breccia. Ultimately it became an impressive feature above the lava plain, rising to well over 11,000 feet, and probably 15 miles in diameter.

7. Eruption of the four shield-type volcanoes.

Raker and Prospect Peaks, Red Mountain, and Mount Harkness erupted during the late stages of Brokeoff Cone's development. These volcanoes, built mainly of andesite, rose at the four corners of the Central Plateau area.

8. Late Pleistocene decline and collapse of Brokeoff Cone; eruption of pre-Lassen dacites.

Near the present location of Lassen Peak, on the subsiding northeast flank of Brokeoff Cone, a new vent opened and extruded flows, of fluid, black, glassy dacite, 1,500 feet in thickness. These flows hardened as the columnar lavas that now encircle the base of Lassen Peak.

About this time Brokeoff Cone collapsed, forming the huge caldera. The pre-Lassen dacites had erupted during an interglacial period; but when the glaciers returned, the soft, decaying rocks of the old volcano were heavily eroded by the ice, thus enlarging the caldera.

9. Eruption of Lassen Peak about 11,000 years ago.

Long after Brokeoff Cone had been reduced to ruins, a large volcanic plug of partly solid, partly viscous, dacite lava began to rise through the northeastern vent on the old cone's flank. The rock enclosing the vent shattered and cracked as the lava kept on rising, producing an extensive covering of talus on the steep slopes of the emerging volcanic dome. Within a comparatively brief time, Lassen Peak attained its present height. The eruption occurred close to the end of the Pleistocene Epoch at a time when a large ice sheet had melted away. This explains why the talus accumulations are relatively undisturbed. Later on, during a return of colder climate, valley glaciers reformed and eroded shallow cirques on the north and northeast sides of the Peak. Small moraines mark the extent of these glaciers.

Smaller, satellite domes of dacite also pushed up, first south and then north of Lassen Peak. Some of these eruptions were accompanied by pyroclastic explosions of tuff and pumice. The most recent of these eruptions were the explosions that produced Chaos Jumbles.

10. Construction of Cinder Cone by pyroclastic eruptions.

Cinder Cone began several centuries ago with violent pyroclastic eruptions that spread ash over 30 square miles. After the cone was built, several basalt flows of unusual composition (quartz basalt) erupted from a vent at the base. The most recent pyroclastic activity occurred in 1851.

11. Eruptive activity of Lassen Peak, 1914–1921.

Before the eruptions began, the pre-1914 crater contained a fairly deep lake. This crater was totally destroyed and the depression filled in by the explosions and subsequent lava flows of the active period. Two craters at the summit can be seen today (fig. 18.5).

Nearly 400 outbursts, a few major but most of them minor, were recorded by observers during the eruptive period. Scientific data collected at that time and subsequent studies of Lassen's activity have been helpful to geologists who study the Cascade volcanoes and try to anticipate what they will do next. Monitoring of Lassen Volcanic Park's volcanic features by seismometers, tiltmeters, and inclinometers is carried on by the U.S. Geological Survey in cooperation with the National Park Service.

Geologic Maps and Cross Sections

American Association of Petroleum Geologists. 1968. *Geological highway map no. 3-Pacific Southwest region.* Tulsa, Oklahoma: American Association of Petroleum Geologists.

Schultz, P. E. 1957. *Geology of Lassen's landscape.* Mineral, California: Loomis Museum Association.

Williams, H. 1932. *Geology of the Lassen Volcanic National Park, California.* University of California Publications in Geological Sciences, volume 21.

Bibliography

Boiling Springs Lake, Lassen Volcanic National Park. 1973. Loomis Museum Association. Leaflet.

Bumpass Hell nature trail, Lassen Volcanic National Park. 1972. Revision. Loomis Museum Association Leaflet.

Cinder Cone nature trail, Lassen Volcanic National Park. 1971. Loomis Museum Association. Leaflet.

Crandell, D. R.1972. *Glaciation near Lassen Peak, northern California.* U.S. Geological Survey Professional Paper 800-C pp. C179–C188.

Diller, J. S. 1889. *Geology of Lassen Peak District.* U.S. Geological Survey Annual Report 8 pp. 395–432.

———. 1895. *Geologic atlas of the United States, Lassen Peak folio, California.* No. 15.

Harbaugh, J. W. 1975. *Northern California,* K/H Geology Field Guide Series. Dubuque, Iowa: Kendall/Hunt Publishing Company.

Harris, S. 1981. *Fire and ice, the Cascade volcanoes,* rev. ed. Seattle, Washington: The Mountaineers.

Holden, J. C. 1980. *Mount St. Helens: the geophysics of volcanoes. Science 80* 1 (6): 49–50, September/October.

Lassen Peak trail, Lassen Volcanic National Park. 1970. Mineral, California: Loomis Museum Association. Leaflet.

Lily Pond nature trail, Lassen Volcanic National Park. 1970. Rev. ed. Mineral, California: Loomis Museum Association.

Loomis, B. F. 1971. *Eruptions of Lassen Peak-pictorial history of Lassen.* 3rd rev. ed. Mineral, California: Loomis Museum Association.

———. 1966. *Pictorial history of the Lassen volcano.* Published in cooperation with the National Park Service and Loomis Museum Association.

Oakeshott, G. B. 1978. *California's changing landscapes.* 2nd ed. New York: McGraw-Hill, Inc.

Schulz, P. E. 1959. *Geology of Lassen's landscape.* 2nd ed., rev. Mineral, California: Loomis Museum Association.

———. 1974. *Road guide to Lassen Volcanic National Park.* Mineral, California: Loomis Museum Association.

Stoiber, R. E., Williams, S. N., and Malinconico, L. L. 1980. Mount St. Helens, Washington, 1980 volcanic eruption: magmatic gas component during the first 16 days. *Science* 208: 1258–9, June 13.

Warren, H. C. 1971. *Lassen's geology.* National Park Service. Mimeographed report.

Williams, H. 1932. *Geology of Lassen Volcanic National Park.* University of California Press, Publications in Geological Sciences, vol. 21.

Address

Lassen Volcanic National Park
Mineral, California 96063

19
Crater Lake National Park

Location: Southwest Oregon
Size: 160,290.33 acres; 250.45 square miles
Established: May 22, 1902

Figure 19.1 Crater Lake with Wizard Island in the foreground, Llao Rock jutting out from the caldera rim in the middle distance, and Mount Thielsen in far background. Photograph by Oregon Department of Transportation.

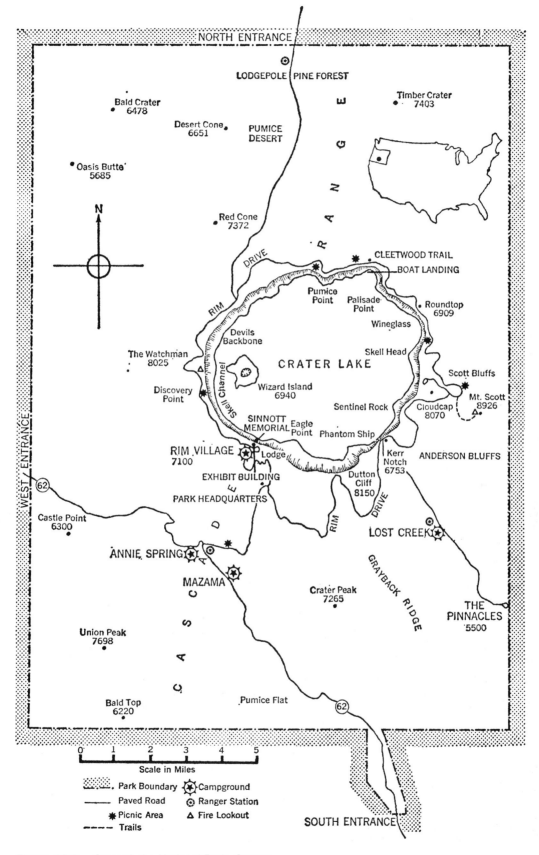

Figure 19.2   Crater Lake National Park, Oregon.

Table 19.1 Geologic Column, Crater Lake National Park

Era	Period	Epoch	Rock Units	Geologic Events
Cenozoic	Quaternary	Holocene		Filling of caldera with water, forming Crater Lake
				Cooling of caldera
			Dacitic scoria	Build-up of cinder cones inside caldera
			Andesitic lavas, scoria	Construction of Wizard Island
			Minor till deposits	Minor glacial advance
			Dacitic ash, pumice, scoria	Collapse of summit; forming of caldera
				Emptying of magma chamber by glowing avalanches, explosive eruptions, ash falls
		Pleistocene	Dacitic pumice; welded tuff	Explosive eruptions; glowing avalanches
			Basaltic scoria; vesicular lavas	Growth of numerous small cinder cones
			Dacitic lavas and pumice	Dacite tuff and pumice eruptions; dacite flows; small volcanic domes
			Andesitic lavas, pyroclastics	Parasitic cones, flank eruptions
			Till deposits and some outwash interbedded with volcanics throughout epoch	Episodes of glaciation before, during, and after eruptions
				Flank eruptions; construction of satellite cone, Mount Scott
			Andesitic lavas, pyroclastics	Construction of Mount Mazama
	Tertiary / Neogene	Pliocene	Some pyroclasts	Construction of Union Peak, Mounts Bailey and Thielsen (composite cones)
			Basaltic and andesitic lavas	Build-up of lava plateau

Souce: modified after Williams 1942

Local History

If you have seen Crater Lake, you can probably imagine how astonishing and mysterious it must have seemed to early visitors to the Oregon Territory. John Wesley Hillman, a young prospector, came upon the place quite by accident on June 12, 1853, as he was following his mule up some ridges in search of a lost mine. When the mule suddenly stopped on the brink of a cliff, Hillman looked down over the edge and saw below the indigo water of the lake in its great caldera, a good five miles from rim to rim. He named the spot where he stood, on the southwest side of the rim, Discovery Point; and he called the lake "Deep Blue Lake," although other members of his party wanted to call it "Mysterious Lake." Several years later gold miners found the lake and referred to it as "Great Blue Lake." Two soldiers from Fort Klamath (a short distance south of the Park) rediscovered the lake in 1865. They wanted to call it "Lake Majesty." But then in 1869, people from Jacksonville, Oregon, a settlement near Medford, decided "Crater Lake" was the ideal name. This is the name it has retained, despite the oft-repeated assertion of teachers, geologists, and park rangers that it really should be called "Caldera Lake"!

The remarkably intense blue color of the lake continues to amaze visitors. The purity of the water and the great depth of the caldera are responsible for the deep indigo shade. Crater Lake's officially measured depth is 1,932 feet, which makes it the deepest lake in the United States. Since the lake has neither inlet nor outlet, all the water in it has accumulated from rain and snow, which accounted for its being devoid of vegetation and aquatic life—up until some 60 or more years ago when fish were introduced. Fresh-water shrimp had to be brought in first so that when the lake was stocked with trout the fish would have something to eat.

Except for minor seasonal changes, the level of the water remains fairly constant from year to year, with intake from precipitation balancing loss from subterranean discharge and evaporation. However, the lake was somewhat higher a few thousand years ago than it is at the present time.

With Crater Lake having so many unusual characteristics, it's no wonder that the legendary folk hero of the logging camps, Paul Bunyan, was given credit for its origin. The story goes that when Paul and Babe (his great Blue Ox) were logging in Oregon Territory, Babe was spooked by the roar of Spokane Falls. Babe took off at a gallop, with the provision sled and the swamp hook crashing and jouncing behind. The hook gouged out the Columbia River Gorge and finally caught fast in the Cascade Mountains. When Paul caught up with Babe and pulled the hook free, water gushed up from the bottom of the hole. Paul threw rocks in to plug up the hole and stop the leak. The last rock he dropped in is Wizard Island, off to one side of the lake. So much for that story!

A Klamath Indian legend is somewhat more in accord with the geological story as we know it, and also suggests tribal memory of a catastrophe many generations before. We know that Klamath Indians were living in the region 6,600 years ago (when Mount Mazama exploded and collapsed, forming the caldera) because their moccasins and other artifacts have been found beneath Mazama ash layers.

The Klamath Indians told of two great volcanoes in mortal combat: Llao, the chief of the Below-World; and Skell, chief of the Above-World. Llao lived in Mount Mazama, and Skell's home was Mount Shasta, about a hundred miles to the south. They thundered at each other, threw barrages of fiery rocks, and sent clouds of flaming ash toward each other. As the two volcanoes fought on, a terrible darkness covered the land for days. (Is it likely that Mazama and Shasta were erupting at approximately the same time, or within a few years of each other? The legend suggests this, but geologic evidence cannot confirm the timing with any exactness.)

In an attempt to pacify the volcano gods and expiate the sins of the tribe, the Klamath medicine men climbed to the top of Mount Mazama and threw themselves into the crater. Impressed by their sacrifice, Skell fought even harder. Finally he drove Llao far down into his cave in the Below-World. Then, with his two fists, Skell, the chief of the Above-World and lord of all the animals, pushed the top of the volcano down to imprison Llao forever.

An Oregonian, Judge William Gladstone Steel, was instrumental in the movement to preserve the Crater Lake region as a national park. As a boy in Kansas, he had read about a mysterious, deep blue lake in Oregon and vowed he would see it some day. When he was 18 years old, his family moved to Oregon, and soon young Steel began searching for the lake, asking all the oldtimers he met where it was. Finally, in 1885, after 13 years of searching, he stood on the caldera's rim and looked down at Crater Lake. Then and there, he decided that it must be protected and preserved. After 17 years of effort, his wish became reality in 1902 when Crater Lake National Park was established. Appointed superintendent in 1913, he

later became the Park commissioner and served in this office until his death in 1934. It was Judge Steel who first put fish into the lake. He hiked, carrying minnows, nearly 50 miles from the Rogue River to Crater Lake.

Highways into the Park lead to a scenic drive around the caldera rim. Probably the best vantage point from which to see a great many features of geologic interest at one time is the top of Mount Scott. It is reached by a trail from the rim road. Looking down one can see both volcanic and glacial landforms around the caldera, as well as an incomparable view of the lake itself. North and south of here rise the High Cascades, topped by snow-capped volcanoes. To the west is "a sea of peaks and ridges"— the Western Cascades and beyond them, the more distant Klamath-Siskiyou Mountains. To the east stretches an immense platform of basalt that goes farther than one can see into Idaho. This lava plateau is cut by canyons, here and there; and a few steep-scarped, tilted, fault-block mountain ranges break the flatness of the surface in the distance.

Volcanic Features and Eruptions

The large caldera (not crater) occupied by Crater Lake and the encircling flanks and ridges are about all that is left of once-mighty Mount Mazama. Mazama's early history was similar to that of the other composite cones of the Cascades. Like its contemporaries—Mount Rainier, Mount Shasta, Mount Hood, and the others—Mazama grew during the Pleistocene, with glaciations advancing and retreating all the while. From its foundation rocks at the base (between 5,000 and 6,000 feet in elevation), Mount Mazama rose another 6,000 feet to an elevation of approximately 12,000 feet above sea level. The cone was broader and less steep than Shasta's or Rainier's because the subsidiary vents and overlapping cones gave the volcano a bulky, irregular shape. The present caldera rim is 7,000 to 8,000 feet above sea level, and geologists have estimated that the original cone probably rose 2,500 to 3,500 feet above that. Consider that the diameter of the caldera at the rim is about 5 miles, and you begin to have a rough idea of the volume of material that was removed by explosion and collapse over 6,000 years ago.

Associated with the construction and destruction of Mount Mazama are a variety of extrusive and intrusive features. Thick pumice and ash deposits, relics of the violent explosive eruptions, lie on the flanks of the mountains, relatively fresh and only mildly weathered by the elements during the short (in geologic terms) time since they fell. Some of the deposits are welded tuffs produced by flaming avalanches, or nuées ardentes. The Pinnacles, about 5 miles southeast of the rim, are erosional remnants of ash deposits 200 to 300 feet thick that probably steamed for years after the hot pyroclastics blanketed the slopes and filled the canyons. Where steam was localized, as in fumaroles, the ash particles became cemented together and were thus more resistant to erosion.

Associated Composite Cones

Mount Bailey and Mount Thielsen, both a short distance north of the Park, probably became inactive early in the Pleistocene, perhaps around the time that Mazama started up. Mount Scott, highest point in the Park (8,926 feet), developed as a parasitic cone from Mazama's eastern base. Mount Scott's very steep cone is made up of layers of andesite lava, closely related chemically to some of the earlier main flows of Mount Mazama. Evidently Mount Scott became extinct before the end of the Pleistocene. A glacial amphitheater (large cirque) on the northwest side indicates severe glacial erosion that was not "repaired" by postglacial volcanic activity. Not much is left of Union Peak in the southwest corner of the Park except its *volcanic plug*, an erosional remnant of lava that solidified to resistant rock in the neck of the volcano.

Shield Volcanoes and Cinder Cones

Basalt lavas and andesitic basalts built a number of satellite cones and craters around Mount Mazama. Crater Peak formed as a shield volcano of andesite and basalt flows and later erupted pyroclastics of andesite and dacite. Another shield volcano, Timber Crater, in the northeast corner of the Park, is also made up of basalt and basaltic andesite flows, but its shield is capped by two cinder cones. None of the cinder cones in the Park are more than a thousand feet high from base to summit. Examples are Desert Cone, Bald Crater, Red Cone, and Diller Cone. Volcanic *bombs* (fragments of lava, or ejecta, usually rounded, ropey, or spiral, that hardened while in the air) can be found on the cinder cones in myriad sizes and shapes. Some of the cinder cones are aligned on what seems to have been a fissure system tapping a magma chamber. Wizard Island is another cinder cone; it formed from eruptions that occurred after Mount Mazama lost its top but before the lake filled the caldera depression.

Igneous Dikes

Most of the large andesite dikes exposed in the Park are old conduits and radial fissure fillings that probably served as feeders for the flank eruptions on Mount Mazama. Devils Backbone, which runs up the cliffs of Crater Lake from the water's edge to the west rim, is the most conspicuous dike (fig. 19.3). A similar dike forms a high vertical wall immediately beneath the Watchman and merges

Devil's Backbone

Figure 19.3 Devils Backbone (on the northwest side of the caldera wall) is an andesite dike that is exposed from the water's edge to the top of the rim. Devils Backbone stands out from the cliff because it is more resistant to weathering and erosion than the volcanics on either side. National Park Service photograph.

upward into the Watchman lava flow (western rim). Dacite dikes, several of which are clustered near Llao Rock (north of Devils Backbone), are smaller and less prominent.

Feeder dikes of Hillman Peak are exposed in the remaining half of the old conduit that was laid bare by the explosions that removed the eastern half of its cone. Hillman Peak, highest spot on the rim, was one of the parasite cones of Mazama and now reveals a nearly perfect cross section of its interior due to the blasting out of the caldera.

Dacite Domes

Dacite flows that began to erupt toward the end of Mount Mazama's development tended to become thicker and more viscous. Some of the earlier flows are overlain by andesite or by glacial drift. The later, more viscous, dacites tended to form domes. Some, such as Lookout, Pothole, and Dry Buttes, are isolated domes. Others, such as the group south and east of Mount Scott, occur in clusters.

The volcanic features briefly described in the foregoing discussion are but a sampling of the numerous landforms and types of volcanic deposits in the Park. What could account for the various volcanic features that we see at Crater Lake? What kind of kettle deep in the earth could spew out so many mixtures of lavas and tephra (pyroclastics) in so many different shapes, sizes, textures, colors, etc.?

It is true that andesite is the dominant rock in the area, but basalts—sometimes fluid, sometimes ejected as cinders—are found interspersed and intermixed. And the later eruptions were more dacitic, or silica-rich, than were the earlier ones. Geologists are not sure what causes this differentiation of magma within a magma chamber, but they believe that length of time is a key factor. In other words, the Mazama magma tended to separate into relatively silica-poor components (that erupted as basalts and andesites) and relatively silica-rich components (that supplied the dacite eruptions). Since the generally more fluid andesites tended to be erupted early, these eruptions were usually quieter and apt to be constructional; i.e., building up the cone. Mount Mazama's eruptions became more violent with time because the increasingly viscous dacite prevented the easy escape of gas.

Glacial Features

Between eruptions, when the mountain was quiet, glaciers rebuilt themselves and carved out deep valleys on the flanks of the volcano. When eruptive activities resumed, the glacial troughs provided convenient channels for the hot lavas, ash flows, lahars, and glowing avalanches. On the west wall of the caldera, Llao Rock, a massive dacite flow, juts up out of a U-shaped glacial trough, showing quite clearly how lavas filled up valleys previously occupied by glaciers. The culminating eruptions and explosions truncated, or "beheaded," both the glacial valley and the hardened lava flow contained therein, leaving this remarkable feature exposed in the caldera rim. Bedded pumice deposits lie on top of the Llao flow; and in some places younger till overlies the pumice. What this shows is that glaciation preceded and followed the Llao dacite flow.

Most of the Mazama glaciers extended from the high slopes to elevations lower than the present caldera rim. That is why the gaping notches of beheaded U-shaped valleys can be seen along the rim. Kerr Valley and Sun Notch on the southeast side, are the two most spectacular U-shaped troughs beheaded by the explosions and now exposed in the precipitous caldera wall. Also exposed in the cliffs rising from the lake are horizons of glacial debris interspersed with layers of lava and pumice. The grooves of glacial striations come to the very brink of the cliffs; and where layers of andesite jut out slightly from the cliff walls, more striations can be discerned on the angular tops. This indicates that after a lava flow hardened, glaciers had time to abrade the new surface (fig. 19.4).

The largest glacier on Mount Mazama flowed down Munson Valley on the southwestern flank, into Annie Creek Valley, and on down beyond the Park's southern boundary, perhaps as far as Fort Klamath. Meltwater streams from the extensive glaciers of the southern and eastern slopes carried large quantities of outwash into the Klamath Marsh. On the west side, several glaciers joined and went down the valley of the Rogue River; while on the north slopes, the Mazama glaciers joined those of Mounts Bailey and Thielsen.

Moraines of varying ages and types are common in the Park. Some are almost completely buried by pumice and by the ash of the culminating explosions. The youngest moraines in the Park were deposited *after* the explosions had ceased, indicating an interval of colder climate sometime within the 6,000 years. No glaciers exist in Crater Lake National Park at the present time, but heavy snows fall each winter—the area receives about 50 feet between September and May.

Geologic History

1. Early Cenozoic uplift and volcanism.

In Oregon, just to the west of the High Cascades and their volcanoes (of which Mount Mazama was one), lies a belt of older lavas and pyroclastics that make up the Western Cascades. These mountains were folded and

Figure 19.4   Southeast section of the caldera rim above Crater Lake. Before the summit of Mount Mazama was removed by explosion and collapse, this side of the volcano was heavily glaciated and was, during maximum glaciation, entirely covered by a thick sheet of ice. Sun Notch is a U-shaped glacial trough, beheaded by the explosion of Mazama. Phantom Ship is an erosional remnant of an old cone that may have been a forerunner of Mount Mazama. National Park Service photograph.

deeply eroded long before the Cascade volcanoes began to erupt. Both the Western Cascades and the younger High Cascades are the result of the continuing subduction of an oceanic plate beneath the westward-moving North American plate.* None of the Tertiary volcanics are exposed in Crater Lake National Park; however, some of these rocks may underlie the foundation rocks of Mount Mazama.

2. Miocene uplift and Pliocene volcanism.

By the end of the Miocene, the High Cascades had risen nearly as high as they are now and had become a barrier to the moisture-carrying winds from the Pacific. Glaciers had probably begun to form at the higher elevations. Pliocene lava flows, mostly basalts and andesites,

*See box 17.1, p. 221.

built up lava plateaus and the older volcanoes near Crater Lake, such as Mounts Bailey and Thielsen and Union Peak.

3. Construction of Mount Mazama beginning in early Pleistocene time.

Mount Mazama began to grow on top of earlier Pliocene eruptions. A smaller cone, that may be the forerunner of Mazama, is partially preserved in Phantom Ship, a small island with triangular "sails" that rises from the lake, close to the rim on the southeast side. Like the other Cascade composite volcanoes, Mount Mazama was built up by alternating lava flows and pyroclastic eruptions, mostly of andesite (fig. 19.5A). Mazama was unique in that it had a great number of parasitic cones and numerous flank eruptions from the gently sloping sides of the volcano. Mount Scott is the largest surviving satellite cone.

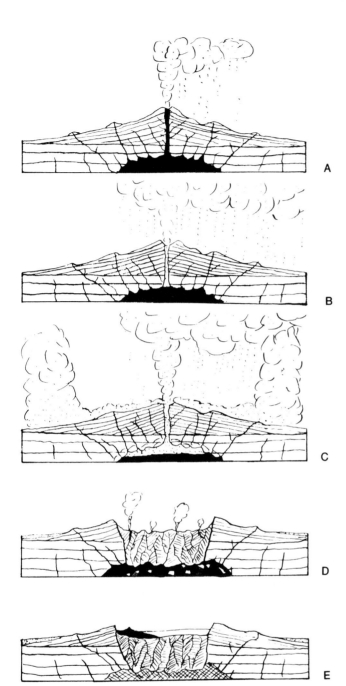

Figure 19.5 The development of Crater Lake. From *Geology of Oregon* (3rd ed.) by E. M. Baldwin. © 1981 by Kendall/Hunt Publishing Company (modified from Williams 1942). *A.* Mount Mazama begins its destructive eruptions about 6,600 years ago. *B.* As eruptions become more violent, the crater enlarges and dacite ash and pumice are ejected and blown over the region (fig. 19.6). *C.* Blocks of rock are thrown from the vent and glowing avalanches of hot gas, ash, and lava pour down the sides of the volcano. The summit begins to collapse. *D.* Collapse of the cone creates a large caldera, five miles wide and 4,000 feet deep. More volcanic material falls back into the caldera than flows out or is blown away during the explosive eruptions. *E.* Subsequent smaller eruptions form Wizard Island and partially fill in the caldera, smoothing the basin floor. Rainwater, meltwater, and surface runoff gradually fill the caldera, creating Crater Lake.

4. Pleistocene glaciations before, during, and after the eruptions.

Glaciers were probably present in the area of the Park even before Mount Mazama began to develop; and throughout much of its history, Mazama's summit was probably ice-covered. In the caldera wall, glacial horizons (layers of glacial deposits) are interbedded with lavas down to the water line, and probably even older horizons exist below the water surface. From this record we can see that glaciers advanced and retreated a number of times, sometimes because of climatic variations and sometimes because of destruction by volcanic activity. Since all of the evidence higher up was destroyed, however, the cross section in the caldera wall cannot be used for correlation of regional advances and retreats. Furthermore, the successions of glacial and volcanic deposits exposed around the caldera wall have no uniformity. For example, note the contrasts between the sequences at Glacier Point on the southeast side of the rim and Pumice Point across the lake, almost directly to the north:

Glacier Point	Pumice Point
Pumice	Pumice and scoria from final eruptions
Andesite	Youngest till
Till	Coarse and fine layers of compacted pink, buff, and white pumice; many angular blocks of andesite throughout
Andesite	Till and glacial sand
Pumice and scoria	Uncompacted lump pumice
Till	Till
Andesite	Andesite

When Mount Mazama was at its maximum height, high peaks and ridges stuck up through the glacial ice. The absence of till at the top of Hillman Peak and Watchman suggests that these peaks were never entirely covered by ice. Hillman Peak probably acted as a "cleaver," or arête, similar to Mount Rainier's cleavers that divide glaciers coming down from the summit.

Apparently long periods of quiescence followed the times of active eruptions. Bands of soil containing charred vegetation at the top of till deposits, later covered by lava flows and pumice, record such quiet times when reforestation of the slopes could take place. The discovery of carbonized stumps of trees that had been growing in till in a glacial col near Discovery Point implies that a long period of quiescence preceded the last catastrophic explosions. The pumice that buried the trees can be correlated with those explosions. The presence of trees on these lower slopes also suggests that at the time of the explosions, the glaciers had receded to higher elevations on the mountain.

It seems likely that the time of maximum glaciation was near the end of the andesitic eruptions and after the early dacitic flows from several fissures on the mountain's south flank. Subsequently dacite eruptions shifted more to the north side, which may be the reason why glacial erosion was more severe on the south side. The large U-shaped valleys on the south side, Munson, Kerr, and Sun, were deeply excavated during this time. Glacial ice was probably a thousand feet thick in the troughs.

5. Differentiation of magma and eruptions of increasing violence.

After Mount Mazama had reached its maximum height, subsequent eruptions tended to be more siliceous. Some dacite domes were protruded from the flanks. As eruptions became more violent, large amounts of dacitic pumice were ejected (fig. 19.5.B).

Eruptions shifted northward from the central vent to a group of new vents that formed an arcuate pattern as they opened. This arc, referred to as the Northern Arc of Vents, may have been the surface expression of a major fissure extending into the magma chamber. It was this curving line that later became the north wall of the caldera. First andesite flows and then great masses of dacite lava flowed from the northern vents, at least six of which are now exposed on the caldera wall. These are, from west to east, the vent of the Watchman andesite, the largely andesitic Hillman Peak composite cone, the vents of the Llao Rock, Cleetwood and Redcloud dacite flows, and the Sentinel Rock andesite.

During this time, in addition to the dacite flows from the northern vents, more dacite domes erupted, some in a cluster on the east side near Scott Mountain. A large number of parasitic cinder cones grew, some to the north of the volcano, some to the south. A few cinder cones even sprouted on the east in the midst of the small dacite domes, indicating that both basaltic and dacitic lavas apparently erupted from neighboring vents in quick succession. All this eruptive activity, much of it violent, must have helped to drain the underlying magma chamber.

In the interval that followed, the mountain remained quiet. Glaciers advanced again, becoming widespread and continuing their work of sculpturing the mountain scenery.

6. The beginning of the end of Mazama.

By the time the magma began to rise again deep within the volcano, glaciers had begun to withdraw from the lower slopes. Another series of violent, explosive eruptions of dacite pumice took place, mostly from the vents on the northern slope. Blocks of old andesite from inside the summit were thrown out with the pumice. Then several glowing avalanches swept down the northeast side of the volcano, leaving behind a thick layer of welded tuff, a

part of which can be seen on the "brim" of the Wineglass, a cliff in the northeast caldera wall (fig. 19.5C). Another quiet time ensued, along with a readvance of glaciers. Then, as the Pleistocene drew to a close, the ice retreated, leaving valley glaciers only on the south flanks of the mountain. Small glaciers also were left at the summit and in deep canyons. Thick forests grew around the base of the volcano, but the upper slopes were largely barren of soil and vegetation.

7. The culminating explosions 6,600 years ago.

The last years of Mazama's existence were at first relatively quiet. There were rumbles and steam emissions, followed by light ash falls and the ejection of pyroclasts. But the explosions increased in intensity. Then the culminating explosions began, similar to but much greater than the May 1980, eruption of Mount St. Helens. At the base of Mazama, some 20 feet of ash accumulated. Ash deposits of various thicknesses spread over the entire region—as far away as Alberta, Canada, to the north; Nevada and northern California to the south; Idaho and Montana to the east, even to the northwest corner of Yellowstone National Park in Wyoming. Fine ash particles were carried far to the east by the prevailing winds (fig. 19.6).

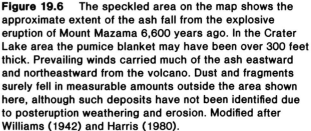

Figure 19.6 The speckled area on the map shows the approximate extent of the ash fall from the explosive eruption of Mount Mazama 6,600 years ago. In the Crater Lake area the pumice blanket may have been over 300 feet thick. Prevailing winds carried much of the ash eastward and northeastward from the volcano. Dust and fragments surely fell in measurable amounts outside the area shown here, although such deposits have not been identified due to posteruption weathering and erosion. Modified after Williams (1942) and Harris (1980).

Radiocarbon dating of the Mazama ash layer places these eruptions at about 6,600 years ago. Wherever a Mazama ash deposit is found, it provides a significant date of reference for geologists and archaeologists. Indian artifacts, such as mats, points, and scrapers, have been found at sites preserved by Mazama ash many miles to the east of Crater Lake. Geologists have been able to determine average rates of sedimentation in a number of lakes in the Pacific Northwest since the time of the Mazama ash falls. They can also tell which glacial moraines in the Cascades are younger than the Mazama ash and which ones are older.

The tremendous explosions of ash and ejecta were followed by a series of large glowing avalanches, or nuées ardentes, that boiled over the crater rim and sped down the mountain flanks, branching into the glacial valleys and spreading destruction everywhere. The whole mountain was set ablaze as trees and anything else combustible were incinerated. One fiery avalanche traveled 40 miles down the Rogue River valley. Another went north in the valley between Mounts Bailey and Theilsen, crossing Diamond Lake and then dumping pumice bombs into the valley of the North Umpqua River. Flows that went east carried large boulders of pumice many miles across the flats east of the volcano. To the south some avalanches ended up in the Klamath Marsh. Floating chunks of pumice were later washed down into the Klamath Lakes.

The last avalanches apparently came from very deep in the magma chamber. Dark-colored scoria was erupted and covered some of the light-colored pumice of the earlier avalanches. On the north side, the dark scoria created the Pumice Desert; and on the south, a scoria flow nearly reached the Klamath Marsh. The light and dark pumice layers are especially noticeable in the Pinnacles.

After the eruptions had ended, the air was full of choking, acrid volcanic smog and dust. Heavy rains fell and clouds of steam and fine dust continued to billow upward. Pumice-filled canyons remained hot and smoking for years. When the dust finally settled and the air cleared, Mount Mazama's summit was gone. Only a great pit was left, a caldera 5 miles wide and 4,000 feet deep. Part of the walls of the old summit had been blown out with the explosions; what was left collapsed into the emptied magma chamber. How much of Mount Mazama's previous bulk disappeared due to explosion and collapse? The generally accepted estimate is 15 to 17 cubic miles (fig. 19.5D).

But this great volume of material is by no means the most astonishing figure. The amount of magma (in the form of pyroclastics) thrown out of the volcano plus some liquid lava that was erupted totaled much more than the part of the cone that was lost. In fact, recent estimates are

that nearly three times as much new rock was spewed out of the volcano's insides as the volume of old rock that had made up the summit. With such an emptying out of the magma chamber, what was left of the hollowed-out volcano fell in on itself. (S. Harris 1980).

8. The aftermath.

Minor activity continued for awhile deep in the caldera. Liquid lava boiled and bubbled out over the caldera floor. Andesitic lava flows built up the area around Wizard Island, and then a cinder cone grew upward from the lava base. Wizard Island is actually about 2,600 feet high from base to top, but only 774 feet of the cone shows above the waterline. Two more, smaller cinder cones (one of them dacitic) grew up from the caldera floor but were later submerged by water rising in the lake. After the caldera finally cooled down, rain and snow accumulated and gradually filled the pit. Eventually the deep blue lake came to exist in the quiet caldera (fig. 19.5E).

Around the caldera, the flanks were attacked again by the forces of erosion—landsliding, mud flows, etc. A brief recurrence of small glaciers left drift patches on top of pumice here and there. Running water reestablished a radial drainage pattern as it tried to clear out the debris-choked valleys. After a time, trees grew around the base and up the slopes, hiding some of the scars.

Will Mazama resume its post-caldera eruptive activity? Will more cinder cones be built up from the lake bottom, or will the present ones be enlarged? Will the old pattern of flank eruptions begin again? After all, Wizard Island is only a few thousand years old. No geologist is ready to say at the present time that Mazama is extinct.

Geologic Map and Cross Section

Williams, H. 1942. *The geology of Crater Lake National Park, Oregon*. Washington, D.C.: Carnegie Institution of Washington Publication 540.

Bibliography

Baldwin, E. M. 1981. *Geology of Oregon.* 3rd ed. Dubuque, Iowa: Kendall/Hunt Publishing Company.

Harris, S. L. 1980. *Fire and ice, the Cascade volcanoes.* (rev. ed.) Seattle, Washington: The Mountaineers.

McKee, Bates. 1972. *Cascadia.* New York: McGraw-Hill Book Company.

Ruhle, G. R. 1964. *Along Crater Lake roads.* Crater Lake Natural History Association, Crater Lake National Park.

Tilden, F. 1979. *The national parks* (rev. ed.). New York: Alfred A. Knopf.

Williams, H. 1941. *Crater Lake-the story of its origin.* University of California Press.

―――. 1942. *The geology of Crater Lake National Park, Oregon.* Washington, D.C.: Carnegie Institution of Washington Publication 540.

―――. 1957. *Crater Lake National Park, Oregon, map.* Geological Survey. Text on back of map.

Address

Crater Lake National Park
P.O. Box 7
Crater Lake, Oregon 97604

20
Hawaii Volcanoes National Park

Location: Island of Hawaii, Hawaii
Size: 229,177.03 acres; 358.09 square miles
Established: August 1, 1916

Figure 20.1 Wave action has created sea arches in lava that flowed into the ocean on the southeast side of Hawaii. Hawaii Volcanoes National Park. National Park Service photograph.

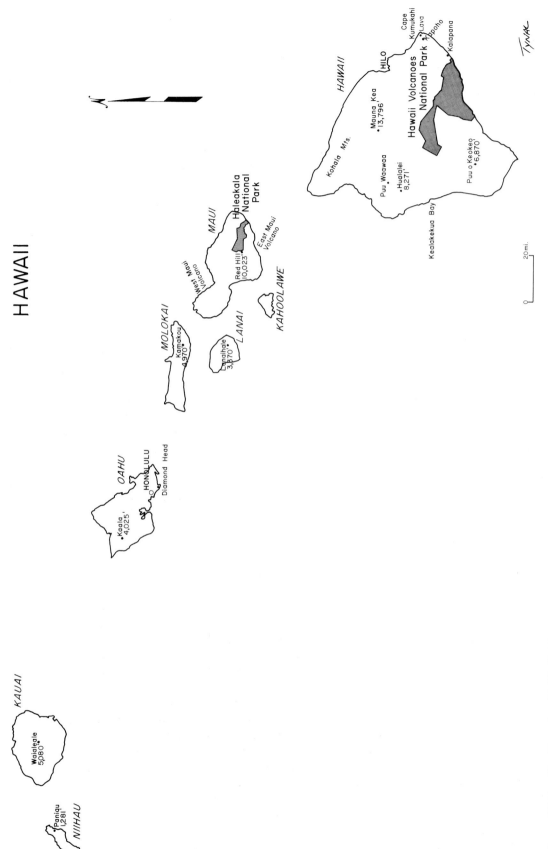

Figure 20.2 The Hawaiian Archipelago.

Box 20.1
The Geography and Geology of the Hawaiian Archipelago
(An Introduction to Chapters 20 and 21)

The Hawaiian Archipelago in the Pacific Ocean is a group of shoals, atolls, and islands, stretching 1,700 miles on a northwest-southeast linear trend, a little over 2,400 miles west of Los Angeles. At the southeast end of the chain, eight main islands and numerous islets make up the state of Hawaii (fig. 20.2). These islands are, beginning at the southeast: Hawaii, largest island and location of Hawaii Volcanoes National Park (chapter 20); Maui, site of Haleakala National Park (chapter 21); Kahoolawe; Lanai; Molokai; Oahu, where Honolulu, the state capital, is located; Kauai; and Niihau. Beyond these islands, extending toward the northwest, are small, mostly uninhabited islands, atolls, and shoals. Midway Island and Kure, at the far end of the chain, are some 1,100 miles northwest of Honolulu.

The two southeasternmost islands, Hawaii and Maui, are the largest and also have the highest volcanic peaks. Hawaii is 4,030 square miles in area and has a maximum elevation (Mauna Kea) of 13,784 feet. Maui, 728 square miles in area, rises to an elevation of 10,023 feet at the summit of Haleakala.

Climate. The subtropical climate is pleasant and mild with air temperatures ranging from below freezing at high elevations to the high 90s on the lee coasts. Because the Hawaiian Islands lie in the path of the northeast trade winds, rainfall is heavy on the windward sides of the islands. Throughout most of the year, the trade winds blow across the islands; but occasionally in winter, winds from the south or southwest ("kona winds") bring heavy storms from that direction and may dump as much as 24 inches of rain in four hours. Maximum precipitation on the islands occurs between 2,000 and 6,000 feet above sea level on windward slopes (i.e., facing northeast), but rain shadow zones reduce precipitation on parts of some islands. Thus annual rainfall ranges from about 10 inches on lee slopes to about 450 inches in the wettest belts. Stream densities are higher and stream erosion more significant on the windward sides of the islands.

Geologic Development. The islands in this long chain are volcanoes built by the eruption of basaltic lavas that rose from beneath the ocean floor. Repeated eruptions of fluid, relatively thin lava flows (only two or three feet thick) brought the volcanoes from water depths of some two and a half to four miles up to sea level in less than a million years. Above sea level, their evolution as volcanic islands has followed a fairly consistent pattern. Some reached greater heights than others; but regardless of how high above sea level the volcanoes grew, each one evolved into an "alkalic cap stage"—that is, the basaltic lavas became poorer in silica and richer in sodium and potassium. Eruptions were less frequent and more explosive during this stage and flows of lava became much thicker.

As volcanic activity waned, erosion began to cut deep valleys into the slopes. On some of the islands—Niihau, Kauai, Oahu, Molokai, Maui, and Kahoolawe—erosion was interrupted by renewed volcanism of even more alkalic and explosive lavas. Diamond Head on Oahu, for example, is the result of a period of renewed volcanism on that island. As each volcanic island aged and was eroded, it sank as the oceanic crust beneath adjusted to the rapid growth of newer volcanoes in the chain. Coral reefs, such as those on Midway, formed on the fringes of some of the older volcanoes. With continued sinking and wave erosion of extinct volcanoes, the reefs were reduced to atolls; that is all that remained above sea level.

Radiogenic dates derived from rock samples taken at various places along the Hawaiian chain clearly show that the volcanoes are progressively younger from northwest to southeast, the youngest rocks being those forming the island of Hawaii.

Moreover, the oceanic crust in that part of the Pacific is much older (Cretaceous) than the lava flows that constructed the Hawaiian volcanoes (most of them younger than Miocene). What caused the Hawaiian chain to form in this linear pattern across the ocean floor with the oldest volcanoes in the northwest and the youngest at the southeast end?

Since the volcanoes are erupting in the middle of an oceanic plate, the volcanism cannot be associated with a subduction zone or with tectonic activity occurring at plate margins. The most likely hypothesis suggests the existence of a "hot spot," or thermal high (perhaps as much as 100 miles in diameter), in the upper mantle beneath the oceanic crust. The source of the heat may be a

mantle plume, a narrow column of hot mantle rock that rises toward the crust and spreads radially outward. The magma produced accumulates in a pocket that bulges up and eventually breaks through the crust, releasing great quantities of basaltic lava and building up shield volcanoes on the ocean floor. As the Pacific plate rides slowly northwestward over the heat source, successive volcanoes form over the hot spot, and then are rafted away on the moving plate and become extinct. The hot spot theory thus explains the basic pattern of a northwest-trending chain of volcanoes with an active volcano on the southeast end and volcanoes becoming progressively older toward the northwest.

Time Units			Rock Units and Geologic Events	
Era	Period	Epoch	Mauna Loa	Kilauea
Cenozoic	Quaternary	"Historic Time" / "Prehistoric" and Holocene / ? / Pleistocene	*Younger Ka'u Formation* Summit eruption, 1975 Numerous summit and rift zone eruptions, 1832–1950 Continuing shield construction by flows of both aa and pahoehoe Enlargement of caldera by pit craters *Older Ka'u Formation* Shield construction by aa and pahoehoe; cinder cones, spatter cones, etc. Development and collapse of Mokuaweoweo caldera; pit craters *Kahuku Formation,* capped by Pahala ash Building up of shield by thick accumulations of aa and pahoehoe Great erosional unconformity Fluctuations of sea level and submergence; intense erosion of shield; canyon cutting *Ninole Formation* Construction of Ninole volcano Volcanic "basement" rocks (not exposed at surface; age unknown	*Younger Puna Formation* Construction of satellite shield Mauna Ulu, 1968–1974 Numerous summit and east and west rift zone eruptions, 1790–1979 Continuing shield construction by flows of aa and pahoehoe Explosive eruptions; 1790, 1924 Appearance, disappearance, reappearance of lava lakes; enlargement of caldera *Older Puna Formation* Shield construction by aa and pahoehoe flows; ash falls, cinder cones, spatter cones Development and collapse of Halemaumau in summit caldera; pit craters *Hilina Formation,* capped by Pahala ash Beginning of shield construction by many flows, both aa and pahoehoe, building up thick accumulations; some ash layers, indicating past explosive eruptions Volcanic "basement" rocks (not exposed at surface); age unknown

Source: modified after Macdonald and Abbott 1970; Easton and Garcia 1981

Table 20.1. Geologic Column, Hawaii Volcanoes National Park

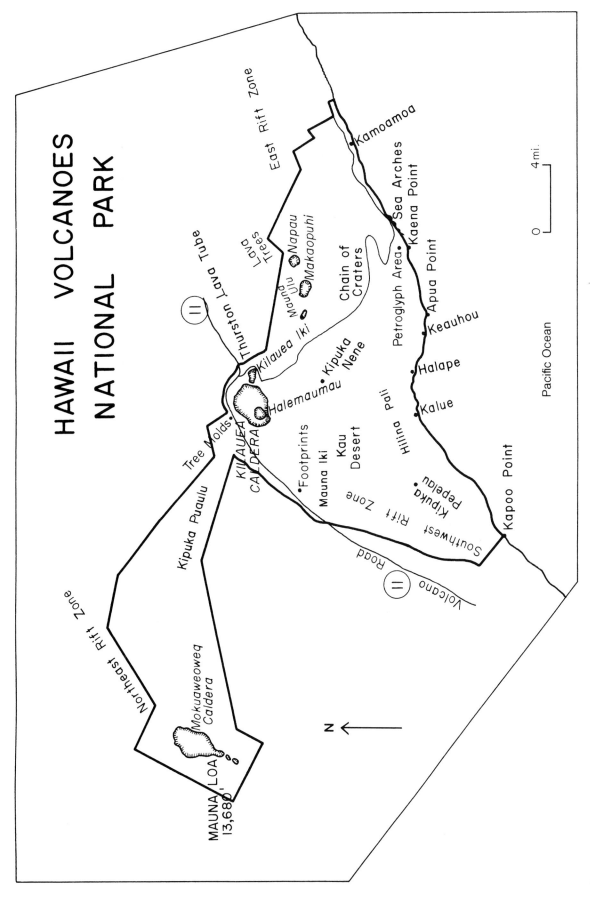

Figure 20.3 Hawaii Volcanoes National Park, Hawaii.

Local History

Polynesian voyagers from the South Pacific were the first known settlers of the Hawaiian Islands. These courageous, seafaring people began arriving at the islands about 2,000 years ago, their final migration being around A.D. 750. Volcanoes were central to their culture and religion as they were believed to be the homes of deities. Especially sacred were Maui and Hawaii, on which many temples (*heiaus*) were built. Pele, the goddess of all volcanoes, lived in Halemaumau, the fire pit in Kilauea caldera on Hawaii, but she was apt to turn up at other volcanoes from time to time. Because she was known to be extremely dangerous when aroused, human sacrifices were made to allay her wrath.

Captain James Cook discovered the islands in 1778 on his way to the Pacific Northwest. He named them the Sandwich Islands for the Earl of Sandwich, who was then First Lord of the British Admiralty. The name was not retained and is now used only for a group of islands in the South Atlantic near Antarctica.

Because Captain Cook arrived at the time of a religious ceremony, the Hawaiians thought he was a god; but when he visited the islands again on his return voyage, they realized that he was not a god and relations were less pleasant. At Kealakekua Bay on Hawaii, Cook got into an argument with some Hawaiians about a rowboat they had taken from his ship, and they killed him.

These were troubled times in the islands, which were then ruled by warring kings. In 1810, after many battles, Kamehameha I, the king of Hawaii, became ruler of all the islands by conquering the other kings. Encouraged by Kamehameha I, American traders and whalers began arriving at the islands in increasing numbers, and they were followed by missionaries. Unfortunately, the white men brought with them infectious diseases which reduced the Hawaiian population and brought about the decline of native cultural traditions.

The Americans encouraged Kamehameha III, who became king in 1825, to establish a constitutional monarchy; and G. P. Judd, an American missionary, became prime minister in 1842. In that year, the United States recognized Hawaii's independence, although France and Great Britain still claimed part of the islands for some years. American influence increased from then on. The last Hawaiian monarch, Queen Liliuokalani, was overthrown in 1893, and seven years later Hawaii became an American territory. Statehood was attained in 1959.

From Captain Cook on, visitors to the islands have been fascinated by the Hawaiian volcanoes. In 1823, a missionary, the Reverend William Ellis, was the first white man to descend into the crater of Kilauea. He described the boiling lake of lava in Halemaumau and recorded eruptions that occurred. Travelers from all over the world came to see this amazing spectacle; among them was Mark Twain, who said that, by comparison, Vesuvius was a "soup kettle"!

The movement to protect and preserve the land around the volcanoes on Maui and Hawaii Islands was led by L. A. Thurston, editor of the *Honolulu Advertiser*. Hawaii National Park, including land on both islands, was established by Congress in 1916. Haleakala National Park on the island of Maui was set aside as a separate park in 1960. Hawaii Volcanoes National Park was designated in 1961.

As founder of the Hawaiian Volcano Research Association, Thurston was also active in helping to establish the Hawaii Volcano Observatory. It was formally organized in 1911 under the auspices of the Massachusetts Institute of Technology and the Carnegie Institute of Washington. Professor T. A. Jagger, a geologist and volcanologist at Massachusetts Institute of Technology, chose Kilauea as the site for the proposed observatory because of the constancy of the volcano's activity and because investigators could get close to active lava without exposing themselves to too much danger. Jagger was the director of the observatory for the first thirty years of its existence. Continuous recording of seismic and volcanic activity has been carried on at the Hawaii Volcano Observatory since 1912 under the successive sponsorship of the Massachusetts Institute of Technology, the U.S. Weather Bureau, the National Park Service, and (since 1956) the U.S. Geological Survey. The Observatory buildings are located in the Park on the rim of Kilauea Crater, overlooking Halemaumau.

Included in the Park is the summit area of Mauna Loa, a strip of land between Mauna Loa and Kilauea, and most of the southern flank of Kilauea, from the caldera to the coast. Mauna Loa and Kilauea make ideal natural laboratories for studying volcanic activity. They erupt frequently but seldom explosively; they emit enormous amounts of lava; and they have large, relatively simple structures (fig. 20.3).

Geologic Features

Five shield volcanoes make up the island of Hawaii. Besides Mauna Loa and Kilauea, which are both active, there are Kohala at the northwestern end of the island, Haulalai on the west coast, and Mauna Kea on the north side of the island. Mauna Kea, the highest cone (elevation 13,784 feet), is also the world's highest peak on an island. At 13,680 feet, Mauna Loa is a close second in elevation. Kohala is considered extinct. Haulalai has erupted only once (1800–1801) in historic time. Mauna Kea is extinct; its most recent dated eruption, based on C^{14}, was approximately 4,000 years ago.

Approaching the island of Hawaii, whether on board ship or in a plane, travelers can see the summits of Mauna Loa and Mauna Kea from great distances if the weather

Figure 20.4   Air view of the shield volcanoes Kilauea (foreground) and Mauna Loa (background) in Hawaii Volcanoes National Park. Note Kilauea's large caldera on the summit. U.S. Geological Survey photograph.

is clear. The closer one gets, the more impressive their bulks appear as they rise from the ocean. Yet much more of their great size is hidden beneath the water; the ocean floor on which the volcanoes stand is 18,000 feet below. Thus Mauna Loa's true height is 32,000 feet, or more than 6 miles! Kilauea, whose highest point is 4,090 feet above sea level, is nearly 10,000 feet lower than Mauna Loa and Mauna Kea. All of the rock above sea level on Hawaii Island has been erupted during the past million years, which gives us an idea of the rapidity with which this tremendous volume of lava has been extruded.

Because Mauna Loa and Kilauea have been continuously active during historic time, many changes in their features have been documented. During the past century and a half, Mauna Loa has produced a lava flow on the average of every seven years and has exhibited minor eruptive activity in between. Its shield-shaped dome is about 70 miles long and 30 miles wide. Even more impressive is its volume of about 10,000 cubic miles, which makes it the world's largest active volcano. At the summit the Mokuaweoweo caldera coalesces with adjacent pit craters. Formed by collapse rather than explosion, the caldera has changed its size, shape, and depth frequently during historic time as lava flows have spread out over the caldera floor or breached the rim (fig. 20.4).

Rising from the southeastern flank of Mauna Loa is Kilauea, which has a shield dome about 50 miles long and 14 miles wide. Its caldera is 3 miles long and 2 miles wide. Within the caldera is the fire pit (or active crater) of Halemaumau that from time to time contains a lake of liquid lava. A *lava lake* forms when magma rises within the volcano and floods the pit in the floor of the caldera. Sometimes a lava lake lasts for years, bubbling and glowing.*

If the level of the molten rock rises high enough, lava spills out over the rim of the fire pit and spreads over the caldera floor. More often, lava breaks out through the flanks of the volcano above or below sea level. As lava pours out of the flanks, the level of the lake falls and the lake may disappear, perhaps for years at a time. If the whole plumbing system of the volcano is drained, then the caldera floor collapses. After a time, more magma rises into the volcano reservoir, or magma chamber, pushes up through the conduit, and the whole process—with modifications—may be repeated.

*In *Roughing It*, Mark Twain describes in awed terms seeing the lava lake after dark: "The greater part of the vast floor of the desert [i.e., caldera] under us was as black as ink and apparently smooth and level; but over a mile square of it was ringed and streaked and striped with a thousand branching streams of liquid fire! . . . Imagine a coal-black sky shivered into a tangled network of angry fire." Mark Twain visited Halemaumau in 1866. The fiery "branching streams" he described were lava fountains.

Volcanic Rocks that Form from the Lavas

The Hawaiian basalts are of two general types, or suites—*tholeiitic* and *alkalic*, both believed to be derived from the upper mantle, but perhaps from different levels. The tholeiitic basalts, sometimes called "primitive basalts," make up the bulk of the shields (99%). The alkalic basalts (1%) are found mainly on the caps of the shields. The predominant phenocrysts in both types of Hawaiian basalts are olivine. Tholeiitic basalts commonly contain low- and high-calcium pyroxenes, both poor in aluminum. Pyroxene phenocrysts are rare in these rocks. Alkalic basalts contain only one pyroxene, a high-calcium and moderately high-aluminum variety that occurs commonly as phenocrysts as well as groundmass grains.

Types of Lava and Lava Features

Since most of the lava is quite fluid, it flows out from vents relatively quietly. Lava in a quiet flow that congeals with a smooth, billowy or ropy surface is called *pahoehoe* (pah-hóay-hoay).* Typically lava erupts in a series of pulses that build up layers of pahoehoe, one on top of the other, during an eruption.

Another common type of lava flow is termed *aa* (ah'-ah). These flows have a rough, jagged, spinose, or clinkery surface. The basalt in the interior of aa flows is dense and thick with many stretched or deflated vesicles left by escaping gas. The front of the flow moves along as a red-hot, clinkery wall. Very massive aa flows may show columnar jointing. Sometimes after gas is dissipated and the lava is cooling, a pahoehoe flow changes to aa; but an aa flow never reverts to a pahoehoe. However, a fresh pahoehoe flow may be erupted on top of an earlier aa flow.

When a lava lake is active or a flank vent is first erupting, gases bubbling up through the magma produce *lava fountains* that may spout from a few inches to 30 feet in height, rising and falling on the hot lava surface. Occasionally lava fountains shoot up much higher, as happened at Kilauea Iki in 1959 when a lava fountain reached more than 1,600 feet in height! As tiny strands or droplets of fountaining lava are caught by the wind, they harden in the air and fall. These glassy pellets and strings, called *Pele's tears* and *Pele's hair,* are fragile and short-lived.

Lava tubes and *lava tunnels* beneath the congealing surface of a pahoehoe flow convey fresh, hot lava on out to the moving front. Later, when the whole mass has cooled, the tunnels and tubes remain. A lava tube serves as a large conduit for the movement of lava down a slope far from the vent. The tube's outer crust acts as an insulator, keeping lava from cooling. Some tubes, such as the Thurston lava tube on the east side of Kilauea, are 20 feet high and 22 feet wide in places. The Thurston lava

tube was formed by the crusting over of a pahoehoe flow from one of the Twin Craters. The molten lava within then drained down the slope. Some lava tubes contain *lava stalactites* which were produced as liquid lava hardened while dripping from the tube roof.

Tree molds and *lava trees* may form in pahoehoe lava when a flow inundates a forest. Hot lava, chilled by its initial contact with a tree, either sticks up against the bark, hardening quickly, or flows around the trunk. After the charred tree rots away, the lava "trees" are left standing. Pail-like or well-like depressions—the tree molds—show where the tree trunks were. Tree molds are numerous in the Park and can also be seen in Lava Tree State Park, just east of the National Park boundary.

Mounds or hillocks on the surface of pahoehoe lava are called *tumuli* (singular, *tumulus*). They tend to form parallel to the flow front in places where the lava is confined, especially in craters or calderas. Solid crust at the top of the flow is dragged along and domed up by the faster-moving fluid lava underneath. Sometimes fresh lava dribbles or spatters out, hardening into *driblet spires* of fantastic shapes.

Lava balls are accretionary features characteristic of aa flows and range in size from a few inches to 10 feet or more. They form when a fragment of solidified lava rolls along, enlarging as it picks up sticky lava. (Children in snowy regions make snowballs large in much the same way by rolling them in sticky snow until they're too big to push.)

An Hawaiian term, *kipuka*, meaning "opening," refers to islandlike areas of older land surrounded by more recently erupted lava. Kipukas, ranging in size from a few square feet to several square miles, are usually the result of irregularities in the land surface. A kipuka surface may be higher or lower than the level of the surrounding lava, the latter case being more common. New lava is sterile and barren of vegetation, but some kipukas have weathered long enough to have soil that can support plant life, making them islands of greenery in a landscape of bare volcanic rock. Several kipukas are in the Park, the best known being Kipuka Puaulu, or Bird Park, northwest of Hawaii Volcano Observatory. Kipuka Puaulu has an arboretum with some 40 varieties of trees. Nearby are tree molds where lava flowed through a forest, incinerating the trees.

Ash Explosions and Pyroclastics

Kilauea has been an ash producer, probably because more moisture is trapped on its windward slopes. Ash explosions are commonly associated with the sudden mixing of ground (phreatic) water and hot lava or magma. The most violent explosions have occurred when water drained down into newly emptied vents or conduits and reached magma. The result has been a *phreatomagmatic explosion.* In 1790 such an explosion killed a whole company of Hawaiian

---

*Varieties of pahoehoe flow are identified by such picturesque terms as *corded, elephant-hide, entrail, festooned, filamented, sharkskin, slab,* etc.

warriors who were marching across the Ka'u desert (a rain shadow zone on the southwest slope of Kilauea). Despite years of weathering, the warriors' footprints are still faintly visible on an ash surface close to the Park's western boundary.

Phreatic explosions occur any time that hot lava encounters ground water and turns it to steam. The most explosive steam explosions in historic time occurred during the 1924 eruptions from Kilauea when large blocks of old lava (that can be seen scattered around the summit area) were ejected from the volcano's main vent. More typically, phreatic explosions produce ash, cinder cones, and pyroclastics of various sizes during eruptions, especially at or near the shore where moisture is plentiful. Most cones, such as those along Kilauea's east rift zone, are a combination of cinders and spatters.

Rift Zones on the Volcanoes

Shield volcanoes are built up not only by summit eruptions but also by eruptions from the *rift zones* on the flanks. Rift zones radiate from the summits of the Hawaiian volcanoes at obtuse angles. Mauna Loa and Kilauea each have two principal rift zones. (Mauna Loa also has a minor rift zone that trends northwest.) Mauna Loa's main growth has been along the two principal rift zones that extend from the caldera to the southwest and to the northeast. Kilauea has built up its flanks along a southwest rift zone and an east rift zone. All of the rift zones extend into the ocean except Mauna Loa's northeast rift.

The rift zones are scored at the surface by many cracks, some of which have served as vents for erupting lava. Below the surface are many dikes of congealed lava filling fissures. The rift zones apparently extend downward over the volcano's magma reservoir, believed to be several miles beneath the summit. Each upwelling of magma from depth causes swelling of the summit and flanks which creates new stresses on the rift zones, and opens new cracks and fissures or reopens old ones. An eruption usually relieves the pressure from the magma reservoir, and the volcano then deflates.

Fault Systems

A fault system between Mauna Loa and Kilauea appears to act as an "expansion joint" between the two volcanoes, which seldom erupt at the same time. Earthquakes in a fault system along Mauna Loa's southeastern base suggest that fault movements are making the necessary adjustments. Step faulting, by which the mountain side has moved downward in response to gravity, has produced a series of fault scarps facing southeast.

Similar step faulting has taken place in the Hilina fault system along Kilauea's southern slope. Here the fault blocks form a series of steps down to the coast, with the trend being parallel to the shore. Some lava flows have cascaded down over the fault scarps and into the ocean.

Shore Features in the Park

Lava flows that reach the ocean produce cinders and ash when the hot lava mixes with the sea water. Cinder cones near Kalapana were formed when aa lava entered the ocean. Waves eroded and transported sediment from the cinder cones, depositing "black sands" on a nearby beach. Since the source of supply (i.e., the cinder cones) is limited and some sand is washed out to sea each year, the black sands may last only two or three centuries.

Sea cliffs and sea arches made up of lava flows (fig. 20.1) are undergoing wave erosion and are retreating. In some places along the shore, recent lava flows protect the cliffs and have halted their retreat temporarily.

Box 20.2
Predicting Eruptions

Crisscrossing the island of Hawaii is a network of scientific instruments for monitoring the volcanoes. Collecting and analyzing the data, as well as making direct observations, is the responsibility of volcanologists at the Hawaii Volcano Observatory and the University of Hawaii. Tied in with this network are seismic stations and investigators in other parts of Hawaii, Japan, New Zealand, and many locations throughout the world.

What are the kinds of changes or signs that scientists look at? What combinations of factors indicate an impending eruption? In suggesting the lines of investigation outlined below, Macdonald (1972) emphasizes that "no single tool is adequate to predict an eruption except in the most general way" (p. 418).

1. *Periodicity.* What is the eruptive history of the volcano? How long are its quiet periods? Such

information is useful in a general way even though eruptive patterns of volcanoes often change unexpectedly.

2. *Changes in the behavior of fumaroles and changes in the output, composition, and temperature of gas.* Steam and gas emissions are often affected by weather (e.g., temperature and humidity), but, nevertheless, observations in this category are very useful. On Kilauea, for example, an increase in the hydrogen content of gas emissions usually signals the onset of an eruption.

3. *Shifts in the orientation of the volcano's magnetic field and reduction of magnetic attraction of rocks due to increasing heat.* Changes in magnetism are difficult to interpret, but they appear to indicate movement of magma within the interior of a volcano. Often an eruption does follow such movement, but sometimes a volcano quiets down without an intrusive event occurring.

4. *Infrared photographs and remote sensing.* Increases in heat and changes in the locations of hot areas show up when these techniques are used. Any eruptive activity is invariably preceded by an increase in heat on the volcano's surface.

5. *Geoelectrical changes.* Variations in electrical currents in a volcano have been observed just before explosions or eruptions, suggesting that geoelectricity may be useful as an indicator.

6. *Tumescence and tilting.* Kilauea and Mauna Loa have histories of swelling and tilting before eruptions, but occasionally swelling goes down without an eruption occurring. Land deformation on any active volcano is considered an ominous sign. The cause of bulging, rising of the summit, tilting, and the like is assumed to be magma filling the volcano reservoir and rising in the conduits. Tiltmeters on Kilauea and Mauna Loa continuously monitor land surface deformation as well as migration of the center of a tumescent area.

7. *Changes in seismicity.* From earliest times people who lived near volcanoes associated earthquakes with eruptions. On Hawaii Island, certain kinds of earth tremors and rumblings were called "Pele's thunder." These were regarded as warning signals of her anger and were not ignored!

Earthquakes are frequent on the island of Hawaii; many of them are so slight that they are imperceptible to residents. Most of the earthquakes are shallow and local and indicate a shifting around of rocks in response to stresses in the volcanic structures. However, severe earthquakes do occur from time to time (for example, a 7.5 quake* on the south coast of Hawaii in 1975), and some quakes originate at depth (25 miles or more below the earth's surface). On Maui and the other Hawaiian Islands, earthquakes are less common and are probably caused by isostatic adjustments of the oceanic crust to the load of overlying volcanoes.

Seismic data collected on Hawaii have been studied for more than sixty years by volcano-watchers seeking clues as to the imminence of eruptions. Several significant patterns have shown up, but they must be combined with other observations to be meaningful. Earthquake "swarms," consisting of thousands of little earthquakes, are recorded on seismographs before eruptions occur on Mauna Loa and Kilauea. These quakes, called *harmonic tremors,* are probably the result of minor faulting that accompanies tumescence or the inflation and pushing up of the volcano top. Fissures and troughs may develop along the rift zones. After the magma has drained and the summit starts to shrink, quakes begin again, sometimes stronger than those that accompany the swelling. Evidence from tiltmeters may confirm whether the volcano is shrinking or, possibly, swelling again. This is important in Hawaii where eruptive periods may last for weeks, months, even years.

Similar techniques for collecting data are used for Mount St. Helens and the other Cascade volcanoes (see box 14.1, p. 174). Predictions regarding possible eruptions are necessarily based on several lines of evidence and are expressed in probabilistic terms. Nevertheless, even though the prediction of volcano behavior remains an inexact science, forecasts can be made that can help to prevent loss of lives when dangerous areas need to be evacuated.

Because the active Hawaiian volcanoes are seldom explosive and because more is known about their behavior, they are considered less hazardous than most other active volcanoes. The Cascade volcanoes, as we have seen in previous chapters, tend to be more erratic, more explosive, and much less predictable, in part because not much data on their infrequent (in human terms) eruptions are available.

*Richter scale

Geologic History

1. Initial eruptions on the ocean floor at the southeastern end of the Hawaiian Archipelago, probably early in Pleistocene time.

The age of the rock at the base of the volcanic pile that is now Hawaii Island has not been determined. One problem on Hawaii Island is that putting down drill holes in order to get core samples for age determination would be a most difficult and costly operation. Moreover, in some locations drill bits might hit magma that would melt them.

Like the other islands in the chain, Hawaii was built up from the ocean floor by many thousands of layers of lava. Magma accumulated in pockets below the ocean crust, worked its way to the surface, and erupted as lava—a process that has continued into the present.

2. Construction above sea level of Hawaii Island, beginning a little more than a million years ago.

Five shield volcanoes, each with its own magma reservoir but all having the same magma source in the upper mantle, built up the island during Pleistocene time. The oldest is Kohala, probably now extinct, at the northwestern end of the island. Kilauea, the youngest volcano, is at the eastern end of the island. South of Hawaii is Loihi seamount, which is regarded as the new surface expression of the hot spot beneath the oceanic crust.

3. The development of Mauna Loa.

Rock units. Mauna Loa's rocks are mainly olivine basalts. The oldest that are exposed are the rocks of the Ninole Formation that make up the core of the mountain. Ninole rocks represent an earlier cone that was deeply eroded before Mauna Loa's eruptions buried it. Ninole beds can be seen in low hills on Mauna Loa's southeast slope. Deep canyons were cut into Ninole lavas during the erosional period that produced an unconformity between Ninole beds and the overlying lavas of the Kahuku Formation. The Kahuku lavas partially filled the canyons and built up the slopes of Mauna Loa. Pahala ash, 5 to 50 feet thick, mantles the Kahuku rocks. Most of the Pahala ash in this area came from Kilauea and was probably the result of violent phreatomagmatic explosions that accompanied the collapse of Kilauea's caldera. Much of the Pahala ash has been covered by younger lavas of the Ka'u Formation. The fairly fresh and relatively unweathered rocks of the Ka'u series include lavas from eruptions during historic time. Mauna Loa had probably attained its present height by the end of the Pleistocene Epoch.

Eruptions in historic time. Mauna Loa's eruptions since early in the 19th century have been mainly of two types: (1) summit eruptions without flank flows and (2) the more typical flank flows from the rift zones that may have been preceded or accompanied by brief summit eruptions. Eruptions generally begin with local earthquakes and the release of clouds of steam and gas that may rise several miles above the summit. Lava fountains may shoot up many feet in Mokuaweoweo, the summit caldera. Strands of Pele's hair are blown out and away by the uprushing winds. During a summit eruption, lava pours from cracks in the caldera wall and spreads over the floor. Pit craters may develop or enlarge in and around the caldera.

When eruptions begin in the rift zones, lava and hot rising gases shoot up, making a "curtain of fire" along the length of a fissure opening. Clots, blobs of pumice, and threads of volcanic glass harden in the air and then fall glistening to the ground. During the first few days, very fluid pahoehoe pours down over the slopes like a slow-moving river (10 to 25 miles per hour). Later some flows change to aa.

The 1859 eruption, which lasted for 10 months, produced a 33-mile flow, the longest recorded in historic time. Where the pahoehoe flow entered the ocean, pillow lava formed beneath the water surface.

Several eruptions from the northeast rift zone have endangered the city of Hilo and its harbor on the eastern coast of the island, the closest call being a pahoehoe flow 29 miles long in 1881 that stopped on Hilo's outskirts. In 1935 and again in 1942, planes dropped bombs on flows that appeared headed for Hilo. The flows did stop well before reaching Hilo, but it is likely that natural causes were responsible. Geologists now believe that the flows stopped simply because the supplies of fresh lava were exhausted.

A major eruption of Mauna Loa took place in June 1950. Lava first began erupting near the caldera at the top of the southwest rift zone. In minutes the erupting fissures had opened up the southwest rift zone for several miles down the slope. Within 24 hours lava fountains blazed in a curtain of fire down the mountainside for 8 miles. On both sides of the rift ridge, rivers of lava poured down the side slopes and on down the mountain. Three flows reached the ocean. The volume of lava (600,000,000 cubic yards) that was extruded during the three weeks of the 1950 eruption equaled the amount of lava that erupted in the 10-month eruption of 1859! In terms of volume, these two eruptions are the largest that have been recorded on the island of Hawaii.

Although Mauna Loa has been mostly quiet since 1950, major eruptions might still occur. In 1975 lava from a summit eruption spread over part of the caldera floor.

4. The development of Kilauea.

Rock units. Eruptions of the Hilina Formation lavas from the southeast flank of Mauna Loa began building up Kilauea in about the middle of Pleistocene time. The Hilina lavas are presumed to be roughly equivalent in age to the Kahuku Formation on Mauna Loa. Hilina rocks, consisting of basalts, including olivine basalts, and some oceanite, are exposed in the fault scarps along the southern coast. Flows of both pahoehoe and aa can be distinguished.

Pahala ash overlies the Hilina Formation. Above the Pahala ash are rocks of the younger Puna Formation. Puna lavas were erupting at about the same time as the Ka'u lavas on Mauna Loa, and the flows interfinger in the area between the two volcanoes. Puna flows were erupted largely from vents in Kilauea's caldera and from the rift zones. Spatter and cinder cones formed on both the southwestern and eastern rift zones, with those on the eastern rift zone being more numerous.

Kilauea is believed to have its own magma reservoir and conduit system without direct connections to Mauna Loa. However, both volcanoes may draw magma from a common source at depth (i.e., below the oceanic crust).

Eruptions in Historic Time. The types of eruptions that have occurred have been (1) flank flows, more often than not preceded by summit activity or eruption; (2) rising and falling of lava lakes; (3) brief summit eruptions; and (4) violent explosive eruptions.

The lava lake in the fire pit on Kilauea's summit apparently filled not long after the violent explosion of 1790. Throughout the 1800s and the first quarter of this century, the lava lake was almost continuously active; although some sinking, followed by refilling, occurred from time to time after lava had been drained off by flank eruptions below sea level. The lava lake was at a high level at the beginning of 1924 when a series of earthquakes began. They first occurred at the summit and then shifted outward along the eastern rift zone, suggesting that magma was moving in that direction. Meanwhile, the level of the lake was dropping several hundred feet. The explosions began in May and they lasted for about three weeks. Ash-filled clouds of steam and gas shot up. Sulfur fumes killed vegetation over a wide area. Blocks of caldera rock were thrown a mile into the air. Local earthquakes, lightning, and heavy downpours of rain accompanied the explosions. Afterwards Halemaumau had increased in diameter from 500 feet to over 3,000 feet, and the floor of the caldera had sunk 11 feet. Even the caldera rim was lowered a few feet. What caused the explosions and collapse was ground water pouring into the hot vents as lava drained away. Instantaneous boiling resulted, and then the violent escape of the steam that was generated. The explosions were not magmatic since no fresh lava was ejected with volcanic debris (fig. 20.5).

Since 1924, the lava lake that had bubbled and flamed for so many years has refilled partially, but only for very brief intervals. The fire pit is now quiet except for wisps of steam. Summit eruptions have been short and infrequent. Flank eruptions, on the other hand, have been more numerous and sustained.

Before the lava lake disappeared, the southwest rift zone was generally more active than the eastern rift zone. Mauna Iki is a secondary lava cone southwest of the caldera that was built by an eruption in 1920. Fresh lava flowed out near Mauna Iki in 1971. The Great Crack, extending from Mauna Iki to the ocean, was the outlet for very fluid lava that erupted in 1823, flowing so fast that canoes on the shore were burned. Lava tree molds are preserved in this flow.

When the center of volcanic activity on Kilauea shifts from the summit to the flanks, particularly to the east rift zone, a number of changes ensue. This rift zone has some rather unusual characteristics. From the caldera, the east rift zone goes southeastward and then takes a sharp bend almost due east to Cape Kumukahi and on out into the

Figure 20.5   The summit and caldera of Kilauea. The pit crater Halemaumau is in the caldera floor. National Park Service photograph.

ocean. The Chain of Craters, on the upper part of the eastern rift zone, used to be a row of 13 pit craters coming down from the caldera. A Park highway went through the area from the Crater Rim road southeastward to the coast. The road was buried by lava in 1969 when an entirely new vent opened up and Mauna Ulu, a satellite cone, began to grow. (The Chain of Craters road was reopened in 1979.) Mauna Ulu continued to erupt until 1974, many of the flows reaching the ocean. Next to Mauna Ulu is the Alae shield, which was built up over an old pit crater that was 440 feet deep. Other nearby craters were partly filled or buried, and some reopened and began pouring out lava or filling up and forming lava lakes. In 1959, Kilauea Iki, a large pit crater close to the caldera, had a major eruption. The lava fountains that shot up were the highest on record, and cinders and pumice devastated a nearby forest. A 365-foot-thick lava lake formed in the crater. The lake has now crusted over, but it is still hot.

The following year, in 1960, lava erupted 28 miles away from the summit near Cape Kumukahi at the extreme eastern tip of the island. Lava flowing from the eastern rift zone destroyed the village of Kapoho and created some 500 acres of new land where the flow went into the ocean. (The question of who owns this new land is still in litigation.)

A 1977 eruption from the eastern rift zone nearly destroyed the coastal village of Kalapana, close to the Park's eastern entrance. This vent, also, opened far from the caldera and very fluid lava erupted. Because the flow changed to very thick aa at the far end, it stopped just before reaching Kalapana.

5. Glaciation.

A small icecap existed on Mauna Kea, Hawaii's highest volcano, during the Pleistocene. Glaciers may have been present on Mauna Loa at the same time, but if they were, all evidence of them has been destroyed or buried by recent lava flows. In today's climate, some snow falls every winter on Mauna Kea and Mauna Loa, but it does not last long.

6. Changes in sea level.

Old shorelines, some higher than the ocean's present stand and some lower, testify to the fluctuating ocean level around the Hawaiian Islands during their comparatively brief geologic history. The most significant long-term trend is that of land submergence; that is, as the volcanic pile has gotten higher and heavier, the sea floor has tended to sink. However, older sea levels are not consistent from place to place because local tilting and warping have occurred. Thus tectonic changes have affected the relative position of land and ocean. At the same time, the islands have been affected by worldwide changes in sea level that occurred during Pleistocene time. Some submerged shorelines can be roughly correlated with lower sea stands

resulting from the withdrawal of ocean water as huge quantities were stored on land in glacial ice. The high stands are probably related to release of water by glacial melting.

The cliffs on the southeastern coast of Hawaii Island along the section owned by the Park are largely the result of tectonic activity. At Puu Kapukapu, for example, a fault scarp 1,000 feet high plunges directly into the ocean. The fault is part of the Hilina fault system that parallels the coast.

Geologic Maps and Cross Sections

American Association of Petroleum Geologists. 1974. *Geological highway map of the state of Hawaii.* Map No. 8. Tulsa, Oklahoma: American Association of Petroleum Geologists.

Macdonald, G. A. 1949. *Petrography of the Island of Hawaii.* U.S. Geological Survey Professional Paper 214-D. Plate 12.

————. 1971. *Geologic map of the Mauna Loa quadrangle, Hawaii.* U.S. Geological Survey Map GQ-897.

Peterson, D.W. 1967. *Geologic map of the Kilauea Crater quadrangle, Hawaii.* U.S. Geological Survey Map GQ 667.

Stearns, H. T., and Macdonald, G. A. 1946. *Geology and ground water resources of the Island of Hawaii.* Hawaii Division of Hydrography Bulletin 9. Fig. 1.

Bibliography

Dalrymple, G. B., Silver, E. A., and Jackson, E. D. 1973. Origin of the Hawaiian Islands. *American scientist* 61 (May-June):294–308.

Easton, R. M., and Garcia, M. O. 1981. Petrology of the Hilina Formation, Kilauea Volcano, Hawaii. *Bulletin Volcanologique* (Macdonald issue), in press.

Frazier, K. 1979. Kilauea, window to nature's power. *Science News* 116 (September 8):170–2.

Herbert, D.; and Bardossi, F. 1968. *Kilauea: case history of a volcano.* New York: Harper & Row Publishers.

Kaye, G. 1976. *Hawaii Volcanoes, the story behind the scenery.* Las Vegas, Nevada: KC Publications.

Macdonald, G. A. 1949. *Petrography of the Island of Hawaii.* U.S. Geological Survey Professional Paper 214-D.

————. 1972. *Volcanoes.* Englewood Cliffs, New Jersey: Prentice-Hall, Inc.

————, and Abbott, A. T. 1970. *Volcanoes in the sea.* Honolulu: University of Hawaii Press.

————, and Hubbard, D. H. 1972. *Volcanoes of the national parks in Hawaii.* 6th ed. Hawaii Natural History Association.

Stearns, H. T. 1966. *Geology of the state of Hawaii.* Palo Alto, California: Pacific Books.

————. 1978. *Road guide to points of geologic interest in the Hawaiian Islands.* 2nd ed. Palo Alto, California: Pacific Press.

————, and Macdonald, G. A. 1946. *Geology and ground water resources of the Island of Hawaii.* Hawaii Division of Hydrography Bulletin 9.

Swanson, D. A., and Peterson, D. W. 1972. *Partial draining and crustal subsidence of Alae lava lake, Kilauea Volcano, Hawaii.* U.S. Geological Survey Professional Paper 800-C.

Wentworth, C. K., and Macdonald, G. A. 1953. *Structure and forms of basaltic rocks in Hawaii.* U.S. Geological Survey Bulletin 994.

Address

Hawaii National Park
Hawaii 96718

21
Haleakala National Park

Location: Island of Maui, Hawaii
Size: 28,655.25 acres; 44.77 square miles
Established: July 1, 1961

Figure 21.1   Haleakala Crater in Haleakala National Park
is neither a true crater nor a caldera; it is the result of
stream action and headward erosion. This is the view from
the observatory on the west rim of Haleakala. National Park
Service photograph.

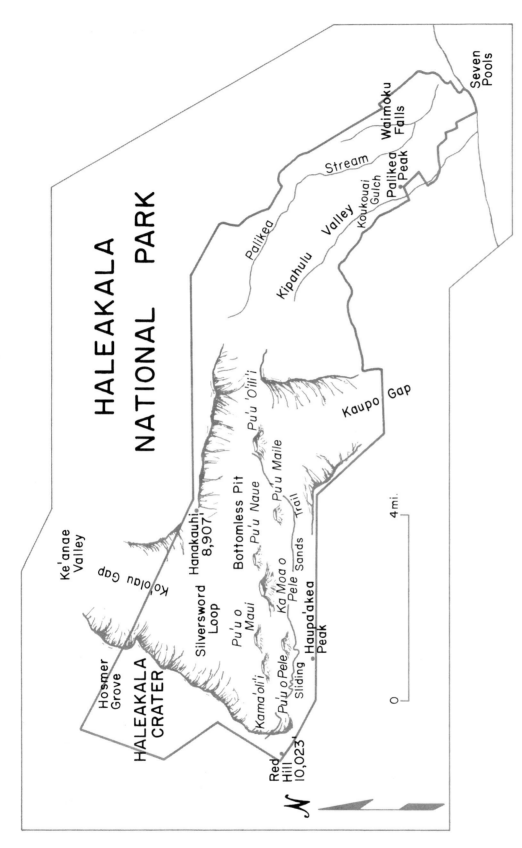

Figure 21.2 Haleakala National Park, Hawaii.

Table 21.1. Geologic Column, Haleakala National Park

Era	Period	Epoch	Rock Units	Geologic Events
Cenozoic	Quaternary	"Historic time" Holocene —?—	Hana Formation	Small lava flows c. 1790
				Valley-filling lavas, crater lavas, ash, cinder cones, spatter cones
			Mudflow debris	Kaupo mudflow
		Pleistocene	Accumulations of sediment	Landsliding, canyon-cutting; development of amphitheaters and Haleakala Crater by stream action, frost action, mass wasting
			Great erosional unconformity	Fluctuation of sea level and submergence
	?—?	?—	Kula Formation	Cinder cones, ash; alkalic cap construction Extensive aa and pahoehoe flows
	Tertiary / Neogene	?— Pliocene	Kumuiliahi Formation	
			Honomanu Formation	Build-up of shield by accumulations of pahoehoe and aa flows
			Volcanic "basement" rocks (not exposed at surface)	Lava flows building up from the ocean floor

Source: modified after Stearns 1966; Macdonald 1978

Local History

Haleakala National Park is on Maui, the next island northwest of Hawaii Island (see box 20.1 p. 255). Maui was a religious and cultural center for the Polynesian people who settled in the Hawaiian Islands beginning some 2,000 years ago. Within Haleakala Crater, their sacred place, are remnants of a road of hand-laid stone and the graves of ancient Hawaiian kings.

In the Hawaiian language, *Haleakala* (ha-lay-áh-ka-lah) means "House of the Sun." According to legend, the Hawaiian god Maui climbed to the top of the mountain and captured the sun, holding him fast by tying down his rays with ropes. When the sun pleaded for his freedom, Maui made the sun promise to move more slowly across the sky so that Hina, his mother, would have more daylight hours in which to do her work.

Captain Cook, who discovered the Hawaiian Islands in 1778, was the first white man to visit Maui. After him came whalers, traders, and missionaries. Lahaina, a port on the west side of Maui, is the site of the first white settlement in the Hawaiian Islands and was the capital until 1845 when the seat of government was moved to Honolulu. Lahaina was an important whaling center in the 19th century, but now people come to the old port to watch whales instead of catch them. Humpback whales spend the winter months in the channel waters between Maui and the neighboring islands of Lanai, Kahoolawe, and Molokai.

Sugar cane and pineapples are grown on parts of Maui today, but these crops are less important to the island's economy than tourism. The park on Maui was originally part of Hawaii National Park, established in 1916, but Haleakala National Park was redesignated as a separate park in 1960. Most of the Park is contained within the Haleakala Crater, but in 1969 the Kipahulu Valley was added which connected the crater section with the ocean. This narrow corridor was given to the Park so that the fragile ecosystem of Kipahulu Valley with its rare and endangered species of plants and birds could be protected. At least 25 bird species that lived on Maui before 1900 became extinct after mosquitoes carrying bird malaria were introduced to the island by a ship from Central America. The heavily forested slopes of the Kipahulu Valley, where a new bird species was recently discovered, are not open to the public but are kept in a natural state for scientific study only.

Dense forests grow on the lower and middle slopes of the Haleakala volcano. The upper slopes and the crater are nearly bare of vegetation, but among the rocks in the crater and on the rim grows the silversword, a rare, yuccalike plant, native to Hawaii, that has a mass of silvery, saberlike leaves. At maturity when the plant is perhaps six feet tall, it produces hundreds of purple flowers; and as soon as the seeds ripen, the plant dies. A hundred years ago the silversword covered acres of the crater floor, but the rare plant nearly became extinct before coming under Park Service protection.

Also making a comeback in the Park is a flock of nenes (naý-nays), or Hawaiian geese, that almost became extinct early in the century. The nene is the state bird of Hawaii.

Geologic Features

Today the island of Maui consists of two eroded volcanoes, but it was once part of a large volcanic massif made up of six major volcanoes. When sea level was lower, the six volcanoes formed one island. Now there are four separate islands—Maui, Kahoolawe, Lanai, and Molokai.

Maui's volcanoes are older and more deeply eroded than the great volcanic shield complex making up the island of Hawaii. In map view, Maui's appearance is like an asymmetrical dumbbell, with the smaller West Maui volcano on one side and the larger East Maui volcano on the other side. Joining the two volcanoes is a low isthmus constructed of younger lavas (from Haleakala) and sediment.

West Maui is only 5,788 feet high and 18 miles across. By contrast, on East Maui, the rim of Haleakala Crater rises to over 10,000 feet and the volcano is about 33 miles across. The so-called crater, the dominant topographic feature of the island, covers 19 square miles and is several thousand feet deep. Its inward-facing slopes are much steeper than its outer flanks. Because it looks like a crater, or actually more like a caldera, this impressive landform will probably always be called "Haleakala Crater," even though by geologic definition it is neither a crater nor a caldera (fig. 21.1).

To be sure, there was a crater of some sort on Haleakala, and the mountain was once much higher. But after the volcano became inactive, very rapid stream erosion began removing the youngest volcanic deposits, which included lava flows, ash, and cindery material. Because rainfall is heaviest on north- and east-facing slopes due to the trade winds, streams draining those areas were able to erode faster and capture the headwaters of lesser streams. Several master valleys developed and two of them, by cutting headward into the mountain's core, finally merged, creating a huge depression on the summit. A low dividing ridge across the floor of this calderalike feature separated the two drainage basins, the Kaupo Valley on the southeast and Keanae valley toward the northeast.

Volcanic activity resumed after the erosional period. The old rift zones of the volcano reopened. Lava pouring from fissures on the sides of the surrounding walls and from the crater floor flowed down the eroded stream valleys. An enormous mudflow, along with cinder blocks and lava, went through the Kaupo Gap and down to the sea. Cinder cones, constructed of lava, pumice, cinders, blocks, ash, and obsidian, built up on the floor of the depression,

with some cones rising as much as 600 feet in height (fig. 21.3). Bizarre spatter cones formed, such as Pele's Pigpen and Bottomless Pit, the latter named for its deep vent. Lava tubes, such as Bubble Cave, were left when pahoehoe lava cooled on the outside while the lava below that was still molten drained away.

One of the striking features of Haleakala Crater is that the rocks in the steep walls and in many of the features of the floor are multicolored. One reason for this is that when the sun's rays strike volcanic glass (obsidian), the light is refracted as with a prism. Sunlight strikes the rocks at different angles during the day, of course, so the colors vary as the sun's position changes. Also affecting the surface coloring are the volcanic gases that are continually escaping; they leave varicolored crusts on the rocks due to precipitation and sublimation. Chemical weathering of the basaltic rocks adds still more colors to the landscape as iron-bearing minerals are oxidized and hydrated. The prevailing warm temperatures and frequent mists and rains intensify the effects of weathering processes on these easily weathered rocks.

"Climatic layering" on Maui produces unique weather patterns and tends to control the prevalence of or lack of vegetation. Maximum precipitation is greatest on northeastern slopes between 2,000 and 6,000 feet, where the trade winds drop more than 300 inches of rain per year. This precipitation and the clouds and fog that often envelop the middle-level slopes of Haleakala are responsible for the luxuriant growth in the rain forests, the Kipahulu Valley being a notable example. Above 6,000 feet and on the leeward slopes, the environment tends to be semiarid. Thus from the air, one sees heavy forests on the lower and middle flanks of the volcano and barrenness at the higher elevations around the summit and crater.

An additional factor causing this contrast is the extreme porosity of the volcanic rocks in the crater. Rain water drains away so fast that not much soil develops and vegetation is sparse. Runoff moves loose pyroclastic material down the inner slopes of the crater, depositing gravel and ash in stream valleys. Wind action piles up dunes, as in the Sliding Sands area on the south side of the crater.

Some of the streams on Maui flow only during and right after rains. These are *ephemeral streams*. On the windward slopes, however, streams cascading down over volcanic cliffs and ledges have created beautiful pools and waterfalls that are cut in the valley-filling lavas. Hikers and swimmers who visit the pools of 'Ohe'o stream and Wailua Cove, for example, in the Kipahulu section of the Park on the eastern shore of Maui, can see not only interesting lava outcrops but also rare birds and flowers. (Sometimes the water rises very suddenly and the stream becomes a raging torrent, so caution is advised.)

Figure 21.3 On the floor of Haleakala Crater are small cinder cones erupted from the old rift zones that reopened after the summit of the volcano had been extensively eroded. National Park Service photograph.

In the rugged sea cliffs on this section of the shore, older and younger lavas are exposed, as well as partially buried cinder cones and other volcanic features.

Geologic History

1. Construction of the volcanic massif.

During late Pliocene time, the shields of six volcanoes were being built up above sea level from a basaltic lava foundation on the ocean floor. At first they were separate islands, but as the volcanoes continued to erupt, they coalesced as a great volcanic complex.

2. Eruption of lavas of the Wailuku Formation (West Maui) and the Honomanu Formation (East Maui).

In earliest Pleistocene time, flows of the Honomanu Formation, many thousands of feet thick, constructed the bulk of the East Maui volcano. West Maui, although it had begun earlier, was being built up by similar outpourings of the lavas of the Wailuku Formation. Rocks of both formations consist of tholeiitic basalts and olivine basalts that hardened from pahoehoe and aa flows. (The rocks that contain about 35 percent olivine phenocrysts are called oceanite.) Honomanu rocks above sea level were buried by the flows of later lavas and are now exposed only in the sea cliff along the northeast side of the island and in the walls of the deeper stream valleys.

3. Eruption of lavas of the Kumuiliahi Formation in Pleistocene time.

At the base of the south wall of Haleakala Crater (north and northeast of Kumuiliahi Peak) and in a small knob at the east edge of the Koolau Gap, are thin-bedded pahoehoe and aa lavas formerly assigned to the Honomanu Formation but now regarded as a discrete rock unit on the basis of chemical composition. Kumuiliahi rocks are too alkalic to be considered tholeiitic, but are less alkalic and also younger than the overlying rocks of the Kula Formation (Macdonald 1978).

4. The Kula Formation of Pleistocene age.

Kula lavas erupted largely along the two rift zones that trend to the southwest and to the east from the summit of Haleakala volcano. Minor eruptions occurred along the third rift zone that extends northward from the summit. These outpourings of lava buried the shield formed by the Honomanu lavas and formed a thick cap (2,500 feet) over the summit, raising the top of the volcano some 2,000 to 3,000 feet above its present elevation. The rocks of the Kula Formation include alkalic basalts, but hawaiite,* an andesitic rock, is the predominant type.

The Kula flows, mainly aa, contain interbedded layers of ash and pyroclastics. Since many of the eruptions were explosive, large cinder cones formed, most of them astride the volcano's rift zones. The presence of lenses of stream gravel between some Kula lavas indicates that quiet intervals occurred between eruptive periods. Quiescent times apparently lengthened toward the end of the Kula eruptive phase. Exposed in the walls of Haleakala Crater are a few dikes of Kula age that parallel the rift zones. The dike rocks, which have a higher proportion of feldspar minerals, tend to be more resistant than the surrounding lavas. The ancient Hawaiians made adzes from an even more resistant, fine-grained, igneous rock exposed in a few intrusive masses emplaced in the west wall of the crater (Macdonald 1970).

5. The great submergence and time of weathering and erosion.

As volcanism subsided, sea level rose until at times the ocean met the shore at levels more than 100 feet higher than at present. East and West Maui were separated by a channel. All through the Hawaiian Islands, extensive beach deposits and deltas were formed, graded to these high sea stands, which are associated with Pleistocene melting of the great continental glaciers. On Maui, most of the erosional and depositional features produced by waves were subsequently eroded by stream action or buried under younger lava flows.

As was the case elsewhere in the Northern Hemisphere during the Pleistocene, more rain and snow fell than usual and cooler climates prevailed. Glaciers probably did not develop on Maui, but frost action may have been severe; some snow accumulated in winter; stream erosion and mass wasting accelerated.

During the time the Maui volcanoes were quiet and erosion was intense, the height of Haleakala was reduced, perhaps by more than 2,000 feet. The south slope of Haleakala became steeper and shorter, probably because of seaward slip faults that triggered huge landslides on that side of the mountain. On the northeast flanks, where precipitation is heavier due to the trade winds, the larger stream valleys cut downward several thousand feet. Because of fluctuations of sea level, the mouths of some stream valleys were scoured out below present sea level and later partially filled with gravel and mudflow debris. In the summit area, intensive headward erosion, downcutting, and the rapid removal of rock and debris produced large, amphitheater-shaped basins at the heads of the major stream valleys.

6. Development of Haleakala Crater.

Harold T. Stearns, who for many years was District Geologist for the U.S. Geological Survey in the Hawaiian Islands and the Pacific area, developed the idea that the calderalike depression at the summit of Haleakala was mainly the result of coalescence of amphitheaters of the Keanae and Kaupo valleys. The Keanae valley trends northeastward from the summit area, and the Kaupo valley runs southeastward and southerly. The large size of Haleakala Crater is due to the valley heads being offset, rather than joining in a straight line across the top of the mountain. Throughout the crater area, the evidence is clear that a great deal of erosion took place, although some of the erosional features have been buried under younger volcanic lavas and mudflows.

7. Late Pleistocene to Holocene volcanic activity.

The lava flows, ash falls, and cinder cones of the Hana Formation that erupted after the erosional period are conspicuous on the floor of the crater, on its western rim, and on parts of the volcano's flanks. The surface of the most recent flows in the crater area is fresh and only slightly weathered; this indicates prehistoric eruptions, perhaps a few hundred years ago. The earliest Hana flows on Haleakala may be as old as 200,000 years but are probably younger. The Hana rocks are alkalic basalt and, like the Kula rocks, distinctly lighter in color than the dark tholeiitic basalts that built up the primitive shield.

Many of the Hana flows traveled much farther from their vents than Kula lavas did; this suggests that, in general, the Hana lavas were more fluid. Some Hana lavas went down the Keanae and Kaupo valleys to the ocean, and some erosional valleys were nearly filled. The Kipahulu valley, for example, was virtually filled with Hana lava; the stream cut a new gorge along the north side of the filled valley; and then fresh Hana lava partly refilled the canyon again. During one eruptive episode, an enormous mudflow came down the Kaupo valley, reaching the ocean.

8. Volcanic activity during historic time.

Carbon 14 ages for charcoal associated with unweathered Hana lavas from the southwest rift zone of Haleakala have been determined as A.D. 1070 ($\pm$ 170 years), A.D. 1310 ($\pm$ 140 years), and A.D. 1370 ($\pm$ 120 years). The most recent dated eruption from the southwest rift zone apparently took place around 1790 (Reber 1959; Macdonald 1978).

At the present time, Haleakala is considered dormant, although there is potential for a renewal of eruptive activity in the future. Earthquakes that occur from time to time on Maui are probably due to subsidence of the crust from the weight of the volcanoes.

---

*Not to be confused with the mineral hawaiite, a gem variety of olivine.

Geologic Maps and Cross Sections

American Association of Petroleum Geologists. 1974. *Geological highway map of the state of Hawaii and the state of Alaska.* Map No. 8. Tulsa, Oklahoma: American Association of Petroleum Geologists.

Macdonald, G. A. 1978. *Geologic map of the crater section of Halaeakala National Park, Maui, Hawaii.* U.S. Geological Survey Map I-1088.

Bibliography

Kyselka, W. and Lanterman, R. 1980. *Maui, how it came to be.* Honolulu: University Press of Hawaii.

Macdonald, G. A. 1978. Summary of the geology of Haleakala. Accompanies U.S. Geological Survey Map I-1088. See above.

———, and Abbott, A. T. 1970. *Volcanoes in the sea.* Honolulu: University of Hawaii Press.

Mack, J. 1979. *Haleakala, the story behind the scenery.* Las Vegas, Nevada: KC Publications.

Reber, G. 1959. Age of lava flows on Haleakala, Hawaii. *Geological Society of America bulletin* 70:1245–6, September.

Ruhle, G. 1968. *Haleakala guide.* Hawaii Natural History Association.

Stearns, H. T. 1942. Origin of Haleakala Crater, Island of Maui, Hawaii. *Geological Society of America bulletin* 53:1–14, January.

———. 1966. *Geology of the state of Hawaii.* Palo Alto, California: Pacific Books.

———. 1978. *Road guide to points of geologic interest in the Hawaiian Islands.* 2nd ed. Palo Alto, California: Pacific Books.

Wallace, Robert. 1973. *Hawaii,* American Wilderness Series. Alexandria, Virginia: Time-Life Books.

Address

Haleakala National Park
Box 537
Makawao, Hawaii 96768

22
Voyageurs National Park

Location: Northeast Minnesota
Area: 219,128 acres; 342.39 square miles
Established: April 8, 1975

Figure 22.1 This air view of part of Voyageurs National Park overlooks Kabetogama Lake. The landscape of waterways and islands developed on eroded Precambrian rocks on the southern part of the Canadian Shield. National Park Service photograph.

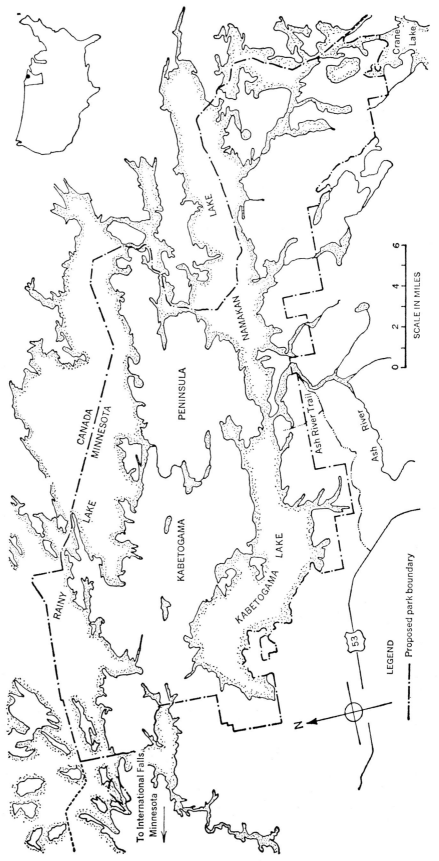

Figure 22.2 Voyageurs National Park, Minnesota.

Table 22.1. Geologic Column, Voyageurs National Park

Time Units			Rock Units	Geologic Events
Era	Period	Epoch		
Cenozoic	Quaternary	Holocene	Alluvium in streams, lakes, bogs, etc.	Continuing glacial rebound; some lakes drained or filled in.
		Pleistocene	Glacial drift	Minor morainal and outwash deposition Severe glacial erosion by four major ice sheets
Paleozoic/ Mesozoic			/////////	Prolonged erosion of the shield
Proterozoic X (Middle Precambrian time)			Mafic intrusive rock	Intrusion of mafic dike
Archean (Early Precambrian time)			Kenoran (Algoman) granitic intrusives	Kenoran (Algoman) orogeny; mountain-building and uplift; metamorphism
			Pre-Kenoran (pre-Algoman) formations (position in column uncertain):	
			Timiskaming (Windigokan)	Rapid deposition of coarse conglomerates, graywackes, silt, etc.
			Laurentian granites	Small granitic intrusions
			Keewatin lavas	Submarine extrusion of mafic lavas; Occasional pyroclastic eruptions
			Couchiching	Marine deposition of coarse and fine sands, silts, etc.

Source: modified from Pye 1968

Geographic Setting

Roads do not exist beyond the entry points of Voyageurs National Park. It is located in the borderland of islands and waterways between Minnesota and Ontario, Canada, Visitors hike, ski, and fly into Voyageurs National Park, but the main means of reaching the Park is by canoe or boat. This is an area of water, woods, wildlife, and very old igneous and metamorphic rocks. Protected in the Park is the only example of a southern boreal forest in the National Park System; that is, a forest characteristic of the transition zone between the vegetation of the Arctic and the Northern Temperate region.

The Park is about 40 miles long, extending from International Falls on the west to the Boundary Waters Canoe Area on the east, and it varies in width from about 3 to 15 miles. The landscape was shaped by the ice sheets of the Pleistocene Epoch. As the great continental glaciers advanced and retreated, they scraped, gouged, and pushed away the original trees, rock, and soil, converting the region of huge forests and a few rivers to a barren land of sparse soil, patches of bogs, grassy meadows, and numerous lakes interconnected by a network of streams. The boreal forests that covered the land when the Indians, voyageurs, and fur traders were here grew in the thin soil left by melting glaciers or deposited in temporary lakes after the last ice sheet melted back about 10,000 years ago. Most of this virgin timber was logged off in the years before and after the turn of the century, but some tracts of the boreal forest remain in Voyageurs National Park, the Boundary Waters Canoe Area, and Quetico Provincial Park in Ontario.

Winters are long and cold in the north woods, and daylight hours are few. Snow may cover the ground from October until May, and ice in the lakes and streams seldom breaks up before the middle of May. Although the summers are relatively brief, temperatures are comfortable and the days are long. Spring and fall are times of intense activity in the wilderness; migratory species return to the northern breeding grounds as days lengthen, and then head south again as winter approaches.

Local History

Prehistoric people came into the north woods about 8,000 years ago. A few pictographs, painted on rock cliffs with red ocher, and artifacts such as slate knives and arrowheads are relics of their Stone Age culture. The Chippewas lived in the region when the French Canadian trappers and voyageurs traveled back and forth along the waterways. These white men learned Indian tongues and Indian ways of adapting to the wilderness, made birch-bark canoes for transport, and intermarried with the Indians in some cases. Presumably, Indian guides taught the voyageurs the safest and most direct routes among the myriad streams and lakes of the region.

It was the voyageurs for whom the Park was named, and their well established waterway between Lake Superior and Lake of the Woods was made the international boundary by the 1783 treaty ending the American Revolution. Fifty-six miles of that boundary, passing through the waters north of the Kabetogama Peninsula, also mark the northern limits of Voyageurs National Park.

In the 18th and 19th centuries the voyageurs plied their trade, paddling and portaging thousands of miles through the wilderness waterways and bringing out many tons of furs. They worked for the North West Company (later the Hudson Bay Company), which made a great deal of money trading English goods for pelts, mostly beaver. The voyageurs' 3,000-mile route between Montreal and Lake Athabasca (northeastern Alberta) was divided into eastern and western stages, each with its own type of birch-bark canoes and group of men. The exchange point was first at Grand Portage (Minnesota), on the western shore of Lake Superior; but after the American Revolution, the English company (wishing to avoid American taxes) moved the rendezvous site to Thunder Bay in Canadian territory.

The larger, 36-foot Montreal canoes, paddled by ten men, transported goods and furs between Montreal and Grand Portage (or Thunder Bay). At the exchange station, cargoes were split up and reloaded into the lighter, 25-foot north canoes, each with a crew of five or six men, who traveled between Grand Portage and the western end of the line at Fort Chipewyan on Lake Athabasca. Outward bound in early spring, the Montreal canoes were loaded with iron stoves, cooking pots, blankets, flour, salt, and other supplies. Meanwhile, *les hommes du nord* ("the men of the north"), as they called themselves, loaded their north canoes with furs and left the remote trading posts

of Alberta and Saskatchewan (where they had spent the winter) as soon as the ice began to break up. The whole trip—getting the furs to the rendezvous, picking up the trade goods from Montreal, and returning to Lake Athabasca—had to be completed between spring thaw and autumn freeze-up. To accomplish this feat, each 25-foot north canoe carried a standard load of a ton and a half, made up in 90-pound bundles that the men carried on their backs over the portages. In making the 2,000-mile journey between Grand Portage and Lake Athabasca, the voyageurs paddled and portaged 15 to 18 hours a day. At first light (three to four a.m. at that latitude), they got up, paddled several hours, and then stopped for a breakfast of mush made from dried peas, corn, and pemmican or salt pork. At brief rest stops during the day, they smoked their clay pipes but did not eat. At dusk (seven to nine p.m.), they stopped for the night, repaired the canoes, and ate their second meal of the day, again mush, occasionally supplemented by fresh game or birds' eggs.

These brave and hardy men, most of them French Canadians, were noted not only for their strength and endurance but also for their songs, their humor, and gusto. They wore deerskin moccasins and leggings tied with thongs to blue breeches, bright scarves and sashes, and blue woolen shirts or jackets. Their red knitted *toques* (caps), each with a tassel hanging down (worn for dress-up occasions), completed their colorful dress.

By the middle of the 19th century, the fur trade had begun to decline (beavers nearly became extinct), and by the 1870s and 1880s, lumbering and mining were assuming more importance. Large-scale logging of the virgin forests went on for about 50 years, and in the peak year of 1905 nearly two billion board feet were cut. Like the trappers and voyageurs, the lumbermen, too, had their legendary heroes, exemplified by the figure of Paul Bunyan.

The Kabetogama Peninsula was the scene of a gold-mining boom between 1893 and 1898. The boom town of Rainy Lake City (northwest corner of the Park) filled up with prospectors and miners but then declined from a settlement of 500 to a ghost town in five years.

Since the closest transportation center was 100 miles away, getting supplies and equipment to the mines and hauling out the ore was not easy. The chief producer was the Little American Mine, located on the island of the same name near the south shore of Rainy Lake. About 500 tons of ore were extracted and crushed, yielding gold that was worth over $4,600 in 1894 and 1895. The gold (of hydrothermal origin) was disseminated in a composite quartz vein, 4 to 6 feet wide, in a belt of sheared chlorite and biotite schist. The open cut of the mine (a 10-foot-wide by 44-foot-deep shaft) was flooded when the water level of Rainy Lake was raised for the purpose of generating hydroelectric power. The Little American Mine, like the other gold mines opened in the area, was close to (i.e., within a mile of) the Vermilion batholith (p. 280).

The Big American Mine, the first mine dug in the area, was on Dryweed Island on the homestead of George Davis, the man who discovered the gold. The pit, 5 feet square and only 5 feet deep, was sunk on a quartz vein. Lyle Mine, on the north side of Dryweed Island, had a shaft 22 feet deep, but Old Soldier Mine, on a very small island off Dryweed, was only a shallow pit. Two other abandoned mine shafts are on the east end of Dryweed.

At Bushyhead Mine on Bushyhead Island, an adit (tunnel) penetrated a shear zone of massive pyrite and quartz just above water level. Gold Harbor and Holman Mines, on a blunt point of the mainland, are shallow pits in a quartz-pyrite vein in a schistose area.

The voyageurs left the wilderness pretty much as they found it, but the mining and lumbering industries—although they brought a measure of prosperity to the region—drastically altered much of the natural environment. Lumbering, in particular, devastated large areas. However, on both sides of the international boundary people who loved the northern lake country felt that the forests and waterways should be preserved and protected. As early as 1891, the Minnesota legislature sought to have a national park established on the Kabetogama Peninsula. Through the years similar efforts were being made in Ontario, resulting in the establishment of Quetico Provincial

Box 22.1
Why Dating and Correlating the Earth's Precambrian Rocks Is Difficult

Precambrian rocks, wherever they are found, represent geologic activities that took place during more than three-fourths of the entire time of the earth's geologic history. The rocks are "samples of time" that provide tantalizing clues about the origin of continents, early tectonics, and the development of the earth as we know it (fig. 22.3).

However, except for radiometric age, geologists have not found common factors that can be used consistently to correlate Precambrian

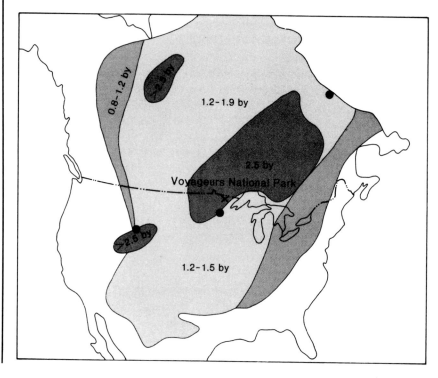

Figure 22.3 Generalized map of radiometric ages of Precambrian rocks of the North American craton. The black dots show locations where dates older than 3.2 billion years (Archean time) have been obtained. Note location of Voyageurs National Park. From *Historical Geology of North America* (2nd ed.) by Morris S. Petersen and J. Keith Rigby. © 1973, 1980 Wm. C. Brown Company Publishers, Dubuque, Iowa. Reprinted by permission.

rocks from one region to another. Outcrops are scattered; index fossils are lacking; some rock units are too deeply buried for cores to be obtained; all have been subjected to repeated metamorphism and extensive erosion; and each geographic district has its own geologic history with different events having occurred in various sequences. Logical breaks in Precambrian chronology show up on every continent, but where these interrruptions come varies so much from place to place that using them for purposes of correlation is not feasible. The result has been that scientists on each continent have had to develop their own system for classifying Precambrian rocks and events.

For example, geologists are in general agreement that Precambrian time can be divided into two eons, Archean and Proterozoic. (An *eon* is the longest geologic-time unit, next in magnitude above an *era*.) Because the Archean was a time of high heat flow, the earth's primitive continental crust probably developed fairly rapidly. In Proterozoic time, the amount of heat generated by radioactive decay dropped substantially; this slowed down the melting of magma. Continents grew more gradually; tectonic mechanisms similar to those of the present began; and primitive life forms developed. According to Windley (1977), "The Archean-Proterozoic boundary can be regarded as a threshold in the geothermal/tectonic evolution of the continents." (p. 337)

But where in global geochronology is this boundary (actually a transition zone) to be placed? On the Baltic Shield, the end of the Archean Eon has been designated as 1,900 million years ago, but in the Soviet Union the division between Archean and Proterozoic is drawn at 2,600 million years ago.

Lacking a uniform system, geologists in different parts of the world who study Precambrian

rocks have experienced difficulties of communication when trying to work together on investigations of mutual interest. The U.S. Geological Survey's approach to Precambrian classification is to use the one common factor that does exist; that is, radiometric dating. The Archean-Proterozoic boundary has been set at 2,500 million years ago, and the letters X, Y, and Z have been assigned to time units in the Proterozoic Eon. Although these time terms are regarding as being "without specific rank," they are useful for comparing and classifying groups of Precambrian rocks that yield dates falling within these informal divisions. Archean rocks were, until recently, designated "Precambrian W," but this time term is no longer used by the U.S. Geological Survey.

Table 22.2 compares the U.S. Geological Survey's Precambrian chronology with the Canadian classification. Although time boundaries in the two systems differ slightly, the two classifications are more similar to each other than either one is to Precambrian chronologies developed for the other continents of the world. Both systems place the end of Precambrian time at about 570 million years ago.

Research now underway on Precambrian life forms may eventually yield information that will be helpful in correlating Precambrian rocks in different parts of the world. In 1954 the first discovery of Precambrian microfossils was made near Thunder Bay on Lake Superior. The Gunflint Chert in which the fossils were found is Proterozoic X in age. Microfossils have not been found in rock units in Voyageurs National Park. However, about 60 miles northeast of the Park, near Atikokan, Ontario, algae fossils, 2.7 billion years old (Archean or early Precambrian), have been found in marine sedimentary rocks of the Steep Rock Formation of the Timiskaming group.

Table 22.2 Comparison of Precambrian subdivisions used in the United States and Canada. (Geologic Names Committee, U.S.G.S., 1980)

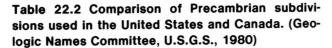

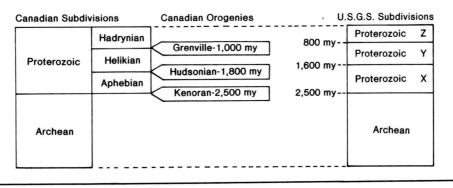

Canadian Subdivisions		Canadian Orogenies		U.S.G.S. Subdivisions	
Proterozoic	Hadrynian		800 my	Proterozoic	Z
		Grenville-1,000 my		Proterozoic	Y
	Helikian		1,600 my		
		Hudsonian-1,800 my		Proterozoic	X
	Aphebian		2,500 my		
		Kenoran-2,500 my			
Archean				Archean	

Park. In Minnesota, the author-naturalist Sigurd Olson was probably the person who worked longest and hardest for the establishment of Voyageurs National Park, and it finally came about in 1975. Both Canada and the United States have designated the boundary lands and waters in this region "to be kept in a state of nature as far as possible."

Geologic Features

The Precambrian Rocks

All of the bedrock exposed in Voyageurs National Park and the surrounding region is Precambrian in age. These ancient rocks are part of a large area of exposed basement rocks called the Canadian Shield, which is, in turn, a part of the North American craton. The oldest shield rocks are thought to be the core or nucleus of early continental crust that formed as a cluster of islands and probably later became a single large continent.

Exposures of the oldest rocks, classified as Archean, have been identified in shields in other parts of the world, such as the Fenno-Scandinavian region of northern Europe and parts of Asia, Australia, South America, Antarctica, and Africa. The fact that the shield rocks are surrounded by platforms made up of somewhat less ancient Precambrian rocks (and early Paleozoic rocks) suggests that the early continents may have grown by accretion. Also the platforms have some characteristics of once-active mobile belts. Both the shield rocks and the surrounding platform rocks are part of the cratons (basement rocks) that have been stable and undeformed on every continent for a prolonged period. Erosion for countless millenia has brought about their characteristic low relief.

Clusters of radiometric dates show that each Precambrian shield area is divisible into subregions, or structural provinces, based on rock ages. Within each province, the trend of tectonic folds and the style of deformation are distinctive from those of adjacent areas. Thus metamorphic boundaries and fault boundaries, as well as radiometric dates, delineate the various provinces. Both Voyageurs National Park and Isle Royale National Park (chapter 23) are in the Superior Province, a large Precambrian subregion with many bedrock exposures. The province extends over a large part of south central Canada and dips slightly into the United States in northern Minnesota, Wisconsin, and Michigan. Because of its mineral resources, which include major occurrences of iron, gold, copper, and nickel, the Superior Province has been studied by Canadian and American geologists for many years.

The Archean rocks of the Voyageurs region are associated with the Kenoran (Algoman) orogeny, which is regarded as the earliest datable orogeny in North America. Two contrasting assemblages are present in these ancient rocks—the granitoid gneisses, which are highly metamorphosed, and the greenstone belts, which are less highly metamorphosed. The granitoid gneisses were derived from high-silica rocks, possibly of volcanic origin, and sandstones.

The greenstone belts occur in large, elongated downwarps between the more extensive granitoid gneiss accumulations. Dark greenish in color, the greenstones owe their name and appearance to the constituent minerals—chlorite, epidote, actinolite, etc.—that were produced by metamorphism. The presence of pillow lava structures in the greenstone belts indicates that the lavas were extruded in the waters of an ancient sea. Although most of the pillow structures in the Voyageurs area have been destroyed by metamorphism, a few "pillows" up to three feet long have been found on a small islet off Steamboat Island (north of Cranberry Bay). Some poorly sorted mixtures of clay, mica, and quartz grains (graywackes) that were derived from the decomposition of pyroclastic debris (tuff, ash, etc.) suggest that some of the volcanic activity was explosive. Some of the clastic material also has characteristics of turbidite deposition; that is, rapid deposition by turbidity currents, possibly in an ocean trench.

The greenstone belt in Voyageurs National Park consists of schistose (i.e., finely foliated) and sheared lavas. In general, the schistosity is so extreme that the composition and characteristics of the original lavas are masked. Major islands in and near the northwest corner of the Park, such as Dryweed, Grassy, and Grindstone Islands, are made up of greenstone, as are sections of Neil Point (across the Narrows) and part of Rainy Lake's south shore between Jackfish Bay and Black Bay.

Between the greenstone tracts are large areas of metamorphosed biotite and mica schists (originally sedimentary and granitic rocks). The main portion of the Kabetogama Peninsula and the Black Bay area are in part of the schist belt, and a contact zone between the greenstone belt and the schist belt can be traced in the Rainy Lake area.

The Vermilion batholith, which is associated with the Kenoran (Algoman) orogeny, underlies the Sullivan Bay and Blind Ash Bay area on the south side of Kabetogama Lake. This large intrusion may have been the source of the gold-bearing quartz solutions that crystallized in veins along an east-west fault zone that extended to Neil Point. In addition to gold, the veins contain other minerals, such as tourmaline, ankerite, pyrite, and quartz.

Box 22.2
How Did the Early Precambrian Greenstone Belts Originate?

The early Precambrian greenstone belts in the Superior Province and elsewhere in the world have been studied for many years because they are "host rocks" for valuable metallic ore deposits. However, despite such interest and numerous investigations in various parts of the world, the question of how the greenstone belts originated and evolved is still a matter of controversy. Some geologists prefer a uniformitarian approach that emphasizes similarities between Precambrian greenstone belts and those of more recent origin, such as the Mesozoic greenstone belts. Other geologists take a nonuniformitarian approach, stressing the significant differences between the Precambrian greenstones and those of younger age. The former approach assumes that plate tectonic mechanisms operated throughout much of Precambrian time in ways quite similar to tectonic activity in the present and relatively recent geologic past. The uniformitarians theorize that Precambrian greenstone belts developed in a "back-arc spreading center" in a marginal basin near a continent-ocean converging plate boundary. For a modern analog, they point to the volcanic and tectonic activity in southern Chile.

The nonuniformitarians, on the other hand, feel that global environmental conditions in the early Precambrian world were so different that the combination of circumstances that produced the greenstone belts at that time has not existed on the earth since they were formed.

Nonuniformitarian models postulate, for example, parallel troughs in unstable, thin, primitive sialic or granitic crust that sagged progressively as they filled with denser volcanic and sedimentary material. Eventually, thickening of the surrounding granitic crust squeezed the troughs, metamorphosing the basaltic material to greenstones. Another nonuniformitarian suggestion is that greenstone belts may have formed as a result of meteorite impacts on the earth's primitive crust. (Meteorite bombardments, such as those that struck the moon, producing lunar maria (impact craters) are believed to have been fairly frequent during the very early geologic history of the earth.)*

The significance of the question of the origin of Precambrian greenstone belts is not that the problem is insoluble; it is likely that at some time in the future new data, perhaps provided by more sophisticated techniques, will yield a hypothesis that will be acceptable to the majority of scientists working in these areas. What *is* important for students to realize is that all of the models were developed *from the same body of data.* In other words, in geology, as in many areas of science, it is possible for several logical but quite different explanations to be formulated based on knowledge that is available at a given time.

*For an interesting, thought-provoking discussion of various models proposed for the origin of Precambrian greenstone belts, see chapters 2 and 3 in Windley (1977).

Effects of Glacial Rebound

The end of Precambrian time brought to a close the long series of tectonic events that created the basement complex of the Canadian Shield. Due to erosion, only the roots of the ancient mountains remained when the great continental glaciers began to form in Pleistocene time. The weight of the huge ice sheets moving southward depressed the crust, as well as scraping and gouging the bedrock. After the ice retreated, lakes filled the basins and spread over the lowlands. Relieved of the weight of ice, the crust probably regained its normal elevation after each major episode of glaciation, a phenomenon called *glacial rebound.* Since the last retreat of ice occurred less than 10,000 years ago, gradual rebound of the Canadian Shield is still going on. As the land has risen, some lakes have drained away; in others water levels have dropped. Some have become bogs and swamps, filled in by vegetation and sediment. Since rebound of the crust has been uneven, some shorelines have been tilted slightly, resulting in uneven slopes along old beach lines and lake terraces.

Geologic History

1. Early Precambrian (Archean) basement complex.

Most of the rocks in the area have been dated as 2,600 million years old, or older. Commonly referred to as "pre-Kenoran" (or "pre-Algoman"), they are intensely metamorphosed rocks of sedimentary, volcanic, and intrusive igneous origin.

The Couchiching rocks (which may or may not be the oldest formation) are foliated gneisses and schists, rich in mica. Originally they may have been graywackes or "dirty sandstones."

A group of similar rocks of sedimentary origin, designated as Timiskaming (or Windigokan), are thought to be younger, mainly because the conglomerates contain large cobbles of granitic material, suggestive of rapid erosion and deposition, perhaps as turbidites.

The volcanic greenstones, or Keewatin lavas, are largely basaltic but include some rhyolitic and clastic material as well.

Associated with the volcanic and sedimentary rocks are small intrusive bodies of fairly dark, granitic rocks.

2. Kenoran (Algoman) orogeny at the end of Archean time and the beginning of Proterozoic X time.

The emplacement of large granitic batholiths in the Superior Province brought about intense folding and uplift of the earlier rocks and the forming of what were probably mountain ranges of considerable bulk and height. The severe deformation that accompanied the orogeny produced some migmatites, similar to those found in North Cascades National Park (p. 175). The subsequent erosion of the mountain ranges has exposed a number of ore-bearing structures in the region. The gold veins found in the Kabetogama Peninsula area (that may have emanated from the Vermilion batholith) are an example.

3. Erosion of the mountains; minor intrusive activity.

The youngest bedrock in Voyageurs National Park is exposed near State Park in a mafic dike that is about 2,100 million years old (Proterozoic X).

Figure 22.4 A forest cover has become reestablished on the thin soil overlying glacially scoured bedrock in Voyageurs National Park. Boulders left by the last ice sheet are being broken up by frost action and other weathering processes. National Park Service photograph.

4. Prolonged erosion throughout Paleozoic, Mesozoic, and most of Cenozoic time.

From Cambrian time about 600 million years ago until the beginning of the Pleistocene Epoch about 2 million years ago, erosion was the dominant process in the region. Probably a fairly uniform soil cover, supporting extensive forests, blanketed the ancient bedrock before the glaciers came. Relief was low.

5. Pleistocene glaciation.

Four times a huge ice sheet advanced southward, pushing and scraping away virtually all of the old soil cover and gouging out some of the bedrock as well. The present thin soil cover, largely of glacial origin, consists of morainal material, outwash gravels, and lakebed sediment. The glacial features are related to Wisconsin glaciation (the last major episode), which obliterated traces of the earlier glaciations.

6. The postglacial, constructional landscape.

Unconsolidated materials, Pleistocene and Holocene in age, have been superimposed on some of the oldest bedrock in North America. Because relief in the area is low, streams have been slow in reworking and sorting the glacial debris. Lakes occupy the large basins; and small ponds, bogs, and wetlands, interconnected by waterways, are all about. These are indications that drainage in the area is still disorganized.

Geologic Maps and Cross Sections

American Association of Petroleum Geologists. 1982. *Northern Great Plains region, geological highway map no. 12.* Tulsa, Oklahoma: American Association of Petroleum Geologists.

Sims, P. K. 1970. Geologic map of Minnesota. In *Geology of Minnesota: a centennial volume,* ed. P. K. Sims and G. B. Morey, plate 1. St. Paul, Minnesota: Minnesota Geological Survey.

Bibliography

Eicher, D. L.; and McAlester, A. L. 1980. *History of the earth.* Englewood Cliffs, New Jersey: Prentice-Hall, Inc. pp. 73–90.

Hofmann, H. J. 1971. *Precambrian fossils, pseudofossils, and problematica in Canada.* Geological Survey of Canada Bulletin 189.

Knauth, P. 1972. *The north woods,* American Wilderness series. Alexandria, Virginia: Time-Life Books.

Minnesota Geological Survey. 1969. *An evaluation of the mineral potential of the proposed Voyageurs National Park, with map.* St. Paul, Minnesota: University of Minnesota.

National Park Service. 1979. *Voyageurs National Park.* Map and brochure.

Ojakangas, R. W. 1972. Rainy Lake area. In *Geology of Minnesota: a centennial volume,* ed. P. K. Sims and G. B. Morey, pp. 163–70. St. Paul, Minnesota: Minnesota Geological Survey.

Petersen, M. S.; Rigby, J. K.; and Hintze, L. F. 1980. *Historical geology of North America.* 2nd ed. Dubuque, Iowa: Wm. C. Brown Company Publishers.

Pye, E. G. 1968. *Geology and scenery, Rainy Lake and east to Lake Superior.* Geological Guide Book no. 1. Ontario Department of Mines.

Schwartz, G. M.; and Thiel, G. A. 1954. *Minnesota's rocks and waters.* Minneapolis: University of Minnesota Press.

Southwick, D. L. 1972. Vermilion granite-migmatite massif. In *Geology of Minnesota: a centennial volume,* ed. P K. Sims and G. B. Morey, pp. 108–118. St. Paul: Minnesota Geological Survey.

Windley, B. F. 1977. *The evolving continents.* New York: John Wiley & Sons.

Wright, H. E., Jr. 1972. Physiography of Minnesota. In *Geology of Minnesota: a centennial volume,* ed. P. K. Sims and G. B. Morey, pp. 561–78. St. Paul, Minnesota: Minnesota Geological Survey.

Address

Voyageurs National Park
P.O. Box 50
International Falls, Minnesota 56649

23
Isle Royale National Park

Location: Northern Michigan, in Lake Superior
Area: 571,796.18 acres; 893.43 square miles
Established: April 3, 1940

Figure 23.1 Tilted beds of ancient, resistant lava form a rugged shoreline in Isle Royale National Park. National Park Service photograph.

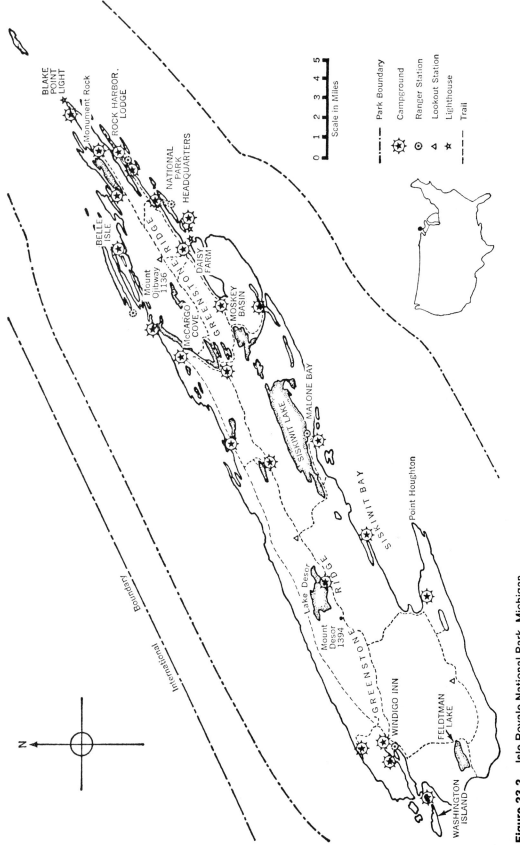

Figure 23.2 Isle Royale National Park, Michigan.

Table 23.1 Geologic Column, Isle Royale National Park

Time Units			Rock Units	Geologic Events
Era	Period	Epoch		
Cenozoic	Quaternary	Holocene	Beach deposits Alluvium	Development of Lake Superior Glacial rebound (Lake Nipissing shore features) Stream erosion; wave action

		Pleistocene	Beach deposits
			Glacial outwash and debris
			Till
				Shorelines formed at various lake levels by wave action
				  (Glacial Lake Minong shore features)
				Drumlins, ice-margin deposits left by Valdes Stade (last
				  Wisconsin advance)
				Severe erosion by repeated glaciations

Mesozoic Paleozoic			/////////	Prolonged erosional interval

Proterozoic Y
(Late Precambrian)
		Middle Keweenawan	Copper Harbor Conglomerate
			
			Portage Lake Volcanics
			
			Marker flows:
			Scoville Point (porphyrite)
			Edwards Is. (trap)
			Middle Point (porphyrite)
			Long Is. (trap)
			Tobin Harbor (porphyrite)
			Washington Is. (ophite)
			Greenstone (ophite)
			Grace Is. (porphyrite)
			Minong (trap)
			Huginnin (porphyrite)
			Hill Point (ophite)
			Amygdaloid Is. (trap)
				Isle Royale fault; uplift
				
				Deposition of volcanic and clastic sediments
				
				Emplacement of native copper and other minerals; mild
				  metamorphism
				
				Continuing subsidence of Lake Superior syncline
				
				Extrusion of mainly basaltic lavas interbedded with
				  sedimentary deposits

Source: modified after Huber 1973, 1973a; Wolff and Huber 1973

Local History

In the northern bulge of Lake Superior, where the international boundary between the United States and Canada bends southwest toward Grand Portage on the Minnesota shore, the border follows a water passage between Ontario and a long narrow island which is in the state of Michigan. This piece of land is Isle Royale. At 45 miles long and 5 to 8 miles wide, it is the largest island in a lake in the United States. The main island, along with its 200 surrounding islets and the waters 4.5 miles out from the shoreline, make up Isle Royale National Park. The Park is accessible only by boat or floatplane. An irregular shoreline, long parallel ridges and intervening valleys with lakes and bogs are features of the Isle Royale landscape.

Isle Royale has long been famous for its native copper; the metallic element copper is found uncombined in a natural state on the island. Indians called the island *Minong,* which means "a good place to get copper." But the Chippewas who lived in the region when the French explorers came did not know how to mine copper or make tools from it. They valued the pieces of "float copper" that they found in loose gravel or on beaches, and they used the raw copper fragments for charms or for crude cutting and scraping tools. Even so, the Chippewas knew nothing about the ancient pit mines that prospectors found on Isle Royale and the Keweenaw Peninsula on the Michigan mainland. Tribal legends did not even preserve the memory of the Pre-Columbian people who for many centuries excavated and removed Lake Superior native copper.

Archeologists investigating some of the thousands of mine pits found charcoal that yielded radiometric dates of 3800 years B.P. and older. Ornaments of delicate design made from Isle Royale copper have been discovered in Hopewell mounds in southern Ohio and at many other sites, proving that the people of the Copper Culture knew how to soften and shape the nearly pure copper. Artisans developed methods of heating and hammering the copper (pure copper is malleable and ductile) in order to make arrowheads, knives, and other weapons and tools. Many of the copper artifacts found in North America are believed to have been made by prehistoric craftsmen using Lake Superior metal.

The method of mining was primitive, time-consuming, and arduous. The Indians built a fire on top of a vein that was to be worked. When the rock was hot enough, they dashed cold water on it, causing the rock to shatter so that they could pound out chunks of copper with stone mauls. Thousands of the stone cobbles that were used for this purpose have been found around the mines. More of the stone mauls found on the Keweenaw Peninsula are grooved for attaching handles than are ungrooved. Most of the mauls found on Isle Royale, however, are rounded beach cobbles that were hand-held. This may mean that the Indians found and excavated copper veins on the island before the veins on the mainland were located. Perhaps the bedrock surface on Isle Royale was more exposed by glacial scouring, making the copper veins easier to find. Although Indian mine pits were left on all parts of the island, the highest concentration is along McCargo Cove on the north side.

When the French occupied the Lake Superior region in the 1600s, they were more interested in fur trading than in mining. Ironically, about the time they began to organize mining companies a century later, the Indian wars started, and subsequently the English gained control of the region. The English government sent a mining expert to Mackinac, who brought back a favorable report, but, again, efforts to get mines operating in the 1770s were unsuccessful.

Although Isle Royale was given to the United States by the 1783 treaty with England, the English remained in control of it until after the War of 1812. The island was important in the fur trade during that time, with both English and American companies maintaining trading posts there. Meanwhile, the Chippewas considered the island to be their territory. Not until after the Chippewas ceded Isle Royale to the United States government in 1843 was the island overrun by prospectors and miners.

A report by Douglass Houghton, Michigan's first state geologist, set off the copper boom in the 1840s, both on the mainland and the island. The famous Ontonagon copper boulder, a glacial erratic long owned by the Chippewas, was the object of a complicated legal wrangle during those years. At last the War Department confiscated the huge specimen, which weighed 3,708 pounds, and removed it to Washington where it was placed (and still remains) on display in the Smithsonian Institution.

All during the second half of the 19th century, mining was carried on at Isle Royale. In order to find ore outcrops more easily, the prospectors and miners burned the forests or cut them down for building settlements. When a promising deposit was found—a contact vein, a dike, or a fissure filled with copper—the miners blasted and excavated. If the site proved out, a shaft was sunk. As a result, in addition to the worked-out mines and the ancient Indian pits, hundreds of exploration pits are scattered over the island. Many of the old mines and pits are overgrown with blueberry thickets or have been filled in with dirt. Because of the severity of the winters, the miners could work on the island only during the short summer season; they had to return to the mainland each fall. In this way they followed the pattern of their prehistoric predecessors, who also spent summers extracting copper but established no permanent settlements on the island. Where these Indians came from and where they returned to is still one of the mysteries of Isle Royale.

The 19th century copper miners found three classes of ore on Isle Royale and the Keweenaw Peninsula, which they called "moss," "barrel-work," and "stamp." Moss copper was made up of large sheets of pure copper that weighed anywhere from a few hundred pounds to several tons. Barrel-work was smaller pieces of copper enclosed in amygdules at the top of brecciated flows. An *amygdule* is a gas cavity, or vesicle, in igneous rock that has been filled with a secondary mineral—native copper, in this case. Stamp copper consisted of pebble-sized pieces bound in a rock matrix. This type of copper was freed by pulverizing the ore in a stamp mill. (A mill was actually a series of heavy metal shoes operated by a cam shaft powered by water or steam.) Occasionally the miners came across large masses or chunks of native copper weighing several thousand pounds. One mass exhibited at the 1876 Centennial Exposition in Philadelphia weighed 5,720 pounds.

The principal mines on Isle Royale (operated at different times between 1843 and 1899) were the Smithwick Mine at Rock Harbor, Siskowit (which produced 190,000 pounds of copper between 1847 and 1855), Minong Mine, and Windigo. Minong Mine, located on Minong Ridge in the midst of ancient mines, was the largest and best known. The Windigo mine site is on a Park Service trail near the Windigo Information Center at the west end of the island. The other mines mentioned are also points of interest along trails in the Park. The Isle Royale mines were abandoned as the ore was exhausted and excavations were no longer profitable. By the end of the century only a few fishermen and their families were left on the island.

A new era began for the Isle Royale when it became a summer resort in the early 1900s. The U.S. Army Corps of Engineers completed the first survey of Isle Royale, and land was put up for sale at $10 an acre. A few "inholdings" are still occupied seasonally by families who bought land and put up summer homes, but these tracts will revert to the Park eventually. The movement to make Isle Royale part of the National Park System was spearheaded by Albert Stoll, Jr., a Detroit newspaperman. Legislation authorizing land acquisition for a national park was passed in 1931, and the Park was established in 1940.

In 1980 Isle Royale National Park was designated a United Nations International Biosphere Reserve, which makes it part of a worldwide data-gathering system that monitors the impact of global pollution. Biosphere Reserves are selected on the basis of their uniqueness as ecosystems.

The Predator-Prey Relationship of Timber Wolves and Moose

Its location and its geologic history have made Isle Royale biologically as well as geologically interesting. The long ridges that make up the island are the result of upfaulting and erosion of late Precambrian volcanics in which the copper was emplaced. After prolonged erosion had worn down the surface, Pleistocene ice sheets overrode the whole region and the weight of the ice depressed the land. Isle Royale was under water (probably more than once) when the ice melted back. Glacial rebound and the lowering of water levels of postglacial lakes exposed the island surface again late in Pleistocene time.

After the final retreat of ice, most of the island was scraped bare of soil and weathered rock, although some glacial till and outwash was left on the western part of Isle Royale. How did plants and animals reestablish themselves on the island? In particular, how did small animals get across the 15 miles of open water between Isle Royale and the Canadian mainland? These are speculative questions to which biologists do not have definitive answers. However, they do know things about the ecological changes that have come about during historic time. Caribou and lynx formerly lived on the island but apparently left not long after the primeval forests were cut off or burned. The trees and undergrowth that replaced the original stand continue to evolve as a mixture of transitional hardwoods and subarctic conifers.

Sometime early in this century, moose arrived on Isle Royale. Presumably they swam over from Canada. No predators were on the island, food was plentiful, and the moose population zoomed to several thousand animals, far more than the island could support. By the 1930s, as the water plants, twigs, and leaves on which moose browse were depleted, hundreds of moose starved. A fire in 1936

burned over a quarter of the island, eliminating more vegetation, and more moose died. However, as new growth, stimulated by the fire, renewed the food supply, the moose population increased rapidly, followed, in about ten years, by another starvation cycle.

In the extremely cold winter of 1948–49, Lake Superior froze over, a rare event that usually occurs only a few times in a century. A pack of eastern timber wolves from Canada crossed over on the ice and established themselves on Isle Royale. Since that time three more packs, either new immigrants or offshoots of the original group, have settled in different parts of the island. The number of wolves on Isle Royale now is between 25 and 40. Each pack kills, on the average, one moose about every three days, the most likely prey being an animal in the herd that is very young, very old, sick, or weak. This culling process keeps the moose population healthy and stable, at a ratio of about 30 moose per wolf. The fact that the moose herd on Isle Royale has a high proportion of twin calves born each year is a further indication that the stock is productive and vigorous.

A third mammal that helps to maintain the balance of wolves and moose on the island is the beaver. Its role is to provide dams for ponds and beaver meadows that supply browse for the moose, as well as food for other animals in the ecosystem.

A cautionary note for hikers and campers who visit Isle Royale is added here. Because moose are infected with hydatid tapeworms, *all surface water* on the island is contaminated with the eggs of this organism. Since the eggs are not killed by chemical purification, all water must be filtered or boiled for at least five minutes before it can safely be used for cooking or drinking.

Geologic Features

Like Voyageurs National Park (chapter 22), Isle Royale National Park is part of the Canadian Shield and is included in the Precambrian Superior Province. However, the Precambrian bedrock exposed at Isle Royale is younger in age than the bedrock of Voyageurs National Park. Thus the events that took place in late Precambrian time (Proterozoic Y) in the Lake Superior region are a continuation of the ancient geologic history that is recorded in the early and middle Precambrian rocks of Voyageurs National Park to the west. (Precambrian chronology is explained in box 22.1, p. 278.)

Ancient basaltic lava flows, mineralized and altered by metamorphism, are a feature of both parks, but distinct differences—in addition to the age factor—are evident between the two series of eruptive events. The Keweenawan lavas of the Lake Superior region poured forth as flood basalts from a rift of continental proportions in the craton. Time and time again throughout late Precambrian time, sheets of molten basalt (from the lower

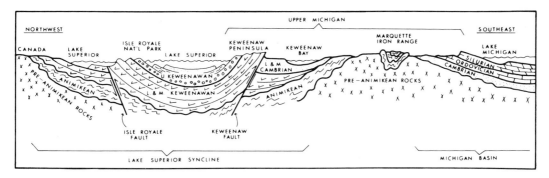

Figure 23.3 Schematic northwest-southeast cross section showing the relationship of Isle Royale, the Keweenawan lava flows, the Keweenaw Peninsula, and major faults to regional structural features. The heavy, wavy lines are major unconformities. From *Geology of Wisconsin and Upper Michigan* by R. K. Paull and R. A. Paull. © 1977 Kendall/Hunt Publishing Company; used by permission.

crust or upper mantle) spread over thousands of square miles of the ancient landscape. Geologists believe that the fissure from which the lavas emanated originally extended from the Lake Superior region all the way to what is now Kansas. However, south of the Michigan, northern Wisconsin, and Minnesota outcrops, the Keweenawan lavas lie in a subsurface belt under many hundreds of feet of younger rock and glacial drift. The trace of the ancient rift has been mapped by geophysical techniques that detect gravity anomalies beneath the surface. The high densities of the buried Keweenawan basalts register large positive anomalies on gravity meters carried or flown over suspected areas.

The Keweenawan volcanics that crop out on Isle Royale on the north side of Lake Superior and their counterparts on the Keweenaw Peninsula to the south are part of an escarpment of resistant basaltic rock that nearly encloses Lake Superior, and the volcanic sequence is continuous beneath the lake basin (fig. 23.3). The present Lake Superior basin was formed by glacial and preglacial erosion of a large structural trough, called the Lake Superior syncline, that may have existed even before the Keweenawan eruptions began. Subsidence of this trough apparently continued during and after the eruptions. After volcanic activity ceased, sediments were deposited over the lavas. Reverse faulting, probably early in Paleozoic time, along both the north and south limbs of the synclinal trough, uplifted the Keweenawan rocks. Subsequently, erosion of less resistant sedimentary rocks left the escarpment now exposed at Isle Royale and on the Keweenaw Peninsula. The lava flows in both the north and south arms of the escarpment dip toward the center of the Lake Superior syncline, but the flows on the Keweenaw Peninsula dip much more steeply than the corresponding flows on Isle Royale. Thus the synclinal structure is somewhat asymmetrical.

The Precambrian Rocks in the Park

Visitors to Isle Royale National Park can find exposures of volcanic and sedimentary rocks in a rich variety of textures and types, in spite of the dense ground cover of bushes and trees. The volcanic flows that were originally stacked more or less horizontally one on top of the other have been tilted by subsidence and faulting so that they are now exposed in strips or layers on ridge tops and along the shore. Thin sedimentary layers between some of the flows represent quiet intervals during which weathering of the flow surfaces and minor erosion and deposition could occur. Individual flows within the volcanic sequence have characteristic textures reflecting different rates of cooling and varying mineral compositions. Most flows are basaltic, but some are andesitic, and a few are dacitic or rhyolitic. They range, that is, from mafic to felsic (see box 12.1, p. 142).

The dark, massive, fine-grained flows, called "trap rock," tend to make up the resistant ridges, such as the Minong Ridge. Coarse-grained volcanic rocks, which solidified more slowly, are easily distinguished by their *ophitic texture*; that is, pyroxene crystals surround and enclose laths of plagioclase, giving the rock a mottled appearance. Weathered outcrops often have a knobby surface. The flows with *porphyritic texture* have well defined plagioclase crystals (slow cooling) scattered throughout a fine-grained groundmass (more rapid cooling). Some of the volcanic rocks of unusually coarse texture are pegmatites.

The native copper in the Isle Royale volcanic sequences may have ascended with the basaltic lavas or via channelways between flows. (Its origin and mode of deposition are uncertain.) The three classes of deposits in which the native copper was found correspond to the categories the 19th-century miners used: 1) Copper in a matrix of conglomerate interbedded with lava flows was what the miners called "stamp." 2) Amygdules of copper filling voids in the brecciated tops of lava flows was "barrel-work." 3) "Moss" was copper found in veins, fissure fillings, fractures, and the like. A mottled-green, semiprecious stone called *chlorastrolite* (meaning "green star") is also found in amygdules in amygdaloidal basalt on Isle Royale. This small, unusual gemstone, which fades in sunlight, is seldom found outside the Lake Superior region. Crystals of it can be seen in bedrock at Scoville Point, Rock Harbor, and Todd Harbor, but they may not be collected within the Park boundaries.

Some of the volcanic flows are topped by pyroclastic material, and pyroclastic debris also appears to be a component of the interbedded sediment. Irregularly shaped, pinkish-red agates occur in tuff on the east end of the island.

Sedimentary units in the volcanic sequences are mainly conglomerates and sandstones that appear to be fluvially deposited. Pebbles in the conglomerates consist of volcanic rock, mostly mafic but some felsic, reflecting the composition of the source rock. Accessible outcrops are in the Chippewa Harbor area, on the south side of Siskewit Lake, and on the north side of Conglomerate Bay. Old mine dumps, such as those at Island Mine, also have accumulations of broken sedimentary rock.

The Copper Harbor Conglomerate. On the south side of Isle Royale National Park, the Portage Lake Volcanics dip conformably and gently beneath the coarse sandstones and conglomerates of a somewhat younger formation, the Copper Harbor Conglomerate. About a fifth of the bedrock of the Isle Royale archipelago is made up of this sedimentary unit; the other four-fifths consists of the Portage Lake Volcanics. Exposures of the Copper Harbor Conglomerate crop out along the southwest shore of the main island, on the small islands offshore, and on the higher parts of Feldtmann Ridge and Houghton Ridge, which parallel the southwest shoreline of the main island. Surficial deposits, mainly of glacial origin, cover a good deal of the Copper Harbor Conglomerate. On the opposite shore of Lake Superior, 40 miles away, the Copper Harbor Conglomerate crops out again, lapping up on the Portage Lake Volcanics in the Keweenaw Peninsula. However, the Copper Harbor sedimentary units may not be continuous beneath the lake due to erosive activity since their deposition in late Precambrian time.

Effects of Glaciation on the Preglacial Topography

The topographic "grain" of the Isle Royale archipelago reflects the structures of the bedrock. The main island and nearly all of the small islands are elongate with a northeast-southwest trend. The parallel ridges with valleys or water-filled troughs in between are the result of differential erosion on the uplifted fault-block. In general, massive basalt flows make up the resistant ridges; less resistant amygdaloidal basalts underlie the valleys and troughs, producing a "corrugated" preglacial topography that has been more sharply delineated by the continental ice sheets that moved over the Lake Superior region (fig. 23.4).

Joints (fractures) and minor faults in the bedrock also affected preglacial and glacial erosion. Ravines developed along joints and fault zones that cut across the main trend of the ridges. McCargo Cove, a northeast-trending inlet on the north side of the island, is an example of a linear topographic feature that was eroded from shattered rock in a transverse fault zone.

The preglacial topography was accentuated by glacial erosion because glacial quarrying and plucking were more effective on the less resistant bedrock underlying the stream valleys. Elongate lakes, such as Angleworm Lake, Sargent Lake, and Hatchet Lake, occupy basins excavated by glacial ice that followed the valley trends. Quarrying also deepened fiordlike inlets along the shore and some of the channels between the elongate islands. Some of the crosscutting ravines, especially those on the north side of the island were noticeably gouged out by glacial quarrying.

On the ridge tops glacial abrasion smoothed and polished the resistant massive basalts, removing weathered bedrock from all but protected locations. Glacial striations are evident on many outcrops.

The last ice advance (which largely obliterated the traces of previous ice sheets) moved southwestward over the eastern half of the island, following the trend of the parallel ridges and valleys. However, the ice veered more to the west over the western end of the island and flowed across the ridges. The change in direction of ice movement is probably the reason for the fact that bedrock on the island's eastern half is more severely eroded, while a number of depositional (constructional) glacial features were left on the western half of the island. Glacial deposits are lacking on the eastern part of Isle Royale except in the bottoms of valleys or small depressions. Glacial erratics are scattered and few in number. However, on the west end of the island, till mantles much of the bedrock, and erratics are abundant.

Constructional features such as drumlins and ice-margin deposits are characteristic of continental glaciation. (Drumlins are seldom formed by valley glaciers in mountainous regions.) A *drumlin* is an elongate hill or ridge made up of compact glacial till. The Isle Royale drumlins, all on the western end of the main island, are of the "crag-and-tail" variety. The *stoss* end (facing the

Figure 23.4 This high-altitude, infrared photograph of the western end of Isle Royale shows the topographic lineation or grain of the bedrock structures that has been accentuated by glacial erosion. National Park Service photograph.

direction from which the ice advanced) is a steep, ice-smoothed knob of rock (the crag) that obstructed the flow of ice. The *lee* end (the tail), consisting of till, is streamlined and tapering. The alignment of the drumlins shows that on this end of the island the ice moved from east to west.

The ice-margin deposits are of irregular sizes and shapes and consist of coarse sand, gravel, and glacial debris. They are strung out across the island in a more or less northward direction west of Lake Desor. They mark the stillstand of an ice front that paused for some time in its retreat.

Abandoned shorelines and old beaches can be discerned in glacial drift on the western and southwestern parts of the main island. Such features record former lake levels when the water was higher (or the land was depressed) as the last ice sheet retreated. On the east end of Isle Royale, where glacial drift is lacking, former shorelines are marked by wave-cut benches in bedrock, and by stacks, arches, and cliffs. Monument Rock (about two miles west of Blake Point) is a stack carved by wave erosion when the water level was about 80 feet higher than it is now.

Geologic History

1. Middle Keweenawan eruptions during late Precambrian time.

Radiometric and paleomagnetic data collected from the Isle Royale lava flows places them with the Portage Lake Volcanics in the Middle Keweenawan rocks of Proterozoic Y time. All the Middle Keweenawan rocks have *normal polarity*; that is, when the lavas solidified, the earth's magnetic field was the same as it is today. Older lavas (Lower Keweenawan) that crop out in the Lake Superior region (and which may be buried beneath the Isle Royale flows) have *reverse polarity*, as do some younger intrusive rocks (Upper Keweenawan) outside Park boundaries. During times of reverse polarity, the direction of the earth's magnetic field was *opposite* to what it is now. (See p. 202 for an explanation of natural remanent magnetism in rocks and global polarity.) The changes in polarity recorded in the rocks of the region are another indication of the length of time during which the rift in the craton was releasing tremendous volumes of largely mafic lava.

Twelve of the Portage Lake lava flows exposed on Isle Royale have been named and designated as *marker flows* (table 23.1), which means that they have distinctive characteristics and can be traced from place to place. In between the marker flows are unnamed lavas and sedimentary units that are less readily identifiable. Six of the marker flows, from oldest to youngest, are described briefly below because of their geologic significance and because outcrops are accessible from trails or along the shoreline (from Huber 1973a).

Amygdaloid Island Flow. This trap rock, an andesite, is the most felsic volcanic rock of the group. Like the other trap rocks, it is resistant and forms the ridge that runs the length of Amygdaloid Island. A wave-cut arch, eroded when the lake was at a higher level, forms a natural bridge near the middle of the island. Pink agate amygdules, some with quartz centers, are a distinctive feature of this rock.

Hill Point Flow. The thickness and coarse grain of this ophitic rock make it a useful marker. Near the west end of the main island, the flow is 158 feet thick. Between Huginnin Cove (west end of the island) and McCargo Cove along the north shore are imposing cliffs eroded from the flow. The *type locality* outcrop is at Hill Point near the eastern end of Isle Royale. (the type locality is the place where a stratigraphic unit is typically displayed.)

Huginnin Flow. The prophyritic texture with large, tabular, plagioclase crystals scattered throughout a fine-grained groundmass, makes this flow distinctive. The rock is easily recognized in beach pebbles and cobbles as well as in bedrock exposures, such as those near Huginnin Cove.

Minong Flow. This flow, consisting of dark, fine-grained trap rock, is massive, thick, and resistant. It runs the entire length of the main island, forming the Minong Ridge. Outcrops can be seen at the shoreline on both the eastern and western ends of the island, as well as at many places along the Minong Ridge trail.

Greenstone Flow. The Greenstone Flow is referred to as the "backbone" of Isle Royale because it forms the high resistant ridge that runs from east to west through the interior of Isle Royale. Mount Desor, the highest point (1,394 feet) on the island, is on Greenstone Ridge, as are Sugar Mountain (1,362 feet), Ishpeming Point (1,377 feet), and Mount Siskiwit (1,205 feet). The thickest part of the Greenstone Flow (about 800 feet) is in the middle of the island, but the flow thins westward to about 100 feet on Washington Island. The rock is mainly ophitic in texture but has some pegmatite zones. Outcrops are numerous along the Greenstone Ridge trail that goes from the Windigo Information Center on Washington Harbor to Lookout Louise near Tobin Harbor.

Scoville Point Flow. This flow is the most prominent of the "Tobin porphyrites," so-called because of their outcrops in the Tobin Harbor area. Scoville Point (where the flow is about 100 feet thick) is the type locality for the rock. In the western part of the island, the Scoville Point Flow caps Red Oak Ridge, which is nearly as high as Greenstone Ridge because of the porphyritic rock's resistance. Fine plagioclase crystals uniformly distributed in a fine-grained matrix give the rock its characteristic appearance.

2. Mineralization and metamorphism in the Portage Lake Volcanics.

The native copper, plus small amounts of silver and other minerals that are found in the volcanic sequence probably rose concurrently with the lavas. Some metamorphism occurred but not to any great extent.

3. Deposition of the Copper Harbor Conglomerate, completing the Middle Keweenawan sequence.

As volcanism in the Lake Superior region diminished, erosional processes stripped off sediments that were then deposited over the area by streams. The direction of sediment transport is assumed to have been from west to east because of the following indications in the Copper Harbor Conglomerates: (1) the orientation of cross-bedding, ripple marks, and the like preserved in the rock; (2) thickening of the formation and decreasing coarseness eastward; and (3) the composition of boulders and cobbles that appear to have come from older Keweenawan volcanics on the Minnesota shore. Nearly all of the coarser sediments in the Copper Harbor Conglomerate are of volcanic origin. This sedimentary unit was not a host rock for the native copper, which suggests that volcanic activity had probably ceased prior to deposition of these sediments.

4. Faulting and uplift, probably in early Paleozoic time.

During the long subsidence of the Lake Superior synclinal trough, more than 40,000 feet of lavas, sandstones, and conglomerates accumulated in its center (fig. 23.3). Then a series of fault movements, possibly beginning before the end of Precambrian time, raised the Keweenawan rocks. The Isle Royale and Keweenaw faults, which are associated with the escarpment around Lake Superior (fig. 23.3), are two of the four major faults that developed along the north and south edges of the syncline during this time.

5. Erosion during Paleozoic, Mesozoic, and Cenozoic time and the development of the Lake Superior basin.

Tectonic activity ceased, followed by many millions of years of erosion. The more resistant lavas tended to form escarpments, while most of the sedimentary rocks, being less resistant, were removed. In this way a basin formed, drained by streams that probably flowed across the area in a northeasterly direction. Lake Superior occupied the basin after Pleistocene glaciation had scoured and deepened it. Because of the synclinal structure of its basin, Lake Superior is the deepest of the Great Lakes. From Isle Royale, on the northwest rim of the lake, the bottom slopes off to a depth of more than 1,000 feet.

6. Pleistocene glaciation.

Although Isle Royale was glaciated repeatedly and severely during Pleistocene time, most of the evidence of earlier ice sheets was destroyed by the last phase of Wisconsin glaciation; that is, the Valdes Stade, a substage about 11,000 years ago during which a partial readvance of ice occurred. This readvance brought ice back over most of the Lake Superior basin and removed almost all of the surficial material that had previously been deposited on Isle Royale. As the Valdes ice began to retreat, the west end of the island was uncovered first. A pause in the retreat allowed the accumulation of the ice-margin deposits around tongues of stagnant ice. Shoreline features associated with a glacial lake that existed at that time developed on the western end of the main island. Water level of the lake was about 200 feet higher than the present level of Lake Superior.

The final retreat of the ice margin was rapid and left only a small amount of glacial debris on the central and eastern parts of Isle Royale. Glacial Lake Minong formed when the ice margin had melted back to the northern shore of Lake Superior. Lake Minong, which filled the whole basin, existed long enough for well developed beaches to form on the island, especially in the glacial debris on the southwestern shore. Monument Rock (fig. 23.2) is a Lake Minong shoreline feature.

7. Glacial rebound and the development of the present lake.

Water levels continued to lower for a time but then began to rise again as glacial rebound elevated the outlets. Because of the tilting that accompanied rebound, abandoned shorelines are difficult to correlate, especially in the Isle Royale area and the northern shore of the basin where upwarping was irregular and uneven. However, during the Lake Nipissing stage about 5,000 years ago, strong wave action produced a number of well defined shoreline features, among them the wave-cut arch on Amygdaloid Island. Nipissing was the largest of all the Great Lakes.

Continued uplift of the land to the north and the downcutting of the Port Huron outlet to the south brought the Nipissing stage to a close. Then the uplift of Lake Superior's outlet separated that basin from Lake Michigan and Lake Huron, forming the Lake Superior of today. The exact time at which this occurred is not known (Hough 1958). Because glacial rebound is still causing gradual upwarping and tilting of the lake basin slightly toward the

south, the shorelines in the vicinity of Isle Royale are rising, while those on the Keweenaw Peninsula on the opposite shore are being very slowly submerged.

Geologic Maps and Cross Sections

American Association of Petroleum Geologists. *Geological highway map of the Great Lakes region.* Map no. 11. Tulsa, Oklahoma: American Association of Petroleum Geologists.

Huber, N. K. 1973. *Geologic map of Isle Royale National Park, Keweenaw County, Michigan.* U.S. Geological Survey, Miscellaneous Geologic Investigations, Map I-796.

Bibliography

Books, K. L. 1972. *Paleomagnetism of some Lake Superior Keweenawan rocks.* U.S. Geological Survey Professional Paper 760.

Dorr, J. A., Jr.; and Eschman, D. F. 1970. *Geology of Michigan.* Ann Arbor, Michigan: University of Michigan Press.

Dott, R. H., Jr.; and Batten, R. L. 1981. *Evolution of the earth.* 3rd ed. New York: McGraw-Hill Book Company.

Hough, J. L. 1958. *Geology of the Great Lakes.* Urbana, Illinois: University of Illinois Press.

Huber, N. K. 1973. *Glacial and postglacial geologic history of Isle Royale National Park, Michigan.* U.S. Geological Survey Professional Paper 754A.

———. 1973a. *The Portage Lake Volcanics (Middle Keweenawan) on Isle Royale, Michigan.* U.S. Geological Survey Professional Paper 754C.

Leskinen, C.; and Leskinen L. 1980. *Copper country history.* Park Falls, Wisconsin: F. A. Weber & Sons, Inc.

National Park Service. 1979. *Isle Royale National Park.* Map and informational brochure.

Paull, R. K.; and Paull, R. A. 1977. *Geology of Wisconsin and Upper Michigan, including parts of adjacent states.* Dubuque, Iowa: Kendall/Hunt Publishing Company.

———. 1980. *Wisconsin and Upper Michigan,* K/H Geology Field Guide Series. Dubuque, Iowa: Kendall/Hunt Publishing Company.

Rakestraw, L. 1965. *Historic mining on Isle Royale.* Isle Royale Natural History Association.

Salop, L. J. 1977. *Precambrian of the Northern Hemisphere and general features of early geological evolution.* New York: Elsevier Scientific Publishing Company.

Wilson, M. E. 1965. The Precambrian of Canada: the Canadian Shield. In *The Precambrian,* ed. K. Rankama, vol. 2, p. 378. New York: Interscience.

Wolff, R. G., and Huber, N. K. 1973. *The Copper Harbor Conglomerate (Middle Keweenawan) on Isle Royale, Michigan.* U.S. Geological Survey Professional Paper 754B.

Address

Isle Royale National Park
87 North Ripley Street
Houghton, Michigan 49931

24

Yellowstone National Park

by Sherwood D. Tuttle
University of Iowa

Location: Northwest Wyoming, Eastern Idaho,
Southern Montana
Area: 2,219,822.70 acres; 3,468.47 square miles
Authorized: March 1, 1872

Figure 24.1   Castle Geyser erupting hot water and steam in Yellowstone National Park. Geyserite (silica) accumulating around the vent has formed a large cone.

Castle Geyser is located in Upper Geyser Basin. Photograph by W. H. Parsons.

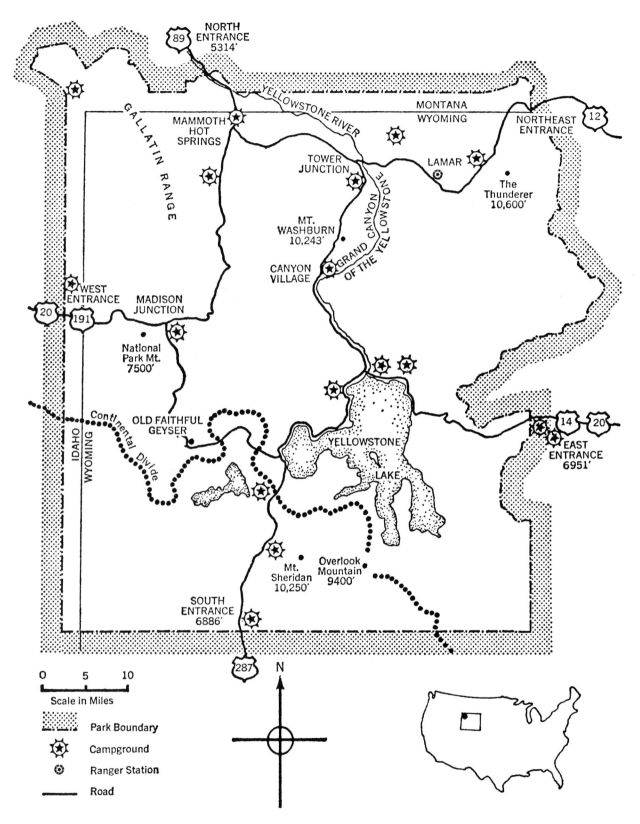

Figure 24.2   Yellowstone National Park, Wyoming,
Montana, Idaho.

Table 24.1. Generalized Geologic Column, Yellowstone National Park.

Time Units (Era)	Period	Epoch	Rock Units	Geologic Events
Cenozoic	Quaternary	Holocene	Alluvial, landslide, glacial, and hot spring deposits	Earthquakes; landsliding; Geothermal activity, hot springs, geysers; Neoglaciation; Regional uplift; doming in center of Yellowstone Plateau; Erosion of Grand Canyon of Yellowstone River
Cenozoic	Quaternary	Pleistocene	•Yellowstone Group (about 20 formations and members): (Plateau rhyolites and basalts), (Lava Creek Tuff), (Mesa Falls Tuff not exposed in Park), (Huckleberry Ridge Tuff)	Pinedale glaciation; Post-caldera lava flows; Obsidian Cliff; Bull Lake glaciation; Pre-Bull Lake glaciation; Climactic eruptions and caldera collapse of third volcanic cycle; rhyolitic ash falls, mudflows, welded tuff, breccias, etc. Second volcanic cycle; Island Park caldera and western part of Yellowstone Plateau; First volcanic cycle; caldera collapse in Island Park area; ash falls, pyroclastics, breccias, etc.; rhyolitic eruptions
Cenozoic	Tertiary (Neogene)	Pliocene / Miocene	/////	Erosion of ranges, canyon cutting, sedimentation in basins; Regional uplift; block-faulting
Cenozoic	Tertiary (Paleogene)	Oligocene / Eocene / Paleocene	•Absaroka Volcanic Supergroup (9 formations)	Repeated eruptions of andesitic pyroclastics and lavas; burial of fossil forests; Basaltic breccias and flows; Post-orogenic faulting, uplift and erosion
Paleozoic/Mesozoic	(Cretaceous)		•Over 40 formations, some in northern and some in southern sections of Park	Laramide orogeny; intense folding and thrust-faulting; uplifting of Rocky Mountains; Extensive deposition of marine and nonmarine sedimentary rocks as seas advanced and retreated over downwarped (geosynclinal) area
Paleozoic/Mesozoic			/////	Major angular unconformity
Precambrian time			Gneisses and schists	Orogenic cycles; intrusions, metamorphism; sedimentation and erosion

* For details of the stratigraphic record, see Keefer (1971, pp. 9–11), Parsons (1978, p. 214), Christiansen and Blank (1972, p. B6), Love and Keefer (1975, p. D9), Ruppel (1972, p. A8), Smedes and Prostka (1972, p. C7).

Yellowstone and the National Park Image

As the oldest and best known (and until recently, the largest) national park, Yellowstone symbolizes to many Americans all that is awe-inspiring, beautiful, and wonderful about our national parks. What are the elements that make up its image as a very special place, the ultimate park of parks?

Some of the Yellowstone mystique is in its location—"out where the West begins." To reach the Park one travels through wide open spaces and high mountain passes.

Old Faithful has fascinated millions of visitors. Every day during the summer, thousands of people gather to watch its hourly eruptions. Parking lots fill, empty, and then refill as this world-famous geyser puts on performance after performance.

Mammoth Hot Springs, its great colorful terraces ever growing and changing, impresses visitors with its size. How could one have imagined that it would be so large!

In the Grand Canyon of the Yellowstone, that deep gash in the Plateau, are the strange "yellow rocks" for which the Park was named.

In the great coniferous forests, people can hike, camp, fish, and relax. And in this sanctuary live the large woodland animals—black bear, grizzlies, bighorns, moose, elk, mule deer, and in the glades, buffalo (or more properly, bison).

And there are the Park Rangers in their green uniforms and broad-brimmed hats—protectors of people and animals, guardians of the environment—who answer questions and patiently give directions.

Stories about Yellowstone's wonders are told over and over by those who have come to look and marvel. And if you think the stories you hear have been overblown through countless retellings since the days of John Colter and Jim Bridger, we can only say, as they did, "Go and see for yourself!"

Discovery and Exploration

Bands of Indians roamed the Yellowstone region for thousands of years before white men came. The Indians hunted and fished in the area and camped near the hot springs and geyers. Something else that drew them was the plentiful obsidian, from which they made tools and weapons, for their own use and for bartering. Large piles of obsidian chips near Obsidian Cliff and the finding of arrowpoints made from Yellowstone obsidian in the Great Plains and as far away as the Mississippi Valley are evidence that the obsidian trade was very important to the Indians.

Like today's visitors, most of the Indians summered in the Park but did not stay through the winters because of the heavy snows and severe cold. One group who did live in Yellowstone the year round were the Wyoming Shoshones. Because they subsisted on the bighorn sheep that lived in the mountains, these Indians were called the "Sheepeaters." The Shoshones called the geysers "water-that-keeps-on-coming-out." Perhaps it was they who built a pile of logs above the mouth of one of the geysers, but no one is certain why this was done.

John Colter, a guide on the Lewis and Clark expedition, was probably the first white man to visit Yellowstone. He left the expedition on the return trip in 1806 to join a party of trappers. The following year, alone and on foot, he crossed the Wind River Mountains and the Teton Range and explored the Yellowstone region, making contact with the various Indian tribes. Although wounded in a battle between Crow and Blackfoot Indians, he managed to make his way to the trading post at the mouth of the Bighorn river (south central Montana). There his description of a place of "fire and brimstone" was dismissed as incredible and doubtless the result of delirium. Later when other trappers and fur traders went into the region and saw what he had seen, they called it "Colter's Hell."

Other explorers passed through Yellowstone, among them the missionary, Father DeSmet, and Captain John Mullen in the early 1850s. But little was known about the region until the famous mountain man Jim Bridger, after returning from an 1857 expedition, began spreading tales of spouting water, boiling springs, yellow rock, and a mountain of glass. Since Bridger was known as a "spinner of yarns," his reports were laughed at and regarded as preposterous. Nevertheless, enough information had been brought out to arouse the interest of influential men, among them the geologist and western explorer, F. V. Hayden. In 1859, he and an Army surveyor, W. F. Raynolds, began a two-year reconnaissance of the Upper Missouri region with Bridger as the party guide. They reached the approaches to Yellowstone but were unable to get through the mountain passes because of very heavy snows. The outbreak of the Civil War put a stop to their explorations. Hayden, who was a medical doctor before becoming a geologist, served as a battlefield surgeon with Union troops. Eleven years passed before he was able to fulfill his dream of making a scientific study of Yellowstone.

Meanwhile, intrigued by the reports of miners who had traveled up the Yellowstone River in an unsuccessful search for gold, a group of Montanans organized the Washburn-Langford-Doane expedition. Headed by Henry Washburn, the surveyor-general of Montana Territory, the party included Cornelius Hedges, a judge, and Nathaniel P. Langford (who later became known as "National Park" Langford). An Army detachment, commanded by Lt. Gustavus Doane, accompanied them. They spent nearly a month exploring the Yellowstone region, naming features, and collecting specimens. As they were gathered around a campfire (near what is now Madison Junction) on their last night in the Park area, they talked about what should be done with this strange and marvelous region. Out of this discussion came the idea of making it a preserve, not open to settlement or mining, under the protection of the government. All agreed to work toward getting the land set aside.

The next year, in 1871, F. V. Hayden led a large, government-sponsored expedition into Yellowstone. His comprehensive report, supplemented by the excellent photographs of W. H. Jackson and paintings by Thomas Moran, helped to convince the Congress to withdraw this unique area from public sale. On March 1, 1872, President Grant signed a bill authorizing Yellowstone National Park. The new law provided that

> ". . . a tract of land in the territories of Montana and Wyoming, lying near the headwaters of the Yellowstone River is hereby reserved and withdrawn from settlement, occupancy, sale under the laws of the United States, and dedicated and set apart as a public park or pleasuring ground for the benefit and enjoyment of the people. . . ."

N. P. Langford, the first superintendent of the Park, served for five years without pay, as did several succeeding superintendents. Unfortunately, Congress did not appropriate funds to implement what was a noble intention. Finally, in 1886, the responsibility for protecting and managing the Park was given to the Army. After the National Park Service was organized in 1916, a corps of rangers, under the leadership of a civilian superintendent, took over the care and development of the Park, continuing the public works and improvements that the Army engineers had begun and supervising the facilities operated by concessionaires.

The administrative structure developed at Yellowstone National Park has served as a model for national parks throughout this country, and in other parts of the world as well. Many individuals contributed to the development of the national park idea, and no more worthy place could have been chosen to work out the concept and put it into effect than Yellowstone.

The Geographic Setting

Yellowstone National Park is astride the Continental Divide, the topographic "ridgepole" of North America that separates Atlantic and Pacific drainages. The Divide "enters" the Park near the southeast corner, crosses between Yellowstone Lake and Shoshone Lake, and then goes into Idaho south of the West Yellowstone entrance to the Park (fig. 24.3).

The Yellowstone River, which rises south of the Park, collects water from a number of tributaries as it flows northward across the Park and eventually joins the Missouri River in Montana. The Missouri, in turn, joins the Mississippi, which empties into the Gulf of Mexico. The drainage from the southwestern third of the Park finds its way into the Snake River, which passes through Grand Teton National Park (chapter 25) and then flows west to join the Columbia River on its way to the Pacific Ocean.

Just outside the southern boundary of the Park is a small, shallow body of water called Two-Ocean Lake. When this lake is unusually full, it discharges water in two directions, some to the Gulf of Mexico and some to the Pacific. At normal levels, the lake's drainage is westward to the Pacific.

The Park is on a high plateau, averaging about 8,000 feet in elevation and nearly surrounded by ranges of the Middle Rockies, 10,000 to nearly 14,000 feet high. To the southwest, the Yellowstone Plateau slopes gently down to the Snake River Plains, which are in one of the Intermontane Provinces.

The ranges that encircle the Yellowstone Plateau are, beginning on the west and going clockwise, the Madison Range; to the north, the Gallatin Range and the Beartooth Mountains; on the east, the Absaroka Mountains; and to the south, the Teton Range and Jackson Hole in Grand Teton National Park.

Geologic Features

Geologically, as well as topographically, the two national parks in the Middle Rockies—Yellowstone and Grand Teton (chapter 25)—are related in origin to Rocky Mountain National Park (chapter 12) in the Southern Rockies and to Glacier National Park (chapter 10) in the Northern Rockies. In general, all the ranges of the Rocky Mountain chain evolved through the same time span of geologic history, were subjected to Laramide tectonic forces of great intensity, and have rocks and structures of similar age. However, the differences, as well as the similarities, among these great ranges continue to challenge geologists. Many questions about the geologic development of the Rocky Mountains are incompletely or only tentatively answered. Why, for example, have the Middle Rockies undergone extensive and repeated episodes of volcanism, while the Northern Rockies have few volcanic rocks?

The Surrounding Ranges

Many of the features of the Yellowstone landscape are associated with volcanic events that took place in relatively recent geologic time; but long before Cenozoic volcanism began, the structures of the ranges were evolving. Thus the story begins not on the Yellowstone Plateau but in the mountains that rise above it (fig. 24.3).

Madison and Gallatin Ranges. These mountains, some of which are in the northwest section of the Park, are elongated topographic highs bounded by major rivers. The Gallatin River flows north between the two ranges.

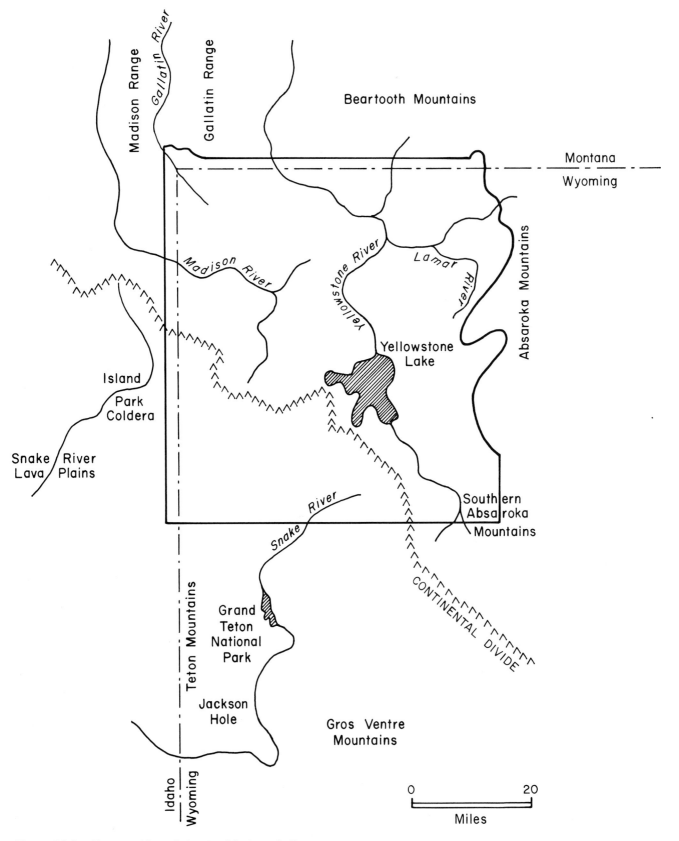

Figure 24.3 Topographic and structural features in the vicinity of Yellowstone National Park.

The Madison River crosses the Park's western boundary and then heads northward along the western flank of the Madison Range. The eastern margin of the Gallatin Range is outlined by the course of the Yellowstone River, which flows out of the Park near North Entrance at Gardiner.

Like many other Rocky Mountain ranges, the Madison and the Gallatin have a core of Precambrian granites and gneisses with flanking Paleozoic and Mesozoic sedimentary rocks, some of which are overlain by volcanics. The northwest-southeast structural trend of the two ranges is characteristic of the general trend of the Rocky Mountains. Typical, also, is evidence of several episodes of mountain-building and uplift, with the Laramide orogeny being the most significant. The present mountains are large blocks bounded by post-Laramide faults that (in Cenozoic time) cut obliquely across older structural trends in a nearly north-south direction. Thus the Madison and Gallatin Ranges can be described as fault-block mountains with the north-south river valleys following downfaulted blocks.

The Beartooth Mountains. Uplift and vertical faulting have raised the Precambrian granites and gneisses of the Beartooth massif high into the air. From its southern tip in Yellowstone Park, the massif rises to Montana's highest point, Granite Peak, elevation 12,799 feet. At the top of the range are a few outliers of a once extensive sedimentary cover, Beartooth Butte being an example. Other Paleozoic rocks, mainly limestones, lap up on the flanks. In these intensely glaciated mountains, U-shaped valleys, 3,000 to 4,000 feet deep, dissect the mountain landscape. The valleys were cut first by streams and then gouged out by alpine glaciers. Mountain uplands, some forested and some scraped bare by ice, are high and inaccessible. The bold eastern front of the Beartooths rises 4,000 to 5,000 feet above the Great Plains. U.S. Highway 212, which leaves the northeast corner of Yellowstone National Park and winds eastward and then north to Red Lodge, Montana, goes through this spectacular mountain terrain.

Absaroka Mountains. This large range, trending northwest-southeast, occupies the eastern part of Yellowstone Park and extends for some distance into western Wyoming. From the southwest edge of the Beartooths, the Absaroka Range stretches down to the Owl Creek Mountains of the Wyoming basin. Absaroka peaks, rising to elevations of 10,000 to 12,000 feet, have been intensely eroded by streams and glaciers and form a jumble of steep ridges above fairly broad valleys. Unlike the other ranges of the Middle Rocky Mountains, the Absarokas are made up of over 10,000 feet of mainly andesitic, pyroclastic volcanics. Dated Eocene in age, these rocks form what is called the Absaroka volcanic field. Eruptions came from numerous centers.

The tuffs, breccias, and lavas of the Absaroka Volcanic Supergroup overlie older metamorphic and sedimentary rock sequences of the Middle Rocky Mountains. What the volcanic rocks represent is an extensive series of volcanic events that began in the Eocene Epoch, and subsided in the Oligocene. The Absaroka volcanic rocks have been somewhat uplifted and warped, but are only mildly deformed.

Jackson Hole and the Teton Range. The northern boundary of Grand Teton National Park lies a few miles south of the Yellowstone National Park southern boundary. (The strip in between is part of the Teton National Forest.) In this area south of Yellowstone are Jackson Hole and the Teton Range. ("Hole" in this context means a sheltered valley, or a place to "hole up" for the winter.) The Teton Range is a tilted and uplifted fault-block, and Jackson Hole is a down-faulted block. These features are described in the next chapter.

Island Park Caldera and the Snake River Plains. The Island Park caldera is a plateau ringed with low hills. Its volcanic rocks are similar to those of the Yellowstone Plateau. Beyond the caldera, the Snake River Plains of southern Idaho slope gently to the southwest, with many flows of young basaltic lava covering the surface of the plains. In Craters of the Moon National Monument (about 120 air miles southwest of Yellowstone), some basalt flows erupted as recently as 2,000 years ago. The volcanism of the Island Park caldera and the Snake River Plains is an integral part of the volcanic history of the Yellowstone Plateau.

The Yellowstone Plateau

Most of the Yellowstone Plateau is occupied by a huge collapsed caldera that has been nearly filled with volcanic material. In general, the rocks of the Plateau are younger than those of the surrounding ranges. Only a few hills and mountains of older rock stick up through the relatively flat surface of the Plateau, the most prominent being Mount Washburn, 10,243 feet high. A depression in the caldera contains Yellowstone Lake, which has an elevation of 7,331 feet. With an area of 139 square miles, it is the largest high-altitude lake in North America.

The Yellowstone Plateau is located at the intersection of three tectonic trends, the expression of which can be traced in the faulting, tilting, uplift, and downwarping of the surrounding mountains and basins. Present earthquake activity in the region indicates that many faults are still active. The stresses and fracturing in the crust are associated with the volcanic activity that is responsible for much of Yellowstone's landscape. On the Plateau are found geyser basins, hot springs, and thousands of cubic miles of young volcanic rocks. The origin and development of these remarkable features is discussed in the sections that follow.

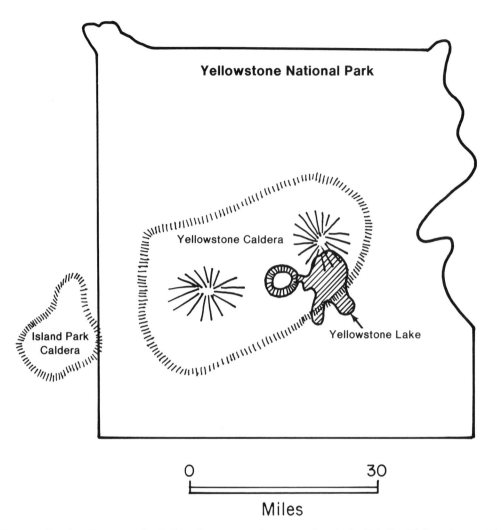

Figure 24.4   Diagram showing the approximate location and extent of Island Park caldera and the Yellowstone caldera, both Pleistocene in age. Two currently expanding domes are located within the Yellowstone caldera. A smaller crater now occupied by West Thumb of Yellowstone Lake is a late Pleistocene feature.

The Yellowstone Caldera

The explosion-collapse caldera of Crater Lake (chapter 19) has an area of barely 20 square miles. What we are describing at Yellowstone is a multiple-eruption caldera with an area of about 1,500 square miles! By means of extensive field investigations and with the aid of satellite imagery, the outlines of this mega-caldera and its extent have been fairly accurately determined. About 40 miles long and 30 miles wide, it reaches from the Madison River (near Madison Junction) to the eastern shore of Yellowstone Lake and from Lewis Falls north to the foot of Mount Washburn (fig. 24.4).

In the volcanic history of the Yellowstone Plateau, three phases, or volcanic cycles, have been recognized, all of which took place during the last two and a half million years. Rocks erupted in the first volcanic cycle are exposed in the Island Park caldera and in southwestern Yellowstone. Identifiable outcrops of the second volcanic cycle have been found near the edge of the Island Park caldera, but most of the second cycle rocks have been destroyed or buried by those of the third volcanic cycle. The eruptive materials of the third cycle make up most of the Yellowstone Plateau. Through careful field mapping and study of satellite photos, ruins of the calderas associated with the three volcanic cycles have been identified in the topography of today—despite the masking effects of later eruptions of lava.

Studies of the predominantly rhyolitic extrusive rocks of the three volcanic cycles in the Island Park-Yellowstone area suggest that during each episode great volumes of magma were brought to the surface and erupted, building up thousands of feet of ash and tuff as well as layers of lava and breccia. Moreover, small lava eruptions, related to the third cycle, persisted until about 70,000 years ago. It was during the third volcanic cycle that the Yellowstone caldera was formed (fig. 24.5).

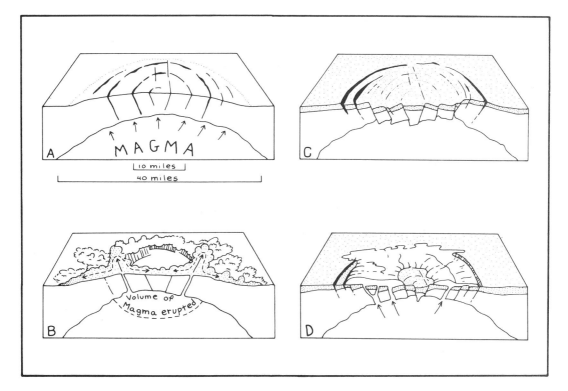

Figure 24.5 Schematic diagrams showing the development of the Yellowstone caldera during the most recent volcanic cycle (600,000 years ago). *A.* Upwelling magma arches the roof rocks, causing concentric ring fracture to form that extend downward toward the magma chamber. *B.* As the ring fractures tap the magma chamber, the release of pressure allows escape of gases that were dissolved in the magma. Catastrophic eruptions of pumice, tuff, and ash flows pour out from the ring fractures. Fine particles, borne by the wind, spread over thousands of square miles. *C.* Normal faulting along the ring fractures collapses the roof rocks, forming an immense caldera. *D.* Magma, depleted of gas, domes up the caldera and flows out from the ring fractures over the caldera floor. From *Middle Rockies and Yellowstone* by W. H. Parsons. © 1978 by Kendall/Hunt Publishing Company (modified from Keefer 1971 and others).

Any explanation for such great volumes of extrusive material requires the presence of a fairly large reservoir of magma, presumably originating in the upper crust and extending fairly close to the surface. As more and more magma was withdrawn by eruptions through fractures and weak zones, the roof rocks over the chamber stretched and thinned, finally collapsing of their own weight into the liquid magma below. Since no identifiable roof rocks or old volcanic cones have been found, we can assume that the missing surface material was either incorporated into the melt or expelled with the ash. The cataclysmic explosions were of a magnitude not known in recorded history and perhaps unequaled in geologic history.

For thousands of years following the climactic eruptions, lava continued to flow periodically until the tremendous depression was nearly filled. Then, beginning about 150,000 years ago, the floor of the caldera began to bulge up again. This doming exposed rocks and structures that have enabled geologists to develop a sequence of events to account for this remarkable feature. Although the details of the geological activities that produced the Yellowstone caldera are technical, the significant lines of evidence can be reviewed and summarized as follows:

Heat Flow. The earth's crust gets warmer with depth. This rate of heat increase with depth is called the *geothermal gradient*. Heat energy tends to move from hotter areas to colder areas; and despite the fact that rocks are poor conductors, the earth's internal heat seems to be moving from great depths toward the surface. Amounts of heat flow can be measured both at the surface and in drill holes. Heat flow values taken from a number of points on the Yellowstone Plateau are abnormally high, many times greater than the continental average. In the bottom of wells drilled by the U.S. Geological Survey, the average temperature is over 400°F. at depths of 500 to 600 feet below the Plateau surface.

Curie Depths. It can be demonstrated experimentally that most rocks lose their magnetism when they are heated beyond the *Curie point*, a threshold temperature of about 560°C. Thus information about the temperature of rocks below the surface can be inferred from the magnetic properties of rock. Using geophysical techniques, investigators have determined that beneath the surface of the Yellowstone Plateau, Curie depth is about 6 miles down (10 kilometers). By contrast, Curie depths under many other parts of the continent are between 10 and 20 miles (15 to 30 kilometers) below the surface. What this means is that rock under Yellowstone reaches the Curie point (c. 560°C) at a considerably shallower depth. Furthermore, *under the caldera* in some spots, Curie depth is less than 4 miles (6 kilometers) (Smith and Christiansen 1980).

Electrical Conductivity. Earth materials at high temperatures tend to be good conductors of electricity. Again, by means of geophysical techniques, the measurements obtained show that electrical conductivity at Curie depth under Yellowstone is good; therefore, the patterns of magnetic and electrical data are in agreement.

Seismic Wave Velocities. For many years seismologists have known that earthquake waves increase or decrease in speed as they travel through the rocks of the crust, depending upon the density of the materials they encounter. Seismic *P* waves, for example, travel faster through the higher-density basalts of oceanic crust and then slow down when passing through granitic continental crust, which has lower density. Seismograph monitoring by the U.S. Geological Survey in the Yellowstone caldera and vicinity has recorded travel-time delays of *P* waves from distant earthquakes. Mathematical modeling of the travel-time delays suggests the presence of a body of low-density material close to the surface and extending downward to the upper mantle.

Gravity Data. A *gravity survey* of an area consists of measuring the gravitational field at a number of different locations with a gravimeter and then plotting the results on a map. The object is to associate gravity variations with differences in the distribution of densities and hence of rock types. If the crust consisted of homogeneous materials, the variations would be due solely to differences in surface elevation. Thus interpretation of gravity data can yield two sorts of evidence. First, the location of rock masses of different types can be determined. Then, after a survey has been repeated at intervals, say five or ten years, the changes in elevation that have occurred can be calculated and the results compared.

Gravity measurements derived from 900 locations on the Yellowstone caldera and the surrounding region have indicated the presence of low-density material, both under the caldera and also in the subvolcanic basement, with the latter showing even lower readings than the material nearer the surface. What this implies is that the presumed shallow chamber contains magma which is high in silica, and that the silica content in rocks above the chamber is less high, perhaps because the silica has been somewhat dispersed. Rhyolites, which are plentiful in Yellowstone, are high in silica and have a density of about 2.5 to 2.6.

Mafic rocks, such as basalts, which are much less common in Yellowstone, have somewhat higher densities, around 3.0.

Gravity data have also confirmed geodetic surveys that indicate the caldera area is rising at the geologically rapid rate of about a half-inch per year.

Depth of Foci and Earthquake Frequency. The *focus* (pl. *foci*) of an earthquake is the point within the earth that is the center of an earthquake. The focus is the initial rupture point in the earth from which seismic waves emanate. The *epicenter* is a point on the earth's surface *directly above* the focus of an earthquake. A seismologist can pinpoint the location of an earthquake and the depth of focus by comparing the seismic wave data recorded on the seismograph at his station with similar information from other stations and then analyzing the results.

During the past century, reports of earthquakes in the Yellowstone region have been more frequent than those from elsewhere in the Northern and Middle Rocky Mountains. The strongest earthquake recorded in the Yellowstone area was the August 1959 quake, which caused extensive landsliding and had a magnitude of 7.1 on the Richter Scale. The epicenter of this major earthquake was at Hebgen Lake, just outside the northwest corner of the Park. Monitoring of many earthquakes in the Yellowstone area, most of which have registered between 1 and 6 on the Richter Scale, has revealed that the average depth of focus is around 9 miles; this is considered fairly shallow. This means that the rocks are solid to that depth, because only solid rocks can fracture and produce earthquake waves. *Earthquake swarms* (that is, series of minor earthquakes) occurring in Yellowstone are similar to those in other areas of active volcanism and suggest movement of magma below the surface.

Ages of the Caldera Rocks. U.S. Geological Survey scientists have ascertained absolute ages for volcanic rocks from the three volcanic cycles using the potassium-argon method of radiometric dating (box 13.1, p. 159). By combining the dates obtained with information from stratigraphic studies and geologic mapping, the following sequence was developed: The first volcanic cycle began about 2.2 million years ago in the Island Park area and lasted until about 1.6 million years ago, with the climactic eruptions and caldera collapse occurring about 2 million years ago. The climactic eruptions of the second volcanic cycle, which also began in the Island Park area, were about 1.2 million years ago. Overall, a smaller volume of material was erupted during the second cycle. The third volcanic cycle, which was marked by a shift in volcanic activity from the Island Park area to the center of the Yellowstone Plateau, began about 1.2 million years ago and lasted until about 70,000 years ago. The climactic eruptions that produced the Yellowstone caldera took place about 600,000 years ago.

The Composition of the Lavas. While most of the volcanic rocks produced by the eruptions from the Yellowstone caldera are rhyolitic, some basalt flows were extruded late in the third volcanic cycle. As noted earlier, basaltic lava flows also followed the rhyolitic flows of the first and second volcanic cycles. The pattern seems to be one of extensive rhyolitic volcanism, with the center slowly moving from southwest to northeast, followed periodically by episodes of basaltic volcanism along the same general trend.

Chemically and mineralogically these two rock types are quite different. Rhyolite is high in silica and low in iron and magnesium, while basalt is high in iron and magnesium and low in silica. The predominant minerals in rhyolite are potassium feldspar and quartz; in basalt, they are usually calcium plagioclase feldspar, olivine, augite, and hornblende. Studies have shown that minerals that form basalt crystallize out from a melt or magma at considerably higher temperatures than do the minerals that form rhyolite. Thus the eruption of rhyolitic lavas and then basaltic lavas from the same vents with no intermediate lavas (e.g., andesites) appearing in the sequence calls for a special explanation. The volcanoes described in the previous chapters (Mount Rainier, for example, chapter 17) changed the composition of their lavas as eruptions ran their course, but there was usually a more gradual transition from one type of lava to another.

It seems logical to theorize that at Yellowstone two different source materials melted in order to form two distinct liquids, one rhyolitic and the other basaltic (Smith and Christiansen 1980). Basaltic magma may be formed by the partial melting of the upper mantle, while rhyolitic magma may be derived from partial melting of metamorphic rocks (such as gneisses and schists that are high in silica) in the earth's lower crust. Long-continued uplift and extension, or stretching, of the crust might cause upward displacement of the mantle. This, in turn, causes reduction of confining pressure that allows materials with low melting points to be liquified. As heat radiates upward in the crust, additional high-silica rocks are incorporated into the melt because they tend to melt first. In this way a high-silica magma, having lower density and greater viscosity and being somewhat more buoyant, begins to rise faster and collect in the upper crust. In the late phase of a volcanic cycle, after most of rhyolitic magmas have been ejected, basaltic material from the upper mantle begins to melt and rise upward under conditions of reduced pressure.

Expanding Domes. The geophysical and geochemical evidence summarized above forms the basis for the idea that beneath the Yellowstone region there has existed for the past several million years a body of very hot, low-density, high-viscosity material capable of generating and extruding tremendous volumes of volcanic material.

What happens as the magma nears the surface? The overlying roof of surface rocks bulges, stretches, thins, and cracks; this produces ringlike fractures, which further weaken the roof rocks, causing more fractures to open until, quite literally, the roof falls in. Very likely, ground water, plentifully supplied by rain and melting snow, plays a role in triggering the catastrophic explosions. The gaseous steam, suddenly released, propels enormous fiery clouds of ash high into the air. Some ash is carried hundreds of miles downwind; some blankets the ground over a large area. Nuées ardentes ("glowing clouds") roar down the slopes, devasting the landscape, cooling finally as welded tuff.

Within the Yellowstone caldera at the present time are two small resurgent domes that are bulging up the center of the plateau at a faster rate (one-half inch per year) than the general uplift still going on in the Yellowstone region. What all these phenomena suggest is that Yellowstone is still an area of potentially active volcanism.

The Mantle Plume, or Hot Spot, Theory

What is the source of the heat that drives the Yellowstone thermal and volcanic activity? And why have the eruption centers shifted over time along a trend from southwest to northeast? Geologists theorize that a hot spot in the mantle, comparable to the one that produced the Hawaiian Island chain (chapter 20) may be causing the episodes of volcanic eruptions. The high heat flow may be coming from a plume rising from a hot spot in the mantle and melting its way through the crust. The eruptions in Yellowstone have been more episodic than the Hawaiian ones, which have tended to be more or less continuous. The differences may be due to the fact that the Pacific plume rises under thin oceanic crust, while thick continental crust overlies the Yellowstone plume.

At the present time, the North American tectonic plate is moving about 4.5 centimeters per year in a southwesterly direction over the hot spot. Since the assumption is that the heat plume has remained stationary in relation to the mantle, the centers of volcanic activity shift northeastward as the crust drifts toward the southwest. In general, the ages of the volcanic rocks tend to fit this model, with the older rocks being in the vicinity of Island Park and the younger rocks in the central part of the Yellowstone Plateau. The young basaltic rocks of the Snake River Plains, which largely covered much earlier rhyolitic eruptions, may thus represent partial melting of upper mantle material that occurred after the emptying of rhyolitic magma from an older chamber. Geologists have suggested that if this sequence continues, which seems probable, a fourth caldera may form in the northeast corner of the Park within the next several hundred thousand years.

Hot Springs and Geysers

When it comes to hydrothermal features, no place in the world can compare to Yellowstone, which has about 62 percent of all known geysers, plus thousands of hot springs, mud pots, and fumaroles. Even the surface waters draining away from the geyser basins and hot springs stay warm for some distance; this is vividly suggested by the name, Firehole River, given to the stream that wanders through the main geyser basins of the Park (fig. 24.6).

A *hot spring* is, by definition, any thermal spring whose temperature is above that of the human body. A *geyser* is an intermittent hot spring that regularly or irregularly erupts jets of hot water and steam, the result of ground water being heated enough by hot rock to create steam under conditions preventing free circulation. In

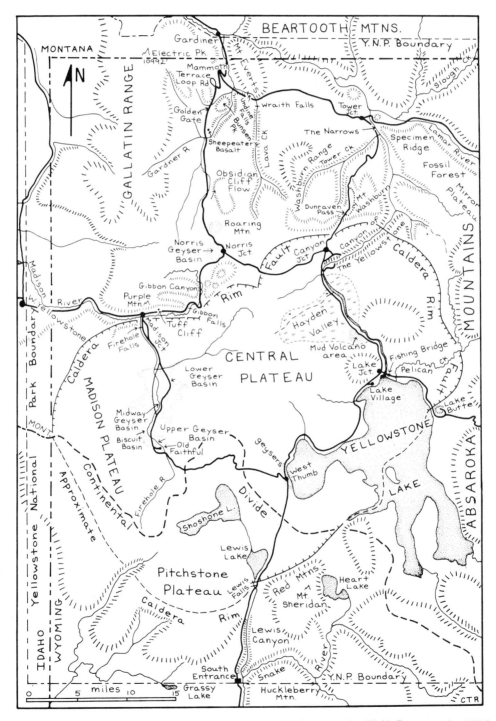

Figure 24.6 Index map locating points of geologic interest in Yellowstone National Park. From *Middle Rockies* *and Yellowstone* by W. H. Parsons. © 1978 by Kendall/Hunt Publishing Company.

other words, confined steam builds up pressure underground until it attains enough force to be forcibly ejected at the surface.

What was so astonishing to the first white visitors to the Yellowstone area was not the hot springs; these thermal features occur in many parts of the world and were well known to the Europeans who settled North America. But none of the early explorers had ever seen a geyser before and had doubtless never heard of Geysir in Iceland, the namesake for all the world's geysers. Geyser fields are rare, and most of them are located in remote areas of the world such as Iceland, New Zealand, and Indonesia, places seldom visited by 19th-century Americans and Europeans. To come upon an area in the wilderness where great jets of steam and scalding water suddenly spouted from the ground, shooting 100—even 200— feet into the air,

only to stop and then, inexplicably, after an interval, begin again, must have been unnerving, as well as awesome to the early hunters and trappers. That people in the more "civilized" parts of the country did not believe the stories that were brought back is not surprising.

The source of the geysers' heat and the origin and operation of their "plumbing systems" have been subjects of scientific speculation for more than a hundred years, but a generally accepted theory of how they work has been developed. For geysers to function, three conditions must be present: (1) a powerful heat source, usually volcanism; (2) a plentiful supply of water; and (3) a fairly watertight plumbing system capable of holding a large volume of water and withstanding the tremendous pressure of steam and superheated water (fig. 24.7).

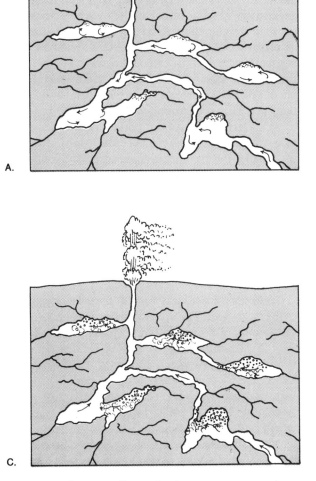

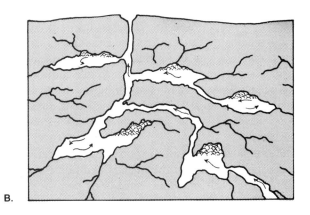

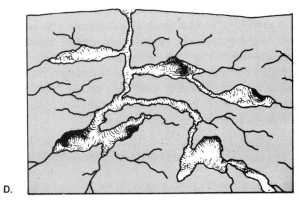

Figure 24.7 Diagrams illustrating how geysers operate. From *The Earth's Dynamic Systems* (3rd ed.) by W. K. Hamblin. © 1982 by Burgess Publishing Company; used by permission. *A.* Ground water circulates through hot rocks, collecting in open spaces and irregular fractures. Steam bubbles tend to clog restricted parts of the geyser plumbing. *B.* Bubbles of steam force water upward until it discharges at the surface vent. *C.* The flow of water out of

the vent reduces the hydrostatic pressure on the water at depth. With the reduction of pressure at the surface, superheated water at depth flashes to steam and expands rapidly. Additional water from side chambers and fractures also flashes into steam, and water and steam are forcibly erupted from the geyser vent. *D.* After the eruption is over, openings in the rock begin filling again. As the water heats up, the eruption cycle is repeated.

Geysers do not have large open underground chambers for accumulating water. The tube or vent at the surface leads down to a fairly shallow plumbing system consisting of intricate conduits that connect small openings in the rock and lenses of porous sediment capable of storing water. Normally, several geysers in a geyser basin are supplied by an interconnecting fracture system. The passageways in the system are lined with *sinter*, which is a chemical sedimentary rock, mainly siliceous, deposited as a hard incrustation by precipitation. The sinter deposits help to keep the system somewhat watertight and also tend to produce constrictions that prevent free circulation of water. Silica-rich rhyolitic rocks are the source of the sinter. Very hot water, deep underground, can dissolve silica from the rocks and carry it in solution toward the surface. Most of the world's geyser fields are, in fact, formed in areas where rhyolite is unusually abundant.

Rainwater (and meltwater from snow) infiltrates the ground, percolating slowly downward through cracks and pores in the rock. Some of the water seeps into the geyser plumbing systems; but much of it goes several thousand feet lower, where it is heated by contact with the hot rocks over the magma chamber to a temperature of more than 400° F (fig. 24.8). Due to hydrostatic pressure, the water cannot boil; but it begins to expand and rise toward the surface. Steam bubbles, rapidly forming in the rising water, heat the cooler water that is filling the lower part of the plumbing system. This filling and mixing process

goes on for awhile until the system is full of water that is ready to boil over. The average boiling point at the surface of the geyser basins in Yellowstone (elevation about 7,500 feet) is about 199° F. (or 93° C).

The steaming water flows out of the hot springs in the geyser basins, because the hot springs have larger openings and less restricted underground circulation. In the geysers, however, the rapidly rising steam bubbles clog up constricted passageways, and the weight of the water above causes a pressure-cooker effect. As the water temperature rises, it reaches a critical point, which varies from geyser to geyser. The upward pressure of the steam bubbles forces out some of the water; this lowers the pressure enough so that the water below flashes to vapor. The vapor then blasts out of the vent with such force that the whole system is nearly emptied out as boiling water and steam shoot into the air. After that, the whole cycle begins again.

The silica that was brought up in solution precipitates both above and below ground as the temperature of the water drops. Sometimes sinter builds up enough to seal off diversionary passages in the plumbing system, thus changing the eruptive behavior of the geysers. Some geysers build large cones or mounds of sinter around the vents. Hot springs may have fantastically shaped mounds of sinter over which the water flows out. Most of the Yellowstone sinter is light colored, but some is tinted various hues.

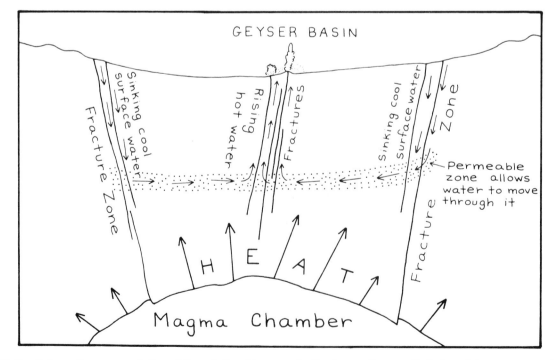

Figure 24.8 Schematic representation of the relationship between a magma chamber at depth and the thermal system below a geyser and hot spring basin. Ground water infiltrating from the surface may percolate downward to a depth of more than a mile before it becomes very hot and then begins to rise. From *Middle Rockies and Yellowstone* by W. H. Parsons. © 1978 by Kendall/Hunt Publishing Company. (Modified from Keefer 1971 and others).

The Geyser Basins. Upper Basin has the highest concentration of geysers and hot springs in the world, the most famous of them being Old Faithful. Its eruptions are watched, photographed, and cheered by the hourly assemblies of visitors who come during the summer season to view the spectacle. Old Faithful overflows for a few minutes at the beginning of the eruption, as if to clear its throat. Then with a roar, the mighty stream shoots up, 100 to 150 feet into the air. After a few minutes, the height of the jet slowly drops and then sinks down. Some puffs of steam drift out, and Old Faithful settles back to fill again.

Some Yellowstone geysers spout higher than Old Faithful and some erupt larger volumes of water, but Old Faithful's predictability is unique, even though its regular timing is not really so precise as many people think. However, the rangers are able to set the "Old Faithful clock" (that gives the time of the next eruption) by timing the length of the previous eruption. If the eruption is short, say a couple of minutes, the interval before the next one will be about about 45 to 56 minutes. If the play of water lasts 5 or 6 minutes, the next eruption will not take place for 75 to 80 minutes. The average time between eruptions is 64.5 minutes. Old Faithful's reliability is probably due to the fact that its hydrothermal system operates in dense rock that is relatively insoluble, even in very hot water. Moreover, Old Faithful's system is apparently unconnected to the underground plumbing of its neighbors. Its operation and volume have changed very little since it was discovered and named by the Washburn-Langford-Doane expedition more than a hundred years ago.

In Lower Basin, which is south of Upper Basin, the groups of geysers and hot springs are more scattered. The Fountain Paint Pots, a cluster of multicolored mud pots, are in this basin amid a large geyser complex. *Mud pots* are hot springs that contain boiling mud and dissolved compounds that the hot water has brought up from the rocks below. Mud pots do not have enough water to flush out sediment as it accumulates (fig. 24.9).

Figure 24.9   Steam rises from Black Dragons Cauldron, a constantly boiling mud pot in Lower Geyser Basin. National Park Service photograph.

Fumaroles, or steam vents, can be readily observed in both Upper and Lower Basin. Some fumaroles become spouters or even small geysers when the water table is high.

Norris Basin, located near the northwest edge of the Yellowstone caldera, is the hottest geyser basin in the Park, presumably because magma is closest to the surface here. Another difference in Norris Basin is that the water is acidic rather than alkaline, as it is in the other geyser basins in the Park. The largest geyser in the world, Steamboat, is located in Norris, but its eruptions are few and far between.

Six other geyser basins are in Yellowstone, several of them in less accessible areas of the Park.

Although most of the Park's hot springs and hot pools are in the geyser basins, another hot spring area, quite different and most spectacular, is in the northwest corner of the Park. This is Mammoth Hot Springs, a mountain of travertine terraces covering nearly a square mile. *Travertine* is a dense, massive or concretionary limestone, formed by rapid chemical precipitation of calcium carbonate. At Mammoth Hot Springs the build-up of travertine deposits by precipitation is much more rapid than the accumulation of sinter deposits around the hot springs and geysers in the geyser basins (fig. 24.10).

The explanation for this is that thick beds of marine limestone underlie the Mammoth area. Since limestone is highly soluble, the rising hot water dissolves and brings to the surface enormous quantities of calcium carbonate—over two tons of it per day, in fact. As the warm, mineralized water flows from the springs and down over the terraces, particles of calcium carbonate precipitate out as growing icicles and spreading sheets of travertine. The millions of algae that live in the warm pools tint the rock shades of brown, red, orange, and green.

Springs and terraces change rapidly. New outlets form as old ones are closed over, and active springs become inactive in a year's time. Terrace Mountain, still under construction by Mammoth Hot Springs, continues to grow, change, and enlarge.

Seismic Activity and the Thermal Features

Earthquakes are apt to cause disruption or changes in the hydrothermal systems of Yellowstone. After the West Yellowstone quake of August 1959 (magnitude of 7.1 on the Richter Scale) many changes in geyser behavior were observed for several years. Immediately after the main shock, hundreds of springs in Yellowstone began to spout jets of water. Clepsydra Geyser erupted wildly for more than three years after the quake and then became dormant. Sapphire Pool, which used to be a placid spring with occasional gentle eruptions 3 to 6 feet high, began erupting violently and voluminously after the quake. Hundreds of tons of siliceous sinter were broken up and washed away. This pattern went on for several years, but gradually Sapphire's eruptive power declined. New fumaroles opened near the Firehole River, one becoming a hot spring after a time. Mud pots spouted, and several springs that had been crystal clear became muddy. Seismic Geyser in Upper Basin is the direct result of a crack in the earth that opened during the 1959 quake.

Figure 24.10   Behind the main terrace deposits at Mammoth Hot Springs, New Highland Spring, which became active in 1952, is depositing terraces of travertine in a forested valley. Trees are killed by the hot water and then buried by the encroaching rock. Photograph by W. H. Parsons.*

*The Parsons photographs that appear in this chapter were previously published in his field guide *Middle Rockies and Yellowstone* (© 1978 by Kendall/Hunt Publishing Company) and are reproduced here with his permission.

Glacial Features

Evidence of early Pleistocene glacial episodes is scattered and incomplete in Yellowstone National Park, because the violent eruptive events, as well as later glaciations, buried or destroyed most of the older features. Most of the extensive erosional and depositional glacial features in the Park and the surrounding area can be identified as belonging to the Pinedale substage of Wisconsin time, which ended in Yellowstone about 8,000 years ago. No glaciers are present on the plateau now, but at higher elevations in the encircling mountains, small glaciers assigned to Neoglacial time (post-Wisconsin) continue to advance and retreat.

Winters on the high Yellowstone Plateau are severe, and snowfalls are heavy. In the colder and wetter Pleistocene climates, valley glaciers started high in the mountains. As they grew and flowed down to lower elevations, they coalesced to form piedmont glaciers. The Lamar Glacier, which originated in the Beartooth Mountains, flowed down from the northeast; the Upper Yellowstone glacier, advancing from the southeast, was fed by valley glaciers in the southern part of the Absaroka Range. Small icecaps developed during the coldest and most prolonged episodes, and covered the plateau and the foothills. Mount Washburn and other high peaks stuck up above the ice as nunataks most of the time but may have been covered during glacial maxima. At its greatest extent, the Pinedale icecap flowed radially outward from about the present location of Yellowstone Lake. Eastward extensions of glacial ice pushed over divides in the Absaroka Range. To the south and southwest, ice flowed over the Continental Divide in the Two Ocean Plateau area and down into Jackson Hole, leaving large end moraines (chapter 25).

Cirques and other erosional features in the high mountains were sculptured by glacial ice. Other evidences of glaciation are erratics, patches of till, and outwash features. A "calling card" of Pinedale glaciation, that gives some idea of its transporting power, is a large glacial erratic of Precambrian gneiss near Inspiration Point, a short distance from the western rim of the Grand Canyon of the Yellowstone River. Weighing more than 500 tons, the erratic measures about 24 by 20 by 18 feet. To get from its source to where it is now, the erratic must have been transported at least 15 miles.

The Grand Canyon of the Yellowstone was not occupied by a valley glacier; the icecap was so thick it simply filled the valley and flowed over the top.

The Grand Canyon of the Yellowstone River

The Yellowstone River, which rises in the Absaroka Range southeast of the Park, carries most of the Park runoff from the eastern side of the Continental Divide. The river flows into the southern end of Yellowstone Lake, out through the northern end, and then northward across the Park into Montana. A principal tributary is the Lamar River that drains the northeastern part of the Park.

The remarkable feature that has made the river famous is the Grand Canyon of the Yellowstone. This narrow, steep-walled canyon evolved through several stages of cutting and filling during the past half-million years. An earlier Yellowstone River valley drained into the Lamar River and then went north. Ash and other volcanic debris filled this valley. Canyon cutting went on between several more episodes of filling by volcanic material. Then advances of Bull Lake glaciers (early Wisconsin time) blocked the Lamar-Yellowstone drainage system. Water backed up, creating a long lake in the canyon, and layers of lake-bed sediment accumulated. After glacial retreat, rapid eriosion cleared the canyon, but the Pinedale icecap overrode the canyon and again blocked the system. About 12,000 years ago the last ice in the canyon melted away. Since that time, the Yellowstone River has eroded from its canyon most of the volcanic debris and the glacial and lake-bed sediment.

The canyon walls have been cut mainly in rhyolitic lava and tuff. Intense chemical weathering from hot ground water and hot springs has changed the color of the rock from an unweathered grayish black to a startling array of reds, browns, and yellows—producing the strange "yellow rocks" for which Yellowstone was named. Near the bottom of the canyon, hot springs bubble out and fumaroles emit steam, continuing the process of hydrothermal alteration, softening and dissolving the rock and thus changing its composition and physical properties.

At the head of the deepest and most spectacular part of the gorge is Lower Falls, 308 feet high (fig. 24.11). Here the river flows over the edge of an unaltered, resistant rhyolite flow that erodes more slowly than the tuff and softened lava below. Above Lower Falls is Upper Falls, which is 109 feet high. The Grand Canyon of the Yellowstone ends near Tower Falls, which flows over the rim of a hanging valley and drops the water of Tower Creek 132 feet into the Yellowstone River below.

Columnar Jointing

Good examples of columnar jointing can be seen in some of the lava flows in the Park. Columnar jointing is clearly visible in basalt flows that crop out in the walls of the Grand Canyon of the Yellowstone and at Sheepeaters Cliff along the Gardner River. The characteristic parallel columns, polygonal in cross section, form by contraction as a lava flow cools (fig. 24.12).

Obsidian

The Yellowstone obsidian, which was of great value to the Indians, is a dark volcanic glass. Obsidian Cliff, along the road from Norris to Mammoth Hot Springs, is presumed

Figure 24.11   Lower Falls and the Grand Canyon of the Yellowstone River. The river is cutting down through the young lavas and pyroclastics that make up the central part of the Yellowstone plateau. Lower Falls, 308 feet high, is formed where the river tumbles over a ledge of unaltered, resistant rhyolite. Hot waters, rising through fractures in the rock, have accelerated processes of chemical weathering in the canyon walls, softening the rock and changing its color to shades of white, yellow, and red. Photograph by W. H. Parsons.

Figure 24.12   Prominent columnar jointing is displayed in the basalt flow that forms Sheepeaters Cliff in northern Yellowstone National Park near the Gardner River. As the lava cooled and contracted, intersecting joints produced the hexagonal-sided columns. Photograph by W. H. Parsons.

to be the "Glass Mountain" that Jim Bridger talked about. Obsidian, or natural glass, forms when lavas (especially felsic lavas) cool and harden before the mineral grains have had time to crystallize.

Buried Forests

At Specimen Ridge, on the south side of the Lamar River valley in the northeast corner of the Park, are petrified tree trunks (some with even the root systems preserved)

sticking up through volcanic material in lifelike positions. Erosion of volcanic material on the slopes has exposed these fossil forests, buried long ago by Eocene ash flows and lahars (mudflows). The trees were subsequently lithified as silica-enriched ground water infiltrated and preserved the woody structure (fig. 24.13). Another, smaller buried forest area is at the tip of the Gallatin Range where it extends into the northwest corner of Yellowstone.

Figure 24.13 Petrified stumps, composed of chalcedony and opal, stick up through a blanket of weathering volcanic breccia on Specimen Ridge. These fossil trees, which grew in Eocene time, were buried by an eruption and subsequently petrified by silica-enriched, percolating ground water. The hard silica that filled the woody cells is more resistant to weathering and erosion than the volcanic breccia that originally covered them completely. Photograph by W. H. Parsons.

At Specimen Ridge, 20 or more successive forests (the exact number is uncertain) were buried, one above the other from bottom to top, during Eocene time. Perhaps as much as 2,000 years passed between burials, as again and again violent eruptions related to the Absaroka volcanic field sent flows of ash and mud rolling down slopes, smashing and devastating the forests.

The tree varieties most commonly identified here are sycamore, walnut, magnolia, chestnut, oak, redwood, maple, and dogwood. Apparently the climate throughout this time remained mild and humid; it was probably similar to that of southeastern United States today.

Geologic History

1. Precambrian erosion, sedimentation, and tectonism.

In the northern part of the Park, mainly in the Gallatin and Beartooth Ranges, are outcrops of contorted metamorphic rock (schists and gneisses), some of which have been dated as being over 2.7 billion years old. Representing fragments of the geologic history of Precambrian time, the constituents of these rocks may have been moved not only up and down in the earth's crust many times, but they probably have also drifted on tectonic plates thousands of miles across the face of the earth. The schists were probably volcanics, sandstones, and shales deposited in downwarps in primitive crust. The gneisses may have been derived from granitic intrusions in ancient mobile belts. Presumably, the rock materials have been "recycled" several times by the processes of erosion, deposition, and tectonism.

2. Paleozoic and Mesozoic deposition.

Over 40 separate rock formations of Paleozoic and Mesozoic age have been identified in Yellowstone National Park. They crop out across the northern section of the Park in the Gallatin and Beartooth Ranges, as well as across the southern margin, where the rock units are found in the Two Ocean Plateau and in the Pitchstone Plateau. Many of the Mesozoic and Paleozoic beds were stripped off during uplift or lie buried under many feet of volcanic rock. Little correlation exists between the northern and southern sedimentary rock units in the Park region.

Predominantly marine, the rocks are an assortment of conglomerates, sandstones, shales, and limestones. Approximately 3,000 feet of Paleozoic rocks and 6,000 feet of Mesozoic rocks were deposited, for the most part in shallow seas that repeatedly advanced and withdrew over what is now western North America. Some beds of terrestrial deposits on beaches and flood plains are interspersed in the Mesozoic formations, but for several hundred million years, this was a relatively quiet region of low elevation and low relief.

3. The Laramide orogeny in late Cretaceous and early Tertiary time; faulting and erosion.

The compression and folding of rock layers, the mountain-building, and rapid increase in sedimentation that began in late Cretaceous and Paleocene time was the result of plate convergence. The Pacific plate (made up of oceanic crust) was subducted, underthrusting the North American plate and bringing about "crustal shortening." The great Rocky Mountain chain was the result of these tectonic forces which continued for some 30 million years.

In south central Yellowstone, the Paleozoic and Mesozoic sedimentary beds were tightly folded into anticlines (trending northwest-southeast), separated by synclines and faults. Movement along one reverse fault in this area was locally more than 10,000 feet. Reverse faulting of this magnitude also occurred in the northern part of Yellowstone. Fault blocks, created by the tremendous stresses, were raised, dropped, or tilted differentially.

An immediate result of the compression and uplift of crustal blocks was intensified erosion. Sediment stripped from the heights was deposited in nearby downwarped areas and basins. The Harebell Formation, for example, in south central Yellowstone, of latest Cretaceous age, received more than 8,000 feet of sediment.

4. Eocene andesitic volcanism; the Absaroka volcanic field.

Tremendous outpourings and explosions of andesitic lava and pyroclastics built up the Absaroka and Washburn Ranges, as well as part of the Gallatins, during the Eocene Epoch. Most of the Yellowstone area was at least partially covered by these rocks. The eruptions of viscous lava, ash, and breccia came from large vents, one of which was in Mount Washburn. The fossil forests in the Park were buried by these eruptions (fig. 24.13).

The composite thickness of the Absaroka Volcanic Supergroup is estimated to be over 12,000 feet. Intrusive equivalents (i.e., having the same age and composition) of the Absaroka volcanics were emplaced as stocks, dikes, and sills. Interbedded with the volcanics are sedimentary beds of reworked ash and volcanic debris. Many deposits show the effects of extensive mass wasting—landsliding, slump, and the like.

5. Oligocene and Miocene erosion.

Large volumes of volcanic sediment were eroded and transported out of the Yellowstone area during this time. No rock units of Oligocene or Miocene age have been found in the Park. If any were deposited, they have since been eroded or buried.

6. Late Tertiary (Pliocene) uplift, faulting, and erosion.

Regional uplift raised Yellowstone and the Middle Rockies several thousand feet. Mountain blocks were further elevated by faulting, while downthrown blocks became basins. The Gallatin Range was uplifted to its present height. To the south, the Teton Range was raised and Jackson Hole depressed, the offset on this steeply dipping normal fault being about 30,000 feet. With these tectonic events, the landscape of the Yellowstone region began to assume its present framework.

Uplift rejuvenated the streams so that they began eroding rapidly, cutting deep canyons and removing more of the Absaroka andesitic volcanics. The remnants exposed in the Park (such as the Mount Washburn eruption center) represent only a small amount of the great volume of Absaroka rocks that buried most of the Yellowstone surface in Eocene time. About 40 million years were to elapse before the next period of volcanism, and that would be quite unrelated to the Absaroka episodes.

7. The Pleistocene rhyolitic eruptions; volcanic cycles and caldera collapse.

The largely rhyolitic rocks that built up the Yellowstone Plateau during the three Pleistocene volcanic cycles are designated the Yellowstone Group (fig. 24.14). The first volcanic cycle (2.5 to 1.5 million years ago) produced a large dome, quantities of welded tuff, and eventually, a collapsed caldera, approximately 30 by 50 miles in size and several thousand feet deep. The Huckleberry Ridge

Tuff (Yellowstone Group) is associated with the explosive eruptions that occurred as the caldera collapsed (fig. 24.15).

The second volcanic cycle, similar in pattern but shorter in duration, was not so widespread or productive of lava as the first cycle. The second cycle reached its climax about 1.2 million years ago with the explosive eruptions that formed the Mesa Falls Tuff, and again, the caldera collapsed.

About 1.2 million years ago, the third, and most recent, volcanic cycle began with a definite shift in activity from the Island Park area to the Yellowstone Plateau. Ring fractures opened around the rising bulge, and through these fissures, gas, viscous rhyolitic lava, and ash were ejected until the roof rocks over the magma chamber became seriously weakened. This led to the climactic eruptions of about 600,000 years ago, during which more than 1,000 cubic kilometers of magma were expelled. Clouds of fiery gas and ash shot into the air and foaming lava poured from the vents, devastating the surroundings. Ash borne on prevailing winds blanketed the midcontinent area. Collapse of the chamber roof produced a caldera 45 by 25 miles in extent and a half-mile deep.

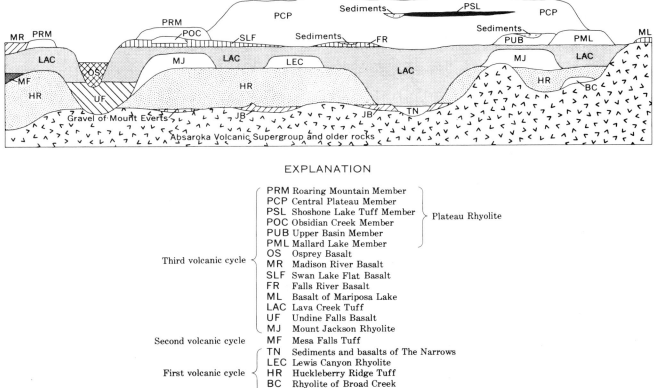

EXPLANATION

PRM Roaring Mountain Member
PCP Central Plateau Member
PSL Shoshone Lake Tuff Member
POC Obsidian Creek Member   } Plateau Rhyolite
PUB Upper Basin Member
PML Mallard Lake Member
OS   Osprey Basalt
MR   Madison River Basalt
Third volcanic cycle {
SLF Swan Lake Flat Basalt
FR   Falls River Basalt
ML   Basalt of Mariposa Lake
LAC Lava Creek Tuff
UF   Undine Falls Basalt
MJ   Mount Jackson Rhyolite
Second volcanic cycle MF   Mesa Falls Tuff
TN   Sediments and basalts of The Narrows
LEC Lewis Canyon Rhyolite
First volcanic cycle { HR   Huckleberry Ridge Tuff
BC   Rhyolite of Broad Creek
JB   Junction Butte Basalt

Figure 24.14 The names and stratigraphic relationships of volcanic rock units recognized in the caldera area. From

Christiansen and Blank 1972, U.S. Geological Survey Professional Paper 729-B, p. B7.

Figure 24.15   Cliffs of welded tuff (Huckleberry Ridge Tuff) at Golden Gate in northwestern Yellowstone National Park. The dense, resistant welded tuff formed from a great ash flow during the first volcanic cycle about 2 million years ago. Photograph by W. H. Parsons.

Many of the features of the Yellowstone landscape were formed at this time: the welded tuff deposits of the Lava Creek Tuff, the depression now holding Yellowstone Lake, and the blurred but still visible caldera rim are perhaps the most notable.

A resurgence of fairly strong volcanic activity took place about 160,000 years ago in the depression now occupied by the West Thumb of Yellowstone Lake. After dome expansion, explosive eruptions, and collapse, a small caldera, about the size of Crater Lake in Oregon (chapter 19), was left.

For thousands of years after the collapse of the main caldera, viscous lava periodically poured out over the caldera floor, sometimes in flows 1,000 feet thick. Some of these eruptions were as recent as 60,000 years ago. Most flows stayed within the caldera, but some of them spilled over the rim. Lava occasionally came up through the ring fissures. A prominent, but small, flow of this last type is the one exposed at Obsidian Cliff. Minor basalt flows were extruded during the sequence.

8. Glaciation and canyon-cutting.

The Yellowstone Plateau was glaciated at least three times during the Pleistocene. A pre-Wisconsin glaciation occurred between 300,000 and 180,000 years ago. Bull Lake glaciation (between 125,000 and 45,000 years ago)

and Pinedale glaciation (between 25,000 and 8,000 years ago) were both Wisconsin in age. The pre-Bull Lake and Bull Lake deposits are scattered and discontinuous, having been buried or obliterated by volcanic activity and Pinedale glaciation. Icecaps that formed during these glaciations covered perhaps 90 percent of the Park area. At various times glacial ice blocked major river valleys, diverted streams, and dammed Yellowstone Lake. During Pinedale glaciation, Yellowstone Lake was approximately 140 feet higher than it is now. In some places in the Park, glacial drift and alluvium are interbedded with volcanic debris, indicating that eruptions occurred when glaciers were in existence.

9. Post-Pleistocene uplift and erosion; Neoglaciation; geothermal activity.

The Yellowstone Plateau and the caldera have undergone uplifting in Holocene time, the latter at a more rapid rate. This development appears to be continuing at the present time. Two small resurgent domes rising in the caldera may presage a resumption of volcanic activity, possibly within 100,000 years.

After the disappearance of Pinedale glaciers during the postglacial optimum (between 6000 and 2000 B.C.),

Figure 24.16   View across The Narrows of the Yellowstone River near Tower Falls. Two massive, resistant lava flows with columnar jointing are exposed in the canyon wall. Between the two flows are a number of layers of stream gravel deposits. The upper flow is overlain by a blanket of glacial till. Photograph by W. H. Parsons.

the Yellowstone River completed clearing out and re-eroding its valley. Erosion continues in the Grand Canyon of the Yellowstone at the present time by the processes of weathering, mass wasting, and stream action (fig. 24.16).

In Neoglacial time, brief intervals of cooler climate allowed accumulation of perennial snowfields in the high mountains and the beginning of small valley glaciers. Since the end of the "Little Ice Age" about 200 years ago, the remaining glaciers have shown evidence of rapid retreat, although this trend may be slowing at present.

Geysers and hot springs have probably been active on the Yellowstone Plateau ever since the end of Pinedale glaciation, and they may have been intermittently active during interglacial periods. Location of the hydrothermal features is believed to be related to the faults and fractures in the hot rock over the magma chamber. Changes in the behavior of the geysers and springs after earthquakes seem to indicate that a direct relationship does exist between fault movements and the passageways by which hot water and steam reach the surface.

Rates of heat flow on the Yellowstone Plateau, the frequency of shallow earthquakes, and the inferred presence of a body of hot, low-density material below the surface all point toward the possibility of eventual new eruptions. Investigators have noted that a major caldera has formed in the Yellowstone region about every 600,000 to 900,000 years in the past two million years. Inasmuch as the present Yellowstone caldera collapsed about 600,000 years ago, we can speculate that the next one may be in preparation!

Geologic Maps and Cross Sections

American Association of Petroleum Geologists. 1972. *Geological highway map no. 5, northern Rocky Mountain region.* Tulsa, Oklahoma: American Association of Petroleum Geologists.

Keefer, W. R. 1971. (See below)

U.S. Geological Survey. 1972. *Geologic map of Yellowstone National Park.* Miscellaneous Geological Investigations, Map I-711.

U.S. Geological Survey. 1972. *Surficial geologic map of Yellowstone National Park.* Miscellaneous Geological Investigations, Map I-710.

Bibliography

Baker, R. D. 1976. *Late Quaternary vegetation history of the Yellowstone Lake basin, Wyoming.* U.S. Geological Survey Professional Paper 729E.

Bartlett, R. D. 1962. *Great surveys of the American West.* Norman, Oklahoma: University of Oklahoma Press.

Boyd, R. D. 1961. Welded tuffs and flows in the rhyolite plateau of Yellowstone Park, Wyoming. *Geological Society of America bulletin* 72:387–436.

Bryan, T. S. 1979. *The geysers of Yellowstone.* Boulder, Colorado: Colorado Associated University Press.

Christiansen, R. L., and Blank, H. R. 1972. *Volcanic stratigraphy of the Quaternary rhyolite plateau in Yellowstone National Park.* U.S. Geological Survey Professional Paper 729-B.

Crandall, H. 1977. *Yellowstone, the story behind the scenery.* Las Vegas, Nevada: KC Publications.

Hammond, A. L. 1980. The Yellowstone bulge, *Science 80* 1:68–73.

Howard, A. D. 1937. *History of the Grand Canyon of the Yellowstone.* Geological Society of America Special Paper 6.

Keefer, W. R. 1971. *The geologic story of Yellowstone National Park.* U.S. Geological Survey Bulletin 1347.

Love, J. D., and Keefer, W. R. 1975. *Geology of sedimentary rocks in southern Yellowstone National Park.* U.S. Geological Survey Professional Paper 729D.

Marler, G. D. 1964. *Studies of geysers and hot springs along Firehole River, Yellowstone National Park.* Yellowstone Library and Museum Association.

Parsons, W. H. 1978. *Middle Rockies and Yellowstone,* K/H Geology Field Guide Series. Dubuque, Iowa: Kendall/Hunt Publishing Company.

Pierce, K. L. 1979. *History and dynamics of glaciation in the northern Yellowstone Park area.* U.S. Geological Survey Professional Paper 729-F.

Ruppel, E. T. 1972. *Geology of the pre-Tertiary rocks in the northern part of Yellowstone National Park.* U.S. Geological Survey Professional Paper 729-A.

Smedes, H. W., and Prostka, H. J. 1972. *Stratigraphic framework of the Absaroka Volcanic Supergroup in the Yellowstone National Park region.* U.S. Geological Survey Professional Paper 729-C.

Smith, R. B., and Christiansen, R. L. 1980. Yellowstone Park as a window on the earth's interior. *Scientific American* 242:104–117.

Thornbury, W. D. 1965. *Regional geomorphology of the United States.* New York: John Wiley and Son.

Walker, B. S. 1973. *The Great Divide,* American Wilderness Series. Alexandria, Virginia: Time-Life Books.

White, D. E.; Fournier, R. O.; Muttler, L. J. P.; and Truesdell, A. H. 1975. Physical results of research drilling in thermal areas of Yellowstone National Park, Wyoming. U.S. Geological Survey Professional Paper 892.

Note: Bryan (1979), Keefer (1971), and Parsons (1978) are especially useful sources for Park visitors who are interested in the geology of Yellowstone to take along when traveling in the Yellowstone region.

The Yellowstone Institute, formed in 1976 in response to public interest in the Yellowstone ecosystem, offers summer seminars, one to six days in length, on the natural history of the Park. College credit is available for courses in wildlife management, Park geology, botany, and so on. For information, write Yellowstone Institute, Box 515, Yellowstone National Park, Wyoming 82190.

Address

Yellowstone National Park
P.O. Box 168
Yellowstone National Park, Wyoming 82190

Part **IV**
Mountain Building and Uplift

© by G. Shimer

Complex rocks and structures underlie a mountain landscape of grandeur and remoteness in the high ranges of the Southern Appalachians. Great Smoky Mountains National Park, Tennessee and North Carolina.

Internal geologic processes on a grand scale are chiefly responsible for the diverse landscapes of the national parks included in Part IV. Although streams have modified the landforms, as have glaciers in three parks, and waves in three others, the controls have been tectonic. Most of the activity has occurred in mobile belts along plate boundaries. Mountains can be built both by compressive forces that shorten and crumple the crust and by tensional forces that pull the crust apart, allowing some blocks to rise and others to sink. Bodies of molten rock (magma) have intruded the mountain roots and then cooled at depth; rocks have been altered by heat and pressure; and lavas have poured out from volcanoes and from submarine vents. Earthquakes in some park regions warn us that tectonic activity is ongoing; in other parks, the processes of erosion have become dominant in landscape development as tectonism has subsided.

25

Grand Teton National Park

by Sherwood D. Tuttle
University of Iowa

Location: Northwest Wyoming
Area: 310,515.93 acres; 485.18 square miles
Established: February 26, 1929

Figure 25.1 A high altitude view, looking north, at Grand Teton (foreground), the tallest peak (13,770 feet) in the Teton Range. Mount Owen, the peak on the right, is 12,928 feet in elevation. The declivity behind the peaks is Cascade Canyon, a glacial trough. The steep slopes, jagged surfaces, and pointed peaks are typical of alpine topography. National Park Service photograph.

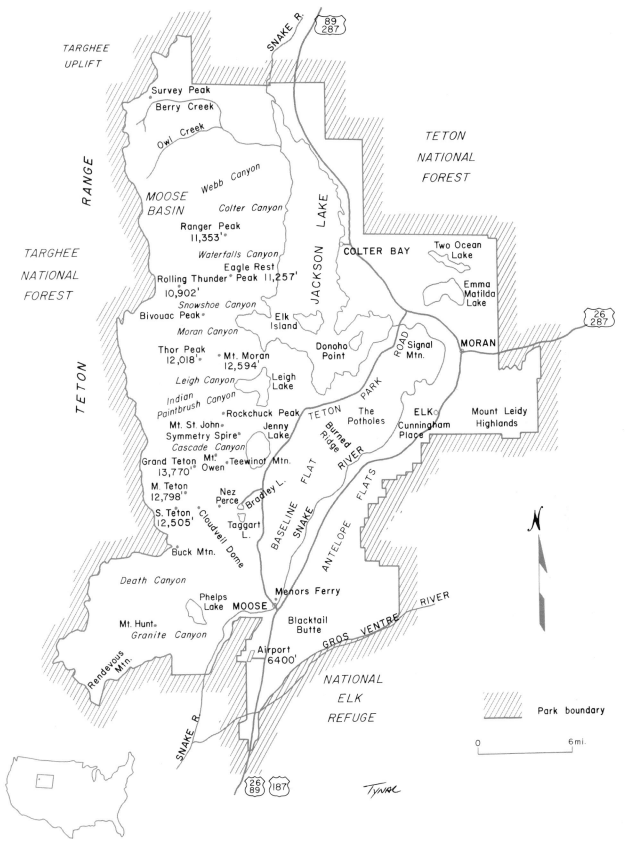

Figure 25.2   Grand Teton National Park, Wyoming.

Table 25.1. Generalized Geologic Column, Grand Teton National Park

Time Units			Rock Units	Geologic Events
Era	Period	Epoch		
Cenozoic	Quaternary	Holocene	Alluvium, talus, landslide deposits; minor glacial deposits	Weathering, erosion, mass wasting, Neoglaciation, Continued faulting
		Pleistocene	Moraine and outwash deposits	Pinedale glaciation (morainal lakes, outwash plain, terrace cutting, etc.) Bull Lake glaciation
			Loess deposits (interglacial)	
			Outwash deposits	Buffalo glaciation, extensive glacial erosion and deposition; Preglacial lakes in Jackson Hole; Volcanic eruptions in southern Jackson Hole
	Tertiary — Neogene	Pliocene	Bivouac Formation	Welded tuff eruptions from Yellowstone area
			Teewinot Formation	Teewinot Lake on subsiding Jackson Hole block; lakebed deposits, ashfalls
			Conglomerates, tuff	Development of Teton fault system; uplift of Teton Range; increasing erosion
		Miocene	Volcanic conglomerates, tuff, sandstones, etc.	Volcanic eruptions in Jackson Hole; basin filling
		Oligocene		
	Tertiary — Paleogene	Eocene	Volcanic tuffs, breccias, conglomerates, sandstones, claystone, shale, coalbeds	Extensive volcanism in the Absarokas
		Paleocene	Pinyon Conglomerate	Buck Mountain fault, uplift; Laramide orogeny; Sediment accumulation
Mesozoic	Cretaceous		Harebell Formation	Extensive nonmarine sedimentation
			About 15 nonmarine and marine formations in Jackson Hole region (for details, see p. 80, Love and Reed 1971)	Upwarps and subsidence, intermittent shallow sea; marine, transitional, and nonmarine sedimentation
	Jurassic			
	Triassic			
Paleozoic			About 9 major formations, limestones, dolomites, shales, sandstones (for details, see p. 68, Love and Reed 1971)	Intermittent deposition in shallow seas between episodes of upwarp and erosion
			/////	Uplift and erosion
Precambrian time			Diabase dikes, Light-colored granites, pegmatites, Dark granites, Gneisses	Intrusions, Orogenies, Metamorphism
				Accumulation of marine and volcanic sediment in crustal trough

Source: modified after Love and Reed 1971, Behrendt et al. 1968

Local History

To the early inhabitants of western North America, Jackson Hole in the Grand Teton region was as attractive a place to come to as it is to people of today, although, to be sure, prehistoric Indians had different motivations for their travels than we do. We know that groups of ancient Indians often camped at Jackson Hole because spearpoints, identified by archeologists as of a type in use 8,000 to 12,000 years ago, were discovered at an old campsite beside Jackson Lake. In their roamings over the land, following the seasons and the migrations of animals, these hunters and gatherers must have looked for the distinctive skyline of the Teton Mountains, just as today we scan the horizon, waiting for those peaks to come into view, whether we travel by plane or by car. The Grand Tetons are like a beacon; and once seen, their outline against the sky is never forgotten. And who has stood on the floor of Jackson Hole, face tilted upward, and regarded the formidable east face of the Teton Range without a sense of awe?

For the primitive peoples, reaching Jackson Hole may have meant survival. Here was a well watered place where food was at hand, a sheltered valley where the clan was protected from sudden spring blizzards and the searing droughts of summer. The later Indians, those tribes who occupied the western regions when Europeans and Americans first came, regularly camped in the valley. It was on several routes of Indian travel and trade. They could follow the Snake River north into Yellowstone for obsidian, or go south to the Colorado Plateau or east to the Great Plains. Teton Pass, at the southern end of the range, was a shortcut to the Pacific Northwest. The Blackfeet, Crow, Flathead, Gros Ventre, Nez Perce, Bannock, Shoshone, and the Utes knew the area well and often summered in Jackson Hole, sometimes peacefully and at other times warring with each other. They tended to come into the valley from either the north or the south, leaving a relatively unoccupied zone in the middle. Perhaps that was their way of avoiding conflict. The Shoshone Sheepeaters from Yellowstone kept up the custom of summer camping in the valley the longest, until nearly the end of the 19th century. Apparently they were able to do this because they lived in small groups, stuck to the old ways, and avoided conflicts with whites.

For 18th-century trappers and traders, the valley and the lake, ringed by mountains, provided a safe place where they could "hole up" for the winter. One of the trappers was David E. Jackson, and it was for him that Jackson Hole and Jackson Lake were named. The early French voyageurs gave the name *Les Trois Tetons* to the three peaks now called Grand Teton, Middle Teton, and South Teton. The common translation used by the English and American mountain men was "The Three Paps." These easily distinguished peaks, which can be seen from both western and eastern approaches, were the landmarks that guided the trappers, who followed trails earlier established by Indians. At the height of the fur trade in the 1830s, Jackson Hole was a crossroads through which the major fur expeditions passed (DeVoto 1947). By cutting through Teton Pass, trappers could reach mountain streams rich in beavers—those draining the western slopes of the Tetons—as well as the wild rivers to the north.

Around 1840, the fur trade ended abruptly as men of fashion began wearing tall silk hats instead of stovepipe hats made of beaver pelts. Except for a few prospectors drifting through, who panned for gold but found nothing worth staying for, Jackson Hole became an isolated valley; it was nearly forgotten for some twenty years.

F. V. Hayden, the geologist who led the first government-sponsored, surveying and scientific expedition into Yellowstone in the summer of 1871, decided to also map the Teton valley that he had visited in 1860 with the Raynolds expedition. Therefore, in the spring of 1872, Hayden split his group into two parties. He led one team into Yellowstone from the north. The geologist James Stevenson led the second party from Utah north to Fort Hall in Idaho and then east across Teton Pass into Jackson Hole. The two parties met late in the summer at Firehole Geyser Basin in Yellowstone. With Stevenson was N. P. Langford, who had been appointed superintendent of the newly established Yellowstone National Park, the photographer William Henry Jackson, the artist-geologist William Henry Holmes, several other geologists and topographers, packers, a cook, and a mountain man (called "Beaver Dick") as guide. As they approached the jagged pinnacles, looming up against the sky, the scientists vowed to climb the highest peak, Grand Teton (elevation, 13,770 feet). Beaver Dick swore it never had been done and never could be done. (Even today, the climb should be attempted only by experienced mountaineers and then only with Park Service permission.) Nevertheless, Stevenson and Langford did get to the top of a summit they thought was Grand Teton. Their claim was discredited by a later surveyor, William O. Owen, who made the first officially affirmed ascent of Grand Teton in 1898.

Among the more lasting accomplishments of that summer and fall of 1872 were the remarkable photographs of Teton country taken by William Henry Jackson; the superb drawings of William Henry Holmes, including one of his famous panoramas showing the Teton Range; topographic maps; scientific reports; fossil and rock collections; and plant and animal specimens, including birds and insects. Among the features the explorers named are Jenny Lake and Leigh Lake, named after Beaver Dick and his Indian wife.

Two more expeditions in 1877 and 1878 completed the mapping of most of Wyoming, Idaho, and Montana and furnished material for the voluminous Hayden Survey reports covering those years. European scholars, as well as distinguished American scientists, contributed their insights to this record of the natural history of a region. By the 1880s this part of the American West was no longer unknown territory to the civilized world.

When the Jackson Hole area was opened to homesteaders after the completion of the surveys, not very many settlers joined John Holland, the first homesteader, who came in 1884. The growing season was too short for most crops, and the valley was isolated for weeks each winter when snow blocked the passes. The population remained sparse well into the 20th century, even after Teton County was organized in 1921.

Meanwhile, a movement had begun to protect the timber south of Yellowstone, and in 1929, part of Teton National Forest was designated Grand Teton National Park. This act by Congress protected the Teton Range and some of the lakes, but most of the floor of Jackson Hole, including Jackson Lake, remained in private hands. Sheep ranchers and cattlemen strongly opposed proposals to incorporate more acreage into the national park.

Nevertheless, many interested citizens throughout the country realized that historically, culturally, and geologically, the Teton Range and the valley floor below the range were two halves of a natural environment that needed to be protected and managed as an integrated unit. John D. Rockefeller, Jr., believed in this concept and had the means to act on his belief. He formed a land company in 1927 and began buying up tracts of land in the Jackson Hole vicinity, eventually accumulating 33,562 acres, to be held until such time as it could be administered by the National Park Service. President Franklin Roosevelt took the next step in 1943 by proclaiming an area of about 223,000 acres (much of which was already public land) as Jackson Hole National Monument. The Rockefeller land was accepted as an addition to the monument in 1949. Finally, after considerable litigation and political maneuvering, Congress established the enlarged area as Grand Teton National Park in 1950, with part of the former Jackson Hole National Monument being absorbed by the National Elk Refuge south of the Park. In recognition of Rockefeller's vision and generosity, the scenic highway that goes through Grand Teton National Park, from its south entrance into Yellowstone National Park, just beyond the northern boundary, was named John D. Rockefeller Memorial Parkway.

Geologic Features

Photographs cannot capture the grand scale of the landforms that the human eye sees in Grand Teton National Park. When you drive along the highway on the floor of Jackson Hole, the massive, dark-gray mountain front looms up, almost within reach. Snow clings to the scarred, roughhewn, lofty slopes; and small glaciers can be seen high up in narrow, steep-walled valleys. In the canyons incising the lower slopes, waterfalls drop in silvery white lines, easily visible from many points along the highway. But the overriding question is, what tremendous force split the ancient crust, raising the mountain range and dropping the valley floor?

Grand Teton National Park includes the major portion of two great landforms: (1) the Teton Range, an elongate, upfaulted block tilted to the west and about 45 miles long; and (2) Jackson Hole, a narrow, intermontane basin, about the same length and 6 to 12 miles wide. Its remarkably flat floor varies in altitude from 6,000 to 7,000 feet. At least seven of the Teton peaks exceed 12,000 feet in elevation, and one, Grand Teton, is over 13,000 feet. The precipitous drop that we see today, from peaks to valley floor, might be even greater were it not for the intense erosion that has crenulated and dissected the mountain range and the thousands of feet of deposited sediment that cover the bedrock of the valley floor (fig. 25.3).

The Paleozoic and Mesozoic sedimentary rocks that once covered the mountain block have been stripped from the top of the range, exposing Precambrian crystalline rock. Remnants of sedimentary rock flank the western slopes of the range. Similar units of sedimentary rock underlie Jackson Hole, buried by alluvial and glacial deposits. In geomorphic terms, the mountain features are erosional landforms and the valley features are depositional landforms.

The Crystalline Rocks

Layered gneisses, granitic gneisses, granites, pegmatites, and some diabase dikes make up the Precambrian rocks that dominate the Teton Range. Strontium-rubidium dating places the age of the granites at about 2.5 billion years. Most of the other Precambrian rocks in the Park are older than the granites, except for the black diabase dikes, which have been dated as 1.3 billion years old.

The banded nature of some of the gneisses suggests that they were formed from sedimentary and volcanic rocks that accumulated in a primitive mobile belt. The massive gneisses were once granitic magma that was intruded and later metamorphosed.

Extending vertically up the face of Mount Moran from base to summit is a black band which is the exposed edge of one of the diabase dikes that cut through the older rocks. The Mount Moran dike is about 150 feet thick and can be traced westward for about 7 miles. *Diabase* is a dense, black, igneous rock studded with light-colored plagioclase crystals, resembling small laths, within the pyroxene crystals.

Figure 25.3   Looking southwest across the Snake River valley in Jackson Hole toward the eroded fault scarp of the Teton Range. The north-south-trending Teton fault system lies along the base of the mountain front. The irregular skyline, with its glacially carved horns, cols, and canyons, is characteristic of alpine topography. The Snake River (flowing away from the viewer) has cut several levels of terraces in the thick glacial outwash deposits that cover most of Jackson Hole. National Park Service photograph.

Some of the gneisses contain dark crystals of garnet set in white "halos," producing the appearance of eyes peering out from the rockface. Geologists call these rocks "bright-eyed gneiss." Pods of heavy dark green or black serpentine (soapstone) are also scattered throughout the gneisses. Indians carved bowls from this smooth, relatively soft, rock. Pebbles of serpentine found in stream gravels *outside the Park* can be polished and used as gems. This stone, called "Teton jade," is softer and less lustrous than true jade.

Paleozoic Sedimentary Rocks

On the western flanks of the Teton Range and at the northern and southern ends of the range, gently dipping sequences of Paleozoic marine rocks top the ridge crests and crop out on the valley walls. These sedimentary units lie unconformably over the Precambrian rocks. Projections updip of the rock layers indicate that they formerly extended eastward clear across the Teton Range and were eroded as the mountain block was uplifted and tilted. Breaks in the rock record, shown by erosional unconformities, indicate many interruptions of deposition. Within the sequence are limestones, dolomites, shales, and sandstones. Careful study of the sedimentary units yields details of source areas, directions of sediment movement, positions of the Paleozoic seas, and evolving marine life forms. Drilling and subsurface geophysical exploration have made possible the identification of the same Paleozoic sequences deeply buried beneath Jackson Hole. Some of the rock units, such as the Flathead Sandstone, Bighorn Dolomite, Madison Limestone, and the Phosphoria Formation, are widely exposed throughout much of the Middle Rocky Mountain region (fig. 25.4).

Figure 25.4   High altitude view looking westward across Grand Teton peak, the Teton Canyon, and in the far distance the Teton Basin. Most of the area in the picture lies outside the western boundary of Grand Teton National Park. The layered rocks in the middle distance are Paleozoic beds, dipping gently to the west on the back of the Teton Range. In the right foreground, are beds of upturned Paleozoic rock sloping away from the Buck Mountain fault, which trends north-south along the western side of the range crest. National Park Service photograph.

Mesozoic Sedimentary Rocks

Mesozoic sedimentary rocks, both marine and nonmarine, were deposited on top of the Paleozoic sequence. Like the Paleozoic rocks they covered, the Mesozoic rocks probably extended eastward across the location of the present Teton Range. Subsequent uplift and erosion stripped off most of the Mesozoic rocks, exposing the Precambrian and Paleozoic rocks. Outcrops of Mesozoic rocks are in the northern, eastern, and southern parts of the Teton region, mostly outside the Park. Among the rock units that are more widespread throughout other parts of the Middle Rockies are the Chugwater, Sundance, Morrison, and Frontier Formations. These units were deposited by interconnecting Mesozoic seas that drained away as the crust was upwarped and deformed by mountain-building (Laramide orogeny) near the close of the Mesozoic.

Cenozoic Sedimentary Rocks and Unconsolidated Sediments

Exposures of Tertiary (or Paleogene) rocks are found along the eastern margin of Jackson Hole and at the north end of the Teton Range, but most of the Tertiary units are buried beneath even younger deposits that cover the floor of Jackson Hole. The Tertiary sequence of conglomerates, sandstones, shales, volcanic breccias, and tuffs have an aggregate total thickness of about 6 miles! No other region in the United States contains a more complete nonmarine Tertiary section than the Teton region. The rock units reflect the tectonic activity of the Laramide orogeny, the Eocene volcanism that produced the Absaroka lavas and ashfalls, and the movement of the fault blocks. The uplift of the mountain blocks accelerated erosion, producing vast amounts of debris; while the downfaulted basins provided places for the sediments to accumulate. A resumption of volcanic activity in Oligocene and Miocene time contributed additional material. Then in late Pliocene or early Pleistocene time, the great volcanic events occurring in the Yellowstone area (chapter 24) spread rhyolitic debris and tuff over the northern end of the Teton Range and over Jackson Hole.

All of the surface materials in the Jackson Hole section of the Park, with the exception of outliers of older rock in Signal Mountain and Blacktail Butte, are Pleistocene or Holocene (Quaternary) in age. These young deposits, which consist of volcanics, lake-bed sediment, glacial drift, and alluvial materials, have covered the thick sections of Paleozoic, Mesozoic, and Tertiary rocks that fill the Jackson Hole basin.

Fossils

Many species of plants and animals thrive under National Park Service protection in Grand Teton today. A logical assumption is that the region also supported life assemblages of richness and variety in the past. However, in comparison to similar rock sequences in the American West, the Teton rocks are not abundantly fossiliferous. The best fossil record is in the Paleozoic sedimentary rocks of the Alaska Basin, just outside the Park's western boundary. The limestones and dolomites contain brachiopods, trilobites, bryozoans, and corals that once grew in Paleozoic seas.

In the Harebell Formation of late Cretaceous time (east of the Park) dinosaur bones have been found. Scattered bones, pollens, and snail shells in the lake-bed sediment document the presence of Cenozoic plants and animals that lived in the Teton region prior to the present.

The Fault Blocks

The two major topographic units in Grand Teton National Park, the mountain range and the valley, are the result of a fault, or break, in the earth's crust. Differential movement along the fault produced two large crustal blocks, one tilted up and the other tilted down. Separating the huge blocks is the Teton fault system, roughly 40 miles long. The trace of the fault goes along the base of the Teton Range, approximately at the break in slope between the mountain front and the western side of Jackson Hole (fig. 25.5).

In structural terms, the Teton fault is a steeply dipping *normal fault*. By definition, the hanging wall of a normal fault has moved downward in relation to the footwall. The Teton fault dips to the east. Therefore, the hanging wall side, the Jackson Hole block, moved downward in relation to the Teton Range block, which moved up. This up and down movement of the blocks along the fault surface caused a tilting action so that both blocks were tilted to the west.

How did geologists know the fault was there? How did they determine when the movement occurred? How much did the blocks move and in what direction? What do geologists know about the rate of movement, and what do they think caused this displacement of crustal units? Most of the geologic evidence is in the rocks that were briefly described in the preceding section.

1. Geologists suspect faulting when rocks of quite different types and ages are next to each other and the stratigraphic and geomorphic conditions make an unconformity unlikely. Beneath the floor of Jackson Hole, the Precambrian crystalline rocks of the Teton Range butt up against much younger rocks. Although the fault contact is obscured by glacial and stream deposits, geologists can locate the fault system by drilling and seismic prospecting, and by projecting data based on isolated outcrops.
2. The relatively straight and smooth eastern front of the Teton Range is suggestive of faulting. Not many geologic processes other than faulting produce linear features. Because the mountain front

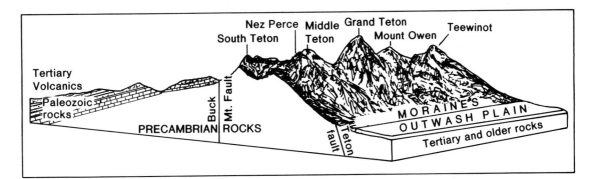

Figure 25.5 Block diagram of the north-south-trending Teton Range. The dissected fault scarp rises from the outwash and moraines on the floor of Jackson Hole. The high central part of the range, cut into horns and cols by Pleistocene glaciers, is bounded by the Teton fault on the east and the Buck Mountain fault on the west. Remnants of the Paleozoic rock units that once covered the entire range lie unconformably on the western slopes over the Precambrian crystalline rocks. (After Horberg 1938; reproduced by permission of Augustana Library Publications.)

has been eroded and cut by canyons, it does not represent the original fault scarp; but it does suggest the approximate slope and location of that scarp.

3. The absence of foothills is significant. This indicates that the mountains were not the result of erosion.

4. The most likely cause of the asymmetry of the range is faulting and tilting.

5. A few small, triangular facets, remnants of the original fault scarp, are spaced along the base of the range.

6. Small, fresh fault scarps in the vicinity of the Teton fault system are regarded as evidence of postglacial faulting and tilting that is likely to continue.

With the presence of the Teton fault inductively confirmed, the next step is to determine which side moved up and how much movement occurred by matching rock types (stratigraphic correlation) across the fault. In matching one side with the other, the best marker is the contact between the Paleozoic rocks and the underlying Precambrian rocks. On the Teton block, we project the westward-dipping Precambrian-Paleozoic contact toward the east until it intersects the steeply dipping plane of the fault. On the Jackson Hole block, the Paleozoic-Precambrian contact can be located far below the surface by deep drilling and by geophysical techniques. The matching of contacts shows that the western block appears to have moved up relative to the eastern block a distance of roughly 30,000 feet (fig. 25.6).

The shape of the Teton block suggests that the Teton fault dips steeply eastward. Data from geophysical studies support this conclusion. Therefore, the Jackson Hole block (the hanging wall) moved downward in relation to the Teton block (the footwall), making the Teton fault a normal fault.

According to the principle of cross-cutting relationships, a fault is younger than the youngest rock that it cuts and older than the oldest rocks that are *not* displaced or cut by fault movement. Applying this rule to the Teton fault, we find that it apparently truncated the youngest sedimentary rocks in the basin, that it began about 9 million years ago, and that it is still active. Most of the movement, which was spasmodic rather than continuous, occurred during late Tertiary time before the coming of the major Pleistocene glaciers. The average rate of movement (derived by dividing the amount of "throw" by the number of years) was about one foot for every 300 years.

In summary, the Teton fault system is a north-south-trending, active, normal fault that dips steeply to the east and has undergone several miles of movement during the last 9 million years.

Several other faults have significantly affected the Teton landscape. The highest mountains in the central part of the Teton Range were elevated by a reverse fault that is older than the Teton fault. The Buck Mountain fault, a reverse fault of moderate displacement, can be traced for a distance of about 10 miles in a north-south direction on the western side of the Teton summit area (figs. 25.5, 25.7). (A *reverse fault* is one where the hanging wall moves

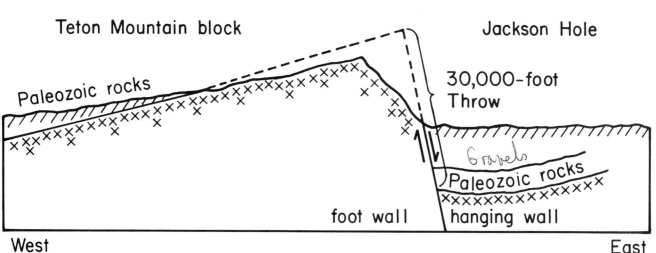

Figure 25.6   Cross section, illustrating the lateral tracing (correlation) of the Precambrian-Paleozoic contact across the Teton fault, shows projections and how the amount of throw is determined. The Teton fault, shown here as a single plane, is actually a fault system made up of a series of individual fault surfaces that form a zone a few feet to a few yards wide. (Buck Mountain fault not shown.)

Figure 25.7 Location map for Buck Mountain fault (left), Teton fault (right), and other significant geologic features in and close to Grand Teton National Park. The Jackson Lake recessional moraine holds in Jackson Lake. The Burned Ridge moraine marks the terminus of the same large glacier (of Pinedale age). Glacial outwash covers the Potholes area and spreads southward over the rest of Jackson Hole. The Snake River has cut terraces (not shown) in the outwash.

Location key
Mountain peaks (from north to south): MO = Mount Owen, GT = Grand Teton, MT = Middle Teton, ST = South Teton, BM = Buck Mountain.
Small fault blocks on the floor of Jackson Hole (from north to south): SM = Signal Mountain, BB = Blacktail Butte, EGVB = East Gros Ventre Butte, WGVB = West Gros Ventre Butte.
Moraine-dammed lakes at the base of the Teton Range (from north to south): L = Leigh, S = String, J = Jenny, B = Bradley, T = Taggart, P = Phelps.
Pre-Pinedale morainal lakes: TO = Two Ocean, EM = Emma Matilda.
Landslide topography: GVS = Gros Ventre Slide.

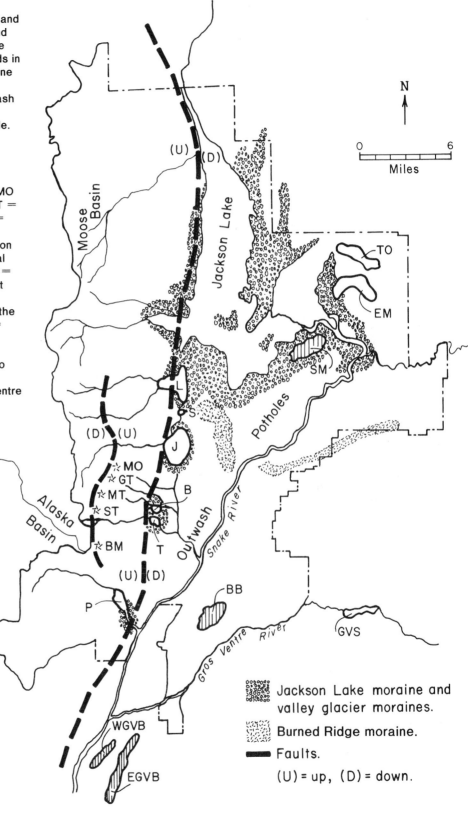

up in relation to the footwall.) West of the Buck Mountain fault line, the Paleozoic sedimentary rocks are bent sharply upward, and on the narrow, uplifted block on the east side of the fault are the dominant peaks of the range—the Three Tetons and several others—in a fairly straight alignment. A logical assumption is that the Buck Mountain reverse fault was a result of orogenic activity that preceded the normal Teton fault of late Tertiary time. Thus the crest area that had been raised by reverse faulting was later raised again by normal faulting to its present height.

On the eastern side of the Jackson Hole block is a nearly vertical normal fault with similar relative movement; that is, west side upthrown, east side downdropped. Other faults have been mapped at both the northern and southern ends of the valley.

The continued tilting of the Jackson Hole block has affected stream drainage and caused some slopes to be susceptible to landsliding. Engineers have erected levees to keep the Snake River from shifting its course more to the western side of the valley, in response to the tilting.

The ultimate cause of the faulting in the Teton region is related, in ways that are not wholly understood, to the geologic history of the surrounding region. Fractures, or faults, in the earth's crust are responses of rock to stresses that may originate in the subcrustal zone of the earth. Normal faulting, looked at mechanically and geometrically, can be explained by stretching or extension of the crust. Such forces are described as *tensional*. Reverse faulting, on the other hand, is usually the result of *compression* of the crust. Deductively it would seem that plate movements may cause both crustal compression and crustal extension at different times. During the Laramide orogeny, parts of the crust in western North America were folded, uplifted, and subjected to reverse faulting and thrust faulting on a grand scale. Then, during late Tertiary time, tectonic stretching and uplift produced many fault-block ranges and basins. However, the relationship of regional tectonic events to the unique geologic history of the Tetons is neither obvious nor simple. Perhaps the key to the puzzle may come eventually from a better understanding of the causes of the volcanism that has played an important role in the development of the Yellowstone-Teton landscape.

How Surface Processes Modified the Teton Landscape

Faulting has been the single most significant geologic process at work in developing the landscape in Grand Teton National Park. On the structurally controlled framework, running water and glacial ice, with the help of mass wasting processes, have sharpened peaks, excavated canyons, and covered the floor of Jackson Hole. The most profound surficial changes were those wrought by Pleistocene glaciers.

The Teton Range is a world-famous example of *alpine topography*; it is certainly as spectacular as the European Alps and the New Zealand Alps. The Teton skyline, as seen from below, is a series of concave curves passing from peak to peak with lower passes in between. The serrated ridge, the conical horns, and the great cirques are all evidence of severe glacial erosion. The nine highest peaks in the center of the range (nearly all of them over 12,000 feet in elevation) are the focus of this grand scenic display.

Despite the fact that the Precambrian rocks are extremely hard and resistant to erosion, the high peaks, which remained above the ice as glaciers grew and shrank, were badly shattered and cracked by frost action on all exposed surfaces. Below the rough, jagged rock of the peaks are smooth ledges that have been scoured and scraped by ice. In the bedrock are shallow, ice-eroded depressions occupied by small lakes. Many of the cirques and rock basins in the glacial troughs also contain lakes.

The large cirques and groups of cirques that head the major canyons are like amphitheaters, with sheer walls hundreds of feet high. The canyons, or glacial troughs, dissect the eastern face of the range. Excavated several thousand feet deep by ice, they have broad floors and sheer, oversteepened walls, forming the U-shape typical of glacial valleys. Tributary glaciers carved hanging valleys, now occupied by waterfalls, in the canyon walls. Some canyons have glacial staircases with falls and cascades rushing down the slopes.

Several times during the Pleistocene, valley glaciers became so large that they coalesced and spilled out over the floor of Jackson Hole. Here they joined great tongues of ice from the north and east that filled the basin. Moraines built at the mouths of the canyons mark the extent of the last Pleistocene glaciation. These moraines now hold a series of lakes (Leigh, String, Jenny, Bradley, Taggart, and Phelps) close against the foot of the mountains (fig. 25.7). Several of the lakes are surprisingly deep, such as Jenny Lake (226 feet) and Leigh Lake (250 feet). Meadows in the moraines have replaced former, more shallow lakes that have been filled in or have drained away. Large glacial boulders, carried down by the ice dot the moraines and rim the lakes.

Jackson Lake, the largest body of water in the valley, occupies the site of the last major glacier from the north that combined with ice from the northern canyons and covered part of Jackson Hole. A large recessional moraine that was built across the valley floor holds in Jackson Lake. South of this moraine is a pitted outwash plain. The small depressions, some filled with water, are *kettleholes*, or *kettles* that were left when stagnant, debris-covered ice blocks melted away. The area is named "The Potholes," although the hollows in the outwash are not potholes in the geologic sense, of course.

Farther down the valley, the Burned Ridge moraine marks the terminus of the last glacial advance; and beyond this moraine, a thick outwash plain, built of glacial gravel, extends for some 10 miles down the valley, its surface sloping gently southward (fig. 25.7).

During times of glacial advance, more debris is put into the meltwater streams than they can carry, and they rapidly aggrade their beds, building up thick accumulations of sand and gravel. When the glaciers retreat, streams are swollen with meltwater but they carry much less debris. This is when they begin cutting trenches into the accumulated outwash, thereby deepening their channels and producing terraces. Terrace tops are built up during glacial advances, and terrace faces are cut during times of glacial retreat.

In Jackson Hole, the Snake River has cut several levels of prominent terraces (fig. 25.3) that can be related to moraines of various ages. The amount of work the Snake has done resorting and redepositing the unconsolidated materials is impressive. With its rapid changes in volume, high velocity, and shifting channels, the Snake deserved the name "Mad River" given it by the mountain men. Even today, despite the damming of the Jackson Lake outlet and the building of levees, the Snake is sometimes difficult to control. For the most part, however, the modern Snake River twists around gravelly bars and quietly reworks loose material in its braided channel.

From its source in Yellowstone, the Snake River flows into Jackson Lake, goes out the southeast end, and then runs on south through Jackson Hole, collecting runoff from the front slopes of the Teton Range and from the highlands on the east side of the valley. The tributaries deliver sediment to the valley floor, and much of it eventually reaches the Snake. The streams from the less steep mountains on the east tend to build alluvial fans where they enter the valley and then flow directly west, some of them joining the Snake almost at right angles. By contrast, the tributaries on the west side of the valley have much steeper gradients but do not develop extensive alluvial fans. Instead, the streams leave the canyons, flow east a short distance, and then bend south, flowing along the mountain front and joining the Snake at the south end of Jackson Hole. Several of the streams flow into interconnected morainal lakes at the base of the Teton Range on their way to the Snake River. This "lopsided drainage," which is due to the tilting of the Jackson Hole block, is another indication of tectonic control of topography.

Great quantities of broken rock and sediment were carried down the mountain sides by glaciers and meltwater during Pleistocene time, but the dumping of rock debris in Jackson Hole did not stop with the retreat of the large-scale glaciers. The work of water and ice goes on, aided by the force of gravity, although at the present time only small glaciers in shaded locations occupy cirques and the upper parts of glacial troughs. Talus accumulates at the bottom of cliffs and slopes, but it only rests there temporarily. The journey of rock debris downslope can be fast or slow or by fits and starts. Some talus, mixed with water, snow, and ice, flows down slowly in rock glaciers. Good examples of active rock glaciers can be seen in Granite Canyon (southern part of the Park) and in a north-trending valley near Eagles Rest (a peak in the northern part of the range). Some talus moves by the process of creep, very slowly (p. 70). Some is brought down rapidly by snow avalanches, especially in late winter when avalanches are fairly frequent.

The vegetation patterns in Jackson Hole reflect in an interesting way the geological pattern of the glacial and alluvial deposits. The evergreen forests that grow on the mountain slopes below the timberline extend out onto the valley floor, but only along the moraines. On the outwash plains and where outwash deposits have buried moraines, only sagebrush, coarse grasses, and a few other hardy plants can survive. Lodgepole pines grow well on the moraines because the unsorted glacial debris is more compact and retains moisture better. Also, the soil has a higher clay content and is richer in minerals that plants need for growth. The coarse, quartzite gravels of the outwash plains, on the other hand, are dry and infertile. Moisture rapidly drains away to the nearest streambed. Along streambeds, below the parched and barren outwash plains, aspens, cottonwoods, and willows thrive on the available moisture and nutriments. Some irregular patches and lenses of loess, blown into the valley between earlier glacial episodes, provide the only good soil in Jackson Hole (except for small amounts of silt). Where loess crops out on banks, badgers and coyotes dig burrows because the loess is easy to dig in and does not collapse.

The Gros Ventre Landslide

In June of 1925, after an unusually wet spring season, a large rockslide moved several cubic miles of rock and debris down Sheep Mountain and across the Gros Ventre River valley in about a minute's time. This occurred about three miles upstream from the present southeastern boundary of the Park. The slide dammed the river, backing up a lake 200 feet deep and five miles long. What caused the slide was the combination of saturation of sedimentary rock layers on Sheep Mountain and undercutting of the slope by the Gros Ventre River that left the base of the slope unsupported. Triggered by minor earth tremors, a rain-soaked slab of Tensleep Sandstone (of Pennsylvanian age) broke loose and slid downdip over soft shale underneath. Two years later the debris dam blocking the river gave way due to the pressure of high water from melting snow. The resulting flood washed out the town of Kelly five miles downstream, and six people were drowned. Now, more than half a century later, the jumbled landslide topography in the valley and landslide scar 200 feet above the river on Sheep Mountain are still clearly visible.

Geologic History

1. Accumulation of sedimentary and volcanic rocks in a Precambrian trough.

The oldest rocks in the Park were probably deposited about 2.8 billion years ago as sandstones, limestones, and shales in a marine environment. Volcanic rocks, possibly extruded from a chain of volcanic islands, were interbedded with sedimentary rocks.

2. Precambrian orogenies and intrusions.

The sedimentary and volcanic rocks were folded and metamorphosed, probably more than once, and the resulting gneisses were intruded by granites more than 2.5 billion years ago. Subsequently, dikes of lighter-colored granite (quartz monzonite) and pegmatite dikes were emplaced.

3. Intrusion of diabase dikes late in Precambrian time.

About 1.3 billion years ago, black diabase dikes, ranging from 5 to 200 feet in thickness, cut through the older Precambrian rocks.

4. Late Precambrian uplift and erosion.

During the last 700 million years of Precambrian time, the region was uplifted and extensively eroded. Finally a new trough, perhaps a geosyncline, began to develop, and an ocean encroached on the land.

5. Deposition of Paleozoic sedimentary rocks.

About nine major formations, representing all periods of the Paleozoic Era except the Silurian, were deposited in Paleozoic oceans that covered the Teton region. The sedimentary units have a composite thickness of about 4,000 feet and include sandstones, shales, limestones, and dolomites. Deposits are discontinuous, and periods of erosion during the Paleozoic may have been longer than periods of sedimentation. However, no significant deformation occurred. Remnants of the Paleozoic rocks are exposed in the Park, but more complete sequences are found outside Park boundaries on the north, south, and west (fig. 25.4).

6. Mesozoic deposition.

Although Mesozoic sedimentary rocks lie conformably over the Paleozoic rock units, the depositional environments of the Mesozoic Era were more varied, shifting from marine to transitional to continental, and alternating back and forth as crustal conditions changed in the Teton region. In general, sedimentation became more rapid throughout the era, especially toward the close. More

than 15 formations with a composite thickness of between 10,000 and 15,000 feet accumulated during Mesozoic time. The Cretaceous units, which are the most numerous and have the greatest thickness, record the last transition from marine to continental rocks. Much of the sediment came from the mountains rising to the west that preceded the mountain-building that was soon to take place in the Teton region. Beds of bentonite between marine sands and shales represent ashfalls (from western volcanoes) that drifted down over a shallow sea. At the present time slides caused by the swelling of wet bentonite clay sometimes block access roads in the northern part of the Park.

As the seas withdrew, coal swamps developed along the old coastline. Some of these coal beds are visible in abandoned mine workings near the Park's eastern margin. A broad, general uplift and the beginning of folding brought the Mesozoic to a close. At this time the 5,000 feet of conglomerates and sandstones of the Harebell Formation (northern and northeastern parts of the Park) were deposited.

7. Laramide orogeny.

The Rocky Mountain phase of the Laramide orogeny began late in the Cretaceous Period. Tectonic activity produced crustal buckling along a northwest-southeast trend and broad uplift throughout the region. From the Targhee uplift, northwest of the Teton region, rapidly eroding rivers brought large quantities of sand and gravel and eventually quartzite cobbles and pebbles, which were spread south and east over the Teton region. These were the sediments that made up the Harebell Formation. The ancestral Teton-Gros Ventre arch, a little to the west of the present Teton Range, was elevated.

Other mountains in the region were raised even higher as the orogeny intensified, with compressive forces causing more folding and faulting. The climax came in early Eocene time as great thrust faults and reverse faults produced mountain blocks and basins. The Buck Mountain fault probably formed at this time.

8. Tertiary nonmarine sedimentation.

In Jackson Hole and the immediate vicinity, vast quantities of clastic sediment that came from the uplifted adjacent areas were spread over the land during Tertiary time. The Pinyon Conglomerate, which overlies the Harebell Formation, is of Paleocene age. On top of coalbeds and claystone that make up the base of the Pinyon are thick zones of well rounded quartzite pebbles derived from the Targhee uplift, like those in the Harebell Formation. The quartzite fragments increase in size toward the northwest and many of them show percussion scars from being pounded during transport by swift mountain streams with steep gradients (Love and Reed 1971). The nature of the tectonic upwarp and subsidence of that time can be inferred from the rock record. Uplift was rapid in the mountain areas that supplied the coarse sediment. At the same time, Jackson Hole was subsiding more and more, thus providing a site for the thick accumulations of eroded sediment. Such a relationship is described as tectonic control of sedimentation.

9. Volcanic activity in early and middle Tertiary time.

As the intensity of the Laramide mountain-building diminished in Eocene time, volcanoes began erupting in the Yellowstone-Absaroka area, producing the tremendous volume of volcanic rocks that make up the Absaroka Volcanic Supergroup. Tuff from these eruptions was deposited in Jackson Hole. In Oligocene and Miocene time volcanoes on the eastern margin of Jackson Hole produced more volcanic debris that filled the basin areas and buried some of the mountains. The Colter Formation, for example, filled a downwarped trough in Jackson Hole with about 7,000 feet of volcanic conglomerate, tuff, and other debris.

10. The Pliocene Teewinot Lake in Jackson Hole.

A fault that trended east-west across the floor of Jackson Hole, near the present south boundary of the Park, impounded the first large freshwater lake in the valley. The water in the Teewinot Lake, dammed by the fault scarp, backed up toward the north, filling the depression over the down-faulted block. For about 5 million years, sediment (including some tuff from nearby volcanoes) accumulated in the lake that covered most of Jackson Hole. The lake persisted because apparently the rate of subsidence kept up with the rate of sedimentation. The presence of fossil snails, clams, and small mammal bones in the sediment indicates that the lake remained fairly shallow.

11. Pliocene volcanics.

Rhyolitic lava, erupted from volcanoes in Yellowstone, flowed down into the northern end of Teewinot Lake and solidified as obsidian, which has been dated by the potassium-argon method as being 9 million years old. Indians used this obsidian for cutting tools and for spears and arrowpoints. Several million years later, after the Teewinot Lake had dried up, nuées ardentes poured down from Yellowstone, spreading welded tuff over the old lake bed. The Bivouac Formation, exposed on the flanks of Signal Mountain, was the result of one of these great catastrophic eruptions. The welded tuff juts out as a ledge on the mountainside with conglomerates above and below.

12. Teton block-faulting over the last 9 million years.

The long-continued upfaulting that elevated the Teton Range and tilted it westward caused the erosion of the Mesozoic and Paleozoic rocks that once covered the summits. However, the range still stands high, both because faulting is still active and because of the resistance of the Precambrian rocks now exposed at the higher elevations (fig. 25.1). In Jackson Hole, sedimentation has kept up with subsidence so that the surface elevation has remained fairly constant. However, local faulting, upwarping, and downwarping in Jackson Hole caused the impounding of two successive preglacial lakes in the south and southwestern part of the valley. The development of the lakes followed a series of volcanic eruptions that occurred near the southern boundary of the Park. Remnants of the volcanic rocks cap East and West Gros Ventre Buttes. These two buttes, as well as Blacktail Butte and Signal Mountain, are small fault blocks that stick up above the flat surface of Jackson Hole (fig. 25.7).

13. Pleistocene glaciation.

The Cenozoic climates, subtropical and humid to begin with, gradually cooled until by Pliocene time the Teton region had perhaps about the same climate as today. Pollen preserved in the Teewinot lakebed sediment came from plants and trees such as fir, spruce, pine, and sage, not unlike those growing in the Park now. With the onset of the much colder climates of the Pleistocene, glacial ice accumulated in the Teton Range, in the Absaroka Mountains to the north, and in the Wind River Range and the Gros Ventre Mountains to the east. At least three times during the Pleistocene, glaciers from these highlands flowed down over Jackson Hole.

During Buffalo glaciation, the first and greatest invasion of ice, glaciers from the north converged and flowed south along the eastern front of the Teton Range. Glaciers from the east flowed down the Gros Ventre River valley and joined the main ice stream near the Gros Ventre Campground. Signal Mountain and the three buttes in the valley were overridden and severely abraded by this great mass of ice which was 2,000 feet thick in many places. At its maximum extent, the ice flowed on down the Snake River valley and into eastern Idaho. The ice melted during the warm interglacial period that followed, and neither of the two later glaciations reached that far. Jackson Hole was left a wasteland. All of the accumulated soil had been scraped or washed away, and the floor of the valley was covered with quartzite boulders and cobbles. The parts of the valley not affected by the later glaciations are still in this barren state; that is, lacking in soil and too stony for all but the hardiest plant varieties to grow.

Bull Lake, the second glaciation, was much less extensive than Buffalo glaciation. Some of the moraines and outwash deposits of Bull Lake glaciation were covered by wind-deposited loess after the glaciers melted back and before vegetation was reestablished. Later glaciation scraped off or covered much of the loess.

The glacial features associated with Pinedale glaciation, especially those in Jackson Hole, are familiar elements of the Park landscape because this was the last major glacial episode. Beginning about 20,000 years ago and lasting until about 9,000 years ago, Pinedale glaciers left behind most of the lake basins, moraines, and gravelly outwash plains that we see today. In terms of volume and erosive power, this last glaciation was the least extensive of the three. The terminal moraine of Pinedale ice is the Burned Ridge moraine, a series of low, hummocky, tree-covered ridges that slant southward across the middle part of Jackson Hole. Between this moraine and the recessional moraine that dams Jackson Lake is the Potholes area described earlier. In the Teton Range, it was the valley glaciers of Pinedale time that put the finishing touches on the sculpturing of the horns, cols, and cirques that we see along the skyline. The string of morainal lakes at the foot of the range are also the result of Pinedale valley glaciers. (The basins of Two-Ocean Lake and Emma Matilda Lake, however, were left by Bull Lake glaciation.) (fig. 25.7)

14. Holocene and historic time.

Pinedale glaciers probably melted away completely during the warm period at the end of the Pleistocene, but small Neoglaciers formed in the Teton Range during the colder climate of the Little Ice Age. These glaciers have not done much eroding, but they are picturesque and are regarded with caution and respect by mountain-climbers.

By building a dam at the south end of Jackson Lake, engineers have continued the work begun by Pinedale glaciers. The purpose of the dam is to prevent excessive raising and lowering of the lake level and thus keep the Snake River from going on a rampage every spring. But sometimes in dry seasons the needs of the Idaho ranchers downstream for more water from the Snake conflicts with the needs of people in the Park who want the lake level kept high for recreation purposes.

South of Jackson Lake the Snake River keeps trying to cut a more westerly channel because of continued slight tilting of the Jackson Hole block. Simultaneously, engineers keep reenforcing levees in order to force the river to remain where it is.

No severe earthquakes have occurred in the valley since it was settled, but one could occur any time. Meanwhile, the usual rockslides, earthflows, avalanches, and so on, interfere with human activities from time to time.

The continual battle on the earth's surface goes on between the "wearing-down forces" and the "resisting framework." In Grand Teton National Park, this age-old conflict produces an interesting, diverse, and ever-changing landscape.

Geologic Maps and Cross Sections

American Association of Petroleum Geologists. 1972. *Geological highway map of the Northern Rocky Mountains region, map no. 5.* Tulsa, Oklahoma: American Association of Petroleum Geologists.

Love, J. D.; and Reed, J. C., Jr. 1971. *Creation of the Teton landscape.* Moose, Wyoming: Grand Teton Natural History Association.

Love, J. D.; Reed, J. C., Jr.; Christiansen, R. L.; and Stacy, J. R. 1972. *Geologic block diagram and tectonic history of the Teton region, Wyoming-Idaho.* U.S. Geological Survey miscellaneous geologic investigations, Map I-730.

Bibliography

Bartlett, R. A. 1962. *Great surveys of the American West.* Norman, Oklahoma: University of Oklahoma Press.

Behrendt, J. C.; Tibbetts, B. L.; Bonini, W. E.; Laven, P. M.; Love, J. D.; and Reed, J. C. 1968. *A geophysical study in Grand Teton National Park and vicinity, Teton County, Wyoming.* U.S. Geological Survey Professional Paper 516E.

Crandall, Hugh. 1978. *Grand Teton, the story behind the scenery.* Las Vegas, Nevada: KC Publications.

Christopherson, E. 1961. *Behold the Grand Tetons.* Missoula, Montana: Teton Book.

DeVoto, B. 1974. *Across the wide Missouri.* Boston: Houghton Mifflin Company.

Fryxell, F. 1946. *The Tetons, interpretations of a mountain landscape.* Berkeley, California: University of California Press.

Horberg, L. 1938. *The structural geology and physiography of the Teton Pass area, Wyoming.* Publication No. 16. Rock Island, Illinois: Augustana Library Publications.

Love, J. D.; and Reed, J. C., Jr. 1971. *Creation of the Teton landscape.* Moose, Wyoming: Grand Teton Natural History Association.

Parsons, W. H. 1978. *Middle Rockies and Yellowstone,* K/H Geology Field Guide Series. Dubuque, Iowa: Kendall/Hunt Publishing Company.

Stegner, Wallace. 1962. *Beyond the hundredth meridian.* Boston: Houghton Mifflin Company

Wyoming Geological Association. 1956. *Jackson Hole Field Conference guidebook.* Casper, Wyoming: Wyoming Geological Association.

Note: Love and Reed (1971) is a well illustrated, nontechnical book about Grand Teton geology that explains the features visitors can see in the Park and is a handy size for tucking into a backpack. Much of the material in this chapter is based on Love and Reed's book.

Address

Grand Teton National Park
Box 170
Moose, Wyoming 83012

26
Redwood National Park

by Lisa A. Rossbacher
Whittier College

Location: Northern California
Area: 109,026.53 acres; 107.35 square miles
Established: October 2, 1968

Figure 26.1 The tallest trees on earth are the coastal redwoods, which may reach over 300 feet in height. They grow only in a narrow belt along the northern coast of California. National Park Service photograph.

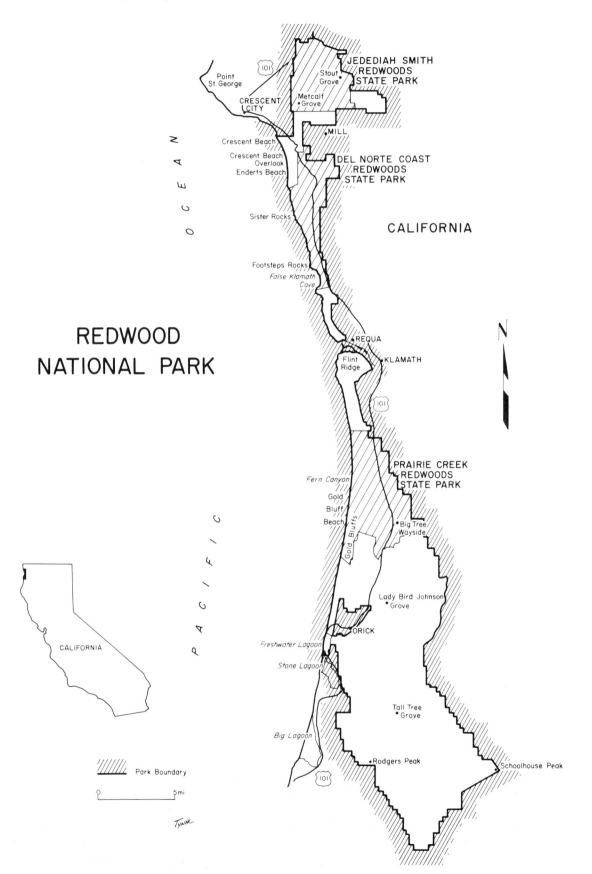

REDWOOD
NATIONAL PARK

Figure 26.2   Redwood National Park, California.

Table 26.1. Generalized Geologic Column, Redwood National Park

Time Units			Rock Units	Geologic Events
Era	Period	Epoch		
Cenozoic	Quaternary	Holocene	Alluvial and mass-wasting deposits	Coastal erosion and spit deposition; accelerated erosion on land; landsliding; continued faulting and coastal subsidence
Cenozoic	Quaternary	Pleistocene		Continuing uplift and erosion; formation of marine terraces in middle to late Pleistocene
Cenozoic	Tertiary (Neogene)	Pliocene	Stream gravels	Coast Range orogeny extending to the south; thick stream gravels deposited in valleys (especially south of Park)
Cenozoic	Tertiary (Neogene)	Miocene	Marine sediments, gravels	Basin forming and sedimentation in Coast Ranges; some volcanism, predominantly submarine
Cenozoic	Tertiary (Paleogene)	Oligocene / Eocene / Paleocene	Marine sediments, some quartzose sandstone	Decreasing depositional rates as seaways narrow and become smaller; weathered clays and clean sandstones indicate nearly tropical climate in Eocene. Thrusting over Franciscan rocks
Mesozoic	Cretaceous / Jurassic		Franciscan rock assemblage (sandstone with quartz, plagioclase feldspar, and chlorite mica), shale, limestone, radiolarian cherts, ultrabasic intrusions	Extensive deposition of marine sediments (over 50,000 feet), followed and accompanied by subduction beneath continental margin, tectonic mixing, and intrusion of ultrabasic peridotite. Metamorphism of sediments and older rocks
Mesozoic	Triassic		(Not exposed in Park)	
Paleozoic			(Not exposed in Park)	Long accumulation of sediments in Paleozoic seas (No Precambrian or early Paleozoic rocks have been identified in the Coast Ranges; if they exist, they have been severely metamorphosed.)

Source: modified after Norris and Webb 1976

337

Local History

The region around Redwood National Park is rich in legends of the Klamath Indians. The Klamaths made dugout canoes from the redwood trees, as did more recent tribes including the Yurok, Tolawa, and Chilula. These later Indians also constructed homes and sweathouses out of redwood planks.

Portola, a Spanish explorer, led an expedition through this area in 1769. In a notebook, Father Juan Crespi, a Franciscan missionary on the expedition, referred to the *palo colorado* or "red trees." The redwoods were officially named in 1823, when A. B. Lambert designated them *Taxodium sempervirens*, or evergreen cypress. Later, a German botanist realized that these trees did not belong to the bald cypress group, and he renamed them *Sequoia sempervirens*. Both the coast redwood and the giant sequoia of the Sierra belong to the family Taxodiaceae, or swamp cypress family.

In a sense, the stately grandeur of the redwoods has been a factor in their destruction. The attributes that make them so impressive and beautiful in the forest are the same qualities that make them attractive for construction purposes; their bark and wood are resistant to decay, fire, and insects. This means that man is the trees' only real enemy. Cutting the trees is the ultimate destruction, but even the stress of people walking on their shallow root systems can bruise them.

This part of northern California was once dense with redwoods. When the explorer Jedediah Smith brought a group into the forests in 1828, he and his company were able to travel only about a mile per day, winding their way through the huge trees. The pioneers and early settlers cut many trees in the mid- to late 1880s for buildings in San Francisco and northern California towns. The advent of motorized equipment made more remote areas accessible to loggers, and more trees were felled. By 1964, only 15 percent of the original stand of redwoods remained.

In order to avert the total destruction of the trees, the Save-the-Redwoods League was established in 1918 to promote the creation of Redwood National Park; it took 50 years to accomplish the League's goal. Feasibility studies were made in the 1920s and 1930s, but no legislative action was taken. In 1946, the Douglas Bill was proposed for the establishment of a national memorial forest that would include 2.5 million acres, about 180,000 of which would be administered according to National Park Service regulations. The Douglas Bill never passed.

Aided by funds from the National Geographic Society, the National Park Service surveyed the redwood area in 1963 and 1964; this study revealed some dreadful statistics. Of the original forest area of nearly 2 million acres, only 15 percent remained—and only 50,000 acres (2.5 percent) were protected in state parks! If the logging rate continued unchecked, all of the unprotected trees would disappear before the year 2000. The study also found a grove of the earth's tallest trees, including one reaching 367 feet high, on private—and unprotected—land.

Redwood National Park was finally established by Congress in 1968, guaranteeing protection of these endangered trees. The Park links three California state parks; from north to south, these are Jedediah Smith, Del Norte, and Prairie Creek State Parks. In addition to the virgin redwood forest, the Park includes stands of maple, cedar, and oak; private homes and small businesses; some cutover land; and more than 40 miles of coastline. Some of the most impressive redwood stands are located in the three adjacent state parks. Although the state and national parks have some cooperative programs, they are administered as completely separate units.

Establishment of the Park assured protection of the surviving trees from logging, but other problems remained. Commercial logging operations outside of the Park boundaries threatened to cause serious erosion of the famous Redwood Creek watershed, which includes the tallest trees. Rich topsoil eroded from drainage basins at higher elevations outside the Park was being deposited in the stream channels within the Park, and this was causing the streams to rise and literally drown the redwood trees. To help solve this problem, Congress extended the Park boundaries in 1978, increasing the total area by over 80 percent, to protect a total of 109,026 acres.

The added Park area is primarily cutover lands, including 300 miles of logging roads, 3,000 miles of skid trails, and countless charred stumps marking former locations of stately redwoods. The Park Service started a program to reclaim and restore this area, and the reclamation has been very successful so far. The techniques being used in the cutover parts of the Park, which include slope stabilization and revegetation, may aid in reclamation of damaged land in other parts of the country.

The Tall Trees

The redwoods include the tallest trees on earth; many of them are over 200 feet tall, and some of them stand more than 300 feet high. The tallest coast redwood was discovered by the National Geographic Society in 1963, near the southern end of the Park in what is now the Tall Trees Grove. This huge tree stands 367.8 feet high and the trunk is 14 feet in diameter near its base. The ages of the tallest trees average only 400 to 800 years, although a tree that was cut in 1934 had reached an age of 2,000 years.

The trees are called redwood because the color of their bark is a dull reddish brown. The bark is dense and fibrous, and it may be as much as a foot thick on a very old tree. The leaves are usually less than an inch long, sharply pointed, and evergreen. Redwood cones are egg-shaped and about 1.5 inches in diameter.

The coast redwood, like its cousin the giant sequoia (chapters 27, 28), grew throughout much of North America when the dinosaurs were alive. Pleistocene glaciation, beginning about 2 million years ago and ending only 10,000 years ago, caused major changes in the soil and climatic conditions, and resulted in a much narrower distribution of the Sequoia family. The coast redwoods are now restricted to a band approximately 35 miles wide and 500 miles in length along the Pacific coast. Fossils of redwood trees have been found in North and South Dakota, Montana, and Yellowstone National Park in Wyoming (chapter 24).

Favorable conditions for redwood trees include moderate temperatures, rainy winters (50 to 100 inches per year), dry summers, and year-round fog. An old saying is that the redwoods grow "as far inland as the fog flows," and that is about 30 miles here.

Geologic History and Features

Redwood National Park is entirely within the Coast Ranges of California, which extend from the Oregon border south to Santa Barbara. This region has provided some important insights into the theory of plate tectonics. The northern Coast Ranges lie east of the San Andreas fault zone, which crosses the coast and goes offshore about 50 miles south of the Park. The moist climate that makes this region so favorable for redwoods and other vegetation also makes the geology very difficult to see; thick soils and dense forests make for few exposures. However, the sea cliffs provide an excellent view of the geology, for anyone willing to brave the rocky beach.

1. **Deep-water, continental, and volcanic deposits in a subducting trough: the Jurassic-Cretaceous Franciscan formation.**

The oldest rocks found in the northern Coast Ranges are part of the Franciscan rock assemblage, which is also called a *mélange*, reflecting the mixture of diverse materials it contains. These rocks are the product of deep-water sedimentation on the continental slope and on oceanic crust in deeper trenches offshore. These sands and muds also included extrusive and intrusive submarine volcanics, radiolarian oozes (mud composed of tiny siliceous animal shells), and sediments eroded from land. This mixture of sediments—carried along on an oceanic plate—was subducted beneath the continental margin about 100 million years ago. Some of this tectonic activity probably occurred at the same time as deposition, so the unconsolidated sediments were severely disturbed. Some outcrops of the Franciscan have rocks that are so contorted that they look as if they had been stirred with a spoon. To complicate the situation further, ultrabasic rocks intruded the Franciscan beds during subduction.

The subduction ended relatively suddenly, and the complex Franciscan assemblage was uplifted and exposed at the surface, probably about 70 million years ago. Thus, its characteristic minerals reflect the low-temperature and high-pressure conditions of their origin.

The Franciscan assemblage is composed primarily of grayish-green graywackes. These sandstones contain relatively high proportions of plagioclase feldspar, as well as quartz; and a chlorite matrix gives the Franciscan its distinctive greenish color. The intruded volcanics were altered to serpentinite by metamorphism, and these metavolcanics are also green.

The Franciscan is a thick and widespread rock unit in the Coast Ranges. A conservative estimate of its volume is 350,000 cubic miles—which could cover all of the lower 48 states to a depth of 600 feet!

2. **Basin-forming and sedimentation in the Miocene.**

The sea advanced over the Coast Range area in the early Miocene, creating a number of small, nearly isolated marine basins. Volcanism, predominantly submarine, also contributed to the accumulation of deposits during this time.

In the northern Coast Ranges, the late Miocene brought further tectonism. Downwarped areas formed basins, defined by the intervening anticlinal arches trending to the northwest. The area where Redwood National Park is now located was surrounded by two of these depressed and flooded embayments, one at Crescent City to the north and the other along the Mad River near Eureka. A few downfaulted remnants of the basin deposits are all that have survived late Cenozoic erosion to show that marine rocks were formerly widespread here.

3. **Thick stream gravels in the Pliocene.**

The orogenic activity that began in the northern Coast Ranges during the late Miocene extended southward throughout the Coast Ranges in the Pliocene. By the late Pliocene, most of the ranges that exist today were exposed as dry land. Thick layers of stream gravels were deposited in valleys, sometimes nearly burying the surrounding hills. Deposition was most extensive in the southern Coast Ranges, but a few patches of gravel have been found north of Eureka. One place where Pliocene-Pleistocene deposits are exposed is along the beach at Gold Bluff, where the sand does contain small flecks of gold.

4. **Quaternary erosion and sea-level changes.**

Because most of the land surfaces in the northern Coast Ranges have been exposed since the Pliocene (about the last 2 million years), the geologic record has been slowly destroyed. Only a few coastal Quaternary deposits cover the Franciscan in this area. The only major Quaternary activity recorded in the northern Coast Ranges is the development of raised marine terraces. Some of this terrace formation was caused by worldwide sea-level changes, or *eustatic* changes, due to glaciation. However, most of the terraces in northern California are wave-cut platforms that have been eroded in bedrock and then raised by localized uplift. Marine terraces are visible in the Crescent City area, at the northern end of the Park.

Tectonic activity has been going on in the Pacific, off the northern California coast, at least since the early Cenozoic. At Cape Mendocino, about 50 miles south of the Park, the San Andreas fault zone goes offshore and merges with the east-west trending Mendocino ridge and fracture zone. About 50 miles offshore, the Gorda Plate is being subducted under the North American Plate (fig. 26.3). This subduction zone was responsible for the November 8, 1980, earthquake in this area; essentially the quake was caused by the friction between these two plates. The magnitude of the tremor was 7.0, which made it the largest earthquake in California in 28 years! Northern coastal California has had at least twenty earthquakes with magnitudes of 6.0 or greater in the last century.

5. Landslides and accelerated erosion today.

The Franciscan rocks are extremely fractured and sheared, and that makes them highly susceptible to mass wasting. Landslides are so abundant in this area that they cause more downslope movement of material than any other geologic process—including stream action! Water is often an important agent in initiating landslides, and the added rock fracturing in fault zones also decreases slope stability. Construction on steep slopes is another significant cause of landslides. Each year, millions of dollars' worth of property damage is done by landslides in the northern Coast Ranges.

Stream activity is important in the Park area today. Stream deposits support some of the largest stands of tall trees. Both the Tall Trees Grove on Redwood Creek and Stout Grove on the Smith River grow on alluvial flats.

The area around Redwood National Park currently has coastal lagoons. These are the drowned mouths of stream valleys that have sand spits separating them from the sea. The coastal lagoons are evidence that this part of

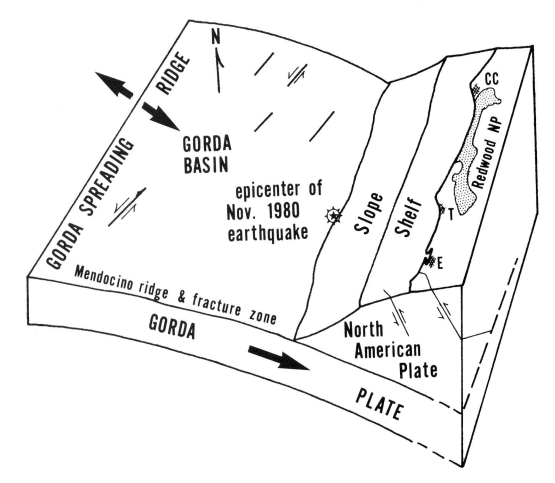

Figure 26.3 This generalized block diagram shows the coast and offshore area of northern California. The San Andreas fault swings west and merges with the Mendocino ridge and fracture zone south of Eureka. (CC = Crescent City, T = Trinidad, and E = Eureka)

Figure 26.4 The view looks south at the drowned mouth of the Klamath River in Redwood National Park. Ocean water moved into the river valley as crustal downwarping caused local subsidence of the coast. The spit was built by longshore drift (p. 211). River and tidal currents keep the mouth of the spit open. The lagoon behind the spit is filling with sediment. National Park Service photograph.

the California coastline is sinking. The reason for this subsidence is not known, but it may be related to subduction offshore.

A recent noteworthy geologic event in the Crescent City area, at the northern end of the Park, was the *tsunami* (seismic sea wave) caused by the 1964 Good Friday earthquake in Alaska.* Sudden changes in the sea floor, caused by earthquakes or large submarine landslides, can start ocean swells that move across the ocean and hit coastlines with destructive violence. Although tsunamis occur with some frequency in the Hawaiian Islands, Crescent City is the only place in the conterminous United States to have recorded serious damage from such an event.

Erosion within the Park has been accelerated by the activities of man. Extensive lumbering and clear-cutting have nearly eliminated the soil protection in some areas. Gully formation, the loss of rich topsoil, and streams choked with debris are all consequences of man's thoughtless exploitation of the redwoods as a natural resource. Redwood National Park was established to save the area from further damage, but it will take many years and considerable work to repair the existing damage. The National Park Service's reclamation project, intended to restore the cutover land purchased in 1978, is expected to involve at least a decade of work to recover the damaged areas. The Park Service is still making decisions about how to develop the newly added park areas. A unique opportunity exists for interested citizens to affect use of national park land, because the superintendent of Redwood National Park has invited public comment on development plans for the Park.

Geologic Map and Cross Section

American Association of Petroleum Geologists. 1968. *Pacific Southwest region, geological highway map no. 3.* Tulsa, Oklahoma: American Association of Petroleum Geologists.

Bibliography

Alt, D.; and Hyndman, D. W. 1975. *Roadside geology of northern California.* Missoula, Montana: Mountain Press Publishing Company.

Bailey, Edgar, ed. 1966. *Geology of northern California.* California Division of Mines and Geology Bulletin 190.

*See chapter 42, p. 512.

Norris, R. M.; and Webb, R. W. 1976. *Geology of California*. New York: John Wiley & Sons.

Oakeshott, G. B. 1978. *California's changing landscapes*. New York: McGraw-Hill Book Company.

Sunset Books editorial staff. 1969. *Redwood country and the Big Trees of the Sierra*. Menlo Park, California: Lane Books.

William, R. L. 1976. *The loggers*, The Old West Series. New York: Time-Life Books.

Zahl, P. A. 1964. Finding the Mt. Everest of all living things, *National geographic* 126 (1) :10–51, July.

Address

Redwood National Park
Drawer N
Crescent City, California 95531

27
Sequoia National Park

Location: East Central California
Area: 403,023 acres; 629.72 square miles
Established: September 25, 1890

Figure 27.1 Giant sequoias growing in Crescent Meadow
(near Giant Forest) in Sequoia National Park. National Park
Service photograph.

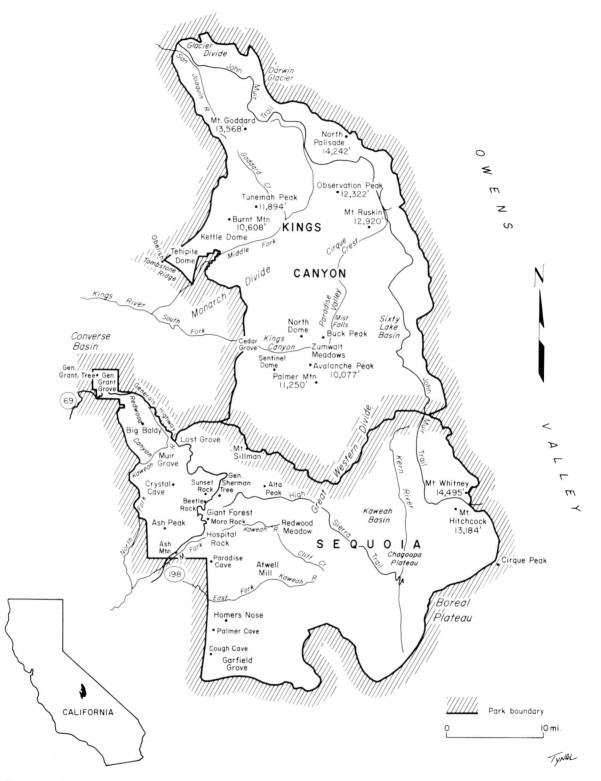

Figure 27.2   Sequoia-Kings Canyon National Parks, California.

Table 27.1. Generalized Geologic Column, Sequoia and Kings Canyon National Parks

Time Units			Rock Units	Geologic Events
Era	Period	Epoch		
Cenozoic	Quaternary	Holocene	Alluvial, glacial, and mass wasting deposits	Stream erosion; Neoglacial activity; landslides, rockfalls Earthquakes, faulting, continuing uplift
		Pleistocene	Till, moraines, glacial outwash of various ages Avalanche deposits	At least three major glaciations, including substages; intense glacial erosion, avalanche chutes, ice-sculptured landscapes, glacial deposition Faulting, relatively rapid uplift
	Tertiary (Neogene)	Pliocene		Probable onset of glaciation Southern Sierra volcanism (outside Parks) Canyon-cutting, intense erosion Beginning of faulting and uplift Westward tilting of Sierra Nevada block
		Miocene		
	Tertiary (Paleogene)	Oligocene Eocene Paleocene		Erosion Pause in tectonic activity Probable beginning of cave development
Mesozoic		Cretaceous Jurassic Triassic	Quartz diorites, granodiorites, quartz monzonites Limestones, shales, sandstones, interbedded with volcanics	Emplacement of plutons in composite batholith Metamorphism of country rock Nevadan orogeny Marine sedimentation and island arc volcanism Subduction and plate convergence
Paleozoic			(not exposed in Parks)	Long accumulation of sediments in Paleozoic seas

Source: modified after Hill 1975; Bateman 1974; Bateman and Wahrhaftig 1966

Local History

Sequoia National Park was established in 1890, just in time to save its groves of giant sequoia trees from destruction by the lumbering industry. The "Big Trees" grow only in scattered patches on the moist western slopes of the Sierra Nevada at elevations between 4,000 and 8,000 feet, from Yosemite National Park (chapter 13) on the north to Sequoia on the south. In between is Kings Canyon National Park (chapter 28), which adjoins the northern boundary of Sequoia National Park. These two Parks are administered as a unit, and their geology is very similar.

About 13,000 giant sequoias (*Sequoiadendron giantea*) grow in the High Sierras, the largest single grove being Giant Forest in Sequoia National Park. Giant Forest has about 3,500 Big Trees, among them General Sherman, the largest living tree in the world. The circumference of its trunk is 102.6 feet, and it is almost 275 feet tall. The mature giant sequoias in the Parks are between 3,000 and 4,000 years old.

The giant sequoias and their close relatives, the coast redwoods (*Sequoia sempervirens*) which grow in a narrow belt along the Pacific coast, are the surviving members of a once large family of huge conifers that were widespread throughout the temperate regions of North America during the Mesozoic Era. Fossilized sequoia stumps have been identified in Yellowstone National Park (chapter 24) and in Petrified Forest National Park (chapter 38). Most of the sequoia species became extinct during the Ice Age, which began about 2 million years ago and ended about 10,000 years ago. Only the two Pacific coast species survived the Pleistocene glaciations and were able to reestablish themselves in the limited environments in which they now grow (see chapter 26, Redwood National Park). The mountain ranges paralleling the Pacific coast provide an ideal environment for the two species: (1) their elevation causes the moisture-bearing clouds from the Pacific Ocean to precipitate large amounts of rain and snow; and (2) the fresh, sandy soils required by sequoias are derived from glacial erosion and the weathering of the mountain rocks.

But perhaps the most important reason for the survival of the giant sequoias and redwoods to the present day has been the recognition by Americans that these living monuments are irreplaceable. This was not apparent at first. A hunter, A. T. Dowd, is credited with discovering Giant Forest in 1852, although earlier hunters and explorers had found giant sequoias in the Sierra region. News spread about the "mammoth trees," and in the 1860s and 1870s claims for land were filed and sawmills were erected. The loggers of the day regarded felling the Big Trees as the ultimate challenge to their skill and bravery. The wood from sequoias is of exceptionally high quality, straight-grained and naturally resistant to decay. A single giant sequoia might yield 300,000 to 400,000 board feet of lumber! Reminders of these early logging days can be seen in Big Stump Basin in Kings Canyon National Park near Generals Highway; where a grove of giant sequoias once flourished, only huge stumps remain.

John Muir came to California in 1868 as a young man and settled in the Yosemite Valley. From that time on, he dedicated his life to saving the Big Trees and preserving the Sierra wilderness. On foot and often alone, he walked and climbed the entire length of the range many times, thoroughly observing and documenting its natural wonders and its plants and animals. Appalled by the devastation of the natural environment by loggers and miners, he enlisted the help of politicians, newspaper editors, and other citizens in the campaign to save the Sierra landscape from exploitation. He led the group that founded the Sierra Club and became its first president. A tireless naturalist and conservationist, he spent many years of his long and remarkable life traveling throughout the world, studying nature and spreading his message of wilderness preservation for the well-being of humankind. But he always returned to his home in the Sierras. The magnificent John Muir Trail, established in his memory, follows the crest of the Sierras, from the Tuolomne Meadows in Yosemite to Mount Whitney in Sequoia, winding through

three national parks and three national forests on its 218-mile route. This is a part of the Pacific Crest Trail, which extends along the highest ranges of California, Oregon, and Washington from the Mexican border to Canada.

The California Geological Survey, under the direction of geologist Josiah D. Whitney, undertook the mapping of the High Sierra in the 1860s. Their primary mission was to locate occurrences of gold and other minerals with economic value. Despite insufficient funds, this group of scientists produced outstanding reports and maps that aided the conservation movement and laid the groundwork for subsequent scientific investigations. The survey party included William Brewer, a botanist who played an important role in the development of soil science; Charles Hoffman, an engineer who became known as the father of modern topographic surveying; and Clarence King, who later was appointed first director of the U.S. Geological Survey.

The Development of the Sierra Nevada

In broad outline, the geologic story of Sequoia and Kings Canyon National Parks is like that of Yosemite (chapter 13). All three Parks lie along the crest and western slopes of the Sierra Nevada, a huge crustal block that has been lifted and tilted westward. Thus, the mountain range is asymmetrical; it has steep slopes on the fault-bounded eastern margin and a more gentle regional slope to the west. Along the eastern foot of the range is the Sierra Nevada normal fault system, which trends north-northwest and marks the break between the bold mountain front and the Owens Valley. With an elevation of 14,494 feet, Mount Whitney is the highest summit in the United States outside Alaska. The floor of the Owens Valley lies nearly 11,000 feet lower. Death Valley, some 80 miles to the east, contains the lowest point in North America, 282 feet below sea level—these figures give an indication of the magnitude of uplift and downfaulting that has occurred in this part of California. The vertical movement along the Sierra Nevada fault system has been at least 15,000 feet.

The composite batholith now exposed in the High Sierra is made up of granites (quartz diorites, quartz monzonites, and granodiorites). This batholith separates the rocks to the west which, according to plate tectonic theory, were formerly oceanic crust, from the rocks east of the fault system that were probably continental crust before mountain-building began in Mesozoic time.

The tectonic events followed a long period of marine deposition during Paleozoic time and were initiated by the subduction of the eastward-moving Pacific plate beneath the westward-moving North American plate. Lava generated by the heat and pressure of subduction erupted offshore, forming a chain of island-arc volcanoes. As plate convergence continued, the rocks along the two plate boundaries became severely compressed and deformed. Portions of oceanic crust were scraped off the descending plate and mixed with mangled pieces of continental crust. Meanwhile, far below the surface, granitic magma was being generated. Gradually it began to rise like a huge blob, pushing up and further deforming the roof rocks overhead. Some of the country rock enclosing the hot body of magma was metamorphosed; some of it dropped down and was melted; and some chunks were left as large discrete masses near the top of the batholith. As the whole crustal mass thickened and bowed upward, faulting began, raising the great Sierran block.

Numerous recent earthquakes on the Sierra fault system are evidence that the mountain-building process continues to the present day. The largest earthquake in historic time occurred on March 26, 1872, near Lone Pine on the floor of the Owens Valley. (The epicenter was about 15 miles east of Mount Whitney.) The quake is estimated to have had a Richter magnitude of at least 8.0. Vertical movement on the fault was as much as 28 feet, and horizontal displacement was about 20 feet. The ground was broken for nearly 100 miles along the fault trace. Nearly all the buildings in Lone Pine were destroyed, and 27 people were killed (Hill 1975).

Geologic Features

As the mountain block rose, erosion intensified, eventually exposing the igneous rocks of the composite batholith. Estimates are that a layer of material probably more than 7 miles thick was stripped from the top of the Sierra Nevada! The uncovering of the batholith also exposed roof pendants of metamorphosed marine sedimentary rock. (*Roof pendants* are downward projections of country rock into the batholith.)

Both Sequoia and Kings Canyon National Parks have roof pendants. In the Kaweah Basin, in the western part of Sequoia, marble in a roof pendant has been partially dissolved by ground water to form under-ground caverns such as Crystal Cave, Clough Cave, Palmer Cave, and Paradise Cave. The metamorphism that changed limestone to marble destroyed any fossils that may have existed, so the age of the rock unit is not known (although it is assumed to be early or middle Mesozoic). As these carbonate rocks were being raised above sea level, ground water began percolating along fractures and bedding planes, enlarging openings by removing more of the material. When the openings were raised above the the water table by continued uplift, ground water then began to deposit as well as to dissolve calcite. Typical dripstone features formed on the ceilings, walls, and floors of the underground chambers. (Cave features and solution processes are explained in more detail in chapters 39, 40, and 41.)

Figure 27.3   Moro Rock, an immense exfoliation dome of massive granite, rises to an elevation of 6,725 feet above the canyon of the Kaweah River. National Park Service photograph.

The shape and appearance of the landforms in Sequoia and Kings Canyon are strongly influenced by the batholith structure and by the weathering characteristics of the intrusions. Because some of the granites are massive and have few fractures (or joints), exfoliation processes have shaped many of the exposed bedrock surfaces. Exfoliation domes, similar to those in Yosemite National Park (chapter 13), are prominent features. Examples of exfoliation in Sequoia National Park are Moro Rock, Alla Peak, Beetle Rock, and Sunset Rock (fig. 27.3). The Watchtower is a granite pinnacle that stands 2,000 feet high.

In well-fractured granites, especially at high altitudes, frost action shatters rock into angular blocks and fragments of various sizes. Vertical cliffs with sharply incised gashes are formed by frost action and mass wasting. At moderate altitudes, where more moisture is available, cascading streamlets have cut slotlike gulches along fractures in canyon walls.

Avalanche chutes are numerous on steep slopes, cirque walls, and canyon walls in the High Sierra. An *avalanche chute* is the steep, smooth track, or trough, left on a slope after an avalanche, carrying ice, rocks, and trees, has descended. Most of the avalanche chutes probably formed during the severe conditions of the Pleistocene, but some of them are still active. In the upper canyon of Cliff Creek, on the face of Mount Hitchcock, and at other places where parallel vertical fractures are closely spaced, the avalanche chutes which developed on the fractures give the slope faces a fluted appearance. On the west flank of Mount Whitney, branching avalanche chutes follow the fractures; some of the chutes are more than 100 feet deep. In the massive granites (most notable on slope faces of the Great Western Divide), large, semicircular, almost perfectly smooth avalanche chutes have developed. They have been altered very little since they were repeatedly abraded and polished by sliding ice and rock during Pleistocene time.

At lower altitudes where the vegetation is heavy and moist soil was undisturbed by glaciation, intense chemical weathering has decomposed some of the older granites. This "rotten granite," which is loose and granular, can be seen in road cuts along the Generals Highway. In some places, the granite is so soft that blasting was not required during highway construction and the road cuts were excavated by power shovels.

Climatic Zones and Precipitation

In many respects the weathering and erosional processes of the High Sierra can be related to climatic zoning, which is fairly well defined on the slopes. The highest parts of Sequoia and Kings Canyon National Parks are a barren, windswept, rocky upland in an Arctic-Alpine climatic (or life) zone. Rainfall is limited, and snow accumulates only in scattered patches because of the constant wind. Flat surfaces (such as Diamond Mesa, the Boreal Plateau, and parts of the summits of Mount Whitney and Cirque Peak) were never glaciated because snow was blown away before it could accumulate. The relative smoothness of these areas is probably due to the slow weathering of the massive, resistant granites. In general, relief is fairly subdued in the high uplands. Only the hardiest plants survive in this environment.

In the transitional Subalpine zone, thin patches of soil support a few hardy shrubs and low plants. Peaks and basins show evidence of past glaciation. The sides of some of the flat uplands are scalloped by small cirques, and small lakes occupy some of the abraded rock basins. Near the timberline, the mountain landscape becomes more irregular. Ridges are serrated with horns and cols. A short distance below the main Sierra crest on the eastern side are the saw-toothed peaks that stand against the skyline so

dramatically when viewed from the Owens Valley far below. Prominent cirques are carved out below peaks and arêtes. Many of the avalanche chutes, described above, extend down into the forested zones.

In the Canadian life zone, which reaches higher on south-facing slopes because plants receive more sunlight than on north-facing slopes, the firs and lodgepole pines resemble those of forests in more northerly latitudes. Heavy snows cover the ground from November to May. Some of the streams flow in U-shaped valleys that were formerly occupied by glaciers. Mountain meadows and marshes mark the locations of old lake basins that were originally eroded by glacial ice and are now filled in with sediment.

On the western slopes, precipitation is sufficient to support forests between elevations of roughly 2,000 feet and 11,000 feet. The zone of maximum precipitation, approximately between the 7,000-foot and the 8,000-foot levels, is about where the Canadian zone blends into the Transition life zone; that is, where the lodgepole pines give way to the taller ponderosa pines and the groves of giant sequoias. Each winter the snowpack at various elevations in the Sierra is closely monitored by snow-survey crews from the California Department of Water Resources. Their estimates of the water content of the snow are vital to the well-being of residents in Southern California cities (especially in the Los Angeles area), who depend on the meltwater for their water supply. The farmers in the Owens and Central Valleys also use this water for irrigation. These conflicting needs have caused political problems in the area since the early 1900s.

Drainage Systems and Divides

The Great Western Divide is a north-south range of rugged mountains located a short distance west of the main crest of the Sierra. This high divide separates the westward-flowing Kaweah Basin drainage from the southward-flowing Kern River drainage (fig. 27.4). The Keweah River is the southernmost of the major streams that flow generally southwest, following the regional slope of the Sierra Nevada. The western part of Sequoia, which is drained by the Kaweah River, is rugged mountain country, deeply dissected by branching canyons and ridges. The main canyon of the Kaweah at Ash Mountain Park Headquarters is more than 4,000 feet deep. Several of the branch canyons, such as East Fork and Middle Fork, are even deeper. However, some plateaulike uplands are found in the Kaweah Basin. Giant Forest occupies one of these gently sloping platforms. The relatively flat surface is apparently due to the resistant nature of the underlying massive granite bedrock.

The Kern River rises in the upper Kern Basin in the eastern part of the Park, between the Great Western Divide and the main crest of the Sierra. The Kern River is

Figure 27.4 The glacially eroded peaks of the Great Western Divide trend generally north-south across the boundary between Sequoia National Park and Kings Canyon National Park. National Park Service photograph.

less deeply entrenched throughout its course than the Kaweah. The Kern's modified U-shaped valley is incised some 2,000 feet in the high upland surface, between the Boreal Plateau on the east and Chagoopa Plateau on the west. Tributaries in hanging valleys at right angles to the main stream are notched in the canyon walls. The Kern River's southward course generally follows the trend of a north-south fault in the Sierra block. The fracture zone of the fault provided a less erosion-resistant course for this subsequent stream (p. 71) as the crustal block was uplifted. The canyon becomes shallower as it approaches the southern boundary of the Park, and thick glacial and alluvial deposits cover the bedrock floor (fig. 27.5).

Geologic History of Sequoia and Kings Canyon National Parks

1. Paleozoic deposition.

Not much is known about the marine sedimentary rocks that were deposited in Paleozoic seas that spread over the region where the Sierra Nevada now stands. Some Paleozoic rocks crop out east of the range, and a few scattered patches have been found in the mountains some distance north of Sequoia and Kings Canyon National Parks.

2. Mesozoic deposition and mountain-building.

Metamorphosed volcanic and marine sedimentary rocks of Triassic and Jurassic age have been preserved as roof pendants and inclusions in the Sierra Nevada batholith. The severe deformation and metamorphism that accompanied the emplacement of the batholith destroyed most of the marine fossils in the rocks. Most of the rocks are marbles (such as those exposed in the caves), schists, and cherts, with intermixed meta-volcanics. In the western parts of Sequoia and Kings Canyon, the metamorphic rocks are well exposed in the canyons of the Kings River and Kaweah River drainages.

The plutons of the batholith, which are Cretaceous in age, were emplaced in a series of pulses; the older granites are found in the western parts of the range and the younger granites are near the crest. In general, the older granites are more mafic in composition (quartz diorites, for example), and the younger granites are more felsic (granodiorites, quartz monzonites).

The Nevadan orogeny that uplifted the Sierra Nevada is regarded as the first wave of the Laramide mountain-building episodes that created the great North American Cordillera. The tectonic activity began about 150 million years ago with the subduction of the Pacific plate. The extensive erosion that accompanied the orogeny stripped off many thousands of feet of Paleozoic and Mesozoic deposition.

3. Cenozoic faulting and uplift.

Erosion continued during the beginning of the Cenozoic Period, but tectonic activity had slowed. Then the Sierra Nevada began to tilt westward. Faulting continued as the mountain block rose higher and higher, and the range began to assume its present position and form. This extensive faulting, uplift, and tilting probably began 2.5 million years ago. Sporadic earthquakes in the southern Sierras are an indication that movement and uplift are continuing. Late in the Cenozoic, some volcanism occurred in the southern Sierra, but those lava flows are outside Park boundaries.

Figure 27.5 Looking south down the U-shaped glacial valley of the Kern River from near the mouth of Whitney Creek. The broad uplands that slope gently to the top of the canyon walls are about 2,000 feet above the river. The stream valley was probably incised in the high upland surface during uplift of the Sierra Nevada fault block and later modified by glacial erosion. U.S. Geological Survey low-altitude, oblique photograph.

4. Pleistocene and Holocene glaciation.

Episodes of glaciation in the southern Sierra were less severe and less extensive than those in the northern Sierra. Pleistocene climates were less harsh in the southern Sierra, and less precipitation was available. Overall, however, the patterns of advance and retreat of ice were similar. Alpine glaciation began early in the Sierra Nevada, probably late in the Pliocene and long before the large continental glaciers accumulated. In Sequoia and Kings Canyon, valley glaciers moved down major stream valleys, enlarging and deepening them. Glaciers were largest and most numerous in the zones of high precipitation on the western slopes.

The valley glaciers on the east side of the crest strongly eroded the peaks but did not extend down into the Owens Valley.

At its maximum, the Kern Glacier, the most southerly of the Sierra Nevada trunk glaciers, nearly reached the southern boundary of Sequoia National Park. In the Kaweah Basin, glaciers partly filled many of the converging canyons, but few of them overtopped the divides except in the upper reaches of the basin. Even at their maximum, the largest glaciers in the northeast part of the basin probably did not extend much below the 5,000-foot level.

Glacier systems were more extensive in Kings Canyon National Park, especially toward the north. Valley glaciers occupied the canyons of the Kings River and its tributaries, as well as the upper reaches of the San Joaquin River basin at the northern tip of the Park. Glacial erosion was severe.

Three major episodes of glaciation occurred during the Pleistocene, with substages involving advances and retreats of ice during these generally colder climatic periods. Till patches and moraines of varying ages mark glacial recessions. The Pleistocene glaciers did not last through the post-Pleistocene climatic optimum.

In Kings Canyon National Park, a few glaciers began occupying the cirques of earlier glaciers during the Little Ice Age (A.D. 1450–1850). These small valley glaciers are in the vicinity of Mount Goddard in the northern part of the Park; others are in the Palisades region, just east of the Park's northeastern boundary.

Prospects for the Future

More than one hundred years have passed since John Muir began his fight to preserve the California wilderness. Yet today, the national parks and national forests of the High Sierra are seriously threatened by throngs of visitors. The crowds have created problems of overuse that tend to damage the fragile environment, especially in the Alpine and Subalpine life zones. Because Sequoia and Kings Canyon are the national parks closest to the Los Angeles metropolitan area, they are particularly endangered.

Ironically, a recent ten-year study by the National Park service indicated that the giant sequoias may also be threatened by *overprotection*. Periodic thinning by fires is needed to remove undergrowth and allow seeds to reach the soil and root in the sunlight. A program of "controlled burning" has been started by the Park Service to help ensure new generations of Sequoias.

How to balance the needs of a large urban population for enjoyment of the parklands and at the same time protect the park environment from the damaging effects of too many people and vehicles, while avoiding overprotection, is a difficult challenge. For the National Park Service, the task involves careful management in the present, long-range planning for the future, and, above all, cooperation and understanding on the part of park visitors.

Geologic Maps and Cross Sections

American Association of Petroleum Geologists. 1962. *Geological highway map no. 3, Pacific Southwest region.* Tulsa, Oklahoma: American Association of Petroleum Geologists.

Jennings, C. W. *Geologic map of California.* California Division of Mines and Geology. Sheets covering southern Sierra Nevada.

Ross, D. C. 1958. *Igneous and metamorphic rocks of parts of Sequoia and Kings Canyon National Parks, California.* California Division of Mines, Department of Natural Resources Special Report 53.

Bibliography

Bateman, P. C. 1974. Model for the origin of Sierra granites. *California geology* 27(1) :3–5.

Bateman, P. C., and Wahrhaftig, C. 1966. Geology of the Sierra Nevada. In *Geology of northern California*, ed. E. H. Bailey, pp. 107–172. California Division of Mines and Geology Bulletin 190.

Bowen, E. 1976. *The High Sierra*, American Wilderness Series. Rev. ed. New York: Time-Life Books.

Colby, W. E., ed. 1960. *John Muir's studies in the Sierra.* San Francisco: Sierra Club.

Frome, M. 1980. *National Park guide.* 16th ed. Chicago: Rand McNally & Company.

Gray, W. R. 1975. *The Pacific crest trail.* Washington, D.C.: National Geographic Society.

Hamilton, W. 1969. Mesozoic California and the underflow of Pacific mantle. *Geological Society of America Bulletin* 80: 2409–2430.

Hartesvelde, R. J. 1973. *Field guidebook for Sequoia and Kings Canyon National Parks; the natural history, ecology, and management of the giant Sequoias.* American Association of Stratigraphic Palynologists, 6th annual meeting, Anaheim, California, October 16–20.

Hill, M. 1975. *Geology of the Sierra Nevada.* Berkeley: University of California Press.

Matthes, F. E. (ed. F. Fryxell). 1950. *Sequoia National Park, a geological album.* Berkeley: University of California Press.

———. (ed. F. Fryxell). 1965. *Glacial reconnaissance of Sequoia National Park, California.* U.S. Geological Survey Professional Paper 504-A.

Oakeshott, G. B. 1978. *California's changing landscapes.* 2nd ed. New York: McGraw-Hill Book Company.

Oberhansley, F. R. 1974. *Crystal Cave in Sequoia National Park.* Sequoia Natural History Association.

Ross, D. C. 1958. *Igneous and metamorphic rocks of parts of Sequoia and Kings Canyon National Parks, California.* California Division of Mines, Department of Natural Resources Special Report 53.

Sharp, R. P. 1976. *Geology field guide to Southern California*, K/H Geology Field Guide Series. Rev. ed. Dubuque, Iowa: Kendall/Hunt Publishing Company.

Wahrhaftig, C. 1965. Stepped topography of the southern Sierra Nevada, California. *Geological Society of America bulletin* 76:1165–89.

Wheelock, W., and Condon T. 1974. *Climbing Mount Whitney*. 3rd ed. Glendale, California: La Siesta Press.

Address

Sequoia National Park
Three Rivers, California 93271

28

Kings Canyon National Park

Location: East Central California
Area: 460,136.19 acres; 718.96 square miles
Established: March 4, 1940

Figure 28.1 Rugged mountain terrain in Kings Canyon National Park developed on massive plutonic rock. The canyon of the Middle Fork of the Kings River can be seen in the middle distance. National Park Service photograph.

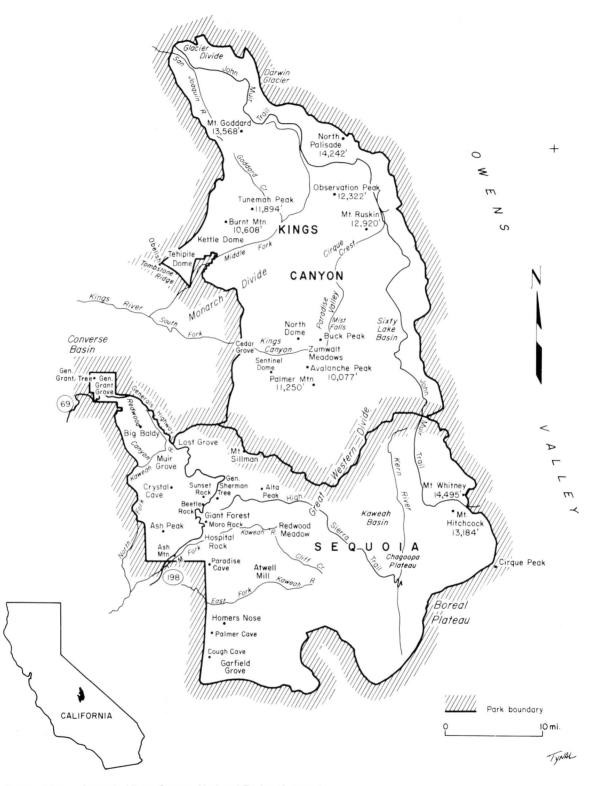

Figure 28.2   Sequoia-Kings Canyon National Parks, California.

Table 28.1. Generalized Geologic Column, Sequoia and Kings Canyon National Parks

Era	Period	Epoch	Rock Units	Geologic Events
		Holocene	Alluvial, glacial, and mass wasting deposits	Stream erosion; Neoglacial activity; landslides, rockfalls Earthquakes, faulting, continuing uplift
	Quaternary	Pleistocene	Till, moraines, glacial outwash of various ages Avalanche deposits	At least three major glaciations, including substages; intense glacial erosion, avalanche chutes, ice-sculptured landscapes, glacial deposition Faulting, relatively rapid uplift
Cenozoic		Pliocene		Probable onset of glaciation Southern Sierra volcanism (outside Parks) Canyon-cutting, intense erosion Beginning of faulting and uplift Westward tilting of Sierra Nevada block
	Tertiary	Miocene		
		Oligocene		Erosion
		Eocene		Pause in tectonic activity
		Paleocene		Probable beginning of cave development
Mesozoic		Cretaceous	Quartz diorites, granodiorites, quartz monzonites	Emplacement of plutons in composite batholith Metamorphism of country rock Nevadan orogeny
		Jurassic		Marine sedimentation and island arc volcanism
		Triassic	Limestones, shales, sandstones, interbedded with volcanics	Subduction and plate convergence
Paleozoic			(not exposed in Parks)	Long accumulation of sediments in Paleozoic seas

Source: modified after Hill 1975; Bateman 1974; Bateman and Wahrhaftig 1966

Local History

General Grant Grove was set aside as a national park in the same year Sequoia National Park (chapter 27) was established. The area was expanded and redesignated in 1940 when a large tract of land extending northward from Sequoia was combined with General Grant Grove to form Kings Canyon National Park. Later the Redwood Canyon area was added. The two Parks, Kings Canyon and Sequoia, are administered as a single unit.

In 1805 Spanish explorers crossed the Kings River (probably near the western foothills of the Sierra) and named it *El Rio de los Santos Reyes*, or "the river of the holy kings." Americans later shortened the name to Kings River. The canyon of the South Fork of the Kings River is 7,000 feet deep—deeper than the Grand Canyon— and it gives the Park its name. The Middle Fork of the river has also cut a deep canyon.

The giant sequoias in General Grant Grove were discovered in 1862 by Joseph Thomas. Five years later, a Mrs. Baker of Visalia, California, visited the grove and named the largest tree in honor of General Ulysses S. Grant, who was soon to become eighteenth president of the United States. She sent small branches of the tree to General Grant when she informed him of the giant sequoia. Congress declared the tree a national shrine and a memorial to veterans in 1956. General Grant is the second largest tree in the world, after General Sherman in Giant Forest in Sequoia National Park.

A giant sequoia that was felled in 1875 was cut into segments and shipped to Philadelphia for the 1876 Centennial Exposition. It was reassembled and exhibited, but Exposition visitors thought it was a fake! The huge stump of the tree still stands in the Park and is called Centennial Stump. This and other reminders of logging days can be seen in Big Stump Basin, where young sequoias are beginning to rise above the sawed-off trunks of the fallen giants.

The Generals Highway, which enters Sequoia National Park from the south, goes through General Grant Grove, then follows the canyon of the Kings River across

Sequoia National Forest, and reenters Kings Canyon National Park where South Fork crosses the boundary. The highway ends at Copper Creek, about ten miles inside the Park. From this point, only horseback and hiking trails lead into the mountain wilderness. In fact, about four-fifths of the area of both Sequoia and Kings Canyon National Parks is accessible only by trails.

Redwood Canyon, between General Grant Grove and Sequoia National Park, has a large grove of sequoias on the flanks of Redwood Mountain. Tungsten and copper ores were taken from mines in Redwood Canyon and along the North Fork of the Kaweah River in the early 1940s, before the area was placed under national park jurisdiction.

Geologic Features

As in Sequoia National Park, exfoliation domes and glacial features are prominent in Kings Canyon National Park. Near the Middle Fork of the Kings River are a group of conspicuous exfoliation domes, including Tehipite and Kettle Domes and the Obelisk. Two large domes that can be seen from the Generals Highway are Sentinel Dome and North Dome. These domes, over 8,500 feet in elevation, are formed in massive granite.

Pleistocene glaciers scoured some of the domes and excavated U-shaped valleys. The largest and longest of the valley glaciers was the one in Kings Canyon; it was a river of ice 44 miles in length that reached a relatively low altitude of 2,500 feet on the western Sierra slope. Hanging valleys, some with beautiful waterfalls such as Roaring River Falls and Mist Falls, once held tributary glaciers that fed Kings Glacier. A few rock glaciers still exist at high altitudes. Sixty-Lake Basin, near the Sierra Crest, is a glacial staircase. Many small rock-basin lakes dot the broad Monarch Divide (elevation 10,000 to 12,000 feet) that separates the Middle Fork from the South Fork.

The floor of Kings Canyon (South Fork) is from one-fourth to one-third of a mile wide and covered with outwash and glacial debris. Steep granite walls rise on either side, giving the canyon the typical U-shape of a glacial trough. By contrast, much of the canyon of the Middle Fork of the Kings River is V-shaped, although the upper reaches are U-shaped (fig. 28.3). The junction of the two Kings River forks in the Sequoia National Forest west of the Park can be seen from an overlook on the Generals Highway.

Figure 28.3   Glaciers did not extend to this part of the canyon of the Middle Fork, Kings River. This view looks northeast from a point near Hume Lake. The tops of

Tehipite Dome and Kettle Dome are visible in the far distance. Both are exfoliation domes of massive granite. National Park Service photograph.

A few modern glaciers that formed in Neoglacial time, such as the Darwin Glacier, cling to cirques on the mountains along the northeast boundary of the Park. Their activity is minor compared to the severe glacial erosion and extensive deposition of their Pleistocene predecessors.

Many of the topographic features of Kings Canyon National Park are grand in size and simple in form. Relief is very high. Drops of several thousand feet between ridge top and canyon bottom are characteristic of the coarse drainage texture in this part of the Sierra. *Coarse-textured drainage* means that the number of streams and valleys per square mile (or given area) is relatively low. On the massive granites of the tilted Sierra block, only a few major streams developed. The reason is that as the streams cut deep valleys during the rapid uplift, the larger rivers collected (or captured) the waters of the smaller streams. Moreover, the drainage networks were greatly modified by several cycles of glaciation during uplift. Like the Kaweah drainage in Sequoia, the main tributaries of the Kings drainage have a dendritic pattern and follow the trend of the regional slope.

Roof Pendants. Extensions of the roof pendants on the western side of Sequoia crop out in Kings Canyon National Park. The mines in Redwood Canyon and along the North Fork of the Kaweah River are in *tactite*, a complex calcareous metamorphic rock formed as a result of hydrothermal and contact metamorphism. During the emplacement of the adjacent plutons, chemically active solutions from the hot magma penetrated the country rock and replaced some of its constituents with new minerals. Contained in the tactite are garnet, pyroxene, quartz, scheelite, chalcopyrite, and bornite. Scheelite is an ore of tungsten. Chalcopyrite and bornite are sources of copper.

In the northern part of the Park is a large roof pendant, trending northwest-southeast, that continues beyond the Park's northern boundary. This is the Goddard Pendant, roughly two miles wide and more than 20 miles long, that is composed primarily of Jurassic metavolcanics. Intrusive breccias crop out along the margins of the pendant, and many dikes, both felsic and mafic, are emplaced in the metamorphic rock and the sheared granites associated with the pendant structure.

A smaller roof pendant, the Bishop Creek Pendant, which contains Paleozoic marbles and other calcareous rocks, is located outside the Park, close to the northeast boundary.

Geologic History

See chapter 27, Sequoia National Park.

Geologic Maps

Bateman, P. C. 1965. *Geologic map of the Black Cap Mountain quandrangle, Fresno County, California.* U.S. Geological Survey GQ Map 428.

Bateman, P. C., and Moore, J. G. 1965. *Geologic map of the Mt. Goddard quandrangle, Fresno and Inyo Counties, California.* U.S. Geological Survey GQ Map 429.

See also map and cross section references in chapter 27.

Bibliography

See chapter 27, Sequoia National Park.

Address

Kings Canyon National Park
Three Rivers, California 93271

29
Channel Islands National Park

by Donald F. Palmer
Kent State University

Location: Southern California (offshore)
Area: 246,870 acres; 385.73 square miles (includes
122,870 acres of submerged land; 124,115 acres
above sea level)
Proclaimed a National Monument: April 26, 1938
Designated a National Park: March 5, 1980

Figure 29.1 This view from behind the beach at Cuyler Harbor on San Miguel looks north toward the barren rock mass of Prince Island. The sand dunes on the backshore are partially stabilized by sparse vegetation. Photograph from Channel Islands National Park Collection.

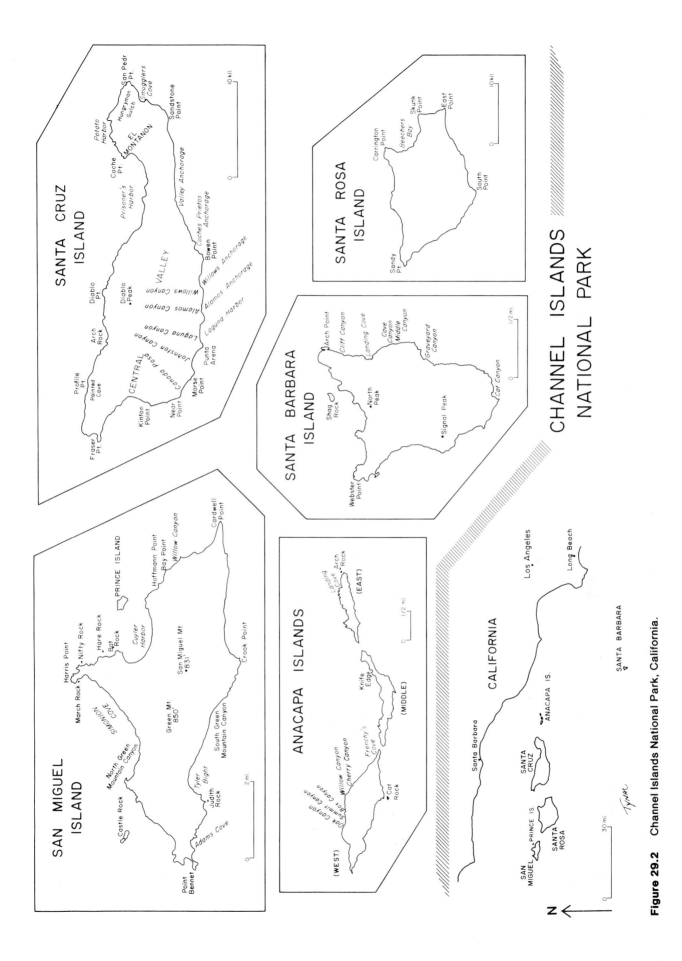

Figure 29.2 Channel Islands National Park, California.

Table 29.1. Geologic Column, Channel Islands National Park

Era	Period	Epoch	Formation	Geologic Events
Cenozoic	Quaternary	Holocene	Alluvium, stream and landslide deposits	Continued uplift and sea level changes
		Pleistocene	Terrace gravels	Repeated uplift and development of marine terraces
			Middle Anchorage	Sedimentation of conglomerate
			alluvium	
			Potato Harbor Formation	Marine transgression, late volcanic activity
	Tertiary — Neogene	Miocene	Monterey and Blanca Formations	Uplift, faulting, and exposure of the basement complex associated with development of the Transverse ranges, movement and rotation due to plate tectonics.
			Santa Cruz Island Volcanics	Rapid accumulation of volcanic lavas, breccias, and pyroclastic debris. Erosion and reworking of older sediments and deposition under increasing depths of water.
			San Miguel clastic and volcanic units	
	Tertiary — Paleogene		San Onofre Breccia	
		Oligocene	Rincon Formation	Deposition of marine and nonmarine clastic sediments mixed with volcanics. Uplift in source areas to the north and east.
			Vaqueros Formation	
			Sespe Formation	
		Eocene	Cozy Dell Formation	Inception of coarse clastic sedimentation due to uplifts in surrounding area. Deep water sedimentation
			South Point and Jolla Vieja Formations	
			Cañada Formation	Continuous deposition in a continental slope environment with a gradually transgressive sea.
		Paleocene	Pozo Formation	
Mesozoic	Cretaceous (Upper)		Jalama Formation	Marine sedimentation and the Laramide orogeny
	Jurassic		Alamos Pluton Willows Plutonic Complex Santa Cruz Schist	Volcanic activity and sedimentation, intrusion of plutons and regional metamorphism.

Sources: modified after Weaver 1969; Howell 1976.
Note: ∿∿∿ erosional events.

Geographic Setting and Local History of the Islands

Channel Islands National Park, created in 1980, includes the six islands of Santa Barbara, Anacapa, Santa Cruz, Santa Rosa, San Miguel, and Prince Island. Of these, Santa Barbara (one of the Southern Channel Islands), Anacapa (one of five Northern Channel Islands), and San Miguel (a Northern Channel Island owned by the United States Navy) previously constituted a National Monument. The inclusion of the three remaining Northern Channel Islands expands the Park as a haven for endangered species of plants and wildlife, many of which are unique to the islands, and also preserves rookeries for sea birds and marine animals such as fur seals, elephant seals, and sea lions. Santa Cruz, for example, is the home of nine endangered plant species as well as of the endangered California brown pelican. San Miguel has the only marine mammal rookery in the world inhabited by six species of seals and sea lions. The unique endangered island fox, much smaller than mainland species, is found on four of the islands. In addition to its rookery, Santa Barbara has a treelike sunflower of the genus *Coreopsis*.

Before the Park was created, the fragile ecosystems of the islands and their surrounding waters were threatened by such human activities as ranching, trapping, oil spills, offshore oil exploration, a proposed pipeline compressor facility for crude oil, and residential development.

Early Inhabitants. The Channel Islands had long supported prehistoric cultures prior to the arrival of the Spanish explorers in the 16th century. Recently discovered evidence indicates that Santa Rosa Island may have had human inhabitants as far back as 30,000 or more years ago. Firepits on Santa Rosa and Santa Cruz have been dated at 37,000 B.P. (Before Present). Roasting pits with dwarf mammoth bones are dated as old as 29,750 B.P.. Definite human skeletal remains have been dated as 11,000 B.P., which was approximately the end of Pleistocene time. The climate then was more moist; the drier climate of today began around 7,500 B.P.

During late Pleistocene time, when sea levels were lower, the Channel Islands occupied a larger land area as part of the Transverse Ranges of Southern California. Pleistocene mammoths that had migrated to the southwestern tip of the area were cut off from the mainland when sea level rose, creating the islands. Once isolated, this mammoth evolved into a pigmy variety that was only four feet high at the shoulder. Indian hunters may have been responsible for their extinction.

The people of the Dune Dweller Culture (c. 7,500 B.P.) depended on sea life for food. Highland Culture Indians (c. 5,000 B.P.) lived in the Catalina cherry and oak forests, and gave way to the seafaring Chumas or Canalino Indians (c. 3,000 B.P.). The Canalinos used tar from oil seeps on San Miguel and the mainland coast as caulking material for their canoes, for mending bowls, and as adhesive for attaching arrow points and other weapon heads to shafts. It was the Canalino Indians who greeted Rodriguez Cabrillo, the discoverer of California, when he stopped at San Miguel in 1542. Cabrillo, a Portuguese mercenary in the service of Spain, led an expedition for the Spanish viceroy to find a sea route to China. After sailing north along the California coast beyond San Francisco, Cabrillo returned to San Miguel to spend the winter. He died on the island in January of 1543.

Sebastian Vizcaino, a Spanish explorer and merchant, led a second expedition up the California coast in 1602. He discovered and named Monterey Bay and drew the first chart that showed Santa Cruz and other Channel Islands.

Santa Cruz, the largest of the Channel Islands, had a population of some 3,000 Indians when the Spaniards first visited it. The island was given its present name in 1769 by Juan Perez, a Spanish naval officer, after the Indians found an iron crucifix that had been lost by the expedition's chaplain. To commemorate the event, Perez struck out earlier names on the charts and declared that henceforth the island would be called *Santa Cruz* ("holy cross").

Diseases introduced by Europeans reduced the native population of the islands in the 17th and 18th centuries. The Canalino Indians abandoned Santa Rosa after a severe earthquake in 1812, and the last Indian was removed from Santa Cruz in 1813. By 1834, no Indians remained on any of the Channel Islands.

Ever since the beginnings of European settlements, the Channel Islands have had separate histories based on the different ownerships and types of land use.

Santa Rosa Island. In 1843 Santa Rosa was awarded to the Carrillo family as a land grant from the government of Mexico, and for more than fifty years the Carrillos and their descendants maintained a prosperous sheep ranch on the island. W. L. Vail and J. V. Vickers bought Santa Rosa in 1902, converted the ranch to a cattle operation, and removed all sheep from the island. The Vail and Vickers Company still owns the island and uses it as a cattle ranch. Eventually the island will be purchased by the federal government.

Santa Cruz Island. The Mexican government had a penal colony on Santa Cruz for a short time in 1829, just long enough to give Prisoners Harbor its name. Andres Castillero was given a land grant for the island in 1838 by the governor of Mexican California but apparently did not develop the property. Sheep and other domestic animals were first brought to the island by an English company that acquired the island in 1859. Ten years later the island was sold to Justinian Caire and his California associates. Caire operated a sheep and cattle ranch and produced wine for a time under the Santa Cruz Island label.

His descendants still own 6,000 acres on the eastern end of Santa Cruz where they operate a sheep ranch. The rest of the island (about nine-tenths of the total acreage) is owned by the Santa Cruz Island Company. Most of the sheep have been removed from the island and those that remain are confined by fences. The grazing of cattle is also restricted to parts of the island that are less subject to damage by erosion.

A University of California research facility, the Channel Islands Field Station, was established on the island in the 1960s in cooperation with the Santa Cruz Island Company. The field station office is located on the University of California–Santa Barbara campus. Eventually the entire island will be owned and managed as a preserve by the Nature Conservancy in cooperation with the National Park Service.

San Miguel Island. The United States government acquired San Miguel when Mexico ceded California territory to this country in 1848. George Nidever, who brought sheep, cattle, pigs, and horses to the island in the 1850s, was the first leaseholder to live on San Miguel. The adobe house that he constructed from local clay is still standing. Prior to the introduction of domestic animals, the island had a luxuriant ground cover, but a combination of overgrazing and repeated droughts destroyed most of the vegetation, and sand dunes blown in by the prevailing northwest winds prevented regrowth. The severe drought of 1863–64 killed 85 percent of Nidever's stock (5,000 sheep, 180 cattle, and 30 horses). An 1870 drought caused the remaining sheep to dig as deeply as two or three feet into the ground for roots; and what the sheep didn't kill, the drifting sand dunes did. Subsequent droughts in 1877, 1893–94, and 1923–24 changed the landscape to near desert conditions which are able to support only a few grasses and wildflowers. The Lester family in the 1920s and 1930s were the island's last residents. The few remaining sheep were removed in the 1960s. The U.S. Navy owns the island and administers it in cooperation with the National Park Service.

Anacapa Island. Anacapa, which is actually three islets (East, Middle, and West) linked by sand bars, was acquired by the United States from Mexico at the same time as San Miguel. In the late 19th century and early 20th century, a few sheep were pastured on the island, but all were removed when Anacapa became part of the National Monument in 1938.

Arch Rock (also called Cabrillo's Arch), located at the eastern tip of East Island, is a well known landmark on Anacapa. James Whistler, the noted 19th-century American painter and etcher, made an etching of this 40-foot high sea arch when, as a young man, he worked as a map engraver for the U.S. Coast and Geodetic Survey.

An automated, 600,000-candlepower beam and foghorn were installed in the lighthouse on East Island in 1966. Prior to that time, Coast Guard personnel had manned the lighthouse, which was constructed in 1932. An acetylene beacon had given warning to coastal shipping for many years before that. In spite of the warning system, wrecks and groundings have occurred on Anacapa's rocky shores, usually because of dense fog conditions. A tuna clipper sank off Middle Island near a commercial fishing camp in 1949. Perhaps the most famous of the shipwrecks was that of the paddle-wheel steamer "Winfield Scott." When it sank in 1853, all 250 passengers got off safely, but they spent several anxious days on Anacapa before being rescued by the S.S. "California." The wreck's remains can still be seen offshore in about 20 feet of water.

Frenchys Cove on the northeast side of West Anacapa was named for a hermit, Raymond LeDreau, who lived alone on the island from 1928 to 1956. Cat Rock, a sea arch on the south side, is supposedly where Frenchy's cats were stranded at one time.

The name Anacapa is from an Indian word, *eneepah*, which means "deception, or mirage"; from a distance the Indians' view of the three islets was distorted. Under certain conditions of weather and light, this distortion can still be seen—probably because, when it is viewed from the north, the marked difference in slope between the north and south sides of the islets causes Anacapa to appear to be a larger landmass than it actually is.

Santa Barbara Island. Sometimes referred to as "Santa Barbara Rock" because of its small size (approximately a square mile), this island is the only one of the Southern Channel Islands to be included in Channel Islands National Park. Archeological evidence indicates occasional extended visits here by prehistoric people but no permanent camps, probably because of the lack of fresh water on the island. A few seeps, such as the one in Indian Water Cave, are the only natural sources of fresh water.

Originally the island was well covered with vegetation and supported a number of bird species, small animals, and marine shore species. However, a century or more of human occupation has drastically changed the island's natural environment, and many of the changes are irreversible. Farming and livestock-raising were carried on for many years. During World War II, the U.S. Navy maintained an early-warning outpost on the island, and in the 1950s, a missile tracking station.

The changes in the island's character can be summarized as follows (with many of the items applying to other Channel Islands as well):

1) The introduction of cats eliminated several species of sea birds.
2) Domestic sheep, rabbits, goats, etc., reduced or eliminated many native plants.
3) Clearing of land for farming removed much of the original vegetation. Native species are left only in isolated, protected areas.

4) Fires, both deliberate and accidental (the worst one in 1959), have burned most of the vegetation to the water's edge. They have destroyed so much of the island's habitat that the Santa Barbara song sparrow has become extinct.

5) The introduction of grasses and mainland weeds forced out native vegetation.

6) The planting of grains brought about overpopulation of native deer mice who, in turn, are decimating the land mollusks and the eggs of sea birds.

7) The introduction of the "ice plant" (*Mesembryanthemum crystallinum*), a fleshy-leaved plant with glistening encrustations, is crowding out native species.

8) The combined effect of burning, exotic plants, and domestic animals has caused considerable erosion in some areas.

Present Use of the Islands in the Park

Primitive camping is permitted on Santa Barbara and Anacapa Islands, and National Park Service rangers live on both islands. The remaining islands may be visited for one day only with permission from the National Park Service or property owners. Tourists on San Miguel must be accompanied at all times by Park Service personnel. Sensitive areas, defined as habitats of endangered plant or animal species, are closed to the public. Qualified scientists, however, are permitted to carry on environmental research projects on several of the islands.

The Channel Islands have generally cool, foggy summers and rainy winters (about 12 inches of rain annually). In winter, during the migration season, the islands are some of the best areas along the California coast for watching whales.

Geologic and Other Natural Features of the Channel Islands

In the 1870s and 1880s, the city of Santa Barbara on the California coast opposite the Channel Islands was a popular health resort because of the numerous natural seeps of tar, oil, and gas that were active along the shoreline there. At that time, natural oil slicks, spread over the coastal waters by winds and currents, were believed to purify the air, making infectious diseases and chronic illnesses less prevalent. Most of the seeps emanated from shale in the Monterey Formation, which crops out in some of the Channel Islands as well as on the mainland. San Miguel has two oil seeps from the Monterey Formation, one near Castle Rock, off the island's northwest shore, and the other near Tyler Bight, off the south shore. Another seep has been burning at the eastern end of Santa Cruz Island near Chinese Harbor since before written records existed.

As noted earlier, the Channel Islands are in two geographically separate groups. The Northern Channel Islands (San Miguel, Prince, Santa Rosa, Santa Cruz, and Anacapa), all of which are in the Park, lie across the Santa Barbara Channel from the mainland. The Southern Channel Islands, which include Santa Barbara (the only one in the Park), San Nicholas, Santa Catalina, and San Clemente, are separated from the mainland by the San Pedro Channel and the Gulf of Catalina. In spite of the fact that the islands are relatively close together, each one has its own history. Thus it is advantageous to look at the islands both together and separately.

First of all, if the ocean water were to disappear so that only land could be seen, it would be obvious that the islands stand as high ridges on the continental shelf and that they are separated by a series of basins. Marine terraces on the islands indicate that intermittent uplift of the islands or lowering of sea level, or both, has occurred in the geologic past.

The islands are located in a transitional region between northern and southern marine fauna, with Point Conception (located on the mainland 23 miles due north of San Miguel) being the dividing point. Therefore, the westernmost islands—San Miguel, Prince, and San Nicholas (not in the Park)—support mainly northern marine fauna since they are in the path of the cold water of the south-flowing California Current. Because of the warm water countercurrent from the south, the easternmost islands—San Clemente, Santa Catalina, and Santa Barbara—support primarily southern species. Anacapa, Santa Cruz, and Santa Rosa, which lie between the two currents, have mixtures of southern and northern species. Waters surrounding all the islands are rich in marine algae and kelp, which are food for marine organisms of many types. About 20 species of whales and porpoises abound in the area, along with six species of sea lions and harbor seals and many species of sea birds.

Some of the islands lack trees, but the vegetation on most of the islands is sage scrub and grasses along the shorelines, with oak woodlands mixed in on the larger islands. Islands that have (or have had) sheep are nearly stripped of vegetation, except where sheep have been restricted to fenced areas, as on Santa Cruz.

Each island has endemic species and subspecies of plants and animals that live only on one island or, in some cases, adjacent islands; examples are unique snails, deer mice, island foxes, and many kinds of plants. Both native and feral animals are found on the islands at the present time. The feral animals—rabbits, cats, sheep, pigs, etc., that have reverted to the wild—have altered the environment and in some instances caused the disappearance of some native species.

Among the many threats still endangering the environments of these peaceful islands are overharvesting of marine resources, such as anchovies (main food of the California brown pelican) and kelp; overgrazing; oil spills from passing tankers; development of offshore oil and gas; sonic booms that disturb seal and bird rookeries; and the introduction of exotic species.

San Miguel Island

Structurally, San Miguel, 14 square miles in area, is the northern flank of a faulted anticline. Its axis trends northwest, which is also the strike of the beds that dip northeast. The ages of the bedrock range from Cretaceous to Miocene. The rocks are mostly sedimentary, and the beds form a homoclinal structure.

Several levels of beaches encircle the island. Because of overgrazing and the loss of vegetation cover, the island slopes are severely gullied and the western two-thirds of the island is covered by sand dunes. Rainwater trapped by the dunes is the source of fresh water springs, making San Miguel one of the few Channel Islands with potable (drinkable, although poor tasting) water. Some of the dunes are being stabilized by regrowth of vegetation and by cementation due to ground water action.

Since most of San Miguel's land area is a plateau (400 to 500 feet high) with two rounded hills, it is lower in relief and more subject to wind erosion than the other islands (fig. 29.3). Prevailing northwest winds have blown away most of the topsoil, exposing caliche and dune rock (eolianite) at the surface. *Caliche*—composed of crusts of soluble calcium salts, mixed with sand, silt, etc.—is formed mainly by capillary action drawing mineral-rich ground water to the surface where it evaporates in the wind and sun, leaving the salts. Caliche also tends to from in a layer just below the surface when not enough moisture is available to flush the calcium salts from the soil. Both surface and subsurface caliche deposits occur over most of the island. *Dune rock*, an *eolianite* made up largely of dune sand, is formed from wind-deposited clastic sediment that has been cemented below the surface by ground water and calcite.

On San Miguel the dune rock is interbedded with layers of soil that represent periods when conditions were stable long enough to permit soil formation by weathering of bedrock. Unstable (and probably drier) periods brought about the resumption of dune growth, and dunes buried the vegetation. Concretions of caliche formed as nodules,

Figure 29.3   Dunes extend parallel to the direction of the strong, prevailing northwest winds on the exposed western end of San Miguel Island on land formerly covered by vegetation. Photograph from Channel Islands National Park Collection.

tubes, and casts around roots and tree trunks that were covered by the sand dunes. Gradually the dunes were consolidated into a dune rock layer. A caliche-encased log, 30 feet long and over 2 feet in diameter, was found on the island. Trunk casts of other logs have also been uncovered by erosion.

Case-hardened caliche is common on the northern and western parts of the island. Here the caliche deposits were exposed to rain and sea spray often enough so that the calcium salts on the surface were dissolved and reprecipitated many times, making the crusts thick and hard. Slabs and large chunks of caliche are scattered over the gound, and the sand dunes have become semi-indurated.

Of interest to geologists are the "elevator rocks" of San Miguel. They are found as a pebble and cobble pavement on old marine terraces that have been raised above sea level and then buried by sediment washed down from still higher terraces or slopes. When enough moisture is present, soil churning forces the coarser gravel and cobbles to the surface, and the finer grains are blown away by wind or washed away by water.

There are many paleontological and archeological sites on the island, including over 500 Indian sites and a large accumulation of pigmy mammoth bones at Running Springs (near Dry Lake on the west side of the island). Another pigmy mammoth site is about three miles west of Cardwell Point on the southeast shore of the island. The bones at Running Springs have been well preserved by calcareous tufa deposited around the springs, which have a high dissolved load of calcite.

The pigmy mammoth *(Mammuthus exilis)* was only four feet high at the shoulder. Carbon-14 dating of mammoth remains yields ages such as $15,630 \pm 460$ years and $16,520 \pm 150$. Burned and calcined mammoth bones, charcoal made of cypress, and mammoth tusks and teeth all suggest that prehistoric people were also present at the sites. The evidence strongly indicates that early man lived on the island and hunted the pigmy mammoth. There is geologic evidence on San Miguel of overgrazing by the pigmy mammoths, periodic forest fires set by man and by lightning, and a 20-year drought cycle.

San Miguel is noted for its unique plant and animal species, which are now under the protection of the National Park Service. The island is on the migration pathway of the humpback, blue, finback, and sei whales. Six species of seal (five of which breed here) frequent the seal rookery at Point Bennett on the western tip of San Miguel. These species are the harbor seal, California sea lion, northern elephant seal, Steller sea lion, northern fur seal, and the Guadalupe fur seal (nonbreeding at this rookery). Eight kinds of plants are found only on San Miguel.

Prince Island

Tiny Prince Island (also called Gull Island) lies a short distance from the northeast shore of San Miguel (fig. 29.1). The highest elevation on Prince Island is only 296 feet, and its dimensions are about 1,000 by 1,500 feet. The island is a major sea bird rookery and it is also in the whale migration pathway. Thin, shallow soils mantle the rocky slopes of this foggy, windswept piece of land. The bedrock, consisting of massive dacite intrusives, flows, and clastics, is Oligocene in age and is part of the Upper Member of the San Miguel Volcanics.

Santa Rosa Island

Santa Rosa, at 84 square miles in extent, is the second largest island in Channel Islands National Park. The island has about 45 miles of shoreline and is approximately 15 miles long and 10 miles wide. Geologically the island is a series of anticlines and synclines that have been faulted. A major fault, the Santa Rosa fault, trends east-west across the island. Movement along this fault that produced both vertical and horizontal displacement has caused distinct differences in topography between the northern and southern parts of the island. The northern half of the island, the downthrown side, has broad, plateaulike terraces into which streams have cut steep canyons. High bluffs face the northern shore. White sandy beaches stretch along the northern and western coast, most of them flanked by dunes up to 400 feet high. South of the Santa Rosa fault, the land is higher and more rugged, the highest point (1,589 feet) being on a mountainous ridge that follows the east-west trend of the fault. The shales, sandstones, and volcanics on the south side of the island are more resistant to erosion than the shales (Monterey Formation) exposed on the north side.

Many archeological and paleontological sites have been located on Santa Rosa, including places where there are remains of pigmy mammoths. Some of the sites have been studied, but much archeological work still needs to be done. A rare subspecies of Channel Islands fox is found only on this island, as is an endemic spotted skunk. The Torrey pine grows only here and in Torrey Pines State Park in San Diego County on the mainland. Twenty-six other plants that do not grow on the mainland have been identified on Santa Rosa. The profuse kelp beds along the shore are a refuge for sea otters and a breeding ground for the California sea lion. Whales, especially the humpback, blue, and sei, swim by the island during their migrations.

Santa Cruz Island

With 96 square miles of land surface, Santa Cruz is the largest Channel Island. It has 65 miles of shoreline, and is 24 miles long and between one and 6.5 miles in width.

Topographically Santa Cruz is the most rugged of the islands with its steep shoreline and occasional pocket beaches. Sea caves (such as Painted Cave) have been eroded in the cliffs along the shore by wave action. Unlike the other Northern Channel Islands, which have relatively few safe anchorages, Santa Cruz's irregular shoreline has a number of protected coves, especially on the south side.

A major east-west-trending fault cuts across Santa Cruz, separating northern and southern mountain ranges, the principal rock types, and island's drainage networks. The floor of the intervening valley, eroded along the fault zone by streams, is about the only flat land on the island. Most of the Central Valley is drained by a stream that flows into Prisoner Harbor on the northeastern side of the island. A smaller stream drains to the west. The fault zone is the result of strike-slip movement along the East Santa Cruz Basin fault that brought together the northern and southern halves of the island during Miocene time.

The northern mountain range has an elevation of over 2,400 feet at Diablo Peak, its highest point. Sharp ridges and steep canyons leading down to the coast dissect the northern range. The somewhat less rugged southern mountain range rises 1,523 feet above sea level. Canyons on the south side, which are broader with gentler slopes, provide easier access to the island's interior.

Some of the sedimentary rocks on Santa Cruz are a continuation of those in neighboring Santa Rosa; but the geology on Santa Cruz is more complex, with exposures of numerous rock types, including volcanic and metamorphic rocks of various ages. Prehistoric inhabitants of the Channel Islands found many excellent raw materials for stone tools among this rich variety. Especially prized was a chert of high quality in beds of the Monterey Formation cropping out on the eastern end of the island. Points and tools chipped from this chert were bartered for goods from Indians on the mainland. Some of the volcanic rocks on the island were also used for chipping and grinding weapons and tools.

Among the endemic plants and animals on Santa Cruz are the great island pine, the Santa Cruz pine, spotted skunk, a species of salamander, and the dwarf fox. About 500 native plant species and subspecies grow on the island, 7 of which are endemic and 31 of which are not found on the mainland. Resident bird species number 44 and include the endangered California brown pelican. Kelp beds surrounding the island provide food for seal and bird rookeries, as well as for fish, shellfish, etc.

Anacapa Island

The three Anacapa islets are segments of a volcanic ridge made up of lava flows, breccias, ash, and cinders erupted from the ocean floor in Miocene time. Many of the lava flows have amygdules of olivine and chalcedony, and a large vein of chalcedony is exposed near Frenchys Cove on West Island. Cobbles and pebbles weathered from the chalcedony are scattered over the cove beach. This beach is one of the few places on Anacapa where boats can land, since sheer cliffs and rocks make up most of the shoreline. The cliffs are particularly high and almost vertical from the ridge crest to the water along the island's south side. The slopes on the north side of the ridge are somewhat less steep.

The highest point of the volcanic ridge is 930 feet on West Island (Vela Peak), and the ridge gradually lowers to 250 feet above sea level, the highest point on East Island. From west to east, Anacapa is 5 miles long, and it is only about a half-mile wide. The largest islet, West Island, has about half of Anacapa's total acreage of a little over one square mile.

Wave-erosional features, such as sea arches, sea caves, stacks, black sand and cobbles, wave-cut platforms, surge channels, and blowholes, are prominent on Anacapa's coast (fig. 29.4). The sea caves, arches, and surge channels tend to form in fracture zones of the bedrock. A well developed surge channel and blowhole is near Cat Rock on the south side of West Island. When a wave swell moves into the narrow channel, air that is trapped and compressed within the fissures forces out sea water explosively, spraying the surrounding area. Raised beaches and marine terraces on parts of Anacapa are indications of changing sea levels in the fairly recent geologic past.

Santa Barbara Island

Santa Barbara (the only Southern Channel Island in the Park) is geologically related to Anacapa, since both islands are almost all volcanic, primarily basalt. Geologic features of Santa Barbara include volcanic cinders, pillow basalts mixed with sedimentary marine beds, marine and nonmarine terraces, eolianites, sea caves, landslides, and evidence of churning and slickensides in the soil, caused by differential expansion and contraction of soil materials.

As pointed out earlier, Santa Barbara has severe erosional problems resulting from overgrazing, burning, and so on. Many of the cliffs and slopes are unstable and subject to landsliding, especially on the north side of the island. Another area of severe erosion is near the base of Signal Peak on its eastern side. Signal Peak (elevation 635 feet), the highest point on the island, overlooks the southwest coast.

Marine terraces, cut by canyons, are prominent on the northern and eastern parts of the island. Probably the entire island was submerged during the highest sea stands of early Pleistocene time. Faunal remains have been found in the terraces, as have numerous archeological sites indicating the frequent presence of prehistoric people on the

Figure 29.4   Low tide at Anacapa Island exposes abalone clinging to sheer rock above the wave-cut platform. Blocks of rock, loosened along joint planes, lie beside the tidal pool. In the foreground is a piece of heavy anchor chain washed up by the surf. Photograph from Channel Islands National Park Collection.

island. The people probably camped on Santa Cruz to fish, hunt, or stop over when canoeing from the Southern Channel Islands to the Northern Channel Islands.

Although Santa Barbara is a small island, it is important for many reasons: for the protection of certain native plants, such as the giant *coreopsis* which attains a height of eight feet and has bright yellow, daisylike flowers in the spring, the live-forever (*Crassulaceae*), and other plants not found on the mainland; for the protection of an endemic night lizard and certain species of land snails; because of rookeries of the California sea lion, elephant seal, harbor seal, and species of sea birds; and as the habitat of a few endemic mammals, such as a small bat and deer mice.

Geologic History

Channel Islands National Park is located on the west side of the boundary zone between the North American Plate and the Pacific Plate, along which movement has been occurring for several million years.* During this time, lateral movement on the San Andreas and other parallel faults has acted to: 1) open the Gulf of California by separating the Baja peninsula from the mainland of Mexico; 2) move Baja California and that part of California west of the San Andreas fault to the northwest; and 3) cause large-scale displacements and rotations of large crustal

*The national parks in the Sierra Nevada—Yosemite (chapter 13), Sequoia (chapter 27), and Kings Canyon (chapter 28)—are in the part of southern California that is *east* of the San Andreas fault system and the plate boundary.

blocks near the California borderland. Such large-scale tectonic changes, when superimposed upon an already complex geologic history, have made the interpretation of the geologic history of the Channel Islands and adjacent mainland very difficult indeed. Continued movement along the San Andreas fault system will lead to greater displacement and perhaps further rotation of the Channel Islands as time goes on.

The Channel Islands are geologically associated with the Transverse Ranges section of the California Coast Ranges, which are included in the Pacific Border Province. In the Transverse Ranges, the basins, ranges, and most other structures trend east-west, or "transverse," to the dominant southeasterly trend of the adjoining areas (the Sierra Nevada and the Great Valley, for example). The aligned peaks of the Northern Channel Islands are, in fact, the tops of the partially submerged, westward extension of the Santa Monica Mountains on the mainland; while the Santa Barbara Channel (north and east of the islands) is the submerged extension of the Ventura Basin, a broad valley north of, and approximately parallel to, the Santa Monica Mountains (fig. 29.5).

The geologic history of the Channel Islands is told by four kinds of evidence: 1) the sedimentary rocks of the area that reveal the depositional and uplift history of the islands in the Park, as well as the surrounding region; 2) igneous rocks and metamorphism and deformation of pre-existing rocks, showing the effects of tectonic processes; 3) paleomagnetic data that demonstrate wholesale clockwise rotation of the Channel Islands and adjacent Transverse Ranges relative to the rest of California, which occurred in post-Miocene time; and 4) Pleistocene fossils and topographic features that help to explain recent history.

1. Mesozoic volcanic activity, intrusion, and metamorphism.

The earliest rocks in Channel Islands National Park are a mixture of metamorphosed volcanic and sedimentary rocks called the Santa Cruz Schist. They formed over 160 million years ago as a volcanic pile made up of basaltic and rhyolitic lavas interlayered with clastic sediments of volcanic origin. Later intrusions by the Willows Plutonic Complex and the Alamos Pluton occurred at about 160 and 140 million years respectively. Then in Jurassic and Cretaceous time the area underwent regional dynamic metamorphism that recrystallized and deformed the rocks.

2. Deposition and deformation during the Cretaceous and early Tertiary.

The Cretaceous ushered in a marine environment with a tropic ocean that extended north of the Channel Islands area. Over 6,300 feet of shales, siltstones, and sandstones were deposited as the Jalama Formation. Deposition continued into the early Tertiary and is marked by coarsening of some sediments as a function of uplift of the adjacent continent and the nearness of the source terrain. During early Tertiary time, 5,000 feet of clastic sediments consisting of cobbles, sands, and shales were deposited to form the Pozo, Cañada, Jolla Vieja, South Point, and Cozy Dell Formations. These formations were deposited as turbidites (p. 204), the sediment coming from a continental mass to the south and southwest of the present islands.

During the Oligocene Epoch, terrestrial sediments of the Sespe, Alegria, Vaqueros, and Rincon Formations were deposited to a thickness of 3,400 feet over a wide area from Santa Cruz Island across the Santa Barbara Channel to the Santa Monica Mountains.

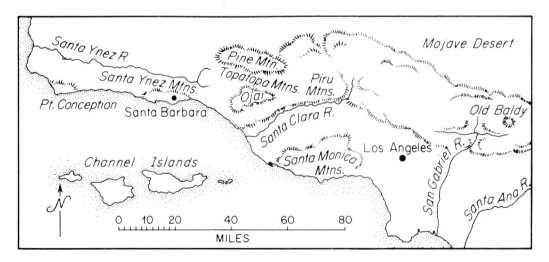

Figure 29.5 Map showing the relationship of the Northern Channel Islands to the east-west trending Santa Monica Mountains and other Transverse Ranges. Modified

from R. P. Sharpe *Southern California Geology Field Guide* © 1976 Kendall/Hunt Publishing Company.

Sedimentation in the early Tertiary was controlled by the Laramide orogeny which caused the general uplift of the western United States and affected specific ranges such as the Santa Ynez Mountains (one of the Transverse Ranges) and other ranges in southern California.

3. Late Oligocene-Early Miocene deformation and continued sedimentation.

Deformation at the end of the Oligocene involved folding and faulting and was followed by renewed volcanic activity in the early Miocene. Volcanic activity was initiated by mafic flows and intrusions and was followed by more felsic flows, breccias, and pyroclastic debris. This 1,400 feet of volcanic rock is known as the Blanca Formation, which, in turn, was eroded with earlier sedimentary rocks and the basement complex. The resulting coarse clastics were deposited as the Beecher Member of the Monterey Formation. Most of the Monterey Formation is a siliceous deposit formed on the continental shelf in warm seas.

4. Clockwise rotation of the Channel Islands and adjacent Transverse Ranges.

Paleomagnetic evidence on Miocene igneous rocks shows that the Channel Islands were rotated between 70° and 80° clockwise since Miocene time. This rotation, which affected much of the Transverse Ranges, appears to be due to "rolling" of the Transverse Range block between the moving north Pacific and North American plates during the late Tertiary. This shearing also created the major offsets along the San Andreas and parallel faults in southern California. The sequence of these events as reconstructed from the paleomagnetic evidence is shown in figure 29.6. The rotation does much to explain the sources of sediment and the location of the seas described earlier. The ocean to the north of the Channel Islands and the south and southwest sources of Tertiary sediments are rotated, upon reconstruction to pre-Miocene time, to a western ocean and eastern sources of sediment, such as we find today.

5. Emergence of the Channel Islands during early Pleistocene time.

Uplift of the Transverse Ranges and Channel Islands formed an area of high relief. Initial uplift formed a major land extension called the Cabrillo Peninsula that stretched from the Santa Monica Mountains out along the present Channel Islands. It was along this peninsula that Pleistocene land mammals migrated, only to become isolated when the land bridge between the Channel Islands and the mainland subsided. Marine terraces found at various levels on San Miguel Island, and as high as 1,800 feet on Santa Cruz Island, attest to the dynamic uplift of parts

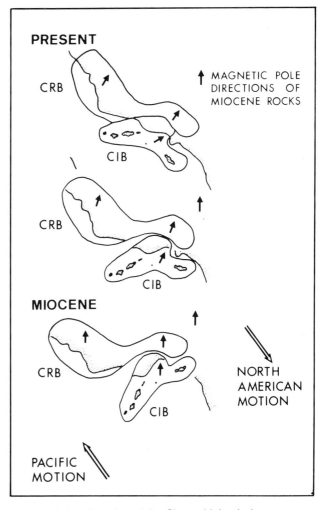

Figure 29.6 Rotation of the Channel Islands from Miocene to the present. The diagram shows a schematic representation of the rotations of the Channel Islands Block (CIB) and the Coast Range Block (CRB) relative to the rest of North America. Evidence for this rotation is found in the paleomagnetic directions to the pole found in Miocene igneous rocks (shown by black arrows on the different blocks) which point in different directions at the present time. These directions were parallel in Miocene time when the igneous rocks cooled in the Miocene magnetic field (bottom of figure). Subsequent rotation in a clockwise direction led to the situation seen at present (top of figure). This rotation was caused by the relative motion of the Pacific and North American Plates (long open arrows) which effectively "rolled" both the Channel Islands and Coast Range Blocks as small fragments caught between the two larger plates. In the figure the borders of the blocks are only approximate and the land areas (stippled line) are given as a reference only.

of the island chain during the Pleistocene. Such uplifts and areas of subsidence have caused the isolation of the Channel Islands, led to the distinctive biological heritage of many of the islands, and produced the impressive topographic features of today.

Geologic Maps and Cross Sections

Howell, D. G., ed. 1976. *Aspects of the geologic history of the California continental borderland.* Pacific Section, American Association of Petroleum Geologists Miscellaneous Publication 24.

Weaver, D. W. 1969. *Geology of the northern Channel Islands.* Special Publication. Pacific Sections, American Association of Petroleum Geologists and Society of Economic Paleontologists and Mineralogists.

Bibliography

Glassow, M. 1977. *An archeological overview of the northern Channel Islands, California, including Santa Barbara Island.* Prepared for the Western Archeological Center, National Park Service, Tucson, Arizona. National Technical Information Service No. 298855.

Hall, C. A., Jr. 1981. San Luis Obispo transform fault and Middle Miocene rotation of the western transverse ranges, California. *Journal of geophysical research* 86:1015–1031.

Howell, D. G., ed. 1976. *Aspects of the geologic history of the California continental borderland.* Pacific Section, American Association of Petroleum Geologists, Miscellaneous Publication 24. (Reports by various investigators on the basement rocks, stratigraphy and paleontology, sedimentology, and tectonics presented at the California Borderland Symposium, Santa Cruz Island, 1975.)

Howorth, P. C. 1982. *Channel Islands, the story behind the scenery.* Las Vegas, Nevada: KC Publications.

Johnson, D. L. 1967. *Caliche on the Channel Islands.* Mineral Information Service, Natural Resources Program, Naval Undersea Center, San Diego, California.

Kamerling, M. J., and Luyenduk, B. P.. 1979. Tectonic rotations of the Santa Monica Mountains region, western transverse ranges, California, suggested by paleomagnetic vectors, *Geological Society of America bulletin* 90: 331–337.

Luyenduk, B. P., Kamerling, M. J., and Terres, R. 1980. Geometric model for Neogene crustal rotations in Southern California, *Geological Society of America bulletin* 91:211–217.

Power, D. M., ed. 1980. *The California Islands: Proceedings of a multidisciplinary symposium.* Santa Barbara, California: Santa Barbara Museum of Natural History.

Resource Management Workshop, Channel Islands, November 1–3, 1978. Mimeographed sheets.

Sharp, R. P. 1976, *Field guide to southern California,* K/H Geology Field Guide Series. Dubuque, Iowa: Kendall/Hunt Publishing Company.

Tilden, F. 1979. *The national parks.* Rev. ed. New York: Alfred A. Knopf.

U.S. Dept. of the Interior. April 1980. *Channel Islands National Monument, California legislative background.*

Weaver, D. W. 1969. *Geology of the northern Channel Islands.* Special Publication. Pacific Sections, American Association of Petroleum Geologists; Society of Economic Paleontologists and Mineralogists.

Address

Channel Islands National Park
1901 Spinnaker Drive
Ventura, California 93001

30
Shenandoah National Park

Location: Northwest Virginia
Area: 194,327.81 acres; 303.64 square miles
Authorized: March 2, 1926
Established: December 26, 1935

Figure 30.1 The two climbers are standing on an outcrop of massive Catoctin lava near the summit of Little Stony Man (elevation approximately 3,500 feet) in the north central section of Shenandoah National Park. They are looking to the west over a winter landscape of mountains and foothills descending to the Shenandoah Valley. The rounded mountain ridges in the middle distance have been weathered and eroded from bedrock of the ancient Pedlar Formation. The convex-upward summit profiles are typical of unglaciated mountains formed from massive, resistant rock. National Park Service photograph.

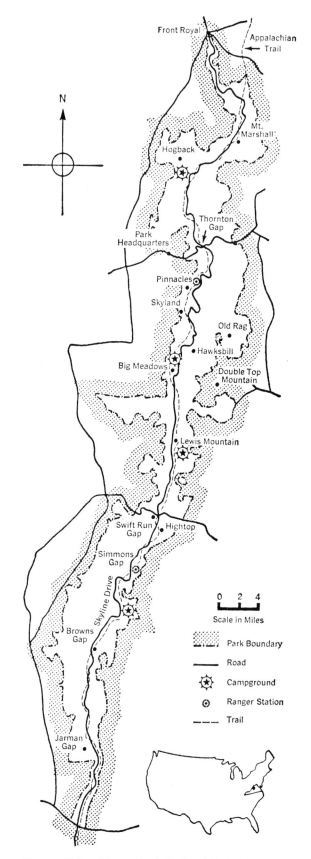

Figure 30.2   Shenandoah National Park, Virginia.

Table 30.1. Geologic Column, Shenandoah National Park

Time Units			Rock Units		Geologic Events
Era	Periods	Epochs			
Cenozoic	Quaternary	Holocene	Alluvium		Frost action, stream erosion, mass wasting
		Pleistocene	Periglacial talus		Intensified frost action and stream erosion
	Tertiary (or Paleogene/ Neogene)				Erosion Evolution of modern drainage Isostatic upwarp
Mesozoic	Cretaceous Jurassic Triassic		Diabase dikes		Erosion Uplift Intrusion of dikes Tensional rifting and faulting
Paleozoic	Permian Pennsylvanian Mississippian — — — —				Extensive thrust faulting Reverse faulting Folding, uplift, metamorphism Allegheny orogeny
	Devonian				Erosion and deposition
	Silurian — — — — Ordovician				Uplifting of Blue Ridge rocks Taconic orogeny

	Cambrian		Chilhowee Group	Erwin (Antietam) Formation
				Hampton (Harper) Formation
				Weverton Formation
					Prolonged subsidence and sedimentation;
					    Quartzites (with *Skolithos* tubes) some interbedded
					        shales
					Shales, siltstones, interbedded quartzites
					
					Conglomeratic quartzites, shales

Proterozoic Z (late Precambrian)			Catoctin Formation Swift Run Formation		Dark greenstones (originally outpourings of basaltic lava and dike intrusions plus ash and tuff) Volcano-clastic deposits
Proterozoic Y			Old Rag Granite Pedlar Formation		Metamorphosed, light gray plutonic rock Metamorphosed greenish-gray gneissic rocks

Source: modified after Gathright 1976; Reed 1969

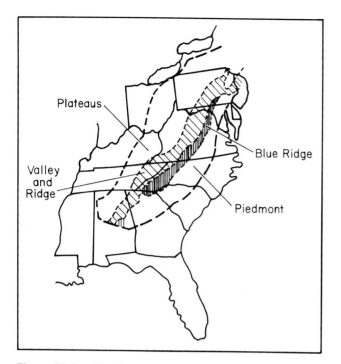

Figure 30.3 Physiographic provinces of the Southern Appalachian region.

Geographic Setting

South of New England, the Appalachian Mountain belt divides into four geologic provinces that trend south to Georgia in roughly parallel bands. From east to west, these divisions are the Piedmont (or foothills), the Blue Ridge, the Valley and Ridge, and the Appalachian Plateau (fig. 30.3). In the Blue Ridge province, which has the highest elevations in the Appalachian system, are two national parks—Shenandoah in Virginia and Great Smoky Mountains (chapter 31) in Tennessee and North Carolina.

The Blue Ridge Mountains, which are the easternmost high range of the Appalachians, begin in southern Pennsylvania as a single narrow ridge, only about 10 miles wide. Trending southwestward, the mountains become closely spaced ridges of greater height and width, ending finally in a broad bulge of complex ridges extending over parts of North and South Carolina, Tennessee, and Georgia. Great Smoky Mountains National Park occupies the highest part of the southern end of the Blue Ridge province.

Shenandoah National Park, near the northern end of the Blue Ridge, is long and narrow. The Park follows the crest of the Blue Ridge for about 80 miles from Front Royal, Virginia (about 60 miles west of Washington, D.C.), to the southern entrance of the Park near Waynesboro. Elevations in Shenandoah National Park range from around 600 feet at Front Royal to 4,049 feet at the top of Hawksbill Mountain.

Skyline Drive, a scenic highway famous for its overlooks, runs the length of the Park and then connects with the Blue Ridge Parkway, which continues southward to Great Smoky Mountains National Park. The Appalachian Trail goes through both parks as it follows the crest of the Appalachians from Maine to Georgia.

The Blue Ridge Mountains, as well as the other ranges of the southern Appalachians, tend to be more rounded and less angular than the mountains of western United States. Both the greater age of the Appalachians and the more humid climate are reasons for this aspect of the landscape. Much of the bedrock of the Appalachians is mantled by residual soil that accumulated through millions of years of weathering. This soil and ample precipitation distributed evenly throughout the year support dense hardwood forests on the mountain slopes. Evergreens that have persisted at higher elevations are probably remnants of the coniferous forests that migrated down from the north during cold Pleistocene climates. Glaciers did not reach Virginia, but periglacial conditions prevailed at high altitudes and must have intensified weathering and erosional processes during that time. The large boulder fields on many of the slopes of the Blue Ridge are probably the result of periglacial activity. Even in today's climate, snow usually blankets the higher levels of the Blue Ridge for at least part of the winter season, and frost-wedging is a significant factor in the weathering of rocks.

Major streams in the area are the Potomac River, which cuts through the Blue Ridge at Harpers Ferry (north of Shenandoah National Park) and the James River, which has carved scenic gorges for its course through the Blue Ridge south of the Park. West of the Park, in the Shenandoah Valley, the Shenandoah River flows north, paralleling the Blue Ridge, and joins the Potomac at Harpers Ferry. The Blue Ridge itself is dissected by numerous tributary streams, many with waterfalls, that drain the mountain slopes. Streams have been the dominant agent of erosion in the region since the mountains were uplifted in Paleozoic time. During heavy rains quantities of loose sediment muddy the rivers, showing that erosion is fairly intense in spite of the protective cover of vegetation. When storms are prolonged or extremely heavy, mudslides occur on steep or denuded slopes.

Local History

Stone burial mounds, arrowheads, spear points, and other artifacts that have been found both within and outside the boundaries of Shenandoah National Park show that native tribes knew the area well for thousands of years before Europeans came to Virginia. Apparently the early inhabitants lived in the lowland river valleys during the winter season and camped in the Blue Ridge during the summer and fall to gather berries, hickory nuts, and chestnuts, and to hunt game. Elk, deer, and bison grazed in the high open meadows and clearings in the forest. Originally the fires that kept the meadows free of trees may have been started by lightning, but as time went on, the Indians burned over the clearings deliberately. This was a method of driving game into the range of spears and arrows. Fire also encouraged the growth of berry bushes.

The Indians who occupied the region when the first white people came belonged to small tribes of the eastern Siouan group, but not much is known about them. Supposedly these tribes were absorbed or annihilated by larger groups, such as the Cherokee, not long after white settlers established themselves on the Virginia coast.

For about a hundred years after the English first settled in the Tidewater region, the Piedmont (the foothills of the Blue Ridge), the Blue Ridge, and the Shenandoah Valley on the far side remained unsettled and largely unexplored. A German, John Lederer, may have been the first European to reach the crest of the Blue Ridge and gaze out over the Shenandoah Valley and the Allegheny Mountains to the west. His first expedition was in 1669, and he made three more trips to the Blue Ridge before returning to his native land. The reports of his explorations apparently didn't receive much attention at the time.

Alexander Spotswood, an able and energetic royal governor of the Virginia colony in the early 1700s, saw the need for expansion as the colony's population grew. He encouraged settlement of the outlying frontiers by exempting settlers from quitrents (a form of taxation). In 1716 he led an expedition to the Blue Ridge and the Shenandoah Valley and a few years later negotiated a treaty by which the Iroquois Indians agreed to stay north of the Potomac and the Blue Ridge. From that time on, settlers began establishing farms and villages on both sides of the Blue Ridge and in the Shenandoah Valley. Some came from eastern Virginia and others moved south from Pennsylvania.

Settlements in the Shenandoah Valley and the Piedmont became increasingly prosperous as time went on, but the mountain people who tried to farm in the upland valleys and summit areas of the Blue Ridge became ever poorer and more isolated. They were limited in their choice of favorable sites not only by the rugged topography but also by the nature of the bedrock. Although they may not have understood the reason, it was obvious to them that the well watered, fertile soils of the broad summit areas and the gentler eastern slopes—underlain mostly by granites and basaltic greenstones—were more suitable for homesteads than the rocky and steep western ridges that had developed thin, infertile soils on resistant quartzites and phyllites. Also, on the west side springs were fewer and intermittent. Unfortunately, what good soils there were in the uplands soon became depleted as crops were planted year after year in the same small fields. Erosion was severe, especially after timber cutting removed much of the protective vegetation from the slopes. Then a blight exterminated the native chestnut trees, which had been one of the finest hardwoods of the region.

The discovery of small deposits of iron, manganese, and copper in the Blue Ridge helped the region for a time during the 19th century, but as the ores were worked out, the mines were abandoned. (Some manganese mining was, however, carried on as late as World War II.) In the western foothills of the Park and just outside the western boundary are more than 40 old mine sites.

As times became ever harder, the mountain people and their unique way of life seemed doomed. But toward the end of the 19th century, the development of mountain resorts high in the Blue Ridge brought new money, new people, and a revival of interest in the region's potential. The beauty of the mountains and the benefits of pure mountain air were now accessible to the large populations of the nation's capital and the cities of the mid-Atlantic seaboard, areas that often sweltered in summer heat and humidity. The idea of establishing a national park in the Virginia Blue Ridge country began to grow. George Freeman Pollock, who operated Skyland Resort on the mountaintop and controlled some 5,000 acres along the crest, became enthusiastic about the project. After Congress authorized the Park's acquisition in 1926, he helped to raise over a million dollars for land purchases. (With no public lands in the area, all the territory for the Park had to be purchased or acquired by donation from private owners. President Herbert Hoover, for example, gave his land and fishing camp on the Rapidan River.) After the Virginia legislature had appropriated another million dollars for the compensation and relocation of landowners, and the state had acquired clear title to 250 square miles of land (most of it along the crest), the tract was turned over to the federal government for the establishment, in 1935, of Shenandoah National Park. Later additions to the Park brought its total area to over 300 square miles.

Skyline Drive with its beautiful panoramas to the east and west goes through the Park along the "spine" of the Blue Ridge, but most of the rest of the land has been left undeveloped or allowed to revert to a wilderness state (fig. 30.4). Some of the ancient clearings and forest openings, such as Big Meadows, have been kept open as campgrounds, but new forests have reclaimed the mountain farms and logged-over areas. By the time land acquisition for the Park began, only a few hundred families were subsisting on small farms and orchards in the mountains.

At the present time, more than three million people visit this exceedingly popular national park each year.

Figure 30.4   View of Stony Man from Stony Man Mountain Overlook on Skyline Drive. The "face" is eroded from the ancient lavas of the Catoctin Formation. Photo courtesy National Park Service.

Bedrock and Geologic Structures

In Shenandoah National Park, hardwood forests and other vegetation mantle bedrock and structures. Most outcrops show evidence of deep weathering; ledges may be shattered or largely covered by broken and crumbling talus. Fresh exposures of bedrock are rare. Nevertheless, the controlling influence of bedrock and structure on landscape development is plain to see.

Consider Big Meadows, for example. Here is a plateau of low relief, about 3,550 feet in elevation and roughly five square miles in extent, that is part of a broad summit area on the crest of the Blue Ridge. Elsewhere along the northern Blue Ridge, the crest is relatively narrow and deeply dissected. Two intermittent swamps have long existed in shallow basins in Big Meadows, fed by springs and runoff from higher slopes. Indians hunted game that grazed in the meadows in prehistoric time, and settlers pastured their herds there until early in the 20th century.

What lies beneath Big Meadows are at least 1,800 feet of massive Precambrian lava flows (Catoctin Formation), originally basaltic but later metamorphosed to greenstones. These highly resistant greenstones cap the highest ridges and peaks of the Park (Hawksbill Mountain, for example). In the Big Meadows area the top of the formation and overlying sedimentary rocks have been removed by erosion, producing the gently rolling topography of this part of the Blue Ridge crest.

Prominent cliffs of columnar-jointed lava, such as Franklin Cliffs and Crescent Rocks can be seen from Skyline Drive (fig. 30.5). (The columnar jointing in the Catoctin lavas indicates that they were erupted on land.) Most of the waterfalls in the Park cascade over cliffs and benches of resistant greenstone. Dark Hollow Falls, which drains much of the Big Meadows area, is an example.

Some of the Precambrian plutonic rocks that are older than the Catoctin Formation have been cut by greenstone feeder dikes that represent intrusions of molten basalt that did not reach the surface. Because the ancient granitic rocks are even more resistant than the greenstones, differential erosion has produced deep notches of varying widths where the dikes were emplaced. Along the Ridge Trail on Old Rag Mountain, a flight of natural stairs has been weathered out between vertical granite walls. The stairs on the dike surface weathered from columnar joint blocks.

The Old Rag Granite exposed at the summit of Old Rag Mountain (and elsewhere in the Park) is easily recognized by its very light color and coarse grain. Because this metamorphosed granitic bedrock is very resistant, weathering is mostly along the joints (fractures) where water and ice can attack the rock. Between the joints, large spheroidal boulders (some as much as 10 feet in diameter) have weathered in place (fig. 30.6).

Figure 30.5   Crescent Rock, near Crescent Rock Overlook on Skyline Drive, is an example of columnar jointing in lava of the Catoctin Formation. Photo courtesy National Park Service.

Figure 30.6   This huge block of Old Rag Granite, perched on the crest of Old Rag Mountain, is not a glacial erratic. The boulder was weathered in place along joint planes in the very resistant, massive, granitic bedrock exposed at the summit. Frost action and spheroidal weathering are the dominant processes attacking the rock. Photo courtesy National Park Service.

Box 30.1
Global Tectonics and the Geologic History of the Southern Appalachians

For nearly two centuries the Appalachian Mountains have been the inspiration and the testing ground for many of the grand ideas and concepts of American geology. For a long time the majority of American geologists were trained on Appalachian rocks and structures because most centers of learning were in the eastern part of the country; and it was in dealing with the complexities of these ancient mountains that the American geologic sciences came of age. Yet still today, questions regarding the origin and development of the Appalachians remain unanswered.

Almost from the beginning, geologists working in North America recognized that masses of rock exposed in the higher Appalachians appeared to be unrelated—or geologically incongruent—to rock units in the foothills and lowlands on either side. Another puzzling observation was the great thickness of early Paleozoic sandstones and carbonates in the Appalachians as compared to thinner rock sequences of similar age and lithology in the central lowlands west of the mountain ranges.

The concept of the geosyncline, developed by American and Canadian scientists from studies of Appalachian geology, was conceived as an explanation of these phenomena. The geosyncline was the first major contribution to geologic thought based wholly on North American work rather than on investigations by European geologists. The original hypothesis (presented by James Hall in 1857 in his presidential address to the American Association for the Advancement of Science) postulated an extensive belt of down-sinking crust in which thousands of feet of sediment accumulated during prolonged subsidence, gradually becoming consolidated as layered rock. The height of the mountains, Hall thought, was directly related to the amount of accumulated sediment.

In the years that followed, other American geologists elaborated on Hall's concept, suggesting possible ways in which mountain-building occurred. The bottom of the geosynclinal trough, weakened by strain, gave way and buckled, producing compressive force that folded and metamorphosed the sedimentary rocks. The linear mountains of the Valley and Ridge Province and the Blue Ridge (on which Shenandoah and Great Smoky National Parks are located) were thought to have been formed in this way.

Because geologists recognized that in general Paleozoic deposits thickened and became coarser toward the east, they assumed that the source of most of the geosynclinal sediment lay in that direction. Paleographic maps were drawn depicting a great landmass, named "Appalachia," located in the Atlantic Ocean east of North America. The mountains and borderlands of eastern United States and Canada were thought to be remnants of this "lost continent." Presumably, the ancient landmass finally sank below the ocean surface after countless years of erosion had transferred most of its rock material to the geosyncline.

Meanwhile, European geologists set about adapting the American geosyncline concepts to the Alps and other mountain belts with which they were familiar. The Appalachian model seemed to fit some mountains fairly well, but it was not very useful in explaining the origin of some other mountain chains. Not all mountain chains are located on continental margins as in North America; nor do all have such thick accumulations of sedimentary material. Furthermore, the model did not account satisfactorily for the underlying cause of mountain-building. How could great lateral contractions, as seen in folded mountain belts such as the Appalachians and the Alps, apparently due to immense horizontal compression, be caused by presumably vertical forces or movements associated with instability in the root of a geosyncline? And what was the geophysical justification for this vague instability?

The idea of "mobile belts" as being areas of orogeny, or mountain-building, both in the present and in the past, brought about further modifications of the geosyncline concept. Again, the Appalachian region was one of the first mobile belts to be studied in detail. The geosyncline or subsiding area was visualized as made up of two parallel subtroughs: (1) an inner *miogeosyncline*, mainly in the continental shelf area, where great thicknesses of shallow-water deposits of shale, sandstone, and carbonate rocks built up; and (2) an outer *eugeosyncline* that filled with highly siliceous volcanic materials from an island arc. In the Southern Appalachians, miogeosynclinal deposits are represented by thick sequences of Cambrian sandstones and Ordovician and Silurian calcareous rocks (limestones and dolomites). Eugeosynclinal deposits are conspicuous in the Northern Appalachians but are less so in the Southern Appalachians. No longer was the geosyncline viewed as the *cause* of mountain belts; rather the development of a geosyncline was regarded as the *result* of mountain-building processes in *some, but not all*, mobile belts.

The advent of plate tectonic theory with its concept of moving crustal plates removed the need for having a fixed landmass as a sediment source for filling developing geosynclines. Instead, continents and continental fragments (microcontinents) were visualized as coming together and breaking apart with intervening oceans closing or opening during the process. When crustal plates come together in such a way that structural downbuckling occurs, an area of subsidence is formed, which may eventually become filled with sediment—in other words, become a geosyncline, according to the modern definition of the term.

Present approaches to the evolution of the Appalachian mobile belt are based on reconstructions of former positions and movements of landmasses during Paleozoic and Mesozoic time. Data derived primarily from studies of magnetic anomalies on the ocean floors have enabled geologists to reconstruct with reasonable accuracy the positions of continents since the beginning of the Mesozoic Era. Correlative data indicating paleolatitudes of landmasses have come from investigations of ancient magnetism in continental rocks and from depositional belts of evaporites, coal, and life forms.

The evidence for changing continental positions during Paleozoic time is less clear, but movement of continents was apparently extensive. Reconstructions of Paleozoic tectonic activity are derived from studies of rocks and structures of old mobile belts, such as the Appalachian Mountains.

The modeling involves paleomagnetic data, the orientation of fold belts, the distribution of sedimentary facies, and geophysical techniques, such as deep seismic reflection profiling—which was originally developed for oil exploration by the petroleum industry. Appalachian studies have provided a key model for plate tectonic mechanisms before Mesozoic time. The outline that follows summarizes a tectonic pattern that many geologists feel helps to explain the development of the major geologic features of the Southern Appalachians.

Outline of the Tectonic Evolution of the Southern Appalachians

1. Precambrian and earliest Cambrian time.

Scientists theorize that the basic mechanism for the movement of lithospheric plates developed at least 2.5 billion years ago, or about the beginning of Proterozoic time (see table 22.2, p. 279). The origin of the ancient granites, gneisses, and schists exposed in the Blue Ridge, some more than a billion years old, is obscure, but they appear to have undergone intense deformation, presumably more than once during Precambrian time. They may have been part of a mountain belt on a Proterozoic supercontinent.

Rifting of this supercontinent in late Precambrian time brought about the initial opening of an ocean basin, the proto-Atlantic Ocean, and produced a subsiding belt that eventually became the Appalachian geosyncline. Some large islands, or microcontinents, that broke off from the cratonic crust during the rifting process moved away from the main continental landmass.

As the Precambrian highlands were eroded, clastic sediments that washed into the ocean were deposited on the shelf areas. This was the beginning of the miogeosynclinal portion of the geosyncline. The early Cambrian sandstones and quartzites now exposed in the northwestern part of the Great Smoky Mountains were deposited during this phase.

2. Cambrian and early Ordovician time.

Continued erosion lowered the land surface, and the ocean began to spread over the continental margins. Carbonates deposited in the shallow marine environment covered the Cambrian sandstones. Fossil remains of early Paleozoic life forms, such as brachiopods and stromatolites, indicate that the climate was tropical or subtropical. In deeper water to the east, a eugeosynclinal environment began to develop, rimmed by volcanic islands. Deep-water sediments and volcanic debris accumulated in the trough, along with shale beds containing graptolites. The eugeosynclinal rocks, now metamorphosed and deformed, crop out in the Piedmont province east of the Blue Ridge.

3. Middle to late Ordovician time.

New oceanic crust spreading outward from a ridge in the Ordovician ocean basin caused compression and subduction on both sides of the proto-Atlantic Ocean. The Appalachian mobile belt was formed by this tectonic episode, which is called the Taconic orogeny. (The Taconic was the first of the Appalachian orogenies that took place in Paleozoic time.) As subduction continued, the North American continental shelf was downfaulted and the eugeosynclinal area was uplifted, resulting in collision of the Blue Ridge and the Piedmont rocks. During the orogeny, granitic intrusions were emplaced in both the Blue Ridge and the Piedmont. From the uplifted mountains, great volumes of eroded sediment were transported westward and deposited in a vast, shallow continental sea that had spread over the central craton during Cambrian time.

4. Silurian and Devonian time.

As tectonic activity subsided in Silurian time, the processes of erosion became dominant. In late Silurian and early Devonian time, the proto-Atlantic Ocean was closing as the European and North American plates moved closer together. The Acadian orogeny (chapter 15) in middle and late Devonian time resulted from the collision of these landmasses. Most of the Acadian tectonic activity took place in the Northern Appalachian region because this was where the ocean closed first.

5. Late Paleozoic time.

While Europe and the northern part of North America were being fused together, the African plate was moving toward what is now southeastern United States. The ensuing collision, which took place over a 50-million-year interval in Pennsylvanian and Permian time, destroyed the proto-Atlantic Ocean and ended deposition in the Appalachian geosyncline. The force of this tectonic episode, called the Allegheny orogeny, caused extensive folding, metamorphism, and thrusting in the Appalachian region south of New England.

In the Southern Appalachians, thrust sheets are the dominant structural features. The ancient crystalline rocks of the Blue Ridge province were thrust westward over the younger sedimentary rocks of the former continental margin. New data derived from deep seismic reflection profiling suggest that the thrusting (especially at the southern end of the Blue Ridge) was on a much greater scale than had been previously estimated. The crystalline rocks of the thrust sheet may have been pushed over the continental margin, in a series of episodes, for as much as 150 miles relative to the younger rocks now buried many feet below the surface!

The creation of the supercontinent Pangea was the eventual result of the Paleozoic continental collisions and orogenies—which also included northward movement of the South American plate against what is now southern United States (chapters 34 and 37).

6. Mesozoic time.

The eroded mountains of the Appalachian belt that we see today probably looked more like the Himalayas at the beginning of the Mesozoic Era. At that time the Southern Appalachians stretched across the interior of Pangea in a humid tropical zone south of the equator.

When Pangea began to split up along a new rift east of the Appalachians, the present Atlantic Ocean basin began to form between the drifting plates. However, the new continents and basins did not have the same margins as the Paleozoic crustal plates that had come together to form Pangea. Small segments of crust that had formerly been part of the European and African (or South American) plates remained attached to what is now northeastern and southeastern United States. Thus the continental margin of eastern North America began to assume its present form as Pangea was breaking up early in the Mesozoic, and the active development of the Appalachian mobile belt came to a close. Meanwhile, to the west on the other side of the continent, tectonic activity was intensifying and new mountains were rising.

References

Ben-Avraham, Zvi. 1981. The movement of continents, *American scientist* 69(3): 291-299, May-June.

Bobyarchich, A. R. 1981. The Eastern Piedmont fault system and its relationship to Alleghenian tectonics in the Southern Appalachians, *Journal of geology* 89(3): 335-345, May.

Dott, R. H., Jr. 1979. The geosyncline—first major geological concept "made in America." *In* C. J. Schneer, editor. *Two hundred years of geology in America*, Proceedings New Hampshire Bicentennial Conference on the History of Geology. Hanover, New Hampshire: University Press of New England, pp. 239–264.

Dott, R. H., Jr. and Batten, R. L. 1981. *Evolution of the earth* (3rd ed.) New York: McGraw-Hill Book Company.

Oliver, Jack. 1980. Exploring the basement of the North American continent, *American scientist* 68(6): 676–683, November-December.

Pique, Alain. 1981. Northwestern Africa and the Avalonian plate relations during late Precambrian and late Paleozoic time, *Geology* 9(7): 319–321, July.

Windley, B. F. 1977. *The evolving continents*. New York: John Wiley & Sons.

The resistance of the gneissic Pedlar Formation relates to the character of the bedrock exposures, that is, whether they are massive or strongly foliated. Excellent outcrops of strongly foliated Pedlar gneisses can be seen along Skyline Drive, especially at the Tunnel Parking and Buck Hollow Overlooks. The Pedlar and Old Rag rock units are the oldest rocks in Shenandoah National Park (table 30.1).

In general, the ancient greenstones and granites are the rocks that form the crest and the gentler eastern slopes of the Virginia Blue Ridge; that is, the areas favored by the Indians and later by the settlers. On the western side are the Cambrian clastic rocks of the Chilhowee Group that have been metamorphosed to quartzites, phyllites, metaconglomerates, and so forth. Steep, sharp ridges, V-shaped hollows, and slopes covered by coarse rock debris characterize the topography formed on these rocks (fig. 30.7).

The Cambrian rocks vary in resistance, with the quartzites (such as those of the Weverton Formation) being the most resistant to erosion. The thin, sandy, and

Figure 30.7 Outcrops of Chilhowee Group rocks in the southern part of Shenandoah National Park sometimes assume unusual shapes while undergoing erosion. This natural chimney near Bull Run Overlook (Skyline Drive) is made up of phyllite metamorphosed from the shales and siltstones that made up the Hampton Formation originally. Photo courtesy National Park Service.

shaly soils that developed on the metasedimentary rocks support forests, but the trees tend to be less diverse, smaller, and less luxuriant than those that grow in the more fertile soil on the plutonic and volcanic rocks. However, part of this difference may be due to lower moisture content in the soils weathered from the sedimentary rocks.

Cleavage and Joints

In greater or lesser degree, as can be seen on exposures throughout the Park, the ancient bedrock is cut by thousands of cleavage planes. (The one exception is the Triassic diabase dikes, which were emplaced after the long series of tectonic events that took place in Paleozoic time.) Clearly, cleavage is the dominant and most obvious feature of the rocks. It is displayed best on the very fine-grained metasedimentary rocks and may completely obscure the original bedding. Cleavage is pervasive in even the massive plutonic rocks and quartzite beds. It is the result of regional metamorphism and compressive deformation during tectonic uplifting and thrusting that transported the Blue Ridge rocks westward over a segment of younger crust.

Joints, or fractures, are also conspicuous in the bedrock with most outcrops displaying fracture trends in several directions. Weathering operates on both cleavage and joints, continuously breaking up the bedrock into boulders, blocks, and rubble. Talus slopes tend to reflect the particular characteristics of the cleavage and joint systems of the bedrock source. For example, below ledges of massive, resistant quartzite are accumulations of white angular blocks that look from a distance like fresh rockslides because vegetation cannot grow over them. Soluble minerals available for soil enrichment that may have been present in the original sandstone have long since been removed, and the metamorphosed quartzite is extremely resistant to chemical weathering. Talus slopes of basaltic blocks, on the other hand, are difficult to see from a distance because trees have grown over the rocks in soil developed by weathering. Talus slopes made up of granite blocks are notable for their range of sizes; they range from small to enormous, depending upon the fractures in their bedrock sources. Below ledges and cliffs of the layered metasedimentary rocks, the talus is usually made up of many small blocks of near-uniform sizes because bedding planes and joints are normally closely spaced.

Folds and Faults

Asymmetric folds, both large and small, are visible on some exposures of Chilhowee rocks, especially in the southern section of the Park. Some of the light-colored, resistant beds of quartzite are easy to trace along hillside slopes. Most folds trend northeast-southwest, parallel to the long axis of the Blue Ridge.

A geologic map of Shenandoah National Park shows that faults are numerous throughout the Park area, but often the trace of a fault has been obscured by talus, gravel or alluvium. However, faults have strongly influenced landform development. The low- to high-angle reverse and thrust faults are the most important structural features in the Park. They are laterally extensive and dip generally to the southeast. Displacements produced by this faulting moved the rock mass of the Blue Ridge over younger Paleozoic rocks.

Also significant are the high-angle transverse faults that trend northwestward across the Blue Ridge. Movement on these faults was mainly vertical and generally less than 500 feet. Uneven movement on these two types of faults produced an irregular lobe and embayment pattern of rock distribution along the eastern edge of the Shenandoah Valley and the uneven summits of the Blue Ridge.

In the massive crystalline rocks, linear shear zones generally indicate the presence of faults. More spectacular are the zones of tectonic breccia that mark reverse faults in the Erwin quartzites. Red iron oxide cement has filled all the voids in the fault zone, binding the brecciated quartzite into very strong, resistant rock.

Water Gaps and Wind Gaps

A *water gap*—for example, the James River gorge, south of Shenandoah National Park—is a deep pass in a mountain ridge, through which a stream flows. A *wind gap* is usually a former water gap that has been abandoned by the stream that formed it. Within the Park, rivers in water gaps no longer cut through the resistant rocks of the Blue Ridge, but there are a number of wind gaps, some of them more than a thousand feet deep. Typical examples are Compton Gap, Thornton Gap, Swift Run Gap, Powell Gap, Jarman Gap, and Rockfish Gap (near the Park's south entrance). Tranverse valleys of this type are called wind gaps not because of any notion that they are the result of wind erosion, but simply because strong winds blow through them. Their shape tends to funnel the summit winds.

Looking at a wind gap, you might ask, what happened to the stream that eroded it? And why does a river in a water gap flow *across* a large topographic and structural feature instead of going around it? A plausible explanation for evolution of drainage on the northern Blue Ridge might be as follows:

1. Streams developing on uplifted structures flowed from a drainage divide to the west (in the Appalachian Plateau province) southeastward across the region to the Atlantic Ocean. These streams were *consequent*; that is, their course and direction were controlled by the slope of the existing land surface.

2. As the streams cut down, they were lowered onto the rocks and structures of the Blue Ridge; and water gaps were eroded transverse to the trend of the ridge. In this way erosion removed a considerable volume of rock material, and the original dendritic stream pattern was *superimposed* (or *superposed*) on the structures and rock units of the region.

3. Because of lithologic and structural variations in the bedrock, some streams had an advantage in downcutting and maintaining their courses (the Potomac and James Rivers, for example). These favored streams and their tributaries were able to establish lower local base levels and erode headward faster than the streams without such advantages.

4. The Shenandoah River, a tributary of the Potomac, cut headward up (south) the Shenandoah Valley, which is underlain by limestone, much more erodible in a humid climate than the crystalline rocks of the Blue Ridge. Thus the Shenandoah River was able to *capture* the headwaters of the consequent streams that flowed across the Blue Ridge between the Potomac and the James Rivers. These water gaps became wind gaps, usually with two small streams flowing away from the top of the gaps, one to the east and the other to the west. The Shenandoah River had become a *subsequent* stream; that is, its development was now controlled by the bedrock structure, in this case a belt of nonresistant limestone. By the capture of the tributaries, the drainage in the Shenandoah Valley became *adjusted* to the structure. In other words, *adjustment of drainage* had taken place.

5. Differential resistance of bedrock and structures continues to control the development of the topography today. During isostatic upwarping in Cenozoic time, streams have maintained their courses by downcutting and adjusting to rocks and structures.

Geologic History

1. The oldest Precambrian rocks in Shenandoah National Park (Proterozoic Y).

Radiometric dating procedures have yielded an approximate age of 1,100 million years ago for the crystallization and metamorphism of the Pedlar Formation and the Old Rag Granite, which make up the core of the Blue Ridge. The ancient, intensely foliated gneisses of the Pedlar Formation may have been intruded by the Old Rag Granite, but both were metamorphosed at depth under conditions of high temperatures and pressures. In some exposures, the darker gneissic rocks of the Pedlar Formation can be seen in gradational contact with the Old Rag Granite. These ancient plutonic rocks are resistant to weathering and form many of the ridges and peaks in the Park. The record of Precambrian tectonic events which presumably brought about the metamorphism and deformation of the Pedlar and Old Rag rock units has been obscured by Paleozoic tectonic activities (box 30.1).

2. Deposition of the Swift Run Formation in late Precambrian time (Proterozoic Z).

Following a long erosional interval, represented by an unconformity in the rock record, a discontinuous sequence of conglomerates, sandstones, and pyroclastics was deposited by streams eroding the Precambrian mountains and building up valley floors.

3. Eruption of the Catoctin lavas through crustal fissures.

Plateau basalts, pouring out over this ancient landscape, filled the valleys and eventually covered the eroded mountains. The rifting which allowed the molten basaltic magma to well up represented the initial opening of the proto-Atlantic Ocean late in Precambrian time. Ash falls interlayered with the lava flows indicate that some of the eruptions were explosive. Stream-deposited clastic materials are also interbedded with the ash and lava. The uppermost layer in the volcanic sequence is a purplish slate (representing an extensive ash fall) that marks the boundary between the Catoctin Formation and the overlying Cambrian beds.

4. Deposition of marine Cambrian sedimentary rocks of the Chilhowee Group.

As the Cambrian seas encroached on the subsiding continental margin, great quantities of clastic (or detrital) sediments were washed down from the cratonic highlands to the west and north. The Weverton Formation, which lies unconformably on the Catoctin rocks and, in some places, on the Pedlar Formation, is made up of 100 to 500 feet of conglomerates, sandstones, and shales that have been metamorphosed to quartzites and phyllites. The absence of the mineral epidote in the sequence distinguishes the Weverton rocks from the older clastics at and near the top of the Catoctin. The epidote in the Catoctin sediments was derived from the volcanic rock. Since the Weverton beds lack epidote, the sediments may have been transported from another source, presumably the older cratonic highlands.

The earliest traces of life forms appear in the quartzites of the Hampton Formation. In these beds are tracks, trails, and *Skolithos* tubes, or burrows, left by a wormlike, marine organism. The Hampton Formation is composed of 1,800 to 2,200 feet of clastic rocks—thin-bedded sandstones and siltstones (now quartzites and phyllites). In the overlying Erwin Formation, *Skolithos* tubes are more abundant and varied, especially in the quartzites.

Figure 30.8 Chimney Rock is a prominent landmark on a trail near the southern boundary of Shenandoah National Park. This outcrop of light-colored, resistant quartzite of the Erwin Formation (Chilhowee Group) is being eroded by weathering and mass wasting along fracture planes. Notice that "root pry" is one of the weathering processes that is breaking up the rock. The tree roots enlarge cracks by physical weathering, while chemical weathering from plant acids penetrates ever deeper into the rock. Photo courtesy National Park Service.

The resistant, massive beds of Erwin quartzite form sharp peaks, flatirons, and ridges along the western flanks of the Blue Ridge. The Erwin Formation is 700 to 1,000 feet in thickness. Most of the manganese and iron deposits in the area have been found in the topmost Erwin beds (fig. 30.8).

5. Continuing deposition and intermittent, increasingly intense tectonic activity, Ordovician through Permian time.

Elsewhere in the Appalachian area thousands of feet of marine sediments accumulated. Probably some of these beds were deposited in the Blue Ridge area but were subsequently removed by erosion or buried by large-scale crustal movements during the various Appalachian orogenies. This long series of tectonic episodes, culminating in the Allegheny orogeny and the ending of Appalachian mobile belt development in late Paleozoic time, greatly altered the rocks of the Blue Ridge (box 30.1). The folds that are oversteepened to the northwest, the thrust faults that pushed from southeast to northwest, and the cleavage and joint systems oriented to compressive stresses coming from the southeast were all produced by continuing plate convergence and eventual collision.

6. Intrusion of Triassic dikes, erosion, and long slow uplift in Mesozoic time.

The diabase dikes that rose through fractures in the Blue Ridge bedrock in Triassic time are probably associated with the break-up of Pangea and the opening of the Atlantic Ocean (box 30.1). Only a few of these dikes are exposed in the Park, most of them in the central and southern sections. McCormick Gap, near the Park's southern entrance, is eroded along the northwest-southeast trend of an extensive diabase dike. The dike rock is less resistant than the Catoctin lavas that it intruded.

The long, slow uplift during the Mesozoic was accompanied by tensional faulting and active erosion that removed great volumes of rock from the uplands. The crustal movements were largely vertical and probably were the result of isostatic adjustment that took place following the late Paleozoic orogeny.

7. Upwarping in Cenozoic time and the development of the modern landscape by erosional processes.

The upwarping of the Appalachian region in Cenozoic time (which may be continuing today at a very slow rate) was due to isostatic compensation of the crust that accompanied the gradual widening of the Atlantic Ocean.

During Pleistocene time, erosion intensified because of increased frost action and precipitation. The wide alluvial fans along the foothills, sloping westward to the Shenandoah Valley, were probably built up in the Pleistocene. The fans are now being eroded by the same streams that deposited them. The talus slopes described earlier are also mainly the result of intensified Pleistocene weathering.

Geologic Maps and Cross Sections

American Association of Petroleum Geologists. 1970. *Mid-Atlantic Region Geological Highway Map No. 4.* Tulsa, Oklahoma: American Association of Petroleum Geologists.

Gathright, T. M., II. 1976. *Geology of the Shenandoah National Park, Virginia.* Virginia Division of Mineral Resources Bulletin 86. Plates 1, 2, 3.

Reed, J. D., Jr. 1969. *Ancient lavas in Shenandoah National Park near Luray, Virginia.* U.S. Geological Survey Bulletin 1265. Plate 1.

Bibliography

Crandall, H. 1975. *Shenandoah, the story behind the scenery*. Las Vegas, Nevada: KC Publications.

Espenshade, G. H. 1970. Geology of the northern part of the Blue Ridge anticlinorium. In *Appalachian geology*, ed. G. W. Fisher et al., pp. 199–211. New York: Interscience Publishers.

Gathright, T. M., II. 1976. *Geology of the Shenandoah National Park, Virginia*. Virginia Division of Mineral Resources Bulletin 86.

Hack, J. T. 1965. *Geomorphology of the Shenandoah Valley*. U.S. Geological Survey Professional Paper 484.

Heatwole, H. T. 1978. *Guide to Skyline Drive and Shenandoah National Park*. Shenandoah Natural History Association.

Hunt, C. B. 1974. *Natural regions of the United States and Canada*. San Francisco: W. H. Freeman and Company.

Reed, J. C., Jr. 1955. Catoctin Formation near Luray, Virginia. *Geological Society of America Bulletin 66: 871–896.*

————. *1969. Ancient lavas in Shenandoah National Park near Luray, Virginia*. U.S. Geological Survey Bulletin 1265.

Rogers, J. 1970. *The tectonics of the Appalachians*. New York: John Wiley & Sons, Inc.

Petersen, M S.; Rigby, J. K.; and Hintze, L. F. 1980. *Historical geology of North America*. 2nd ed. Dubuque, Iowa: Wm. C. Brown Company Publishers.

Shenandoah National Park guide. 1981. Shenandoah Natural History Association.

Tuttle, S. D. 1980. *Landforms and landscapes*. 3rd ed. Dubuque, Iowa: Wm. C. Brown Company Publishers.

Address

Shenandoah National Park
Route 4, Box 292
Luray, Virginia 22835

31
Great Smoky Mountains National Park

Location: Tennessee-North Carolina boundary
Area: 517,368.15 acres; 808.39 square miles
Authorized: May 22, 1926
Established for full development: June 15, 1934

Figure 31.1 Deciduous trees and evergreens of many varieties cover even the steeper slopes in Great Smoky Mountains National Park. In this photograph, taken from Newfound Gap, the Transmountain Highway is visible next to the right margin and Mount LeConte can be seen in the far distance. National Park Service photograph.

GREAT SMOKY MOUNTAINS

NATIONAL PARK

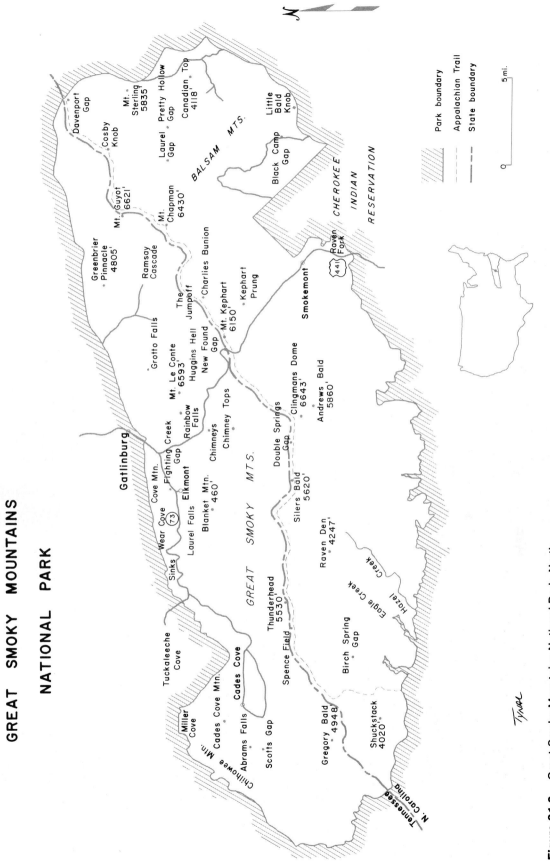

Figure 31.2 Great Smoky Mountains National Park, North
Carolina and Tennessee.

Table 31.1. Geologic Column, Great Smoky Mountains National Park

Time Units			Rock Units		Geologic Events
Era	Periods	Epochs			
Cenozoic	Quaternary	Holocene	Alluvium		Stream erosion, mass wasting
		Pleistocene	Talus slopes Alluvium		Increased frost action, weathering, stream erosion during climatic fluctuations
	Tertiary (or Paleogene/Neogene)				Long erosion, establishment of modern drainage patterns Sporadic uplift
Mesozoic					Prolonged erosion Uplift due to isostatic adjustments
Paleozoic	Permian				Gatlinburg faults (?) Great Smoky thrust faults Allegheny orogeny, folding, regional metamorphism
	Pennsylvanian				
	Mississippian		(outside Park)		Deposition in vicinity of Great Smoky Mountains
	Devonian				
	Silurian		///////		Erosional interval
	Ordovician	(Middle)	(outside Park)		Deposition in vicinity of Great Smoky Mountains Taconic orogeny, folding, Greenbrier thrust faults
			///////		Unconformity
		(Lower)	Knox Group		Carbonates (exposed in coves)
	Cambrian	(Lower)	Chilhowee Group		Thick deposits, mostly sands, on continental shelf Exposed in valleys and ridges north and west of Park
	?				Disconformity (?)
Proterozoic Z (late Precambrian)			Ocoee Series	Walden Creek Group	Slates, shales, siltstones
				Unclassified formations	Sandstones
				Great Smoky Group	Mainly thick-bedded sandstones
				Snowbird Group	Sandstones, siltstones
			///////		Erosional interval
Proterozoic Y			Basement complex		Accumulated sediments and intrusives now highly metamorphosed and deformed

Source: modified after King, Neuman, and Hadley 1968

Geographic Setting

Great Smoky Mountains National Park, which straddles for some 60 miles the boundary between Tennessee and North Carolina in the southern Blue Ridge province, is in the highest part of eastern United States. Only Mount Mitchell (6,684 feet in elevation) in the Black Mountains northeast of the Park is higher than the major peaks of the Great Smokies. Clingmans Dome, highest point in the Park, is 6,642 feet in elevation; and the second highest peak, Mount Guyot, is 6,621 feet. More than 20 other summits in the Great Smokies are over 6,000 feet in height. The two highest mountains in the Park were named for geologists T. L. Clingman and Arnold Guyot, who made the first scientific reconnaissance of the range in the late 1860s.

The crest of the southern Blue Ridge in Great Smoky Mountains National Park is more than 2,000 feet higher than the Blue Ridge crest in Shenandoah National Park (chapter 30), some 400 miles to the north. Moreover, in Shenandoah the spine of the Blue Ridge is relatively narrow, with spurs jutting out on either side; while in Tennessee, North Carolina, and Georgia—at the southern end of the province—the Blue Ridge becomes a sprawling complex of ridges lacking the distinctive lineation of the mountains to the north. What is locally called "the Blue Ridge" in North Carolina is a prominent frontal scarp that rises from the Piedmont along the eastern edge of the Blue Ridge province.

Another striking difference between the northern and southern parts of the province is the fact that no rivers cross the Blue Ridge south of Roanoke, Virginia. South of the Roanoke River, the drainage divide follows the crest of the Blue Ridge province. Streams on the eastern slopes and the Piedmont drain to the Atlantic, but most of the water from the southern Blue Ridge flows west or northwest and then south to the Gulf of Mexico.

The southern highlands are also more humid because warm winds from the Gulf of Mexico discharge quantities of moisture on the mountain slopes. Clingmans Dome and the other high summits in the Great Smokies usually receive 85 inches or more of precipitation annually, making these mountains almost as wet as those of the Pacific Northwest. Even the valleys in the southern Blue Ridge receive about 50 inches of rainfall per year. The abundance of moisture and a thick mantle of residual soil in valleys and foothills has supported for millions of years a luxuriant and diversified plant life, with many species attaining extraordinary size. Recently Great Smoky Mountains National Park was selected for International Biosphere Reserve status because of its significant ecological resources.

Droplets of mist in the air, vapors rising from the lush vegetation, and wisps of fog are what cause the "smoke" that gives the Great Smokies their name. The gently swirling rise and fall of the haze under varying conditions of temperature and light produces an effect of bluish-gray smoke around the summits and in the valleys (fig. 31.3).

Great Smoky Mountains National Park is visited by more than 9 million persons annually, partly because a major highway (US 441) crosses the mountains through the Park. Here also is the southern terminus of the scenic Blue Ridge Parkway (chapter 30). Yet despite the fact that this Park has more visitors than any other in the National Park system, about 75 percent of its 800 square miles is preserved as a wilderness or is being allowed to revert to a wilderness state. Although two-thirds of the total area had been logged over or burned before coming under national park protection, establishment of Great Smoky National Park saved the largest remaining virgin forest in eastern United States.

Local History

The Southern Appalachians, in particular the Great Smoky Mountains, are regarded by the Cherokee Indians as sacred and as their ancestral home. In the high *Shaconage,* the Cherokees' name for the "mountains of the blue smoke," dwelt the tribal spirits, the "little people" who were the guardians of Cherokee tradition. Each autumn the elders of the tribe communed with the spirits in an encampment near the jagged peaks of Chimney Tops. At the foot of the lofty mountains, where Deep Creek campground now is, was Kituwha, their principal village.

The Cherokee had developed a settled and advanced culture long before the coming of Europeans to North American shores. The Cherokee system of government was democratic and their way of life generally peaceful. Their early contact with white settlers was friendly although the Indians fought when they were provoked. After a war with Carolina settlers ended in the 1760s, the Cherokee withdrew to their lands in the mountains and foothills of the Blue Ridge and concentrated on developing their own nation. Their great leader Sequoyah created an alphabet and written language and taught thousands of his people to read in the early 1800s. They kept tribal records and even published their own newspapers. Their prosperous and successful way of life ended, however, after gold was discovered on Cherokee land. In 1838 most of the Cherokee were forcibly deported from their ancient homeland and escorted by U.S. Army troops to Oklahoma Territory. Thousands died from starvation and exposure during the march, referred to as the "Trail of Tears." Some Cherokee escaped deportation by hiding for years in their sacred places high in the mountains between Clingmans

Figure 31.3   The "smoke" (haze) that hangs over the valleys and wreathes the summits in Great Smoky Mountains National Park is due to droplets of mist borne on moisture-laden winds from the southwest. This view from Cave Creek Gap looks southward over North Carolina. National Park Service photograph.

Dome and Mount Guyot. Eventually the federal government permitted Cherokee who wished to do so to return from Oklahoma. About 4,500 descendants of the Cherokee who resisted removal or returned now live in the Qualla Indian Reservation in North Carolina on land adjoining Great Smoky Mountains National Park.

In the late 18th and early 19th centuries, white settlers, mainly emigrants from England, Scotland, and Ireland, began to occupy land in the foothills and valleys on both the Tennessee and North Carolina sides of the Great Smokies. After the fertile bottom lands in the valleys were taken, families sought farm sites higher on the slopes. Those who were fortunate found hidden "coves" in the mountains and "balds" on the ridges. Cove is the local term for an open, flat, or gently sloping, valley underlain by limestone bedrock and protected from wind by high ridges. (Their geologic origin is discussed later in the chapter.) The balds are treeless openings in the forest on slopes at higher altitudes. Some are grassy, but most are covered by a nearly impenetrable maze of closely intertwined rhododendrons and mountain laurel shrubs. The settlers called these spaces "heath balds" or "laurel slicks."

Their origin is uncertain, but they may be long-weathered scars of old mudslides or natural fires. Settlers laboriously cleared some balds and used them for pasture. In the sheltered coves they built homes and cultivated fields.

Cades Cove, first settled in 1818, is maintained as a historical and cultural preserve by the National Park Service. In this isolated valley, deep in the Great Smoky Mountains, descendants of some of the early pioneers live and work their farms, keeping up the traditions of their forebears. Exhibits, museums, a blacksmith shop, a working grist mill, and the like, tell the story of life as it was in the southern highlands for well over a century—self-sufficient, independent, and shut off from the outside world. The overshot water wheel that powers the grist mill, for example, is one of dozens of small water wheels the settlers built on mountain creeks for grinding their grain. With high gradients and high velocities, the streams were a reliable and convenient energy source.

In the early 1900s, the mountain people became less isolated as lumbering companies acquired vast tracts of land in the Great Smoky Mountains. Railroads and towns were built and sawmills set up. Forests centuries old were leveled, and mountain streams were used as sluiceways for the fallen logs. Splash dams were constructed on creeks so that the water could be released in a flood that would wash the logs downstream. The Sinks, a series of cascades on the Little River, carried logs by the thousands from the slopes of Clingmans Dome.

Slash was burned with little regard for the steep slopes that were now exposed to erosion. Landslides occurred with every heavy rain. Charlies Bunion, a bare slate knob on the crest of the range (near Newfound Gap), was completely denuded of soil and vegetation by a fire and subsequent flash flood in the 1920s. Only a few hardy plants clinging to crevices in the jagged bedrock have been able to grow back.

People in Tennessee and North Carolina who deplored the destruction of the forests and the pollution of the streams began to organize a conservation movement for the purpose of establishing a national park in the Smokies. Others throughout the nation, realizing that the mountains were a botanical treasure house, joined in the effort. After Congress authorized the Park in 1926, the difficult task of raising money to buy up the land from thousands of individual owners began. Citizen groups in Tennessee and North Carolina gathered contributions, and the two state legislatures each appropriated two million dollars. Still more money was needed, and an appeal was made to John D. Rockefeller, Jr., long a supporter of the national park movement. His donation of five million dollars, plus a million and a half from the federal government, finally brought the undertaking to successful completion. Congress established the Great Smoky Mountains National Park for full development in 1934, and it was dedicated in 1940.

An ecological problem causing concern at the present time is the recent increase in "acid rain," a result of air pollution. The situation is being monitored carefully, but the degree of damage to the Park's environment has not been completely determined.

Geologic Features

"*What* geology?" the ranger naturalist said with a chuckle to an enquiring tourist. "We don't have any geology in this park!"

What he meant, of course, is that the bedrock geology in Great Smoky Mountains National Park is not obvious. Outcrops are scanty, and geologic structures are obscured by the forest cover. These difficulties, as field geologists agree, are "compounded by the enigmatic nature of the rocks themselves" (King and others 1968, p. 1).

However, some of the geomorphic aspects of the mountain landscape can be related to geologic processes and climatic factors by even casual observers. Summits and ridges in the Park are more rounded than sharp. In profile they tend to be convex upward. Glaciers have not sculptured the skyline of the Southern Appalachians. The outward appearance of these magnificent and imposing mountains is, rather, the result of millions of years of weathering and erosion in a humid, predominantly warm to temperate climate.

The climatic fluctuations of Pleistocene time mainly affected the plant and animal life of the region, but effects can be seen in the Park of intense frost action during the ice advances over the North American continent. Although they are now fairly stable and masked by tree growth, the bouldery talus slopes in the steep draws and gullies at higher altitudes probably developed during the colder times of the Pleistocene. At lower altitudes patches of deeply weathered soils may reflect the increased warmth and humidity of interglacial episodes.

The results of mass wasting processes are evident, especially on the steeper, higher slopes where the soil layer over the bedrock is thin and fragile. Slopes where regrowth has not become well established since lumbering days are particularly vulnerable to landsliding during heavy or prolonged rains. In September 1951, when a thunderstorm dumped four inches of rain in one hour on Mount LeConte, more than 40 slides and avalanches, carrying rocks, trees, and debris, roared down the mountainside. At the present time, some of the larger landslide scars are still clearly visible from Newfound Gap Road (fig. 31.4).

Throughout the Park, because of the high precipitation, many streams, large and small, race down the slopes through stone-filled channels and over waterfalls. Several of the Park's scenic highways, such as Little River and Laurel Creek Roads, follow winding streams along their valleys at lower elevations. Hiking trails, from easy to strenuous, lead up to waterfalls that plunge over bedrock ledges (figs. 31.5, 31.6).

Bedrock and Geologic Structures

Because they were formed under similar geologic conditions, the rocks in Great Smoky Mountains National Park are in many (but not all) respects like those of its neighbor to the north, Shenandoah National Park (chapter 30). In general, the regional geologic and tectonic events summarized in box 30.1 (p. 378) apply to both parks. A significant difference is that deformation was more severe and the extent of thrust faulting was considerably greater in the southern section of the Blue Ridge province than in the northern part.

Figure 31.4   The landslide scars on Mount LeConte (ridge in background) are the result of slides that occurred in September, 1951, during a downpour of four inches of rain that fell in less than an hour. The slides stripped trees and soil from the steep slope, exposing the bedrock. This view looks northwest over Anakeesta Ridge (foreground) from Chimney Tops. National Park Service photograph.

Much has been learned in recent years about the origin and evolution of the complex rocks and structures at the southern end of the Southern Appalachians, and ongoing investigations continue to fit together pieces in the geologic story—particularly in regard to the deep subsurface relationships of the rocks. However, a generally acceptable synthesis has yet to be proposed that can satisfactorily account for the geometry, history, and lithologies of the very old, considerably altered bedrock of the Great Smokies.

In gross terms, three groups of rocks make up the Great Smokies and the surrounding foothills. In chronological and roughly geographical order, from southeast to northwest across the mountains, these groups are as follows:

1) The basement complex. These are rocks of middle and late Precambrian age that have been deeply buried and have undergone repeated episodes of metamorphism.

2) The Ocoee Series. This is a thick sequence of metamorphosed sedimentary rocks of late Precambrian age that form the main bulk of the Great Smokies.

3) The Chilhowee Group and other early Paleozoic rock units. These rocks crop out on the lower ridges, in the coves, and in the Appalachian Valley, mostly outside the Park.

The Basement Complex

In the southeastern foothills on the North Carolina side of Great Smoky Mountains National Park are exposures of ancient, much altered crystalline rocks that underlie the rocks and structures of the region. Some of the layered gneisses and schists were probably formed from sand, silt, and volcanic sediments. The granitic gneisses may originally have been plutonic rocks. All of these basement rocks have undergone so many changes that their predeformation positions and age relationships cannot be determined.

Figure 31.5 Laurel Branch (a stream) tumbles over resistant ledges of Precambrian Thunderhead Sandstone to form Laurel Falls in Great Smoky Mountains National Park. Laurel Falls is accessible by trail from Little River Road. National Park Service photograph.

Figure 31.6 LeConte Creek drops over a resistant outcrop of Thunderhead Sandstone, forming Rainbow Falls. The horizontal fractures in the bedrock are probably the result of regional metamorphism and do not represent the original bedding of the rock. The large blocks in the foreground have broken off from the ledge along these fracture planes. A trail from Cherokee Orchard Road leads to Rainbow Falls. National Park Service photograph.

Deep Creek Campground and the Oconaluftee Visitor Center in the southeast corner of the Park are underlain by Precambrian basement complex rocks.

The Ocoee Series

Great thicknesses of Ocoee rocks not only make up most of the mountain mass in the Park but also extend for many miles along the trend of the southern ranges, both northeast and southwest of the Great Smokies. These are the rocks of the great thrust sheet that was pushed northwestward over younger Paleozoic rocks in a series of movements in Paleozoic time.

Included in the Ocoee Series are the Snowbird Group, which lies unconformably over the basement complex; the Great Smoky Group, by far the thickest and most massive; some unclassified sandstone formations; and the Walden Creek Group, which crops out mostly in the foothills on the western and northern sides of the Great Smokies. All are nonfossiliferous, clastic marine sedimentary rocks that have been metamorphosed in varying degrees. The Walden Creek rocks contain large amounts of shaly and silty clastics and lenses of conglomerate, but the other Ocoee rock units consist mainly of many layers of fine- to coarse-grained sandstones.

The most prominent are the formations of the Great Smoky Group, which are exposed on the summits in the Park, in numerous roadcuts along the transmountain highway (US 441), and in most of the Park's waterfalls. The Elkmont Sandstone, consisting of relatively fine-grained sandstones, is at the base of the Great Smoky Group. The coarse-grained Thunderhead Sandstone, which is thick-bedded and quite resistant, forms ledges and cliffs on some of the mountain slopes, as well as the summit areas of Clingmans Dome and Mount Guyot, the highest mountains in the Park. Overlying the Thunderhead Sandstone on some other parts of the crest, such as Mount LeConte, Chimney Tops, and The Sawteeth, are the dark, slaty rocks of the Anakeesta Formation. The craggy pinnacles and steep-sided ridges eroded from the Anakeesta rocks form prominent landmarks along the Appalachian Trail.

The rock units of the Great Smoky Group are mostly massive and were initially remarkably homogeneous. The present variations in their lithologies are due mainly to the several metamorphic facies that reflect differing degrees of heat and pressure when the rocks were undergoing progressive regional metamorphism. The more intensely metamorphosed rocks (amphibolite facies) are in the southeastern part of the mountains, while rocks of the same formations to the northwest show decreasing grades of metamorphism (greenschist facies). The chemical and mineralogical variations in the rocks of the thrust sheet indicate that on the southeast side metamorphism was probably more prolonged and increasingly intense. The presence of abundant water in the buried rocks was also a significant factor affecting their alteration by metamorphic processes (Allen and Ragland 1972).

The Paleozoic Rocks

Some of the older Cambrian beds of the Chilhowee Group are exposed in roadcuts along the Foothills Parkway that skirts the western edge of Great Smoky Mountains National Park. Primitive fossil shells have been found in these rocks. However, the only significant outcrops of Paleozoic rocks within Park boundaries are Lower Ordovician limestones of the Knox Group that underlie the coves nestled among the western foothills.

Igneous Rocks

A geologic map of the Great Smokies region shows extensive sequences of the dominant sedimentary and metamorphic rocks, but only a very few minor bodies of igneous rocks—although the basement complex is assumed to contain components that were originally extrusive and intrusive rocks. Some narrow metadiorite sills of Paleozoic age crop out between beds of the Thunderhead Sandstone near the top of Clingmans Dome. The sills extend southwestward down the mountain and are probably associated with some old copper mines along Hazel Creek in the southern part of the Park.

Folds and Faults

Folds large and small are numerous in the layered rocks of Great Smoky Mountains National Park. Some are broad, open folds with long stretches of uniform dip; others are small and sharp. In the southeastern area, where metamorphism was more intense, folds of several ages intersect, producing complex relationships. The majority of folds align with the dominant regional trend, northeast-southwest. They are asymmetric in that the northwest limbs dip more steeply than the southeast limbs.

Three major thrust fault systems dominate the mountain structures of the region: the Greenbrier faults, the Great Smoky faults, and the Gatlinburg faults. All three trend generally northeast-southwest and are the result of strong compressive tectonic forces from the southeast that eventually carried Precambrian Ocoee rocks many miles northwestward over younger rocks. The fact that the fault systems formed at different times during the Paleozoic Era suggests that they are linked to successive mountain-building episodes in Southern Appalachian tectonic history (box 30.1).

Oldest of the three are the faults of the Greenbrier system. Prominent exposures of the faults are in the southeast boundary area of the Park between the Oconaluftee section and the Cherokee Indian Reservation. Greenbrier faults also loop around the northeast corner of the Park and extend westward, where they are broken up and obscured by younger faults. The low-angle Greenbrier thrust faults carried rocks of the Great Smoky Group over the underlying rocks of the Snowbird Group, perhaps for a distance of 15 miles. Basement complex rocks in the southeast were also involved in Greenbrier faulting. In the Straight Fork area they overrode younger Snowbird rocks that are now exposed in a window, or fenster. (*Window*, when used as a tectonic term, refers to an eroded area of a thrust sheet that displays the rocks beneath the thrust sheet. The term *fenster*, a German word, is synonymous.)

Unlike the Greenbrier fault system, which is within the Great Smokies district and is mainly local in scope, the Great Smoky fault system is only one segment of an extensive displacement involving Southern Appalachian structures from Virginia to Alabama. The Great Smoky fault and its associated faults, all low-angle thrust faults, moved Precambrian Ocoee rocks in the Great Smokies area for many miles over Ordovician and occasionally younger Paleozoic rocks. Most exposures of the Great Smoky fault system are between the foothills and the Appalachian Valley, west and north of the Park, but some are within the Park, as, for example, fault segments along the coves. The main thrust sheet of the Great Smoky fault is split on the western side of the mountains into several fault blocks. The coves—that is, the windows in the thrust sheet—are located in these breaks and openings. From the leading edge of the thrust fault to where the fault surface is exposed in windows is a distance of nine miles, but this represents only a small part of the overall movement of the thrust sheet. The tectonic movement along the Great Smoky fault system occurred after the Ocoee rocks had undergone regional metamorphism.

The Gatlinburg faults, believed to be younger than the Great Smoky faults, dip much more steeply and cut both Great Smoky and Greenbrier faults in some places. Most of the Gatlinburg faults trend east-northeast and lie outside the Park's western boundary; but the Oconaluftee fault, which is related to the Gatlinburg system, branches off toward the southeast from the vicinity of Meigs Mountain, crosses the crest of the Great Smokies near Newfound Gap, and goes on down the valley of the Oconaluftee River. Gatlinburg faults have had a controlling effect on stream drainages where streams have eroded shattered rock along the fault traces. Some of these faults cross from one stream valley to the next through notches in intervening ridges.

How the Coves Developed

Cades Cove, Crib Gap, and White Sink in the western part of the Park are windows in the thrust sheet of Precambrian Ocoee rocks where younger Paleozoic rocks of the Knox Group (Ordovician) are exposed. Tuckaleechee and Wear Coves, just west of the Park boundary, are windows of larger size. The coves have been described as "Appalachian Valleys in miniature" (King and others 1968) because they are low, fertile, slightly rolling patches of lowland underlain by the same sedimentary, unmetamorphosed carbonate rocks as the long, narrow lowland stretches of eastern Tennessee's Appalachian Valley farming country.

The coves developed among the western foothills of the Great Smokies where the thrust sheet becomes thinner and broken up along its leading edge. Because the limestones beneath the thrust sheet were much less resistant to erosion than the metamorphosed Ocoee rocks, the breaks in the thrust sheet became enlarged. Runoff from the plentiful rains was channeled into the openings as the mountains were uplifted, and quantities of sediment built alluvial fans and spread over the bottoms. This eventually produced the rounded and roughly oval coves surrounded by high ridges. A few very small, remote coves that were not cleared by settlers or lumbermen remain in their natural state. The rich soils of these protected, well-watered coves support a unique community of plants, the "cove hardwoods" (yellow birches, basswoods, buckeyes, sugar maples, etc.), and a profusion of wild flowers.

Geologic History

1. The Precambrian basement complex.

The rocks of the basement complex accumulated and were metamorphosed, probably more than once, in ancient mobile belts. The oldest isotopic dates derived from minerals in these rocks indicate that plutonic and metamorphic episodes were occurring in the region about 1.0 billion years ago, or Proterozoic Y time. A scattering of later dates in these rocks suggests that "radioactive clocks" were reset by recrystallization that took place during Paleozoic metamorphism. The basement complex rocks were deeply eroded before younger sediments were deposited on them.

2. Deposition of the Ocoee Series in late Precambrian time (Proterozoic Z).

For millions of years an ancient ocean received the clastic sediments that make up the Ocoee rocks. The sediments, mostly sand and having a cratonic source, were deposited on a broad continental shelf that gradually sank as the ocean encroached on the landmass. Paleogeographers speculate that the ocean and its adjacent lands lay well south of the equator during that time. Although fossils have not been found in the rocks, numerous changes in the grain size and in the bedding are indications of changing environments of deposition over time; for example, varying depths of water, ocean currents, types of sediment (pebbly, sandy, muddy), and the like. Approximately 50,000 feet of sediment accumulated overall in the Ocoee Series. Of this thickness, the Snowbird Group made up about 13,000 feet; the Great Smoky Group, 25,000 (in some places); unclassified formations, 3,000 to 4,000; and the Walden Creek Group, 8,000. The stratigraphic units of the Ocoee Series were disrupted by late Paleozoic faulting so that correlations are uncertain between some units that are now north of the Greenbrier fault and those south of the fault (King and others 1968).

3. Deposition of Paleozoic rocks.

In the region outside the Park, to the north and west, a variety of marine sedimentary rocks were deposited, ranging in age from Cambrian to Mississippian. The Cambrian through Lower Ordovician rock units, like the older sedimentary rocks, were deposited on the continental shelf; but as sea levels continued to rise worldwide, sedimentation in the region slowed, and the quartz sands gradually changed to carbonates. Some of the rock units became part of the thrust sheet produced by the Great Smoky fault, as, for example, the Chilhowee rocks exposed on the northwest face of Chilhowee Mountain (outside the Park). Other units were overridden by the thrust sheet and are now exposed in valleys and coves, some inside the Park and some outside.

4. Paleozoic tectonic episodes.

An unconformity between Lower and Middle Ordovician stratigraphic units marks the beginning of the mobile belt in the Southern Appalachian region. Subduction and compression resulting from plate convergence during the Taconic orogeny caused the land to the southeast to rise. Clastic sediments from the uplifted regions washed into the Appalachian geosyncline and into the shallow continental sea to the west. The Greenbrier fault is associated with the orogenic activities of early Paleozoic time.

After a relatively quiet period of erosion and deposition, tectonic activity resumed in the Southern Appalachians with increasing intensity during the Allegheny orogeny. The Ocoee and Cambrian rocks were folded and metamorphosed. The thrust sheet, by a series of movements, was pushed farther to the northwest as the Great Smoky fault and its related faults developed. This occurred late in Paleozoic time after the Mississippian rocks (northwest of the Park) had been deposited. The Gatlinburg faults, believed to be younger than the Great Smoky faults, were probably active near the end of Paleozoic time. According to plate tectonic theory, the collision of plates that occurred in late Paleozoic time destroyed the proto-Atlantic Ocean and produced the supercontinent Pangea. The Great Smokies were part of a great mountain belt that stretched across much of Pangea (box 30.1).

5. **Mesozoic isostatic adjustments and erosions.**

Although tectonic activity in the Appalachian mobile belt came to an end, isostatic adjustments caused continued uplift of the Great Smoky Mountains. Meanwhile, erosion shaped the ridges and valleys and uncovered many of the older rocks. To the south and east, new rifts opened and a spreading center developed. The basin of the present Atlantic Ocean began to form, and Africa and Europe drifted away from North America. The climate of the Southern Appalachian region became less tropical and more temperate as the North American plate moved westward and northward.

6. **Continued erosion and sculpturing of the Great Smoky Mountains throughout Cenozoic time.**

Sporadic uplift may have continued into early Cenozoic time. Modern drainage patterns developed partly in the direction of regional slopes and partly by differential erosion of rock units. In Pleistocene time periglacial processes affected rocks and slopes at higher elevations although the mountains were not glaciated. Climatic oscillations not only caused increased erosion but also brought about great diversity in the plant and animal life of the region. Some of the unusual plant species found in the Great Smoky Mountains National Park vicinity may have had a much wider range throughout North America prior to the Ice Ages, which put great stress on both plant and animal life. In the Southern Appalachian region fossil remains have been found of extinct giant mammals, such as mammoths, saber-toothed tigers, and the giant sloth. Their bones have yielded carbon-14 dates of about 11,000 to 12,000 years ago, which means that these creatures managed to survive the extremely cold times, probably by sheltering in protected valleys and caves. Scientists theorize that the giant mammals were wiped out by early hunters who migrated into the region after Pleistocene time ended.

Geologic Maps and Cross Sections

American Association of Petroleum Geologists. 1970. *Mid-Atlantic region geological highway map, no. 4.* Tulsa, Oklahoma: American Association of Petroleum Geologists.

King, P. B., Neuman, R. B., and Hadley, J. B. 1968. *Geology of the Great Smoky Mountains National Park, Tennessee and North Carolina.* U.S. Geological Survey Professional Paper 587. Geologic map and structure sections are in the pocket.

Bibliography

Allen, G. C., and Ragland, P. C. 1972. Chemical and mineralogical variations during prograde metamorphism, Great Smoky Mountains, North Carolina and Tennessee. *Geological Society of America bulletin* 83: 1285–1298 (May).

Cantú, R. 1979. *Great Smoky Mountains, the story behind the scenery.* Las Vegas, Nevada: KC Publications.

Doolittle, J. 1975. *The Southern Appalachians,* American Wilderness Series. Alexandria, Virginia: Time-Life Books.

Dott, R. H., Jr.; and Batten, R. L. 1981. *Evolution of the earth.* 3rd ed. New York: McGraw-Hill Book Company.

King, P. B., Neuman, R. B., and Hadley, J. B. 1968. *Geology of the Great Smoky Mountains National Park, Tennessee and North Carolina.* U.S. Geological Survey Professional Paper 587.

Locke, J. (n.d.) *Great Smoky Mountains National Park.* Charlotte, North Carolina: Aerial Photography Services, Inc.

Address

Great Smoky Mountains National Park
Gatlinburg, Tennessee 37738

32

Big Bend National Park

by Sherwood D. Tuttle
University of Iowa

Location: West Texas
Area: 708,118.40 acres; 1106.44 square miles
Established: June 12, 1944

Figure 32.1 Rio Grande (flowing to the left) in Boquillas Canyon in Big Bend National Park. The land in the foreground is in Texas, and the cliffs across the river are in Mexico. Here the river crosses the Sierra del Carmen, a range that extends southward into Mexico for many miles. Exposed in the cliff face are dipping beds of the Santa Elena Limestone, Cretaceous in age. National Park Service photograph.

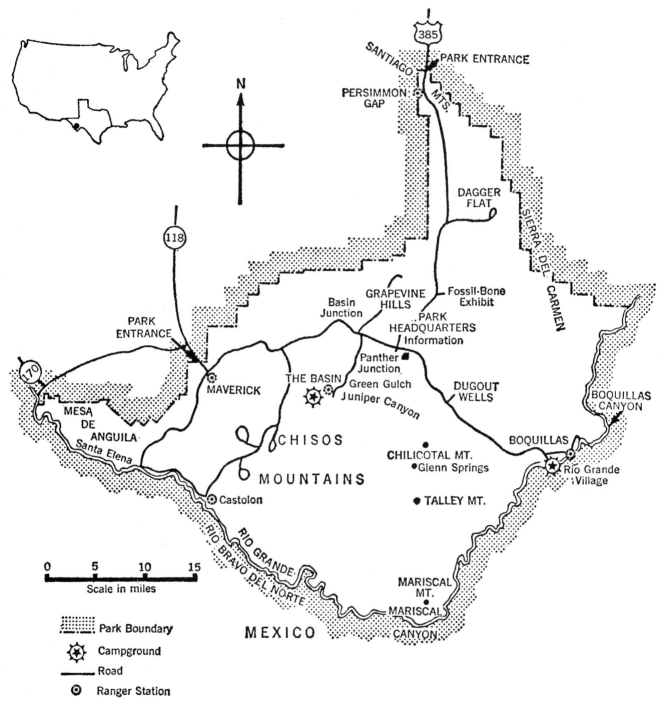

Figure 32.2 Big Bend National Park, Texas.

Table 32.1. Generalized Geologic Column, Big Bend National Park

Era	Periods	Epochs	Rock Units	Geologic Events
Cenozoic	Quaternary	Holocene / Pleistocene	Younger gravels and terraces	Erosion and deposition Erosion, deposition, major canyon-cutting Uplift, renewed faulting
Cenozoic	Tertiary (Neogene)	Pliocene / Miocene	Older gravels, pediment gravels, older terraces	Basin filling
Cenozoic	Tertiary (Neogene)		/////////////	Block faulting
Cenozoic	Tertiary (Paleogene)	Oligocene ---?---	South Rim Formation	Volcanic eruptions Hydrothermal mineralization Igneous intrusions
Cenozoic	Tertiary (Paleogene)	Eocene	Chisos Formation	Igneous intrusions
Cenozoic	Tertiary (Paleogene)	Eocene	Canoe Formation	Block faulting Mixed clastics and ash
Cenozoic	Tertiary (Paleogene)	Paleocene	Hannold Hill Formation Black Peak Formation	Nonvolcanic deposits (mammal bones)
Mesozoic	Cretaceous		/////////////	Laramide orogeny, folding, thrust-faulting
Mesozoic	Cretaceous	(Upper)	Javelina Formation	Nonmarine deposition (dinosaur bones, silicified wood)
Mesozoic	Cretaceous	(Upper)	Aguja Formation	
Mesozoic	Cretaceous	(Upper)	Pen Formation	Transitional and shallow marine deposition in transgressing and withdrawing Cretaceous seas
Mesozoic	Cretaceous	(Upper)	Boquillas Formation	
Mesozoic	Cretaceous	(Lower)	Buda Limestone	
Mesozoic	Cretaceous	(Lower)	Del Rio Clay	
Mesozoic	Cretaceous	(Lower)	Santa Elena Limestone	
Mesozoic	Cretaceous	(Lower)	Sue Peaks Formation	
Mesozoic	Cretaceous	(Lower)	Del Carmen Limestone	
Mesozoic	Cretaceous	(Lower)	Telephone Canyon Formation	
Mesozoic	Cretaceous	(Lower)	Glen Rose Limestone	
Paleozoic	(Ordovician to Pennsylvanian)		/////////////	
Paleozoic	(Ordovician to Pennsylvanian)			Ouachita orogeny
Paleozoic	(Ordovician to Pennsylvanian)		Deformed shales, sandstones, novaculite, conglomerates, probably Tesnus, Caballos, and Maravillas Formations	Shelf deposits; deep-ocean deposits

Source: modified after Maxwell 1968

Local History and Geographic Setting

The first superintendent of Big Bend National Park was Ross Maxwell, a geologist who worked on the geology of the Big Bend country for many years. In his guidebook (Maxwell 1968), he relates an Indian legend to show that prehistoric people recognized the rock formations and landforms as being strange and unusual. Like modern visitors to Big Bend, the earliest inhabitants wondered about the origins of the features they saw. According to the Indian story, when the Great Spirit had finished making the earth, the stars, and all living things, he had a lot of scraps and stony trash left over that he couldn't decide what to do with. Finally he just piled it all up in a heap and left it. That was how Big Bend country came to be.

Big Bend country is a triangular-shaped region formed where the Rio Grande runs south and then north in a great loop. At the bottom, or elbow, of the bulge is Big Bend National Park (fig. 32.2). Mexico lies across the river, and the Chihuahuan Desert extends for many miles both north and south of the international boundary. Big Bend National Park is far from a settlement of any size. The town of Marathon, 70 miles north of the Park, is on the Southern Pacific Railroad as well as the nearest east-west highway. The U.S. Customs facility for the area is also in Marathon. Interstate 10, the main highway across southern Texas, is an additional 50 miles north of Marathon. Nevertheless, Big Bend is not the isolated region it once was because a good highway (US 385) goes from Marathon to Persimmon Gap (the north entrance of the Park), and a second highway goes north from the western part of the Park to Alpine, Texas.

Although Big Bend National Park occupies only about a third of what is thought of as Big Bend country in west Texas, it is still one of the larger national parks in the United States. Within its bounds are the spectacular canyons of the Rio Grande, the jagged igneous peaks of the Chisos Mountains, barren highlands of sedimentary rock, superb desert scenery, unusual fossils, vestiges of prehistoric cultures, and rare plants and animals. Because of the generally sparse vegetation in the arid climate, angular landforms and rock units exposed in all their starkness dominate the landscape. It is interesting to note that in 1963 and 1964, in preparation for moon landings, astronauts studied rocks and practiced moving about on the rocky terrain in Big Bend National Park.

Despite the generally harsh climate, the region has been inhabited for at least 12,000 years. The nomadic Basketmakers took shelter in caves in the canyons and later built pit houses. They utilized yucca and other native plants for making sandals, mats, and baskets; and they left hearths, grinding pits, burial sites, petroglyphs, and numerous artifacts in the Park area. The Chisos Indians succeeded the Basketmakers and were there when white people came to the region. The Chisos lived along the river and tilled small fields of corn, beans, and pumpkins until they were forced to retreat to the Chisos Mountains by marauding Mescalero Apaches. The Comanche Trail goes through Big Bend. Every autumn the fierce Comanches rode from Oklahoma across Texas and into Mexico to raid the ranches and steal horses and cattle. Both the Apaches and the Comanches resisted for many years the incursions of white settlers or other intruders into their territory.

The Spanish conquistador, Alvar Nunez Cabeza de Vaca, was the first white man to visit Big Bend country. He was one of the survivors of the de Narvaez expedition that was shipwrecked on an island off the Texas coast in 1528. Made a slave by the Indian inhabitants, Cabeza eventually escaped and made his way overland with two companions. After wandering through the desert wilderness for eight years, they finally reached the Spanish settlements in Mexico. In Cabeza's account of his travels and hardships is the first description of Big Bend country.

During the next three centuries, the Spaniards and later the Mexicans tried to control the Indians along the Rio Grande, but neither missionaries nor soldiers had much success in subduing the tribes.

After the Mexican War in the 1840s, Major W. H. Emory surveyed the Rio Grande and the Big Bend area for the International Boundary Commission. His 1852 report was the first scientific description of the region. Emory Peak (7,835 feet), the highest point in the Park, is named for him.

A few years later, the United States War Department sent an expedition to Big Bend to locate the Comanche Trail and try to devise ways of protecting the Texas settlements from Indians. The expedition was also for the purpose of determining the capabilities of camels imported for military transport in the arid southwest. A young Army lieutenant named Echols led the camel caravan along the base of the Santiago Mountains, entering territory that is now Big Bend National Park near Persimmon Gap and Dog Canyon. (The latter was originally named Camel Pass by Lt. Echols.) They crossed the difficult desert terrain and followed Terlingua Creek to the mouth of Santa Elena Canyon on the Rio Grande, a trip that took more than a month (fig. 32.3). For several days in the desert, death from thirst seemed imminent, but finally enough water was found to enable them to complete their journey to the border. Lt. Echols reported that the twenty camels stood the trip better than the mules and even better than the men. However, with the beginning of the Civil War, the camel experiment ended. The camels were turned loose in the desert and they disappeared.

Mining activities began in Big Bend in the 1890s with the opening of quicksilver (mercury) mines near Terlingua, now a ghost town outside the Park's western boundary. The Spaniards had earlier hunted, probably

Figure 32.3 This aerial view, looking southwestward toward the mouth of Santa Elena Canyon, is one of the most spectacular landscapes in Big Bend National Park. The Rio Grande flows through the canyon, toward the foreground of the photograph, and then bends left, or southeastward, continuing on downstream. The cliffs on the right side of the river and the land in the foreground are in the United States. The cliffs on the left side of the river are in Mexico. The broad wash in the right foreground, filled with alluvium, is the bed of Terlingua Creek, which flows down from the north to join the Rio Grande. Santa Elena Canyon extends upstream for about 18 miles, cutting across the Mesa de Anguila, a large fault block (called Sierra Ponce on the Mexican side). For the first 7 miles upstream, the gorge is as steep and narrow as the opening shown here. In some places the bottom of the canyon narrows to 30 feet and the walls are sheer vertical cliffs. The canyon's maximum depth is over 1,500 feet.

The imposing fault scarp of the uplifted Mesa de Anguila block stretches across the photograph, with the trace of the Terlingua fault running along the base. The foreground area is part of the downdropped block. Faulting with a throw of about 3,000 feet began in late Cretaceous time, with renewed movement on the fault occurring in late Tertiary time. Approximately 1,600 feet of sedimentary beds are exposed at the mouth of the canyon. The prominent cliff-forming rock unit near the top is the Santa Elena Limestone; below is a shaly formation (Sue Peaks) with talus-covered slopes; below that another limestone, the Del Carmen. Two more formations crop out at the base of the escarpment. All are Lower Cretaceous. (See fig. 32.5.)

Note the large slump block jutting out on the left side of the canyon mouth. Another example of mass wasting, a short distance upstream, is a rock slide, 250 feet high and nearly a quarter of a mile long that partially dams the Rio Grande and creates dangerous water conditions for rafts and riverboats. National Park Service photograph.

unsuccessfully, for gold and silver in the mountains, but tales of their "lost mines" persist to this day. Some quicksilver was also produced at the Mariscal mine in the southern part of the Park (fig. 32.4). Maxwell (1968) reports that the quicksilver mined and smelted in the Terlingua district from 1896 until the 1940s amounted to "one-fourth of the total mercury production in the United States" (p. 115).

A few hardy ranchers and homesteaders settled in the wider parts of the canyon bottoms along the Rio Grande in the early 1900s. The settlers were able to grow cotton and livestock feed by drawing irrigation water from the river. At the site of some warm springs near the river, a small village and a bathhouse were erected. Invalids came there to soak in tubs of the spring water. Big Bend was

Figure 32.4 Mariscal Mountain, a plunging anticlinal fold. This aerial view, looking south toward Mexico in the far distance, shows a relatively rare landform, an anticlinal mountain eroded from a plunging fold. The axis of the fold runs from the lower right of the photograph to upper left along the mountain crest. Hogback ridges (formed by differential erosion), running along the base of the mountain on both sides, converge in the photograph's foreground; and elevation of the mountain crest (the trace of the fold axis) decreases toward the foreground. Thus the anticline plunges toward the lower right in this view. The hogbacks on the right dip about 40°, and those on the left, 15° to 20°; indicating that the fold is asymmetrical with a steeper right limb. Small streams flowing from the crest area have cut gullies through the hogbacks and are depositing sediment in the long valleys on either side of the mountain. The notch in the far distance on the mountain's right slope is the entrance to Mariscal Canyon, which was cut through Mariscal Mountain by the Rio Grande.

The rocks here include almost the entire Cretaceous section exposed in Big Bend National Park. The oldest rock, cropping out in the center of the fold is the Santa Elena Limestone. Strips of the younger Pen Formation are found along both sides of the mountain. A small quicksilver mine, now abandoned, is visible in the vicinity of the road junction in center at the bottom of the photograph. The cinnabar ore was recovered from the Boquillas Formation, a clayey limestone (p. 405). National Park Service photograph.

still the "wild west," however; it was the last frontier of the great American West. Raiders, gangs of outlaws, and Mexican revolutionaries crossed the border to plunder ranches and settlements. U.S. Army outposts were established to protect the settlers and patrol the border, but the raids persisted until about the time of World War I.

The movement to establish a park in the Big Bend began in 1933 when Texas Canyons State Park was set aside by an act of the Texas legislature, and the process of acquiring additional land was started. Two years later the United States Congress authorized the inclusion of the area in the national park system, pending completion of land acquisition. Through donations and state appropriations during the next ten years, the work went on. Finally in 1944, the large tract of mountains, desert, and 107 miles of international boundary was transferred to the federal government, thus establishing Big Bend National Park.

Unique species of plants, birds, and animals live in the Park, especially in the Chisos Mountains, where there is a little more moisture at the higher elevations. A few species of colder-climate trees, such as ponderosa pine, cypress, and aspen, have survived among the peaks since the Ice Age. When the climate was cooler and moister during the Pleistocene, these species grew over a wider area, but as conditions became drier and hotter, they were replaced by desert vegetation except at the high altitudes.

Throughout Big Bend country rainfall averages less than 10 inches per year. When it does come, rain is usually local, in a cloudburst (perhaps as much as an inch in an hour), and during late summer or early fall. At these times, erosion is intense, although brief. Temperatures are warm the year round; cooler in winter and at higher elevations, and very hot in summer. NEVER hike or drive in Big Bend National Park without taking along drinking water.

Geologic Features

The Landforms

The locations and shapes of the infinite variety of landforms in Big Bend National Park can be explained in terms of geologic structure and the degree of resistance to erosion of specific rock units. For example, the three great canyons of Big Bend through which the Rio Grande flows are located at places where geologic structures have produced mountains. In the southeastern part of the Park, Boquillas Canyon (longest of the three) cuts across the Sierra del Carmen, a range of mountains that trends north-south along the Park's eastern boundary (fig. 32.1). Upstream, at the southern tip of the Park, is Mariscal Canyon, where the Rio Grande crosses the structure of Mariscal Mountain (fig. 32.4). At the western end of the Park, Santa Elena Canyon, the most spectacular of the gorges, was eroded by the river through Mesa De Anguila, a fault-block mountain, probably during uplift (fig. 32.3). In all these mountains the principal bedrock is in the formations of Cretaceous limestones that have been folded and faulted.

Seemingly unrelated are the volcanic and plutonic rocks that make up the Chisos Mountains in the center of Big Bend National Park (fig. 32.6). The Chisos, which are several thousand feet higher in elevation than the features along the Rio Grande, were built by igneous emplacements and extrusions, Oligocene in age or younger, and are the youngest bedrock in Big Bend.

The oldest rocks in the Park are Paleozoic units exposed in Persimmon Gap at the Park's northern entrance. Persimmon Gap, which is an opening in the Santiago Mountains, is a wind gap left by stream capture that probably took place early in Cenozoic time.

People have given vivid and colorful names to oddly shaped natural landmarks in the Park. Most of these features were produced by the processes of weathering and erosion on rocks of differing resistance. Cowboy Boot, for example, which looks like a giant inverted boot, is a rock pillar eroded along joints in volcanic rock. Two resistant dikes, side by side and looking like a mule's ears sticking up on a mountaintop, are the result of differential erosion. The landform, of course, is named Mule Ear Peaks. Elephant Tusk, a steep-sided peak with a pointed top, was eroded from a resistant igneous intrusion. Alsate's Face is an eroded ridgetop that resembles the head in profile of a recumbent Indian gazing at the sky. (Alsate was an Apache chieftain of local renown.) Casa Grande ("big house") is a lava-capped mesa with nearly vertical sides and flat top. A small cave eroded in a cliff near the top of Lost Mine Peak is named the Watchman's House. The cave, according to legend, is occupied by the ghost of an Indian who was left to guard a mine once worked by the Spaniards. These are but a few of many interesting landforms that can be seen in the Big Bend.

A large open valley (altitude, 5,401 feet) nestled among the peaks of the Chisos Mountains was thought by some early investigators to be either a large cirque or a caldera; but further study showed that the feature was the result of erosion by running water. (No glaciers formed in the Chisos Mountains in Pleistocene time, but erosion was intensified.) This landform, called the Basin, is the site of the Chisos Mountains Lodge and Campgrounds. The Window, a narrow gorge cut in the Basin's western rim, opens onto a 75-foot dry fall. The Window is the outlet for all the drainage from the Basin.

Evidence of deposition is as widespread throughout Big Bend as the features of erosion, although the sheet-wash gravels, filled basins, terrace gravels, and gravel-strewn pediments may not have the dramatic shapes of the erosional landforms. Great quantities of sand and gravel have been washed down from the mountains and deposited in valleys and basins, with deposition being especially heavy in Pleistocene and post-Pleistocene time.

Because of the absence of vegetation, Big Bend is a place where many of the features plotted on geologic maps (anticlines, fold axes, fault traces, fault blocks, etc.) can be readily seen. Noteworthy examples are Cow Heaven anticline and the Mariscal anticline (fig. 32.4).

In learning about such a multiplicity of structural features and landforms, it is helpful to remember that the limestones and other sedimentary units are mainly Cretaceous; the igneous rocks are early Cenozoic in age; and the major structures are related to Laramide tectonic activity in latest Cretaceous and early Cenozoic time. Block-faulting and mountain uplift occurred in Tertiary time along with intense erosion and deposition. This part of Texas is included in the Basin and Range province.

Box 32.1
The Rio Grande

All of the Big Bend drainage is collected by the Rio Grande, which forms the southern boundary of the Park for more than a hundred miles. One of the great rivers of the continent, the Rio Grande rises in the San Juan Mountains of southwest Colorado in the Southern Rockies, flows through part of the Basin and Range province in New Mexico and west Texas, skirts the southern tip of the Great Plains, and empties into the Gulf of Mexico by crossing the Texas portion of the Coastal Plains province. On its 1,885-mile journey from the mountains to the ocean, the river crosses many diverse rocks and structures.

Although the Rio Grande forms a natural boundary between the United States and Mexico from El Paso to Brownsville, its shifting channel has been the cause of many disputes over the exact location of that boundary. An International Boundary Commission mediates these disputes. However, a 114-year controversy over the location of the boundary at El Paso was not settled until 1968 when the stream was diverted into a concrete channel. In this century a number of dams and reservoirs have been constructed on the river to control its flow and supply water for irrigation. A 1945 treaty with Mexico provides for sharing the water of the Rio Grande. Despite its volume and length, the Rio Grande is navigable for only a short distance inland from the coast.

Throughout most of its length, it appears that the Rio Grande is younger than the geologic structures over which it flows and that it developed its course by flowing across basin fills and fault blocks. This is clearly evident in the Big Bend boundary area where the Rio Grande is a *transverse stream*; that is, it flows in a crosswise

direction over the faults and fault blocks. The river is also *superimposed*, which means that it was let down from above by erosion and that it cut through the formations on which it originally developed and through unconformities beneath. During this process, the Rio Grande and its tributaries removed quantities of sand and gravel (that once filled the basins) and exposed the bedrock of the upfaulted blocks.

From the size of the valley of the Rio Grande and the large amount of fairly recently deposited sediment on the flood plain, we can conclude that the volume of the river has fluctuated considerably in late Cenozoic time. During the Pleistocene, the Rio Grande was fed by valley glaciers that grew and melted in the Southern Rockies, greatly increasing the river's volume. Periods of pluvial climate brought greater precipitation to the whole southwest region; but at other times, the regional climate was even drier than at present. Such climatic fluctuations profoundly affected the river's regimen, especially its cycles of flooding, and also caused changes in the processes of weathering, mass wasting, and runoff in the Big Bend region and elsewhere.

Detailed studies of river gravels, terraces, and pediments in the Rio Grande valley (mainly in New Mexico) show the occurrence of alternating periods of *aggradation* (the building up of the land surface by deposition) and *degradation* (the lowering of the land surface and general reduction of relief by stream action). This sort of cut-and-fill went on in the canyons as well. In some places on the less steep sidewalls, pediments cutting across both bedrock and gravel were left high above the present valley floor.

Vertebrate Fossils

Beside the road that goes from Persimmon Gap at the northern entrance of Big Bend National Park to Park Headquarters at Panther Junction, is an exhibit shelter displaying mammal and dinosaur bones discovered in sedimentary rocks in the Park. The remains of very ancient, primitive mammals were found in Paleocene beds near the roadside exhibit. These extinct mammals, the oldest that

have been found in Texas, were small, forest-dwelling carnivores and herbivores having no direct relationship to modern species. (Such mammals may have coexisted earlier with late Mesozoic dinosaurs, whose fossilized remains have also been found in Big Bend.)

Eocene beds overlying the Paleocene rocks near the exhibit contain the bones of somewhat more advanced mammals. These animals were not direct ancestors of living mammals either, but they had some anatomical similarity to later species. The remains include the earliest "horse" (*Hyracotherium*, or eohippus), an early cat, piglike and hippopotamuslike animals, and a rhinoceroslike species. In younger rocks elsewhere in Big Bend, doglike mammals, a rabbit, camels, a small deer, and sheeplike animals have been identified.

Dinosaur fossils in Big Bend have been found in shallow marine and nonmarine deposits, late Cretaceous in age, that accumulated in a warm, swampy, coastal environment. In addition to dinosaur bones, fossils in these rock units include petrified wood, turtle shells, fish scales, shark teeth, crocodile bones, and so on.

By far the most remarkable finds have been wing bones and other fragments from pterosaurs (flying reptiles). In 1971 a student at the University of Texas, Douglas Lawson, discovered wing bones (including a large forearm bone) of a giant pterosaur, which has become known as the "Big Bend Pterodactyl." Diligent later searches turned up fossil fragments of other individuals of the same species, probably the largest creatures ever to fly. The Big Bend Pterodactyl had an estimated wing span of 38 feet! The mechanics of pterosaur flight are puzzling. How were such enormous wings controlled and maneuvered? How did the flying reptiles take off and how did they land? Did they "ride" updrafts and air currents as hang gliders do? Did the huge creatures hunt over land as well as water? How did they get around when they were on the ground? (For a fascinating discussion of such questions, see Langston 1981, cited in the bibliography for this chapter.)

At the end of the Cretaceous Period, the pterosaurs, who had dominated the skies for millions of years, died out—along with all the other dinosaurs on earth. The birds, who took over the realm of the sky, developed from another form of reptilian life, unrelated to the pterosaurs.

Mining in the Big Bend Area

Except for the quicksilver mines mentioned earlier, a hundred years of prospecting in and around the mountains of the Park failed to turn up significant deposits of other ore metals, despite legends of gold and silver from "lost mines" worked by the early Spaniards. However, fluospar (fluorite) ore is mined in Coahuila, Mexico, across the Rio Grande from Boquillas. Concentrate from the mines is trucked to the railroad at Marathon, north of Big Bend National Park.

The quicksilver that was mined for many years in Big Bend country came from cinnabar ore hydrothermally deposited, probably during Tertiary igneous activity. The mineral *cinnabar* (mercury sulfide) was brought upward in hydrothermal solutions that were then caught in a stratigraphic trap under the impervious Del Rio Clay and deposited along with calcite in the Santa Elena limestone,

just below the clay. The stratigraphic trap was formed by a faulted anticline. Most of the ore was found within 50 feet of the bottom of the Del Rio Clay, but in some places the ore extended downward as much as 200 feet into the Santa Elena beds. Some of the richest ore was discovered in collapsed sinks and pipes.

In the Terlingua district (outside the Park's western boundary), the cinnabar ore occurs in an east-west belt about 16 miles long and 6 miles wide, extending from Study Butte to Lajitas on the Rio Grande. Scattered occurrences of cinnabar have also been reported at sites in the lower Big Bend area. A mine was operated for a few years in the Mariscal Mountain area (fig. 32.4). Here the ore is associated with hydrothermal activity during the emplacement of sills in the Boquillas Formation.

Cinnabar ore from the Big Bend mines was smelted locally and the recovered quicksilver was shipped out in iron flasks. Obtaining fuel for the furnaces was always a problem. Most of the forests on the lower slopes of the Chisos Mountains were cut down for fuel. Because of the soil erosion that resulted and the arid climate, regrowth of trees was not possible and the slopes have remained barren. Eventually deposits of subbituminous coal were located northeast of the mines, and for some years the coal was used in the smelters.

Geologic History

1. Sedimentation and mountain-building during the Paleozoic Era.

The oldest rocks in Big Bend National Park are deformed Paleozoic units exposed in thrust slices at Persimmon Gap. Here in the northernmost section of the Park are the Santiago Mountains, which trend northwestward into the Marathon Basin and the Glass Mountains of west Texas. Although undifferentiated, the rocks at Persimmon Gap are probably part of the Maravillas, Caballos, and Tesnus Formations, which include shales, limestones, chert, and novaculite. On the basis of age and lithology, these rocks are related to the Paleozoic sequences of the Ouachita fold belt in Oklahoma and Arkansas. The novaculite, a dense, light-colored, siliceous rock, is similar to that found in Hot Springs National Park (chapter 37).

The Paleozoic rocks were deposited in an ancient ocean between a continental plate that eventually became the North American landmass and another continent to the south. In late Paleozoic time the ocean closed and the two continental plates collided. Folding and faulting occurred and some of the rock units from the deep part of the ocean basin were thrust over units along the continental margin.

Except for the outcrops at Persimmon Gap, the Paleozoic rocks of the Big Bend area have been stripped off or buried by much younger rocks. However, in Mexico, across the Rio Grande from Boquillas, some older metamorphic Paleozoic rocks are exposed in the Sierra del Carmen escarpment.

2. Accumulation of Cretaceous sediments.

The absence of Triassic and Jurassic rocks in the Big Bend sequence probably indicates a long period of erosion. Throughout Cretaceous time, the Big Bend region was part of a coastal plain and a shallow marine shelf along the margin of a continental landmass that sloped toward an ancestral Gulf of Mexico. Unconformities in the sedimentary sequence mark several advances and withdrawals of the ocean during the 90 million years of the Cretaceous Period. The final draining away of seawater occurred when the region was uplifted late in Cretaceous time.

Massive, limey units, such as the Glen Rose Formation, the Del Carmen Limestone, and the Santa Elena Limestone, accumulated during the higher sea stands (fig. 32.5). Clastic sediment from the north and northwest spread over the ocean floor when sea level was lower or the land higher. Units interbedded with the limestones include claystone and mudstone (some of it bentonitic), shales, calcareous sandstones, thin sandstone beds, stringers of silt and coal, gypsiferous marls, chalk, cherty limestone, and basal conglomerates. Careful study of these sequences allows geologists to document sedimentary facies changes that show transgressions and withdrawals of the sea, episodes of erosion, and a wide variety of sedimentary environments.

Most of the rocks are fossiliferous and contain molluscan remains typical of marine Cretaceous sequences. In the transitional and nonmarine units (the Aguja and Javelina Formations) that accumulated late in Cretaceous time are fossilized logs and various vertebrate remains, including dinosaur bones.

3. Deformation of Mesozoic rocks in latest Cretaceous and early Tertiary time; Laramide orogeny.

The Big Bend region is near the eastern edge of the vast area of western North America that was profoundly affected by the complex and long-lasting tectonic events associated with the Laramide orogeny. Thus deformation was moderate in Big Bend in comparison to that occurring in regions farther north and west. Compressive forces produced tight folds and overthrusting along the eastern and western margins of the Big Bend and broader anticlinal folds in between. Some of the older folds and structures, such as those at Persimmon Gap, were realigned and refolded with the overlying Cretaceous beds. If you draw a line on a geologic map from Santa Elena Canyon to the northeast corner of Big Bend National Park, you will note that the line crosses more than a dozen folds and several major faults.

Figure 32.5 Vertical cliff showing effects of weathering and erosion in Santa Elena Canyon. Only the lower part of the canyon wall is shown. Exposed here from bottom to top are the Glen Rose Formation (bedded rocks just above the waterline), the Telegraph Canyon Formation (covered with talus), and the Del Carmen Limestone, a cliff-former. Rock units are Lower Cretaceous in age. This view is directed downstream (or north-northeast) across the Rio Grande. National Park Service photograph.

On the mountainous topography that developed as a result of tectonic activity, the processes of erosion went to work. The structural trends controlled the patterns of valleys and ridges that evolved. However, not much evidence of these landforms can be found in the present landscape of Big Bend because they were destroyed or buried by subsequent geologic events.

4. Paleocene and Eocene nonvolcanic deposition.

Erosion of the Cretaceous rocks and structures produced clastic sediments that accumulated between folds and fault blocks, forming nonvolcanic rock units, Lower Tertiary in age and approximately 3,000 feet in thickness. These units lie unconformably over the older Cretaceous rocks. Lithologically the rock units are composed of conglomerates, sandstones (including channel sandstones), and shales. In the Canoe Formation nonvolcanic beds grade into mixtures of clastics, tuff, and ash, interbedded with lava flows, signifying the onset of volcanic activity. Because fairly intense erosion continued on the mountain slopes and on the nonvolcanic sequences that had accumulated in the valleys, local unconformities cut across all three Lower Tertiary formations and some of the uppermost Cretaceous units as well.

5. Extrusion and intrusion of igneous rocks, Eocene through Oligocene time.

In early Tertiary time, after the slackening of Laramide compressive forces, magma began rising into the crust. Some of the molten rock reached the surface and was extruded in the form of lava or as pyroclastics. Some magma cooled below the surface. The intrusions include dikes, sills, sheets, laccoliths, plugs, and other discrete bodies of irregular shape. Many were covered by lavas and pyroclastics and some may have been feeders for volcanic eruptions. Because most of the intrusions are fine grained, they probably cooled at depths of only a few thousand feet or less. Subsequent erosion in Big Bend National Park and adjacent areas has exposed many of these intrusions. They are largest and most abundant in a broad, northwest-trending belt centering in the Chisos Mountains. In terms of composition, the intrusive rocks range from felsic rhyolites (the riebeckites) through intermediate rocks, such as trachytes and andesites, to mafic basalts. They range in age from latest Cretaceous to late Oligocene (fig. 32.6).

A number of volcanic centers opened in the Big Bend area and from these vents, lavas and pyroclastics were erupted. In basins and valleys volcanic rocks are interbedded with sandstone and other clastics. Some beds that were formerly continuous have been interrupted by faulting and erosion. The sequence shows that between eruptions, streams reworked and redeposited loose debris that was later buried by new volcanic material.

Figure 32.6 Shown here is the south side of Vernon Bailey Peak in the Chisos Mountains. This domelike landform, located on the northwest side of the Basin, was formed from resistant igneous rock (a felsic quartz-riebeckite) emplaced in a small stock and subsequently exposed by erosion. The massive appearance of the feature, with relatively few joints, is characteristic of small intrusive bodies. The igneous rock is believed to be the intrusive equivalent of the Burro Mesa Riebeckite Rhyolite, which is the youngest eruptive rock in Big Bend National Park and has been dated as Oligocene and younger. National Park Service photograph.

The Chisos and South Rim Formations are made up of 3,000 to 5,000 feet of massive lava flows, flow breccias, conglomerates, tuffaceous sandstones, mudstones, tuffaceous clays, and ash beds. Chisos rocks have been dated as mainly late Eocene in age, and the South Rim Formation as probably Oligocene or younger. Potassium-argon dates taken from basalts in the lower part of the Chisos Formation yield dates of 38 to 42 million years before the present. Dates from riebeckite rhyolite near the top of the South Rim Formation record eruptions that occurred about 30 million years ago.

6. Tertiary block-faulting.

The Big Bend landscape of the present day was formed by the erosion of upfaulted blocks and deposition in down-faulted basins. Toward the close of Laramide tectonism, segments of crustal material that had been contorted and squeezed together by compression began to readjust isostatically in order to attain equilibrium. Tensional forces, due to sagging and stretching, broke the crust into blocks that moved up or down. In the Big Bend area evidence based on cross-cutting relationships indicates that throughout Tertiary time high-angle faulting occurred intermittently, producing a series of northwest-trending fault blocks. Faulting was most intensive, however, during early and middle Tertiary time. Most geologists believe that the igneous activity was associated with the faulting, although the causal relationship (if it exists) is not entirely clear.

7. Extensive late Tertiary erosion and deposition.

The unconsolidated sediments shown as "older gravels" on geologic maps of the Big Bend are the remnants of a sedimentary cover that was once much more widespread. Indirect evidence suggests that eroded clastic and volcanic debris filled the fault basins and probably covered some of the fault-block ranges. It was also during this time that the present drainage systems (graded to the Rio Grande) were organized.

8. Pleistocene and Holocene faulting and erosion.

Probably because of complex changes in rainfall and runoff patterns, a large part of the basin fill and other Tertiary deposits was eroded and carried away by swollen Pleistocene streams. Renewed uplift on some of the Tertiary faults elevated the highlands, at the same time increasing the erosive power of the streams. The Rio Grande (box 32.1) and some of its larger tributaries cut canyons and widened and deepened valleys. Landforms referred to as "older terraces" developed at this time, and cliff recession produced pediments that truncated bedrock and older gravels.

In post-Pleistocene time as the regional climate became increasingly arid, streams continued to erode and deposit but at a slower rate. Young gravels have accumulated in valley bottoms and on narrow flood plains. Pediments, fans, and terraces are being dissected by ongoing erosion.

Geologic Maps and Cross Sections

American Association of Petroleum Geologists. 1973. *Geological highway map of Texas, map no. 7.* Tulsa, Oklahoma: American Association of Petroleum Geologists.

Maxwell, R. A., and Lonsdale, J. T. 1966. *Geologic map of the Big Bend National Park, Brewster County, Texas.* University of Texas Bureau of Economic Geology (See below: Maxwell and Dietrich 1965; Maxwell et al. 1967).

Bibliography

Jackson, D. D.; and Wood, P. 1975. *The Sierra Madre,* American Wilderness Series. Alexandria, Virginia: Time-Life Books.

Langston, W., Jr. 1981. Pterosaurs. *Scientific American* 244 (2) :122–136.

Maxwell, R. A. 1941. *Fall field trip, Big Bend Park area, Brewster County, Texas.* West Texas Geological Society.

————. 1968. *The Big Bend of the Rio Grande,* a guide to the rocks, landscape, geologic history, and settlers of the area of Big Bend National Park. Guidebook no. 7. Austin, Texas: University of Texas Bureau of Economic Geology.

————; and Dietrich, J. W. 1965. *Geology of the Big Bend area, Texas*; Field trip guidebook with road log and papers on the natural history of the area. West Texas Geological Society Publication 65–51.

————; Lonsdale, J. T.; Hazzard, R. T.; and Wilson, J. A. 1967. *Geology of the Big Bend National Park, Texas.* Publication no. 6711. Austin, Texas: University of Texas Bureau of Economic Geology.

Sheldon, R. A. 1979. *Roadside geology of Texas.* Missoula, Montana: Mountain Press Publishing Company.

Tilden, F. 1979. *The National Parks.* Rev. ed. New York: Alfred A. Knopf, Inc., pp. 137–147.

Address

Big Bend National Park
Big Bend National Park, Texas 79834

33

Guadalupe Mountains National Park

by Ann Budd Foster
University of Iowa

Location: West Texas
Area: 76,293.06 acres; 119.21 square miles
Authorized: October 15, 1966
Established: September 30, 1972

Figure 33.1   The resistant Capitan limestone, which caps most of the Guadalupe Mountains, represents the ancient reef that encircled the Delaware Basin during the Permian Period. Weathering and mass wasting in an arid climate have produced the talus slopes below the cliffs. This view is from Pine Spring Canyon in Guadalupe Mountains National Park. National Park Service photograph.

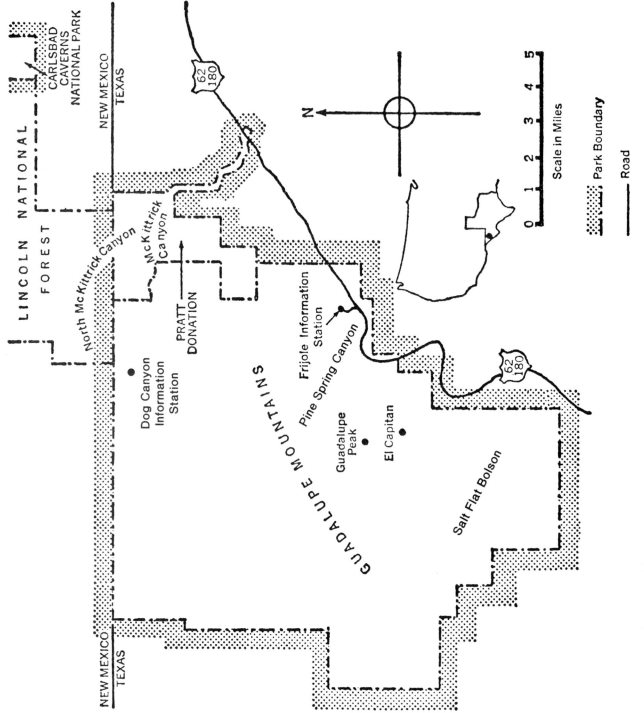

Figure 33.2 Guadalupe Mountains National Park, Texas.

Table 33.1. Geologic Column of Guadalupe Mountains National Park

Era	Period	Epoch	Back reef, lagoon	Reef crust, fore reef	Offshore basin	Geologic Events
			Time Units	**Rock Units**		**Geologic Events**
Paleozoic	Permian	Ochoan?	Rustler Formation			Progradation of terrigenous red sandstones and siltstones across evaporites
			Salado Formation	Salado Formation		Widespread deposition of evaporites, filling up the lagoon and back reef areas
				Castile Formation		
	Upper	Quadalupian	Carlsbad Group (Seven Rivers, Yates, Tansill Formations)	Capitan "Reef"	Bell Canyon Formation	Reduced basin subsidence and tectonic stabilization Rapid reef growth outward over basin sediments
			Queen Formation	Goat Seep "reef"	Cherry Canyon Formation	Increased subsidence of deeper offshore basin during which three distinct reef facies (nearshore dolomites, small banks, offshore basin deposits), developed; reef increased in size but most growth was directed upward
			Grayburg Formation	Getaway "Bank"		
	Lower	Leonardian	San Andres Formation	Brushy Canyon Formation		Slow subsidence of offshore basin; reef development continued
			Yeso Formation	Bone Spring Limestone		Infilling of the basin with carbonates and shale lenses to form a gently sloping basin and shelf margin; development of small patch reefs.

Source: modified after Newell et al. 1953

Local History

Studies of pictographs in caves, artifacts associated with roasting pits, and other archeological finds suggest that as far back as 12,000 years ago, Indians inhabited caves in Guadalupe Mountains National Park. They did not farm but moved with the seasons from the valley floors to the high canyon rims collecting plants and hunting animals, some of which are now extinct. Spanish conquistadores were in the vicinity in the 1700s, but they usually followed the river beds so it is doubtful that they entered the area that is now the Park. Throughout much of the 1700s and 1800s, the Guadalupe Mountains were inhabited by the Mescalero Apache Indians who discouraged travel or settlement in the area by other peoples.

Although the term *Guadalupe* was used on a map as early as 1828, the Park area was not extensively explored by people of European descent until 1849 when a series of surveys investigated possible wagon routes. The best known, the Ford-Neighbor Survey, explored two potential routes between Waco and El Paso, Texas. Also in 1849, a wagon train proceeded through the area for the first time.

It was escorted by the U.S. Army Fifth Infantry led by Captain Randolph Marcy and moved around the southern end of Guadalupe Peak on its way from Fort Smith, Arkansas, to Santa Fe, New Mexico. In 1850, John R. Bartlett (Commissioner of the Mexican Boundary Survey) traveled through the Park region and wrote detailed descriptions of his observations. Captain John Pope attempted to establish a post in the lower Pecos Valley in 1855 and the Butterfield Stage Line tried to set up a station at Pine Springs on the east flank of Guadalupe Peak in 1849, but both were abandoned in less than a year due to lack of water and personnel and to Indian trouble. In the following years ranchers began to settle the area but many Indian attacks ensued. In 1869 a Lieutenant Cushing moved against the Indians and destroyed their camp. Numerous encounters followed, culminating in the defeat of the Indians. In 1879 the remaining Indians were moved to a reservation. Battles also occurred during this time between the U.S. and Mexico, especially over the salt flats. An uneasy peace was finally brought about between the two countries in 1878.

The Park was authorized in 1966 after Wallace Pratt, an oil geologist and vice-president of the Standard Oil Company of New Jersey, donated the North McKittrick Canyon area to the federal government for use as a national park (a total of 5,632 acres). An additional 72,000 acres was then purchased by the federal government from J. C. Hunter, Jr., of Abilene, Texas, for 1.5 million dollars (approximately $21 per acre). Grazing has always been restricted in the Park and hunting has been strictly limited. Rigorous conservation practices have enabled the fragile terrain and vegetation to be preserved.

The highest point in the state of Texas is Guadalupe Peak (elevation, 8,751 feet), which is about in the center of the Park and directly north of El Capitan (elevation, 8,078 feet).

Geologic Features

The Capitan Reef

The Guadalupe Mountains are a portion of one of the largest fossil "reefs" in the world, the Capitan Reef. The reeflike deposits are exposed only for 40 miles along the northwest margin of the feature (specifically in the Guadalupe Mountains, Carlsbad Caverns National Park (chapter 39), the Glass Mountains, and Apache Mountains); but subsurface drilling shows that the so-called reef actually extends to the southeast for a distance of approximately 300 miles. It forms the shape of a horseshoe around what was in Permian time a deeper offshore marine basin, called the Delaware Basin. The most striking feature of Guadalupe Mountains National Park is the 1,000-foot-high El Capitan Cliff composed exclusively of "reef" limestone (fig. 33.1).

By definition, a *reef* is a submerged resistant mound or ridge formed by the accumulation of plant and animal skeletons. Although reefs can form in deeper, cooler regions, the largest living reefs occur in shallow (not over 425 feet) marine, tropical seas (temperatures over 68°F) where deep sea upwellings provide plentiful nutrients for organic growth. The seawater must also be clear, have normal marine salt content (27–40 parts per thousand), and be free of sediment or freshwater carried in by nearby streams. Reefs generally form in areas with heavy wave activity so that the upper surface of the reef, the *reef crest*, is subjected to the constant pounding of breaking waves. The sediments associated with reefs are quite different from those found in many other depositional settings. Reef sediments are largely composed of calcium carbonate ("lime") containing varied amounts of magnesium, and they consist exclusively of broken fragments of the skeletons of plants and animals that live on or near the reef. These fragments are irregular in size and shape, and are poorly sorted. They therefore show little evidence of being transported great distances.

Types of Reefs and Their Characteristics

There are four major types of reefs, which differ in shape and their relationship to nearby land masses: (1) barrier reefs, (2) fringing reefs, (3) atolls, and (4) patch reefs. *Barrier reefs* are long, narrow ridges that form parallel to a shoreline and are separated from the shore by a deep (>5m) basin of water called a lagoon. *Fringing reefs* are small, linear reefs that also form parallel to a shoreline but lack a lagoon separating them from the shore. (A shallow channel, however, may exist between reefs and shore.) *Atolls* are circular reefs formed far from any landmass. With an atoll a lagoon is formed in the middle of the reef. *Patch reefs* are small reefs of irregular shape that may form in the lagoon of a barrier reef or atoll. Of these four major types, the Capitan reef of the Guadalupe Mountains could be considered a barrier reef. The reef is horseshoe-shaped but, as evidenced by the nearby land-derived and lagoonal sediments, it formed near the shore and was separated from the shore by a lagoon.

Different portions of a reef are subjected to different environmental factors during its construction. Relatively high wave activity on the seaward-facing section of the reef (the fore reef) causes both large and small blocks of living and dead skeletal material to roll down the fore reef slope into the deep sea basin and thereby form a slope of debris or talus at the base of the reef. Before their fall, these blocks are severely weakened by bio-eroders (organisms) that rasp and bore through the dead skeletal material. Relatively high light intensity on the top of the reef (the reef crest) fosters a rapid upward growth of plants and animals that form the solid framework of the reef. There is relatively low wave or current activity on the shoreward-facing section of the reef (the back reef); therefore, only fine sediment is carried back into this area, and the water itself is often stagnant and muddy. The salinity in this region is especially high in areas lacking inflowing streams.

With each part of a reef having a characteristic marine environment, it is not surprising that the *biota* (that is, the living organisms) of the reef area are highly diverse and also different from one zone of the reef to the next. For instance, plants and animals with more robust skeletons tend to inhabit the reef crest and fore reef areas, whereas those that burrow or thrive in muddy places tend to live in the back reef. The fastest growing plants and animals usually occupy the shallow fore reef and reef crest where competition for space is especially intense. Because of the differences in the physical environment and in the constituent biota whose skeletons form the reef, distinctive sediment types are found in the fore reef, reef crest, and back reef areas. Each of these sediment types is called a *facies* (different rock types deposited in different environments at the same time). The sediments accumulate gradually as more plants and animals live and die, and their skeletons are washed over the lagoon and fore reef basins. As they are buried, the skeletal remains and the sediments created from them are compacted and recrystallized, becoming lithified as limestones. Sediment deposited in stagnant back reef or lagoon waters often contains high amounts of magnesium. This magnesium combines with the lime sediment to form the rock dolomite. In especially salty water, the salt alone often forms rocks as the water evaporates. These rocks are called evaporites. Three types of evaporites are associated with the Permian West Texas reefs: anhydrite, gypsum, and halite. As the salt content of the water increases during evaporation, first anhydrite (calcium sulfate, commonly as nodules or nodular layers) forms, then gypsum (hydrous calcium sulphate that tends to form long slender crystals), and last halite (common table salt, or sodium chloride).

During late Permian time (230-280 million years ago) in West Texas and New Mexico, three reef facies formed synchronously along the south-central portion of the North American craton in an inland sea comparable in size to today's Black Sea (fig. 33.3). To the east in the Appalachian region, orogeny took place as the African and North American plates collided. To the west along the edge of the craton through Nevada and Idaho, subduction was taking place producing volcanoes. The reef facies formed

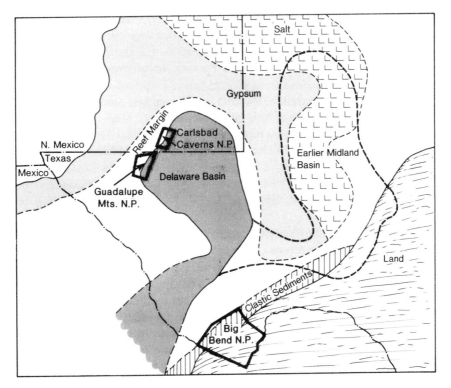

Figure 33.3 Location and extent of the Delaware Basin and the earlier Midland Basin in Permian time, shown superimposed on a present-day map of parts of New Mexico and Texas. The southern part of what is now the North American plate was close to the equator throughout the Permian Period. Paleogeographic reconstructions indicate that the Permian paleomagnetic equator crossed northeastward through southern Texas. Modified from *Historical Geology of North America* (2nd ed.) by Morris S. Petersen and J. Keith Rigby. © 1973, 1980 Wm. C. Brown Company Publishers, Dubuque, Iowa. Reprinted by permission.

in a dry, tropical region south of the Permian equator. They consisted of: (1) a fore reef talus facies and deeper sea basin facies formed in the so-called Delaware Basin, (2) a reef crest and shallow fore reef facies formed in the surrounding El Capitan Reef Complex, and (3) a back reef and lagoon facies formed in the northeasterly Midland Basin. The rocks of the Delaware Basin consist largely of black limestones with lesser amounts of quartz sandstone. The black color of the limestone is believed to have been caused by high organic content similar to that of restricted marine environments of today where the bodies of floating microscopic animals fall to the bottom and are mixed with sediment. Such basin bottoms presumably contain little or no oxygen and thus few bacteria to consume the dead organic material. The El Capitan Reef Complex consists of light-colored, massive fossiliferous limestones; massive primarily because they were formed by the growth and accumulation of skeletons rather than by deposition of sediment from moving water. Such skeletons were stabilized by encrusting organisms that grew over and thereby cemented the solid reef-rock. They accumulated predominantly in shallow water where ample supplies of light existed to provide the energy necessary

for algal photosynthesis. The Midland Basin is composed of bedded dolomites and evaporites. Both of these rock types form in areas where circulation is poor and the salt content of the water is high.

A walk through McKittrick Canyon (fig. 33.4) proceeds through all three major reef facies. The walk begins in the deep basin facies (the Bell Canyon Formation) and moves into the talus slope at the base of the fore reef (the Capitan talus). The Bell Canyon Formation contains numerous beds of sedimentary breccia (a rock composed of nonsorted, nonlayered, angular, and often large rock fragments) surrounded by fine-grained basin deposits. The breccias are believed to have been deposited when submarine slides rolled down the steep face of the fore reef and talus slope. The Capitan talus consists almost exclusively of steeply sloping thick beds of this breccia. On top of the Capitan talus lies the shallow fore reef and reef facies (the Capitan Formation) and then the back reef/lagoon facies (the Carlsbad Group). The reef limestone itself is composed largely of sponges, algae, and bryozoans (tiny colonial animals). Brachiopods (shelled animals), crinoids (sea lilies), and fusulinids (large, single-celled animals) are common in the fore reef and back reef facies. In general, reefs of Permian age are unusual because they

Figure 33.4   View of McKittrick Canyon in Guadalupe Mountains National Park. The three major reef facies— deep basin (fore reef) reef, and back reef—are exposed in the canyon walls. National Park Service photograph.

were not built primarily by corals, the principal reef-formers of the middle Paleozoic Era and of the Mesozoic and Cenozoic Eras. The back reef facies show a progressive change from carbonates (limestones and dolomites) through evaporites to clastic terrigenous (land-derived) rocks. Several specialized rock types can be found: (1) *coquina* ("shell hash"), deposited in the lagoon in water having normal marine salt content; (2) lenses of *pisolites* (spherical particles containing concentric layers, sometimes called "peastone"), deposited by algae in intertidal areas along the shore; (3) fine-grained dolomite, deposited in intertidal and supertidal areas along the shore in salty water; and (4) red siltstone and sandstone mixed with anhydrite, deposited in supratidal areas along the shore in salty ponds.

Recently, controversy has developed over whether the Permian Capitan Reef Complex was actually a wave-resistant structure. Some workers have questioned if the intertwined algal fronds forming the "reef" could have been strong enough to withstand wave impact. Others have claimed that the large volume of detrital (i.e., composed of broken-up fragments) sediments in all three major facies suggests that the reef and the organisms forming the reef played only a minor role in controlling the sedimentation patterns of the area. Many geologists, therefore, have reinterpreted the Capitan Reef to represent merely a bank of skeletal debris below wave base. Despite this re-evaluation, most workers agree: that the Delaware Basin sediments were deposited in an offshore enclosed marine basin; that the Capitan Reef Complex was formed on the edge of the basin in shallow, marine conditions; and that the Midland Basin sediments were formed shoreward of the skeletal accumulations on an intertidal-supratidal, hypersaline (excessively salty) platform.

Geologic History

1. **Formation of the Delaware Basin in West Texas during early Permian time (Wolfcampian Epoch).**

    The Delaware Basin formed by earliest Permian time as an inland basin with a narrow outlet to the open ocean (fig. 33.5). More or less oval in shape, the body of water covered an area of around 10,000 square miles. Although the basin was largely enclosed, some ocean water entered periodically to replace water lost by evaporation. The circulation patterns within the basin are believed to have been similar to those in the present-day Mediterranean Sea. During this time, 1,600 to 2,200 feet of limestones and interbedded dark shale were deposited in the shallow subsiding basin.

2. **Initial formation of small banks along the margin of the basin during later early Permian time (Leonardian Epoch).**

    By the beginning of mid-Permian time, the basin temporarily ceased to subside. Alternating beds of gypsum, sandstone, and dolomitic limestone (the Yeso Formation, the Victoria Peak Member) were deposited in nearshore areas; while thin-bedded black limestones (the Bone Spring Limestone) accumulated in the stagnant deeper portion of the basin. Small discontinuous patch reefs formed in shallow water along the rim of the basin.

3. **Continued patch reef development during the early Upper Permian (Guadalupian Epoch).**

    By mid-Permian time, the basin began to subside again and many larger patch reefs formed along the basin margin. Cherty dolomites (San Andres Formation) built up close to shore, while quartz sandstones (Brushy Canyon Formation), containing scattered patch reefs, formed in and around the basin.

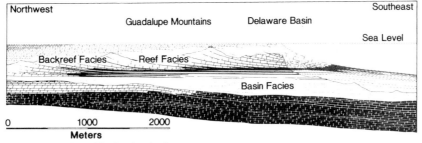

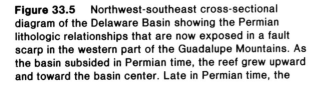

Figure 33.5 Northwest-southeast cross-sectional diagram of the Delaware Basin showing the Permian lithologic relationships that are now exposed in a fault scarp in the western part of the Guadalupe Mountains. As the basin subsided in Permian time, the reef grew upward and toward the basin center. Late in Permian time, the basin filled with evaporites as the shallow equatorial sea dried up. From *Historical Geology of North America* (2nd ed.) by Morris S. Petersen and J. Keith Rigby. © 1973, 1980 Wm. C. Brown Company Publishers, Dubuque, Iowa. Reprinted by permission.

4. Formation of the first barrier reef during the Upper Permian (Middle Guadalupian Epoch).

After the patch reefs began to develop along the basin margins during mid-Permian time, the Delaware Basin itself began to subside rapidly. This subsidence may have been related to compressive folding resulting from the colliding North American and African plates. As the basin subsided, reef growth intensified predominantly in an upward direction, and the first barrier reef formed, the Goat Seep reef. Three distinct facies developed: (1) a lagoon and back reef facies composed of sandstones and dolomites (the Queen Formation, the Grayburg Formation), (2) a reef crest and shallow fore reef facies containing dolomitized sponge and algae skeletons (the Goat Seep reef, the Getaway Bank), and (3) an offshore basin facies composed largely of quartz sandstones (the Cherry Canyon Formation) (fig. 33.5).

5. Formation of the most extensive barrier reef during the Upper Permian (Late Guadalupian Epoch).

By the end of mid-Permian time, the basin slowly ceased to subside and achieved a stable position. As this position was reached, the largest reef, the Capitan Formation, expanded rapidly 350 miles around the rim of the basin and gradually prograded toward the center of the basin. The three facies differentiated further. Limestones and sandstones (the Bell Canyon Formation) were deposited offshore in the basin; massive skeletal limestones (the Capitan Formation) accumulated in shallow areas encircling the basin; and dolomites and fine-grained sandstones (the Carlsbad Group) were formed in nearshore areas (fig. 33.5).

6. Infilling of the basin with evaporites during the Upper Permian (Ochoan Epoch).

Continued sedimentation in the stable basin and possibly a drop in sea level caused the gradual shallowing and ultimate drying up of the basin late in Permian time. As the basin dried up, extensive accumulations of nearshore evaporite deposits covered the lagoon and reef and eventually the basin itself. Similar deposits, called *sabkha facies,* accumulate today in such arid areas as the Persian Gulf. Laminae* (bands of layers 1/16 inch in thickness) of gray gypsum and brown calcite (the Castile Formation) precipitated first in the basin. Halite and potassium-rich salts (the Salado Formation) formed synchronously in nearshore areas and around the basin margin. Eventually the salts were deposited exclusively throughout the area.

---

*A *lamina* is the thinnest recognizable unit layer of deposition in a sediment or sedimentary rock. The brown calcite laminae in the Castile Formation are believed to have formed by algal growth, periodically interrupted by storms. The gray gypsum laminae formed during storms as sediment washed over a layer of algae covering the sea floor. The algae re-established themselves after the storm by growing up through the storm deposits.

7. Migration of terrigenous red beds over evaporite deposits during the Upper Permian (Ochoan Epoch).

At the very end of Permian time, terrigenous red siltstones and sandstone (the Rustler Formation, the Dewey Lake Formation) were deposited on the evaporites as rivers began to migrate across the area.

8. Uplift and tilting of the area during late Mesozoic and early Cenozoic time.

A series of tectonic movements created a fault on the west side of the Park; and a graben (a downfaulted block) at Salt Flat Bolson tilted the entire region to the northeast. These movements occurred during the Laramide orogeny in Cretaceous and early Cenozoic time as sedimentary rocks deposited in the Rocky Mountain region were thrust eastward onto the craton. The fault on the west side of the Park occurs within a fault-line valley bringing Cretaceous beds in the valley in contact with Permian beds that form the nearby hills.

9. Erosion during the Cenozoic Era

Stream erosion has removed much of the softer sediment and has leveled the region to its present state. Ground water erosion formed numerous caves throughout the limestone units, the largest being the Carlsbad Caverns (chapter 39). Because of the steep relief and lack of vegetation, landslides have eroded some of the hillsides, a process of mass wasting that has continued into the present.

Geologic Maps and Cross Sections

American Association of Petroleum Geologists. 1973. *Geological highway map of Texas, map no. 7.* Tulsa, Oklahoma: American Association of Petroleum Geologists.

Roswell Geological Society. 1964. *Geology of the Capitan Reef complex of the Guadalupe Mountains, field trip guidebook.* Roswell, New Mexico: Roswell Geological Society, p. 68.

Bibliography

Barnett, J. (n.d.) *Guadalupe Mountains National Park—its story and scenery.* Carlsbad, New Mexico: Carlsbad Caverns Natural History Association.

Christiansen, P. W., and Kottlowski, F. E. 1967. Mosaic of New Mexico's scenery, rocks, and history. In *Scenic trips to the geologic past,* no. 8. 2nd ed. New Mexico Bureau of Mines and Mineral Resources.

Heckel, P. H. 1974. Carbonate buildups in the geologic record: a review. In *Reefs in time and space*, ed. L. F. Laporte, pp. 90–154, Special Publication 18. Tulsa, Oklahoma: Society of Economic Paleontologists and Mineralogists.

Kendall, A. C. 1979. Continental and supratidal (sabkha) evaporites (#13) *and* Subaqueous evaporites (#14). In *Facies models*, ed. R. G. Walker, pp. 145–74. Ontario: Geological Association of Canada Publications.

Newell, N. D.; Rigby, J. K.; Fischer, A. G.; Whiteman, A. J.; Hickox, J. E.; and Bradley, J. S. 1953. *The Permian reef complex of the Guadalupe Mountains region, Texas and New Mexico.* San Francisco: W. H. Freeman & Co.

Roswell Geological Society. 1964. Geology of the Capitan Reef complex of the Guadalupe Mountains, Culberson County, Texas, and Eddy County. In *New Mexico field trip guidebook.* Roswell, New Mexico: Roswell Geological Society.

Selley, R. C. 1978. *Ancient sedimentary environments.* 2nd ed. Ithaca, New York: Cornell University Press.

West Texas Geological Society. 1960. *Geology of the Delaware Basin and field trip guidebook.*

Wilson, J. L. 1975. *Carbonate facies in geologic history.* New York: Springer-Verlag.

Address

Guadalupe Mountains National Park
3225 National Parks Highway
Carlsbad, New Mexico 88220

34

Virgin Islands National Park

by Ann Budd Foster
University of Iowa

Location: St. John Island (Virgin Islands)
Area: 14,708.71 acres, 22.98 square miles
Established: August 2, 1956

Figure 34.1 Little Cinnamon Bay·on the north side of St. John Island is formed mainly of the Louisenhoj Formation which weathers to a reddish color. The vegetation is second growth consisting of breadfruit, mahogany, palms, mangroves, soursop, and kapok.

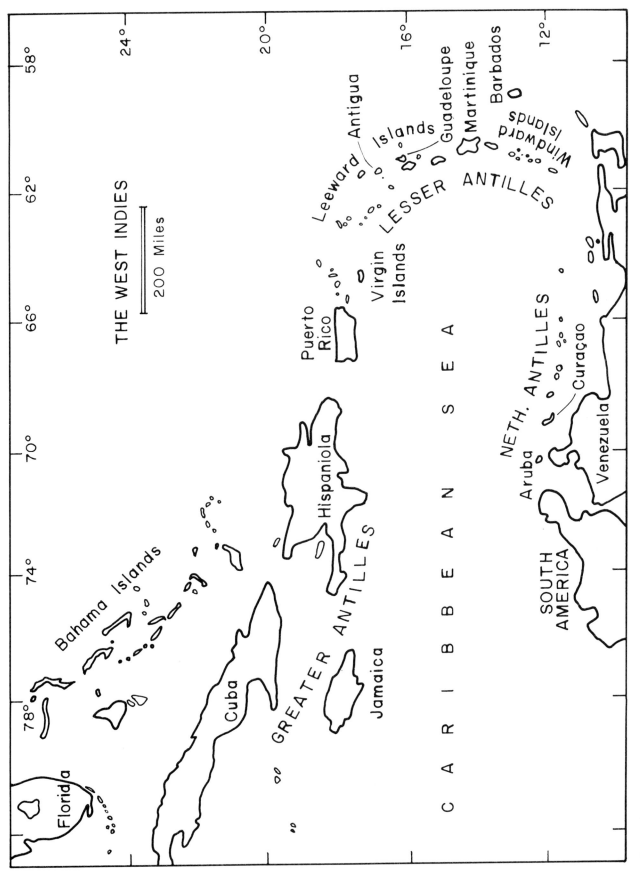

Figure 34.2 Location map showing the major islands of the Caribbean Sea.

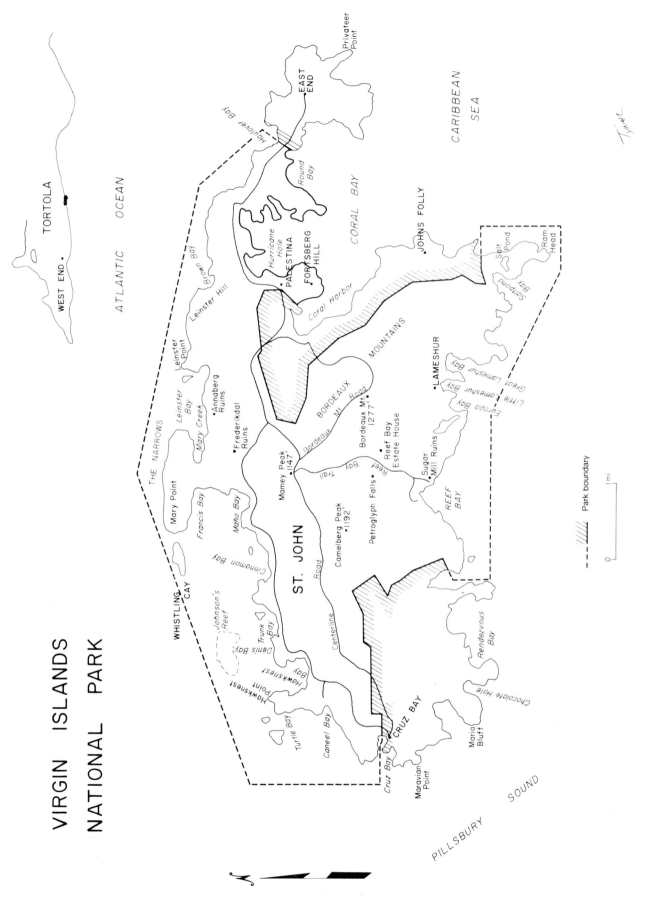

Figure 34.3 St. John Island, Virgin Islands National Park.

Table 34.1. Geologic Column of the Virgin Islands National Park

Time Units			Rock Units	Geologic Events
Era	Period	Epoch		
Cenozoic	Quaternary		Alluvium	Erosion by streams and ground water; formation of submarine limestone terraces as sea level periodically rose and fell during glacial and interglacial stages.
	Tertiary / Paleogene	Eocene	Scattered small dikes and plugs	Emplacement of dikes. Widespread faulting; emplacement of dioritic plutons; localized metamorphism.
			Hans Lollik Formation	Renewed volcanic activity; Folding and strike-slip faulting.
		--?--		
Mesozoic	Cretaceous	Albian	Tutu Formation	Renewed fault movement deposition of terrigenous material primarily as turbidites.
			Outer Brass Limestone	Deposition of deep water limestones during a period of volcanic quiescence and sea bottom subsidence.
			Louisenhoj Formation	Development of red soil horizon and subaerial volcanic activity. Initial formation of deep ocean trench north and east of the Lesser Antilles.
		Neocomian?	--unconformity--	--subaerial erosion--
			Water Island Formation	Submarine extrusion of volcanic material along an east-west high-angle fault onto a relatively flat sea bottom. Gradual drop in sea level and/or tectonic uplift.

Local History

The first inhabitants of St. John Island were a peaceful group of Indians, the Arawaks and the Ciboneys, who had fled in their canoes from aggressive neighboring Indians along the north coast of South America. The Arawaks first reached the Virgin Islands around A.D. 300 and lived there more than 1,000 years. Small groups settled along the north and northwest shores of St. John in places such as Cinnamon Bay, Turtle Bay, Caneel Bay, and Cruz Bay. They were farmers who primarily used the slash-and-burn method of agriculture to raise *manioc* (a starchy root, also called *cassava*) which they had brought with them from South America. They supplemented their diet with birds, fish, shellfish, and sea turtle. During the 1300s, the Carib Indians migrated from the Amazon River Basin and either absorbed or exterminated the Arawaks. The Caribs were good hunters, swimmers, canoeists, and fighters, and they also practiced cannibalism.

Christopher Columbus officially discovered the Virgin Islands in 1493 during his second voyage to America. Columbus called the archipelago *Las Virgenes*, or "The Virgins," in honor of the martyred St. Ursula of Cologne and her 11,000 virgin followers. During the century after

Columbus's discovery, explorers came from the Netherlands, England, France, Denmark, and Spain. Then in the 17th century, pirates such as Captain Kidd and Bluebeard may have visited the area, for it was part of the Spanish Main.

The Danes officially claimed the Virgin Islands in 1687. However, St. John Island was not actually settled by Europeans until 1716 when Peter Durlieu and other Danish and Dutch planters who lived on St. Thomas moved over there and built plantations. Soon afterwards, other planters also decided to settle on St. John and the island was divided into estates. The planters raised sugar and cotton and brought in slaves (predominantly West African blacks) for labor. By 1726 all potentially usable land was being farmed. A drought triggered a slave revolt in 1733 which resulted in heavy loss of life. The revolt was finally put down by French soldiers from Martinique, after which the plantations were re-established and the planters prospered once again. In 1848, however, there was a second slave revolt, and the Danish governor soon emancipated the slaves. Shortly thereafter, many planters and freed slaves left St. John because farming without slave labor was unprofitable. The population of the island was reduced from 2,000 to about 700, and the land went back to bush. After 1875 some of the plantations were revived when a new variety of sugar was introduced. However, competition from sugar beets, poor farming methods, and depleted soils all contributed to the ultimate collapse of the sugar farming industry in the early 1900s. Rum is still distilled in the Islands at present, and cattle-raising is also important to the economy.

During World War I, the United States bought the Virgin Islands from Denmark for 25 million dollars in order to prevent possible German occupation of the islands. With his own money and funds from the Jackson Hole Preserve Corporation, Laurance Rockefeller gradually bought up most of the land on St. John Island. In 1956 he presented 5,000 acres to the United States as a national park. In 1960, 1962, and 1978, additional land and offshore areas were incorporated into the Park to protect the coral reefs from collection and commercial exploitation. The island itself is only five miles wide by nine miles long, and most of the topgraphy is mountainous. Bordeaux Mountain, the highest peak, is 1,277 feet above sea level. The climate is pleasantly mild year round with Fahrenheit temperatures seldom in the high 90s because the northeast trade winds and the ocean moderate the tropical heat. Moisture is also supplied by the trade winds; and rainfall, collected in cisterns, is the chief source of fresh water.

Tectonic Activity in the Caribbean Region

There are four major groups of island chains or archipelagoes in the Caribbean Sea (fig. 34.2): (1) the Bahama Islands (running northwest to southeast, southeast of Miami); (2) the Greater Antilles including Cuba, Jamaica, Hispaniola, and Puerto Rico (running west to east, south and southwest of the Bahamas); (3) the Lesser Antilles including Antigua, Guadeloupe, Martinique, and Barbados (running north to south, southeast of Puerto Rico); and (4) the Netherland Antilles including Curacao and Aruba (running west to east along the north coast of Venezuela). The Virgin Islands (about 100 in all) lie at the extreme eastern end of the Greater Antilles between

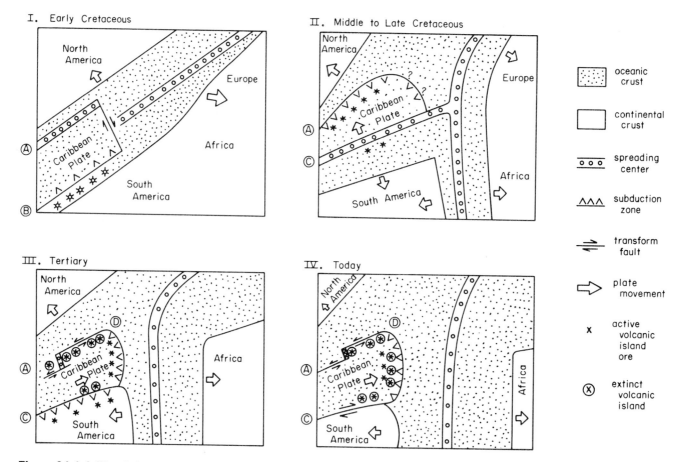

Figure 34.4, I–IV   Schematic representations of major plate movements in the Caribbean Sea during the past 130 million years. Encircled letters indicate zones of significant tectonic activity. See explanation in text. (Simplified and redrawn from diagrams and data in Mattson, 1977.)

that archipelago and the Lesser Antilles bending southward. The U.S. Virgin Islands include St. Thomas, St. John, St. Croix, and other islands on the west side of the group. The British Virgin Islands, in the northeast part of the archipelago, comprise about 40 islands, including Tortola, Virgin Gorda, and Anegada.

The Caribbean Sea formed during the early part of the Cretaceous Period as Europe and Africa moved away from North and South America (fig. 34.4-I). At that time, the Caribbean plate formed between a spreading center (A) (in the area now occupied by the Greater Antilles) separating it from the North American plate. A subduction zone (B) (in central Venezuela) was active on the boundary between the Caribbean plate and the South American plate (fig. 34.2). Shortly thereafter, the spreading center (A) became a subduction zone (A), and a new spreading center (C) formed in the area now between the Greater Antilles and the northern Venezuela coast (fig. 34.4-II). The core of the Greater Antilles consists of rocks extruded from volcanoes in an island arc that developed along the newly formed subduction zone during early to middle Cretaceous time. The Netherland Antilles, which were probably part of the Greater Antilles island arc, lay on the other side of the new spreading center (C) and therefore, "drifted" away from the Greater Antilles during middle to late Cretaceous time. The older subduction zone (B) through central Venezuela became inactive during the middle Cretaceous. The newer spreading center (C) and subduction zone (A) ceased activity during the late Cretaceous or early Tertiary time, and a *transform fault* (A) formed in the location of the subduction zone (A) (fig. 34.4-III). A transform fault is an extensive strike-slip (i.e., having only horizontal movement) fault that occurs at the boundary between two tectonic plates. In this type of plate boundary, no crust is consumed (as in a subduction zone) or created (as in a spreading center); therefore, the North American and Caribbean plates merely slid by one another. A subduction zone (C) and associated volcanic island arc also formed at the same time near the spreading center (C) along the northern coast of Venezuela. A second subduction zone (D) and island arc became active during the early Tertiary along the line of the modern Lesser Antilles. Subduction along the north coast of Venezuela ceased during late Tertiary time and a transform fault (C) formed along its former position. Subduction of the North American plate beneath the eastern Caribbean plate margin still continues today near the Lesser Antilles (fig. 34.4-IV). Throughout the last of the Mesozoic Era and in early Tertiary time, no land barrier separated the Caribbean Sea from the Pacific Ocean. Late in the Miocene Epoch, the Isthmus of Panama emerged as a result of activity in the subduction zone along the eastern Pacific coast of Central and South America.

This complicated geologic history is reflected in the structure and rock composition of various islands. Generally, the Bahamas consist of more than 10,000 feet of Cretaceous and Tertiary limestones deposited in the miogeosynclinal portion of the geosyncline formed in the Cretaceous Greater Antilles subduction zone (A) (fig. 34.4-II). The Greater Antilles are composed largely of highly deformed igneous and metamorphic rocks of more or less similar ages that formed in the eugeosynclinal portion of the same geosyncline. Volcanic and seismic activity in the Greater Antilles has now largely subsided and the region is thought to be relatively stable tectonically. Some scientists believe that the Greater Antilles are related to the North American Cordillera (on the western side of the North American plate), because the Greater Antilles are structurally more similar to northern Central America than is southern Central America (i.e., the Isthmus). Certainly both the North American Cordillera and the Greater Antilles experienced major deformation during the time of the Laramide orogeny. The Lesser Antilles, however, consist primarily of Tertiary and younger volcanics. The islands are arranged in two arcs: (1) an inner chain of volcanoes (some dormant and some active) consisting largely of late Tertiary and Quaternary volcanic rocks, and (2) an outer chain of early Tertiary volcanoes capped by late Tertiary limestones. In contrast, the Netherlands Antilles consist of Cretaceous volcanic rock overlain by widespread Eocene limestones. As in the Greater Antilles, volcanic and seismic activity in the Netherlands Antilles has largely subsided today.

The rock composition and deformation history of the Virgin Islands is more like that of the Greater Antilles than the Lesser Antilles. Rocks forming St. John Island consist largely of Lower Cretaceous volcanics. These were intruded and deformed during late Cretaceous and early Tertiary time (i.e., when the Laramide orogeny was building the North American Cordillera).

Development of Coral Reefs

Much of the modern offshore topography of St. John (as well as almost all of the Caribbean Islands) was developed during the Pleistocene and Holocene Epochs as fringing reefs grew upward and outward along the windward coast on the eastern side of the islands. Such reefs range today from southern Florida (chapters 35, 36) to the southern margin of the Caribbean Sea, throughout an area where temperatures are relatively warm and sedimentation of fine particles is reduced. Although a few atolls have been described in the western Caribbean and some of the reefs in the Bahamas and Florida could be considered fringing-barrier reefs, throughout the Pleistocene and Holocene Epochs the large majority of reefs in the Caribbean were and are either fringing or patch reefs.

Unlike the Permian algal reefs in Guadalupe Mountains National Park (chapter 33), these Pleistocene and Holocene reefs are constructed primarily of coral skeletons. A *coral* is a cylindrical-shaped marine invertebrate that lives either singly or structurally bound together with others in a colony. (Corals belong to class Anthozoa, phylum Coelenterata.) Before an individual living coral (a *polyp*) matures, it attaches its cup-shaped base to a hard rock substrate where it remains fixed for its entire adult life (sometimes several centuries). Each polyp then grows upward by continuously secreting an external lime skeleton around the lower cup-shaped portion of its body. The polyp(s) thus occupies only the outer surface of a sometimes large (may be more than 5 to 6 feet in diameter) skeleton. Corals feed by actively capturing microorganisms with a ring of tentacles that encircle the outer edge of the upper surface of each polyp. Each polyp then uses its tentacles to insert the captured prey into a mouth located in the center of its tentacular ring. Some corals contain algae (*zooxanthellae*) within their living tissue that aid in their metabolism. Corals containing numerous algae can secrete skeletons and thereby grow; they grow especially rapidly (as much as 3 inches per year) in well lit habitats. Colonial (not solitary) corals are largely responsible for reef-building. They assume a great variety of shapes, each adapted for a specific wave energy, light intensity, or sedimentation rate.

Like the Permian reefs of Guadalupe Mountains National Park, the corals forming the Quaternary reefs of St. John are distinctively zoned parallel to the reef crest. In offshore areas with high wave energy (for example, across Denis and Coral Bays), the reef crest is dominated by the thick-branched "elkhorn" coral *Acropora palmata* and the mound-shaped brain coral *Diploria strigosa*. On either side of the crest lies a second zone dominated by the thick-branched "staghorn" coral *Acropora cervicornis* (fig. 34.5). Still farther from the crest in both back reef and fore reef areas, one finds a third zone dominated by the knobby "finger" coral *Porites porites* and the mound-shaped "star" coral *Montastraea annularis*. In areas where the wave energy is lower (for example, off Annaberg near Mary Creek) the first zone does not occur. Here the second zone lies on the reef crest and is surrounded only by the third zone. These zones of corals occur today and existed throughout the Quaternary Period across the

Figure 34.5 The *Acropora cervicornis* zone in Caneel Bay of the Virgin Islands National Park. This portion of the reef is dominated by the staghorn coral, *A. cervicornis* (lower left corner of the photo). Soft corals (upper right corner of the photo) and fish (center) live among the staghorn coral branches.

entire Caribbean. However, it is important to remember that, although corals form the reef framework, they do not live alone on these reefs. Coral reefs consist of a diverse, well integrated community of marine animals which might also include sea urchins, fish, shrimp, sea fans, sea whips, snails, worms, and sponges.

Reef Terraces

During the Pleistocene Epoch, several *reef terraces* (horizontal topographic plains bounded on either side by steep slopes) formed along the coasts of all the modern Caribbean Islands. Most of St. Johns's reef terraces are now submerged; however, on many of the islands (the most spectacular examples can be seen on Barbados), several terraces lie above sea level. These reef terraces have been cut back about 500 feet and relief between terraces is about 50 feet. They are composed of lithified skeletons of corals that lived and formed reefs immediately prior to terrace formation. It is believed that, throughout the Caribbean, reef development flourished at times when sea level rose in the Pleistocene Epoch. Erosion occurred forming a terrace each time sea level was lowered. Terraces were formed repeatedly during the Pleistocene Epoch because sea level changed frequently at that time. The ocean rose as much as 60 feet above today's sea level when the glaciers melted during interglacial periods, and dropped as much as 300 feet below today's sea level during periods of glacial accumulation. The exact number and timing of Pleistocene terrace formation throughout the Caribbean Islands is not clearly understood because of the complicating effects of vertical tectonic movements and isostatic adjustments that occurred on all the islands during this time. Recently sea level has risen on St. John; therefore upward reef growth is intensified and many of the ancient reef terraces are drowned. One such terrace occurs off the south shore at depths of 30 to 100 feet.

Wave action has formed the beaches of St. John Island, with the reefs and submerged terraces being the main sources of sand. Because no permanent streams exist on the island, the processes of runoff do not supply much sediment to the lagoons and beaches except during heavy tropical storms or occasional hurricanes. However, storm waves and "swells" (large waves that travel across many miles of open ocean before breaking on a shore) erode both reefs and shore rocks, thus producing sand and cobbles. Some of the beaches on the island, such as Europa Bay on the southeast shore, are almost entirely made up of coral rubble; but Little Lameshur, which is nearby, is mainly fine sand. Cobblestones, well rounded by wave action, cover much of Greater Lameshur, the adjacent beach. Each beach on the island's shore has been shaped by local conditions of the coastal environment which include the bottom topography offshore, the character of the reef at that location, tidal currents, wind and wave dynamics, etc. The Park Service's intention is to preserve this unique ecosystem of shore, ocean, and reef as much as possible in its natural state.

Geologic History

1. Submarine extrusion of lava flows and subsequent uplift during the Lower Cretaceous Period (Neocomian Epoch).

The rocks forming the core of the Virgin Islands (the Water Island Formation) consist of more than 15,000 feet of *keratophyre* (a light-colored igneous rock composed predominantly of plagioclase feldspar and quartz) and *spilite* (a mafic green igneous rock composed predominantly of plagioclase feldspar and chlorite). The lavas were extruded over deep sea clay sediments by quiet underwater volcanic flows onto a deep (more than 15,000 feet), relatively flat sea bottom. These rocks are found throughout the entire southeastern half of the island and can be best seen near the Centerline Road and Bordeaux Mountain Road, near Chocolate Hole, and near Reef Bay trail. The lack of pumice and of associated land-derived sediment in these rocks, as well as their chemical composition, suggests that they were formed in deep sea water. The volcanoes that produced the lavas were part of the island arc chain that built up the Greater Antilles Islands. During the Cretaceous Period, these islands lay in the western portion of the "Tethyan" seaway, an ancient ocean that extended around the central tropics from the Middle East and northern India through the Mediterranean Sea across the then-narrow Atlantic Ocean and into the Caribbean Sea.

2. Subaerial erosion; subaerial volcanic eruption of ash and coarse debris during the Lower Cretaceous Period (Albian Epoch).

The Water Island Formation is immediately overlain by a 7,000-foot-thick series of blue andesite ash beds, pyroclastic breccias (containing blocks as large as 4 feet in diameter), and brick red soil layers (the Louisenhoj Formation). These rocks cover the western half of St. John and can be seen near Cruz Bay and Rendezvous Bay. The actual volcanic vent through which the lavas and ash were extruded is probably located in Pillsbury Sound between St. John and St. Thomas Islands. The ash beds and soil layers were formed subaerially after the volcanic island arc had grown so that the volcanoes emerged above sea level. They are separated from the Water Island Formation by a pronounced unconformity which indicates that the Water Island Formation was first subaerially exposed and eroded by running water. The ash and volcanic debris were then exploded violently out of the volcanic craters across the land surface and shallow sea floor; however, much of the pyroclastic material appears to have later been deposited underwater by submarine slides and slumps on the steep slopes that had formed along the volcano flanks.

3. Sea floor subsidence, volcanic quiescence, and limestone deposition during the Lower Cretaceous Period (Albian Epoch).

Following the period of widespread volcanic eruptions, reduced volcanism and gradual sea-floor subsidence took place. During this time, 600 feet of thin-bedded, dark, silica-rich limestones (the Outer Brass Limestone) were deposited on a moderate slope in water depths of a few hundred feet. The rocks can be seen near Mary Creek. The limestones are silica-rich for two reasons: (1) the abundance of silica tests (shells) of microscopic floating animals, and (2) the occurrence of thin, interbedded layers of tuff.

4. Faulting and deposition of submarine slides and turbidites during the Lower Cretaceous Period (Albian Epoch).

Toward the end of the limestone deposition, tectonic activity intensified, causing vertical fault movement which, in turn, created steep submarine slopes and subaerial emergence of portions of the Louisenhoj Formation. Submarine slides and turbidity currents (p. 204) moved material eroded from the Louisenhoj Formation down these steep sea-floor slopes. During this time 6,000 feet of flyschlike (widespread sandy and calcareous shales containing sandstone and conglomerate beds) rock suites composed of volcanic wacke (poorly mixed sediment consisting of volcanic particles) and limestone megabreccia (large angular blocks cemented together) (the Tutu Formation) were deposited. The Tutu Formation can be seen from Maho Bay to Leinster Point. The limestone megabreccia occurs at Mary Point, Leinster Point, and in the Annaberg area.

5. Renewed volcanic activity, faulting, and plutonism during the late Cretaceous through Tertiary Periods (Eocene Epoch).

Although they only crop out on islands northwest of St. John Island, at least 10,000 feet of water-laid andesite pyroclastic rock (the Hans Lollik Formation), similar to the Louisenhoj Formation, overlie the Tutu Formation. This suggests that a third period of intense volcanic activity occurred, possibly as late as the Eocene Epoch. This volcanic activity was accompanied by the intrusion of dioritic stocks and dikes and by contact metamorphism. Much of the British Virgin Islands and three islands south of St. John (Buck Island National Monument, Frenchman Cap, and Flanagan Island) were formed entirely by these dioritic rocks. Outcrops on St. John can be found at Leinster Hill and near Mary Point. Extensive vertical faulting occurred concurrently with the dioritic intrusions. Later in the Eocene Epoch, numerous dikes were

intruded. One can be seen near Caneel Bay on St. John Island. By mid-Eocene time, the Caribbean region had assumed its modern geographic configuration, distinctly separated by the widening Atlantic Ocean from the Mediterranean Sea. Subduction of the North American plate below the Caribbean plate was especially intense during Eocene time, but thereafter movement along the northern margin of the Caribbean was largely horizontal along an east-west transform fault.

6. Alternations in sea level and reef development during the Quaternary Period (Pleistocene and Holocene Epochs).

Reefs formed around St. John Island as seawater warmed when sea level periodically rose in the Pleistocene Epoch. These reefs were later eroded to form terraces during periods when sea level dropped. Over the past 15,000 years sea level has slowly risen and submerged many of the older terraces. Living reefs have become established on the Pleistocene terraces and have grown upward during this last sea level rise.

Geologic Map and Cross Section

Cosner, O. J., and Bogart, D. B. 1972. *Water in St. John, United States Virgin Islands.* Geological Survey, Water Resources Division, Caribbean District, open file report, p. 9.

Donnelly, T. W. 1966. *Geology of St. Thomas and St. John, United States Virgin Islands.* Geological Society of America Memoir 98, pp. 85–176.

Bibliography

Colin, T. J. 1981. The U.S. Virgin Islands. *National geographic* 159(2):225–43 (February). Map supplement included.

Donnelly, T. W. 1966. *Geology of St. Thomas and St. John, U.S. Virgin Islands.* Geological Society of America Memoir 98, pp. 85–176 (also in Mattson 1977).

———. 1968. Caribbean island arcs in light of the sea-floor spreading hypothesis. New York Academy of Science Transactions, Series 2, volume 30, pp. 745–750 (also in Mattson 1977).

Greenberg, J., and Greenberg, I. 1972. *The living reef.* Miami: Seahawk Press.

Khadoley, K. M., and A. A. Meyerhoff. 1971. *Paleogeography and geological history of Greater Antilles.* Geological Society of America Memoir 129.

Mattson, P. H., ed. 1977. *West Indies island arcs.* Stroudsburg, Pennsylvania: Dowden, Hutchinson, and Ross, Inc.

Robinson, A. H. 1974. *Virgin Islands National Park, the story behind the scenery.* Las Vegas, Nevada: KC Publications.

Wood, Peter. 1975. *Caribbean Isles.* New York: Time-Life Books.

Address

Virgin Islands National Park
Box 806
Charlotte Amalie
St. Thomas, Virgin Islands 00801

PART V
Ground-Water Action

© by G. Shimer

Pisolites ("cave pearls"), about the size of cherries, on a cave floor. Carlsbad Caverns National Park, New Mexico.

In regions underlain by soluble rocks, ground water percolates downward, dissolving, transporting, and redepositing rock material in myriad forms, sizes, and shapes. Great subterranean chambers, a maze of interconnecting passages, and speleothems in amazing variety can be viewed in three of the national parks in Part V: Carlsbad Caverns, Mammoth Cave, and Wind Cave. Heated ground water of rare purity returns to the surface at Hot Springs National Park. Ancient logs have been transformed by ground-water action to gem-quality stone in Petrified Forest. In Everglades and Biscayne, surface water, ground water, and ocean water have collaborated in the development of exceedingly beautiful and fragile features both above and below sea level.

35
Everglades National Park

Location: Southwest Florida
Area: 1,398,800 acres; 2,185.63 square miles
Authorized: May 30, 1934
Established: June 20, 1947

Figure 35.1 Taylor Slough in the foreground is a natural drainage channel on the southeast side of Everglades National Park. The slough was probably a tidal channel when sea level was higher. The slightly elevated, tree-covered mound in the background is a hardwood hammock near the Anhinga Trail and the southern tip of the Atlantic Coastal Ridge. National Park Service photograph by Zimmer.

EVERGLADES
NATIONAL PARK

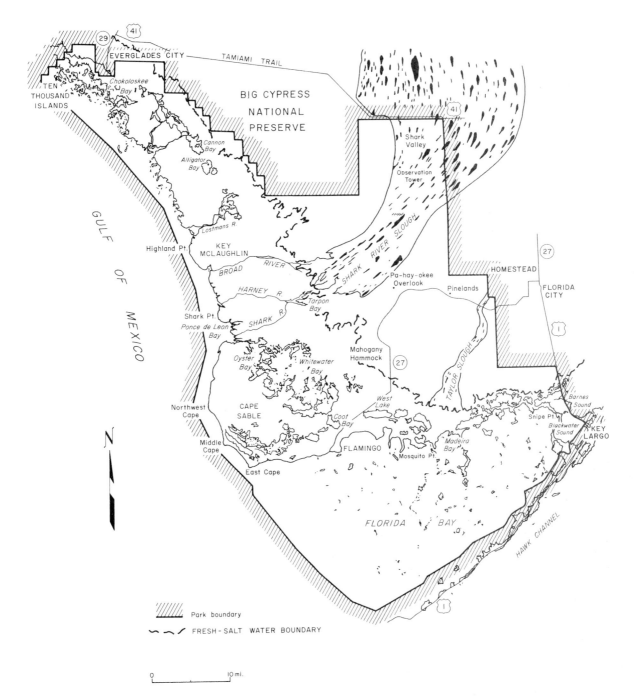

Figure 35.2   Everglades National Park, Florida.

Table 35.1. Geologic Column, Everglades National Park

Time Units			Rock Units	Geologic Events
Era	Periods	Epochs		
Cenozoic	Quaternary	Holocene	Carbonaceous sands in beach ridges; limy muds, mucks, peat, soil; limy marine muds in bays	Nonmarine accumulations of unconsolidated sediment in shallow stream channels and on bedrock surface
		Pleistocene	Miami Limestone (oolitic and bryozoan facies)	Accumulation of ooids and bryozoan masses in shallow seas
			Fort Thompson Formation (in subsurface of Biscayne Aquifer)	Marine and freshwater marls, limestones, sandstones, shell hash, etc.
	Tertiary (Neogene)	Pliocene	////////////////	Slight erosion
		Miocene	Tamiami Formation	Marine limestones deposited in shallow water; silts, shelly sands
	Tertiary (Paleogene)		////////////////	Disconformity
			Underlying marine carbonates of the Coastal Plain (approx. 3 miles in thickness)	Only slight deformation
Mesozoic	Cretaceous			

Source: modified after Hoffmeister 1974

The Wetlands Environment— Past and Present

Everglades National Park is the largest remaining subtropical wilderness in the United States. It includes both freshwater and saltwater areas; offshore islands; dense forests; swamps; and open, grassy prairies, called "glades." Although Everglades National Park and Big Cypress National Preserve (a large adjoining tract north of the Park) occupy most of southwestern Florida, the Everglades *region* is larger still, extending as far north as Lake Okeechobee and originally as far east as the coastal ridge along the eastern shore. Moreover, the Everglades wetlands are ecologically linked with a series of lakes, streams, and swamps that reach north from Lake Okeechobee almost to Orlando in the center of the state. What this means is that the whole lower half of the Florida peninsula is an interdependent hydrologic system, with the Everglades wetlands, both inside and outside Park boundaries, acting as a delicate and vitally necessary balancing mechanism. Only comparatively recently has this function of the Everglades wetlands system been recognized, and much is still to be learned about its intricate workings that protect and preserve the state's water resources. The Everglades is a unique, natural treasure as necessary to the enjoyment of life in southern Florida as the ocean, the beaches, and the benign climate.

The natural rhythms of the Everglades have been evolving for at least 5,000 years, during which time climate and sea level have remained relatively unchanged. The water supply of the wetlands is renewed and restored annually by a rising water table and increased circulation of water during the hot, wet summer season. Most of the year's precipitation comes in brief, intense showers and torrential rains between May and October. The mild winter season is drier; little rain falls between November and April. Most Everglades flora and fauna are not only adapted to, but dependent upon these wet and dry seasons. Their reproductive cycles have become synchronized with times of increasing and decreasing moisture.

In the past, disruptions of the system by periods of drought, lightning-caused fires, and hurricanes were accommodated without much difficulty; in fact, such natural disasters became part of the ecological processes of renewal. Today controlled burning is a part of the National Park Service management policy in the Everglades, because natural communities like the piney woods are restored by this means.

Figure 35.3   This air view of a fresh-water marl prairie in Everglades National Park shows winding, tree-lined waterways and marshes filled with sawgrass. National Park Service photograph.

The ecological crisis that the Everglades wetlands system has *not* been able to absorb and recover from is the loss and diversion of a great deal of its fresh water due to recent residential, commercial, and agricultural development in the surrounding areas. Until about a hundred years ago, water from Lake Okeechobee flowed very slowly, but unhindered, through sloughs and swamps toward the south, emptying at last into Florida Bay and the Gulf of Mexico. The interior of the Florida peninsula south of Orlando can be thought of as a broad shallow basin, tilted just enough toward the south to permit the natural drainage to move across the surface as sheetwash and to flow sluggishly through irregular channels and stretches of open water (fig. 35.3). Decaying vegetation and fine sediment built up layers of peat and muck on the limestone bedrock. The Everglades vegetation thrives in these muck soils, which are enriched annually by saturation and flooding.

For several centuries after Spanish conquistadors discovered the Florida peninsula, the Everglades remained a little-known wilderness, uninhabited except for a few Seminole Indians and runaway slaves who took refuge in the swamps and secluded hammocks in the 19th century. Through the years a few hunters and trappers came to

know the myriad waterways of the Everglades, but settlers seldom penetrated this formidable and mysterious wilderness. They were discouraged not only by its vastness and the difficulties of traversing its depths, but also by the swarms of ravenous mosquitoes and by several varieties of toxic plants. The manchineel tree in particular was feared because of its poisonous "apples" and toxic milky-white sap. It was even believed that a man who rested in its shade would fall asleep and die. The Indians, however, held the manchineel in high esteem. They crushed the bark and tossed it into a pond or stream. When the stunned fish came to the surface, the Indians hauled them in and ate them.

The opening up of the Everglades began in this century. Drainage of the wetlands started around Lake Okeechobee with the digging of the first canal in 1906. Muck lands south and east of the lake were cleared for the growing of fruits and vegetables. This set off a land boom that reached its peak in the 1920s. Levees and dikes around the southeast side of Lake Okeechobee confined the water. Drainage ditches kept the fields well watered but not saturated, and bountiful crops were harvested from the rich muck soils. During several drier than normal years in the early 1920s, land promotion schemes brought thousands of people into southern Florida. Several years of abundant rainfall ensued, including disastrous hurricanes in 1926 and 1928 that caused Lake Okeechobee to overflow its dikes. Reclaimed land was flooded, crops were ruined, and several thousand people perished. The land boom of the Twenties collapsed.

Population pressures around the Everglades slowed during the depression years of the 1930s, but picked up again with the influx of military personnel and defense establishments in the 1940s. More canals and drainage ditches were dug without adequate regional controls, and the cycle of disastrous floods, droughts, and fires began again. Finally, after the havoc wrought by the hurricane of 1947, a regional flood control district was set up by local authorities, working with the U.S. Corps of Engineers and the U.S. Geological Survey, to deal with the problems of too much water and too little water. Recognizing that Lake Okeechobee, with an area of 700 square miles (about one-third the size of Lake Michigan), was the key to control of the system, engineers set about making the lake a huge storage reservoir.

Canals had been dug, both to the Atlantic and the Gulf, to improve drainage of excess water to the ocean. These were integrated into an elaborate network with pumps, gates, and levees designed to retain water needed during droughts. During heavy storms, excess water could be back-pumped from the reclaimed farm lands into the lake. Later, when it was needed, the water would be drawn back for irrigation. Another project involved dredging and channelizing the Kissimmee River, which collects water from the north and carries it into Lake Okeechobee. This shortened the river's length and increased the rate of flow into the lake.

The water management plan worked well for a number of years and brought great benefit to many people. However, serious problems recurred when the great growth in population in the 1960s and 1970s put enormous stresses on the system. The threat of saltwater encroachment, which began when the first canals were dredged to the Atlantic Ocean—thus disrupting the natural drainage to the south—became more serious. During dry periods, saltwater moved into the canals and into the porous bedrock, displacing freshwater. A critical factor here is the level of the water table along the coast, and this level has permanently dropped. Freshwater wells near the coast became brackish, and new wells had to be dug farther inland. Attempts have been made to stabilize the water table by maintaining freshwater pressure high enough to prevent further encroachment of salt water; but whenever rainfall is less than abundant, the water supply problem in the coastal areas becomes acute.

In the agricultural lands reclaimed from the Everglades, the muck soils are drying out, oxidizing, and being blown away, the loss being at the rate of about an inch per year. The only way of renewing and regenerating the muck is to allow tracts of land to be flooded for at least half of the year so that more organic material can decay and fine sediment can be deposited. The fact that about 1,200 square miles of the Everglades agricultural area has already sunk six to seven feet means that the loss of muck is approaching a critical point.

The channelization of the Kissimmee River lowered lake levels in the northern part of the system and caused the drying out of bottom lands exposed to the atmosphere. This resulted in more soil loss. When the Kissimmee meandered over its mile-wide flood plain, more water soaked in, the water table was higher, and during rainy seasons the bedrock aquifer was recharged.

The reduced flow of freshwater into the southern Everglades, including Everglades National Park, has been detrimental to the habitat of a number of plant, animal, and bird species, some of which are now endangered. Sometimes water has been released too late in the breeding season or too early or too much all at once, instead of flowing slowly and continuously through the Everglades during the dry season as in the past. The most serious effect of reduced freshwater on life forms has been the stress on nurseries of shrimp and both freshwater and saltwater fish in the southern Everglades and Florida Bay. The brackish estuaries open to the Bay have become more salty, greatly reducing the fish and shrimp population and resulting in losses to the fishing industry.

The East Everglades wetlands adjoining the Park is still an unspoiled area, but it, too, is threatened by residential development. Any further draining of this area will not only be detrimental to the Park's ecosystem but will also deprive the Biscayne Aquifer (box 35.1) of the fresh water it needs for recharge. This is a vital consideration, because the aquifer is the chief source of water for more than two million people.

Severe droughts in recent years have alarmed many residents of the Florida peninsula and have given renewed impetus to the conservation movement. The Everglades is now seen as the ecological base for repairs to the hydrologic system (because restoration is impossible). With prudent and farsighted management, the water resources of south Florida can be maintained, used carefully, and not wasted.

Years of effort by devoted conservationists in the 1920s and 1930s went into the creation of Everglades National Park. Although the Park was authorized by act of Congress in 1934, land acquisition proceeded slowly due to lack of funds. The disastrous hurricane of 1947 provided the spur to action that finally brought enough land together by donation and purchase for the establishment of Everglades National Park. The second big step in the preservation of the wetlands came in 1974 with the setting aside of Big Cypress Swamp north of the Park as a National Preserve. At the present time the federal government is buying more wetlands acreage in the Big Cypress Swamp area; the National Audubon Society is creating a wet prairie sanctuary on the former flood plain of the Kissimmee River; and tracts of land in East Everglades (on the east side of the Park) are being withdrawn from development. All these measures and more are necessary for the protection of the Everglades wetlands and the preservation of water resources.

A National Park Service brochure that is given to visitors to Everglades National Park explains the wetlands problem plainly and succinctly:

> "Water, fresh water, the life-blood of the Everglades, appears to be everywhere, but . . . despite an apparent lush richness, water supplies are critical. Porous limestone underlies the entire park, and rooting plants have only a thin mantle of marl and peat atop this limestone for their support. If not protected, the Everglades' fragile richness would quickly vanish.

> " . . . Once the benefactor of south Florida's naturally well watered richness, the Everglades now competes at the end of a controlled supply line. To the north, flood prevention, frost protection, pest control, drinking water, and sewage dilution systems siphon off shares. When rainfall is ample, few problems arise. But in drought years arrangements for sharing are required."

The Ecosystems of the Wetlands

Within Everglades National Park are eight zones, or ecosystems, that are dependent upon a balance between wet seasons and dry seasons, between freshwater and saltwater, and, in some cases, flood and fire. Primarily the zones are defined by climatic conditions; elevation in relation to sea level and to the water table; and the geologic aspects of the surface and subsurface materials.

The Marine and Estuarine Zones. Most of Florida Bay is within the boundaries of Everglades National Park. This is a very shallow body of water, roughly triangular in shape lying at the southern end of the peninsula. It is of special interest to many geologists because it is studied as a modern analog for ancient epicontinental seas in which deposition of fine sediment and organic material took place in a brackish environment with little circulation. Sandbars and banks covered by dense growth of mangroves shut off the eastern end of the bay from tidal interchange with Barnes Sound; and the line of Florida Keys from Key Largo to Key West—along the edge of the continental shelf—keeps out most of the water from the ocean. Between the mainland and the Florida Keys, a lacelike pattern of winding, elongated mudbanks and bars, dotted with

hundreds of tiny mangrove islands, spreads over the shallow bottom. Most of the bay is four or five feet deep and the maximum depth is only about nine feet. Even on the west side of the bay, which is open to the Gulf of Mexico, the daily tides have little effect since they penetrate the bay for only a short distance. It takes a full-fledged hurricane to flush out Florida Bay.

Florida Bay is part of the great "carbonate rock factory" (chapter 36) that produces different kinds of limestone in the various marine environments off the Florida coast. By studying the chemical, biological, and environmental conditions under which carbonate rock is forming in Florida Bay, geologists seek to better understand the processes that formed ancient sediments which are sources of petroleum. Much of the research that has been and is being done is supported by petroleum companies.

The fine-grained calcareous mud that is accumulating in Florida Bay (and that may eventually be consolidated into limestone) is less than 5,000 years old and lies on top of the oolitic facies of the Miami Limestone, which is Pleistocene in age. The southern Florida Keys (outside the Park), which rim the southeast side of the bay are also made up of oolitic limestone, which covers an older coral reef. Ooids have not formed in this area since Pleistocene time. (For an explanation of ooid formation, see chapter 36.) The age relationship of the bedrock of the bay and the overlying mud suggests that Florida Bay has been under water a relatively short time.

The estuarine environments of Everglades National Park can best be observed along the course of the Wilderness Waterway, a boat and canoe route that goes from the Visitor Center at Flamingo on Cape Sable northwestward to the Gulf Coast Ranger Station (near Everglades City) in the northwest corner of the Park. This winding, 100-mile inland waterway goes through vast stretches of coastal swamp along an irregular, island-dotted shoreline. The underlying bedrock (rarely exposed because of the thick vegetation) is the bryozoan facies of the Miami Limestone in the southwest and the Tamiami Limestone in the northwest. Saltwater extends several miles up the estuaries.

Both the marine and estuarine zones serve as breeding grounds, spawning grounds, and nurseries for life forms of many kinds, from microscopic organisms on up to large species, such as crocodiles and manatees, both of which are in the endangered category. Crocodiles, which have a shorter, more tapered snout than alligators and are much less numerous, live mainly in the eastern end of Florida Bay and build their mounded nests on the Upper Key Largo shore and along the Bay shore. Manatees, or "sea cows," are large aquatic mammals that feed on vegetation in mangrove-lined creeks.

The Mangrove Zones. Mangrove trees, which grow in shallow, brackish water, establish themselves in the tidal zone of the marine areas and in the transition zone between the freshwater glades and the salty water in the estuaries. The mangroves have an important geologic role because they extend the land as they spread along the coastline and protect it from storm erosion. In the estuaries mangroves line the banks and hold soil in islands (box 36.1, p. 450).

In the Ten Thousand Islands region, part of which is within the northwestern section of Everglades National Park, mangroves have created elongate islands from oyster beds in shallow waters offshore. When the upward growth of oyster beds and accumulated sediment reaches the tidal zone, mangrove seedlings are able to gain a foothold on top of the beds; eventually the trees displace the living oysters, and the oysters reestablish beds in new locations farther offshore. Generations of mangroves have built up thick layers of peat over old oyster beds. These layers, representing successive tidal zones, provide a record of gradually rising sea level along the coast.

The Cypress Zones. A cypress stand develops in solution pits and sinkholes in limestone that have been filled with accumulated muds and marls. Dwarf cypress grow in brackish areas inland from the southern coast and near the edge of the sawgrass glades. The larger pond cypress grow where more freshwater is available near the sloughs. Some build up slightly elevated "domes" of mud and vegetation on top of sinkholes. Mixed with other species, pond cypress also thrive on tree islands (bayheads) in the marshes and sloughs.

The Coastal Prairie Zone. Salt-tolerant plants, such as cactus and yucca, grow among grasses in the thin salty soils of the coastal prairie in the southern tip of the Park on Cape Sable and at the lower end of Taylor Slough. North of West Lake a stand of hardwoods struggles for survival on an old shell mound left by prehistoric coastal Indians. The shell mound does not get the tree roots high enough out of the salt for the trees to do very well. Patches of coastal prairie are also found between the mangrove-lined shore and the coastal swamps on the Gulf side of the Park. Ancient shell mounds mark the sites of camps where Indians feasted on shellfish. One of these at Chokoloskee near Everglades City covers 135 acres. Twenty feet high, this man-made "hill" is the highest point in Everglades National Park. Pots made from marl clay, shark-tooth knives, and other artifacts have been excavated from some of these archeological sites.

The Freshwater Sloughs and the Freshwater Marl Prairie. Shark River Slough and Taylor Slough are the two main channels that move fresh water slowly southward through the Everglades. Both sloughs occupy broad, shallow depressions in the limestone bedrock and were probably free-flowing rivers at one time. Seen from the air, the sloughs are linear complexes of interconnected waterways and ponds, swamps, and sawgrass glades. The sloughs blend into the freshwater marl prairie on either side, and both prairie and sloughs are dotted with innumerable tree islands, called "bayheads" or "hammocks" (fig. 35.3).

Shark River Slough, the larger, central slough, is more than 40 miles wide during the summer rainy season but dwindles to a few narrow streams during the winter droughts. At its lower end, the slough empties into the myriad waterways and lakes of the Shark River estuary in the Cape Sable area at the southwestern end of the peninsula.

Between Shark River Slough and Taylor Slough, the southern tip of the Atlantic coastal ridge forms an interfluve, not more than three feet high, that divides the drainages of the southern Everglades. Before drainage canals were dug in the East Everglades area, Taylor Slough cleaned and stored water for a much larger part of southeast Florida. At the present time, state and federal authorities are attempting to preserve the wetlands between the Park's eastern boundary and the metropolitan area spreading southwest from Miami in order to protect the region's water supply and also keep the feeding and breeding grounds for wildlife in the Park from being further imperiled. In recent years heavily populated Miami and its environs have drawn so much water from the Biscayne Aquifer (box 35.1) that the water table has dropped rapidly, permitting saltwater to seep into the aquifer in increasing amounts.

The freshwater marl prairies, which support many of the more than 100 species of grass growing in the Park, have by far the greatest extent of the Everglades ecosystems. The seemingly endless expanse of grassy glades around the horizon are what inspired the term "everglades," and this became the name for the whole region. Small ponds and "gator holes" are scattered over the prairies. The latter are pits and solution holes in the limestone bedrock that have been cleaned out by alligators during the dry seasons. They thrash their tails around, clearing out mud and vegetation so that water can collect in the depressions. Such waterholes enable many Everglades species to survive droughts.

The famous sawgrass of the glades and sloughs is really a type of sedge that grows as high as 12 feet and is hardier and tougher than most grasses. (Sedges have solid stems; true grasses have hollow stems.) Sawgrass stems have sharp, toothed edges, hence the name. The Everglades is the largest sawgrass marsh in the world (fig. 35.4).

The Pinelands and Hardwood Hammocks. The pinelands in Everglades National Park are on the slightly elevated limestone of the end of the ridge between Taylor

Figure 35.4   Sawgrass covers a marsh dotted with hammocks, which are slightly elevated knobs of limestone on which trees can take root. National Park Service photograph.

Slough and Shark River Slough. Here on mostly bare limestone rock, the pines survive by sending their shallow roots down into pits and crevices in the oolitic rock. The stands of slash pine in the Park are about all that is left of the pine forests that once covered the Atlantic Coastal Ridge for some 50 miles, from north of Fort Lauderdale southward to the end of the ridge in the Everglades. The plant species in the pinelands have adapted to the periodic natural fires that sweep over the ridge during severe droughts. The slash pine, for example, germinates best after a ground fire has burned off low vegetation, releasing ash that enriches the scanty soil; and the thick bark of the mature trees protects them from the fire. Controlled burning of the pinelands is now an established part of the Park's management.

The hammocks of the ridge area are slightly elevated, isolated clumps of tropical trees growing only in spots not burned over by fires (fig. 35.1). Like the pinelands, the hardwood hammocks were once much more extensive on the Atlantic Coastal Ridge. Now they are found only in the Park and a few protected areas outside. The trees and shrubs of the tropical hammocks in south Florida are of West Indian origin, and the taller, broad-leaved trees form a dense canopy that retains moisture for the forest floor. The hammocks also tend to grow on the edges of sinkholes or other sites where the roots can draw more water. The extra moisture has apparently enabled the groves to withstand the ground fires that have raced through the neighboring pinelands.

The hammocks of dense trees and shrubs that make up the tree islands (bayheads) of the sloughs and freshwater prairie may have less than a foot of elevation above the water surface. Many of them are teardrop-shaped, with the points facing into the current, which, although slow and weak, is able to gently mold the island banks. Some hammocks occupy rocky knobs or slight irregularities that jut up above the general flatness of the bedrock surface.

Box 35.1
The Biscayne Aquifer

This important water-bearing rock series is a nonartesian aquifer underlying all of southeast Florida. (An *aquifer* is an rock unit—or units—of moderate to high permeability from which ground water can be pumped.) The Biscayne Aquifer is the principal source of freshwater for the densely populated region from Fort Lauderdale to Key Largo.

The aquifer is supplied by rainwater and surface runoff that seep into the ground and fill the pore spaces in the bedrock. The ground water cannot penetrate deeply because the lower part of the Tamiami Formation is impermeable and thus holds the water in the overlying rocks. The total section of bedrock saturated with water is approximately 10 feet of thickness in western Broward and Dade counties (in the center of the Florida peninsula), increasing to a maximum of about 200 feet along the southeast coast. Throughout most of the area, wells need only to be drilled below the water table in the aquifer in order to tap a supply of potable water.

The best water-bearing unit in the aquifer is the Fort Thompson Formation because of its thickness and high permeability. A good deal of the Fort Thompson is made up of shell hash and contains numerous cavities. The permeable Miami limestone overlying the Fort Thompson is also part of the aquifer. Because of the solution-riddled characteristics of these limestones, the Biscayne is regarded as one of the most permeable aquifers in the world. What this means is that the aquifer

benefits from fairly rapid recharge during abundant rains but also that it is subject to pollution and to encroachment by salt water.

The Biscayne Aquifer has always lost water due to natural causes such as evapotranspiration and ground-water outflow into Biscayne Bay and the ocean. In recent years the amounts of available water have been greatly reduced by (1) the loss of recharge area to industrial, residential, and agricultural development; (2) the extensive system of drainage canals (many of which are highly polluted); and (3) increasing drawdown by pumping from wells.

The pressure of freshwater in the aquifer is what keeps the saltwater out. At the beginning of this century, the peninsular water table was so high and the aquifer was so full that freshwater bubbled up in Biscayne Bay, and freshwater wells could be dug a few feet back from the shore. Now saline water has seeped in along the base of the aquifer and extends inland underneath most of Dade and Broward Counties, which depend on the aquifer for most of their freshwater. Dams have been constructed in canals to maintain high water levels and prevent loss of freshwater. Other control measures have been put into effect; but the water supply situation remains critical. In the summer of 1981, after a decade of drier than normal years, the drought became so severe that the governor of the state declared south Florida a disaster area.

Geologic History

1. **Cretaceous and Tertiary carbonate rocks of the Coastal Plain.**

The surface bedrock of the Everglades is underlain by about three miles of marine sedimentary rocks, mainly carbonates, that were deposited in Cretaceous and Tertiary time and have undergone little deformation. None of these beds crop out at the surface in south Florida.

2. **Miocene deposition of the Tamiami Formation.**

In Miocene time, a blanket of cream, white, and greenish-gray clayey marl, silt, sand, and sandy marl—up to 100 feet in thickness—accumulated in the south

Florida area. These shallow-water sediments were consolidated and cemented into the limestones of the Tamiami Formation, which makes up the bedrock surface of the northwestern part of Everglades National Park. The highly permeable upper part of the Tamiami forms the basal portion of the Biscayne Aquifer. The low permeability of the lower Tamiami members confines the ground water of the aquifer.

3. **Pliocene emergence and slight erosion.**

Pliocene rocks are not present in Everglades National Park, but sandy and shelly marls in the Okeechobee area to the north suggest that the land there may have been at or slightly above sea level.

4. Pleistocene deposition.

The Lower Pleistocene Fort Thompson Formation does not crop out in the Park, but this very permeable unit of marine and freshwater marls, sandstones, shelly limestones, etc., is the main water-bearing component of the Biscayne Aquifer (box 35.1).

When sea level was high during the last Pleistocene interglacial episode (preceding Wisconsin glaciation), several rock units—deposited under different environmental conditions but all about the same age—accumulated in the south Florida area. These rock units make up the surface bedrock of most of south Florida and the Florida Keys, and underlie Biscayne Bay and Florida Bay. The Miami Limestone formed two distinct facies. The oolitic facies makes up the Atlantic Coastal Ridge, the East Everglades region, parts of the southern coast, Florida Bay, and the southern Florida Keys. The oolitic facies interfingers with the bryozoan facies, which makes up the bedrock of southwestern and south central Florida. (The Key Largo Limestone of the same age, which does not crop out in Everglades National Park, is described in chapter 36.)

Examination of a specimen of the Miami Limestone from the oolitic facies shows that its layers consist of billions of tiny spherical grains, or *ooids,* cemented together to make a rocky mass. Ooids are formed by precipitation of calcium carbonate, and the grains accumulate as sedimentary layers (chapter 36). The ooids in the Miami Limestone were cemented by calcite dissolved by rainwater and ground water after sea level lowered, exposing the land to processes of weathering and erosion.

The ooids that make up the Miami Limestone in the Atlantic Coastal Ridge were piled up along a late Pleistocene shoreline by ocean waves. The higher parts of the ridge toward the north, 20 to 40 feet above present coastal elevation, have been exposed to the atmosphere much longer than the southern part of the ridge, which is only about two feet high at its terminus in the southeast part of Everglades National Park.

In the central part of the Park, the byrozoan facies of the Miami Limestone forming the bedrock surface is made up of bryozoan remains as well as ooids. Bryozoans, which are small marine invertebrates living in colonies, secrete calcium to form the tiny cuplike cells in which the individual organisms live. Byrozoan remains, accumulated over many generations, make up about 70 percent of the rock volume of the Miami Limestone. In Florida Bay modern bryozoans continue to grow in knoblike clusters encrusted with seaweed. These small animals have lived in marine environments and produced carbonate material since late in Cambrian time.

5. Holocene mucks, soils, and sediments.

As the great ice sheets of Wisconsin glaciation melted, sea level rose again fairly rapidly, the trend continuing into early Holocene time. Then the rate decreased, and sea level has risen much more slowly in the past 4,000 to 5,000 years. During this time of relative stability, unconsolidated sediments have accumulated on the land surface of southern Florida. Southward-flowing waters spread weathered and eroded material from the Tamiami and Miami Limestones over the saucerlike interior basin now occupied by the Everglades. Mixed with the sediment are abundant organic remains of decayed vegetation. The resulting peat and muck soils are immature but highly fertile, especially under conditions of high moisture and humidity. On land protected by natural vegetation, muck continues to accumulate; but on drained land the surface material becomes powdery during times of drought and is easily eroded by wind. Then when rains return, the loose soil particles are washed away in the drainage ditches.

Box 35.2
The Glacio-Eustatic Theory

The surficial and subsurface geology of Florida's two national parks—Everglades and Biscayne, chapters 35 and 36—show many evidences of sea level changes that came about because of the advances and retreats of the great ice sheets of Pleistocene time. Such variations in sea level are called glacio-eustatic changes. The question arises—how do worldwide changes in sea level relate to the dating of rock units and the geologic history of south Florida? (fig. 35.5)

Not long after the acceptance of the idea of multiple glaciations during the Pleistocene Epoch, geologists began trying to relate global changes in Pleistocene sea levels to the growth and retreat of ice sheets on the continents. At that time, most glaciologists believed that glaciations during the Pleistocene were worldwide and synchronous. In

Figure 35.5 A beach ridge, or berm, on Middle Cape, which is on the Gulf Coast side of Cape Sable. The berm, which is made up of uncemented shell and coral fragments, pebbles, and coral sand, probably accumulated when sea level was somewhat higher and is now being cut back by wave action. National Park Service photograph.

other words, if advances of ice sheets and valley glaciers—as well as retreats and stillstands (i.e., stable periods)—occurred everywhere at the same time, these episodes would be of sufficient magnitude to be correlated with specific rises and falls of sea level.

The concept was simple and logical. As visualized in the hydrologic cycle, the earth's water budget is essentially a closed system in which circulating water may change its form (i.e., liquid, solid, or gas) but none is added nor lost. Small amounts of magmatic water may be added to the system by volcanism, and some water is locked up in hydrated minerals; but nearly all of the global water supply is in the oceans, in the atmosphere, on the land surface (in streams, lakes, swamps, frozen in glaciers), and in the subsurface as ground water.

When continental ice sheets form, the water to make the ice comes from the circulation of the hydrologic system, the result being that glacial growth causes worldwide lowering of the sea level. Calculations based on area of land covered with ice and assumed thickness of ice sheets gave approximate volumes of ice needed to form the great continental glaciers. Subtraction of ice volume (converted to water volume) from the calculated volume of ocean water yielded an estimate for lowering of the ocean surface. The maximum lowering of sea level during major glaciations was estimated to be approximately 300 feet (or 100 meters).

If all of the ice now on the earth melted, sea level would rise about 100 feet (30 meters). Geologists assumed that virtually all of the world's glaciers melted during major interglacial stages of the Pleistocene. Therefore, sea levels may have fluctuated from about 300 feet below present elevation to about 100 feet above several times during the past one or two million years.

Indications of changing sea levels, such as high shorelines and drowned shorelines, are found on all continents. The elevations of ancient sea levels in unglaciated regions were regarded as more useful for purposes of correlation than changes that occurred on glaciated coasts, because the former were not affected by isostatic depression due to the weight of ice and subsequent postglacial rebound. Graphs of sea level curves with elevations plotted against time were developed, based on evidence accumulated from all over the world and covering the entire Pleistocene Epoch. Sea level curves have been used in two different ways. If a date, either absolute or relative, was available for a deposit or feature, the value was put on a sea level curve and an elevation of sea level determined for that particular time. Conversely, when elevation of a former sea level was known—from freshwater peat deposits or a raised beach, for example—that measurement was put on a sea level curve and the age of the feature determined.

Logical as this procedure might seem, the results have been far from satisfactory because locally and regionally other factors have, more often than not, affected age or elevation determinations, or both. Only in the broad sense of major trends of rising or falling sea levels does the concept hold true. Over short periods of time, glaciers made advances and retreats that were in response to local conditions and not synchronous with similar advances and retreats elsewhere. A well documented glacial event in Illinois, for example, cannot be directly correlated with a specific sea level change on the Florida coast.

Moreover, geologic activities other than glaciation affect sea level. Obviously, tectonic changes in the earth's crust have caused parts of the sea floor and the land to rise or subside, thus bringing about sea level changes. Further complicating the correlation of sea level elevations are changes caused by movement of tectonic plates (e.g., sea-floor spreading), isostatic adjustments, sedimentation, oceanic volcanism, and combinations of various factors.

Present-day investigators of eustatic change believe that while the dominant, worldwide trends of sea level elevation can be attributed to additions of water to, or removal of water from, the great continental ice sheets, *specific* changes in sea level cannot be correlated from region to region. Also, individual glacier regimens are so variable that they cannot be matched with fluctuations on a sea level curve plotted for a nonglaciated coast a thousand miles or more away. Use of glacial chronologies based on substages or minor advances or retreats in one region to account for sea level changes in another region implies a precision that is not justified.

In south Florida, where little or no tectonic activity has occurred throughout most of the Cenozoic Era, the elevations of most of the young marine limestones suggest that they were deposited in water slightly above present sea level. Moreover, these same limestones have been eroded and pitted by stream action graded to sea levels slightly below present elevations. The modern landscape is being very gradually submerged by sea levels that have been rising during Holocene time. According to the glacio-eustatic theory of worldwide sea level fluctuations, it appears that the limestones formed during the last Pleistocene interglacial phase, when sea stands were high, and that the surface was eroded during the last major glacial episode (Wisconsin time) when sea level was lower. Several lines of evidence seem to confirm these generalizations, but the suggestion that melting of modern glaciers could account for currently rising sea levels may or may not be true.

On-going research in Florida involves detailed studies of accumulated soils, vegetation, sediments, fossils, pollens, etc., and comparisons of onshore and offshore sites. Gradually more precise dates and elevations are being derived from these interdisciplinary studies. These data are enabling scientists to prepare a more detailed and complete geologic history of south Florida that is not based on glacio-eustatic theory and that includes fluctuations of sea level over the last hundred thousand years.

Geologic Maps and Cross Sections

American Association of Petroleum Geologists. 1975. *Geological highway map of the southeastern region, map no. 9.* Tulsa, Oklahoma: American Association of Petroleum Geologists.

Puri, H. S., and Vernon, R. O. 1964. *Geologic map of Florida.* Plate 2. *Summary of the geology of Florida and a guidebook to the classic exposures.* Florida Geological Survey Special Publication No. 5.

Bibliography

Boyle, R. H., and Mechem, R. M. 1981. There's trouble in paradise. *Sports illustrated* 54 (6) : 82–96, February 9.

Carr, Archie. 1979. *The Everglades,* American Wilderness Series. Rev. ed. Alexandria, Virginia: Time-Life Books.

Davis, J. H., Jr. 1943. *The natural features of southern Florida.* Florida Geological Survey Geological Bulletin No. 25.

Hoffmeister, J. E. 1974. *Land from the sea, the geologic story of south Florida.* Coral Gables, Florida: University of Miami Press.

De Golia, Jack. 1978. *Everglades, the story behind the scenery.* Las Vegas, Nevada: KC Publications.

Fishbein, S. L. 1980. Everglades National Park, Florida, in *The New America's Wonderlands, Our National Parks.* Washington D.C.: National Geographic Society, pp. 327–38.

Multer, H. G. 1977. *Field guide to some carbonate rock environments, Florida Keys and Western Bahamas* (new edition). Dubuque, Iowa: Kendall/ Hunt Publishing Company.

National Park Service. 1978. *Everglades National Park.* Map and brochure.

Address

Everglades National Park
P.O. Box 279
Homestead, Florida 33030

36
Biscayne National Park

by Ann Budd Foster
University of Iowa

Location: Southern Florida
Area: 274.53 sq. miles; 175,700 acres
Proclaimed a National Monument: October 18, 1968
Established as a National Park: June 28, 1980

Figure 36.1 Air view looking north over the islands and channels of Biscayne National Park. The islands, which were built up by a coral reef during Pleistocene times of higher sea level, are thickly wooded. National Park Service photograph.

BISCAYNE
NATIONAL PARK

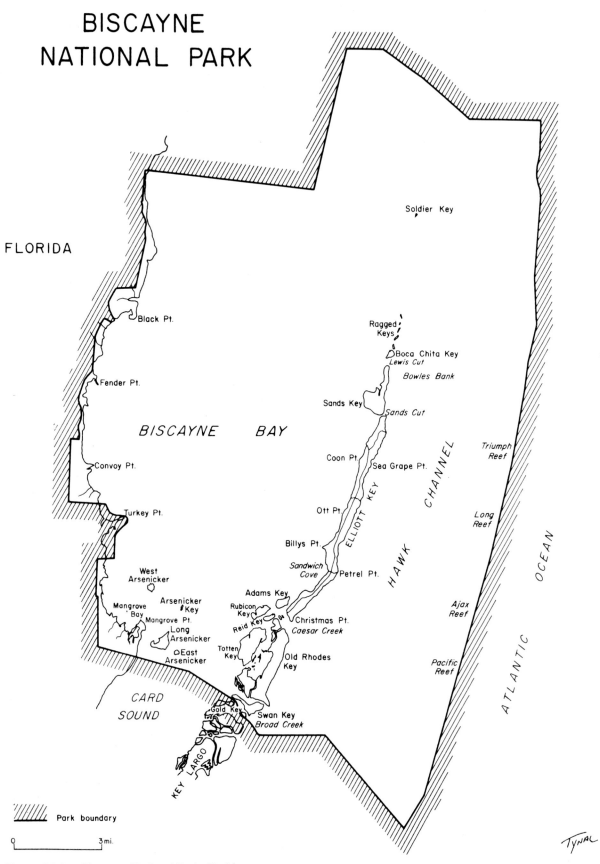

Figure 36.2   Biscayne National Park, Florida.

Table 36.1. Geologic Column, Biscayne National Park

Time Units			Rock Units		Geologic Events
Era	Period	Epoch			
Cenozoic	Quaternary	Holocene	Accumulations of thin soils, marls, peat on land. Accumulations of fine bottom sediments, mainly carbonates, in seawater		Development of fringing mangrove swamps Generally rising sea level Growth of coral reefs
		Pleistocene	Miami Limestone (oolitic facies)	Key Largo Limestone	Fluctuating sea levels, with coral reef growth during high sea stands, erosion during low sea stands Oolitic deposits on mainland and lower keys

Source: modified after Hoffmeister 1974

Local History and Geographic Setting

Biscayne National Park is a large marine preserve south of Miami and offshore from the mainland of southeastern Florida. Within its boundaries, from west to east, are most of shallow Biscayne Bay, including a narrow strip of the Florida shoreline; a north-south chain of 32 small islands, or keys (the northern part of the upper Florida Keys); and a large reef tract, comprising a back reef area and the northern part of an outer reef. This reef tract is the only living coral reef off continental United States. Hawk Channel, a seaway between the islands and the outer reef, is also in Park waters. Biscayne National Park was established to protect three natural ecosystems—(1) the nursery grounds for many fish and shellfish species in Biscayne Bay, (2) the native vegetation of the upper keys, and (3)the marine plants and animals of the coral reefs.

The term "keys," derived from the Spanish word *cayo* meaning "islet," is given to the coral islands that form a line of fossil patch reefs off the southern Florida coast. A topographic and geologic distinction is made between the lower Florida Keys, which extend southwestward from Big Pine Key to Key West, and the upper keys, which run from Big Pine to the Park's northern boundary. The Park's southern boundary is between Old Rhodes Key and Key Largo. A causeway connects Key Largo and the rest of the southern keys with the mainland, but access to the keys in Biscayne National Park is by boat only.

In the 17th and 18th centuries the waters off the tip of Florida were the scene of battles between pirates and sailing ships from many nations. The pirates lay in wait in the Florida Keys for ships sailing through the Florida Straits between Cuba and the mainland. Usually the pirates attacked ships that were crippled by storms. Strong currents, high winds, hurricanes, and sudden violent storms, combined with the hidden reefs and shoals, made these waters dangerous for all sailing vessels. They were frequently blown off course onto the shoals and reefs, where they were easy prey for the pirates. Hawk Channel is marked with the locations of some of the sunken wrecks.

Caesar Creek, a natural passage between Elliott Key, the main island in the Park, and Old Rhodes Key, is named for Black Caesar, one of the cruelest and most notorious of the pirates. He embedded an iron ring in a ledge jutting up above the water (called Black Caesar's Rock), attached his ship to the ring, and heeled it over so that the masts could not be seen by merchant ships passing through Hawk Channel. Then he pretended to be adrift in an open boat. When a ship stopped to pick him up, his crew emerged from behind the rock that was hiding them, attacked the merchant ship, and wrecked it. Most of the officers and seamen were murdered, but some were held in prison camp on one of the keys until they agreed to join the pirates.

Eventually Black Caesar joined up with the infamous Blackbeard (Edward Teach). Their ship was captured in 1718 by an English man-of-war. Blackbeard was killed in the battle, but Black Caesar was captured, taken to Virginia, and hanged.

Other pirates, just as notorious, continued the harassment of shipping for more than a hundred years after Blackbeard and Black Caesar were caught. After Florida was purchased from Spain in 1821, the United States Navy undertook a campaign to rid the West Indies and the Florida Straits of pirates. Commodore David Porter organized the "Mosquito Fleet" and set out from Baltimore in 1823 to establish a base at Key West. His plan was ingenious. A small steamboat and a large storeship (that could later be used as a decoy) were converted to warships and armed with guns. Eight shallow-draft schooners were also armed. Five sturdy, 20-oar barges were outfitted with weapons and crews of 40 men each for working in close to shore. (Appropriately, they were named *Mosquito, Gnat, Midge, Sandfly,* and *Gallinipper.*) All the vessels were light, maneuverable, and well adapted to coastal waters except for the steamship and the storeship. Other American naval ships joined Porter in the West Indies until 25 vessels were under his command.

While buildings and facilities were being constructed on Key West, the commodore kept the vessels on the move. The larger ships provided convoy service to merchantmen and set up blockades. The small vessels were sent out pirate-hunting among the keys and straits. The most intensive searching was done by the sailors and marines in the open barges. Those men sometimes spent weeks at a time away from the mother ship while they examined the inlets, rocks, reefs, and hidden coves along the many miles of wild and irregular coast. After more than a year of this kind of arduous and dangerous work, all of the pirates were captured or driven out, and the Florida Keys and Straits were made secure for commercial shipping—although to this day, smugglers still operate in the area.

The earliest inhabitants of the Florida Keys were Arawak and Carib Indians. Their distinctive pottery with incised designs has been found in Indian mounds. Later these Indians were forced out by Calusa Indians, who were fishermen and mariners. They left huge shell mounds that actually enlarged some of the islands. In the 1830s bands of Seminoles, driven from their homes in northern and central Florida by white settlers, took refuge in the Everglades and in the Florida Keys, where some of their descendants still live.

In recent years the coastal waters, shorelines, and islands of Florida have suffered severe environmental damage from the pressures of rapidly increasing population and from overdevelopment, both commercial and residential. Spurred by conservationists and alarmed by growing problems of water pollution, loss of fisheries, and destruction of wildlife habitat, the Florida legislature passed a Beach and Shore Preservation Act in 1965. This established coastal construction setback lines that have saved some of the state's beaches and inlets from being urbanized. The authorization by the United States Congress of Biscayne National Monument in 1968 stopped the construction of a refinery project on Biscayne Bay, as well as urban development proposed for the northern keys. The monument was expanded and redesignated Biscayne National Park in 1980.

Geologic Features

The narrow strip of shoreline, the keys, the coral reefs, and the ocean floor in Biscayne National Park are all made of carbonate rock, most of it of organic origin. The marine platform on which the keys and the reefs were built is part of the Atlantic coastal shelf, which is at its narrowest here along the southeast coast of Florida. A few miles offshore,

Figure 36.3 On the ocean side of Elliott Key, mangroves have grown almost to the edge of the coral rock shoreline. National Park Service photograph.

just east of the outer reef, the ocean floor drops off steeply into the deep water of the Florida Straits. The Gulf Stream flows through the 90-mile-wide straits, between Florida and Cuba to the south and the Great Bahama Banks to the east.

Landward of the outer reef, the water is shallow and the relief of the keys and the mainland is low. None of the keys rises more than 18 feet above sea level and some are barely above high tide level. In fact, the keys and all of southern Florida were under water during the last Pleistocene interglacial episode when sea level was higher than now. The patch reefs that make up the Florida Keys were built by corals that lived and grew on the marine platform during times of high sea level in the Pleistocene (box 35.2). This means that in the geologic sense all the rocks, soil, and sediment of the region are very young, and the processes that are significant in the geologic story of the Park are still going on.

The "Carbonate Rock Factory"

The marine preserve of Biscayne National Park is part of a large carbonate-producing region that includes the waters around the western Bahamas as well as the Florida Keys. Here in a variety of carbonate environments, a wide range of limestones are accumulating, both organic and inorganic. In the "labor force" are many species of marine plants and animals, with the corals and calcareous algae being the main producers. "Manufacturing" processes include photosynthesis (corals and algae cannot work in darkness), precipitation, evaporation, and supersaturation. By these processes the chemical constituents, or "raw materials," are extracted from seawater and transformed into skeletal parts, crusts, and ooids (or ooliths).

Most *ooids* are tiny spheres of inorganically precipitated calcium carbonate. (Noncalcareous ooids, which are less common, are generally formed by replacement processes.) At the present time ooids are forming only in the Western Bahamas part of the carbonate-producing factory. Strong ocean currents there keep the waters agitated and sloshing back and forth over shallow and deep bottom areas. Intense sunlight raises the temperature of the tidal pools and shallows, and evaporation increases salinity, concentrating the calcite and causing it to precipitate around any fragment (a speck of shell, quartz sand grain, etc.) that acts as a nucleus. Precipitation of successive carbonate layers around such nuclei forms ooids.

Although conditions on the Florida side of the Florida Straits do not now permit ooid production, they accu-

Figure 36.4 The shallow, bay side of Elliott Key attracts many swimming and wading birds, who feed on the marine organisms of the lagoon. National Park Service photograph.

mulated in vast quantities in the past in the waters that covered the lower Florida Keys and southern Florida. Ooids were compacted and cemented to form oolitic limestone, some of which has been eroded to produce sand-sized grains and other sediment. Ooids in the form of sediment, either loose or in sedimentary rock, are referred to as *ooliths* or *oolites*. The small beaches on the Florida Keys and the larger beaches along the extreme southern tip of the Florida mainland consist principally of oolitic and coralline sand rather than quartz sand.

Geologists, marine biologists, and other scientists come from all over the world to study the carbonate-producing environments of the coastal waters of Florida and the Bahamas. Much of the research consists of detailed study of the processes of the present in order to better understand how the vast accumulations of carbonate rock on the earth's surface were produced in the geologic past. Because carbonate rocks are an important source of petroleum, a good deal of the research being carried on has significant economic implications.

The Living Reefs

The corals of the reef tract in Biscayne National Park are building reefs in the same way their ancestors built the older reefs that make up the keys. The fossil corals of the islands are not different from their living counterparts in the surrounding waters. The same coral species, about 50 in all, live in their chosen habitats throughout the Florida reef tract and the Bahamas. The coral groups are also closely related to those living in the reefs off the Virgin Islands (chapter 34). A few species in the Florida reef tract seem able to thrive in any environment conducive to coral growth; but most of them prefer one of three main types of environments, either (1) the outer reef along the open-ocean side of the marine platform, (2) the numerous patch reefs in the back reef zone (between the keys and the outer reef), or (3) the shoals on either side of the keys.

Each of these three habitats is dominated by a characteristic set of corals (fig. 36.5). The outer reef is covered largely by the moosehorn coral, *Acropora palmata*, and to a lesser extent by the staghorn coral, *Acropora cervicornis*. These two corals grow rapidly (100 mm/year) in well lit, agitated environments to form dense thickets of branches that shade and thereby inhibit the growth of other corals. The back reef is composed mostly of mound-shaped corals, including the star coral, *Montastraea annularis*, and the brain coral, *Diploria strigosa*. Single mounds of these corals can attain enormous sizes (more than 5 meters in diameter) after several centuries of growth. The corals found in the inshore shoals include small (5 cm diameter), almost round star corals, such as *Siderastrea radians*, and knob-shaped finger corals, such as *Porites divaricata*. These corals grow slowly but are able to remove sediment easily and to resist extreme or fluctuating temperatures and salinities.

Figure 36.5   The living outer reef is dominated by the large, rapidly growing moosehorn (also called elkhorn) corals, *Acropora palmata*, that are strong enough to withstand the wave motion of the open ocean. The moosehorn coral is one of the major reef builders in Biscayne National Park. Artist's rendering of National Park Service underwater photograph. © 1982 by Kendall/Hunt Publishing Company.

Geologic History

1. Pleistocene reef-building; Key Largo Limestone

In the upper keys (including those in Biscayne National Park), the Key Largo Limestone is the only bedrock that is exposed. During the last interglacial episode of Pleistocene time, when sea levels worldwide were higher than now, a long, arc-shaped coral reef was built up on the marine platform. Data from drilling show that when it was living and growing, the reef had a total length of at least 220 miles and extended from Virginia Bay (part of Greater Miami) to the Dry Tortugas in the Gulf of Mexico, west of Key West. However, the Key Largo Limestone is exposed at the surface only from Soldier Key (northernmost key in the Park) to Big Pine Key (about 35 miles east of Key West), a distance of 110 miles (Hoffmeister 1974). The distinction between the upper keys and lower keys is based on the fact that in the lower keys, from Big Pine Key to Key West, the Key Largo Limestone is overlain by the oolitic facies of the Miami Limestone (chapter 35). The differences between these two types of limestone account for most of the erosional and topographic differences between the upper and lower keys.

Apparently the upper keys were constructed as a series of patch reefs in a back reef environment similar to the one that exists today. The principal frame-builders of both the keys and the living patch reefs are the star corals (*Montastrea annularis*), which usually grow in a protected marine environment with limited wave energy. Species such as the moosehorn coral (*Acropora palmata*), which thrive in the surf of the outer reef, have not been found either as fossils in the Key Largo Limestone or as living corals in the back reef area. What seems likely is that an outer reef was growing during Pleistocene time at the same location as that of the present outer reef.

2. Erosion during the last major Pleistocene glaciation.

During Wisconsin time, as the last advance of the great continental glaciers occurred, sea level dropped more than 300 feet, exposing nearly all of the continental shelf off southern Florida. At the seaward edge of the marine platform, wave erosion lowered the surface of the old outer reef by more than 50 feet. Erosion first by waves and then by streams that began to drain off newly exposed land also stripped off the tops of the patch reefs and the upper parts of the present keys. Channels between the keys, such as Caesar Creek, were widened and deepened. Hawk Channel, which parallels the upper keys, was scoured, and ridges or banks of sedimentary debris were deposited along its sides. (In the back reef area of the present time, tidal currents continue to scour the channels and deposit sediment in shoals and banks.)

Rain and ground water flushed salt from the reef rock of the keys and dissolved some of the calcite. Small pits, sinkholes, and other solution features developed in the rock. Gradually a thin soil developed and the upper keys became thickly wooded.

3. The development of Biscayne Bay.

Biscayne Bay, which lies between the northern part of the upper keys and the mainland, is only about 6 feet deep near Miami Beach but deepens somewhat toward the south. The bay is also quite shallow next to the mainland but slopes off toward the keys to 12 or 15 feet. When the keys were being built up by coral reefs during the time of high sea levels in the last Pleistocene interglacial episode, Biscayne Bay was a large coastal lagoon between the Atlantic Coastal Ridge rimming the Everglades (chapter 35) and the reef complex offshore. When sea level was lowered due to Wisconsin glaciation, the lagoon drained off, exposing the limestone bedrock of the ocean floor. Streams cut valleys in the sides and floor of the troughlike depression. Solution pits developed in the limestone. Freshwater calcitic muds that were deposited below present sea level indicate that for a time a freshwater lake occupied part of the Biscayne basin.

As sea level began to rise again, brackish water entered the lagoon and mangrove swamps developed. Eventually seawater filled the lagoon to its present depth, creating Biscayne Bay. Meanwhile, in the shallows next to the keys, small inshore corals began to grow again. The presence of sediment, constantly stirred up by currents, prevents extensive coral growth in Biscayne Bay, however.

Beneath the bottom sediment, on the bedrock floor of Biscayne Bay, is the contact between the Key Largo Limestone and the oolitic facies of the Miami Limestone, which covers all of southeastern Florida (chapter 35).

Figure 36.6   Coral heads and knobs are exposed above water level on the bay side (near University Dock) in Biscayne National Park. Wave action is producing coral pebbles and sand. National Park Service photograph.

Box 36.1
The Geologic Role of Mangrove Swamps

In Biscayne National Park, as well as elsewhere on the Florida coast, mangroves have had an important role in the development of the present shoreline. Scientists estimate that, despite occasional losses of ground due to hurricanes, the mangrove swamps along the keys, Biscayne Bay, and Florida Bay have in this century added more than 1,500 acres of dense forests where formerly open shoals existed (Carr 1979). In the Park, fringes of mangrove-built shoreline can be seen on both sides of the keys and at Mangrove Point on the mainland.

Among the salt-tolerant trees of the shore zone are the dominant red mangroves *(Rhizaphora),* the only true mangroves; the black mangroves and white mangroves, which belong to two other species although they are called "mangroves"; and the buttonwoods. The buttonwoods grow farther inland than the mangroves and are often found mixed in with mahogany and other tropical hardwoods. The mangroves grow right at the edge of the ocean in the zone between high tide and low tide. In addition to being salt-tolerant, these remarkable trees have an amazing ability to cling to unstable ground. When storms beat in on the coast and hurricanes bring devastating waves and winds, the mangroves usually hold on. Only the most disastrous hurricanes can uproot them. Even then, some of the older, stronger trees that are flattened by the wind manage to send out new anchoring roots; then their trunks bend back upward and start growing toward the sun again. Before the days of advance warning for hurricanes, old-timers often took refuge in firmly anchored small boats in narrow creeks among the mangroves. The protection afforded by the mangroves usually enabled people to ride out the storm.

A full-fleged hurricane always produces coastal changes, but where mangroves grow, such changes are minimized. In fact, if it were not for the mangroves protecting the tidal environment, the Florida coastline would be reshaped by every big storm. Stretches of coastal swamps were drained, filled, and developed as waterfront real estate before the protective function of the

mangrove swamps was understood. When hurricanes struck, as, for example, Donna in 1960 and Betsy in 1965, the damage was appalling.

Another important adaptation of red mangroves is their ability to colonize. Seeds germinate on the mature trees and then fall as seedlings (fig. 36.7). some take root in water beneath the parent trees, but other seedlings are spread by tides and currents along the coast, sometimes traveling considerable distances before stranding and sending out roots. The young trees anchor themselves by strong roots that sprout both from the trunk and from the lower branches. The roots (1) form a complex network of props and braces that hold up the crown, (2) bring up sap, (3) raise the trunk above the water, (4) furnish oxygen which is unavailable in the watery muck,and (5) catch and hold silt and debris. It is this last function that extends the coastline into the water, continually creating new land (fig. 36.8).

At the present time, various lines of evidence show that for several thousand years at least, sea level has been rising very gradually along the

Figure 36.7   Mangrove sapling establishing itself in the tidal zone. National Park Service photograph.

Figure 36.8   The complex roots of a dead mangrove are still holding sediment in place and helping to protect the shoreline. National Park Service photograph.

Florida shorelines. Geologists investigating shoreline changes have found mangrove peat accumulations to be a useful indicator of approximate rates of sea level rise, because mangroves grow only in the tidal zone. In tropical and semitropical latitudes, tidal ranges are low; on Florida coasts the average range between high and low tide levels is only about three feet. Therefore, if sea level were static, mangrove peat accumulations would be about three feet in thickness. However, in protected locations on the shore, peat accumulations have been found to be 12 to 15 feet in depth, which means that sea level must have been rising while successive generations of mangroves grew and peat accumulated (Hoffmeister 1974).

The importance of the mangroves to the nutrient cycles of the coastal environment can hardly be overstated. Besides supporting a large number of rookeries, the mangrove swamps and bordering waters are the habitat of alligators, snakes, turtles, manatees, raccoons, and many other animals. Even the mosquitoes that breed by the millions in the mangrove swamps are a main source of food for many birds, frogs, reptiles, and fish. The tidal inflow and outflow circulate nutrients for the breeding grounds of fish and shellfish in the bays. Nutrients carried out to sea help to supply food for the corals and other organisms living in the reefs.

Geologic Maps and Cross Sections

American Association of Petroleum Geologists. 1975. *Geological highway map of the southeastern region, map no. 9.* Tulsa, Oklahoma: American Association of Petroleum Geologists.

Hoffmeister, J. E. 1974. *Land from the sea, the geologic story of south Florida.* Coral Gables, Florida: University of Miami Press.

Bibliography

Carr, A. 1979. *The Everglades,* American Wilderness series. Rev. ed. Alexandria, Virginia: Time-Life Books.

Hoffmeister, J. E. 1974. *Land from the sea, the geologic story of south Florida.* Coral Gables, Florida: University of Miami Press.

Multer, H. G. 1977. *Field guide to some carbonate rock environments, Florida Keys and Western Bahamas* (new edition). Dubuque, Iowa: Kendall/Hunt Publishing Company.

Wheeler, R. 1969. *In pirate waters.* New York: Thomas Y. Crowell Co.

Address

Biscayne National Park
Box 1247
Homestead, Florida 33030

37
Hot Springs National Park

Location: West Central Arkansas
Size: 5,826.48 acres; 9.10 square miles
Hot Springs Reservation set aside, April 20, 1832
Redesignated Hot Springs National Park, March 4,
*   1921*

Figure 37.1   Hot water flows naturally from the Display Springs in Hot Springs National Park. The water from the other springs in the Park is collected in pipes and kept covered to protect its purity. Photograph courtesy Hot Springs Chamber of Commerce.

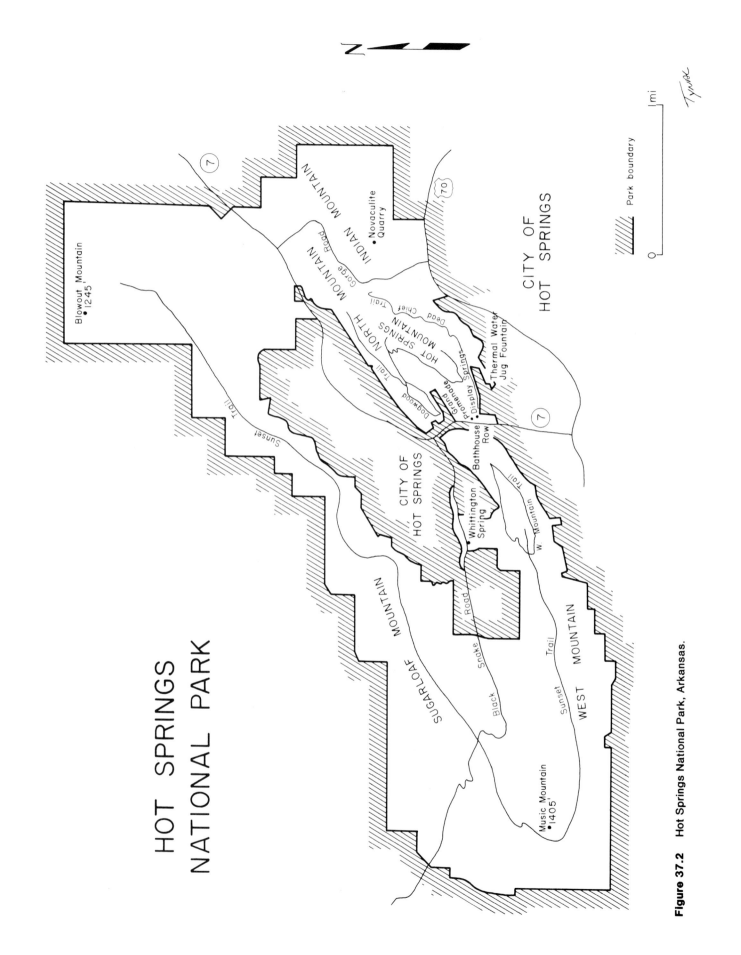

Figure 37.2 Hot Springs National Park, Arkansas.

Table 37.1. Geologic Column, Hot Springs National Park

Time Units			Rock Units		Geologic Events
Era	Period	Epoch			
Cenozoic	Quaternary	Holocene	Alluvium		Erosion and alluviation
		Pleisto-cene	Terrace deposits		Development of hot springs hydrologic/geothermal system Pluvial climate
	Tertiary (or Paleogene/ Neogene)		///////		Long erosion Major unconformity End of uplift
Mesozoic	Cretaceous		///////		Syenite stocks intruded close to Park; small dikes in Park
	Jurassic		///////		Long erosion, major unconformity
	Triassic		///////		Uplift, faulting
Paleozoic	Permian		///////		Ouachita orogeny (plate collision, thrust faulting, intense deformation, uplift)
	Pennsylvanian		(not exposed in Park)		Shales, sandstones deposited in closing seaway.
	Mississippian		(Formation) Stanley Shale	(Members) Chickasaw Crk. Shale Hatton Tuff Hot Springs Sandstone	Flysch deposits in deep trough between closing plates Ash falls Beginning of flysch facies: turbidites, conglomerates, etc.
	——?——		/////// ///////		
	Devonian		Arkansas Novaculite	Upper	Siliceous shales, ash falls
				Lower	"Starved basin" deposition
				Middle	
	Silurian		Missouri Mountain Shale		
			Blaylock Sandstone		
	——?——				
	Ordovician	(late)	Polk Creek Shale		
		(middle)	Bigfork Chert		Slow accumulation in deep-water, stable basin
			Womble Shale		Minor unconformities
		(early)	(not exposed in Park)		
	Cambrian		///////		Opening of ocean; subsidence of sea floor
	Precambrian time		///////		Major unconformity

Source: modified after Bedinger et al. 1979; Viele 1979

Local History

The Indian name for the hot springs was *Tah-ne-co*, meaning "the place of the hot waters." Here was the dwelling place of the Great Spirit, whose warm breath heated the healing waters and vapors of the springs. Many battles were fought among various tribes for possession of the sacred springs. Finally the tribes gathered together and decided that the Great Spirit had meant the springs to be for the benefit of all who were sick. From that time on, the region of the springs became neutral ground and was called "the land of peace." Braves from enemy tribes would lie side by side in the hot mud to ease the pain from rheumatism and old wounds.

When the foundations for the present Quapaw Bathhouse in the Park were dug, a narrow winding cave leading to the springs was discovered. The cave was probably used by the Indians. Former campgrounds and burial mounds of Tunicas, Quapaws, Cherokees, Choctaws, Natchez, Pawnees, and other Indian tribes have been found in the region.

Hernando de Soto, in his search for treasure in the New World, visited the hot springs in 1541 shortly before his death. Chronicles of his ill-fated expedition describe his discovery of "large streams of hot water and a small lake" near a Tunica village when he was journeying between the Mississippi River and what is now Oklahoma. Since there are no other hot springs near the route along the Arkansas River that he followed, the presumption is that he was in the area of Hot Springs National Park.

In 1804, after the Louisiana Purchase, President Jefferson sent William Dunbar and Dr. George Hunter, a chemist, to explore the Ouachita Mountains and to investigate especially the hot springs which Jefferson had heard about. Spending a month in the area, they counted about 70 hot springs and analyzed the water issuing from the rocks. Theirs was the first scientific report to be written about the springs. They found wooden buildings at the site that had been put up by early settlers and trappers who came to soak in the warm, healing water. Other explorers who visited the Arkansas hot springs were U.S. Army officers, Zebulon M. Pike in 1806 and Stephen H. Long in 1818. Two members of Long's expedition, Kearney Swift and Edwin James, the latter a botanist and geologist, studied the springs. They suggested that build-up of *tufa* (calcareous-siliceous deposits around spring openings) blocked the flow from some springs, causing variations from time to time in the total number of springs. In their natural state (i.e., before "improvements" were made), the springs flowed from a number of openings in a belt about a quarter of a mile long and several hundred feet wide along the southwest slope of Hot Springs Mountain.

As time went on, local residents began putting up cabins, bathhouses, and even hotels to accommodate visitors who sought the benefits of thermal bathing. The natural environment of the hot springs became completely altered. Springs were excavated and lined with masonry to increase and control the flow and then covered to protect the water from contamination. To prevent commercial exploitation and ensure that all people would have access to the springs, President Andrew Jackson persuaded Congress to withdraw from public sale four sections of land (four square miles) around the hot springs. This was in 1832, four years before Arkansas became a state.

In 1876, a year after a rail line was extended to the area, the city of Hot Springs was incorporated. Part of the city is now surrounded by the Park, and part has spread out next to the Park's southern boundary. The preserve around the springs was dedicated as a permanent reservation for public use in 1880, and in 1921 it was redesignated a national park. Today in this famous health resort, there are 12 concessionaire bathhouses (five in the city, seven in the Park), all under federal regulations. The former Army and Navy Hospital is now the Arkansas Rehabilitation Center, operated by state and federal authorities. A private facility, the Libbey Memorial Center for Physical Medicine, operates under federal lease.

Of the 47 hot springs in the Park, two—the Display Springs—have been left in a natural state so the Park visitors can see what the springs looked like originally (fig. 37.1). The remaining 45 springs are sealed to prevent contamination and loss of gas. The combined flow from the springs ranges from a high of around 950,000 gallons a day in winter to 750,000 gallons in summer and fall. Water temperature in the springs is about 143° F. The water is carried by insulated pipes from the springs to large underground reservoirs. Hot spring water is mixed with cooled spring water from the reservoirs to reach the desired temperatures. Cooled water is also pumped to drinking fountains and jug fountains. Unlike the water of many natural hot springs, the spring water in Hot Springs National Park does not have an unpleasant taste or odor.

Through the years, the Park Service has acquired additional land in the region of the springs for the purpose of protecting the infiltration and recharge areas from pollution. In map view, the Park land forms a loop with a narrow corridor through the south side that connects the two parts of the city of Hot Springs. The mountains in the Park (between 1,000 and 1,400 feet in elevation) are part of the ZigZag Mountains, a range in the Ouachitas. Music Mountain, in the southwest corner of the Park, is the highest point (1,405 feet) and is at the center of a great horseshoe-shaped ridge, with West Mountain and Sugarloaf Mountain forming the roughly parallel extended

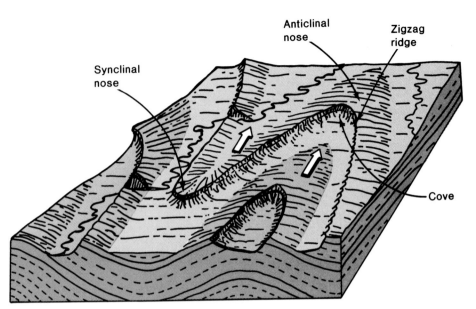

Synclinal nose

Anticlinal nose

Zigzag ridge

Cove

© WARD'S Natural Science Establishment, Inc. 1964.

Figure 37.3   Block diagram illustrating how zigzag ridges may develop due to differential erosion of plunging folds. Less resistant rock layers, usually shales and limestones, tend to erode more rapidly, forming valleys, while the more resistant beds, such as sandstones, form long ridges, called hogbacks. The arrows on top of the block indicate that both the open syncline and adjacent anticline are plunging toward the back of the block. This causes the resistant and less resistant sedimentary beds to develop a zigzag pattern as they are eroded. A rounded bend, or *nose*, forms where a ridge crosses the center (axis) of a fold from one side to the other. The term *cove*, in the sense used here, refers to a dead-end valley (a landform) partially enclosed by a hogback ridge. The folded structures in Hot Springs National Park are tighter and more complex than the folds shown here. © 1964 Ward's Natural Science Establishment, Inc.; used by permission.

"arms" (fig. 37.3). A few roads and many miles of bridle and hiking trails wind through the woodlands and rugged hill country of the Park.

Geologic Features

How Did the Hot Springs Originate?

Hot Springs National Park is in a unique geological setting. The origin and heat source of the hot springs have aroused scientific curiosity since Dunbar and Hunter first made their report to President Jefferson. Scientists (mainly geologists) have argued over whether the water is meteoric, magmatic, juvenile, or of mixed origin; that is, is rain the source of the water, or is it newly risen from magma deep in the earth? If the former is the case, how did the water get so hot? Has the water been stored for a long period of time at considerable depth? How and from where has it moved to its point of issue? Since the mid-1800s, numerous investigations have been made, and hypotheses both simple and complex have been proposed.

The modern explanation, which appears fairly conclusive, draws on several lines of evidence. The seasonal variations in flow—higher in wet seasons and lower in dry periods—suggest rainwater as the most likely source of supply for the hot springs. Reinforcing this idea is the fact that the chemical composition of the hot springs water is similar to that of the local ground water, cold spring water, and well water, after allowances are made for the higher content of dissolved silica contained in the hot water. (Hot water can hold greater concentrations of dissoved compounds than cold water.)

Bedinger and his co-investigators (1979) used mathematical models to test several conceptual models of the hot springs flow system. They concluded that the "geochemical data, flow measurements, and geologic structure of the region support the concept that virtually all the hot springs water is of local, meteoric origin." They further deduced that:

> Recharge to the hot springs artesian flow system is by infiltration in the outcrop areas of the Bigfork Chert and the Arkansas Novaculite (fig. 37.4 A,B). The water moves slowly to depth where it is heated by contact with rocks of high temperature. Highly permeable zones, related to jointing or faulting, collect the heated water in the aquifer and provide avenues for the water to travel rapidly to the surface. (p. C1)

The Paleozoic sedimentary rocks of the Park, which collect the rainwater and from which the hot springs emerge, are not only ancient but they also have been deformed and metamorphosed by episodes of intense folding, faulting, and thrusting. Many complex geologic

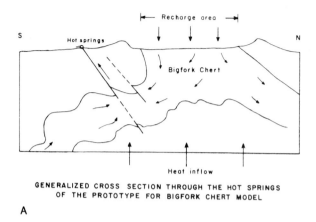

GENERALIZED CROSS SECTION THROUGH THE HOT SPRINGS
OF THE PROTOTYPE FOR BIGFORK CHERT MODEL

A

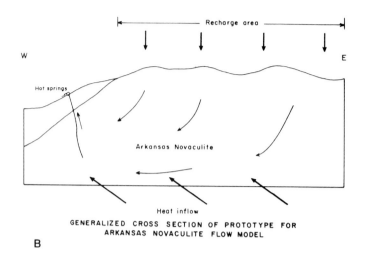

GENERALIZED CROSS SECTION OF PROTOTYPE FOR
ARKANSAS NOVACULITE FLOW MODEL

B

Figure 37.4   Generalized cross-sectional models illustrating directions of ground-water flow in Hot Springs National Park. From U.S. Geological Survey Professional Paper 1044-C, Bedinger et al. 1979. A. North-south cross section approximately through the center of the Bigfork Chert recharge area, about two miles away from the hot springs. Details of the folds and faults are not shown. B. East-west cross section of the Arkansas Novaculite recharge area, less than a mile from the hot springs.

structures have resulted. The ZigZag Mountains basically owe their topographic expression to the resistant ridges of the Arkansas Novaculite Formation, which crop out in a zigzag pattern because of the erosion of tightly compressed plunging folds. Weaker, less resistant rocks underlie the valley floors. Thus the landscape in Hot Springs National Park is a good example of control by geologic structure (fig. 37.5).

The main collectors of rainwater are the permeable ridge crests of novaculite (locally intensely fractured) and the Bigfork Chert, which forms lower ridges and rounded knobs. The shale beds below the permeable rocks tend to be impermeable. On Hot Springs Mountain, between the novaculite on the ridge and the Stanley Shale beds in the valley, the Hot Springs Sandstone crops out. This massive, quartzitic sandstone (which structurally is near the axis of a plunging anticline) contains large joints and fractures. The water of the hot springs emerges from the sandstone between the traces of two thrust faults that parallel the anticlinal axis. This structural situation accounts for the hot springs' alignment in a belt along the mountain slope.

Thus we can visualize the general hydrologic and geothermal flow system as follows:

1. In the Park's higher elevations among the ridges and higher valleys, rainwater infiltrates recharge areas on the Arkansas Novaculite and the Bigfork Chert, with the latter having the larger collecting area.

2. The cold ground water works its way very slowly downward to depths (probably more than a mile below the surface) where, with increasing geothermal gradient, the rocks are very hot.

3. As the cold water sinks, the heated water rises rapidly, following paths of least resistance—in this case the permeable fault zones that bring the water close to the surface.

4. From the fault zones, the water moves along joints in the Hot Springs Sandstone aquifer and, still very hot (143°F.), flows out from openings at the surface. Since the recharge area is at a higher elevation than the discharge area, the springs are artesian; that is, the water has built up a hydrostatic head due to artesian pressure and flows without pumping.

Estimates as to the age of the hot springs water and the travel time involved in the journey from the surface to depth and back to the surface have been derived from tritium and carbon-14 analyses. *Tritium* (H_3), a naturally occurring, radioactive isotope of hydrogen, is formed by cosmic rays in the upper atmosphere and enters the earth's hydrologic cycle via precipitation. Tritium's half-life is about 12-¼ years. *Tritium dating*, involving measuring the rate of radioactive decay in water samples, is useful for calculating the age of a sample and also for tracing subsurface movements and velocities of ground water. The carbon-14 dating of ground water is calculated from dissolved soil gas (CO_2) with corrective adjustments made for carbonates dissolved from rock through which the water travels. (For an explanation of radioactive dating, see p. 159.) The results from many carefully collected and analyzed water samples indicate that water in the hot springs is a mixture of water that is about 4,400 years old with a small amount of water less than 20 years old (Bedinger et al 1979). The inference is that movement of water down through the rocks takes much longer than its rise through the fault zones after being heated.

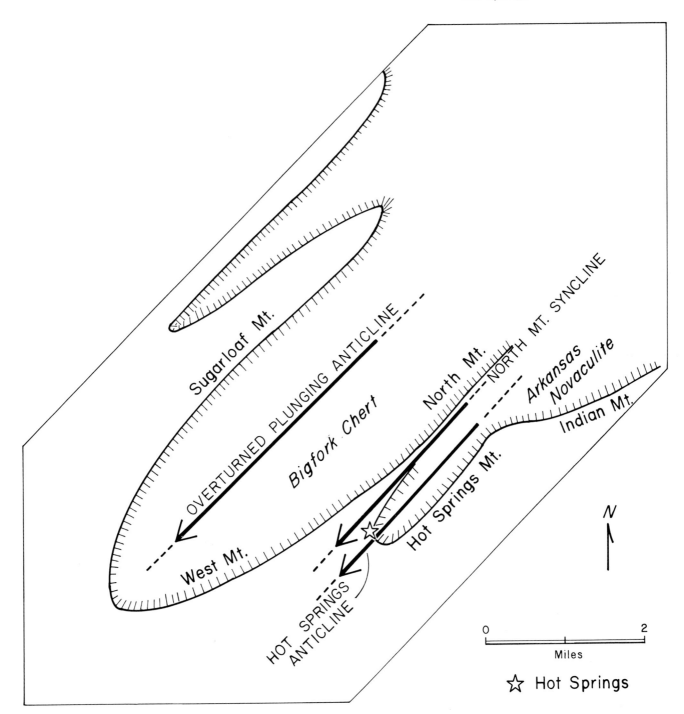

Figure 37.5   Sugarloaf Mountain, West Mountain, North Mountain, Hot Springs Mountain, and Indian Mountain, in and adjacent to Hot Springs National Park, are an example of zigzag topography produced by the erosion of plunging folds (fig. 37.3). The arrows show the northeast-southwest-trending folds plunging to the southwest. All of the folds shown here have overturned southeast limbs, which means that both limbs dip to the northwest. The folds are cut by reverse faults (not shown) that trend parallel to the folds and dip steeply to the northwest.

Water infiltrates into the recharge area of the Bigfork Chert in the middle of the anticline between Sugarloaf Mountain and West and North Mountains. After being heated at depth, the water rises along faults on the northwest limb of Hot Springs anticline, which plunges southwest (fig. 37.4A). Water also infiltrates into the outcrops of the Arkansas Novaculite in the center of the Hot Springs anticline, moves downward and westward, and then rises to the surface along faults on the northwest limb of the Hot Springs anticline (fig. 37.4B). Water from both recharge areas supplies the hot springs, with the greater volume coming from the Bigfork Chert. Simplified from geologic map in Bedinger et al. 1979.

Table 37.2
Chemical Analysis of Hot Springs Water*
(milligrams per liter)

Silica (SiO_2)	42.0	Bicarbonate (HCO_3)	165.0
Calcium (Ca)	45.0	Sulfate (SO_4)	8.0
Magnesium (Mg)	4.8	Chloride (Cl)	1.8
Sodium (Na)	4.0	Fluoride (F)	.2
Potassium (K)	1.5		
Dissolved oxygen (DO)			3.0
Free carbon dioxide (CO_2)			10.0
Radioactivity through radon gas emanation**			0.81
millimicrocurie per liter			

*Analysis by U.S. Geological Survey, 1972
**Analysis by University of Arkansas

As noted earlier, except for the higher temperature and hence higher content of dissolved silica, the hot springs water and the local cold ground water are very similar. The Arkansas hot springs are much less mineralized than the water in most hot springs elsewhere. What this means is that the rocks through which the Arkansas hot springs water travels have fewer mineral ions available for solution. The dissolved constituents are listed in table 37.2.

Minute amounts of radium and its radioactive by-product, radon gas, are present in the hot springs water and have been studied by a number of investigators. The source of the radium is not known, but its presence in trace amounts in the ground water of the region is not unusual. Most of the gas in the hot springs is carbon dioxide (CO_2) with smaller amounts of dissolved oxygen and nitrogen. Analyses of water samples from the different springs and from samples collected over a period of time have shown slight variations in dissolved constituents; but on the whole, the water has remained fairly uniform and consistent. Besides being odorless, the water is remarkably free of bacteria, viruses, or other impurities.

Rocks of Special Interest in the Park

The novaculite found in Hot Springs National Park and vicinity is an unusual and relatively uncommon rock. *Novaculite* is a dense, hard, even-textured, light-colored, siliceous sedimentary rock that resembles chert. Novaculite is cryptocrystalline (i.e., having crystals too small to be distinguished under a microscope), but, unlike chert, true novaculite is made up of more microcrystalline quartz than chalcedony. Novaculite was formerly thought to be only the result of primary deposition of silica; but in its type occurrence in the Lower Paleozoic rocks of the Ouachita Mountains (the region in which the Park is located), true novaculite appears to be a thermally metamorphosed, bedded chert (Keller et al. 1977). In other words, the parent rock, a bedded chert, was recrystallized by heat (and to a lesser extent by confining pressure) and thus altered to novaculite.

In some areas outside the Park, novaculite beds are quarried for use as whetstones, and some specimens that are richly colored are suitable for lapidary work. The Indians prized the rock for making arrowheads and stone tools. On Indian Mountain, in the northeast corner of the Park, are old quarries opened by the Indians; the largest of them is 150 feet in diameter and 25 feet deep. The Indians' method of breaking up and excavating this hard, dense rock was to build a fire to get the rock as hot as possible and then throw cold water onto it. This process fractured the rock into small pieces; but it was also dangerous sometimes when the cold dousing of hot rock caused an explosion that filled the air with sharp projectiles of novaculite.

The Hot Springs region is also noted for beautiful quartz crystals of many shapes and sizes. Some crystals are two or three feet long. Specimens must not be collected within Park boundaries, but may be purchased at mines and rockshops in the vicinity.

Tufa deposits surround the Display Springs and can be seen at other locations where spring water flowed out naturally in the past. Some of the tufa was precipitated by blue-green algae, the only form of life that can exist in water that hot.

Geologic History

1. Late Precambrian and early Paleozoic rifting.

According to plate tectonic theory, the North American plate was much farther south on the face of the globe in late Precambrian time than it is today. At that time, as a result of rifting, or breaking apart of plates, an ocean began to open between the North American plate and a continental plate to the south.

2. Ordovician to Mississippian deposition.

As the continents drifted apart and the ocean floor subsided, sediments accumulated very slowly in the abyssal depths. Fine silts, organic material, siliceous particles, and precipitated silica made up the sediment that eventually became shales, chert, and novaculite in a sedimentary sequence characteristic of "starved basin" deposition. Such a sequence implies that the basin remained stable for a long period of time with the volume of sediment gradually decreasing.

Beds in the sequence now exposed in or close to Hot Springs National Park are the Womble Shale, Bigfork Chert, and Polk Creek Shale of Ordovician age; the Silurian Blaylock Sandstone and Missouri Mountain Shale; and the Arkansas Novaculite Formation, which is Devonian and Lower Mississippian in age. The rocks are not highly fossiliferous, although a few graptolites (extinct colonial marine organisms) have been found in the Polk Creek Shale.

3. Closing of the ocean in Mississippian time.

The appearance of interbedded siliceous shales in the uppermost part of the Arkansas Novaculite Formation is an indication that the ocean was beginning to close. Some geologists believe that siliceous ash, blown from volcanoes on the south side of the narrowing ocean and deposited on the sea floor, was the principal source of sediment for these beds.

As the two continental plates converged, the sea-floor margin of the southward-moving North American plate plunged under the leading edge of the north-moving southern plate and was subducted. The Stanley Shale, a flysch formation, marks the beginning of this tectonic event. A *flysch* is a marine sedimentary facies that is the result of accelerated accumulation of sediment and rapidly changing conditions of deposition. Typically a flysch is deposited in a deep-water trough between approaching plates and consists of thinly bedded, graded deposits of marls and sandy shales interbedded with conglomerates, graywackes, and the like. Beds that show evidence of deposition by turbidity currents (i.e., turbidites) are characteristic.

The Hot Springs Sandstone, the lower member of the Stanley Shale Formation, contains novaculite pebbles in conglomerate layers and shows some interbedding of thin black shale. It is from the Hot Springs Sandstone outcrops on Hot Springs Mountain that the springs emerge.

The Hatton Tuff (of volcanic origin) and the dark shales and graywackes of the upper part of the Stanley Shale are less resistant beds that underlie the wide valleys in the Hot Springs region.

Deposition continued into Pennsylvanian time, with the beds becoming sandier as the ocean became shallower and the plates closed. These rock units are not exposed in the Park.

4. The Ouachita orogeny in late Pennsylvanian and Permian time.

Caught between colliding continental plates, the Paleozoic marine beds were severely deformed by intense folding, thrust-faulting, and uplifting. As the plates locked, some thrusting was to the north, and some was to the south. Plunging anticlines and synclines, some of them overturned, produced complex mountain structures as the whole region buckled and was pushed upward. It was this collision of plates that formed the Ouachita Mountains.

5. Continuing uplift and erosion during Mesozoic time.

The geologic history of the Ouachita Mountains, in post-orogenic time, is somewhat obscure. Most probably, isostatic adjustments were made as crustal blocks rose differentially in trying to regain equilibrium. Apparently a gradual uplift continued in the Ouachitas until early in the Tertiary. The local faults that control the flow of the hot springs water may have developed during this time. During Cretaceous time, intrusions, particularly of syenite, were emplaced in the Ouachita region.

Erosion was rapid as streams stripped off weaker beds, exposing the more resistant rock layers and revealing the fold patterns—as in the ZigZag Mountains.

6. Development of the hot springs hydrologic system.

The hot springs hydrologic system is probably older than the oldest water now carried in the system. The pluvial climate of the Pleistocene, when moisture was plentiful, may have brought about the development of the hot springs. Flow will probably continue as long as the regional climate is temperate and humid. Only a slight lessening of the average flow has been observed since monitoring of the system began. The main concern of Park Service officials is to protect the recharge areas, maintain the quality of the local ground water, and prevent contamination of the hot springs so that people can continue to enjoy and benefit from these unique thermal waters and the beautiful surroundings.

Geologic Maps and Cross Sections

American Association of Petroleum Geologists. 1966. *Geological highway map of the mid-continent region, map no. 1.* Tulsa, Oklahoma: American Association of Petroleum Geologists.

Bedinger, M. S.; Pearson, F. J., Jr.; Reed, J. E.; Sniegocki, R. T.; and Stone, C. G. 1979. *The waters of Hot Springs National Park, Arkansas—their nature and origin.* U.S. Geological Survey Professional Paper 1044-C.

Viele, G. W. 1979. *Geologic map and cross section, Eastern Ouachita Mountains, Arkansas.* Map and Chart Series MC-28F. Geological Society of America. (This cross section is drawn a short distance east of Hot Springs National Park and is helpful in showing the ancient plate margins and explaining the regional tectonic history.)

Bibliography

Bedinger, M. S., et al. 1979. *The waters of Hot Springs National Park, Arkansas—their nature and origin.* (see above).

———. 1974. *Valley of the vapors: Hot Springs National Park.* Philadelphia: Eastern National Park and Monument Association.

Hanor, J. S. 1980. *Fire in the folded rocks: geology of Hot Springs National Park.* Philadelphia: Eastern National Park and Monument Association.

Keller, W. D., Viele, G. W., and Johnson, C. H. 1977. Texture of Arkansas novaculite indicates thermally induced metamorphism. *Journal of sedimentary petrology* 47:834–843.

Stone, C. G.; Hale, B. R.; and Viele, G. W. 1973. *Guidebook to the geology of the Ouachita Mountains, Arkansas.* Arkansas Geological Commission.

Stone, C. G., ed. 1977. *Symposium on the geology of the Ouachita Mountains,* vols. 1, 2. Arkansas Geological Commission.

Viele, G. W. 1979. *Geologic map and cross section, Eastern Ouachita Mountains, Arkansas.* Map and Chart Series MC-28F. Abstract and Discussion. (Plate Margins Working Group, U.S. Geodynamics Committee.) Geological Society of America.

Address

Hot Springs National Park
P.O. Box 1219
Hot Springs National Park, Arkansas 71901

38

Petrified Forest National Park

by Sherwood D. Tuttle
University of Iowa

Location: East Central Arizona
Area: 93,492.57 acres; 146.08 square miles
Established: December 9, 1962

Figure 38.1 Agate Bridge in Petrified Forest National
Park. A ravine has been eroded beneath this great petrified
log, which is over 100 feet long. The National Park Service
has built supports underneath Agate Bridge to prevent its
collapse. National Park Service photograph.

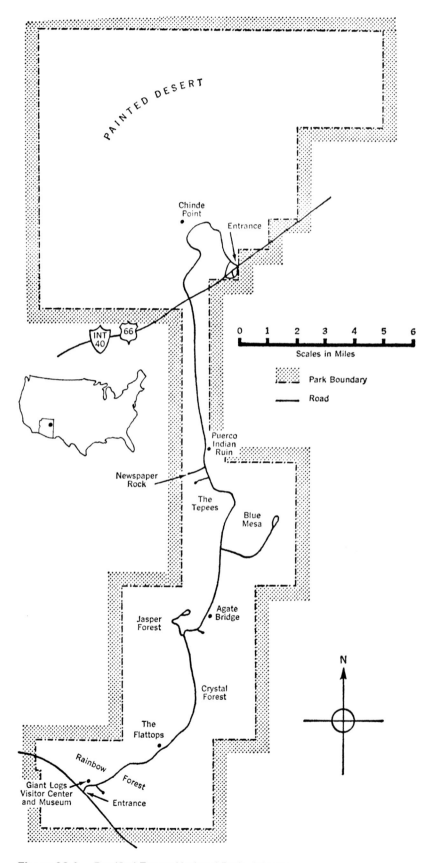

Figure 38.2   Petrified Forest National Park, Arizona.

Table 38.1. Geologic Column, Petrified Forest National Park

Era	Periods	Epochs	Formations	Members	Geologic Events
Cenozoic	Quaternary	Holocene / Pleistocene	Stream, wind, slopewash, and slump deposits		Fluctuating climate, becoming more arid / Weathering, erosion, mass wasting / Development of badland topography / Intensified erosion during pluvials
Cenozoic	Tertiary — Neogene	Pliocene	Bidahochi Formation		Coarse sediment interbedded with and topped by pyroclastics from explosive eruptions northwest of Park
Cenozoic	Tertiary — Neogene	Miocene	(hatched)		Extensive erosion
Cenozoic	Tertiary — Paleogene	Oligocene / Eocene	(hatched)		
Cenozoic	Tertiary — Paleogene	Paleocene			Upwarping (Defiance arch) / Laramide orogeny
	Cretaceous		(rock units no longer exposed in Park)		Extensive marine and nonmarine sedimentation
	Jurassic		(hatched)		Erosion
Mesozoic	Triassic		Chinle Formation	Owl Rock	Gypsiferous clays in playas or lagoons; ashfalls; deposition in interior basin, shales, sandstones
Mesozoic	Triassic		Chinle Formation	Upper Petrified Forest	(Largest accumulation of trees and plant debris in Sonsela)
Mesozoic	Triassic		Chinle Formation	Sonsela Sandstone	Extensive deposition in interior basin
Mesozoic	Triassic		Chinle Formation	Lower Petrified Forest	
Mesozoic	Triassic		Chinle Formation	Shinarump	Basal conglomerate containing Moenkopi fragments
Mesozoic	Triassic		Moenkopi Formation		Nonmarine and shallow marine sedimentation
Paleozoic			(hatched) (rock units not exposed in Park)		Erosion

Source: modified after Billingsley and Breed 1980

Geographic Setting

Petrified Forest National Park is located on the southern part of the Colorado Plateaus, about 120 miles (air distance) southeast of Grand Canyon National Park. The geologic history of Petrified Forest is similar to that of the other national parks of the Colorado Plateaus (chapters 1 through 7), and erosional landforms characteristic of arid regions can be seen in this Park as well as in other parts of the Southwest. However, Petrified Forest National Park has been included in Part V (rather than Part I) because the petrified wood—which the Park was established to protect—is mainly the result of ground-water processes. Scattered over the desert surface are thousands of jewellike logs, broken branches, segments of tree trunks, and sparkling wood chips. All have been changed from wood to stone. Here is an array that, in variety and profusion, is unsurpassed anywhere in the world.

The petrified trees have been eroded from a sequence of Triassic rocks that are a part of the vast Painted Desert of Arizona. Although the American Southwest has many areas of brightly colored rocks referred to as "painted deserts," the geographic entity officially designated as the Painted Desert curves across Arizona from Lees Ferry in Grand Canyon southwestward to Holbrook, then east across Petrified Forest National Park, and on to the New Mexico border.

The Little Colorado River, which, in a general way, skirts along the southern edge of the Painted Desert, flows northwest to join the Colorado River, collecting most of the drainage of the area as it goes. The intermittent Puerco River, which cuts across the middle of the Park in a wide, flat-bottomed arroyo, meets the Little Colorado west of the Park at Holbrook.

Evidence of stream erosion by the Puerco and its tributaries is everywhere about the landscape of the Park, but water is seldom seen on the surface. The arid plateau on which the Park is located (elevation roughly 5,500 feet) is smooth in some places, but elsewhere the landscape is intricately dissected. Mesas, buttes, gullies, arroyos, sinkholes, and other features show that even in the desert, water is the dominant agent of erosion.

The desert climate is variable with wide fluctuations in rainfall and sudden changes in temperature. Mean annual rainfall is 9 to 10 inches, most of which comes in brief, violent thunderstorms during the summer months. Annual temperatures may range 127 degrees Fahrenheit between the highs and lows! Occasionally snow falls in the winter, but does not long remain on the ground. In times of drought, many months may pass without any precipitation. In this harsh climate, rocks are well exposed, but vegetation is sparse. Sagebrush, greasewood, cactus, yucca, and the like are some of the hardy plants that grow here.

Local History

Since prehistoric times, a well traveled, east-west route has followed the valley of the Puerco River through this part of Arizona. For the Indians, the valley was on a trade route that went to Baja California and the Pacific Coast. Explorers and pioneers used the trail. A transcontinental railroad followed the same route, and later a transcontinental highway, Route 66, was built. Today Interstate 40, on its way across the country, goes through the Park, providing convenient access to its features.

Ancient Inhabitants. The oldest ruins in the Park are on The Flattops, an area of sandstone-capped mesas near the southern entrance. Ancestors of the Pueblo people probably lived here in a village of pit houses, which are partially dug into the earth and lined with sandstone slabs. The inhabitants may have raised crops on the flat mesa tops, probably more than 1,500 years ago, when the climate was less arid than now. Nearby (south of the Park Museum) is Agate House, a partially restored pueblo built entirely of petrified wood.

A village at Twin Buttes (west of the Park) was occupied by people of a later, less primitive culture. Pit houses were excavated more deeply and had more rooms and storage chambers. The inhabitants built long, low structures of stone that may have served as windbreaks to protect crops and reduce the movement of sand over the land surface. The Hopi Indians of today still build such walls. Pottery and tools found at the site are more advanced and include artifacts made from Pacific seashells and other materials foreign to this region that were probably received in trade for articles made from petrified wood.

Droughts that began early in the 13th century brought about changes in the Indians' way of life. They began building their villages on the flood plains where the water supply was more dependable. The Puerco Indian Ruin, which may have had a population of more than 200 people at its peak, is a large, rectangular pueblo of uncemented masonry construction on a terrace overlooking the Puerco River. About 150 rooms, many of them interconnected, are laid out around a large courtyard. The Indians grew crops on the level terrace tops.

Around the beginning of the 14th century, precipitation increased, but instead of being evenly distributed throughout the growing season, the rain came in heavy bursts and thunderstorms. Time and time again, the river flooded, washing out fields, eroding topsoil, and ruining crops. Finally the pueblo people (who still farm and live in villages) could no longer maintain their preferred way of life. According to Hopi tradition, their ancestors moved from the Puerco region to the northwest about 80 miles and built new pueblos in what is now the Hopi Indian Reservation.

Exploration and Discovery. When Spanish explorers came through the Puerco region in the 16th century, they found no Indian farmers in the valley although Navajo nomads roamed the area. The Spaniards did not mention the petrified wood in their accounts. Perhaps they were too absorbed in their search for gold and silver to take notice of the gemstones contained in the fossilized wood.

The first official report of the petrified wood was made by Captain Lorenzo Sitgreaves, who led an expedition down the Zuni and Colorado Rivers in 1851. He told of finding large amounts of petrified wood a few miles southeast of the Park area near the junction of the Little Colorado and Zuni Rivers. Five years later, Amiel Whipple, while surveying a route for the transcontinental railroad, found petrified wood in such abundance that he named a tributary of the Puerco River Lithodendron ("stone-tree") Creek. Later geologic investigations throughout the Colorado Plateau revealed that fossilized tree remains are not uncommon in Mesozoic formations. However, nowhere else are they found so well preserved and in such abundance as in Petrified Forest National Park.

Interest in the petrified wood continued to grow. General W. T. Sherman, while on a tour of the western territories, was so impressed by what he saw that he asked that specimens be sent to the Smithsonian Institution in Washington, D.C. An Army detail from Fort Wingate, New Mexico, set out in the spring of 1879 to accomplish the task. Near Bear Springs at the head of Lithodendron Wash, they found huge buried logs that seemed suitable. They dug up two large segments and transported them by wagon back to Fort Wingate. From there, the larger piece was shipped by rail to the Smithsonian, where it remains on display to this day.

Development of the Park. When the transcontinental railroad was extended through Arizona, the railroad company constructed a coaling and water stop in the Puerco River Valley that became the town of Adamana. Train passengers were encouraged to take excursions to "Chalcedony Park," as the area was then called, to see the "petrified forests." Word spread, and from then on, souvenir hunters, gem collectors, and commercial jewelers began hauling away quantities of petrified wood, quartz and amethyst crystals, and Indian relics. Some collectors,

Box 38.1
Silica Can Assume Many Forms

Inasmuch as silicon (Si) and oxygen (O) are the two most abundant chemical elements in the rocks of the earth's crust, it is not surprising that naturally occurring silica—the chemical compound silicon dioxide (SiO_2)—is found in many varied forms, both crystalline and noncrystalline. The silicified logs in Petrified Forest National Park display many interesting examples of this mineral. The variations in color of both crystalline and noncrystalline silica are caused by minute amounts of other compounds that are precipitated with the silica. In the petrified logs the color tones of yellow, red, and brown were produced by iron in the ground water. Copper added tints of blue and green; manganese or carbon blended black in; and so on.

All of the crystalline varieties of precipitated silica are quartz, but they have different names according to their color. The most common are rock crystal (clear), amethyst (lavender or purple), smoky quartz (gray to black), milky quartz (white or cream), and citrine (yellow). If the mineral can

precipitate, or grow, in an unoccupied space, it may develop its regular, distinct crystal form as a six-sided pyramid or prism. If the quartz grows in very small or irregular spaces, it becomes crystalline grains.

When the internal crystalline arrangement of the quartz is so irregular that it has no characteristic external form, or when its internal structure cannot be detected by a petrographic microscope, the specimen is described as amorphous or *cryptocrystalline*. (Extremely fine crystallinity may, however, be demonstrated by the use of an electron microscope.) Such varieties include: *chalcedony*—white, pale blue, gray, brown, or black with a waxlike luster and uniform tint; *jasper*—red chalcedony or chert (a sedimentary rock composed primarily of silica); *carnelian*—a translucent red or orange-red variety of chalcedony; *agate*—a banded or variegated chalcedony, characterized by colors arranged in alternating stripes or bands but occurring in other forms as well.

looking for crystals, used dynamite to break up the logs. Fossil wood was carted away by the wagonload. When a stamp mill was erected at Adamana to crush the logs for the manufacture of abrasives, the local people became alarmed and intensified their efforts to preserve the petrified forests. The Arizona territorial legislature petitioned the United State Congress to set aside Chalcedony Park as a national preserve. Lester Ward, a paleobotanist of the U.S. Geological Survey, investigated the area in 1899 and recommended that the land be withdrawn from homesteading and placed under government protection. In 1906, after Congress passed the Antiquities Act, President Theodore Roosevelt declared the area a national monument. In 1962, after several additions, including a part of the Painted Desert, the monument was redesignated Petrified Forest National Park. A wilderness area of 50,260 acres (where camping is permitted) was placed under Park Service jurisdiction in 1970. Overnight and camping facilities are not available in the Park. *Removal of petrified wood or any unauthorized materials from Petrified Forest National Park is prohibited.*

From the north-south highway through the Park, the petrified forests and other geologic features, as well as Indian ruins, are easily accessible (fig. 38.2). Several overlooks in the northern part of the Park provide sweeping views of the Painted Desert.

Geologic Features

The Petrified Wood

In Petrified Forest National Park, logs and chunks of all sizes lie scattered on the ground or stick out of the soft, shaly bedrock. Examination of the logs shows that many of them are just bare trunks without roots, limbs, or twigs. Much of the bark is scarred and abraded. The silica that infiltrated the wood made the logs very heavy as well as brittle. Some are split lengthwise and some are broken clean across, as if sawed. Probably some logs were broken after petrification—during uplift, possibly by earthquakes, or perhaps when tumbling down slopes as they were uncovered and loosend by erosion (fig. 38.3).

Figure 38.3 Differential weathering and erosion have uncovered segments of petrified logs, which are much more resistant to erosion than the beds of the Triassic Chinle Formation that were removed. The absence of vegetation, except for a few dry clumps of grass, is characteristic of the semiarid climate now prevailing at Petrified Forest National Park. From *Geology of Arizona* by D. Nations and E. Stump. © 1981 Kendall/Hunt Publishing Company.

Box 38.2
How Does Wood Turn to Stone?

Like a number of other geochemical processes, petrifaction is not completely understood. How helpful it would be if wood could be silicified and the process studied in a laboratory under controlled conditions! So far no one has successfully accomplished this, although some investigators have tried. A human lifetime is too short for the completion of experiments to duplicate processes that operate on geologic time in nature.

The logs in Petrified Forest show that the processes of petrifaction are infinitely varied. In some logs the internal woody structure has been preserved with amazing fidelity. Others are *pseudomorphs*; that is, they are recognizable as logs, but the internal woody structure was not retained in the precipitated jasper, chalcedony, or agate. The wood has been entirely replaced by silica. Moreover, in some logs both replacement of woody tissue by silica and preservation of cellular structure have occurred in close proximity.

The process by which the details of cellular structure are preserved is called *permineralization*; it is a type of fossilization whereby mineral matter is deposited within and around the cells or pores without distorting or destroying them. Wood that has been silicified in this way can be immersed in hydrofluoric acid so that the silica is chemically removed, leaving a soft mass of woody tissue from which slides can be made for microscopic study, as with any modern wood.

What probably helps to make this process of fossilization possible under certain natural conditions is the fact that inorganic molecules are much smaller than organic molecules. In other words, the preservation of wood structure in faithful detail by petrifaction is not accomplished by replacement of wood cells by silica ("molecule for molecule," as was once thought to be the case), because there is considerable discrepancy in size and geometry between organic and inorganic molecules. What does happen is that the woody components are penetrated and surrounded by "microcrystalline mineral matter often without visible disturbance of either the tissue or the interpenetrating microscopic crystal patterns" (Schopf 1971, p. 523).

Natural conditions under which permineralization of wood may take place involve ample moisture as well as time. In the swampy lowlands in which the wood was deposited, the water table was close to the surface. Ground water circulated very slowly and the plant material was constantly saturated. Then quantities of fine volcanic ash (from volcanoes to the west) settled over the land and more was washed down from surrounding highlands. (The ash was the main source of the silica.) Ground water, somewhat acidic from decaying plants, began dissolving silica and traces of other mineral matter from the ash. Thus enriched, the ground water solutions gradually permeated the woody tissue, infiltrating the cells, filling the intercellular cavities, and coating the cell walls. We can only speculate as to what chemical or environmental conditions caused some of the precipitating silica to *replace* cells and some to *preserve* cells. We cannot know whether the processes operated concurrently, successively, or what.

The well developed crystals of quartz and amethyst (some of gem quality) that are found in some of the petrified logs grew in hollows or open spaces in the wood that probably existed in the original tree or may have formed soon after deposition in the swamp.

Although plant fossils are by far the most abundant in the Park, teeth, scales, bones, and other remains of extinct species of fish, clams, amphibians, and reptiles record the presence of animal life in the Triassic swamps. Many of these fossils were also petrified by permineralization or pseudomorphic replacement.

However, most of the trees appear to have been up-rooted, transported, tumbled, and swirled about by raging streams and then left in a jumble of woody trash and sediment, probably some distance from where they grew originally. To account for so much wood being washed out, transported, dumped, and then buried, we have to visualize large rivers that during times of high water spread over flood plains and deltas. Perhaps the climate was monsoonal, and winds brought in severe tropical typhoons and seasonal downpours from the coast to the west.

About 40 different species of fossil plants (most of them extinct) have been identified in the Chinle Formation, the principal rock unit exposed in the Park. In addition to the petrified logs, careful searches have uncovered roots, leaves, seeds, and other plant debris preserved in fine-grained shale beds. Included in the Triassic flora are rushes and horsetails, large and small ferns, and large fernlike cycads that are seed-bearing. Such plants thrive on warm, swampy flood plains and in marshes.

Most of the logs are conifers belonging to a group named *Araucarioxylon*. Distant living relatives of these conifers are the monkey puzzle tree (native to Chile) and the Norfolk Island pine (native to Norfolk Island in the South Pacific). Some of the petrified trees were more than 200 feet tall and 10 feet in diameter (Ash and May 1969). Evidently the conifers grew on higher ground or slopes upstream and were brought down and buried in the lowlands by sediment-laden flood waters. Some trees were healthy specimens, perhaps growing on riverbanks undercut during high water; but other trees were dead or dying of natural causes when they toppled down or were blown down. Evidences of insect borings, fungus rot, and charring by forest fires have all been preserved in the petrified remains of the fossil trees. Burial was rapid once the logs came to rest. Not many logs rotted before being covered. The water-saturated muds and sands sealed out air and prevented decomposition.

Petroglyphs and Desert Varnish

Carving petroglyphs in desert varnish on rocks seems to have been a favorite pastime of the prehistoric people who lived in the American Southwest, and there are thousands of their mysterious inscriptions in Petrified Forest National Park. Some of the petroglyphs are drawings of human and animal figures, and others are abstract, geometric patterns, similar to the designs on Indian pottery. Newspaper Rock, near the Puerco Indian Ruin, is a massive sandstone block with many interesting examples of Indian rock art on its surface. The meaning of the petroglyphs is not clear, although some of them are clan symbols and many probably have ceremonial significance.

Archeologists and anthropologists who have studied the rock art in the Park assume that the drawings were done at different times and by individuals from different cultures, but no satisfactory method of dating the petroglyphs has been discovered. The Indians made the drawings by using a sharp tool, probably a piece of petrified wood, to scratch or peck through the dark desert varnish that coated the sandstone, thus exposing the fresh, light-colored rock underneath.

As sandstone weathers in the desert climate, a thin, dark brownish-red or black patina, or *desert varnish*, develops on the rock surface. The varnish consists of iron and manganese oxides with some silica and is probably caused by the repeated evaporation of water drops from the rock surface. As the moisture dries, dissolved mineral matter exuded from the rock is left. Some authorities believe that acid-secreting lichens help to fix the varnish coating to the rock. Blowing sand polishes the varnish, giving it a gloss. Because the rate at which desert varnish accumulates is highly variable, it cannot be used as tool for dating the inscriptions.

Caliche and Other Desert Features

In addition to the desert varnish, other features typical of arid lands are common in the Park. Tubes and nodules of caliche are exposed in ravines and on the flanks of mesas. *Caliche* is a general term for concentrations of soluble calcium salts (such as calcium carbonate) that precipitate from soil moisture at or near the surface. Under arid conditions caliche accumulates because rainfall is insufficient to wash calcium carbonate away or dissolve it. Sometimes caliche forms an impermeable "hardpan" layer beneath desert soil, as well as nodules and tubes. The latter represent roots that have been coated and eventually replaced.

Desert pavement tends to form on flat surfaces, such as on the flood plain of the Puerco River. Wind has removed soil and fine particles and left a closely packed surface of wind-polished pebbles and rock fragments, usually somewhat cemented together by calcium carbonate. Loose rocks, many of them fragments of petrified wood, have been so abraded by wind-blown sand that they have smooth, polished faces and sharp edges. These are *ventifacts*.

Badland Topography and the Painted Desert

To the north along Interstate 40 in this part of Arizona is the panorama of the Painted Desert—a landscape of brilliant, improbable colors against a deep blue sky. The multi-colored rocks of the Chinle Formation are exposed in the myriad shapes of badlands—gullies, ridges, arroyos, washes, mounds, escarpments, and mesas. Like the colors of Badlands National Park in South Dakota (chapter 8)—but much more brilliant—the hues of the Painted Desert change with the hours of the day. Colors are most brilliant in the morning and late afternoon, least intense at midday. Colors change when the surface is wet, and on a partly cloudy day shaded areas and sunlit slopes contrast beautifully. Thus the natural variations in daylight act on the sedimentary rocks to produce the spectacular chromatic effects. The colors of the various shales, marls, sandstones, and limestones of the Chinle are purple, pink, gray, yellow, ash gray, lavender, rose, maroon, sienna, lilac, cream, chocolate, and various other shades of red and brown. Patches of blue, white, and even black are seen. Dominating overall are the reds, pinks, and purples.

Although rainfall is low in this arid land, the relatively nonresistant Triassic rocks are still being eroded, and landforms change with every cloudburst and thunderstorm. Much of the badlands topography, however, was formed when rainfall was somewhat more plentiful than now. Interbedded with the thin layers of poorly cemented layers of gravel, mud, and sand are water-laid units of volcanic ash. Weathering has altered the ash to *bentonite*, a type of soft clay that absorbs water and swells as it becomes wet. When saturated, the clay slumps or may even flow as a fine muck. After it dries, the clay hardens again. Local slopes that develop on the bentonite clay units are rounded and bulbous because of the clay's plasticity (fig. 38.4). *Piping* causes sinkholes to develop in tops of mounds and in gully bottoms. During downpours, water drains through tunnels, or "pipes," inside the mounds and the gullies, washing out loose material and undermining slopes that later collapse as sinks and slumps. New slope profiles and sinkholes may develop in a few hours when flash floods occur.

Figure 38.4 Badland topography typical of the Painted Desert in Petrified Forest National Park. Bulbous slopes and slumps have been eroded in the bentonite clay and shale beds of the brightly colored Chinle Formation. On the ridge in the foreground, the remnant of a more resistant sandstone bed has provided some protection from erosion to the underlying clay and shale. Drawing by Genevieve Shimer from *Field Guide to Landforms in the United States* by J. A. Shimer. The Macmillan Company. © 1972 by J. A. Shimer; used by permission.

Steeper slopes and escarpments develop on the sandstone and conglomerate beds of the Chinle. Some of the more resistant units form caprocks on the buttes and mesas of the landscape.

Geologic History

1. Subsurface Paleozoic rocks.

Paleozoic rock units that are well exposed in many other parts of the Colorado Plateau are not seen in the region of Petrified Forest National Park. Some of the Paleozoic rocks were probably removed by erosion; the rest are buried by younger rocks.

2. Deposition of the Moenkopi Formation in early Triassic time.

At the beginning of Triassic time, the landmass that was to become North America was still part of the supercontinent Pangea, and Arizona was much closer to the equator than it is now. On a low coastal plain that sloped gently westward across northern Arizona toward a geosynclinal trough in Nevada, sandy and muddy beds accumulated, 300 to 600 feet in thickness. Marine and calcareous units interfingered with the continental beds, indicating that at times the sea encroached on the coastal plain. In some places gypsiferous beds formed in shallow lagoons between layers of mud. Sand-filled channels trace the courses of ancient streams. The climate was warm but arid much of the time.

3. Erosion of the Moenkopi rocks.

Only a few beds of the Moenkopi Formation remain in the Petrified Forest area. Most of them were eroded as the sea withdrew and the ancient plain was upwarped. Gradually erosion lowered the relief of the land, leaving extensive coastal plains. Sediment-laden rivers built broad flood plains and complex deltas.

4. Deposition of the Shinarump Member of the Chinle Formation.

In a subsiding basin in north-central and northeastern Arizona, a typical basal conglomerate, containing many Moenkopi pebbles, was deposited unconformably over the eroded Moenkopi Formation. The gravelly, sandy, lower layers are indicative of high source areas and, presumably, streams able to transport the clastic material. The coarse sediment grades upward into finer material. This conglomerate, averaging 35 to 50 feet in thickness, is the Shinarump Member of the Chinle Formation. (In some parts of the Southwest, the Shinarump is a separate formation.)

5. Deposition of Chinle beds during late Triassic time.

The sea that lay to the west continued to regress, leaving an immense alluvial plain in the Colorado Plateaus area. Sediment transported by the rivers became finer. The climate, still warm, became more humid, and savannahs and marshes developed. Aprons of sand, mud, and silt were spread over the northern Arizona lowlands, covering old flood plains, deltas, and backswamp areas. Localized units of gypsiferous clay represent evaporite deposits in lagoons or playa lakes. Perhaps there were seasonal wet and dry periods in the Triassic.

Several hundred feet of sandy, shaly material accumulated during this time to form the Lower Petrified Forest Member and the Upper Petrified Forest Member of the Chinle Formation. These two members are separated in some places by a discontinuous unit, the Sonsela Sandstone Member, which contains more petrified wood than any of the other rock units in the Park. The Owl Rock Member at the top of the section completes the Chinle Formation in Petrified Forest National Park (Billingsley and Breed 1980).

Tectonic activities toward the end of the Triassic began to affect the northern Arizona basin. In the western sea a chain of volcanoes (about where the California-Nevada boundary is now) began erupting large quantities of ash that were carried eastward by winds and dropped on the plains, basins, and highlands of the interior. Heavy rains washed a large quantity of the ash, mixed with mud and other debris, from nearby highlands into the Petrified Forest area. The volcanic ash is believed to be the source of the silica that was deposited in the logs by the ground water.

To the south an ocean trough (the Tethyan Sea) began opening, splitting apart North and South America. Freed from Pangea, North America began a slow drift toward the northwest. From this event, some 200 million years ago, global geography, as we know it today, began to develop.

6. The wood in the Chinle Formation.

The Chinle Formation is famous for its petrified wood. Although the finest and most abundant fossil specimens are in the Petrified Forest area, the same species of conifers have also been found in Chinle beds in eastern Utah and other places where the formation crops out. Obviously the forests were widespread, covering the low hills and uplands of the region. A late Triassic highland extending from central Colorado to northern New Mexico may have been the source of the large conifers washed into the northern Arizona basin. It was there in the swampy lowlands that the ferns and cycads flourished. The fact that their fossil remains are found together, showing little evidence of decay, suggests that the uprooting, transportation, and swamp burial of the conifers occurred in relatively short spans of time, possibly during monsoonal or seasonal floods of considerable magnitude.

7. Silicification of the wood.

Just when and for how long the processes of permineralization and petrifaction went on is not known. Perhaps the elevation of the Colorado Plateau region,

beginning at the end of the Mesozoic Era, terminated the silicifying of the buried wood.

8. Accumulation of Jurassic and Cretaceous sediments.

Several thousand feet of Jurassic and Cretaceous sediments were deposited on top of the Chinle Formation in the Colorado Plateaus region; but in the Petrified Forest area, these rock units were subsequently eroded. The weight of these rock units (although they were later removed) helped to compact the Chinle Formation, and ground water percolating through them probably still affected the trees in the Chinle.

During the last of the Mesozoic and the beginning of the Cenozoic Eras, pulses of tectonic activity, shifting from west to east, culminated in the Laramide phase of the Rocky Mountain orogeny.

9. Tertiary uplift, warping, and erosion.

The Colorado Plateaus (as was described in Chapters 1 through 7) were uplifted and slightly tilted during the great mountain-building episodes of late Cretaceous and early Tertiary time, but the flat-lying rocks remained relatively undeformed. Around the margins of the plateau, however, some broad folding and upwarping occurred. In northeastern Arizona, the upwarping (Defiance uplift) that took place resulted in the shaping of the Park landscape that we see today. Erosion stripped off all the Jurassic and Cretaceous rocks and some of the Chinle beds from the Petrified Forest region. (Some of these units are preserved in downwarped basins only about 50 miles away from the Park.) The valleys of the Little Colorado River and its tributary, the Puerco, that goes through the Park are cut mostly in the soft Triassic shales of the Chinle Formation.

Meanwhile, as the Mesozoic lowlands were slowly elevated, the climate was becoming cooler and increasingly arid. The mountain barriers rising to the west cut off moisture-laden winds from the Pacific.

10. Late Cenozoic sedimentation and volcanism.

In Pliocene time, an irregular blanket (in places several hundred feet thick) of coarse sediment and volcanic debris was spread unconformably across the eroded surface of Mesozoic rocks in this part of Arizona. This is the Bidahochi Formation, but only a few remnants of it are left in the Park, mainly as cap rock on mesas and divides in the Painted Desert. Overlying the Bidahochi sediments (and interbedded with them) are units of basaltic debris derived from volcanic vents northwest of the Park. Large *volcanic bombs* (ellipsoidal fragments that were ejected in a molten state and then hardened in the air) can be seen in roadcuts in the northern part of the Park. The bombs, as well as large angular boulders of basalt in the formation, show that the eruptions were nearby and highly explosive.

11. Pleistocene and Holocene erosion.

During the waxing and waning of the great Pleistocene ice sheets to the north and northeast, regions such as Petrified Forest, outside the glaciated areas, had intervals of *pluvial climate*, that is, periods of high precipitation. In general, the pluvials correspond to transitional or glacial periods in the glaciated regions. Erosion accelerated during the pluvial intervals, and probably more rain fell throughout the year instead of being largely concentrated in summer as at the present time. The nearly complete removal of the Bidahochi sediment and volcanic debris indicates that the intense erosion of Tertiary time continued through the Pleistocene but that erosion is somewhat slower today due to the increasing aridity of the region.

12. Changes in the regimen of the Puerco River early in the 14th century.

In the arid lands of the Southwest, slight changes in weather patterns and annual precipitation that become climatic trends or persist more than a few years have profound effects on plant and animal life and on human habitation. During the long 13th-century droughts, the prehistoric people survived by abandoning their fields on the mesa tops and tilling new areas on the flood plain terraces of the Puerco River. But when rainfall increased again in the early 1300s, the added precipitation came in the form of cloudbursts and heavy summer thunderstorms that caused the river to rise out of its banks and wash out terraces, covering the precious soil of the flood plain with coarse gravels. Because the Indians could not (or would not) adapt to this change in the environment, they moved away. Since the climate in the Petrified Forest region has tended to become more, rather than less, arid since that time, their decision appears to have been wise.

Geologic Maps and Cross Sections

American Association Of Petroleum Geologists. 1967. *Geological highway map of the southern Rocky Mountain region, map no. 2.* Tulsa, Oklahoma: American Association of Petroleum Geologists.

Billingsley, G. H., and Breed, W. J. 1980. *Geologic cross section along Interstate 40 from Kingman to the New Mexico-Arizona state line.* Petrified Forest National Park, Arizona: Petrified Forest Museum Association.

Bibliography

Ash, S. R.; and May, D. D. 1969. *Petrified Forest, the story behind the scenery.* Petrified Forest National Park, Arizona: Petrified Forest Museum Association.

Bezy, J. W.; and Trevena, A. S. 1975. *Guide to twenty geological features at Petrified Forest National Park.* Petrified Forest National Park, Arizona: Petrified Forest Museum Association.

Gregory, H. E. 1917. *Geology of the Navajo country.* U.S. Geological Survey Professional Paper 93.

Mears, B., Jr. 1968. Piping. *In Encyclopedia of geomorphology*, ed. R. W. Fairbridge, p. 849. New York: Reinhold Book Corporation.

Nations, D. and Stump, E. 1981. *Geology of Arizona.* Dubuque, Iowa: Kendall/Hunt Publishing Company.

Petersen, M. S.; Rigby, J. K.; and Hintze, L. F. 1980. *Historical geology of North America.* 2nd ed. Dubuque, Iowa: Wm. C. Brown Company Publishers.

Rigby, J. K. 1977. *Southern Colorado Plateau*, K/H Geology Field Guide Series. Dubuque, Iowa: Kendall/Hunt Publishing Company.

Schopf, J. M. 1971. Notes on plant tissue preservation and mineralization in a Permian deposit of peat from Antarctica. *American journal of science* 271:522–543 (December).

———. 1975. Modes of fossil preservation. *Review of paleobotany and palynology* 20:27–53.

Wagoner, J. J. 1971. *This is Painted Desert.* Petrified Forest National Park, Arizona: Petrified Forest Museum Association.

Wilson, E. D. 1962. *A resume of the geology of Arizona.* Arizona Bureau of Mines Bulletin 171.

Address

Petrified Forest National Park
Arizona 86028

39

Carlsbad Caverns National Park

by Arthur N. Palmer
State University of New York, Oneonta

Location: Southeast New Mexico
Area: 46,755.33 acres; 73.05 square miles
Established: May 14, 1930

Figure 39.1 The semiarid plateau in which Carlsbad Caverns National Park is located consists of reef and back-reef limestones deposited around the perimeter of the Delaware Basin during the Permian Period. Walnut Canyon, shown here, is one of the deeply entrenched valleys of Pleistocene age that dissect the plateau. The prominent bedding of the back-reef limestones is visible in the far wall of the canyon. National Park Service photograph.

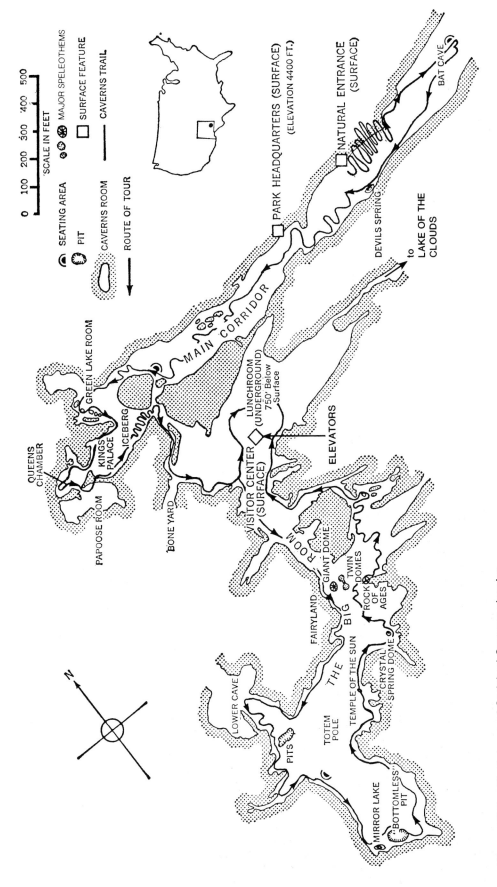

Figure 39.2 Partial map of Carlsbad Caverns showing underground areas open to the public.

475

Table 39.1. Geologic Column, Carlsbad Caverns National Park

Time Units — Era	Period	Epoch	Rock Units	Geologic Events
Cenozoic	Quaternary	Holocene	surficial gravels	Drying of climate; deposition of cave popcorn
		Pleistocene		Deposition of most speleothems in cave
				Draining of cave
				Erosion of Castile Formation exposing reef front; erosional dissection of limestones
			surficial gravels?	Renewed uplift, faulting, tilting; major solutional development of Carlsbad Caverns; deposition of gypsum in cave
	Tertiary (Neogene)	Pliocene		
		Miocene		Slow erosion
	Tertiary (Paleogene)	Oligocene		Early stages of origin of Carlsbad Caverns
		Eocene		
		Paleocene	Surficial gravel and sand?	Laramide orogeny—uplift, faulting, tilting toward northeast
Mesozoic	Cretaceous			
	Jurassic			Minor erosion, low relief
	Triassic			
Paleozoic	Permian	Ochoan	Rustler Formation	Thin, widespread, terrigenous deposits
			Salado Formation	
			SE: Castile Formation	Deposition of evaporites in Delaware Basin
		Guadalupian	NW: Artesia Group — Tansill Fm., Yates Fm., Seven Rivers Fm., Queen Fm., Grayburg Fm.; Capitan Limestone (reef); SE: Bell Canyon Formation	Possible first stage of karst development
			Goat Seep Dolomite; SE: Cherry Canyon Formation	Limestone deposited in subsiding Delaware Basin

Source: modified after Hayes 1957; Hayes and Koogle 1958; Jagnow 1979; and Queen 1981

Regional Setting

Carlsbad Caverns National Park is located in the south-eastern corner of New Mexico along the prominent southeast-facing Guadalupe Escarpment, which extends northeast with diminishing height from Guadalupe Mountains National Park (chapter 33). The two parks share nearly the same geologic setting of massive reef and bedded back-reef limestones of Permian age. This is a region of deeply dissected plateaus, with an average altitude of 4,000 to 6,000 feet, overlooking the gypsum plain of the Delaware Basin 1,000 feet below.

Carlsbad Caverns, the major cave in the Park, matches perfectly the popular image of how a cave ought to look: huge, interconnected rooms with shadowy, vaulted openings leading off in every direction; massive and profuse decorations of stalactites and stalagmites; prodigious length and depth. The cave's largest room alone, the Big Room, is one of the world's largest. It covers 14 acres, and its maximum floor-to-ceiling height is 370 feet at the so-called "Bottomless Pit." In the past geologists treated the cave as a grand example of how limestone caves in general must form. It is surprising, therefore, to learn that in its geology and origin Carlsbad Caverns is actually quite different from most other caves.

Public access to the cave is facilitated by well-lighted trails. The tour is self-guided with an interpretive program keyed to the various locations in the cave through portable radio receivers with headphones. Park Service guides are stationed throughout the cave to answer questions. Exit is provided directly to the surface by a 750-foot elevator, which also affords entry for those who wish a shorter and less strenuous tour. The cave temperature is approximately 56°F year-round, equivalent to that of the mean surface temperature of the region. A light jacket or sweater and sturdy walking shoes are recommended.

Local History

The Guadalupe Mountains were inhabited by humans as early as 12,000 years ago. Although the bones of at least two individuals are reported to have been found in the entrance of Carlsbad Caverns, Indians apparently did not explore the cave because of a sharp drop-off a short distance inside. By the late 19th century the entrance was well known to the local ranchers because of the dense cloud of bats that exits each evening for their nocturnal feeding on insects, and then returns to the cave near daybreak. (These useful animals, Mexican free-tailed bats, numbering in the hundreds of thousands, inhabit a room just inside the cave, called the Bat Cave, which is not open to the public. Each night they consume more than a ton of insects.)

The first person to enter the cave was apparently 12-year-old Rolth Sublett, whose father lowered him into the entrance on a rope in 1883, but he did not explore beyond the limit of daylight. The Bat Cave contains enormous deposits of bat droppings (guano), and the economic potential of the material as a fertilizer did not escape the attention of local speculators. In 1903 Abijah Long filed a claim for the area around the mouth of the cave and established a guano-mining operation. A shaft was constructed from the surface directly into the Bat Cave and the guano was removed in large cable-driven buckets, which also served as the primary means of entry and exit for the miners. The enterprise proved to be an economic failure, however. In fact, over a period of 20 years six different companies tried unsuccessfully to run the operation. In all, more than 100 million pounds of guano were excavated, lowering the cave floor in places as much as 50 feet.

A young miner named James L. White, who worked for five of the guano enterprises, was fascinated by the huge, steeply descending underground corridor that led from the Bat Cave past the natural entrance into the unknown. He made frequent explorations deep into the cave, often alone, and brought back wondrous tales. When the final mining company went bankrupt soon after World War I, Jim White stayed in the area and led private tours into the cave. At first his enthusiastic reports were scoffed at by most people. However, it was not long before others joined him in his praise, including several officials of the federal government who visited the cave, particularly Willis T. Lee of the United States Geological Survey. Lee led a six-month National Geographic Society expedition to study and map the cave, with White serving as a guide. With the size and beauty of the cave substantiated, President Coolidge initiated legislation to establish Carlsbad Caverns National Monument. In 1930 its status was elevated to that of a national park.

Today Carlsbad Caverns is the most popular cave in the country, with more than a million visitors each year. Twenty miles of rooms and passageways have been explored and surveyed, and new discoveries are made almost yearly. More than 50 other caves are known within the Park boundaries, although none approaches the size of Carlsbad Caverns. One of the largest of these, New Cave, is open to the public on an unimproved basis. Park Service guides lead groups through with hand-held lanterns. Although it is not so large as Carlsbad Caverns, New Cave contains travertine formations every bit as grand and bizarre as those of its larger neighbor.

How Caves Are Formed

Virtually all major caves are formed by the dissolving action of underground water within soluble bedrock. Limestone, dolomite, marble, and gypsum are the most common rocks of this type. All but gypsum require that the underground water be at least slightly acidic for the solutional activity to be rapid enough to form sizable caves.

Chemical reactions showing how carbonic acid breaks down calcite:

1. H_2O — water $+$ CO_2 — carbon dioxide gas $\rightleftharpoons$ H_2CO_3 — carbonic acid

2. H_2CO_3 — carbonic acid $\rightleftharpoons$ H^+ — hydrogen ion in solution $+$ HCO_2^- — bicarbonate ion in solution

3. H^+ — hydrogen ion in solution $+$ HCO_3^- — bicarbonate ion in solution $+$ $CaCO_3$ — calcium carbonate (calcite in limestone) $\rightleftharpoons$ Ca^{++} — calcium ion in solution $+$ $2HCO_3^-$ — bicarbonate ions in solution

Limestone, by far the most common cave-forming rock, is attacked mainly by water containing dissolved carbon dioxide, a natural gas present in the atmosphere and in soil, that combines with water to form weak carbonic acid. The chemical reactions that take place are shown above. In this way the carbonic acid breaks down calcite, the principal mineral in limestone, into soluble components that are carried away by water.

Water that enters the ground through openings such as fissures in bedrock or cave entrances is able to dissolve limestone throughout its entire underground flow path (though at a decreasing rate with time), provided the water does not lose its solutional potential by releasing carbon dioxide gas (CO_2). Discrete flows of water, such as those that enter the ground through major depressions, or karst features, at the surface, generally gain carbon dioxide from the subsurface openings rather than losing it, so their ability to dissolve may be sustained for a great length of time.

Water that infiltrates into the ground through soil picks up a great deal of carbon dioxide (from decaying vegetation, etc.), so that a given volume of this water is able to dissolve much more limestone than infiltrating ground water from other sources. However, if this water emerges into an air-filled cave, some of the dissolved carbon dioxide is released into the cave atmosphere. When this happens, the solution process is reversed, causing calcium carbonate ($CaCO_3$) to be precipitated in the form of calcite (limestone). This is how many of the cave decorations, or *speleothems*, are formed.

The dissolving of limestone to form caves may also be caused by sulfuric acid produced by the oxidation of sulfides. Although this acid is comparatively rare in most ground water, recent evidence suggests that the enormous caves of Carlsbad Caverns National Park owe their origin mainly to this reaction. Metallic sulfides, such as pyrite, are typical materials that can oxidize to form sulfuric acid.

The solution reaction, which is *not* generally reversible in the cave environment, follows the steps listed on the page following (shown only in part, for clarity). The calcium and sulfate ions may, if concentrated enough, combine to precipitate gypsum ($CaSO_4 \cdot 2H_2O$) at the same time the limestone is dissolving.

As water infiltrates into the ground, it is affected by gravity and by capillary forces. Except temporarily and locally, water does not entirely fill the available cracks and pores in the ground. Wherever the water flow is substantial enough to overcome the capillary forces, gravity pulls it downward via the steepest available openings until it reaches a zone where all available openings in the ground are filled with water. In this zone the water flows by a combination of gravitational force and hydrostatic pressure, eventually emerging at springs at lower elevations such as river valleys. The zone of descending gravitational water is called the *vadose zone*, and the zone in which all openings are filled with water is called the *phreatic zone*. The top of the phreatic zone is known as the *water table*.

Cave-forming processes can operate in both the vadose and phreatic zones. Vadose cave features show a gravitational orientation (e.g., stalactites and stalgmites, vertical solution pits, and cave passages with consistent downward trends), whereas phreatic features have no consistent gravitational orientation (e.g., irregular, vaulted rooms).

Geologic Features

Bedrock and Geologic Structure

Details of the geologic setting are similar to those of Guadalupe Mountains National Park, described in chapter 33, so only a brief overview is given here. Three rock sequences, all about the same age but of different lithologies, occur in Carlsbad Caverns National Park. The

Chemical reactions in dissolving of calcite by sulfuric acid:

1. S^- or $S^=$ + O_2 + $2H_2O$ $\rightleftharpoons$ $2H^+$ + $SO_4^=$

sulfur ion   oxygen   water

$\underbrace{\qquad\qquad}$ sulfuric acid $\underbrace{\qquad\qquad}$ in solution

2. H^+ + $CaCO_3$ $\rightleftharpoons$ Ca^{++} + HCO_3^-

soluble hydrogen ion   calcium carbonate (calcite in limestone)   calcium ion in solution   bicarbonate ion in solution

Guadalupe Escarpment is formed by an exhumed barrier reef that developed around the perimeter of the Delaware Basin during the Permian Period (fig. 39.3). This reef, called the Capitan Limestone, contains a framework of bryozoan, sponge, and algae fossils in a matrix of unbedded, fine-grained limestone. To the north-northwest of the reef are the back-reef limestones (Artesia Group), which are prominently bedded. They contain a large percentage of silt eroded from nearby land areas, as well as gypsum that accumulated in areas of restricted water circulation. To the southeast, the Capitan reef is bordered by an apron of reef talus that merges with dark, fine-grained limestones deposited in the deeper waters of the Delaware Basin, contemporaneously with the other rock units. The basinal limestones are overlain by gypsum and evaporites, mainly of the Castile Formation, which are also Permian

in age. Figure 39.3 is a diagrammatic representation of the sedimentary facies that make up the bedrock of the Carlsbad region.

As a result of Mesozoic and Cenozoic uplift, the processes of solution and erosion have removed much of the gypsum so that the more resistant limestones now stand out in relief, with nearly the same topographic expression they had while they were being deposited in the inland sea 200 to 250 million years ago. Uplift of these rocks was accompanied by southeasterly tilting and broad folding. Faults with displacements up to several hundred feet are distributed throughout the Park, particularly along the border zone between the reef and basinal facies. Normal faults along the western edge of the Guadalupe Mountains (outside the Park) attain a total displacement of several thousand feet. Prominent joints in the vicinity of Carlsbad Caverns are oriented in two sets, parallel to and

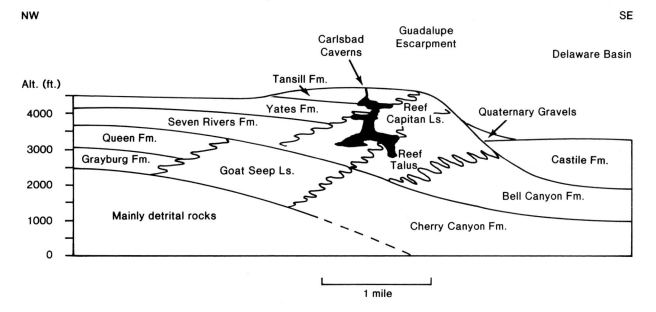

Figure 39.3 Geologic cross section through the Guadalupe Escarpment, showing the location of Carlsbad Caverns.

perpendicular to the reef front. Many of the major joints parallel to the reef front have been filled with shallow-water deposits of land-derived sediment. These deposits now show up in the limestone as clastic dikes of sandstone and siltstone.

Cavern Features

A landscape dominated by solutional features is referred to as *karst topography*. A typical karst region contains many solutional fissures, pits, and funnel-shaped depressions called sinkholes, in which surface water collects and flows underground. Caves show evidence of subsurface solution by underground water. In the semiarid climate of the Guadalupe Mountains, surface karst features are almost entirely absent, so the region is not considered true karst topography, despite its many caves. Karst topography is best developed in a humid climate such as that of Mammoth Cave National Park (chapter 40).

Most caves consist of long, conduitlike passageways dissolved by underground streams. In such a cave it is easy to distinguish the points through which ground-water recharge took place, as well as the springs through which the water emerged at the surface, particularly since many caves are still actively forming today. In contrast, Carlsbad Caverns and most other caves in the Guadalupe Mountains consist of large, interconnected rooms with highly irregular patterns and profiles. Nearly all of these caves are inactive today, having been abandoned by the water that formed them. Clues to the original water sources, patterns of flow, and springs are virtually absent, and attempts to describe Carlsbad Caverns in terms of "normal" circulation of ground water and the carbonic-acid reaction do not explain the unusual character of the cave complex.

Carlsbad Caverns is located mainly in the reef limestone, but it also extends into the adjacent back-reef beds (in the northwestern part of the cave) and the fore-reef talus (at the south end of the Big Room). Nevertheless, the pattern and character of the cave are remarkably uniform, despite the great contrast in rock type. The main chambers of the cave are broad, arched rooms and linear, fissurelike passages whose major trends are controlled by the prominent jointing of the Capitan Limestone. In places the rooms are bordered by intricate zones of *spongework*, where the original pores in the limestone have been enlarged by solution to form an interconnected maze of openings similar to that of Swiss cheese. The Boneyard is one of the best examples. The rooms, fissures, and spongework are the result of solution in the phreatic zone (below the water table). Vadose solution features are limited to a few minor rills and drip pockets.

Many of the largest rooms and passages are grouped at three different elevations below the entrance: Bat Cave at -160 feet; the Big Room and Lunch Room at -660 feet; and the Lower Cave, King's Palace, and neighboring rooms at -740 feet. These levels are not controlled by differences in rock type and may represent separate stages of cave development. Several small lakes are scattered throughout the lowest rooms, including the lowest point in the cave, Lake of the Clouds, at a depth of 1,037 feet below the entrance. These are merely local zones of ponding in the vadose zone and do not represent the water table. The original shape of the cave was modified in various places by the dropping of large rocks from the ceiling. The enormous Iceberg Rock is the largest of these *collapse blocks*. The presence of vadose travertine on some of the blocks shows that the collapse took place late in the developmental history of the cave; yet the ceilings are quite stable today, for no rocks have dropped in the entire time the cave has been known to man.

The deposits within Carlsbad Caverns are unusually extensive and varied (fig. 39.3). The most spectacular are the *speleothems* ("cave formations") composed of travertine, which is calcium carbonate precipitated when the carbon dioxide in underground water is released into the cave atmosphere. The rate at which these features grow can vary a great deal, but maximum rates of about one centimeter per hundred years have been observed in other caves. The great majority of speleothems have been precipitated by vadose water seeping into the cave. Where water runs over a cave wall, sheets or draperies called *flowstone* are formed. Icicle-shaped *stalactites* are deposited by water dripping from the ceiling, and where the drops hit the floor they deposit mounds shaped like totem poles, called *stalagmites*. Where a stalactite and stalagmite are produced by the same source of dripping water and grow large enough to meet, a *column* is formed. If the amount of water entering the cave is too small to form distinct drips, it is held to the growing speleothems by surface tension, and the calcium carbonate is deposited in irregular, twisted fingers called *helictites*.

Where water seeps out of pores in the rock, nodular, grapelike protrusions called *cave popcorn* are precipitated (fig. 39.4). These deposits may form on other speleothems in the same way. Cave popcorn is most abundant in areas where capillary water is drawn from the moist rock by gentle, dry air currents that flow through the cave, as shown by the fact that popcorn growing on stalactites and stalagmites is generally much more profuse on the windward side. However, warm, moist air rising from lakes deeper in the cave has caused water to condense on the colder rock surfaces above. The windward sides of speleothems in such an area are partly dissolved, whereas popcorn is deposited only on the lee sides.

Where water drips into a shallow basin containing small particles such as sand grains, the particles may become coated with calcium carbonate. If the agitation is sufficient to keep the growing particles from being cemented to the floor, rounded *cave pearls* are formed. The

Figure 39.4 The Temple of the Sun is one of the larger travertine columns in the Big Room of Carlsbad Caverns. Many of the profuse stalactites and stalagmites shown here are encrusted with cave popcorn, which gives them a bizarre, knobby appearance. The ceiling height is about 50 feet in this part of the cave. National Park Service photograph.

largest cave pearls in Carlsbad Caverns are more than an inch in diameter.

Certain speleothems are deposited in ponded water. These include calcite crystals such as dogtooth spar, travertine deposits called *rimstone* around the edges of pools, and bulbous, mammillate wall coatings that look like clouds (hence the name "Lake of the Clouds").

Most of the large speleothems are not actively growing today and are in the process of drying out. Their profusion shows that the climate must have been much more humid in the past, presumably during glacial episodes of the Pleistocene Epoch. It was discovered recently that a great deal of moisture was being lost as water vapor through the elevator shaft, so air locks have been placed at the elevators to prevent this accelerated drying of the cave.

Sediments in the Caverns A variety of cave sediments, of both mechanical and chemical origin, provide evidence for the processes that formed Carlsbad Caverns. Most prominent of these are extensive bodies of gypsum deposited in the middle levels of the caverns. The gypsum deposits, which reach a maximum thickness of 20 feet, have been partly removed by solution, so only isolated blocks remain today. Water dripping from the cave ceiling has dissolved vertical "drill holes" through the gypsum in many places.

Most of the gypsum precipitated on the cave floor from sulfate-rich brine that once filled the cave; but in some areas the gypsum has actually replaced the limestone wall rock to a depth of several feet. Recent analysis of the gypsum shows that it is highly enriched in the light sulphur isotope S^{32}, which indicates a biological influence upon its origin. Bacteria probably aided in the production of sulfides in the reducing environment of the Delaware Basin during the Permian Period. These sulfides were later oxidized to sulfate to form the gypsum. The gypsum was not derived from the Castile Formation, at least directly, because the Castile is not enriched in S^{32}.

The presence of the clay mineral endellite, overlain by montmorillonite, gives a further clue to the origin of Carlsbad Caverns. *Endellite* is formed by the alteration of other clay minerals in an acidic, sulfate-rich environment, whereas the *montmorillonite* (also a clay mineral) is a typical insoluble residue from the solution of the limestone. The combined presence of gypsum and endellite suggests that sulfuric acid played a more significant role than carbonic acid in the origin of the cave. Sulfuric acid can be produced by the oxidation of pyrite, which is abundant in the back-reef limestones, but except for a few rare examples the cave patterns show no relationship to the distribution of pyrite. Oxidation of sulfides in rising oil-field brines seems to be a more likely source of the sulfuric acid. Hydrogen sulfide gas (H_2S) occurs within rocks of the Delaware Basin that are rich in petroleum. Hydrogen sulfide oxidizes to sulfur and water, and the sulfur oxidizes further to form sulfuric acid. Remnants of pure sulfur have been found in Carlsbad Caverns and other caves nearby; this supports the view that hydrogen sulfide played a role in the cave's origin. The lack of relationship between the way the cave evolved and the usual patterns of ground-water flow can also be explained. The three levels in the cave may represent average positions of the water table at various times, inasmuch as the rate of oxidation of sulfides would be greatest at and above the water table and would diminish downward within the phreatic zone.

The bedded gypsum overlies silt that is mainly insoluble material released during the solution of the limestone. Other sediments include sparse, rounded limestone pebbles carried in from the surface late in the developmental history of the cave, but they represent only a minor event that has little bearing on the cave's origin.

Geologic History

1. Deposition of limestones in the Delaware Basin during the Permian Period.

The Delaware Basin was a small inland sea 200 to 250 million years ago during the Permian Period. The Capitan reef was one of many that developed around the perimeter of the basin, with impure, bedded limestones and gypsum accumulating on the landward side of the reef and fine-grained limestones in the center of the basin. Details of the sedimentary history are given in chapter 33.

2. First stage of karst development late in the Permian Period.

Late Permian glaciation in the polar regions caused a drop in sea level, exposing the recently deposited limestones to solution by surface runoff and by circulating ground water. Karst features formed at this time have been largely destroyed by later erosion or filled by detrital sediment.

3. Filling of the Delaware Basin with gypsum.

Waning of the late Permian glaciation caused sea level to rise. During and after the polar glaciation, the Delaware Basin was nearly isolated from the ocean, and a high rate of evaporation caused gypsum and other evaporites to fill the basin at least to the height of the Capitan reef.

4. Deposition of detrital sediment near the end of the Permian Period.

In a continental environment, the area was covered by silt and sand derived from nearby higher land areas. This created the Salado and Rustler Formations. Some of the detrital fill in the earlier karst features may date from this episode.

5. Tectonic uplift and deformation during the late Cretaceous Period and early Tertiary Period; early stages of cave development.

As part of the Laramide orogeny, the Guadalupe Mountains were formed by a local uplift, accompanied by faulting and southeastward tilting. Erosion and karst development began in the higher regions, while sediment derived from these processes may have filled any remaining karst features of Permian age at lower elevations. Carlsbad Caverns began to form at this time within the phreatic zone. Mobilization of sulfide-rich brines by the tectonic activity may have occurred at this time, producing sulfuric acid by mixing with oxygenated water that infiltrated from the surface. Some acidity could also have been produced by mixing of descending CO_2-rich water with sulfate brines derived from the Castile Formation. The extent to which Carlsbad Caverns formed at this time is not yet known.

6. Renewed uplift of the Guadalupe Mountains during the late Pliocene Epoch and early Pleistocene Epoch; further enlargement of caves.

The major phase of solutional enlargement of Carlsbad Caverns is thought to date from the time of this uplift. Extensive gypsum deposits in this and many other Guadalupe caves indicate that the solution was caused by sulfuric acid derived from the oxidation of dissolved sulfides released from deeper rocks.

7. Draining of caves during the Pleistocene Epoch; deposition of speleothems.

The major Guadalupe caves gradually emerged into the vadose zone during the Pleistocene Epoch as uplift and erosional dissection continued. Pauses in the emergence, accompanied by continuing solution, may account for the three distinct levels in Carlsbad Caverns. Erosion and solution of the soft Castile gypsum exposed the Capitan reef, which forms the Guadalupe Escarpment today.

Infiltration of CO_2-rich water from the surface deposited travertine speleothems, particularly in the relatively humid climates that accompanied episodes of continental glaciation. Uranium-thorium dating of a speleothem from Ogle Cave, near New Cave, indicates an age of a least 200,000 years, showing that the caves have been drained of water for at least that length of time.

Intersection of the upper cave level by erosion formed the entrance of Carlsbad Caverns. Gradual diminution of infiltrating water and drying of the cave in the semiarid climate since the last glaciation has reduced the rate of speleothem growth. This change has favored the deposition of nodular travertine, such as popcorn, rather than stalactites and stalagmites.

Today Carlsbad Caverns National Park is not only one of the greatest tourist attractions of the Southwest, but it is also the subject of considerable research in the fields of sedimentary processes, climatic history, and the unusual mode of cave origin that has occurred in the region.

Geologic Maps

Hayes, R. T. 1957. *Geology of the Carlsbad Caverns east quadrangle, New Mexico.* Geological Survey map GQ-98.

Hayes, R. T., and Koogle, R. L. 1958. *Geology of the Carlsbad Caverns West Quadrangle, New Mexico.* Geological Survey map GQ-112.

Bibliography

Barnett, J. 1980. *Carlsbad Caverns National Park: silent chambers, timeless beauty* Carlsbad, New Mexico: Carlsbad Caverns Natural History Association.

Bretz, J H. 1949. Carlsbad Caverns and other caves of the Guadalupe Block, New Mexico. *Journal of geology* 57: 447–463.

Davis, D. G. 1980. Cave development in the Guadalupe Mountains, a critical review of recent hypotheses. *National Speleological Society bulletin* 42: 42–48.

Dunham, R. J. 1972. Capitan Reef, New Mexico and Texas: facts and questions to aid interpretation and group discussion. Permian Basin Section, Society of Economic Paleontologists and Mineralogists, Publication 72–14.

Egemeier, S. J. 1973. Cavern development by thermal waters with a possible bearing on ore deposition. Unpublished Ph.D. dissertation, Stanford University.

Good, J. M. 1957. Noncarbonate deposits of Carlsbad Caverns (New Mexico). *National Speological Society bulletin* 19: 11–23.

Halliday, W. R. 1966. *Depths of the earth*. New York: Harper and Row.

Harmon, R. S., and Curl, R. L. 1978. Preliminary results on growth rate and paleoclimate studies of a stalagmite from Ogle Cave, New Mexico. *National Speleological Society bulletin* 40: 25–26.

Hill, C. A. 1981. Speleogenesis of Carlsbad Caverns and other caves of the Guadalupe Mountains. In *Proceedings*, Eighth International Congress of Speleology, pp. 143–144.

Jagnow, D. H. 1979. *Cavern development in the Guadalupe Mountains*. Columbus, Ohio: Cave Research Foundation.

Kirkland, D. W. 1981. Origin of gypsum deposits of Carlsbad Caverns, Eddy County, New Mexico. Pre-publication manuscript. Dallas, Texas: Mobil Research and Development Corporation.

McLean, J. S. 1971. *The microclimate in Carlsbad Caverns, New Mexico*. U.S. Geological Survey Open File Report at Albuquerque, New Mexico.

Morehouse, D. F. 1968. Cave development via the sulfuric acid reaction. *National Speleological Society bulletin*, 30:1–10.

Newell, N. D,; Rigby, J. K.; Fischer, A. G.; Whiteman, A. J.; Hickox, J. E.; and Bradley, J. S. 1953. *The Permian reef complex of the Guadalupe Mountains region, Texas and New Mexico—a study in paleoecology*. San Francisco, California: W. H. Freeman and Co.

Queen, J. M. 1981. A discussion and field guide to the geology of Carlsbad Caverns. Preliminary report to the National Park Service for the Eighth International Congress of Speleology.

Thrailkill, J. V. 1971. Carbonate deposition in Carlsbad Caverns. *Journal of geology* 79: 683–695.

Address

Carlsbad Caverns National Park
3225 National Parks Hwy.
Carlsbad, New Mexico 88220

40

Mammoth Cave National Park

by Arthur N. Palmer
State University of New York, Oneonta

Location: West Central Kentucky
Area: 52,128.92 acres; 81.45 square miles
Established: July 1, 1941

Figure 40.1 Flowstone draperies in Mammoth Cave near the Violet City Entrance. These travertine deposits are formed where water high in dissolved limestone content enters the cave and loses carbon dioxide to the cave air, precipitating calcium carbonate. National Park Service photograph.

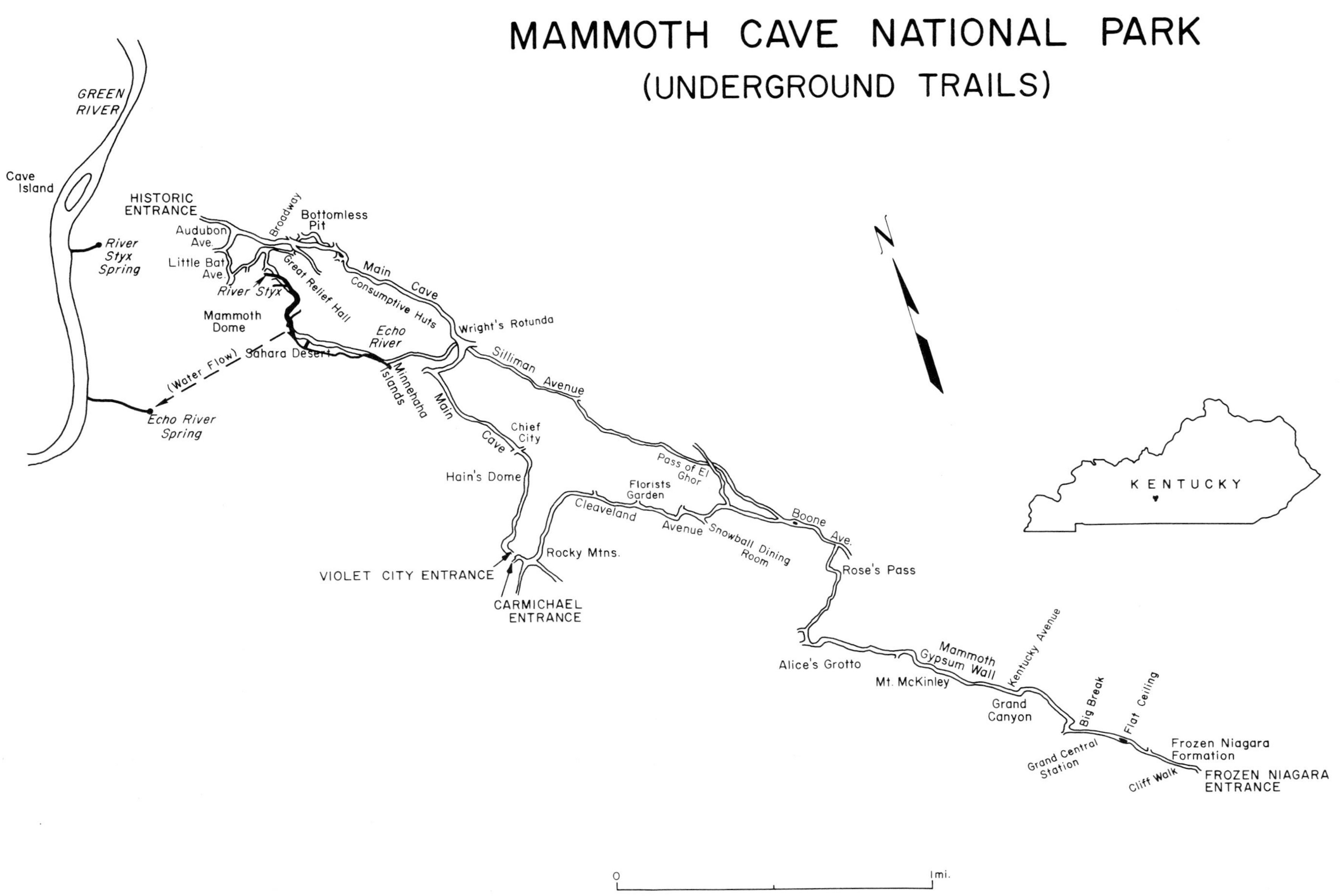

MAMMOTH CAVE NATIONAL PARK
(UNDERGROUND TRAILS)

KENTUCKY

N

GREEN RIVER

Cave Island

River Styx Spring

HISTORIC ENTRANCE

Bottomless Pit

Broadway

Audubon Ave.

Little Bat Ave.

Main Cave

Consumptive Huts

Great Relief Hall

Echo River

River Styx

Mammoth Dome

(Water Flow) Sahara Desert

Echo River Spring

Wright's Rotunda

Silliman Avenue

Minnehaha Islands

Main Cave

Chief City

Hain's Dome

VIOLET CITY ENTRANCE

Pass of El Ghor

Florists Garden

Cleaveland Avenue

Snowball Dining Room

Rocky Mtns.

CARMICHAEL ENTRANCE

Boone Ave.

Rose's Pass

Alice's Grotto

Mammoth Gypsum Wall

Mt. McKinley

Grand Canyon

Kentucky Avenue

Big Break

Grand Central Station

Flat Ceiling

Cliff Walk

Frozen Niagara Formation

FROZEN NIAGARA ENTRANCE

Tyner

0   1 mi.

Figure 40.2 Map of passages in Mammoth Cave that are open to the public. About ten miles of passages are shown here.

Table 40.1. Geologic Column, Mammoth Cave National Park

Era	Period	Epoch	Rock Units	Geologic Events
Cenozoic	Quaternary	Holocene	Alluvium	Alluviation; drowning of lowest levels of Mammoth Cave
Cenozoic	Quaternary	Pleistocene	Alluvium	Rapid entrenchment alternating with minor aggradation; lower levels of Mammoth Cave formed
Cenozoic	Tertiary (Neogene)	Pliocene		Upper levels of Mammoth Cave formed; nearly flat surface of Pennyroyal Plateau formed during period of relatively static base level
Cenozoic	Tertiary (Neogene)	Miocene		
Cenozoic	Tertiary (Paleogene)	Oligocene		Slow erosion, probably interspersed with periods of aggradation
Cenozoic	Tertiary (Paleogene)	Eocene		
Cenozoic	Tertiary (Paleogene)	Paleocene		
Mesozoic				
Paleozoic	Permian			
Paleozoic	Pennsylvanian	Morrowan	Caseyville Formation (conglomerate and sandstone)	Insoluble rocks deposited; gradual emergence of continent above sea level
			⟨ Disconformity ⟩	⟨ Erosion ⟩
Paleozoic	Mississippian	Chesterian	Leitchfield Formation (shale and sandstone)	
Paleozoic	Mississippian	Chesterian	Glen Dean Limestone	
Paleozoic	Mississippian	Chesterian	Hardinsburg Formation (sandstone)	
Paleozoic	Mississippian	Chesterian	Haney Limestone	Mainly insoluble rocks deposited in continental sea
Paleozoic	Mississippian	Chesterian	Big Clifty Formation (sandstone and shale)	
Paleozoic	Mississippian	Chesterian	Girkin Formation (limestone)	
Paleozoic	Mississippian	Meramecian	Ste. Genevieve Limestone	Relatively pure limestones deposited in continental sea
Paleozoic	Mississippian	Meramecian	St. Louis Limestone	
Paleozoic	Mississippian	Meramecian	Salem Limestone	Impure limestones deposited in continental sea
Paleozoic	Mississippian	Meramecian	Harrodsburg Limestone	

Source: Palmer (1981)

Regional Setting

Mammoth Cave National Park is located in west central Kentucky, midway between the cities of Louisville and Nashville, in the heart of America's most extensive limestone region. Solution of the limestone by underground water has created the longest known series of caves in the world. Mammoth Cave, the largest of these, contains more than 230 miles of surveyed passageways. About ten miles of the passageways are open to the public (fig. 40.2).

The Park is in a plateau of limestone bedrock capped by insoluble clastic rocks, mainly sandstones, at the southeastern edge of a regional downwarp, the Illinois Basin (a structural basin). The sedimentary rock layers in the plateau dip gently northwestward. Steep-walled valleys have breached the insoluble rocks, exposing 300 feet of cavernous limestone in irregular ridges. This rugged, hilly region is called the Chester Upland. To the southeast, the insoluble rocks have been removed by erosion, and the limestone is exposed at the surface in an uninterrupted plain, covering thousands of square miles, called the Pennyroyal Plateau. Together, the Chester Upland and the Pennyroyal Plateau comprise one of the world's finest examples of *karst topography*, a landscape dominated by solutional features such as caves, closed surface depressions, sinking streams, and large springs. (fig. 40.3).

Several different guided cave tours are available to Park visitors. The main tours in Mammoth Cave are briefly described here: (1) The Historic Tour emphasizes the pre-Columbian archeology of the cave and its history of saltpeter mining during the early 19th century. It also gives the clearest view of the limestone stratigraphy and of the numerous levels at which the cave has formed.(2) The Half-Day Tour is a five-hour hike through more than four miles of passageways. This tour best illustrates the variety of passage types and their interrelationships, as well as the most typical speleothems ("cave formations"). (3) The Frozen Niagara Tour visits the part of the cave decorated with the best examples of travertine speleothems. That part is also shown on the Half-Day Tour. (4) The Echo River Tour covers the same route as the Historic Tour, but includes a boat ride on a lake in the lowest level of the cave. The tours are moderately strenuous, and hiking shoes are recommended. The cave temperature is about 55–57° F, corresponding to the mean annual temperature of the region, and a light jacket or sweater is recommended.

Local History

The first people to enter Mammoth Cave were prehistoric Indians of the Woodland Culture, who lived in the area roughly 3,000 to 4,000 years ago. They used the larger entrances for shelter and explored as far as several miles into the cave in their quest for minerals used as medicines, for flint used in making tools, and probably also for adventure. For light they used torches made of cane, reeds, and twigs, remnants of which have been found in many of the dry upper-level passages of the cave. Other archeological finds include bare foot prints in mud and dust, woven baskets, fiber slippers, and two Indian "mummies" whose desiccated bodies were preserved naturally in dry parts of the cave.

Mammoth Cave became known to white settlers in the years just prior to 1800. Under private ownership, some of its passages were mined for saltpeter (nitrates) to make

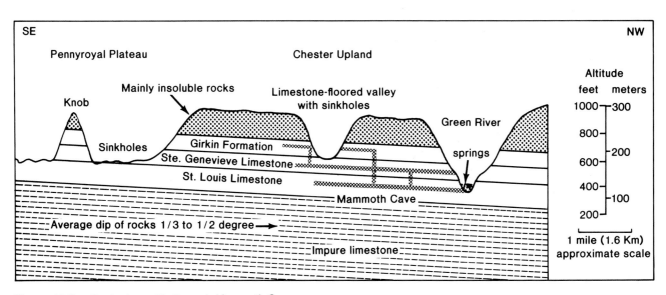

Figure 40.3 Geologic profile through Mammoth Cave National Park, showing the relationship of the cave to local topography and rock units.

gunpowder, particularly during the War of 1812. Well preserved remnants of the saltpeter works are shown on the Historic Tour.

The saltpeter operation closed down after the War of 1812, but in its place grew an increasing tourist trade, spurred by many published articles about the vast cave. One of the early guides, a black slave named Stephen Bishop, became America's first truly great cave explorer. He discovered sections of Mammoth Cave far beyond the previous limits of exploration, including Echo River and some of the passages shown today on the Half-Day Tour.

In 1908 the great size of the cave was verified for the first time when a young German visitor, an engineer named Max Kaemper, surveyed 45 miles of passageways in it. Although he used only a hand-held compass and measured distances by pacing, his map was remarkably accurate.

At that time the known parts of the cave did not extend beyond the flanks of the five-mile-long Mammoth Cave Ridge, one of many flat-topped limestone ridges in the Chester Upland capped by insoluble rocks and separated from neighboring ridges by dry karst valleys. In the late 19th and early 20th century many new caves were discovered and explored in nearby ridges. Some of these were opened for tourists, and a great commercial rivalry developed among the various cave owners. Although Mammoth Cave was still by far the most popular attraction of the region, its tourist trade suffered a blow when new entrances into the same cave were excavated beneath neighboring property and its southeastern end was opened to tours in competition with the "historic" northwestern end.

One of the residents of nearby Flint Ridge, to the northeast of Mammoth Cave Ridge, was an enthusiastic cave explorer named Floyd Collins. He discovered a large cave on his family's property, which he called Great Crystal Cave, and opened it for tourists. Of the many tourist caves in the area, his was one of the least accessible by road, so Floyd spent a great deal of time searching for a new and more conveniently located cave. In 1925, while exploring a tight, unstable cave called Sand Cave, he dislodged a loose rock that fell on his ankle, trapping him. Despite efforts by many people to free him, he died of exposure. Floyd Collins' fate received enormous publicity, bringing even more visitors to the Mammoth Cave area.

Mammoth Cave was authorized as a national park in 1926 and was officially dedicated in 1941. The Park boundaries encompassed many of the neighboring caves that had once been open to the public. Several of the former operators and explorers of Floyd Collins' Great Crystal Cave formed the Cave Research Foundation in 1957, dedicated to the exploration and scientific study of these and other caves within the Park. By 1961 the group had discovered connections between all of the major caves in Flint Ridge, revealing a vast cave system that rivaled Mammoth Cave in length. In 1972 a connection was found between the Flint Ridge Cave and Mammoth Cave through a narrow, wet passageway extending beneath the dry karst valley that separates the two ridges. This joined the two caves into a single enormous system with a combined length more than twice that of any other known cave in the world. Today a total of 230 miles of passageways has been surveyed in Mammoth Cave, and continued exploration will almost certainly reveal connections with other large caves nearby.

Geologic Features

Stratigraphy and Geologic Structure

The rocks of Mammoth Cave National Park originated as sediments on the floor of an extensive shallow sea that covered most of southern North America during the Mississippian Period, approximately 350 to 325 million years ago. Limestone was the major rock type deposited in this continental sea, although sand, silt, and clay were carried in from the land area that lay to the north. The water was very shallow, despite its broad extent, with average depths of only a few tens of feet. The limestones deposited at this time are highly varied in character and possess rather thin and prominent bedding.

Mammoth Cave is located within three different limestone formations. From oldest to youngest, these are the St. Louis Limestone, the Ste. Genevieve Limestone, and the Girkin Formation. The St. Louis is about 200 feet thick, although Mammoth Cave extends into only the upper half. The cave occupies the entire thickness of the Ste. Genevieve (100–120 feet) and the Girkin (135–140 feet). The climate was rather arid while the St. Louis Limestone was forming. As a result, gypsum beds are common at depth below the surface, although the gypsum has been almost entirely removed from the near-surface parts of the formation by ground-water solution. Chert beds and lenses are common, and it is thought that many of these were formed by the replacement of evaporite deposits by silica (SiO_2). The climate grew more humid while the Ste. Genevieve Limestone was accumulating, so this formation contains little or no gypsum. Chert is limited to large rounded nodules and dike-shaped bodies. The upper half of the Ste. Genevieve Limestone contains several thin, impure, silty beds that represent the first stages of encroachment of detrital sediment from the land into the Mammoth Cave area. The Girkin Formation consists of thick limestone members, each about 20 feet thick, interspersed with thin silty limestones and shales.

Above the Girkin Formation lie the younger rock units of the Chester Series, in which the relative proportion of limestone and detrital sediment is reversed (table 40.1). Thick formations of sandstone and shale are interspersed with limestones only 10 to 40 feet thick. The Chester Series forms the resistant cap rock on the limestone ridges of the Chester Upland. This series is truncated by a widespread disconformity with local relief of several hundred feet in places. (The surface of a *disconformity*, representing missing rock layers, is parallel to beds above and below it.) Pennsylvanian conglomerates and other detrital rocks overlie the disconformity in the northern part of Mammoth Cave National Park, but in the southern ridges of the Park, including Mammoth Cave Ridge, the Pennsylvanian rocks have been removed by erosion.

Downwarping of the Illinois Basin occurred while the Paleozoic sedimentary rocks were being deposited, causing the beds to dip gently toward the center of the basin (located in southeastern Illinois). In Mammoth Cave National Park the dip averages one-third to one-half degree toward the northwest, although the direction and amount of dip vary a great deal from place to place because of irregularities in the beds.

Joints are rather small and inconspicuous in the cavernous limestones. Faults with up to 50 feet of displacement occur within the Park, but the largest that have been observed in the cave have displacements of less than a foot. Because the fracturing of the limestone is so slight, individual beds and bedding-plane partings exert the main control over the directions of passages in Mammoth Cave.

Surface Features

Unlike Carlsbad Caverns (chapter 39), Mammoth Cave is intimately related to the hydrology and geomorphic evolution of its surrounding landscape. Almost all the runoff in this humid region sinks underground through karst features in the land surface. Such features include *sinkholes*, which are funnel-shaped depressions where soil has subsided into solutionally enlarged cracks in the limestone. Collapse of bedrock into the underlying openings also plays an important role in forming large sinkholes. In the Mammoth Cave area these features number in the hundreds of thousands, ranging in size up to a thousand feet in diameter and more than a hundred feet deep. Some have open holes in their bottoms or sides that lead directly to caves. Throughout much of the Pennyroyal Plateau the density of sinkholes is so great that there is virtually no surface runoff. Surface streams form where water collects on adjacent insoluble rocks, but where they encounter limestone they disappear underground into sinkholes. These are called *sinking streams*. Despite the large number of karst features, the overall surface of the Pennyroyal Plateau is of remarkably low relief, appearing almost flat when viewed from an overlook. Actually the land slopes gently toward the few entrenched surface rivers that cross the area.

The Chester Upland, in which Mammoth Cave is located, rises abruptly 200 to 300 feet above the Pennyroyal Plateau surface. This steep rise separating the two regions is called the Chester Escarpment. The edge of the Chester Upland is highly irregular, with many erosional embayments and isolated remnants called *knobs*, which are outliers of limestone capped by insoluble rocks. The Chester Upland is a rugged, hilly region rising to altitudes of nearly 1,000 feet. It is extensively dissected by steep-walled valleys, many of which have lost their streams to underground piracy and are now dry, sinkhole-floored karst valleys. The only surface streams in the Chester Upland and Pennyroyal Plateau are major rivers into which ground water emerges through karst springs. The master stream in the Mammoth Cave area is the Green River, which passes westward through both the Pennyroyal Plateau and the Chester Upland.

The Pennyroyal Plateau is lower than the Chester Upland because the erosional removal of the late Mississippian detrital rocks exposed the less resistant limestones beneath. Rocks like sandstone, which make up most of the cap rock of the Chester Upland, are much more resistant than limestone in this humid environment; limestone is attacked by rapid solution as well as by mechanical weathering, whereas sandstone is affected very little by solution. The very low relief of the Pennyroyal Plateau resulted from a prolonged period of weathering at or near base level, during which erosional dissection of the region slowed or stopped. Although the limestones were weathered and eroded almost to a flat surface at that time, the resistant cap rock prevented all but a modest widening of stream valleys in the Chester Upland. Subsequently renewed dissection of the region has caused major rivers to entrench about 200 feet into the Pennyroyal Plateau and to deepen the valleys in the Chester Upland (fig. 40.3).

Underground water in Mammoth Cave National Park emerges at dozens of springs in the Green River valley. Their flow varies greatly with the seasons and responds quickly to rainfall. Most of the largest springs are fed by cave passages that formed when the Green River valley was deeper than it is now. These passages were flooded when river sediment filled the bottom of the valley to depths of at least 30 feet late in Pleistocene time. At present, in order to reach the river, underground water from these passages is forced to rise upward through funnel-shaped openings in the sediment covering the floor of the Green River valley. In addition, many small springs emerge *above* the Green River level in the walls of valleys, particularly in the thin limestone formations sandwiched between detrital rocks in the cap rock overlying the main cavernous limestones.

Streams in the main part of Mammoth Cave discharge through sediment in the floor of the Green River valley at Echo River Spring. During wet periods the underground water overflows through other passages to River Styx Spring, which is located exactly at the level of the Green River. Both of these springs can be viewed by following short surface trails.

Cave Passages

Mammoth Cave consists mostly of long, winding, nearly horizontal tunnels, many of which can be followed for several miles without interrruption. These passages were dissolved by running water that has infiltrated from the land surface, mainly through sinkholes. The various passages join each other as tributaries, forming a crude dendritic pattern. This pattern is not at all apparent from a map of the cave for several reasons:

(1) Subsurface water is strongly controlled by the local geologic structure, as will be shown later. As a result, cave passages have much more erratic patterns than those of dendritic surface streams.

(2) Many different levels of cave development are superimposed upon each other. As the Green River erodes downward, the water in the cave passages diverts to progressively lower routes, forming new passages, and the abandoned passages are left high and dry. However, at any given time during the evolutionary history of the cave, the passages that are actively forming possess a dendritic pattern.

(3) Most passages cannot be followed the entire distance from water input to spring, because they either become too small to explore or are terminated by collapse or sediment fill.

(4) Overflow passages can form where periodic flooding of the lowest cave levels causes water to be forced into alternate routes through the limestone, particularly where a stream passage is partly blocked by collapse material or sediment.

(5) Many passage junctions have resulted from the random intersection of passages, for instance by collapse of the bedrock separating two passages where they cross one another on different levels.

As a result of all these factors, Mammoth Cave appears to the visitor as a confusing three-dimensional maze, rather than an orderly dendritic pattern of stream channels.

All of the passages in Mammoth Cave are well adjusted to the patterns of past or present ground-water circulation in the region. Passages emanate from areas of ground-water recharge, such as the Pennyroyal Plateau and karst valleys in the Chester Upland, and terminate at or near the Green River valley, the only significant area of outflow. The chemical character of this water indicates

that Mammoth Cave formed by the solution of limestone by carbonic acid in the water. Measurements in active stream passages show that the limestone is dissolving away at a maximum rate of slightly less than a millimeter each year. This process, which is responsible for nearly all major caves and karst features, is described in chapter 39 on Carlsbad Caverns National Park. Note that Carlsbad Caverns, which apparently formed by the solutional action of sulfuric acid generated by the oxidation of sulfides, does not possess the close association with the land surface and the present patterns of ground-water circulation that Mammoth Cave shows so well, nor does Carlsbad Caverns contain distinctly stream-carved tunnels.

Passages in Mammoth Cave exhibit a variety of shapes determined by the conditions of water flow that formed them. Infiltrating water flows downward through cracks and pores in the limestone through the *vadose zone*, pulled down mainly by the force of gravity. This water eventually reaches the *phreatic zone*, where all openings in the rock are filled with water. In the phreatic zone the water moves under the combined influence of gravity and pressure along subhorizontal paths to springs in the nearest available river valley.

There is a misconception among many geologists that caves must form in the phreatic zone, at or beneath the water table, and are exposed to vadose conditions only when the water table drops with time. This is not true, because vadose water that enters the ground through sinkholes and sinking streams is able to form caves as readily as phreatic water. In fact, the majority of the passages in Mammoth Cave formed in the vadose zone, above the water table. Many vadose passages are still being enlarged by underground streams. The sinkholes that feed these passages evolved along with the cave, and, except in rare cases, they do not represent points where surface water was diverted underground into pre-existing phreatic caves.

Vadose water descends along the steepest available paths, preferably through vertical joints or faults. However, in the Mammoth Cave area faults are very sparse, and most joints are narrow, inconspicuous, and extend through only a single bed. If no vertical opening is present, vadose water follows the next steepest path. Therefore, the most common passage types in Mammoth Cave are those formed by vadose water that initially flowed not vertically downward, but instead followed the dip of the limestone along bedding-plane partings or favorably jointed beds. When enough solution occurred to allow a discrete stream of water to flow through, such a passage would enlarge by the downward solution and erosion of the floor, thus creating a narrow, sinuous *canyon passage*. A canyon passage looks like the canyon of an arid landscape, except that it is roofed by limestone. Examples include Boone Avenue and Rose's Pass on the Half-Day Tour. Most canyon passages in Mammoth Cave are less

Figure 40.4 Broadway, on the Historic Tour, is a huge canyon partly filled with detrital sediment. This passage is typical of the large upper levels of the cave at an altitude of 600 feet. The wooden pipes are remains of the saltpeter mining operation during the War of 1812. They were used to transport leached nitrates to the Historic Entrance with the aid of large hand-driven pumps. (Photo by A. N. Palmer)

than 10 feet wide, but they range up to 100 feet high. However, some of the largest passages in the cave are broad canyons as much as 80 feet wide, such as Kentucky Avenue, Broadway (fig. 40.4), and Audubon Avenue. The pattern of a canyon passage is intimately controlled by the local dip direction, and minor irregularities in the dip cause a marked sinuosity in the passage. Because the direction of dip is so variable, canyon passages are not all oriented in the direction of the northwesterly regional dip.

Where water is able to flow vertically downward along a fracture, the fracture is widened by solution wherever the flow is greatest, and a *vertical shaft* is formed. These features are shaped like the inside of a silo or well, with vertical, cylindrical walls. From the bottom, a vertical shaft may look like the inside of a high, narrow dome, but from the top it looks like a pit. For this reason they are sometimes called *domepits*. An example is Mammoth Dome on the Historic Tour. Water generally enters laterally through a passage at the ceiling level (usually a canyon passage) and exits through a similar passage that leads out through the shaft wall at the floor level. In the poorly jointed limestones of the area most vertical shafts owe at least part of their vertical extent to the coring action of the descending water upon the limestone floor. It is common for there to be abandoned drain passages at various levels in the shaft walls, representing former floor levels. Vertical shafts in Mammoth Cave range in size up to 30 feet in diameter and 200 feet from the top to bottom. They may intersect other passages as they grow, providing connections between different levels in the cave. Most shafts are still actively forming and contain water flow ranging from small trickles to major waterfalls.

Tubular passages are tunnels with elliptical or lenticular cross sections formed in the phreatic zone at or below the water table. The long dimension of their cross sections is controlled by bedding planes or, less commonly in Mammoth Cave, by joints or faults. They vary in size up to 30 feet high and 100 feet wide. An example is Cleaveland Avenue on the Half-Day Tour (named for a 19th century geologist; the spelling is correct). Tubular passages either form while completely filled by the water flowing through them, or are filled with water only during periods of high flow. In either case the solutional enlargement is not only downward but lateral and upward as well, so that broad, arched ceilings are formed. The overall gradients of these passages are very low, but in places there may be abrupt jogs upward or downward across the bedding for as much as 30 feet. As the Green River on the surface erodes downward, the underground water seeks lower paths through the limestone, abandoning the older phreatic passages as relict features within the vadose zone.

Tubular passages show no consistent relationship to the dip of the strata, because phreatic water has no inherent tendency to flow straight down the dip. Instead they develop along the most efficient flow path to the nearest available spring outlet. Because of the prominent bedding and the decreasing openness of partings and fractures with depth, many tubular passages follow a single stratigraphic horizon for thousands of feet, or even miles, commonly with a nearly strike-oriented trend along the contemporary water table.

Most large passages in Mammoth Cave have vadose sections in their upstream ends and phreatic sections in their downstream ends. In an actively forming passage the transition between the two sections is obvious, but the transition is also rather clear in a dry, abandoned passage. The passage may change from a relatively steep canyon to a nearly horizontal tube, and from a predominantly dip orientation (i.e., controlled by dip of beds) to a direction other than down the dip. A fine example of this transition is where the upper level of Boone Avenue, a canyon, changes to the tubular Cleaveland Avenue.

The major cave passages cluster at several significant levels: 600 feet, 550 feet, and 500 feet above sea level. The Green River is presently at 425 feet, and any cave passages below this elevation are completely flooded. The upper level (600 feet) consists of wide canyons and tubes partly filled with sediment. These are the largest and oldest passages in the cave. Audubon Avenue is an example of a wide canyon at this level that is filled with stream-deposited sand and gravel to more than half its height, so its canyonlike shape has been obscured. Apparently the filling took place soon after the passages at this level formed, while they still carried water. These passages correlate in elevation and probably also in age with the surface of the Pennyroyal Plateau. Their large size and sediment fill indicate slow entrenchment by the major rivers in the area, interrupted by at least one period of aggradation (deposition). The two lower levels are formed mainly by large tubular passages that represent periods of rather static regional base level. It is important to note that the cave levels are not the result of favorable zones caused by differences in bedrock solubility, for each level consists of passages that occupy different sedimentary layers, depending on the dip of the rocks.

By examining abandoned cave passages, it is possible to reconstruct former patterns of ground-water flow in the region. Although specific flow routes have undergone a great amount of capture (piracy) and diversion throughout the history of Mammoth Cave, the overall pattern of flow has remained rather constant. Recent dye tests by the National Park Service show that water passes through Mammoth Cave from sources as far as ten miles away in the Pennyroyal Plateau. With so much water flowing through open conduits in the limestone, great care must be taken to avoid contamination of water supplies in the region by chemical spills and improper waste disposal.

Solutional Features in the Cave

In addition to the cave passages themselves, several minor solutional features provide evidence of past flow conditions.

Scallops are asymmetrical solution hollows formed by turbulent water flow. The bedrock surfaces of most canyons and many tubular passages are covered with a continuous dimpling of scallops. The scallops vary in length from roughly one-half inch to several feet, with the length inversely proportional to the velocity of the water that formed them. Their asymmetry also indicates the direction of flow, as with sand ripples or dunes; the steep faces of the scallops are oriented in the downstream direction. Scallops are easily visible in Rose's Pass on the Half-Day Tour and in River Hall on the Historic Tour.

Flutes are parallel grooves formed in cave walls by water dripping vertically or flowing down a steep slope. They are aligned in the direction of flow and resemble the fluting of the columns of classical architecture. Flutes are very common in vertical shafts, such as Mammoth Dome on the Historic Route and the minor shafts in Boone Avenue on the Half-Day Tour.

Anastomoses are small, winding tubes that interconnect with each other in a braided pattern, normally along bedding-plane partings. The tubes have semicircular to circular cross sections and are dissolved upward into the base of the overlying bed. They range in diameter from about one-fourth inch to a foot. Anastomoses open into the walls of many tubular passages, and in some places they are exposed in the cave ceiling where the underlying bed has dropped or dissolved away. They are usually interpreted as remnants of the initial solution channels from which the cave passages developed. However, recent evidence suggests that they are more readily formed during periodic floods late in the solutional history of a passage, when solutionally active floodwaters fill the passage under pressure, forcing water into partings in the cave walls. Remnants of anastomoses can be seen in Kentucky Avenue on the Half-Day Tour and in Little Bat Avenue on the Historic Tour.

Cave Deposits

Mammoth Cave contains most of the speleothem types that occur in Carlsbad Caverns (chapter 39), but they are concentrated in relatively few places. Nearly all of them are still actively growing today. Various forms of travertine ($CaCO_3$) can be seen on the Frozen Niagara Tour and in a few scattered places throughout the other tours. These include flowstone (fig. 40.1), stalactites, stalagmites, columns, rimstone, and popcorn (chapter 39). The largest of these features is the 75-foot-high flowstone drapery called the Frozen Niagara.

Travertine is deposited in caves only when the water infiltrating from the surface is able to lose carbon dioxide (CO_2) to the cave air. This requires that the water first pass through an overlying soil in which the carbon dioxide content is higher than that of the cave. Vadose seepage passes through the soil, picking up a large amount of carbon dioxide, and readily dissolves the underlying limestone. If the seepage takes place in rather small quantities, it quickly approaches saturation with respect to the limestone. When seepage water loses carbon dioxide to the cave air, it becomes supersaturated and precipitates some of its dissolved carbonate as travertine. Large discrete flows of vadose water, particularly those that enter the ground through sinkholes, do not possess such a high carbon dioxide content and may even gain, rather than lose, carbon dioxide in the cave. They remain undersaturated and account for the solutional development of cave passages.

The insoluble cap rock that overlies most of the ridge in which Mammoth Cave is located severely limits the amount of water infiltrating into the cave. Only around the perimeter of the ridge, where the cap rock has been eroded away, is it possible for large amounts of seepage to enter the cave through soil-covered limestone and to deposit large amounts of travertine.

Beneath the insoluble cap rock, passages that have been abandoned by the water that formed them are generally quite dry. Capillary water is drawn toward them in small quantities from the moist surrounding limestone and evaporates when it emerges into the cave. Many varieties of evaporite minerals are deposited by this water. The most common is gypsum ($CaSO_4 \cdot 2H_2O$), which is derived at least in part from the oxidation of pyrite (FeS_2) in the limestone and overlying sandstone. The process is similar to the sulfuric-acid reaction and gypsum deposition described in chapter 39, except that the capillary nature of the water produces speleothems quite different from the bedded gypsum of Carlsbad Caverns. In general, the gypsum in Mammoth Cave forms a thin white or golden crust on the cave walls up to several inches thick. Where the deposition of gypsum is particularly intense, delicate *gypsum flowers* grow from the walls as curved fronds, forming clusters that resemble lilies or cotton. Many gypsum deposits are seen on the Half-Day Tour, particularly in Cleaveland Avenue.

Mirabilite ($Na_2SO_4 \cdot 10H_2O$) and *epsomite* ($MgSO_4 \cdot 7H_2O$) crystallize as long hairlike tendrils or as cottony tufts, and nitrates are carried into the cave by water that has passed through organic material at the surface. In the saltpeter mining operation of the early 19th century, calcium nitrate was leached out of the cave sediment, and the resulting solution was mixed with wood ashes to produce potassium nitrate, one of the essential ingredients of gunpowder (fig. 40.4).

Deposits in Mammoth Cave resulting from mechanical weathering include sediment carried in by cave streams and breakdown blocks that have fallen from the ceiling and walls. The sediment consists of gravel, sand, silt, and clay derived mainly from weathering of the overlying insoluble rocks, plus a small insoluble residue from solution of the limestone. *Breakdown* is collapse material in the form of blocks, slabs, and chips. It is produced in several ways: by solution along the partings and joints that bound a block of limestone; by the loss of buoyant force on the ceiling when a passage is drained of its water by lowering of the water table; by sagging and peeling of limestone beds in the cave ceiling after the underlying beds have been removed by solution; and by the wedging effect of minerals such as gypsum that grow within fractures in the limestone. Breakdown of a passage ceiling usually progresses upward to the top of a weak rock unit such as shaly limestone, and afterward the passage is relatively stable. Where one passage crosses over another, breakdown may occur between them, producing a junction room. Several entrances to the cave have been formed by the intersection of a passage by surface erosion, but in most such places the incipient entrance is totally obscured by breakdown.

Geologic History

1. Deposition of limestones during the Mississippian Period, 350 to 325 million years ago.

The shallow continental sea, in which the St. Louis Limestone, the Ste. Genevieve Limestone, and the Girkin Formation were deposited, had covered most of the southern part of the North American craton since early in Paleozoic time. Thus many beds of older siltstones and impure limestones (not exposed in Mammoth Cave) underlie the 400 to 500 feet of uninterrupted Mississippian limestones (fig. 40.3). The climate gradually changed, as the limestones accumulated, from relatively arid to humid, with progressively greater amounts of detrital sediment being carried in from land areas to the north. Downwarping of the Illinois Basin occurred simultaneously with this deposition, imparting a gentle northwesterly dip to the rocks.

2. Deposition of predominantly detrital rocks late in Mississippian time and in the Pennsylvanian Period.

Increasing amounts of detrital sediment accumulated in the continental sea, producing the varied rocks of the Chester Series, which today (with the exception of the limestones of the Girkin) form the largely insoluble cap rock overlying the cavernous limestone of Mammoth Cave National Park (table 40.1). A drop in sea level relative to the surface of the continent allowed extensive stream erosion of the Mississippian rocks. With a subsequent rise in sea level early in the Pennsylvanian Period, conglomerate and sandstone were deposited over the erosion surface.

3. Slow subaerial weathering and erosion, beginning in late Paleozoic time.

The Pennsylvanian rocks are the last to have been deposited in the Mammoth Cave region. Since then, several hundred feet of Pennsylvanian and late Mississippian rocks have been removed by erosion, exposing the Mississippian limestones over broad areas. Weathering and erosion were rather slow, probably interrupted by periods of aggradation in river valleys. Karst features began to develop wherever limestone was exposed at the surface. Mammoth Cave belongs to the last stages of this erosional history.

4. The forming of Mammoth Cave late in Tertiary time, perhaps 30 to 5 million years ago.

The earliest passages to form in Mammoth Cave were large tubular and canyon passages now located 570 to 690 feet above sea level. The highest of these (in Crystal Cave, not open to the public) lies above the level of the Pennyroyal Plateau and has an irregular ceiling profile, indicating that it originated below the contemporary elevation of the Green River. Therefore, this passage predates the Pennyroyal surface. Later aggradation filled the passage nearly to the ceiling with stratified silt, sand, and gravel.

The major late Tertiary passages originated during the period of rather slow fluvial entrenchment that allowed the nearly flat Pennyroyal Plateau surface to form. Erosion was interrupted by at least one period of aggradation. Cave passages were formed by sinking streams draining from the Pennyroyal Plateau and from karst valleys in the Chester Upland. Audubon Avenue, Broadway, Main Cave, and Kentucky Avenue were formed at this time as a single enormous passage that has since been segmented by breakdown.

5. Development of lower cave levels during the Pleistocene Epoch.

Pleistocene glacial advances fell short of the Mammoth Cave region, but their influence on the erosional history of the region was profound. Prior to continental glaciation, the Ohio River was not nearly so large as it is today. Drainage from most of the Appalachian Mountains flowed through what is now northern Indiana into the Mississippi River, bypassing the Ohio River entirely. Early Pleistocene glaciers blocked this drainage, diverting it into the Ohio River, which was located approximately along the farthest advance of the ice. The greatly increased discharge caused rapid entrenchment of the Ohio. The Green River, which controls the levels of passage development in Mammoth Cave, is a tributary of the Ohio River, so it also began to entrench rapidly. As a result, the regional water table dropped and streams in the caves were diverted to lower routes. All but the largest surface streams on the Pennyroyal Plateau responded in a similar way, flowing underground as sinking streams.

Several lengthy pauses in fluvial entrenchment occurred throughout the Pleistocene, accompanied by minor interludes of aggradation. Each pause allowed the cave streams to remain stable at the same level for a long period of time, enlarging the cave passages at that level to comparatively great size. The longest pauses resulted in major tubular passage levels at the present altitudes of 550 and 500 feet. Cleaveland Avenue is the best example at the 550-foot level, reaching widths of 50 to 60 feet and a height of 15 to 20 feet. Great Relief Hall is part of the 500-foot level. As the water level dropped in the cave, travertine and evaporite speleothems began to accumulate in dry passages.

Terraces formed in the Green River valley during the periods of relatively static base level, and they correlate in elevation with the cave levels of the same age. Only the 500-foot-level terrace is relatively intact today, as the higher terraces have been severely eroded. Tentative dating of the terraces and cave deposits by various methods, including uranium-thorium radiometric dating of travertine and paleomagnetic measurements of sediment, indicate that the 550-foot level is probably of middle Pleistocene age (approximately 700,000–800,000 years old) and that the 500-foot level is probably of late Pleistocene age (at least 350,000 years old).

One of the latest geomorphic events has been the filling of the Green River valley with 30 to 50 feet of alluvium late in the Pleistocene Epoch, probably during Wisconsin time when the last major ice advance occurred. As a result, the lowest passages in Mammoth Cave are now completely water filled.

Natural entrances to Mammoth Cave have probably existed since the cave first began forming, for instance at springs and at the points where sinking streams entered the ground. However, the present entrances are mainly secondary features created by the random intersection of cave passages by surface erosion. The Historic Entrance is an example of an entrance formed where an upper-level passage was intersected by a minor tributary valley of the Green River valley.

The processes that formed Mammoth Cave continue to operate today. For this reason it is an ideal "underground laboratory" for studying the ways in which caves and their related karst features evolve in a humid environment.

Geologic Maps

Haynes, D. D. 1962. *Geology of the Park City quadrangle, Kentucky*. U.S. Geological Survey Map GQ-183.

————. 1964. *Geology of the Mammoth Cave quadrangle, Kentucky*. U.S. Geological Survey Map GQ-351.

————. 1966. *Geology of the Horse Cave quadrangle, Kentucky*. U.S. Geological survey Map GQ-558.

Klemic, H. 1963. *Geology of the Rhoda quadrangle, Kentucky*. U.S. Geological Survey Map GQ-219.

Bibliography

Brucker, R. W., and Watson, R. A. 1976. *The longest cave*. New York: Alfred A. Knopf.

Harmon, R. S., Thompson, P., Schwarcz, H. P., and Ford, D. C. 1975. Uranium-series dating of speleothems. *National Speleological Society bulletin* 37:21–33

Hess, J. W. 1976. A review of the hydrology of the Central Kentucky Karst. *National Speleological Society bulletin* 38:99–102.

Hill, C. A. 1981. Origin of cave saltpeter. *National Speleological Society bulletin* 43:110–126.

Livesay, A. 1961. *Geology of the Mammoth Cave National Park area*. Revised by P. McGrain. Kentucky Geological Survey, Series 10, Special Publication 7.

McGrain, P., and Walker, F. H. 1954. Geology of the Mammoth Cave region, Barren, Edmonson, and Hart Counties, Kentucky. Kentucky Geological Survey Field Trip Log.

Meloy, Harold. 1971. *Mummies of Mammoth Cave*. Shelbyville, Indiana: Micron.

Miotke, F.-D., and Palmer, A. N. 1972. *Genetic relationship between caves and landforms in the Mammoth Cave National Park area*. Würtzburg, Germany: Böhler Verlag.

Palmer, A. N. 1981. *A geological guide to Mammoth Cave National Park*. Teaneck, N.J.: Zephyrus Press.

Pohl, E. R. 1970. Upper Mississippian deposits in south-central Kentucky. Kentucky Academy of Sciences, *Transactions* 31:1–15.

————., and White, W. B. 1965. Sulfate minerals: their origin in the Central Kentucky Karst. *American mineralogist* 50:1462–1465.

Quinlan, J. F., and Rowe, D. R. 1977. Hydrology and water quality in the Central Kentucky Karst: Phase I. University of Kentucky, Water Resources Research Institute, Research Report No. 101.

Schmidt, V. A. 1982. Magnetostratigraphy of sediments in Mammoth Cave, Kentucky. *Science* 217(4562): 827–829, August 27.

Thrailkill, John. 1968. Chemical and hydrologic factors in the excavation of limestone caves. *Geological Society of America bulletin* 79:19–46.

Watson, P. J., ed. 1974. *The archeology of the Mammoth Cave area*. New York: Academic Press.

White, W. B., Watson, R. A., Pohl, E. R., and Brucker, R. W. 1970. The Central Kentucky Karst. *Geographical review* 60:88–115.

Address

Mammoth Cave National Park
Mammoth Cave, Kentucky 42259

41

Wind Cave National Park

by Rodney M. Feldmann
Kent State University

Location: Southwest South Dakota
Area: 28,292.08 acres; 44.21 square miles
Established: January 9, 1903

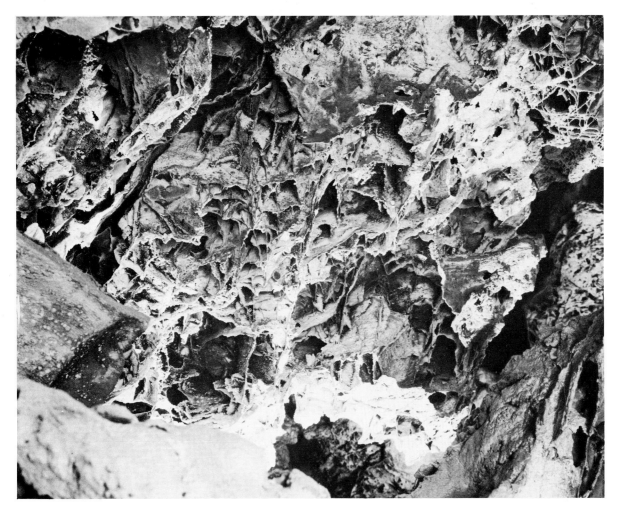

Figure 41.1   Boxwork results when a fracture system is filled in by secondary calcite which is left standing in relief, following solution of the surrounding original bedrock. Wind Cave is noted for this unusual feature. National Park Service photograph.

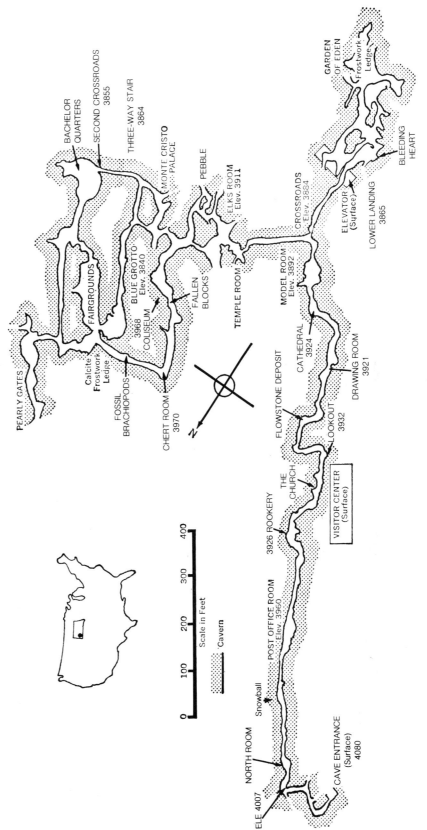

Figure 41.2 Wind Cave National Park, South Dakota.

Table 41.1. Geologic Column, Wind Cave National Park

Time Units		Rock Units	Geologic Events
Era	Period	Formation	
Cenozoic	Quaternary	alluvium	Final erosion of Park landscape; development of solution and precipitation features in Wind Cave.
Cenozoic	Tertiary / Paleo-gene	Brule Formation / Chadron Formation	Final uplift of Black Hills; deposition of terrestrial sediments from erosion of the Hills.
Mesozoic	Cretaceous	Fall River Sandstone / Fuson Shale / Lakota Sandstone	Uplift of Black Hills Late Cretaceous associated with Laramide orogeny; deposition of marine and nonmarine sediments throughout most of the Mesozoic Era.
Mesozoic	Jurassic	Morrison Formation / Unkpapa Sandstone / Sundance Formation	
Mesozoic	Triassic	Spearfish Formation	
Paleozoic	Permian	Minnekahta Limestone / Opeche Formation	Deposition of marine and some nonmarine sediments; formation of karst terrane on Pahasapa surface.
Paleozoic	Pennsylvanian	Minnelusa Formation	
Paleozoic	Mississippian	Pahasapa Limestone / Englewood Formation	Deposition of limestone in which Wind Cave is formed.
Paleozoic	Ordovician	Whitewood Formation	Deposition of marine clastic and carbonate rocks
Paleozoic	Cambrian	Deadwood Formation	
Precambrian		Harney Peak Granite; unnamed igneous and metamorphic rocks	Complex history of deposition of sediments, tectonic activity, metamorphism, and intrusion.

Source: modified from Deike and White 1961

Local History

The Sioux Indians (or Dakotas) called the cave the "Sacred Cave of the Winds." They had several legends about it, one telling that this was the cave through which their ancestors arrived at the Black Hills. Another told that Wokan Tanka (the "Great Mystery"), sent the buffalo out into their hunting grounds from this cave.

The area is filled with legends, even as to how the cave was discovered. One story claims it was discovered in 1877 by "Lame Johnnie," a notorious outlaw and stagecoach robber. Another story claims a cowpuncher was riding by when the wind blew his hat off and into a small hole. When he went to look for it he discovered the cave. Since the opening was very small, he could not retrieve the hat and was forced to leave without it. The next day he rode that way again and the wind brought his hat back to him—and was even nice enough to place it on his head.

The official discoverer was Tom Bingham, one of the pioneer settlers in the Black Hills. As he was deer hunting one day, he heard a strange whistling sound and began to search for the source. He found a small hole eight by twelve inches in size through which the wind was rushing and making noise. Since the natural opening was too small to permit entrance, another was dug nearby to a depth of six feet.

The region around the cave was open to claims. In 1890 the South Dakota Mining Company filed a claim on the cave. Jesse McDonald and his sons (Alvin and Elmer) were hired to take care of the property. Alvin McDonald did most of the exploration. He named many of the features in the cave, and kept a diary of all his mapping and exploration. He is now buried near the cave with a plaque marking the location.

In 1892, Jesse McDonald, Charles Stabler, John Stabler, and several other partners formed a company called Wonder Wind Cave Improvement Company. They took over the property, opened up some passageways, put in stairs, and opened the cave to the public. Magazine stories were printed about the cave in 1900 and 1901 which brought it to the attention of the government. President Theodore Roosevelt signed a bill to create the Park in 1903.

Geologic Features

Origin of Caves

Caves formed in limestone are produced when soluble rock material is jointed, or fractured, to permit passage of ground water along selective pathways. These pathways are then enlarged by solution of adjacent bedrock and subsequent precipitation of material as some form of cave deposit. Variations in rock type, jointing pattern, and ground water chemistry result in a wide diversity of forms and a broad spectrum of caves and cave deposits. Therefore, although Wind Cave is one of a number of caves preserved by the National Park Service, it illustrates features that are unique. For example, although most caves seem to be best known for development of dripstone, stalactites, and stalagmites, these features are not particularly common in Wind Cave. Rather, boxwork, formed very early in the development of the cave, is the dominant precipitate (fig. 41.1).

Primary and Secondary Features

As suggested above, development of caves involves an initial period of growth and enlargement, primarily by solution, followed by a period of infilling by precipitation of minerals from ground water. The formations observable in caves can be classified on the basis of whether they were formed during the enlargement stage, *primary features*, or during the stage of infilling, *secondary features*.

Primary Features. The most important primary feature developed in Wind Cave is the *boxwork*. In addition to the regional joint pattern that dictated the orientation of the major passageways in the cave (fig. 41.2), the Pahasapa Limestone became highly fractured sometime during its history. The orientation of the fractures differs somewhat from the regional joint pattern and is reflected on an entirely different scale. Fractures in which boxwork ultimately formed are separated from one another by inches, or fractions of inches. Calcite was deposited in these fractures as secondary filling by ground water. Later, the primary limestone was removed by solution, leaving the secondary material behind as thin fins that crisscross one another. Wind Cave is one of only two caves in the world in which boxwork developed, and nowhere is it developed in greater volume or wider variety. The other cave with boxwork is located in Czechoslovakia.

Cave earth, a residue of clay particles and other insoluble materials, is found in all caves. In Wind Cave, cave earth ranges from a thin veneer to layers several feet thick. It accumulated either as the insoluble residue from solution of limestone or as material washed into the cave from outside sources. The Pahasapa Limestone consists of about 98 percent soluble material, so that, given the large volume of cave earth in Wind Cave, it is likely that most of the material infiltrated the cave during weathering and erosion of overlying rocks.

Collapse blocks are the portions of the ceiling and walls that have detached themselves from the surface of the cave and are now lying on the floor. It is thought most likely that the blocks loosened and fell when the cave was flooded, because boxwork on the underside of the blocks is undamaged. Had the cave been dry, the impact of the fall would probably have shattered the delicate structures.

Collapse breccia is found in parts of the Pahasapa Limestone where the ceiling of the cave collapsed. This caused portions of the overlying Minnelusa rock to fall with the Pahasapa as a plug into the cavern.

Spongework, a maze of openings or pores ranging in size from a few inches to several feet across, is very common in Wind Cave. The passages thus formed produce a randomly oriented, three-dimensional network large enough, in some places, for a person to crawl through. Spongework appears to form early in the solution process as the most soluble material is selectively removed.

Secondary Features. Wind Cave has a few stalactites and stalagmites, but not very many. They are not discussed here as they have been described earlier (chapters 39 and 40). The same holds true for the other flowstone and dripstone features.

Cave lining was deposited when the cave was completely under water. The deposit formed during, or following, boxwork formation and coated all of the walls. Composition of the lining can be calcite, hematite, or quartz. The cave was drained several times and then resubmerged, with the composition of the material coating the boxwork varying from time to time. This occurred somewhere between 50 million years ago (Eocene) and 10 million years ago (Pliocene). Final draining took place at about the end of the Pliocene (2 or 3 million years ago). Erosional and weathering processes have removed much of the lining in the upper level.

Cave ice features, or calcite rafts, form in bodies of standing water in a cave. They are common in Wind Cave and are also found in Carlsbad Caverns. They are roughly rectangular pieces of calcite so thin that they are supported on top of the water by surface tension. They range in size from one to three inches across. Disturbance of any type causes them to sink, and new ones can form within a year. In the area of the cave referred to as the Calcite Jungle, the floor is covered with these rafts; they formed at a time when the water level was higher.

Popcorn (globulites) is a series of layered nodules (1 to 10 mm in diameter) supported on a very small stalk (1 mm long). If the nodules are covered with a bristlelike cluster of aragonite needles, they are called *frostwork*. This is the second feature for which the cave is famous. At first, popcorn was thought to have formed beneath the level of the water table, but at the present time it appears to be forming adjacent to a wall where dripping water splashes at the boundary between the top of the water and the air. Another possible point of origin is where water is seeping from the walls at the water line. Frostwork may also form as a coating of needles directly on the wall.

The needles may be formed of either aragonite or selenite. *Aragonite* is chemically the same as calcite but belongs to a different crystal system; calcite is hexagonal, whereas aragonite is orthorhombic. *Selenite* is a clear, cleavable form of gypsum. Temperature seems to determine which forms develop. For example, aragonite has a tendency to form in areas where the mean annual temperature is between 56°F and 63°F. Presently, the mean annual temperature of the region is about 50°F and the temperature of the cave is only 47°F, so no aragonite is now forming. However, Wind Cave is situated in an area of the Black Hills with known geothermal activity. For example, hot springs are currently active just south of the Park in the city of Hot Springs, and evidence in many surrounding areas suggests more extensive activity in the past. Therefore, the ground water temperature may have fluctuated through much higher temperatures in the past.

Several other forms of secondary deposits can be identified in the cave, such as *dogtooth spars*, which are well formed calcite crystals, and *hydromagnesite* or "moon milk," a white, earthy magnesium mineral. However, none of these forms occurs in Wind Cave in significant quantity.

Geologic History (Table 41.1)

1. Deposition of Paleozoic and Mesozoic beds on Precambrian basement.

During much of the Paleozoic and Mesozoic Eras, the Black Hills region was either slightly below sea level and receiving marine sediments, or was elevated slightly above sea level. Nearly the entire sequence of Paleozoic and Mesozoic rock units found elsewhere in the Black Hills is also found in Wind Cave National Park. The Pahasapa Limestone, in which the cave is developed, accumulated as a combination of organic fragments and calcareous mud in warm, shallow marine habitats during the Mississippian Period. Following deposition of the Pahasapa, some erosion and development of karst features resulted in a hummocky surface in the southern Black Hills. The overlying Minnekahta Formation was, therefore, deposited on an irregular surface. Preliminary investigation of the upper levels of the cave suggests that the Pahasapa surface may have had a few hundred feet of relief, but this has not been confirmed.

2. Uplift of Black Hills about 70 million years ago.

During Cretaceous and Paleogene time, the Black Hills region was subjected to forces that elevated two crustal blocks of the Precambrian basement and the overlying sedimentary rock sequence. The result was an ovoid dome about 150 miles long and 60 miles wide (fig. 41.3).

Had the uplift occurred all at once with no accompanying erosion, the Black Hills would have been some 7,000 feet higher than they are today. Instead, erosion has stripped sediments from the central part of the Black Hills and redeposited them elsewhere. Some of the sediments were deposited on the flanks of the uplift during Oligocene and Miocene time and are now exposed in the nearby badlands.

3. Formation of Wind Cave in the Mississippian Pahasapa Limestone.

The cave has developed in the upper 100 to 150 feet of the Pahasapa Limestone Formation, which has a total thickness of 300 to 600 feet in the Black Hills area. Perhaps cave development began as early as middle Eocene (50 million years ago), but as late as Pliocene (10 million years ago) is more likely. In either case, Wind Cave is one of the oldest in the world, as most caves have formed only in the last million years.

As is typical of any cave, Wind Cave developed along the regional joint system and passed through a solutional and a depositional phase when it was below the ground water level and above the ground water level, respectively. The regional joint system, developed during the Black Hills uplift, trends northwest-southeast and northeast-southwest. Examination of the plan of the cave shows that the northwest-southeast direction has been the primary compass direction of cave development. A third direction, parallel to the bedding in the Pahasapa, is also significant. Much of the growth of Wind Cave has been lateral; this is reflected in the observation that the cave is confined to the upper portion of the formation, even though jointing extends throughout this and other sedimentary units in the area. This suggests that the special combination of conditions necessary to cave development, including opening of the joint system, position relative to the ground water level, and water chemistry, were met only in the upper portion of the Pahasapa.

4. Development of cave zones.

Since development is mostly lateral, no well defined individual levels have formed and the cave floor slopes gently. Total relief is only 450 feet. The differences in lithology have produced various zones. The upper zone forms in the breccia (angular debris formed in ancient sinkholes). No boxwork exists, nor does cave lining. The upper zone contains mainly crawlways of small and mazelike spongework. The middle zone contains the biggest caverns, well developed boxwork, and cave lining. There has not been complete development of the passages in the lowest zones. The passages are narrow and strongly controlled by the joint system. The Calcite Jungle is located here, and the zone contains dogtooth spar crystals and cave rafts or "ice."

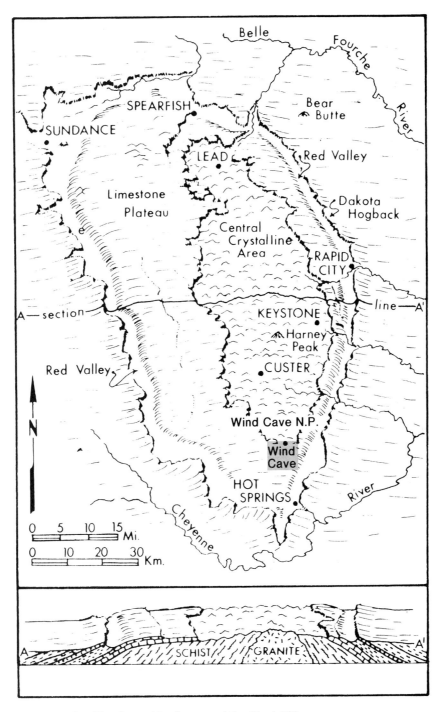

Figure 41.3   Physiographic diagram of the Black Hills
showing the relationship between the broad domal structure
and the erosional expression. From *The Black Hills* by R. M.
Feldmann and R. A. Heimlich. © 1980 Kendall/Hunt
Publishing Company. (**Modified from Thornbury, 1954**).

5. Development of boxwork and other features.

The fracture system that developed was filled with secondary calcite. Then the more soluble limestone was dissolved, leaving calcite fins or boxes. The colors are various shades of yellow, brown, pink, and blue. In some chambers (e.g., Fairground) the calcite is fluorescent and glows blue, green, red, and violet under "black light."

The boxwork has a preferred direction of N 70°–80° E and a second preference of N 10°–20° W. The boxwork even cuts through geodes formed in the limestone, showing that the geodes developed first and the fracture systems later.

6. Erosion and formation of topography.

The climate is semiarid today as is shown by the *intermittent stream* (that is, it does not flow year round) that flows through Wind Cave Canyon and Red Valley. The Wind Cave Canyon Creek may flow through the cave at times when it has sufficient water. Evidence of stream flow through the cave can be found near the Garden of Eden.

Beaver Creek is the only permanent stream in the Park. It cuts across the ridges or hogbacks that are found in the region. The landscape is a combination of steep canyons, hills, and flat upland areas. Origin of these flats (e.g., Bison Flats) is uncertain. They could be structural or erosional.

All in all, Wind Cave is a remarkably ancient cave, with at least 28 miles of passageways that have been mapped, and many other unmapped (but known) ones. Because of stratigraphy the cave can only enlarge along the strike (trend) of the beds, for up and down the dip it intersects either the surface or the water table. It is conceivable that 50 to 100 miles of passageways could exist.

Geologic Map and Cross Section

Map, Wind Cave National Park, Hot Springs, South Dakota.

Robertson, F. *Geologic map portfolio—a laboratory study for historial geology.* Williams & Heintz Pub. Map no. 7.

Bibliography

Darton, N. H. 1909. *Geology and water resources of the northern portion of the Black Hills and adjoining regions in South Dakota and Wyoming.* U.S. Geological Survey Professional Paper 65.

Deike, G. H., III; and White, W. B. 1961. *Preliminary report on the geology and mineralogy of Wind Cave.* National Park Service.

Feldmann, R. M.; and Heimlich, R. A. 1980. *The Black Hills,* K/H Field Guide Series. Dubuque, Iowa: Kendall/Hunt Publishing Co.

National Park Service. n.d. *Subsurface geology of Wind Cave National Park.* (mimeo)

———. n.d. *Surface geology of Wind Cave National Park.* (mimeo)

Palmer, A. N. 1981. *The geology of Wind Cave.* Wind Cave National Park, South Dakota: Wind Cave Natural History Association.

Thornbury, W. D. 1954. *Principles of geomorphology.* New York: John Wiley and Sons, Inc.

Address

Wind Cave National Park
Hot Springs, South Dakota 57747

Part VI
The Last Frontier

© by G. Shimer

Glaciers and alpine topography in the St. Elias Range.
Wrangell-St. Elias National Park and Preserve, Alaska.

Vast areas of arctic and subarctic wilderness, great
mountain ranges, wild rivers, glaciers, icecaps, seashores, active volcanoes, and much more are found in
the new national parks in Alaska. The geology of Alaska
is exceedingly complex, especially that of southern
Alaska, which is one of the most tectonically active regions in the world. Chapter 42 contains capsule descriptions of the geology of the following parklands:

Gates of the Arctic National Park and Preserve
Kobuk Valley National Park
Lake Clark National Park and Preserve
Katmai National Park and Preserve
Kenai Fjords National Park
Wrangell-St. Elias National Park and Preserve
Glacier Bay National Park and Preserve

42

An Introduction to the Geology of the New Alaskan National Parklands

by Sherwood D. Tuttle
University of Iowa

Figure 42.1 Grand Plateau Glacier in Glacier Bay National Park and Preserve extends almost to the Gulf of Alaska (body of water in the foreground). The knobby, forest-covered ridge just in from the ocean shoreline is an end moraine built when the glacier was longer than it is now. The glacier now calves into the lagoon behind the moraine. Debris and vegetation cover most of the glacier's snout. Snow and ice to feed Grand Plateau Glacier accumulate in the Fairweather Mountains in the background. National Park Service photograph by Robert Belous.

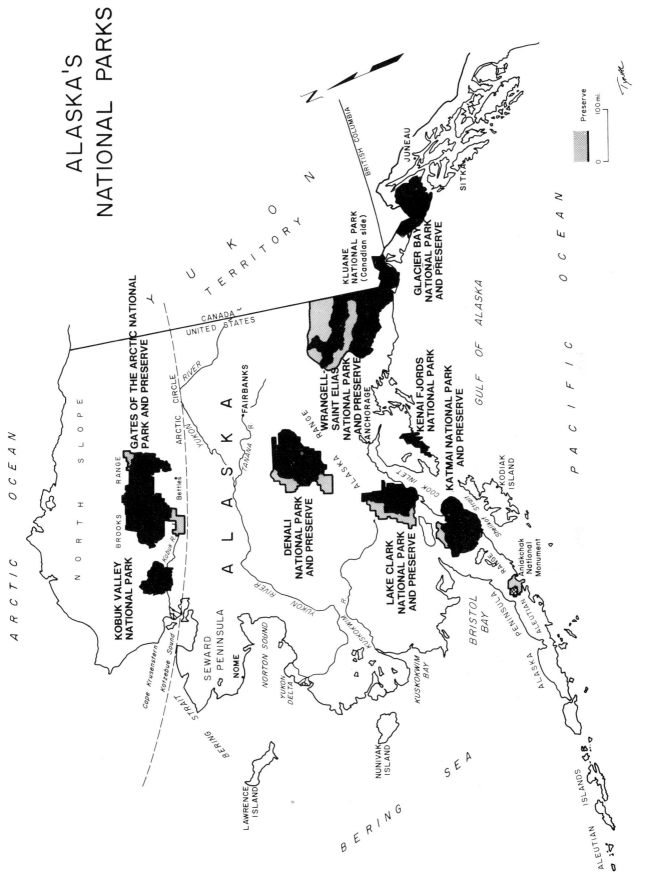

Figure 42.2 Alaskan National Parks and Preserves.

The Alaska National Interest Lands Conservation Act

Passage of this act, commonly called the "Alaska Lands Bill," came about as the result of legislative compromise in the 1980 "lame duck" session of the 96th Congress. President Jimmy Carter signed the act on December 2, 1980. Under the provisions of the new law, an amount of Alaska land greater than that contained in the states of California and Maine combined came under federal control. The new protected areas include 43.6 million acres (68,125 square miles) for national parks, preserves, and monuments; more than double the amount of land in the National Park System formerly. Seven new national parks were established; in addition, Mount McKinley National Park was more than doubled in size and redesignated Denali National Park and Preserve. (The geology of the Denali-McKinley area is the subject of chapter 11.)

Parts of some of the new parks have been designated as "preserves," which means that both subsistence and sport hunting and fishing are permitted. Subsistence hunting and fishing are allowed *only* in specific national park areas. Subsistence is defined in the new law as "the customary and traditional uses by rural Alaska residents of wild, renewable resources for direct personal or family consumption. . . ." The intent of the law is that the native peoples of Alaska be deemed as worthy of protection as geological, archeological, and other resources.

A unique provision in the Alaska Lands Law calls for formal arrangements by which state and federal authorities are to plan for and manage the new parklands. Many of the new areas have no visitor facilities, and access, even by bush aircraft, may be quite limited. Some areas may need to be kept as pristine wilderness, while others may be suitable for at least partial development. Areas under federal control but *not* designated as national parks may at some time in the future have to be exploited for their natural resources, including mineral resources. The problem, for which there are no easy solutions, is this: "On the one hand, we must preserve for future generations what remains of our . . . irreplaceable virgin wilderness. At the same time, we must also find ways of using the natural resources these wilderness areas contain to meet the energy and raw material needs on which the survival of our children depend."* The Alaska Lands Law is a significant and responsible step toward effective conservation and management of natural resources through federal, state, and local cooperation.

Considering the size and diversity of the state of Alaska, the selection of areas remarkable enough to be elevated to full national park status was a formidable task. That an agreement—through long and sometimes heated discussions and difficult compromises—was finally reached

*Congressman James Leach (R-Iowa), personal communication.

must be regarded as a major accomplishment. In this chapter, we attempt to show, from a mainly geological viewpoint, the principal reasons for these selections. The descriptions of the new national parks are necessarily brief and introductory. Investigations of Alaskan geology are ongoing but far from complete. Moreover, new interpretations of the data and observations accumulated in past studies will continue to refine geological concepts regarding the bedrock, structures, tectonics, and landscapes of the 49th state.

The new national parks included in this discussion are as follows: Gates of the Arctic National Park and Preserve, Kobuk Valley National Park, Lake Clark National Park and Preserve, Katmai National Park and Preserve, Kenai Fjords National Park, Wrangell-St. Elias National Park and Preserve, and Glacier Bay National Park and Preserve (fig. 42.2). Five additional areas were brought into the National Park System as monuments or preserves, but they are not described here except for Aniakchak National Monument and Preserve (fig. 42.5), which was given brief mention because of its particular geological significance in Alaskan tectonics.

Gates of the Arctic National Park and Preserve

Location: North Central Alaska
Area: 9,164,000 acres; 14,819 square miles
Proclaimed a National Monument: December 1, 1978
Established as a National Park and Preserve: December 2, 1980

Geographic Setting

This vast wilderness Park and Preserve, entirely north of the Arctic Circle, extends for some 200 miles from east to west along the central part of the Brooks Range. From south to north, Gates of the Arctic ascends from long, broad, glaciated valleys and rounded foothills to the sharp arêtes and turrets of the crest, and then goes down the northern foothills onto the flat tundra of the North Slope and the Arctic Coastal Plain. The region was explored and named about 50 years ago by Robert Marshall of the U.S. Forest Service. As he was following up the North Fork of the Koyukuk, he saw ahead two great, facing peaks framing the valley he was ascending. This was the magnificent landscape that he named the Gates of the Arctic.

The next river to the west, the John River, rises in the vicinity of Anaktuvuk Pass on the crest of the Brooks Range. This pass is a migration route for caribou, which move in great herds each spring from the interior valleys over the range and down to the tundra on the North Slope where they spend the summer. Anaktuvuk village in the pass is the home of a Nunamuit Eskimo tribe who live in the Park and Preserve.

Mount Igikpak, rising to an elevation of 8,510 feet among the Arrigetch Peaks, is the highest point on the Brooks Range (fig. 42.3). The Arctic Divide, which is the continuation of the Continental Divide through Alaska, follows the crest of the Brooks Range. The northern limit of wooded country (the "tree line") passes below the crest of the Brooks Range along the southern slopes. Thus the southern foothills and valleys of the Park and Preserve are in the boreal forest zone. The summit areas of the range are barren rock and frost-broken rubble. Beyond, to the north, there is only the low-growing vegetation of the tundra, stunted shrubs, lichens, mosses, and herbs. *Permafrost,* the permanently frozen ground of the far north, is at or near the surface throughout the Park and Preserve. Here in the Arctic most of the permafrost is "dry"; that is, the frozen ground contains little moisture below that which collects on the surface.

Winters are long and cold in Gates of the Arctic; summers are short and mild. In the lowlands midsummer temperatures occasionally rise into the 80s (F.) during the long daylight hours. Autumn begins in August and is usually rainy. In this greatest remaining wilderness of North America is habitat vital to the Arctic caribou, the grizzly bear, Dall sheep, moose, wolves, and raptors (birds of prey).

Access to Gates of the Arctic is by air from Fairbanks (200 miles to the southeast) and Bettles, a small village outside the southern boundary of the Park and Preserve.

Figure 42.3 Glacial sculpturing on Mount Igikpak and the Arrigetch Peaks in Gates of the Arctic National Park and Preserve. Mount Igikpak (elevation, 8,510 feet) is the highest peak in the Brooks Range. Sharp arêtes lead up to Igikpak's horn, which is capped by twin turrets. Snow blowing off the crest helps to nourish the glaciers in this region of low to moderate precipitation. Occupying the large cirque in the left foreground is a valley glacier, its surface broken by crevasses where it flows over the lip of the cirque. Quantities of frost-loosened rubble, fallen from the precipitous rock walls, form a surface moraine that blackens the top of the moving ice. Note that the lineated pattern of the metasedimentary rocks strongly influences weathering and mass wasting processes. National Park Service photograph by Robert Belous.

The Trans-Alaska Pipeline and private road (which goes from Prudhoe Bay on the Arctic Ocean to Anchorage) skirts the eastern boundary of the Park and Preserve, but this route is not open to the public. No roads or established campgrounds are within the Park and Preserve area, but campsites are available for visitors who wish to backpack or canoe into the wilderness. The long glacial valleys, alpine lakes, and clear-water rivers are the main scenic attractions for visitors. Six major streams that drain the slopes of the Brooks Range have been designated National Wild Rivers; namely, the Alatna, John, Kobuk, Noatak, North Fork of the Koyukuk, and Tinaykuk. Many of the lakes and smaller rivers in the area are as yet unnamed. Fishing is permitted in the Gates of the Arctic; but no sport hunting is allowed at the present time.

Geological Features

The Brooks Range, named for the U.S. Geological Survey geologist Alfred H. Brooks, who explored much of Alaska early in this century, is an east-west-trending range of "fold mountains." The bedded sedimentary and metamorphic rocks of the mountain mass, ranging in age from Silurian to Mississippian, are exposed along the summits of the range. On the flanks are younger sedimentary rock units, Permian through Cretaceous in age. The rocks of the Brooks Range were strongly deformed early in Tertiary time by tight folding, uplift (more in the center of the range than to the east or west), and thrust faulting from south to north. During and following this tectonic episode, the mountains were extensively eroded, but they were uplifted again in late Tertiary time. In terms of age, types of sedimentation, and tectonic history, the rocks of the Brooks Range may be related to the rocks of the Northern Rocky Mountains and the Canadian Rockies. The Brooks Range region has probably been part of the North American plate for many millions of years (see box 42.2).

The landscapes and topography of the Gates of the Arctic are the result of long-continued sculpturing by glaciers, streams, mass wasting, and weathering. Glaciation was not so extensive here in Pleistocene time as it was in the mountains of southern Alaska where precipitation is higher. However, large cirques, horns, arêtes, rock glaciers, and the long, deep, U-shaped valleys all attest to the intensity of Pleistocene glacial episodes. In Gates of the Arctic at the present time, a few cirque glaciers are active among the higher peaks (fig. 42.3).

Address

Gates of the Arctic National Park and Preserve
P.O. Box 74680
Fairbanks, Alaska 99707

Kobuk Valley National Park

Location: Northwest Alaska
Area: 1,710,000 acres; 2,672 square miles
Proclaimed a National Monument: December 1, 1978
Established as a National Park: December 2, 1980

Geographic Setting

The broad valley of the Kobuk River, which empties into Kotzebue Sound, north of Bering Strait, has been a major transportation route into the northern Alaskan interior since prehistoric time. The Kobuk River rises in the Brooks Range on the south side of the Arctic Divide, and then flows nearly due west to the ocean. Kobuk Valley National Park occupies a mountain-rimmed basin, some 75 miles inland from the river's mouth. This wilderness park, entirely within the Arctic Circle, was set aside for the protection of (1) a number of significant archeological sites that reveal more than 10,000 years of human habitation, (2) about 25 square miles of active sand dunes, and (3) habitat and migration routes of many species of arctic wildlife.

The Baird Mountains, 3,500 to 5,000 feet in elevation, enclose the northern side of Kobuk Valley National Park. The northern limit of wooded country runs across the northern part of the Park. Thus the area is in a transition zone between the boreal forests and the Arctic tundra. Bordering the south side of the Kobuk Valley are the low Waring Mountains, 1,000 to 2,000 feet high. The dunefields are between the river and the Waring Mountains on the flat flood plain. The Kobuk River, joined by numerous tributaries coming down from the mountains, meanders over this flood plain, which is three to eight miles wide and covered with vast quantities of glacial drift, outwash, alluvial material, and many acres of dunes, both stabilized and unstabilized (fig. 42.4). Permafrost underlies the whole area, keeping drainage water near the surface in thaw lakes, oxbow lakes, and patches of wet, swampy tundra.

The winter seasons, from September to May, are very cold and dry; summers are short, cool, and moist. The annual mean temperature is about 22° F., although readings may get into the 80s during the longest days. Relict plants, remnants of cold-climate flora that had a much wider range in Pleistocene time, still survive in the Kobuk Valley.

W. C. Mendenhall in the summer of 1901 made the first geological reconnaissance of the Kobuk Valley, traveling by canoe with a mapping party from the U.S. Geological Survey. Today access to the Park is mainly by air from Kotzebue to Eskimo villages outside the Park where boats, kayaks, and float-equipped planes can be chartered. The Park has no roads or campgrounds.

Figure 42.4   Great Kobuk Sand Dunes, a large active dunefield on the flood plain of the Kobuk River in Kobuk Valley National Park. The Baird Mountains in the far distance rim the north side of the valley. Polar easterlies, the prevailing winter winds (in this photo, coming from the right) sweep down the westward-trending valley, shaping and reshaping the dunes. The dunefield is encroaching on the boreal forest in the foreground. National Park Service photograph by Robert Belous.

How the Dunefields Developed

The two large areas of active dunes on the Kobuk River flood plain—the Great Kobuk Sand Dunes and the Little Kobuk Sand Dunes—are about 25 square miles in extent; but they are only one part of a large complex of stabilized and unstabilized dunes that are spread over some 300 square miles of the central Kobuk Valley, both inside and outside the Park boundaries. Because of the low precipitation, the region would be a desert if it were not for the permafrost that keeps surface moisture from draining away.

The dunes began to form early in Pleistocene time. The dune-building winds were then and are still the polar easterlies that are funneled through the westward-trending Kobuk Valley during the cold, dry winter months. These easterly storm winds were stronger and more frequent during glacial periods because of "intensified atmospheric circulation and steep pressure gradients between ice-covered and ice-free areas" (Fernald 1964, p. K22). Glacial erosion and stream erosion furnished the enormous amounts of loose sand grains; the severe climate restricted the growth of vegetation; and the strong, dry prevailing winds completed the combination of conditions favorable to extensive dune formation.

Dune-building slowed and vegetation increased when the climate moderated between Pleistocene glacial episodes and after the last ice advance of Holocene time. The

main sand sources for the dune that are active at the present time are the bare sand bars along the river.

The dune shapes in the active dunefields vary from east to west as the vegetation decreases and the amount of free sand increases. On the east side, where more dunes are stable and there is more vegetation, *parabolic* dunes are dominant; that is, U-shaped dunes with the arms pointing in the direction *opposite* to the prevailing wind. On the west side are large *barchans,* some 100 feet high, U-shaped dunes with the arms pointing in the *same* direction as the prevailing wind. In between, the dunes change from parabolic to large longitudinal ridges and then to long, undulating transverse ridges that finally assume the barchan shape. Both active dunefields have a gently rising wedge of sand on the eastern border (the windward side) and a steep escarpment along the western border (the lee side).

Address

Area Director
National Park Service
540 West 5th Avenue
Anchorage, Alaska 99501

or

Kobuk Valley National Park
P.O. Box 287
Kotzebue, Alaska 99752

Box 42.1
Changing Views of Alaskan Tectonics

For many years the major tectonic units and physiographic provinces of Alaska were thought to be—for lack of a better generalization—continuations of roughly similar units of western North America that are south of Alaska—except that in Alaska the dominant structural trends have been rotated about 90° from those characteristic of the North American Cordillera. For example, the Arctic North Slope was thought to be the counterpart of the Great Plains; the Brooks Range was doubtless an extension of the Rocky Mountains; the Alaska Range was analogous to the Cascade-Sierra chain; the Aleutian and coastal ranges of southern and southeastern Alaska were comparable to the Coast Ranges of California and the Pacific Northwest; and so forth. In broad outline, this concept seemed plausible; but when looked at more closely, the pieces of the puzzle didn't seem to fit together very well. Geologists noted that in a number of Alaskan areas, rock types and fossils appeared to be completely out of place, both in relation to adjacent areas and when compared with their supposed counterparts to the south.

As the combined theories of sea-floor spreading and plate tectonics gained acceptance, some geologists began to hypothesize that the major features of Alaska were the result of the Pacific plate pushing northward against the northwest corner of the American plate. The idea had merit, but it seemed too simplistic to account for the complex relationships of rock groupings in Alaska. More detailed studies suggested that a number of the rock units in question originated in separate and totally different environments of deposition. Regional correlations in terms of age, origin, source areas of sediments, fossils, and so on, did not fit a "logical" pattern. Great sequences of metasediments, clastic wedges, segments of oceanic rock, ancient volcanics, and the like, lay in positions that defied explanation. As is the case with many other phenomena in nature, the more data collected the more confused the issue.

Meanwhile, an increasing accumulation of paleomagnetic and radiometric data was beginning to suggest to some geologists that certain rock masses in Alaska had been formed originally thousands of miles to the south, possibly as far away as the South Pacific. About the same time, paleontologists found fossilized remains of microscopic organisms that are foreign to North America but typical of Asian species. In 1977 U.S. Geological Survey scientists (D. L. Jones, N. J. Silberling, and J. Hillhouse) proposed that a long narrow block, extending from southern Alaska into British Columbia, had been within 15 degrees of the equator in the middle of Mesozoic time (200 million years ago). They named this exotic block (technically referred to as an *allochthon* or as *allochthonous terrane*) Wrangellia, after the Wrangell Mountains (p. 517). Their studies, based on paleomagnetic data and improved techniques for dating microfossils, indicated that Wrangellia may have traveled 9,000 kilometers before docking at its present position of 50° to 60°N.

Further detailed investigations and the refinement of older data have amplified these tentative conclusions. Geologists soon identified more alien blocks, each of them structurally and tectonically distinct. Fifty major allochthonous terranes have now been located and identified, mainly in southern and central Alaska. They range in size from one square mile to several hundred square miles, and they are bounded by faults, along which, presumably, the blocks come together. The terranes and faults are shown on a geologic map of Alaska, published by the U.S. Geological Survey in 1980.

The idea that most of Alaska, especially southern Alaska, consists of fragments of land from distant places is startling. Perhaps even more surprising is the conclusion that these blocks, called *microplates,* drifted to Alaska during Mesozoic and early Tertiary time, which in terms of geologic time implies that plate movement was more rapid than is usually the case. Geophysicists speculate that extensions of some of the Alaskan allochthonous terranes lie beneath continental shelf sediments in the Bering Sea. At the time the microplates arrived, the Aleutian island arc had not yet formed (fig. 42.5).

Figure 42.5 Aniakchak crater and caldera (elevation, 4,448 feet) in Aniakchak National Monument and Preserve on the Alaska Peninsula. The explosion and collapse that produced the caldera probably happened between 500 and 1,000 years ago. Subsequent eruptions have built the cone in the center of the caldera, especially those occurring in 1931 and 1942. Aniakchak, one of the active volcanoes in the Aleutian Range, is about 130 miles southwest of Katmai National Park and Preserve. Both Aniakchak and Katmai volcanoes are on the volcanic arc marking the plate boundary where the Pacific plate is being subducted beneath the North American plate. Aniakchak caldera, approximately 5 by 6 miles in size, is one of the largest calderas in the world. National Park Service photograph by M. Woodbridge Williams.

The question is, how did this assortment of discrete rock masses from low latitudes get jammed up against northern Alaska? Certainly the arrangement does not fit the more usual plate tectonic model of continental accretion, in which sedimentary rocks on a plate being subducted are "scraped off" against a continental margin, thus enlarging the overriding plate. One geologist has described the Alaska fragments as being like "platters stacked side by side in a dish drainer" (West 1981). Did some of the microplates become fused together enroute? Or did they straggle along erratically like ships blown off course in a storm? Are the terranes unmatched fragments of an ancient continent that split apart, scattering chunks on all sides? Whatever the mechanism was that collected the pieces and ferried them to Alaska, no one can say. But perhaps answers can be found in an eventual paleogeographic reconstruction of an ancient Pacific ocean and its landmasses.

Although the geologic history of Alaska is clouded by many unanswered questions, the present tectonic situation is becoming clearer. Without argument, Alaska is seismically one of the most active regions on the globe. Currently, tectonic activity is going on mostly within a wide mobile belt that stretches in a great looping arc from Canada to Siberia. In the far north and northeast, above the mobile belt, some of the Alaska rocks appear to fit rather well with the rest of North America. The land there may not have been part of the original North American craton, but at least it has been with the continent for a very long time. To the south and west of the Alaska mainland, the deep Aleutian trench lies on the open ocean side of the Aleutian Islands chain, a volcanic island arc. The line of active volcanoes that runs from southern Alaska southwestward through the Aleutians marks the subduction zone where the Pacific plate is plunging beneath the Bering Sea and the southwestern Alaska mainland.

In central Alaska, the Denali fault system and associated faults run roughly east-west. Similar major faults trend northwest-southeast in the

Alaska Panhandle. The allochthonous terranes in these sections of Alaska appear to be separated by strike-slip faults that show both vertical and horizontal movements. Horizontal displacement is especially evident along the Denali fault system (see chapter 11). The explanation seems to be that the Pacific plate is underthrusting in a northeastward direction beneath the Aleutians and the Bering Sea and, at the same time, pivoting to the left in southeastern and southern Alaska—in other words, sliding by rather than subducting in the latter area (although subduction may have been extensive in the past).

Earthquakes are frequent all around the edge of the Gulf of Alaska and are sometimes very severe. The 1964 Good Friday earthquake, which centered about 80 miles east of Anchorage, had a Richter magnitude between 8.3 and 8.75, making it one of the strongest ever recorded on the North American continent. The earthquake caused significant changes in land levels and sea floor levels over an area of several hundred square miles in the vicinity of Cook Inlet, the Kenai Peninsula, and Prince William Sound. Such an event indicates movement of rocks along faults in response to tectonic stresses of considerable force.

Virtually all of the landscapes that we see today in Alaska are in a fundamental sense structurally controlled. The surficial processes (glaciation, stream erosion, etc.) continually modify the landforms, but from time to time they are interrupted or altered by tectonic events resulting from internal processes that are changing the structural framework of the crust.

Lake Clark National Park and Preserve

Location: Southwestern Alaska
Area: 3,653,000 acres; 5,708 square miles
Proclaimed a National Monument: December 1, 1978
Established as a National Park: December 2, 1980

Geographic Setting

The remarkable scenic and geological diversity of Lake Clark National Park and Preserve includes jagged mountain crests, great spires, snowfields and glaciers, steep-walled alpine valleys, two symmetrical active volcanoes, many glacial lakes, waterfalls, tundra-covered plains west of the mountains, and marshy lowlands along the coast. Located on the western shore of Cook Inlet, in the Chigmit Mountains, the Park and Preserve is about 100 miles southwest of Anchorage. There are no highways in the Park and Preserve, but the area is conveniently accessible by air from Anchorage.

Except for several private lodges on Lake Clark, near the western boundary of the Park and Preserve, the area is an undeveloped wilderness. The first geological explorations of the region were not carried out until the late 1920s when Stephen R. Capps led several U.S. Geological Survey expeditions into this part of Alaska. He discovered Lake Clark and Lake Chakachamna, which he ranked "among the most beautiful bodies of water in the world" (Wahrhaftig in Williams 1958, p. 50).

Climatic conditions in the interior of the Park and Preserve are quite different from weather on the coast. Temperatures in the interior may be extremely low in the winter (−40° F.), and they may rise into the 80s (F.) in summer. The interior is also sunnier and drier. Winters on the coast are milder, but precipitation is higher there throughout the year. The lowlands on the coast are also in the path of frequent severe storms, especially in winter. Strong winds, dangerously strong in the mountain passes, may occur at any time of year.

Geological Features

Lake Clark National Park and Preserve is in the area of the angular intersection between the Alaska Range that trends roughly east-west across southern Alaska and the Aleutian Range that trends southwestward down the Alaska Peninsula and the Aleutian Islands. The Park and Preserve is about midway between Denali National Park and Preserve (chapter 11) to the northeast, and Katmai National Park and Preserve (fig. 42.2) to the southwest. These three Alaska units in the National Park System plus Kenai Fjords National Park, Wrangell-St. Elias National Park and Preserve, and Glacier Bay National Park and Preserve are all within the active mobile belt discussed in Box 42.1. All are affected periodically by seismic activity, and several of them have active volcanoes.

The two active volcanoes in Lake Clark National Park and Preserve—Redoubt (elevation, 10,197 feet) and Iliamna (elevation, 10,016 feet)—are at the northern end of the Aleutian volcanic chain. Both are composite volcanoes that have erupted explosively several times since the late 1770s (i.e., since records have been kept in Alaska). The volcanoes are near the coast, Redoubt at the north end of the Park and Preserve and Iliamna at the south end. Redoubt erupted explosively from the north side of its caldera four times in the 1960s. Iliamna's most recent explosive eruption was in 1953, but steam emissions were observed in 1978.

The rocks in the Park and Preserve area are similar in stratigraphy, age, and structure to those in the Alaska Range. Fold axes and fault lines run generally northeast to southwest, with overthrusting to the northwest. On strike-slip faults where displacement is occurring, the northwest side is moving toward the southwest.

Figure 42.6 Twin Lakes in Lake Clark National Park and Preserve are the headwaters of the Chilikadrotna and Mulchatna Rivers, which have been designated as National Wild Rivers. The broad, steep-sided valley holding the lakes is a typical glacial trough. On the right-hand side of the valley (partly obscured by the small cloud), glacial ice has removed the lower part of the ridges, forming truncated spurs. In the tributary valleys, meltwater streams, carrying rock flour, silt, and gravel from glaciers at higher elevations, are building deltas out into the lake. The dark smudges in the water in the foreground are bars made of the finer sediment that settles out beyond the deltas. Float planes bringing visitors to these remote mountains land and take off on the numerous glacial lakes in the Park and Preserve. National Park Service photograph by M. Woodbridge Williams.

Although the nonvolcanic mountains in the Lake Clark region are less than 10,000 feet high, snowfields are extensive, and glaciers have remained active in Holocene time because of the fairly high precipitation, especially on the coastal side of the ranges. During Pleistocene time, the whole area, including Cook Inlet, was covered by ice several times. The moraine-dammed lakes in the long, glacially scoured valleys and the ice-sculptured peaks and ridges are the result of intense and repeated glacial episodes (fig. 42.6). Lake Clark Pass and Merrill Pass are steep-walled canyons that cut through the mountains and are lined by many hanging glaciers and cascading waterfalls.

Address

Lake Clark National Park and Preserve
P.O. Box 61
Anchorage, Alaska 99513

Katmai National Park and Preserve

Location: Southwest Alaska
Area: 5,638,126 acres; 8,809.6 square miles
Proclaimed a National Monument: September 24, 1918
Established as a National Park and Preserve: December 2, 1980

Geographic Setting

Located on the Alaska Peninsula, overlooking the Shelikof Strait, Katmai National Park and Preserve is 290 air miles southwest of Anchorage. With its recently added land, the Park and Preserve now occupies most of the upper end of the Alaska Peninsula from Kamishak Bay to the Becharof National Wildlife Refuge (fig. 42.7). The site of one of the most cataclysmic volcanic events of all time is at Katmai. That was the violent eruption of Novarupta in 1912, which created the ash-filled "Valley of Ten Thousand Smokes." Also in the Park and Preserve are several other volcanoes, mountains, glaciers, lakes, streams, forests, tundra, and marshes, plus a scenic coastline of fiords, wave-cut cliffs, and waterfalls. The world's largest carnivores, the Alaska brown bears, thrive in this wilderness. They feed on the red salmon that spawn in the lakes and streams.

Figure 42.7   A hanging waterfall drops over a wave-eroded sea cliff onto a wave-cut platform at Puale Bay on the Shelikof Strait. Wave action is highly effective here because frequent storms beat in against the shore. This stretch of uplifted shoreline is in Becharof National Wildlife Refuge, adjoining Katmai National Park and Preserve. Hanging waterfalls are numerous along this tectonically active coast because the land is rising faster than small streams can cut down their valleys. The dipping layers of sedimentary rock are late Jurassic in age. National Park Service photograph by Robert Belous.

The mountains of the Aleutian Range in this part of the Alaska Peninsula rise over 7,000 feet from the shores of Shelikof Strait, and here the range is about 100 miles wide from east to west. Several hundred miles down the Peninsula, the range becomes broken up into the Aleutian Island chain, which stretches in an arcuate pattern westward for another thousand miles out into the North Pacific. Throughout its length the Aleutian Range has 60 volcanoes, of which 47 are known to have been active since 1760 (fig. 42.5). The volcanoes in the Park and Preserve trend from northeast to southwest, parallel to the coast on the east side of the Alaska Peninsula.

Facilities in the Park and Preserve are open only from June to September, although charter flights over Katmai can be made during the winter. In summer, access to Katmai is by air from Anchorage to King Salmon, a town on the western edge of the Park and Preserve where Park Headquarters is located. Float planes ferry visitors from there to the Brooks River Camp on the shore of Lake Naknek. A four-wheel-drive bus makes daily trips to an overlook on the rim of the Valley of Ten Thousand Smokes. This is the only "road" in the Park and Preserve. Summers in Katmai are cool and moist, with clear skies only about 20 percent of the time. High temperatures in summer average 63° F. and lows average 44° F. Strong winds and williwaws (sudden, gusty rainstorms) frequently sweep the area.

Local History

In the Katmai region, the Aleutian Range was such a difficult barrier to cross that separate native cultures developed on the Gulf of Alaska and Bristol Bay sides of the Alaska Peninsula. Russian fur traders, based on Kodiak Island across the Shelikof Strait from Katmai Bay, were the first Europeans to live in the area. They were succeeded by American traders after the Alaska purchase in 1867. The establishment of salmon canneries on the Bristol Bay side of the Alaska Peninsula and the Nome gold rush brought more people into the area in the late 19th century. In order to avoid the long, stormy passage around the Aleutian Islands, traders and prospectors used the historic Katmai Trail as a shortcut across the mountains of the Alaska Peninsula. This difficult and arduous route went from Katmai Bay, through Katmai Pass, and down to Bristol Bay, where prospectors could take sailing ships to Nome and the goldfields. The 1912 eruption destroyed the trail. Parts of the route have been restored on the west side of Katmai Pass and can be used by backpackers, but the section from Katmai Pass to Katmai Bay has not been opened because of heavy brush, braided streams, quicksand, quickmud, and frequent slides.

The 1912 Eruption

Apparently no lives were lost during this catastrophic event. This was probably due to the sparse population and the fact that a week-long series of earthquakes prior to the explosion on June 6 prompted natives in the vicinity to flee to the coast. Around noon of that day people on the other side of the Gulf of Alaska knew that an extraordinary volcanic event had occurred because they heard loud booms and saw in the distance great clouds of ash. However, because the ashfall lasted for three days in the immediate vicinity of the eruption, wireless communications were disrupted, and no word reached the outside world until June 9. More than a week passed before a rescue party, paddling through chunks of floating pumice, could reach Katmai Bay from Kodiak Island. They found the native village deserted and ash from three to fifteen feet deep over the whole area. Kodiak Island was also covered by a foot of ash and many homes were destroyed there as well. A U.S. revenue cutter happened to be in Kodiak Harbor (100 miles east of the volcano) on June 6, and it took local residents on board in complete darkness during the ashfall and then lay at anchor until the skies cleared and it was safe to go ashore; that was on the morning of the third day after the first explosions.

Although much of the ash and pumice fell in the Gulf of Alaska, over 3,000 square miles of land were covered by about a foot of pyroclastic debris. Ash fell in Juneau on the Alaska Panhandle, in British Columbia, and as far north as Fairbanks. Ash reaching the upper atmosphere was carried around the globe, causing the average annual temperature of the Northern Hemisphere to be reduced by 1.8° F. for more than two years.

Later that summer a geologist, George C. Martin, was sent from Washington, D.C., to Katmai by the U.S. Geological Survey and the National Geographic Society to make the first scientific assessment of the disaster. Over the next six years the National Geographic Society sponsored four more scientific expeditions to the eruption area. As a result of these investigations, Katmai was proclaimed a national monument in 1918 by President Woodrow Wilson. The Society sponsored two subsequent expeditions in 1919 and 1930. All six expeditions were led by Robert F. Griggs, a botanist, who discovered and named Novarupta ("new volcano") in 1917. On previous expeditions he had explored and named the Valley of Ten Thousand Smokes and had discovered the caldera and crater lake on Mount Katmai. Griggs and his fellow scientists accumulated a remarkable body of data and observations under the most arduous and often dangerous field conditions. Their work, and studies done since, have finally enabled scientists to reconstruct a fairly complete account of what happened during the Katmai eruptions and how they came about.

Both Mount Katmai, which was probably at least 7,500 feet high before its summit collapsed, and its lower neighbor, Novarupta (elevation, 2,757 feet), were involved in the sequence of events in a rather unusual way. The first visitors to the area thought that the Katmai vent was the main source of the approximately seven cubic miles of volcanic debris ejected by the eruptions. However, the earthquakes and fault movements that preceded the eruptions apparently caused andesitic magma from Katmai to flow laterally through fissures and connect with conduits containing rhyolitic magma rising beneath Novarupta. As the two magma masses mingled, they frothed up and exploded, ejecting from Novarupta's vent great clouds of fine, white pumice. Fissures opened in the smaller volcano, releasing a colossal nuée ardente that roared westward down the valley. The incandescent mass devastated everything in its path and buried the valley floor with up to 700 feet of ash. Hot gases and steam rose from the flowing mass. (When Griggs first saw this area—the Valley of Ten Thousand Smokes—in 1915, countless fumaroles of steam, some a thousand feet high, were still issuing from the plain of ash, but at the present time only a few active vents remain.)

As the magma chamber emptied, the unsupported summit of Katmai collapsed, forming a crater three miles long, two miles wide, and 4,460 feet deep. Then the rhyolitic lava in the crater of Novarupta hardened and formed a plug that sealed the throat of the volcano. When that vent was closed off, the remaining andesitic lava resurged up through the collapsed debris of Katmai's summit. Another smaller eruption of ash, pumice, and broken rock ensued, and a small cone of andesitic lava was built on the caldera floor. Only minor explosive eruptions have occurred in these volcanoes since 1912, the most recent activity on Katmai being in 1931 and the latest at Novarupta in 1950.

A lake formed in Katmai crater not long after the initial cooling, and later two glaciers developed on the inward wall of the caldera. In recent years the lake level has been rising, and it may soon run over into the caldera. The posteruption elevation of Mount Katmai is 6,715 feet.

Other Volcanoes in Katmai National Park and Preserve

Kaguyak, about 40 miles northeast of Katmai, also has a caldera and a lake but it has been inactive in historic time. Fourpeaked Mountain, farther to the northeast, has had no volcanic activity in historic time and is almost completely covered by an icefield and glaciers. Neighboring Mt. Douglas, also ice-covered except for its 7,000-foot peak, steams intermittently. Between Kaguyak and Katmai are four inactive volcanoes—Devils Desk, Stellar, Denison, and Snowy—and one, Kukak, that is fumarolic. All five peaks are almost covered by an icefield. Mount Griggs, which rises above the Valley of Ten Thousand Smokes, across from Novarupta, has been inactive in historic time but has two sulfurous fumarole fields (*solfataras*). Trident, Mageik, and Mount Martin, which trend southwest from Katmai, have been active in historic time. Trident has had a number of small ash eruptions, the most recent being in 1974 when a dome formed. Mageik's last ash eruption was a minor one in 1953. Mount Martin steams intermittently and had a small ash eruption in 1951.

Types of Bedrock

The volcanic rocks in the Park and Preserve are on top of sedimentary rocks of late Jurassic age that make up the main mass of this part of the Aleutian Range (fig. 42.7). Toward the west, the Aleutian range slopes more gradually down to Bristol Bay. Granitic rocks of the Aleutian batholith, separated by faults from the topographically higher volcanic and sedimentary rocks, are exposed in the western foothills. The igneous rocks were intruded earlier in Jurassic time and have since been uplifted and uncovered by erosion.

Address

Katmai National Park and Preserve
P.O. Box 7
King Salmon, Alaska 99613

Kenai Fjords National Park

Location: South Alaska
Area: 570,000 acres; 890 square miles
Proclaimed a National Monument: December 1, 1978
Established as a National Park: December 2, 1980

Geographic Setting

The Kenai Peninsula juts out into the Gulf of Alaska directly south of Anchorage. Kenai Fjords National Park is on the southeast side of the Peninsula. There are no roads or facilities within the Park itself, but highways go down either side of the Kenai Peninsula to Homer (west of the Park) and Seward, near the eastern edge of the Park. "Flight-seeing" tours over the fiords and the Harding Icefield are available at Seward. A state ferry system connects Prince William Sound on the east side of the Peninsula with Seward, Homer, and Kodiak Island (southwest of Kenai), and charter boats are also available. The Kenai National Moose Range, a wildlife refuge, adjoins the north side of the Park.

The climate of the Kenai region is maritime with high precipitation. Moisture-bearing winds sweeping up the Gulf of Alaska drop abundant rain and snow when they come up against the coastal mountains. Late spring and summer are slightly less wet than fall and winter, and clear, sunny days are exceptional any time of year. Heavy storms frequently beat in on the exposed southern coast. Homer, which is in the rain shadow of the coastal mountains, has slightly less precipitation and overcast than Seward.

Geologic Features

The Harding Icefield, an icecap 800 square miles in extent, covers all but the highest peaks of the Kenai Mountains. From this nearly flat, mile-high icecap, valley glaciers radiate in all directions. Some go down into fiords and become tidewater glaciers that calve (split off icebergs) directly into seawater. The fiords are glacial troughs, excavated during Pleistocene and Holocene time when the glaciers were more extensive (fig. 42.8). As the glaciers receded, saltwater moved into the fiords. The islands offshore (federally owned but not part of the Park) are partly submerged mountains along this drowned coast. Cirques and U-shaped valleys beneath the surface of the water are an indication that this particular section of the coast has been subsiding due to tectonic activity.

On the peninsulas, lower ridges, and islands, wherever the ice has been gone for several hundred years, mature rain forests of spruce and hemlock have grown up. At higher elevations, between the timberline and the ice margin, mountain goats live on the rocky ledges.

The Kenai Mountains, along with the Chugach Mountains to the east and the Kodiak Island Mountains to the southwest, form an arcuate-trending structure made up of folded and faulted Cretaceous and early Tertiary rocks that have been partly metamorphosed. Volcanic and granitic rocks of late Tertiary age are exposed in the coastal cliffs and ridges and the offshore islands. Earthquakes are frequent in this seismically active area. The epicenter of the 1964 Good Friday earthquake (p. 512) was very close to the Kenai Peninsula. Rockslides triggered by the quake plunged into Kenai Lake, producing waves so violent that they peeled bark off of trees around

Figure 42.8 A sea arch produced by wave erosion at Resurrection Bay, the largest fiord in Kenai Fjords National Park. Where steep rock cliffs come right down into the surf zone, wave action may differentially erode parts of the bedrock, forming a sea cave and eventually a sea arch when the waves break through the ridge. Just above water level on the outer edge of the cliff is a wave-cut notch that continues along the base of the arch. National Park Service photograph by M. Woodbridge Williams.

the shore. Earthquake shocks, seismic sea waves 200 feet high, and fires from ruptured oil tanks destroyed most of the town of Seward.

Address

Kenai Fjords National Park
P.O. Box 1727
Seward, Alaska 99664

Wrangell-St. Elias National Park and Preserve

Location: Southeast Alaska
Area: 12,318,000 acres; 19,247 square miles
Proclaimed a National Monument: December 1, 1978
Established as a National Park: December 2, 1980

Geographic Setting

Wrangell-St. Elias National Park and Preserve, the largest unit in the entire National Park System, occupies a vast, mountainous wilderness where the "panhandle" of Alaska is attached to the "pan." Its area extends from the Tetlin lowlands in the north to Yakutat Bay in the south. Across its eastern border, the international boundary, is Kluane National Park in Canada's Yukon Territory. The western boundary follows the Copper River valley nearly to the Gulf of Alaska and then bends southeastward. The coastal section includes the beaches bordering the Malaspina Glacier, Icy Bay, and Yakutat Bay. Within this geologically diverse region are the Wrangell Mountains, the eastern section of the Chugach Mountains, the Alaska part of the St. Elias Mountains, and the broad valley of the Chitina River. The Park and Preserve has the largest assemblage of glaciers on the North American continent and the greatest number of peaks over 16,000 feet in elevation. Mount St. Elias, 18,008 feet in elevation, is surpassed in North America only by Mount McKinley and Canada's Mount Logan in Kluane National Park. The Malaspina Glacier, which covers an area larger than the state of Rhode Island, may be the largest piedmont glacier in the world. Because of their spectacular landscapes and rich diversity of plant and animal life, the two adjoining parks, Wrangell-St. Elias and Kluane, have been designated a World Heritage Site.

Although many parts of the Wrangell-St. Elias Park and Preserve are remote, inaccessible, and even dangerous, visitors can reach a number of interesting sites by ground transportation and by air. From the town of Chitina (on the western boundary), a dirt road (four-wheel-drive vehicles only) follows an abandoned railroad right-of-way for some 65 miles up the Chitina River valley to the old mining towns of McCarthy and Kennicott, where lodges are open in the summer season. Road access into the northern section of the Park and Preserve is from

Slana, which can be reached via the Tok cutoff from the Alaska Highway. A state secondary road goes from Slana about 45 miles into the mountains to the abandoned mining community of Nabesna. All other access is by air. At present, charter boats are not available at Yakutat Bay, perhaps because the Gulf of Alaska coast between Prince William Sound and the waterways of the Alaska Panhandle is regarded as too stormy.

Summer weather in the area is cool, often overcast, and rainy, with July likely to have more sunny, warm days than August. The fall season is short, but may have clear days with few mosquitoes. Winters are long, dark, and cold.

Geologic Features

Prospectors, mining engineers, and geologists were attracted to the formidable mountain ranges of the Wrangell-St. Elias region around the turn of the century by the lure of mineral wealth, especially after gold was discovered in the Klondike in 1896. Also noted as a promising

clue was the fact that Indians in the region possessed tools fashioned from native copper. Intensive searches revealed the sources of the copper in the Copper River valley and the Chitina River drainage. Copper ores and nuggets of native copper were found in the Nikolai greenstones, a series of Permian lava flows and tuffs cropping out on the north side of Chitina Valley. Rich copper ores were also discovered in massive, cliff-forming limestones of Triassic age exposed between McCarthy Creek and the Chitistone River. From the early 1900s until 1938, when the ores ran out and the Kennicott operation closed, the mines were among the richest copper bonanzas in the world. The old mining towns in the area are now tourist attractions.

Some placer gold has been taken from stream valleys in the foothills north and south of the Wrangell Mountains, and widely scattered veins of gold, silver, zinc, and other metals have been found in mountain bedrock. However, commercial mining has not been carried on in the area for some years.

The volcanic rocks that make up the summits of the Wrangells are the andesitic and basaltic Wrangell lavas that began to erupt in Tertiary time (fig. 42.9). Mount

Figure 42.9   Mount Blackburn (elevation, 16,523 feet) and the upper Nabesna Glacier in Wrangell-St. Elias National Park and Preserve. The Nabesna Glacier, one of the longest in Alaska, flows northeastward for roughly 80 miles and ends in the Nabesna River valley outside the Park and Preserve. Shown here is the Nabesna's zone of accumulation where several snowfields feed cirque glaciers that merge to form this large valley glacier. The pattern of crevasses shows how the upper surface of the ice fractures as the main glacier and its tributary glaciers move downslope. The dark stripes are medial and lateral moraines (consisting of talus) that lead back up to bedrock ridges and peaks rising from the cirques. Mount Blackburn is an eroded volcano that became inactive before Pleistocene time. National Park Service photograph by M. Woodbridge Williams.

Wrangell (elevation, 13,950 feet) is the only volcano in the group that has been active in historic time, but only occasional steam clouds and fumaroles have been noted over the summit crater since a small eruption in the last century.

The St. Elias Mountains in the southern part of the Park and Preserve are the highest coastal mountains anywhere in the world. They rise from the sea to elevations of 10,000 to more than 18,000 feet. The St. Elias Mountains are cut by northwest-southeast-trending faults, along which thrusting, strike-slip displacements, and uplift have taken place. Frequent earthquakes in the region are a sign that tectonic activity is continuing. In both structure and bedrock lithology, the St. Elias Mountains are related to the Chugach, Kenai, and Kodiak Mountains that rim the Gulf of Alaska in an arcuate mountain chain. However, the details of the geology of the St. Elias Mountains are little known. Extensive icefields and snowfields cover all but the highest peaks along the top of the range, and on the slopes more than 100 glaciers obscure the bedrock and structures. Even where exposed in cliffs or on the shore, the volcanic and clastic rocks are difficult to correlate because of an absence of fossils and the extreme deformation that has occurred.

Glacial processes are actively modifying the landscapes of the Park and Preserve, especially in the coastal section. Once freed from the confining walls of their mountain valleys, the glaciers broaden into great piedmont lobes along the narrow strip of coastal lowland. The Malaspina Glacier, extending some 40 miles in width between Yakutat Bay and Icy Bay, is the largest piedmont glacier, but the Guyot Glacier on the west side of Icy Bay is a close second. The Malaspina is 2,000 feet thick in some places, and so much glacial debris has accumulated around its margins that forests of alder and spruce grow on top of the soil-covered ice. Meltwater streams that empty into the sea are choked with sediment and form braided patterns over the beaches.

Some studies suggest that the advances of modern St. Elias glaciers have been even more extensive than those of late Pleistocene time. The latest advance apparently ended at the close of the 18th century, and since that time the glaciers have been generally retreating.

Address

Wrangell-St. Elias National Park and Preserve
P.O. Box 29
Glennallen, Alaska 99588

Glacier Bay National Park and Preserve

Location: Southeast Alaska
Area: 3,935,269 acres; 6,149 square miles
Proclaimed a National Monument: February 25, 1925
Established as a National Park and Preserve:
December 2, 1980

Geographic Setting

Glacier Bay National Park and Preserve is a region of tidewater glaciers, fiords, islands, and mountains at the northwest end of the Alexander Archipelago between the Gulf of Alaska and British Columbia. On the ocean side, Icy Point is more appropriately named than Cape Fairweather. Heavy storms, high tides, and gales beat against the outer coast. Behind the coast is the Fairweather Range (fig. 42.1), which is a southward extension of the St. Elias Mountains. (Wrangell-St. Elias National Park and Preserve is about 75 miles to the north.) The highest mountain in the range, Mount Fairweather (elevation, 15,300 feet), and its adjacent neighbor, Mount Quincy Adams (13,560 feet), form corners on the international boundary between the United States and Canada, where it angles from summit to summit at the top of the Alaska Panhandle. Klondike Gold Rush National Historical Park is also in this area near the international boundary.

At the northern tip of the Park and Preserve is the mouth of the Alsek River at Dry Bay. The Alsek is of interest to geologists because it is the only stream that cuts across the St. Elias Range. It rises far to the north in Canada and then flows through a deep, east-west-trending valley through the mountains before reaching the Gulf of Alaska.

Cross Sound, at the southern end of the Park and Preserve, also trends east-west. This embayment connects Glacier Bay, which follows the dominant north-south grain of the region, with the Gulf of Alaska. Most of the narrow bays, inlets, and valleys that open on Cross Sound lead back to glacial snouts and icefields. Glacier Bay and the other valleys invaded by seawater are steep-sided glacial fiords that have been excavated by ice erosion and exposed by glacial retreat.

Access to Glacier Bay National Park and Preserve is by air from Juneau to Gustavus, a town near the mouth of Glacier Bay. Cruise ships and charter boats also bring visitors to the area. Kayak tours are available when weather permits. Summer temperatures seldom exceed 72° F., and periods of rain and mist are to be expected. It is this high precipitation that nourishes the extensive icefields and glaciers in the mountains (fig. 42.10).

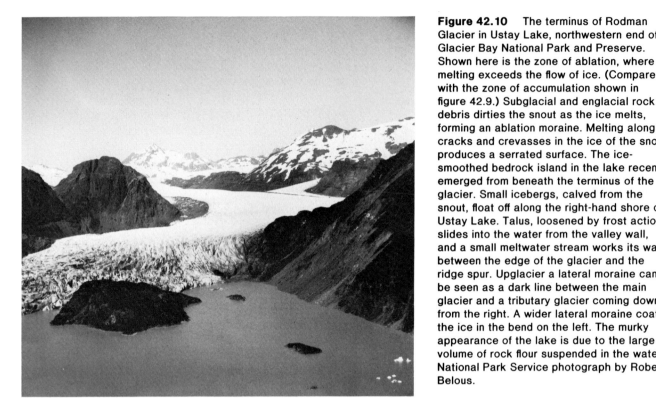

Figure 42.10   The terminus of Rodman Glacier in Ustay Lake, northwestern end of Glacier Bay National Park and Preserve. Shown here is the zone of ablation, where melting exceeds the flow of ice. (Compare with the zone of accumulation shown in figure 42.9.) Subglacial and englacial rock debris dirties the snout as the ice melts, forming an ablation moraine. Melting along cracks and crevasses in the ice of the snout produces a serrated surface. The ice-smoothed bedrock island in the lake recently emerged from beneath the terminus of the glacier. Small icebergs, calved from the snout, float off along the right-hand shore of Ustay Lake. Talus, loosened by frost action, slides into the water from the valley wall, and a small meltwater stream works its way between the edge of the glacier and the ridge spur. Upglacier a lateral moraine can be seen as a dark line between the main glacier and a tributary glacier coming down from the right. A wider lateral moraine coats the ice in the bend on the left. The murky appearance of the lake is due to the large volume of rock flour suspended in the water. National Park Service photograph by Robert Belous.

Geologic Features

The Glacier Bay area was set aside by President Coolidge in 1925 at the urging of scientific and conservation societies because of its uniqueness as a "natural laboratory" for the study of glacial processes and postglacial ecology. The focus of this scientific interest has been on the relatively rapid retreat of the active tidewater glaciers (16 in all) and the rapidity of vegetational change as the land was uncovered.

When the English explorer, George Vancouver, first saw the mouth of Glacier Bay in 1794, it was completely blocked by a high wall of ice that extended from shore to shore and connected with the mountains on either side. But when John Muir, the American naturalist, began his explorations and studies of the region in 1879, he found, instead of an ice sheet, an inlet that extended back to a glacial snout more than 40 miles up the fiord. In tributary valleys other retreating tidewater glaciers, such as Muir Glacier and Geike Glacier, met the invading sea. At the present time, Glacier Bay's main glacier, Grand Pacific, has gone back to Canada. Tarr Inlet washes against the Grand Pacific snout just about on the international boundary, which is roughly 65 miles from the mouth of Glacier Bay. Nowhere else have glaciers retreated so rapidly.

What is even more puzzling is the fact that while some of the tidewater glaciers were melting back at an astonishing rate, others stabilized in the late 1920s and some have even advanced. (Grand Pacific Glacier seems to be fluctuating back and forth over the international boundary.) Johns Hopkins Glacier, in an inlet on the west side of Glacier Bay, appears to be advancing and is one of the most spectacular glaciers in its activity. From its frontal cliffs of ice, 200 feet high, such great blocks of ice are discharged that boats must stay back two miles or risk being swamped by waves.

The glaciers that descend on the Gulf of Alaska side have been more stable, and some are probably about as far down as they were when the region was first explored. LaPerouse Glacier ends in a large piedmont lobe on the shoreline and calves directly into the Pacific Ocean. Another piedmont glacier, Fairweather, is just back from the shore but sticks a finger of ice down to a small coastal inlet on Cape Fairweather.

Brady Glacier, which flows down from the large Brady icefield, used to calve into Cook Sound and then retreated into Taylor Bay. Brady is the largest glacier in the Park and Preserve.

Land uncovered by ice is barren but a short time. Mosses, perennial weeds, herbs, and hardy shrubs spring up on any accumulations of glacial sediment. Thickets of willow, cottonwood, and alder then take over. They are succeeded by spruce trees that are eventually crowded out by taller hemlocks. Muskeg develops in wet or boggy patches. All of these plant communities in various stages of transition can be seen in the Park and Preserve.

In some places, eroding streams have uncovered the buried stumps of older forests that grew and thrived during an earlier retreat, only to be overrun subsequently by

Figure 42.11 An air view looking east up Lituya Bay, a large fiord that opens into the Gulf of Alaska. The spit in the foreground is made up mostly of morainal material being reworked by waves and currents. In the middle of the fiord, a segment of end moraine is plastered over Cenotaph Island. A major fault runs across the mouth of Lituya Bay and a second, parallel fault follows Desolation Valley, passing across the head of the bay. Fault movement caused a severe earthquake in 1958 that shook loose an estimated 90 million tons of rock and ice, which fell 3,000 feet into Lituya Bay. Surging water from the fall created huge waves that washed away the trees on the spurs projecting into the bay. National Park Service photograph by Robert Belous.

advancing ice—ice that is now melting back. Studies show that several glacial advances have occurred in southeastern Alaska during the last 4,000 years. The most extensive advance ended about A.D. 1200 and was followed by several hundred years of retreat. A readvance began before A.D. 1700 and lasted nearly to the end of the 18th century. Recession has been generally dominant since then.

Glacial activity in the Glacier Bay region can be correlated with global climatic changes in only the most general way. Careful records have been kept for a century and many studies have been carried out on the glaciers and their dynamics, but the wide variations in glacier behavior within such a limited area have yet to be satisfactorily explained. The glaciers appear to be affected more by variations in local conditions than by regional climatic trends. The effects of tectonic activities, such as earthquakes, are difficult to evaluate. During a strong earthquake in 1958, millions of tons of rock, weakened by glacial erosion, broke loose from the rock walls above the terminus of Lituya Glacier on the west side of the range. The great rockslide tore off the snout of the glacier, and the whole chaotic mass of rock and ice crashed into Lituya Bay (fig. 42.11). A giant, destructive wave, 1,740 feet high, surged over the ridge on the opposite side, fell back, and then swept down to the ocean. Waves sloshed back and forth in the narrow bay for 20 minutes. When it was over, all the trees in a wide swath along the sides of the bay were gone. Similar rockslides, triggered by earthquakes,

have occurred from time to time in other parts of the Glacier Bay area. The earthquakes are associated with the rapid uplift of the Fairweather Range, which is still rising.

The rocks exposed in the Glacier Bay region are mainly Paleozoic and Mesozoic and vary widely in origin, type, and age. They have had little effect on the development of the landscape. The major faults, on the other hand, have strongly influenced the trend and position of the larger fiords and glacial valleys.

Address

Glacier Bay National Park and Preserve
P.O. Box 1089
Juneau, Alaska 99802

Geologic Maps and Cross Sections

American Association of Petroleum Geologists. 1974. *Geological highway map no. 8, Alaska and Hawaii, circum-Pacific edition.* Tulsa, Oklahoma: American Association of Petroleum Geologists.

U.S. Geological Survey. 1980. Geologic map of Alaska. Compiled by H. M. Beikman in cooperation with State of Alaska Department of Natural Resources.

Bibliography

Alaska Travel Publications Editors. 1974. *Exploring Katmai National Monument and the Valley of Ten Thousand Smokes.* Anchorage, Alaska: Alaska Travel Publications, Inc.

Boehm, W. D. 1975. *Glacier Bay.* Anchorage, Alaska: Alaska Northwest Publishing Company.

Bohn, Dave. 1967. *Glacier Bay, the land and the silence.* Alaska National Parks and Monuments Association.

Brown, Dale. 1976. *Wild Alaska.* Rev. ed. New York: Time-Life Books.

Churkin, M., Jr., Carter, C., and Trexler, J. H., Jr. 1980. Collision-deformed Paleozoic continental margin of Alaska—Foundation for microplate accretion. *Geological Society of America bulletin* 91:648–654.

Clarke, M. 1979. *The life and adventures of John Muir.* San Francisco: Sierra Club Books.

Fernald, A. T. 1964. Surficial geology of the central Kobuk Valley, northwestern Alaska, U.S. Geological Survey Bulletin 1181-K.

Gray, W. R.; Grove, N.; Judge, J.; Kline, F.; and Ramsay, C. R. 1976. *Alaska: High roads to adventure.* Washington, D.C.: National Geographic Society.

McGeary, S. E., and Ben-Avraham, Z. 1981. Allochthonous terranes in Alaska: Implications for the structure and evolution of the Bering Sea shelf. *Geology* 9:608–614, December.

National Parks. 1981. Special issue; Alaska national interest lands. 55 (3) : 5–27, March.

Pett, S., and Loh, J. 1974. The Good Friday Quake of 1964. In *The Cook Inlet collection,* ed. M. Sherwood. Anchorage, Alaska: Alaska Northwest Publishing Company.

Péwé, T. L. 1975. *Quaternary geology of Alaska.* U.S. Geological Survey Professional Paper 835.

Simmerman, N. L. 1982. Our lasting frontier, traveling to the national parks of Alaska. *National Parks* 56(5–6): 16–25, May-June.

Smithsonian Institution. 1981. *Volcanoes of the world.* Stroudsburg, Pennsylvania: Hutchinson Ross Publishing Company.

Walker, Bryce. 1982. *Earthquake,* Planet Earth Series. Alexandria, Virginia: Time-Life Books. (Chapter 1, Alaska 1964: A world gone mad, pp. 18–35.)

West S. 1981. Alaska: The fragmented frontier. *Science news* 119(1) : 10–11, January 3.

Williams, Howel, ed. 1958. *Landscapes of Alaska.* Berkeley: University of California Press.

APPENDIX

U.S. Geological Survey Topographic Maps of National Parks plus selected USGS Topographic Maps, United States Series (scale 1:250,000), showing National Park areas.

Index maps issued by individual states list additional USGS topographic maps of various scales that are available. Because the U.S. Geological Survey is shifting to the metric system, maps for some states are now published in metric scales; e.g. 1:100,000, 1:50,000, and so on. In some states metrication has been delayed in order to maintain continuity—for example, until complete coverage by 1:24,000-scale maps is available. Other states are using a combination of systems with partial metrication.

The U.S. Geological Survey publishes some of the National Park topographic maps in both contour (C) and shaded-relief (SR) editions. On the shaded-relief maps, the shading accentuates the physical features by simulating in color the appearance of sunlight and shadows on the terrain, thereby creating the illusion of a three-dimensional land surface. On contour maps the earth's surface is depicted by contour lines.

The National Park topographic maps and other U.S. Geological Survey maps can be ordered as follows:

Areas west of the Mississippi River

Western Distribution Branch
U.S. Geological Survey
Box 25286, Federal Center
Denver, Colorado 80255

Areas east of the Mississippi River

Eastern Distribution Branch
U.S. Geological Survey
1200 South Eads Street
Arlington, Virginia 22202

Many of the 1:250,000-scale maps are available in raised-relief, plastic editions. Publication and distribution of USGS topographic maps in this format are now handled by Hubbard, P.O. Box 104, Northbrook, Illinois 60062.

USGS Topographic Maps of National Parks* and Scales (available 1981) C = Contour; SR = Shaded Relief	United States Series, Scale 1:250,000, Covering National Park Areas
Acadia National Park and vicinity, ME, 1:50,000(C)	Bangor
Alaska 1980 (shows boundaries of Alaska Parklands) 1:2,500,000 and 1:5,000,000 (C)	
Arches National Park, UT, 1:50,000 (C)	Moab
Badlands National Park and vicinity, SD, 1:50,000 (C, SR)	Hot Springs, Martin
Big Bend National Park, TX, 1:100,000 (C, SR)	Emery Peak
Bryce Canyon National Park, UT, 1:31,680 (C, SR)	Cedar City
Canyonlands National Park and vicinity, UT 1:62,500 (C, SR)	Salina, Moab, Cortez
Channel Islands National Park (Anacapa and Santa Barbara Islands only), CA, 1:24,000 (C)	Long Beach, Los Angeles
Crater Lake National Park and vicinity, OR, 1:62,500 (C, SR)	Roseberg, Crescent, Medford, Klamath Falls

*Maps for National Park areas not in the special series follow this list.

Denali National Park and Preserve, Mount McKinley section, AK, 1:250,000 (C, SR); see also Alaska 1980	Healy, Talkeetna, Tanana
Gates of the Arctic National Park and Preserve, AK, see Alaska 1980	Hughes, Survey Pass, Killik River, Chandalar, Phillip Smith Mtns., Wiseman, Ambler River, Chandler Lake
Glacier Bay National Park and Preserve, AK, see Alaska 1980	Juneau, Mt. Fairweather, Skagway, Yakutat
Glacier National Park, MT, 1:100,000 (C)	Kalispell, Cutbank
Grand Canyon National Park and vicinity, AZ, 1:62,500 (C, SR); former National Monument section 1:48,000 (C)	Grand Canyon, Marble Canyon, Williams, Flagstaff
Grand Teton National Park, WY, 1:62,500 (C, SR)	Ashton, Driggs
Great Smoky Mountains National Park (east half), NC-TN, 1:62,500 (C); Great Smoky Mountains National Park (west half), NC-TN, 1:62,500 (C); Great Smoky Mountains National Park and vicinity, NC-TN, 1:125,000 (C, SR)	Knoxville
Hot Springs National Park (north), AR, 1:24,000 (C)	Little Rock
Isle Royale National Park, MI, 1:50,000 (C, SR)	Thunder Bay, Hancock
Katmai National Park and Preserve, AK, see Alaska 1980	Afognak, Iliamna, Mt. Katmai, Karluk, Naknek
Kenai Fjords National Park, AK, see Alaska 1980	Seldovia, Kenai, Seward, Blying Sound
Kobuk Valley National Park, AK, see Alaska 1980	Misheguk, Howard Pass, Baird Mtn., Amber River
Lake Clark National Park and Preserve, AK, see Alaska 1980	Lime Hill, Lake Clark, Tyoney, Kenai, Iliamna, Seldovia
Lassen Volcanic National Park and vicinity, CA, 1:62,500 (C, SR)	Susanville
Mammoth Cave National Park and vicinity, KY, 1:24,000 (C)	Evansville
Mesa Verde National Park, CO, 1:24,000 (C, SR)	Cortez
Mount Rainier National Park, WA, 1:50,000 (C)	Yakima
North Cascades National Park, WA, 1:100,000 (C)	Concrete, The North Cascades
Olympic National Park and vicinity, WA, 1:125,000 (C, SR)	Cape Flattery, Victoria, Copalis Beach, Seattle
Petrified Forest National Park, AZ, 1:62,500 (C)	Gallup, St. Johns
Rocky Mountain National Park, CO, 1:62,500 (C, SR)	Greeley
Sequoia and Kings Canyon National Parks and vicinity, CA, 1:125,000 (C, SR)	Fresno, Mariposa
Shenandoah National Park (northern section), VA, 1:62,500 (C); Shenandoah National Park (central section), VA, 1:62,500 (C); Shenandoah National Park (southern section), VA, 1:62,500 (C)	Charlottesville
Theodore Roosevelt National Park (north unit), ND, 1:24,000 (C); Theodore National Park (south unit), ND, 1:24,000 (C)	Dickinson, Wartford City
Voyageurs National Park, MN, 1:50,000 (C)	International Falls
Wind Cave National Park and vicinity, SD, 1:24,000 (C, SR)	Hot Springs
Wrangell-St. Elias National Park and Preserve, AK, see Alaska, 1980	Mt. St. Elias, Yakutat, Gulkana, Bering Glacier, Cordova, Nabesna, McCarthy, Valdez
Yellowstone National Park and vicinity, WY-MT-ID, 1:125,000 (C, SR)	Ashton, Cody, WY, Bozeman, Billings, MT
Yosemite National Park and vicinity, CA, 1:125,000 C, SR: Yosemite Valley, CA, 1:24,000 (C, SR)	Walker Lake, Mariposa
Zion National Park (Kolob section), UT, 1:31,680 C, SR; Zion National Park (Zion Canyon section), UT, 1:31,680 (C, SR)	Cedar City

Selected USGS Topographic Maps for National Parks Not Included in the Preceding List

Biscayne National Park, FL, 1:250,000, Miami

Capital Reef National Park, UT, 1:250,000, Salina

Carlsbad Caverns National Park, NM, 1:250,000, Carlsbad; 1:62,500, Carlsbad East; 1:62,500, Carlsbad West

Everglades National Park, FL, 1:250,000, Miami, Key West

Guadalupe Mountains National Park, TX, 1:250,000, Van Horn

Haleakala National Park, HI, 1:250,000 and 1:62,500, Maui

Hawaii Volcanoes National Park, HI, 1:250,000, Hawaii

Redwood National Park, CA, 1:250,000, Eureka, Weed

Virgin Islands National Park, VI, 1:250,000, Puerto Rico and Virgin Islands; 1:24,000, Western St. John

Glossary

aa
A lava flow characterized by a spiny, rubbly surface (see *pahoehoe*).

abrasion
The mechanical wearing, grinding, scraping, or rubbing away of rock by friction or impact. Solid rock particles, transported by wind, ice, waves, running water, or gravity, are the tools of abrasion.

absolute age
Age given in years or some other unit of time.

adjustment of drainage
A condition whereby streams are fitted to structure by headward erosion along outcrops of less resistant rocks and by stream capture.

agate
A banded or variegated chalcedony, characterized by colors arranged in alternating strips or bands but occurring in other forms as well.

aggradation
The building up of a land surface by deposition (see *degradation*).

alkalic basalt
A group (or suite) of basalts that contain only one variety of pyroxene and a considerable amount of olivine; undersaturated in silica (see *tholeiitic basalt*).

allochthon, allochthonous terrane
A mass of rock (exotic in its present location) that has been moved from its site of origin by tectonic forces.

alluvial fan
A large, fan-shaped pile of sediment that usually forms where a stream's velocity decreases as it emerges from a steep canyon onto a flat plain at the foot of a mountain range.

alpine topography
Mountainous topography resembling that of the European Alps, and made up of landforms produced by intensive glacial erosion.

amphibole group
A group of dark, rock-forming, mafic minerals (e.g. *hornblende*).

amygdule
A gas cavity or vesicle in an igneous rock that is filled with a secondary mineral, such as zeolite, calcite, quartz, or chalcedony.

anastomoses (in caves)
Nets of braided, interconnected tubes, commonly confined to a discrete layer of rock (*sing.,* anastomosis).

andesite
Fine-grained, extrusive rock of intermediate composition.

angular unconformity
A type of unconformity in which younger beds of rock overlie an erosional surface on tilted or folded layered rock, so that the bedding planes of the rock units above and below the erosion surface are not parallel (see *nonconformity, disconformity*).

antecedent stream
A stream that maintains its original course or direction despite subsequent deformation or uplift (see *superimposed stream*).

anticline
An upfold opening downward; a fold with older beds in the center (see *syncline*).

aphanitic (texture)
Describing the texture of an igneous rock in which the grains are too small to be seen without a microscope.

aquifer
A rock unit, or units (such as a series of beds), of moderate to high permeability from which ground water can be pumped.

aragonite
A mineral having the same composition as *calcite* but belonging to a different crystal system.

arête
A sharp ridge that separates adjacent, glacially carved valleys.

arkose
A variety of sandstone containing grains of quartz and at least 25 percent feldspar fragments.

atoll
A roughly circular coral reef, enclosing or nearly enclosing a lagoon; formed far from any landmass.

avalanche chute
A track, or trough left on a steep slope after avalanches, carrying ice, rocks, and trees, have descended.

badlands, badlands topography
Intricately dissected topography characterized by a very fine drainage network, high drainage density, short, steep slopes, and narrow interfluves. Badlands develop on a surface with little or no vegetative cover in unconsolidated or poorly cemented clays or silts.

bajada
A broad, gently sloping, depositional surface formed at the base of a mountain range in a dry region by the coalescing of individual alluvial fans; a landform.

balanced rock
A large rock resting more or less precariously on its base, formed by weathering and erosion in place. (In glaciated terrain, balanced rocks may have been left by glaciers—see *erratic*.)

bar (stream)
A ridgelike accumulation of sand, gravel, etc., in a stream channel, along banks, or wherever a decrease in velocity induces deposition.

barchan (dune)
A crescent-shaped dune with the horns of the crescent pointing downwind.

barrier reef
A long, narrow reef that forms parallel to a shoreline and is separated from the shore by a basin of water called a lagoon.

basal conglomerate
A well sorted bed of pebbles or coarse rock fragments that forms the bottom stratigraphic unit of a sedimentary series and that rests on an erosion surface, thereby marking an unconformity.

basalt
A dark, fine-grained, mafic igneous rock composed of calcium-rich feldspar and ferromagnesian minerals.

base level
A theoretical limiting surface below which a stream (or network of streams) cannot erode.

braided stream
A stream that flows in an interlacing, tangled network of rivulets and branches separated from each other by islands and bars.

batholith
A large, generally discordant, plutonic mass that has more than 40 square miles in surface exposure and no known floor.

beach
Strip of sediment (usually sand, but sometimes pebbles, shell fragments, cobbles, or mud) that extends from the low-water line inland to a cliff or zone of permanent vegetation.

beach drifting
The lateral movement of sand and pebbles along the foreshore caused by the oblique arrival of waves against a coast.

bedding plane
A nearly flat surface separating two beds (layers) of sedimentary rock.

beheaded stream
A stream that loses its headwaters due to stream capture.

bentonite
A very fine-grained sedimentary rock derived from igneous material, such as tuff and volcanic ash. Greasy and soaplike to the touch, it absorbs water quickly and swells when it becomes wet. When saturated, it slumps or may even flow as a fine muck. After it dries, it hardens again.

biota
All the living organisms, plant and animal, of an area.

biostrome
A distinctly bedded and widely extensive or broadly lenticular, blanketlike mass of rock, built by and composed mainly of the remains of sedentary organisms.

blanket sand
A blanket deposit of sand or sandstone of unusually wide distribution, typically quartz sandstone deposited by a transgressing sea advancing for a considerable distance over a stable shelf area.

bomb (volcanic)
A fragment of lava, or ejecta, usually rounded, ropy, or spiral in shape, that cooled and hardened while spinning in the air.

boulder
A detached rock mass having a diameter greater than 10 inches (at least the size of a volleyball); the largest size of rock fragments regarded as sediment.

boxwork (cave)
After a network of intersecting blades or plates of calcite was deposited in cracks in highly fractured limestone, the host rock was removed by solution, leaving the intricately crisscrossed fins of the boxwork on cave walls.

breakdown (cave)
The accumulation of debris formed by the collapse of the ceiling or walls of a cave.

breaker
A wave that has become so steep that the crest of the wave topples forward, moving faster than the main body of the wave.

breccia
A coarse-grained clastic rock composed of large (greater than sand-size), angular, and broken rock fragments that are cemented together in a finer-grained matrix; can be of any composition, origin, or mode of accumulation.

brecciated (texture)
Describing rocks formed from an accumulation of angular fragments or broken and crushed particles.

butte
An isolated, steep-sided hill, with or without a flat top, but usually smaller than a mesa (see *mesa*).

calcite
A rock-forming mineral composed of calcium carbonate ($CaCO_3$); the principal constituent of limestone.

caldera
A large, basin-shaped volcanic depression, more or less circular in form, the diameter of which is many times greater than that of the original crater. Most calderas result from the collapse of the roof of a magma chamber due to removal of magma by voluminous pyroclastic eruptions or by subterranean withdrawal of magma.

caliche
A rock composed of crusts of soluble calcium salts mixed with sand, silt, etc.; formed mainly by capillary action drawing mineral-rich ground water to the surface where it evaporates, leaving the salts.

calving
Breaking away of blocks of ice from a glacier front floating in a body of water, forming icebergs.

canyon passage (caves)
An underground stream bed that resembles a chasm.

capture (stream)
The natural diversion of a stream's headwaters into the channel of another stream having greater erosional activity and flowing at a lower level.

carbonates
Sedimentary rocks that contain more than 50 percent by weight of calcite or other carbonate minerals.

carbonation
A type of chemical weathering in which carbon dioxide in air and soil combines with water to form carbonic acid, which reacts with minerals or rocks.

carnelian
A translucent red or orange-red variety of chalcedony.

cave (cavern)
Naturally formed (usually by solution processes) underground chamber.

cave earth
A residue of clay particles and other insoluble materials found in all caves. It accumulates as the insoluble residue left by the solution of limestone or as fine sediment washed into the cave from outside sources.

cave ice, calcite rafts
Thin, rectangular pieces of calcite that form in bodies of standing water in caves.

cave-in lake
A shallow body of water whose basin is produced by collapse of the ground following contraction due to unequal thawing of ground ice in regions underlain by permafrost (see *thaw lake*).

cave pearl
A concretion of calcite formed in subterranean streams when a small particle such as a grain of sand becomes coated with calcium carbonate.

cave popcorn
Nodular, grapelike protrusions formed where water seeps out of pores in the rock of a cave.

cementation
The process by which clastic sediments become consolidated into compact rock, usually through precipitation of minerals in the spaces between sediment grains.

chalcedony
A cryptocrystalline (having no visible crystalline structure) variety of quartz that is sometimes found in chert or as concretions in sedimentary rocks. It possesses *conchoidal fracture*.

chattermarks
A series of small, close, curved cracks made by vibratory chipping of ice-embedded rock fragments against a bedrock surface.

chemical weathering
See *decomposition*.

chert
A hard, fine-grained sedimentary rock formed almost entirely of silica.

cinder cone
A relatively small volcano constructed of loose fragments *(tephra)* ejected from a central vent.

circum-Pacific belt
The great volcanic belt that borders the Pacific Ocean and marks the location of plate boundaries; also called the "Rim of Fire." Earthquakes are frequent along the belt.

cirque
A steep-sided, amphitheaterlike hollow carved into a mountain by glacial erosion at the head of a glacial valley.

cirque glacier
A glacier that occupies a hollow or basin (a cirque) at the head of a mountain valley; usually the remnant of a much larger glacier.

cirque lake
A small, deep, almost circular, glacial lake occupying a cirque; it has no prominent inlet or outlet, being fed by runoff from the surrounding slopes and damned by a lip of bedrock or by a small moraine.

clastic (detrital)
Describing fragments derived from pre-existing rocks or minerals transported as separate particles to their places of deposition by mechanical agents (water, ice, wind, gravity). When consolidated, such sediment (sand, gravel, etc.) becomes *clastic sedimentary rock.*

clastic dike
A sedimentary dike consisting of a variety of clastic materials derived from underlying or overlying beds.

clastic (texture)
A rock texture produced by the lithification of an accumulation of fragments.

clay
Sediment made up of particles less than 4 microns in diameter; i.e., smaller than very fine silt grains.

clay minerals
A group of finely crystalline, hydrous aluminum silicates, formed chiefly by weathering of silicate minerals (feldspars, pyroxenes, etc.); able to absorb substantial amounts of water.

closed drainage basin
A drainage basin that does not have enough streamflow to reach another basin, river or the sea; an area of interior drainage.

coarse-textured drainage
Descriptive of an area where the number of streams and valleys per unit of area (e.g., square mile) is relatively low.

coast
A strip of land of indefinite width that extends from the low-tide line inland to the first major change in landform features.

cobble
A rock fragment, usually somewhat rounded by abrasion, larger than a pebble and smaller than a boulder (i.e., between 2.5 to 10 inches in diameter).

col
A saddle-shaped depression or pass in a ridge resulting when glacial ice breaks down an arête between cirques.

collapse block (in a cave)
A portion of the cave ceiling or wall that has detached itself from the surface of the cave and is now lying on the floor.

collapse breccia
A breccia formed by the collapse of rock overlying an opening, as in a cave.

colloidal dispersion
The condition that exists when submicroscopic particles carry a slight charge of electricity that causes them to repel each other; this keeps the particles in suspension.

column (dripstone)
A dripstone feature formed when a stalactite growing downward and a stalagmite growing upward meet and join.

columnar structure, columnar jointing
Volcanic rock solidified from lava in parallel, vertical columns, polygonal in cross section; formed as a result of contraction during cooling.

composite volcano
Volcano (or volcanic cone) constructed of alternating layers of pyroclastics and rock solidified from lava flows; also called a *stratovolcano.*

compression, compressive force
A force that tends to shorten a body (see *tensional force*).

conchoidal fracture
The characteristic of certain minerals or rocks to break along smoothly curved surfaces.

concordant
Parallel to layering or earlier developed planar structures, such as bedding, joints, etc. (see *discordant*).

concretions
Hard, compact, rounded masses, or nodules, of mineral matter, of varying size and composition, that have been precipitated out from solution, usually around nucleii of some type (such as fossils).

conglomerate
A coarse-grained, clastic sedimentary rock formed by the cementation of rounded gravel.

consequent stream
A stream whose course is controlled by the form and slope of an existing land surface (see *subsequent stream*).

contact metamorphism
A local process of thermal metamorphism related to the intrusion of magma and affecting rocks at or near their contact with an igneous body (see *regional metamorphism*).

continental crust
The thick granitic (sialic) crust that underlies the continents.

continental glacier (ice sheet)
A glacier of considerable thickness completely covering a substantial part of a continent and obscuring the relief of the underlying surface.

continental shelf
A submarine platform at the edge of a continent, inclined gently seaward.

continuous permafrost
A zone of permafrost that, for the most part, is uninterrupted by pockets or patches of unfrozen ground.

converging plate boundary
Boundary between two tectonic plates that are moving toward one another.

coquina
A soft, porous, coarse-grained, off-white limestone made up of loosely aggregated shells and shell fragments.

coral
A cylindrical-shaped marine invertebrate that lives either singly or bound together structurally with other corals in a colony attached to the sea floor; a reef-forming organism.

correlation
Determination of age relationships between rock units (or geologic events) in separate areas.

country rock
Any rock that was older than and intruded by an igneous body.

crater
A basinlike depression over a vent at the summit of a volcanic cone; formed by eruptive action.

craton
A portion of continental crust that has been structurally stable for a prolonged period of time.

creep
The very slow, continuous downslope movement of soil or unconsolidated debris.

crevasse
An open fissure, or crack, in a glacier caused by fracturing of the ice during movement.

crust
The thin, outermost layer of rock that makes up the earth's surface.

cryptocrystalline
Having an internal crystalline structure so fine that it cannot be detected by a petrographic microscope.

crystalline (rock)
Referring to a rock consisting wholly of crystals or crystal fragments, such as an igneous rock that cooled from a molten state or a metamorphic rock that has undergone recrystallization.

cuesta
An asymmetrical ridge that slopes gently on one side, conforming to the gentle dip of resistant beds, and has on the opposite side a steep scarp (or even a cliff) formed by the outcrop of the resistant beds (see *hogback*).

Curie point
The temperature above which thermal agitation prevents spontaneous magnetic ordering. Specifically, the temperature at which there is a transition from ferromagnetism to paramagnetism. Above the Curie point, rocks do not retain their magnetic properties.

dacite
A fine-grained, extrusive igneous rock, intermediate in composition between rhyolite and andesite; a quartz andesite.

daughter product
The isotope of an element produced by radioactive decay.

debris flow
Mass wasting involving rapid flowage of debris of many sizes under a variety of conditions.

decomposition
Weathering processes by which rock is changed due to chemical activity mainly involving water and atmospheric gases; also called *chemical weathering* (see *disintegration*).

deflation
The removal by wind of small sediment particles from a land surface; a form of wind erosion.

degradation
The lowering of a land surface and general reduction of relief by erosion (see *aggradation*).

dendritic drainage pattern
Drainage pattern of a stream and its tributaries that resembles branches of a tree or veins in a leaf.

desert pavement
A closely packed surface of wind-polished pebbles and rock fragments, usually somewhat cemented together by calcium carbonate, which tends to form on flat surfaces in arid regions after wind (or water) has removed the soil and fine particles. Once formed, a desert pavement tends to protect underlying sediment from further deflation (wind erosion).

desert varnish
A thin, dark, brownish-red or black patina that develops on the surface of pebbles, boulders, or bedrock in a desert region; it consists of iron and manganese oxides with some silica, and is probably caused by repeated evaporation of water drops from the rock surface.

detrital (detritus)
See *clastic*.

diabase
A dense, black, intrusive igneous rock (frequently found in dikes) studded with light-colored plagioclase crystals, resembling small laths, within the pyroxene crystals; similar in composition to a *diorite*.

differential erosion (or weathering)
Erosion (or weathering) that occurs at irregular or varying rates, caused by differences in the resistance and hardness of rocks, minerals, or surface materials.

dike
A tabular, discordant intrusive body of igneous rock (see *sill*).

diorite
Coarse-grained, intrusive igneous rock, intermediate in composition; composed of approximately equal amounts of plagioclase feldspar and ferromagnesian minerals.

dip
The vertical angle and direction at which a plane is inclined from the horizontal.

disconformity
A type of unconformity that indicates missing rock layers, but the beds above are parallel to those below (see *nonconformity, angular unconformity*).

discontinuous permafrost
A zone of permafrost containing patches of unfrozen ground; it occurs in an intermediate zone between the *continuous permafrost* and sporadic permafrost.

discordant
Cutting across any layering or earlier developed planar structures, such as bedding, joints, etc. (see *concordant*).

disintegration
The breaking up of rock into smaller pieces by frost action, abrasion, etc. Also called *physical weathering* or *mechanical weathering* (see *decomposition*).

diverging boundary
Boundary separating two tectonic plates that are moving away from each other; a spreading center.

dogtooth spar
A variety of calcite with sharply pointed crystals resembling eyeteeth.

dolomite
A sedimentary rock composed of the mineral dolomite (calcium-magnesium carbonate), commonly formed by replacement of limestone.

dome pit (cave)
An underground vertical passage or high chamber formed by solution. It has a domed ceiling and vertical walls; sometimes called a *vertical shaft*.

drainage basin
The total area drained by a stream and its tributaries.

drainage divide
The boundary between adjacent drainage basins.

driblet cone
A spire-shaped mound resulting from the spattering of lava ejected from an opening in a lava tube; also called *hornito*.

drumlin
A smooth-surfaced, elongate hill or ridge made up of compact glacial till.

dune rock
An eolianite consisting of consolidated dune sand.

earthquake swarm
A series of minor earthquakes occurring in a limited area and time.

ejecta
Material, both solid and liquid, thrown out by a volcano. Sizes may range from ash particles to large blocks of rock and volcanic bombs.

endellite
A clay mineral formed by the alteration of other clay minerals in an acidic, sulfate-rich environment.

end moraine
A ridgelike accumulation of till deposited at the lower or outer end of a glacier.

eolian
Referring to wind action.

eolianite
A consolidated sedimentary rock consisting of clastic mineral deposited by the wind.

eon
The longest geologic-time unit, next in magnitude above an *era;* e.g., the Archean Eon.

ephemeral stream
A stream that flows briefly only in direct response to precipitation in the immediate locality (see *intermittent stream*).

epicenter
The point on the earth's surface directly above the focus of an earthquake.

epoch
A unit of geologic time shorter than a *period;* e.g., the Holocene Epoch of the Quaternary Period.

epsomite
A mineral consisting of native Epsom salts and occurring in colorless, prismatic crystals, masses, and incrustations in gypsum mines or limestone caves, or mineral waters (in solution).

era
A geologic-time unit next in order of magnitude below an *eon* and including two or more *periods;* e.g., the Paleozoic Era.

erosion
The general process (or group of processes) whereby the materials of the earth's crust are loosened, dissolved, or worn away and simultaneously removed by running water, glacial ice, wind, or other geologic agents.

erosion surface
A land surface that has been lowered and smoothed by prolonged erosion; usually nearly level or flat.

erratic
A relatively large rock fragment that has been transported by ice from its place of origin; often different in composition from the underlying bedrock.

eugeosyncline
The volcanic portion of a geosyncline, usually located on the side away from the craton; tends to fill with submarine lava flows, volcanic ash, marine sediment, etc.

eustatic
Pertaining to worldwide changes in sea level that affect all the oceans.

evaporites
Sedimentary rocks that form by crystals precipitating during evaporation of sea water; e.g., rock salt, rock gypsum.

exfoliation (sheet jointing)
The weathering process by which concentric scales, plates, or shells of rock are successively spalled or stripped from the bare surface of a large rock mass; caused by physical and chemical forces producing differential stresses within the rock, *or* by release of confining pressure (i.e., *pressure-release jointing*).

exfoliation dome
A large, rounded landform developed in a massive rock, such as granite, by the process of exfoliation (sheet jointing).

extrusive rock
An igneous rock that has cooled or solidified at the earth's surface, whether it solidified from a lava flow or from fragments of volcanic debris.

facies
See *metamorphic facies, sedimentary facies*.

fault
A fracture in bedrock along which movement has taken place.

fault-block mountain range
A range produced by uplift along normal or vertical faults.

feldspar group
A group comprising the most common minerals of the earth's crust; all feldspars contain silicon, aluminum, and oxygen and may contain potassium, sodium, calcium, or other elements in varying combinations. Feldspars are components of many types of rock and altogether make up about 60 percent of the crust.

felsic
Describing silica-rich igneous rocks with a relatively high content of potassium and sodium (see *mafic*).

ferromagnesium minerals
Generally dark, iron-magnesium-bearing minerals, such as olivine, hornblende, biotite, augite; the *mafic* minerals.

fetch
The extent of open ocean over the surface of which the wind blows with constant speed and direction, thereby creating a wave system.

fin (ridge)
Local term (Colorado Plateau) for a sharp, narrow ridge, usually eroded from sandstone.

fine-textured drainage
Descriptive of an area where the number of streams and valleys per unit of area (e.g., square mile) is relatively high.

fiord
A coastal inlet that is a glacially carved valley, now occupied by an arm of the ocean.

firn
A compacted mass of granular snow, transitional between snow and glacial ice.

firn line
An irregular, shifting line marking the highest level to which the winter snow cover on a glacier is lost during a melting season.

flood plain
That part of a valley floor that is periodically covered by flood waters that leave a blanket of sediment on either side of the stream channel.

flowstone
A general term for any accumulation in caves of calcium carbonate or other mineral matter formed by flowing water on the walls or floor of a cave.

flute (cave)
An incised, vertical or nearly vertical groove developed in a cave wall by dripping water.

flysch
A marine sedimentary facies that is the result of accelerated accumulation of sediment and rapidly changing conditions of deposition; e.g., alternating beds of shales and sandstones.

focus
The point within the earth from which seismic waves originate in an earthquake (see *epicenter*).

fold
A bend in layered bedrock, usually produced by deformation; a type of geologic structure.

foliated (foliation)
A planar arrangement of textural or structural features in any type of rock, but especially in metamorphic rock (see *nonfoliated*).

footwall
The underlying side of a fault, especially the wall rock beneath an inclined fault (see *hanging wall*).

foreshore
The area between mean low water level and mean high water level; the zone that is regularly covered and uncovered by the rise and fall of tides, and upon which waves usually break.

formation
A mappable, lithologically distinct rock unit, usually made up of one or more beds of sedimentary rock, but may be composed of metamorphic or igneous rock units. Contacts separating a formation from adjacent rock units must be recognizable.

fringing reef
A relatively small, linear reef that forms parallel to a shoreline and is directly attached to or borders the shore. There is usually no body of water between the fringing reef and the mainland.

frost action
Mechanical weathering processes caused by cycles of freezing and thawing of water in pores, cracks, and other openings at or near the surface.

frost heaving
The uneven lifting of rock or soil by the expansion of freezing water.

frost splitting, frost wedging
The splitting of rock or soil due to the great pressure exerted by the freezing of water contained in cracks or along bedding planes.

fumarole
A vent, usually volcanic, from which gases and vapors are emitted.

gabbro
A mafic, coarse-grained igneous rock composed mainly of ferromagnesium minerals and calcium-rich plagioclase feldspar.

geanticline
A large anticlinal structure of regional extent that upwarps due to lateral compression in geosynclinal sediments.

geode
A hollow or partly hollow, globular-shaped body occurring in sedimentary rock that weathers out as a discrete nodule or concretion; a type of concretion.

geosyncline
A mobile downwarping of the earth's crust, usually elongate, and hundreds of miles in extent, that acts as a collecting basin for thousands of feet of sedimentary and volcanic rock.

geothermal gradient
The rate of increase of temperature in the earth with depth. The average geothermal gradient in the crust is about 25° C/km of depth.

geyser
An intermittent hot spring that regularly or irregularly erupts jets of hot water and steam.

glacial drift
Unconsolidated rock debris, both stratified and unstratified, that has been transported by glaciers or meltwater streams and deposited directly on land or in the sea.

glacial pavement
A bare bedrock surface that has been planed down, striated, and polished by glacial abrasion until it is relatively smooth.

glacial rebound
Upward movement of the earth's crust after the weight of glacial ice that had depressed the crust is removed; also called *crustal rebound.*

glacier
A large, long-lasting mass of ice, formed on land by the compaction and recrystallization of snow, which moves because of its own weight.

glassy (vitreous) texture
An extrusive igneous rock texture that is similar to that of broken glass and developed as a result of rapid cooling of lava without visible crystallization into individual mineral grains.

glauconite
A dull-green mineral, an iron-potassium silicate of the mica group; its presence in a sediment or sedimentary rock indicates very slow accumulation.

gneiss
A coarse-textured metamorphic rock comprised of bands or lenses of light and dark minerals.

graben
A crustal block between faults that has been dropped relative to adjacent blocks (see *horst*).

granite, granitic rock
Any felsic, coarse-grained, light-colored, intrusive igneous rock containing quartz as an essential component, along with feldspar and mafic minerals.

granular disintegration
A type of mechanical weathering consisting of grain-by-grain breakdown of rock masses. The mineral grains tend to separate from one another along their natural contacts to produce coarse sand or gravel.

gravel
Rounded sediment particles coarser than sand.

gravity survey
The process of measuring the direction and intensity of the earth's gravitational field at a number of different locations, using a gravimeter, and plotting the measurements on a map.

ground moraine
A blanket of till deposited by a glacier or released as glacial ice melts.

ground water (underground water)
The water beneath the ground surface that fills the cracks, crevices, and pore space of rocks.

gypsum
A mineral (hydrous calcium sulfate) which, when lithified, becomes a sedimentary rock, called rock gypsum.

gypsum flower, cave flower
Gypsum that occurs as a curved, elongate deposit from a cave wall and resembles the shape of a flower. (Epsomite also can occur as cave flowers.)

hanging valley
A smaller valley that terminates abruptly above a main valley or a shoreline. The floors of the small valley and the main valley (or shoreline) are discordant in level, usually causing waterfalls to develop.

hanging wall
The overlying surface of, or the surface above, an inclined fault plane (see *footwall*).

harmonic tremors
Earthquake vibrations, associated with volcanic activity, which are harmonious or regular in displacement and arrival time.

headward erosion
The uphill growth of a valley above its original course by gullying, mass wasting, and sheet erosion.

helictite
A speleothem that resembles a stalagmite in origin but that angles or twists erratically.

high-energy coast
A coast exposed to ocean swell and stormy seas and characterized by average breaker heights of greater than 50 cm.

hogback
A long, narrow, sharp-crested ridge formed by differential erosion on the outcropping edges of steeply inclined or highly tilted resistant rocks (see *cuesta*).

horn, matterhorn
A sharp peak produced when cirques cut back into a mountain summit from three or four sides.

hornblende
The commonest mineral of the amphibole group; found in both igneous and metamorphic rocks.

hornfels
A fine-grained, unfoliated metamorphic rock.

horst
A crustal block between faults that has been raised relative to adjacent blocks (see *graben*).

hot spring
A spring with a water temperature warmer than human body temperature.

hydration
A type of chemical weathering in which water combining with minerals in rock causes the mineral grains to expand, thus weakening the rock.

hydromagnesite
A white, earthy magnesium mineral, sometimes called "moon milk."

hydrothermal metamorphism
Alteration of a rock by hot water (usually bearing mineral ions in solution) passing through it.

iceberg
Block of glacier-derived ice floating in water.

icecap
A large glacier covering a relatively small area of land, but not restricted to a valley.

ice cave
An artificial or natural cave in which ice forms and persists throughout all or most of the year.

ice sheet
See *continental glacier*.

ice wedge
A large, wedge-shaped, vertical or inclined sheet, dike, or vein of ice that tapers downward into permafrost. It originates by the freezing of water in a narrow crack produced by thermal contraction of the permafrost. An ice wedge enlarges with subsequent freezing and thawing.

igneous rock
Rock formed from solidification of magma on the earth's surface or within the crust.

impermeable
Not allowing a fluid such as water or petroleum to flow through (see *permeable*).

incised meander
A meander that retains its sinuous curves as it cuts vertically downward below the level at which it originally formed.

inclusion
A rock fragment that is distinct from the igneous rock in which it is enclosed.

intermediate rocks
Igneous rocks that are transitional or intermediate in composition between felsic and mafic rocks.

intermittent stream
A stream that flows only at certain times of the year (see *ephemeral stream*).

intrusion
The upward movement of a body of magma into the crust; also an igneous rock mass so formed within surrounding rock.

intrusive rock
Igneous rock that has cooled within the crust of the earth.

intrusive sheet
A tabular igneous intrusion, usually concordant.

isoclinal fold (isocline)
A fold in which the limbs are parallel to one another, with both limbs having the same amount and direction of dip.

isostasy
The balance of equilibrium, comparable to floating, between adjacent blocks of crust resting on a plastic mantle.

isostatic adjustment
Concept of vertical movement of blocks or sections of the earth's crust in order to attain equilibrium.

jasper
A variety of chert, characteristically reddish in color.

joint
A surface fracture, or parting, in a rock without displacement. The joint surface is usually a plane and often occurs with parallel joints to form a *joint set*.

karst topography
A type of topography that develops due to erosion of limestone or dolomite (occasionally gypsum) by the solution of rock; characterized by closed depressions, sinkholes, caves, underground drainage, etc.

keratophyre
A light-colored volcanic rock, composed predominantly of plagioclase feldspar and quartz, that was extruded from submarine vents.

kettle, kettlehole
A depression caused by the melting of a stagnant block of ice that was surrounded or buried by glacial outwash.

kettle lake
A body of water occupying a kettle in a pitted outwash plain or in a moraine.

kipuka
Islandlike surface area surrounded by a lava flow; highly variable in size (from a Hawaiian word meaning "opening").

klippe
An isolated rock unit that is an erosional remnant or outlier of a formation or a thrust sheet.

knob (topographic)
A protruding mass of resistant rock; any rounded mound or hillock.

lahar
A mudflow on the flank of a volcano, sometimes triggered by glacial meltwater. The debris carried in the flow consists of pyroclastics, volcanic debris, mud, etc.

lamina
The thinnest recognizable unit or layer of deposition in a sediment or sedimentary rock.

landslide, landsliding
A general, inclusive term for slope failure or downslope movement of rock and debris that is sudden, fairly rapid, or slow.

lateral moraine
A low, ridgelike accumulation of till carried along the side of a glacier.

lava
Magma that reaches the surface, either through fissures or volcanic vents.

lava ball
An accretionary feature that forms when a fragment of solidified lava rolls along, growing in size as it picks up sticky lava.

lava fountain
A jet of incandescent lava that shoots into the air as lava is forced up to the surface by hydrostatic pressure and the expansion of gas bubbles.

lava lake
A lake of molten lava (usually basaltic) in a volcanic crater or depression; may be solidified or partly solidified.

lava stalactite
An iciclelike pendant of lava that hardened while dripping from the roof of a lava tube or lava tunnel.

lava tree
A lava tree mold that projects above the surface; a lava tree case (see *tree mold*).

lava tube, lava tunnel
A hollow space, cave, or tunnel beneath the surface of a solidified lava flow, formed by the withdrawal of molten lava after the solidification of the surficial crust.

lee
Said of the side or slope of a hill, knob, or prominent rock located away from which an advancing glacier or ice sheet moved; facing the downstream side of a glacier, and relatively protected or sheltered from its abrasive action (see *stoss*).

lignite
A brownish-black coal that is intermediate in coalification between peat and subbituminous coal.

limb
One side of a fold; each limb is usually part of two adjacent folds.

limestone
A sedimentary rock that may be organic or inorganic in origin and is composed mainly of calcite; a carbonate rock.

lithification
The consolidation of sediment to form sedimentary rock.

loess
A fine-grained deposit of wind-blown dust, composed of angular, unweathered grains of quartz, feldspar, and other minerals—unconsolidated, but sometimes held together by calcareous cement.

longshore current
A moving mass of water that develops parallel to a shoreline, caused by the oblique impingement of waves against a coast. This piles up water, some of which then flows along the shore.

longshore drift
Movement of sediment parallel to the shore carried by longshore currents.

low-energy coast
A sheltered coast protected from strong wave action by headlands, islands, or reefs and characterized by average breaker heights of less than 10 cm.

mafic
Describing silica-poor igneous rocks having a relatively high content of magnesium, iron, and calcium (see *felsic*).

magma
Naturally occurring mobile, melted rock material, generated within the earth, that has not reached the earth's surface.

magma chamber
Space within the earth's crust occupied by a body of magma.

magnetic field (of the earth)
Region of magnetic forces surrounding the earth.

magnetic poles (of the earth)
The points of emergence at the earth's surface where the strength of the magnetic field is greatest. Magnetic lines of force appear to enter at the negative (presently south) pole and leave at the positive (presently north) pole; also called *geomagnetic poles*.

magnetic reversal
A change in the earth's magnetic field between *normal polarity* and *reverse polarity;* also called *geomagnetic reversal.*

mantle
The thick rock shell of the earth's interior below the crust and above the core.

mantle plume
A narrow column of hot, molten mantle rock that rises toward the crust and spreads radially outward; also called *plume.*

marble
A coarse-grained metamorphic rock made up of interlocking calcite crystals, or recrystallized calcite. Some marbles contain dolomite crystals.

marker flow
A lava flow with distinctive characteristics that allow it to serve as a stratigraphic reference or to be traced over a distance.

massive (rock)
Describing a stratified rock that appears to be without internal structures such as minor joints, fissility, and lamination; may also apply to igneous rocks (e.g., granite, diorite) that have fairly homogeneous texture over wide areas.

mass wasting
A group of processes involving loosening and downslope movement of soil and rock in response to gravitational stress. Rates of movement range from creep (very slow) to very rapid falls and slides.

matrix
Finer-grained material (groundmass) enclosing larger grains or particles of a rock or filling the spaces between grains.

meander
A pronounced sinuous curve or wide swing in a stream's course produced by lateral erosion.

mechanical (physical) weathering
See *disintegration.*

medial moraine
A single, long ridge of till riding on top of a glacier, formed by the joining of adjacent lateral moraines of merging glaciers moving downslope.

mesa
A broad, flat-topped hill bounded by cliffs and capped with a resistant rock layer (see *butte*).

metamorphic facies
A group of metamorphic rocks having similar mineral compositions and physical characteristics and hence assumed to have formed under a particular range of pressure and temperature conditions.

metamorphic rock
Rock that has been transformed from pre-existing rock into texturally or mineralogically distinct new rock as a result of high temperature, high pressure, or both, but without the rock melting in the process.

mica
A group of rock-forming minerals (e.g., muscovite, biotite) that commonly occur in flakes, sheets, shreds, or scales. In composition micas are complex silicates. They are characterized by low hardness and perfect basal cleavage.

microplate
A relatively small tectonic plate.

microseisms
Very small earth tremors.

mid-oceanic ridge
A continuous, mostly submarine mountain range that extends through the Atlantic, Indian, and Pacific Oceans. New crustal material forms in the ridge's rift or spreading center.

migmatite
A rock composed of igneous (or igneous-appearing) and metamorphic material.

mineral
A naturally occurring, inorganic, usually crystalline solid with a definite chemical composition and characteristic physical properties that are either uniform or variable within definite limits.

miogeosyncline
The nonvolcanic portion (usually on the cratonic side) of a geosyncline, characterized by thick accumulations of shales, sandstones, and limestones, but generally lacking volcanics (see *eugeosyncline*).

mirabilite
A white or yellow monoclinic mineral ($Na_2SO_4 \cdot 10H_2O$) that occurs as a residue from saline lakes, playas, and springs, and as an efflorescence; also called Glauber's salt.

monadnock
An upstanding rock or hill of resistant rock rising conspicuously above the general level of an erosion surface; a remnant of a former upland remaining after most of the upland rock has been eroded away.

monocline, monoclinal fold
A fold that bends, or flexes, from the horizontal in only one direction.

montmorillonite
A group of clay minerals generally derived from the alteration or weathering of calcic feldspars, ferromagnesian minerals, and volcanic glasses; common in soils, the main constituent of bentonite, and also a typical insoluble residue in cave earth.

moraine
A body of till either being carried on a glacier or left behind after a glacier has receded.

mudball
A spherical mass of mud or mudstone in a sedimentary rock, developed by weathering and breakup of clay deposits. It may be as large as 8 inches in diameter.

mudflow
A debris flow consisting of soil, sand, silt, ash, clay, etc., mixed with enough water so that it moves downslope as a viscous mass; a type of mass wasting.

mud pot
A hot spring that contains boiling mud and dissolved compounds that hot water has brought up from the rocks below.

natural arch
A bridgelike or arch-shaped landform produced by weathering and erosion.

natural levees
Low ridges of flood-deposited sediment, formed on either side of a stream channel, that thin away from the channel.

needle
A narrow, pointed spire of rock formed by weathering and erosion (see *pillar*).

nonclastic
Describing a sediment or rock that formed by chemical precipitation or organic secretion (see *clastic*).

nonconformity
A type of unconformity that developed between sedimentary rocks and older plutonic or metamorphic rocks that were exposed to erosion before the overlying sediments covered them (see *disconformity, angular unconformity*).

nonfoliated
Pertaining to metamorphic rocks (or massive igneous rocks) lacking *foliation*.

normal fault
A type of fault caused by tensional force, in which the hanging wall has moved down in relation to the footwall (see *reverse fault*).

normal polarity
A natural remanent magnetization closely parallel to the present ambient magnetic field direction; i.e., the direction of magnetic north when the rock was formed was approximately the same as it is now (see *reverse polarity*).

novaculite
A dense, hard, even-textured, light-colored, siliceous sedimentary rock that resembles chert.

nuée ardente
An incandescent, turbulent, swiftly flowing cloud of hot ash, dust, and gas erupted from a volcano. The denser, lower part of the cloud, containing pyroclastics, may form a *glowing avalanche*.

obsidian
Volcanic glass, usually black or dark-colored and rhyolitic in composition; characterized by conchoidal fracture.

oceanic crust
That part of the earth's crust that underlies the ocean basins; predominantly basaltic (sima) (see *continental crust*).

oceanic trench
A deep, narrow trough parallel to an island arc or a continental margin (e.g., Aleutian Trench).

offset
Horizontal displacement along a fault. (Note: The *offset* of a stream or ridge by faulting may be lateral, vertical, or oblique.)

olivine
A common, dark-colored, rock-forming, ferromagnesian mineral, $(Fe,Mg)_2SiO_4$.

ooids, ooliths
Tiny, spherical grains about the size of fish roe, usually composed of calcium carbonate, that form by precipitation around a nucleus (shell fragment, sand grain, etc.) in shallow, wave-agitated water. The grains accumulate as sedimentary layers.

oolite
A sedimentary rock, usually a limestone, made up chiefly of ooids, or ooliths, cemented together.

ophitic (texture)
An igneous rock texture (especially in diabase) in which lath-shaped plagioclase crystals are partially or completely included in pyroxene crystals (typically augite).

ore
Naturally occurring material that can be profitably mined.

original horizontality, principle of
Most water-laid sediment is deposited in horizontal or near-horizontal layers that are closely parallel to the earth's surface.

orogeny
The deformational crustal processes by which mountain systems are formed. Metamorphism and igneous activity generally accompany orogeny.

orthoclase feldspar
A light-colored, rock-forming mineral of the alkali feldspar group ($KAlSi_3O_8$).

outwash (glacial)
Glacial drift consisting of stratified detritus (chiefly sand and gravel) removed or "washed out" from a glacier by meltwater streams and deposited in front of or beyond the terminal moraine or the margin of a glacier. The coarser material is deposited nearer to the ice (see *till*).

outwash plain
Smooth, broad, gently sloping plain formed by the deposits of heavily loaded meltwater streams flowing from a glacier margin (see *valley train*).

oxidation
The uniting of oxygen with another element or compound; one of the processes of chemical weathering.

pahoehoe
A basaltic lava flow with a ropy or billowy surface texture (see *aa*).

paleomagnetism
The record of the earth's ancient magnetic fields.

parabolic dune
A deeply curved dune in a region of abundant sand. The horns, or arms, point upwind and may be anchored by vegetation.

patch reef
A relatively small, moundlike organic reef, usually part of a larger reef complex.

patterned ground
More or less symmetrical forms, such as circles, polygons, nets, steps, and stripes that are characteristic of, but not necessarily confined to, surficial material subject to intensive frost action—as in polar and subpolar regions.

pebble
A small, rounded, waterworn stone (a sedimentary particle) ranging in size between a small pea and a tennis ball (1/6 to 2.5 inches).

pediment
A broad, flat (or gently sloping) bedrock erosional surface that develops at the base of a mountain front, usually in a dry region; typically covered with a thin veneer of gravel.

pegmatite
An extremely coarse-grained igneous rock, usually granitic in composition and made up mainly of large quartz and feldspar crystals; occurs principally in dikes.

periglacial
Referring to the processes, conditions, areas, climates, and topographic features at the immediate margins of glaciers and ice sheets, where freezing and thawing occur somewhat continually.

period
A geologic time unit longer than an *epoch* and a subdivision of an *era*; e.g., the Cambrian Period.

permafrost
Ground that remains permanently frozen over a period of many years.

permeability
The capacity of a rock to transmit a fluid such as water or petroleum (see *porosity*).

permineralization
A process of fossilization whereby the original hard parts of an organism have had additional mineral material deposited in their pore spaces.

phaneritic (texture)
Describing the texture of an igneous rock in which the individual components are visible to the unaided eye.

phenocryst
A relatively large, conspicuous crystal in the groundmass of a porphyritic igneous rock.

phreatic eruption
Explosive, steam-propelled ejections of ash and mud caused by heating and expansion of ground water.

phreatic explosion
An explosion of steam, mud, or other material that is not incandescent; it is caused by the heating and consequent expansion of ground water due to an underlying heat source; more violent than a *phreatic eruption*.

phreatic zone
The subsurface zone (below the water table) in which all rock openings are filled with water; also called the *zone of saturation*.

phreatomagmatic explosion
A violent volcanic explosion that extrudes both magmatic gases and steam along with pyroclastics. It is caused by the mixture of magma with ground water or with ocean water.

phyllite
A fine-grained metamorphic rock in which clay minerals have recrystallized into microscopic micas. A silky sheen on corrugated cleavage surfaces is characteristic of phyllite.

phytosaur
A large, extinct aquatic reptile, similar in appearance to the crocodile.

piedmont glacier
A thick, continuous sheet of ice resting on land at the base of a mountain range; formed by the spreading out and coalescing of valley glaciers from the higher elevations of the mountains.

piercement structure, diapir
The piercing or rupturing of domed or uplifted rocks by mobile material pushed up by tectonic stresses.

pillar
A natural, pillar-shaped landform of rock or earth, produced by weathering and erosion. Also, a bedrock support in a cavern remaining after removal of the surrounding rock by solution processes.

pillow lava, pillow structure
Submarine extrusions of basaltic lavas that solidify as pillow-shaped masses, closely fitted together. The "pillows" range in size from a few inches to a few feet in diameter.

piping
Erosion by percolating water in a layer of subsoil, resulting in caving and in the formation of narrow conduits, tunnels, or "pipes" through which soluble or granular soil material is moved.

pisolite
A sedimentary rock, usually a limestone, made up chiefly of small round accretionary bodies cemented together; similar to *oolite*.

placer
A surficial sedimentary deposit, usually on a beach or in a stream bed, containing particles of valuable minerals or native metal (e.g., gold) in unusually concentration due to their greater density (higher specific gravity).

plagioclase feldspar
A group of abundant, rock-forming minerals containing sodium and/or calcium in addition to aluminum, silicon, and oxygen.

plastic
Referring to a substance capable of being molded, bent, or deformed under strain without rupturing (up to the limit of its malleability).

plastic flow zone (glacier)
The bottom part of a glacier where ice is molded or bent under stress so that it flows without fracturing.

plateau
A broad, comparatively flat landform of considerable extent elevated above adjacent countryside; usually bounded by cliffs or an escarpment on at least one side.

plate, tectonic plate
A large, mobile slab of rock making up part of the earth's crust.

plate tectonic theory
The concept that intense geologic interactions occur along boundaries between the rigid, slowly moving plates that make up the earth's surface.

plucking, quarrying
A process of glacial erosion by which large rock fragments, or blocks, are loosened and detached from bedrock by the freezing of water in cracks and joints and then removed by advancing ice.

pluton
A large body of igneous rock that formed at depth in the earth's crust.

pluvial climate
A climate characterized by abundant rainfall (such as during a transitional or glacial period).

polyp
An individual living coral.

porosity
The percentage of a rock's volume that is taken up by openings (see *permeability*).

porphyritic (texture)
An igneous rock texture in which larger crystals (phenocrysts) are set in a finer groundmass which may be crystalline or glassy or both.

pothole
A deep, circular depression eroded into the hard rock of a stream bed by the abrasive action of the stream's sediment load.

Precambrian shield
An extensive area of ancient Precambrian plutonic and metamorphic rocks at or near the surface; part of a craton.

pressure-release jointing
Exfoliation that occurs in once deeply buried rocks that have been unloaded by erosion, thus releasing their confining pressure.

primary structural features
Features in rocks that were formed at the time of origin of the rock, but before its final consolidation.

protalus rampart
A ridge of loose, angular rock fragments that accumulated along the bottom of a large perennial snowbank that lay at the base of a cliff. When the climate warmed, the snowbank melted and left an arcuate ridge of boulders and other coarse debris beyond the toe of the talus slope.

pseudomorph
A mineral whose outward crystal form is that of another mineral species.

pull-apart valley
A valley produced by tensional forces acting on the crustal rocks at the earth's surface.

pumice
A light-colored, vesicular, glassy rock, commonly of rhyolitic composition and often sufficiently bouyant to float on water; usually formed during an explosive eruption when lava contains gas and water vapor in large amounts.

pyroclast
Rock fragment formed by volcanic explosion.

pyroclastic (texture)
An igneous rock texture implying the presence of fragments, such as ash, pumice, and other ejecta.

pyroxene group
Dark, rock-forming silicate minerals (e.g., *augite*) similar in chemical composition to the amphiboles.

quartz
Crystalline silica (SiO_2), an important rock-forming mineral.

quartzite
(a) A metamorphic rock formed by recrystallization of sandstone or chert by regional metamorphism. (b) A very hard but unmetamorphosed sandstone, consisting mainly of quartz grains that have been solidly cemented with secondary silica.

quicksand
A thick bed of fine sand that is usually saturated with water flowing upward through the voids, forming a soft, shifting, semiliquid mobile mass that yields easily to pressure and will not support much weight.

radioactive decay
The spontaneous disintegration or transformation of certain unstable atoms, called the parent material, to daughter products through the gain or loss of nuclear particles.

radiometric dating
Methods for determining the age in years (absolute time) of geologic materials, based on nuclear decay of naturally occurring radioactive isotopes; e.g., carbon-14, potassium-40/argon-40.

rain-shadow zone
A region on the lee (downwind) side of a mountain range that has little rain because most of the precipitation falls on the upwind side of the mountains.

recharge
The addition of water to an aquifer or to the zone of saturation.

recrystallization
The development in a rock of new crystalline grains that may or may not have the same composition but are usually larger than the original grains.

reef
A submerged, resistant mound or ridge formed in the ocean by the accumulation of plant and animal skeletons and other debris.

reef crest
The upper surface of a reef.

reef terrace
A shelflike, eroded surface that may develop at several levels on a reef because of successive episodes of submergence and emergence.

regional metamorphism
Metamorphism affecting an extensive area and involving relatively high temperatures and pressures; associated with orogenic belts.

regression (sea)
The retreat or withdrawal of a sea from land areas, and the consequent evidence of such withdrawal (see *transgression*).

relative time
The sequence in geologic time in which events took place (not measured in time units); time determined by organic evolution or superposition.

relict
(a) Describing a topographic feature developed by erosional or depositional processes no longer operating because of a change in environmental conditions (e.g., climatic change). (b) A relict landform (noun).

relief
The difference in elevation between the highest and lowest points on an area of the earth's surface.

remanent magnetism
The permanent magnetism induced in (a) an igneous rock when it cooled from a liquid to a solid state (i.e., past the *Curie point*), *or* induced in (b) a sedimentary rock by settling of magnetic grains during deposition so that they became aligned with the earth's magnetic field.

reservoir rock
A rock that is sufficiently porous and permeable to store and transmit petroleum.

reversed polarity
A natural remanent magnetization opposite to the present ambient magnetic field direction. When the rock was formed, the direction of magnetic north was approximately 180° different from what it is now (see *normal polarity*).

reverse fault
A fault (caused by compression) in which the hanging-wall block moved up relative to the footwall block (see *normal fault*).

rhyolite
A fine-grained, felsic, extrusive rock composed chiefly of feldspar and quartz.

rift zone
(a) A system of crustal fractures (such as the rift valley along the crest of the mid-oceanic ridge), caused by tensional forces pulling the crust apart. (b) On Hawaiian volcanoes, a linear zone of volcanic features associated with underlying lava conduits.

rigid zone (glacier)
The upper, brittle part of a glacier; does not flow plastically.

rimstone (cave)
A thin, crustlike deposit of calcite that forms a ring around an overflowing basin or pool of water in a cave.

ring dike
A dike that is arcuate or roughly circular in plan and is vertical or inclined away from the center of the arc.

roche moutonnée
An elongate knob or hillock of bedrock that has been smoothed and scoured by moving ice on the upglacier (stoss) side so that the rock is gently inclined and rounded. On the downglacier (lee) side the rock is steep and hackly from glacial plucking or quarrying.

rock
A solid aggregate of one or more minerals (as grains, fragments, precipitates, organic matter, etc.).

rock-basin lake
A lake occupying a depression scoured or excavated in bedrock by glacial action.

rock cycle
A theoretical, graphical way of relating tectonism, erosion, and various rock-forming processes to the common rock types.

rockfall
The sudden free fall of a segment of bedrock from a cliff or steep slope; the fastest form of mass wasting.

rock flour
Fine, powdery glacial sediment consisting of ground-up rock, produced by abrasion underneath a moving glacier.

rock glacier
A mass of poorly sorted, angular boulders and fine materials, cemented by interstitial ice and flowing downslope by means of glacier motion. An inactive or relict rock glacier may descend by creep (mass wasting).

rockslide
Rapid sliding of a large block or mass of bedrock that has become detached from an inclined surface of weakness (such as a bedding plane).

roof pendant
A downward projection of *country rock* into an igneous intrusive body, such as a batholith.

sabkha facies
A type of sedimentary rock facies that consists of an accumulation of evaporites, tidal-flood, and eolian deposits on an arid or semiarid coastal plain, between the land surface and the tidal zone; characterized by an absence of fossils, thin pebble conglomerates, mud cracks, disrupted bedding, etc.

salt-crystal cast
A crystal cast formed by solution of a soluble salt crystal, followed by filling with mud or sand or by crystallization of a pseudomorph.

sandblasting
Abrasion by wind-blown sand.

sand crystal
A large euhedral or subhedral crystal (i.e., bounded or nearly bounded by natural crystal faces) of barite, gypsum, or especially calcite composition that contains inclusions of detrital sand; developed by growth in an incompletely cemented sandstone during cementation.

sandstone
A medium-grained sedimentary rock formed by cementation of sand grains, usually 85 to 90 percent quartz; may be deposited by water or wind.

scallop (cave)
An asymmetrical solution hollow formed by turbulent water flow.

schist
A strongly foliated, crystalline, metamorphic rock with excellent cleavage, characterized by coarse-grained minerals that are oriented approximately parallel to each other.

scoria
A highly vesicular basalt.

sea wall (beach)
A long, steep-faced, natural embankment or berm of shingle or boulders (without gravel), built by powerful storm waves along a seacoast at the high-water mark.

sediment
Unconsolidated, loose, solid fragments or particles that can originate by (1) weathering and erosion of pre-existing rocks; (2) chemical precipitation from solution, usually in water; and (3) secretion by organisms. Sediment particles may be transported and deposited by water, wind, or ice.

sedimentary facies
Significantly different rock types occupying laterally distinct parts of a layered rock unit or formation.

sedimentary rock
Rock that has formed from (1) lithification of any type of sediment, (2) precipitation from solution, (3) consolidation of the remains of plants and/or animals, or (4) combinations of processes.

seismic waves
Oscillations (waves) of energy produced by an earthquake.

selenite
A clear, colorless, cleavable variety of gypsum.

shale
A fine-grained, clastic (mostly clay) sedimentary rock that splits easily into thin slabs parallel to bedding planes.

shatter zone
An area of randomly fissured or cracked rock; may have a network of veins filled with mineral deposits.

sheet jointing
See *exfoliation*.

shield volcano
A broadly convex, gently sloping, volcanic cone built by flows of low-viscosity lava.

sial
The upper layer of the earth's crust, or continental crust, composed of rocks rich in silicon and aluminum and relatively less dense (see *sima*).

sill
A tabular concordant body of intrusive igneous rock (see *dike*).

silt
Fine-grained sediment intermediate in size between clay particles and sand particles.

siltstone
A fine-grained clastic sedimentary rock made of silt, mud, etc., lithified by compaction and cementation; does not split like shale.

sima
The lower layer of the earth's crust, or oceanic crust, underlying the *sial* and composed of rocks rich in silica and magnesium.

sinkhole (karst)
A closed depression found on land surfaces underlain by limestone; formed by solution or when the roof of a cavern collapses.

sinking stream
A surface stream that disappears into an underground channel and that may not reappear in the same or even adjacent drainage basin.

sinter
A chemical sedimentary rock, mainly siliceous, deposited as a hard incrustation by precipitation.

slate
A fine-grained metamorphic rock that splits easily along smooth, parallel planes.

solfatara
A type of fumarole, the gases of which are characteristically sulfurous.

solifluction
The downslope flow of water-saturated debris over impermeable material; a type of mass wasting especially common in permafrost areas.

solution weathering
A type of chemical weathering by which the soluble minerals in rock are dissolved.

source rock (for petroleum)
A sedimentary rock containing organic matter that is converted to petroleum by burial and other postdepositional changes over a period of time.

sparry crystal growth
See *dogtooth spar*.

spatter cone
A small cone formed of globs of lava piled up around a volcanic vent.

speleothem
Any secondary mineral deposit that is formed in a cave by the action of water.

spheroidal weathering
A chemical weathering process that tends to round off a block of rock by the loosening and detachment of successive, concentric shells of decayed rock from the block; similar to *exfoliation*.

spilite
A mafic, green, igneous rock composed predominantly of plagioclase feldspar and chlorite; an altered basalt.

spit
A fingerlike extension of a beach, with the far end terminating in open water (a landform).

spongework (cave)
An entangled pattern or complex of irregular, interconnecting, tubular channels or cavities of various sizes in the walls of limestone caves; separated by amazingly intricate and perforated partitions and remnants of partitions.

spreading center
A crustal rift, such as the crest of the mid-oceanic ridge, where tectonic plates are being pulled apart by tensional forces and new crust is being formed.

stack, sea stack
A pillarlike erosional remnant left detached from a headland as a wave-eroded coast retreats inland.

stalactite
An icicle-shaped pendant of dripstone formed on a cave ceiling.

stalagmite
A cone-shaped mass of dripstone formed on cave floors, generally directly below a stalactite.

stock
A small pluton that has an area of surface exposure of less than 40 square miles (see *batholith*).

stoss
Describing the side or slope of a hill, knob, or prominent rock facing the upstream side of an advancing glacier, and most exposed to its abrasive action (see *lee*).

stream capture
The natural diversion of the headwaters of one stream into the channel of another stream which has greater erosional activity and flows at a lower level; especially diversion effected by a stream eroding headward at a rapid rate so as to tap and lead off the waters of another stream.

striations
Furrows or lines inscribed on rock surfaces as rock fragments embedded in ice are dragged along by a moving glacier.

strike
The compass direction or trend of a line formed by the intersection of an inclined plane (such as a fault plane) with a horizontal plane.

strike-slip fault
A fault in which movement is parallel to the strike of the fault surface. Two blocks slide past each other, resulting in lateral displacement.

stromatolite
A complex sedimentary structure of sediment and organic remains, principally blue-green algae, that occurs in a variety of shapes and forms—horizontal, columnar, branching, etc.

subduction
The sliding down of the leading edge of an oceanic plate beneath a continental margin or an island arc.

subsequent stream
A tributary that has developed its valley (mainly by headward erosion) along a belt of underlying weak rock and is, therefore, adjusted to the regional structure (see *consequent stream*).

superimposed (superposed) stream
A stream that was established on a new surface and maintained its course by cutting across buried rock structures as it eroded downward into underlying rocks (see *antecedent stream*).

surf zone
The strip of shore where the waves break and roll up the beach or beat against a cliff face (see *foreshore*).

syncline
A downfold opening upward; a fold with younger beds in the center (see *anticline*).

tactite
A complex, calcareous metamorphic rock formed as a result of contact metamorphism.

talus
Broken rock that accumulates at the base of a cliff or slope.

tectonic mélange
A body of rock (large enough to be mapped) that includes fragments and blocks of numerous sizes and kinds of rock embedded in a generally sheared matrix or more malleable or tractable rock, such as claystone or shale.

tension, tensional force
Stress that tends to elongate or pull apart a body (see *compressive force*).

tephra
A general term for fragmental volcanic rocks or pyroclastics.

terminal moraine
An end moraine marking the farthest advance of a glacier.

terrace
A general term for a landform that resembles a stairstep with a flat tread (the top) and a steep riser (the face).

texture (rock)
The general physical appearance of a rock, including the size, shape, and arrangement of its components, grains, etc.

thaw lake
A pool of water formed on the surface of a large glacier by accumulation of meltwater; also a cave-in lake in permafrost.

tholeiitic basalt
A group (or suite) of basalts that contain several varieties of pyroxenes and little or no olivine; oversaturated in silica (see *alkalic basalt*).

thrust fault
A reverse fault in which the fault plane has a low angle of dip (see *reverse fault*).

till
Glacial drift consisting of unsorted and unstratified rock debris carried or deposited by a glacier (see *outwash*).

time-transgressive rock unit
A sedimentary formation that becomes younger (or older) when traced across country; e.g., a marine sandstone that becomes younger in the direction in which the shoreline of a sea was advancing.

transform fault
A special variety of strike-slip fault along which the displacement suddenly stops or changes form; associated especially with mid-oceanic ridges where the actual slip is opposite from the apparent displacement across the fault.

transgression (by the sea)
The spread or extension of the sea over land areas, and the consequent evidence of such advance (see *regression*).

transverse stream
A stream that flows across bedrock structures, such as folds, faults, and fault blocks.

travertine
A porous deposit of calcite formed by rapid precipitation of calcium carbonate from surface or ground water *or* by evaporation around the mouth of a spring, especially a hot spring.

tree mold
A cylindrical hollow in a lava flow formed by the envelopment of a tree by the flow, solidification of the lava in contact with the tree, and disappearance of the tree by burning and subsequent removal of the charcoal and ash. The inside of the mold preserves the surficial features of the tree (see *lava tree*).

tritium
A naturally occurring, radioactive isotope of hydrogen (H_3); formed by cosmic rays in the upper atmosphere, it enters the earth's hydrologic cycle via precipitation.

tritium dating
Calculation of an age in years by measuring the concentration of tritium in a substance, usually water. The method also provides a means of tracing subsurface movement of water and determining its velocities.

truncated spur
The faceted lower end of a ridge projecting into a glacial valley that has been eroded by glacial ice.

tsunami
A seismic sea wave produced by any large-scale, short-duration disturbance of the ocean floor, principally by a submarine earthquake.

tubular passage (cave)
A cavern or tunnel with an elliptical or lenticular cross section formed in the phreatic zone.

tufa
A chemical sedimentary rock composed of calcium carbonate, formed by evaporation as an incrustation around the mouth of a hot or cold calcareous spring or seep.

tuff
A type of igneous rock resulting from the solidification of volcanic dust, ash, or pumice.

tumuli
Mounds or hillocks on the surface of pahoehoe lava (sing. *tumulus*).

turbidites
Rocks or sediments that were deposited in a marine environment by settling out from muddy or turbid water that flowed along a sloping ocean bottom.

turbidity current
A flowing mass of sediment-laden water that is heavier than clear water and therefore flows downslope along the bottom of the sea or a lake.

ultramafic rock
A dark, plutonic rock (such as peridotite) composed almost entirely of mafic minerals.

unconformity
A planar surface between rock units that represents a time break in the geologic record, with the rock unit immediately above it being considerably younger than the rock beneath (see *disconformity, nonconformity, angular unconformity*).

uniformitarianism, principle of
The assumption that geological processes that are operating today are the same processes that have operated in the geologic past.

U-shaped valley
The characteristic cross profile of a glacially excavated trough; that is, flat-floored with nearly perpendicular side walls.

vadose zone
The subsurface zone above the water table in which the rock openings are filled partly with water and partly with air; located above the *phreatic zone*. Also called the *zone of aeration*.

valley train
Outwash dropped or spread within the valley of a meltwater stream that flows out from a glacier (see *outwash plain*).

ventifact
A boulder, cobble, or pebble with flat surfaces worn or polished by abrasion of wind-blown sand.

vesicle
A small cavity or opening in volcanic rock caused by gas bubbles trapped in solidifying lava.

vesicular (texture)
An igneous rock texture characterized by abundant cavities formed by the entrapment of gas bubbles during solidification of lava.

viscosity
Resistance to flow. (A thin, runny lava has low viscosity; a thick, sticky lava has high viscosity.)

volcanic dome
A steep-sided, pluglike, or spinelike mass of volcanic rock formed from viscous lava that solidifies in or immediately above a volcanic vent; also called *plug dome* or *lava dome*.

volcanic neck
A vertical, pipelike body of rock that represents the conduit to a former volcanic vent; an erosion remnant of a volcanic cone.

water gap
A deep, narrow, low-level pass penetrating to the base of and across a mountain ridge, and through which a stream flows; especially a narrow gorge or ravine cut through resistant rocks by an antecedent stream.

water table
The surface within the ground below which the rocks are saturated; separates the *phreatic zone* (below) from the *vadose zone* (above).

wave-cut platform, wave-cut terrace
A horizontal bench in the surf zone cut by wave erosion on a retreating coast; usually overspread by beach sediment.

weathering
A group of processes that alter or break down rock at or near the earth's surface (see *decomposition, disintegration*).

weather(ing) pits
Shallow depressions on the flat or gently sloping surface of large exposures of granitic rocks, attributed to strongly localized solvent action of impounded water.

welded tuff
A glass-rich volcanic rock generally composed of silicic pyroclasts welded together.

wind gap
A shallow notch in the crest or upper part of a mountain ridge, usually at a higher level than a water gap. A former water gap, now abandoned (as by capture) by the stream that formed it.

window (landform)
A hole that appears in a fin (or narrow ridge) as the result of differential weathering; also the opening under a natural arch or natural bridge.

window (structure)
An eroded area of a thrust sheet that displays the rocks beneath the thrust sheet; also called *fenster*.

xenocryst
A crystal resembling a phenocryst in igneous rock that is foreign to the body of rock in which it occurs.

zone of accumulation
The portion of a glacier with a perennial snow cover, where more snow accumulates than melts each year.

zone of aeration
See *vadose zone*.

zone of saturation
See *phreatic zone*.

zone of wastage
The portion of a glacier where melting or loss of ice and snow exceeds the accumulation.

General Bibliography

National Parks

Fleisher, P. J. 1975. *Geology of selected national parks and monuments.* Dubuque, Iowa: Kendall/Hunt Publishing Company. 182 p.

Frome, Michael. 1979. The national parks. Chicago: Rand McNally & Co. 151 p.

————. 1982. 16th ann. ed. *National park guide.* Chicago, Illinois: Rand McNally & Company.

Harris, D. V. 1980. 3rd ed. *The geologic story of the national parks and monuments.* New York: John Wiley & Sons. 323 p.

Krell, D. N. (editor) 1980. 3rd ed. *National Parks of the West.* (Sunset Books) Menlo Park, California: Lane Publishing Company. 256 p.

Matthews, W. H., III. 1968. *A guide to the national parks: their landscape and geology.* Volume 1, The western parks, 480 p.; Volume 2, The eastern parks, 287 p. Garden City, New York: The National History Press.

National Geographic Society editors. 1979. *National parks, the best of our land.* Reprinted for National Park Service from *National Geographic magazine* vol. 156, no. 1 (July). Washington, D.C.: National Geographic Society, 152 p.

————. 1980. (rev. ed.) *The new America's wonderlands, our national parks.* Washington, D.C.: National Geographic Society. 463 p.

National Park Service. 1979. *Index, National Park System and related areas.* GPO stock number 024–005–00763–6. (The 1981 *Index* had not been issued when this book went to press.)

Quirk, P. J., and Fise, T. F. 1979. *The complete guide to America's national parks.* Washington, D.C.: National Park Foundation. 292 p.

Rowe, R. C. 1977. 2nd ed. *Geology of our western national parks and monuments.* Portland, Oregon: Binford & Mort, Publishers. 220 p.

Tilden, Freeman. 1979. rev. ed. *The national parks.* New York: Alfred A. Knopf. 486 p.

Journals

National Parks, the magazine of the National Parks & Conservation Association, 1701 Eighteenth Street NW, Washington, D.C. 20009.

Earth Science, published 4 times a year by the American Geological Institute, Falls Church Virginia 22041.

Regional Geology

Eardley, A. J. 1962. 2nd ed. *Structural geology of North America.* New York: Harper & Row. 743 p.

Hunt, C. B. 1974. *Natural regions of the United States and Canada.* San Francisco: W. H. Freeman and Company. 725 p.

K/H Geology field guide series. Dubuque, Iowa: Kendall/Hunt Publishing Company. Individual titles cited in bibliographies for specific parks.

King, P. B. 1977. rev. ed. *The evolution of North America.* Princeton, New Jersey: Princeton University Press. 197 p.

Pirkle, E. C., and Yoho, W. H. 1982, 3rd ed. *Natural regions of the United States.* Dubuque, Iowa: Kendall/Hunt Publishing Company. 384 p.

Shimer, J. A. 1972. *Field guide to landforms in the United States.* New York: The Macmillan Company. 272 p.

Thornbury, W. D. 1965. *Regional geomorphology of the United States.* New York: John Wiley & Sons. 609 p.

Physical and Historical Geology

Allison, I. S., and Palmer, D. F. 1980. 7th ed. *Geology.* New York: McGraw-Hill Book Company. 579 p.

Dott, R. H., Jr., and Batten, R. L. 1981. 3rd ed. *Evolution of the earth.* New York: McGraw-Hill Book Company. 113 p.

Hamblin, W. K. 1982. 3rd ed. *The earth's dynamic systems.* Minneapolis: Burgess Publishing Company.

Levin, H. L. 1983. 2nd ed. *The earth through time.* Philadelphia: CBS College Publishing. 513 p.

MacFall, R. P. 1980. *Rock hunters guide: How to find and identify collectible rocks.* New York: Crowell.

Petersen, M. S., Rigby, J. K., and Hintze, L. F. 1980. 2nd ed. *Historical geology of North America.* Dubuque, Iowa: Wm. C. Brown Company Publishers. 232 p.

Plummer, C. C. and McGeary, D. 1982. 2nd ed. *Physical geology.* Dubuque, Iowa: Wm. C. Brown Company Publishers 500 p.

Press, F. and Siever, R. 1978. 2nd ed. *Earth.* San Francisco: W. H. Freeman and Company. 649 p.

Stearn, C. W. 1979. 3rd ed. *Geological evolution of North America, the methods and problems of historical geology.* New York: John Wiley & Sons. 570 p.

Tuttle, S. D. 1980. 3rd ed. *Landforms and landscapes.* Dubuque, Iowa: Wm. C. Brown Company Publishers. 175 p.

Geologic Terms

American Geological Institute. 1976. rev. ed. *Dictionary of Geological Terms*. Falls Church, Virginia: American Geological Institute. 472 p. paperback.

Bates, R. L. and Jackson, J. A., editors. 1980 (2nd ed.) *Glossary of geology*. Falls Church, Virginia: American Geological Institute.

Fairbridge, R. W., editor. 1968. *The encyclopedia of geomorphology*. New York: Reinhold Book Corporation.

Lapedes, D. N., editor-in-chief. 1978. *McGraw-Hill encyclopedia of the geological sciences*. New York: McGraw-Hill Inc.

Plummer, C. C. and McGeary, D. 1982. (2nd ed.) *Physical geology*. Dubuque, Iowa: Wm. C. Brown Company Publishers.

Films, Videocassettes

The National Park Service Film Collection. National AudioVisual Center, National Archives and Records Service, General Services Administration, Reference Section CH, Washington, DC 20409.

Index